FDA'S DRUG REVIEW PROCESS AND THE PACKAGE LABEL

FDA'S DRUG REVIEW PROCESS AND THE PACKAGE LABEL

Strategies for Writing Successful FDA Submissions

Tom Brody

Academic Press is an imprint of Elsevier
125 London Wall, London EC2Y 5AS, United Kingdom
525 B Street, Suite 1800, San Diego, CA 92101-4495, United States
50 Hampshire Street, 5th Floor, Cambridge, MA 02139, United States
The Boulevard, Langford Lane, Kidlington, Oxford OX5 1GB, United Kingdom

British Library Cataloguing-in-Publication Data
A catalogue record for this book is available from the British Library

Library of Congress Cataloging-in-Publication Data
A catalog record for this book is available from the Library of Congress

ISBN: 978-0-12-814647-7

Publisher: Mica Hayley
Acquisition Editor: Kattie Washington
Editorial Project Manager: Kathy Padilla
Production Project Manager: Mohana Natarajan
Cover Designer: Greg Harris

Typeset by MPS Limited, Chennai, India

To Dawnia and Shideh

Contents

Biography

The author received his PhD from the University of California at Berkeley in 1980, and conducted postdoctoral research at the University of Wisconsin-Madison and also at the University of California at Berkeley. The author has 15 years of pharmaceutical industry medical writing experience, acquired at Schering-Plough, Cerus Corporation, SciClone Pharmaceuticals, and Athena Neurosciences (Elan Pharmaceuticals), and has contributed to FDA submissions for the indications of multiple sclerosis, melanoma, head and neck cancer, hepatocellular carcinoma, idiopathic thrombocytopenic purpura, and hepatitis C virus. The author has published two editions of the textbook, Clinical Trials: Study Design, Endpoints and Biomarkers, Drug Safety, and FDA and ICH Guidelines.[1] This book provides background information for the topics in the present book, but the information in these two books overlaps by only 5%.

The author has 15 years of training and experience in the Code of Federal Regulations as it applies to pharmaceuticals. Shortly before entering the pharmaceutical industry, he published two editions of Nutritional Biochemistry, which describes the clinical features, diagnosis, and treatment of metabolic diseases.[2] The author has 20 research publications from his laboratory bench work on the enzymology and metabolism of folates and related amino acids,[3,4,5,6,7,8,9,10] on the blood-clotting cascade,[11,12] and on the structure of an

[1] Brody T. Clinical Trials study design, endpoints and biomarkers, drug safety, and FDA and ICH guidelines. 2nd ed. New York, NY: Elsevier, Inc.; 2016.

[2] Brody T. Nutritional Biochemistry. 2nd ed. New York, NY: Elsevier, Inc.; 1999.

[3] Brody T, Stokstad ELR. Folate oligoglutamate: amino acid transpeptidase. J. Biol. Chem. 1982;257:14271–9.

[4] Brody T, et al. Folate pentaglutamate and folate hexaglutamate mediated one-carbon metabolism. Biochemistry 1982;21:276–82.

[5] Brody T, Stokstad ELR. Nitrous oxide provokes changes in folylpenta- and hexaglutamates. J. Nutr. 1990;120:71–80.

[6] Brody T, Stokstad ELR. Incorporation of the 2-ring carbon of histidine into folylpolyglutamate coenzymes. J. Nutr. Biochem. 1991;2:492–8.

[7] Brody T, et al. Separation and identification of pteroylpolyglutamates by polyacrylamide gel chromatography. Analyt. Biochem. 1979;92:501–9.

[8] Brody T. The influence of folate binding proteins on 1-carbon metabolism. Pteridines 1989;1:159–65.

[9] Brody T, et al. Rat brain folate identification. J. Neurochem. 1976;27:409–13.

[10] Brody T, Shane B. Folic acid. In: Rucker RB, Suttie JW, McCormack DB, Machlin LJ, editors. Handbook of the vitamins. 3rd ed. New York, NY: Marcel Dekker, 2001. pp. 427–62.

[11] Brody T, Suttie JW. Glutamate carboxylase: assays, occurrence, and specificity. Methods Enzymol. 1984;107:552–63.

[12] Brody T, Suttie JW. Evidence for the glycoprotein nature of vitamin K-dependent carboxylase from rat liver. Biochim. Biophys. Acta. 1987;923:1–7.

antibody drug for multiple sclerosis (natalizumab).[13] The author also conducted laboratory bench research in oncology, where he cloned, sequenced, and expressed an oncogene (XPE gene).[14,15,16]

[13] Brody T. Multistep denaturation and hierarchy of disulfide bond cleavage of a monoclonal antibody. Analyt. Biochem. 1997;247:247–56.

[14] Brody T, et al. Human damage-specific DNA binding protein p48 subunit mRNA. GenBank Accession No. U18299; 1995.

[15] Keeney S, Eker AP, Brody T, et al. Correction of the DNA repair defect in xeroderma pigmentosum group E by injection of a DNA damage-binding protein. Proc. Natl Acad. Sci. 1994;91:40536.

[16] Dualan R, Brody T, Keeney S, et al. Chromosomal localization and cDNA cloning of the genes (DDB1 and DDB2) for the p127 and p48 subunits of a human damage-specific DNA binding protein. Genomics. 1995;29:629.

Abbreviations and Definitions

ADR adverse drug reaction
AE adverse event
agonist An agonist is a drug or naturally occurring ligand that binds to and stimulates (or activates) a receptor. The fact that a drug binds to a receptor, without more, does not imply that it is an agonist or an antagonist.
antagonist An antagonist is a drug or naturally occurring ligand that binds to and inhibits (or blocks) a receptor. The fact that a drug binds to a receptor, without more, does not imply that it is an agonist or an antagonist.
ALT alanine aminotransferase. Also called, serum glutamate pyruvate transaminase (SGPT)
AST aspartate serum transaminase. Also called, serum glutamate oxaloacetate transaminase (SGOT)
AUC area under the curve. AUC is a measure of exposure to a given drug, as determined by analysis of blood concentration over a period of hours, with a multiplicity of blood samples, taken after an oral or intravenous dose.
BCS class Biopharmaceutics Classification System. BCS class is used to designate or to predict the bioavailability of drugs that are taken orally. BCS class takes into account solubility in water and permeability to cell membranes.
BID Twice a day. From the Latin, bis in die.
BLA biologics license application
BMI body mass index
CBER Center for Biologics Evaluation and Research
CDER Center for Drug Evaluation and Research
CFR Code of Federal Regulations. The CFR provides rules, which take the form of "administrative law"
CLcr Creatinine clearance. See FDA. Guidance for Industry. Pharmacokinetics in Patients with Impaired Renal Function—Study Design, Data Analysis, and Impact on Dosing and Labeling; 2010.
Clinical "Clinical," as used in "clinical studies," "clinical effect," or "clinical trials," refers to use with humans, not animals
Clinical hold A Clinical Hold is a notification from FDA to the Sponsor to delay a proposed clinical trial or to suspend an ongoing clinical trial (21 CFR §312.42). Section 312.42 states, for example, "When an ongoing study is placed on Clinical Hold, no new subjects may be recruited to the study ... patients already in the study should be taken off therapy."
ClinPharm Review Clinical Pharmacology Review. These reviews are published on FDA's website, together with the Approval Letter, Medical Reviews, Pharmacology Reviews, and package label, when FDA grants approval to a drug.
CAZ/AVI Ceftazidime/avibactam combination. This is a medication taking the form of a combination of two drugs.
Combination Rule (21 CFR §300.50) "Two or more drugs may be combined in a single dosage form when each component makes a contribution to the claimed effect and the dosing of each component is such that the combination is safe and effective for a significant patient population" (see U.S. Dept. Health and Human Services. FDA. CBER. Guidance for Industry. Chronic Obstructive Pulmonary Disease: Developing Drugs for Treatment, 2007).
COX-2 cyclooxygenase-2
CRDAC Cardiovascular and Renal Drugs Advisory Committee
CRF Case Report Form
CRO Contract Research Organization
CSP Clinical Study Protocol, or "Protocol"
CTCAE Common Terminology Criteria for Adverse Events
CTD Common Technical Document

CYP Cytochrome P-450
DAARP Division of Analgesia, Anesthesia, and Rheumatology Products. DAARP is a division in the U.S. FDA
DDI drug–drug interaction. DDI refers to the influence of the study drug on metabolism of any second drug that is coadministered, and to the influence of any second drug that is coadministered on the study drug's metabolism.
DDI drug–disease interaction. DDI refers to the influence of a disease, such as renal impairment or hepatic impairment, on the study drug's exposure.
DLT dose-limiting toxicity. DLT finds a basis in FDA's Guidance for Industry. See FDA. Guidance for Industry. Estimating the Maximum Safe Starting Dose in Initial Clinical Trials for Therapeutics in Adult Healthy Volunteers; 2005.
DRESS Drug Reaction with Eosinophilia and Systemic Symptoms. Adverse events that are hypersensitivity reactions include DRESS, rashes, anaphylactic reactions, and Stevens-Johnson syndrome (SJS).
eCTD Electronic Common Technical Document
EGFR epidermal growth factor receptor
EMA European Medicines Agency
EMDAC Endocrinologic and Metabolic Drugs Advisory Committee
ESRD end stage renal disease
Exposure In pharmacokinetics studies "exposure" is usually a measure of concentration of a drug in the bloodstream. Exposure is expressed in terms of, AUC, $AUC_{0-24\ hours}$, $AUC_{0-infinity}$, C_{max}, C_{min} (C_{trough}), t_{max}, and so on.
FDA U.S. Food and Drug Administration
FDC fixed dose combination. 21 CFR §300.50 states that a fixed combination drug is, "Two or more drugs ... combined in a single dosage form when each component makes a contribution to the claimed effects." This textbook provides an account of FDA's approvals of these fixed dose combinations: lumacaftor/ivacaftor, ceftazidime/avibactam (AVYCAZ), sumatriptan/naproxen, canagliflozin/metformin, and elvitegravier, cobicistat, emtricitabine, tenofovir alafenamide (E/C/F/TAF).
FGF fibroblast growth factor
FPI Full Prescribing Information. FPI is part of the package label. See FDA. Guidance for Industry. Labeling for Human Prescription Drug and Biological Products—Implementing the PLR Content and Format Requirements; 2013.
GMR Geometric Mean Ratio is a unit used pharmacokinetics analysis. See Davit BM, *et al.* Comparing generic and innovator drugs. Ann. Pharmacother. 2009;43:1583–97.
HAHA human antihuman antibodies. In clinical studies on human subjects treated with a humanized therapeutic antibody, subjects sometimes respond by generating antibodies against the administered antibody. This issue is routinely assessed by FDA in its reviews of BLAs. A more general term is "antidrug antibodies."
HBV hepatitis B virus
HCV hepatitis C virus
HED human equivalent dose
5-HT 5-hydroxytryptamine
ICAC Independent Central Adjudication Committee
ICH International Conference on Harmonization
ITP idopathic thrombocytopenia purpura
ILD interstitial lung disease
IND Investigational New Drug application
Investigator "Investigator means an individual who actually conducts a clinical investigation ... the investigator is the responsible leader of the team" (21 CFR §312.3). The terms "sponsor" and "investigator" have different definitions in the CFR.
IR drug immediate release. IR drugs are distinguished from, for example, timed release drugs
ITT Intent to treat. ITT refers to the population of all of the subjects that were initially enrolled and randomized in a clinical trial. ITT analysis refers to the analysis of safety or efficacy results for the ITT population, without regard to whether a given subject complies with all of the requirements of the Clinical Study Protocol.
IVIVR In vitro–in vivo relationship. This is a term used in food effect studies. See Mathias N, *et al.* Food effect in humans: predicting the risk through in vitro dissolution and in vivo pharmacokinetic models. AAPS J. 2015;17:988–98.

LUM/IVA lumacaftor/ivacaftor combination. This is a medication taking the form of a combination of two different drugs.
MDRD Modification of Diet in Renal Disease. MDRD is an equation used to determine "estimated glomerular filtration rate" (eGFR) which, in turn, is a measure of renal function. See Stevens LA, *et al.* Measured GFR as a confirmatory test for estimated GFR. J. Am. Soc. Nephrol. 2009;20:2305–13.
MedDRA Medical Dictionary for Regulatory Activities
mITT modified intent to treat
MRSD maximum recommended starting dose
MTC medullary thyroid cancer
MTD maximum tolerable dose. The EMA provides a basis for the MTD. See European Medicines Agency. Guideline on the quality, non-clinical and clinical aspects 5 of gene therapy medicinal products; 2015.
NDA new drug application
NME new molecular entity. FDA provides a definition of NME in Lanthier M, *et al.* An improved approach to measuring drug innovation finds steady rates of first-in-class pharmaceuticals. Health Affairs 2013;32:1433–9.
NOAEL no observable adverse effects level. NOAEL is used in animal studies, where the goal is to arrive at an appropriate dose for humans, when initiating clinical studies on the drug.
NSAIDs nonsteroidal antiinflammatory drugs
NSCLC non-small-cell lung cancer
ONJ osteonecrosis of the jaw. ONJ is a type of adverse event, which was found with, for example, denosumab
PBMC peripheral blood mononuclear cell. PBMCs, which are acquired from samples of whole blood, include T cells, B cells, dendritic cells, and NK cells, but do not include red blood cells, platelets, neutrophils or eosinophils.
PD pharmacodynamics. PD refers to the time course of the body's response to an administered drug, for example, in the response of gene induction.
PDGF platelet-derived growth factor
P-gp P-glycoprotein. P-gp is a membrane-bound transporter that mediates efflux of xenobiotics and drugs out of the cell
PK pharmacokinetics
PMR postmarketing requirement. PMR refers to studies that are required as a condition for FDA approval, and that sponsors conduct after approval to gather additional information about a product's safety or efficacy. Where the PMR requires that additional studies be performed, the new data may appear on revised versions of the package label (21 CFR §314.81(2) (iii and vii))
Postmarketing Postmarketing refers, for example, to information gathered about a medication after it is approved by the FDA.
PP per rotocol. Per protocol analysis refers to an analysis of safety or efficacy results, where the analysis is conducted only on subjects that complied with all of the requirements set forth by the Clinical Study Protocol.
Preclinical "Preclinical" as used in "preclinical studies" refers to studies with animals, cultured cells, perfused organs or tissues, purified enzymes, mutagen testing with bacteria, and so on. "Preclinical" refers to studies conducted prior to an associated clinical trial, as well as to studies conducted once the clinical trial is under way. See FDA. Guidance for Industry S6 Preclinical Safety Evaluation of Biotechnology-Derived Pharmaceuticals; 1997.
PT Preferred Term. PT refers to a classification of terms in the MedDRA dictionary
QD Once a day. From the Latin, quaque die
Q2W Once in every 2 weeks
RECIST Response Evaluation Criteria in Solid Tumors. RECIST, which takes into account tumor size and number, is a standard criterion for assessing drug efficacy. See Eisenhauer EA, *et al.* New response evaluation criteria in solid tumours: revised RECIST guideline (version 1.1). Eur. J. Cancer. 2009;45:228–47.
REMS Risk Evaluation and Mitigation Strategy. REMS is part of FDA's drug approval process for some drugs, where the REMS imposes a set of requirements on the Sponsor during marketing of the drug, to be proactive in preventing or reducing drug toxicity
SAE serious adverse event
SEER Surveillance, Epidemiology, and End Results Database. This is a database of the National Cancer Institute. The database is sometimes used as a historic control, where the Sponsor does not have sufficient information from its own clinical trial, to assess association of tumors with the study drug. See the use of SEER data from package label for abatacept (ORENCIA (abatacept) for injection, for intravenous use). June 2017. BLA 125-118.

SMQ Standardised MedDRA Queries
SNOMED-CT Systematized Nomenclature of Medicine-Clinical Terms
SOC System Organ Class
SOF/VEL sofosbuvir/velpatasvir combination. SOF/VEL takes the form of a combination of two different drugs.
Sponsor "Sponsor means a person who takes responsibility for and initiates a clinical investigation. The sponsor may be an individual or pharmaceutical company, governmental agency, academic institution, private organization, or other organization. The sponsor does not actually conduct the investigation unless the sponsor is a sponsor-investigator" (21 CFR §312.3). The terms "sponsor" and "investigator" have different definitions in the CFR.
SSRI serotonin-specific reuptake inhibitor
Subject According to the Code of Federal Regulations, "Subject means a human who participates in an investigation, either as a recipient of the investigational new drug or as a control. A subject may be a healthy human or a patient with a disease." 21 CFR §312.3. Definitions and interpretations.
SUSAR Serious Unexpected Adverse Reaction
TEAE Treatment Emergent Adverse Event
TFLs Tables, figures, and listings. TFLs are source documents that are used for drafting the Clinical Study Report.
TMDD target-mediated drug disposition
TNF tumor necrosis factor
Trough concentration trough concentration refers to the lowest concentration (C_{min}) of a study drug in the bloodstream, where the drug is dosed on a periodic basis, for example, daily or weekly. Trough concentration is determined in blood withdrawn immediately before the next dosing.
ULN upper limit of normal. ULN refers to concentrations of an enzyme or metabolite in the bloodstream, for example, to activity of lactate dehydrogenase.
ULRR upper limit of the reference range
USC United States Code. The USC provides a body of statutes. Note that statutes, as provided by USC, are not the same as rules, which are provided by CFR.
USPI United States Prescribing Information
VEGF vascular endothelial growth factor
VEGFR vascular endothelial growth factor receptor
VHD valvular heart disease

CHAPTER

1

Introduction to Regulated Clinical Trials

I. INTRODUCTION

This book details the Food and Drug Administration's review process for pharmaceuticals, that is, for small molecule drugs and biologicals. The book reproduces data provided by the Sponsor, FDA's analysis and critique of the data, and pathways of logic and reasoning that result in FDA giving recommendations for the package label, and that result in FDA's approval of the drug and of the package label. The author acquired the data that had been submitted to FDA in 100 New Drug Applications (NDAs) and Biological License Applications (BLAs). The data are available on FDA's website in the form of various reviews. Thus, as source material for writing the book, the author read one hundred of FDA's Medical Reviews, one hundred of FDA's Clinical Pharmacology Reviews, and one hundred of FDA's Pharmacology Reviews. These documents are published on FDA's website, along with FDA's Approval Letter, package label, and Risk Evaluation and Mitigation Strategy.

As set forth in Title 21 of the Code of Federal Regulations (CFR), the main goal of FDA-regulated clinical trials is to establish that the study drug is safe and effective in humans.[1] The goal of efficacy is set forth in the Indications and Usage section and in the Dosage and Administration section of the package label, whereas the goal of providing safety is set forth in the Black Box Warning, Contraindications section, Warnings and Precautions section, Adverse Reactions section, and Drug Interactions section of the label. The package label has three large sections: (1) Highlights of Prescribing Information, (2) Full Prescribing Information Table of Contents, and (3) Full Prescribing Information. The first large section takes the form of bullet point information. The third large section reiterates what is in the first section but in the form of detailed paragraphs.

[1] 21 CFR §314.2. *Applications For FDA Approval to Market a New Drug*, requires that, "The purpose of this part is to establish an efficient and thorough drug review process in order to: (a) Facilitate the approval of drugs shown to be **safe and effective**; and (b) ensure the disapproval of drugs not shown to be **safe and effective**."

FDA's Drug Review Process and the Package Label
DOI: https://doi.org/10.1016/B978-0-12-814647-7.00001-4

The third section also contains clinical data and animal data of a neutral nature, that is, not intended as a warning and not intended as an instruction to the physician prescribing the drug. The formatting, organization, and type font of the package label have been outlined.[2]

This is the first book to use FDA's reviews as a source of information for the topics of safety, efficacy, clinical trial design, and package labeling. FDA's reviews that were used in writing this book include:

- Medical Reviews,
- Clinical Pharmacology Reviews,
- Pharmacology Reviews,
- Cross Discipline Team Leader Reviews,
- Summary Reviews, and
- Statistical Reviews.

FDA's recommendations for a given section of the package label are typically dispersed over these various reviews. In other words, FDA's train of logic and reasoning, for example, for arriving at the Warnings and Precautions section, occurs in fragments where each fragment resides in a different review. For this reason, the author needed to explore all of these reviews, especially where any steps of logic were missing in any one review.

The timeline for drug development and for subsequent FDA submissions includes these steps:

Step 1: Basic research is used to characterize the mechanism of action of a given disease, where data on mechanism are acquired from human subjects, animal models of the disease, cell culture models, gene expression data, and so on.
Step 2: When sufficient information on disease mechanism is at hand, the Sponsor designs a drug that acts at one or more steps in the mechanism, where the drug is expected to ameliorate one or more features of the disease.
Step 3: The Sponsor requests permission from FDA to conduct a clinical trial on human subjects, using the preferred drug candidate. Permission is requested by way of an application called Investigational New Drug (IND). As part of the request to conduct the clinical trials, the Sponsor drafts a Clinical Study Protocol. The Clinical Study Protocol is an instruction manual for physicians and other healthcare workers, which teaches all aspects of conducting the clinical trial, such as criteria for selecting study subjects and methods of dosing the investigational drug. The textbook, *Clinical Trials*[3]

[2] Watson KT, Barash PG. The New Food and Drug Administration drug package insert: implications for patient safety and clinical care. Anesth. Analg. 2009;108:211–8.

[3] Brody T. CLINICAL TRIALS Study Design; Endpoints & Biomarkers; Drug Safety; FDA & ICH guidelines, 2nd ed. Elsevier, Inc., New York, NY; 2016.

the CFR (21 CFR §312.21) describes the relationship between the Sponsor's clinical studies and the package label:

> Phase 3 studies are expanded controlled and uncontrolled trials. They are performed after preliminary evidence suggesting effectiveness of the drug has been obtained, and are intended to gather the additional information about effectiveness and safety that is needed to evaluate the overall benefit-risk relationship of the drug and to provide an adequate basis for physician **labeling**. Phase 3 studies usually include from several hundred to several thousand subjects contains about three chapters devoted to each of the topics found in a typical Clinical Study Protocol. These topics include efficacy endpoints, safety endpoints, instructions for randomization and allocation, inclusion criteria, exclusion criteria, dose modification instructions, the study schema, the run-in period as a feature of study design, and nature of the control arm, for example, a placebo control or a comparator drug control.

Before conducting the Phase 3 clinical study, the Sponsor can meet with FDA and discuss, among other things, the proposed package label (21 CFR §312.47). Thus, according to Section 312.47:

> the sponsor should submit background information on the sponsor's plan for Phase 3, including summaries of the Phase 1 and 2 investigations, the specific protocols for Phase 3 clinical studies, plans for any additional nonclinical studies, plans for pediatric studies, including a time line for protocol finalization, enrollment, completion, and data analysis, or information to support any planned request for waiver or deferral of pediatric studies, and, if available, **tentative labeling for the drug**.

The Sponsor's application (NDA; BLA) includes the Sponsor's proposed package label (21 CFR §314.50). Thus, according to Section 314.50:

> Content and format of an application. Applications ... are required to be submitted in the form and contain the information, as appropriate for the particular submission ... [a]n application for a new chemical entity will generally contain an application form, an index, a summary, five or six technical sections, case report tabulations of patient data, case report forms, drug samples, **and labeling**.

When submitting an NDA or BLA, the Sponsor submits the information in a format known as the Common Technical Document (CTD). The CTD includes the outcome of the conducted clinical studies, where this information is in the Sponsor's Clinical Study Reports and other documents. The Sponsor includes a draft package label in Module 1 of the CTD. Regarding Module 1, FDA's Guidance for Industry teaches:

> For the NDA, you should provide a copy of the **proposed labeling text** with annotations directing reviewers to the information in the summaries and other modules that support each statement in the labeling, as described in 21 CFR §314.50(c)(2)(i). The annotated labeling text should include the content of the labeling described under 21 CFR §201.57 and all text, tables, and figures used in the package insert.[4]

[4] US Department of Health and Human Services; Food and Drug Administration; Center for Drug Evaluation and Research (CDER); Center for Biologics Evaluation and Research (CBER). Guidance for Industry Submitting. Marketing Applications According to the ICH-CTD Format—General Considerations; 2001(15 pages).

FDA's Guidance for Industry provides additional information for drafting the package label, in the additional documents cited here.[5,6,7,8]

Step 4: As revealed by FDA's Medical Reviews, Clinical Pharmacology Reviews, and Pharmacology Reviews, FDA assesses the Sponsor's data on efficacy and safety, contemplates the Sponsor's proposed label, and then makes recommendations and requirements regarding the information needed on the label.

Step 5: When FDA grants approval, it issues an Approval Letter, where the letter approves the package label's text, methods for manufacturing the drug, and approval for interstate commerce of the drug. The importance of the package label is revealed by an excerpt from the Approval Letter for **simeprevir**, "We have completed review of this application, as amended. It is approved, effective on the data of this letter, for use as recommended in the enclosed **agreed-upon text.**[9]"

Similarly, the Approval Letter for **pembrolizumab** for melanoma read, "We have approved your BLA for ... pembrolizumab ... effective this date. You are hereby authorized to introduce or deliver for introduction into interstate commerce ... [u]nder this license, you are approved to manufacture Keytruda drug substance ... for use as recommended in the enclosed **agreed-upon text.**[10]"

The term "package label" refers to the information printed on the outside of the package, to the folded-up paper document that resides inside the package (the package insert), and also to the package label available on FDA's website. In these excerpts from FDA's Approval Letters, the phrase "agreed-upon text" refers to the text of the draft package label.

The above steps are summarized by the timeline in the flow chart:

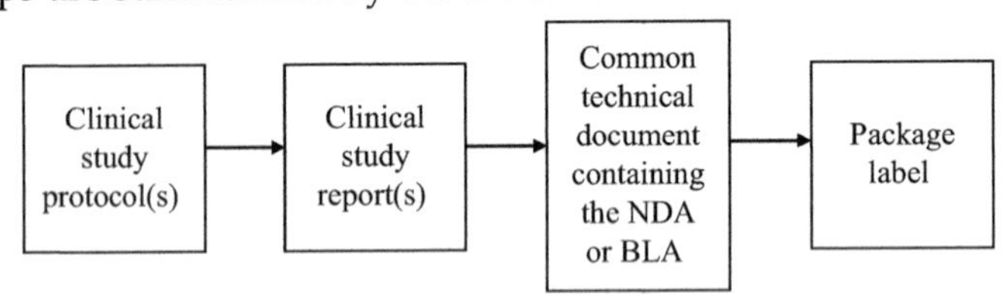

[5] US Department of Health and Human Services; Food and Drug Administration; Center for Drug Evaluation and Research (CDER); Center for Biologics Evaluation and Research (CBER). Guidance for Industry. Warnings and Precautions, Contraindications, and Boxed Warning Sections of Labeling for Human Prescription Drug and Biological Products—Content and Format; 2011 (13 pages).

[6] US Department of Health and Human Services; Food and Drug Administration; Center for Drug Evaluation and Research (CDER); Center for Biologics Evaluation and Research (CBER). Guidance for Industry Adverse Reactions Section of Labeling for Human Prescription Drug and Biological Products—Content and Format; 2006 (16 pages).

[7] US Department of Health and Human Services; Food and Drug Administration; Center for Drug Evaluation and Research (CDER); Center for Biologics Evaluation and Research (CBER). Guidance for Industry Labeling for Human Prescription Drug and Biological Products—Implementing the PLR Content and Format Requirements; 2013 (34 pages).

[8] US Department of Health and Human Services; Food and Drug Administration; Center for Drug Evaluation and Research (CDER). Guidance for Industry Revising ANDA Labeling Following Revision of the RLD Labeling; 2000 (6 pages).

[9] FDA Approval Letter (September 30, 2013) NDA 205-123 for semeprevir (Olysio®).

[10] FDA Approval Letter (September 4, 2014) BLA 125-514 for pembrolizumab (Keytruda®).

The following concerns the terminology used to refer to clinical studies that are conducted by the Sponsor. FDA appears to use the terms "study" and "trial" interchangeably. For example, FDA's Medical Review of osimertinib, the reviewer referred to "**Study** 5160C00001" and also referred to, "5160C00001 ... the secondary objectives of the **trial** included."[11] Further showing interchangeability of terms, FDA's review for lenvatinib referred to, "Tables of Studies/Clinical Trials" and also referred to, "The application is supported by a single major efficacy trial, **Study** E7080-G00-303 ... a randomized, placebo-controlled **trial.**[12]" Similarly, for FDA's Medical Review of pomalidomide, the review referred to, "Pooling of Data Across Studies/Clinical Trials to Estimate and Compare Incidence," to "certification for **study** CC-4047-MM-002," and to "possibility of submitting clinical **trials** CC-4047-MM-002."[13] FDA's Medical Review of pomalidomide, the review referred to, "Pooling of Data Across Studies/Clinical Trials to Estimate and Compare Incidence." The same review referred to, "Clinical **trial** CC-4047-MM-003."[14] FDA's Manual of Policies and Procedures[15] provides distinct definitions for "study" and "trial," but these definitions are not followed in FDA's Medical Reviews for NDAs and BLAs. This book uses the term "trial" when referring to a Phase I, Phase II, or Phase III trial, and uses the term "study" to refer to individual studies, where a given trial comprises a plurality of studies.

II. DETAILS ON FDA SUBMISSIONS

a. Clinical Study Protocol

The Clinical Study Protocol is the instruction manual used by investigators for conducting the clinical study. During the clinical study, data are produced on efficacy and safety. Initially, these data are memorialized on Case Report Forms (CRFs), and eventually the data are collated and analyzed, where the results of the analysis are written in the Clinical Study Report. Typically, the drug development process for any given study drug involves several different clinical studies, with the generation of several different Clinical Study Reports.

The raw data used to draft Clinical Study Reports take the form of tables, figures, and listings (TFLs), which are generated by the investigator. The medical writer then uses the TFLs, together with other documents such as the Clinical Study Protocol, Investigator's Brochures, and medical journal articles for drafting the Clinical Study Report.[16,17]

[11] NDA 208-065. Pages 22 and 33 of 137-page Medical Review.

[12] NDA 206-947. Page 176 of 185-page Medical Review.

[13] NDA 204-026. Page 11 of 101-page Medical Review.

[14] NDA 204-026. Pages 11 and 65 of 101-page Medical Review.

[15] Office of New Drugs; Center for Drug Evaluation and Research (CDER). Manual of Policies and Procedures. MAPP 6010.3 Rev. 1; December 2010 (110 pages).

[16] Miller S, Billiones R. Biostatistics and medical writing: synergy in preparing clinical trials documents. Medical Writing. European Medical Writers Association (EMWA). 2016;25:43–5.

[17] Hamilton S. Effective authoring of clinical study reports: a companion guide. Medical Writing. European Medical Writers Association (EMWA). 2014;23:86–92.

b. Common Technical Document

Eventually, the Sponsor may decide to format the information as an electronic Common Technical Document (eCTD), which can then be submitted to the FDA. For an FDA submission that is an NDA or BLA, the format takes the form of the eCTD.[18]

Fig. 1.1 shows the pyramid diagram used to illustrate the organization of the eCTD and its modules.[19] Module 5 of the pyramid contains Clinical Study Reports, whereas Module 4 contains data on efficacy and toxicology from animal studies. The Clinical Study Report contains many Appendices, where the first Appendix takes the form of the Clinical Study Protocol. In other words, the Clinical Study Protocol is attached to the Clinical Study Report which, in turn, resides in Module 5, which, in turn, is part of the eCTD.

Each Clinical Study Report may contain about 20 appendices, where the first appendix takes the form of the Clinical Study Protocol. As can be seen from the pyramid diagram, Module 4 contains data from studies on animals, cultured cells, tests for mutagens, and so on. Module 3 includes data from routine quality control work on the drug, the drug product (tablet or solution), and any excipients. Quality control work involves testing drug purity and stability with storage. The appendices for any given Clinical Study Report can include those named below in references.[20,21]

c. Finding Package Labels

FDA's website provides package labels and updated package labels that are published in the years following issue of the Approval Letter. The following steps provided access to the package label, FDA's Approval Letter, and to FDA's Medical Reviews, FDA's Clinical Pharmacology Reviews, FDA's Pharmacology Reviews, and other reviews:

1. Access www.fda.gov
2. Click "Drugs" tab

[18] US Department of Health and Human Services; Food and Drug Administration; Center for Drug Evaluation and Research (CDER); Center for Biologics Evaluation and Research (CBER). Guidance for Industry. Providing Regulatory Submissions in Electronic Format—Certain Human Pharmaceutical Product Applications and Related Submissions Using the eCTD Specifications; 2015 (11 pages).

[19] US Department of Health and Human Services; Food and Drug Administration; Center for Drug Evaluation and Research (CDER); Center for Biologics Evaluation and Research (CBER). Guidance for Industry. M4: Organization of the CTD; 2001 (7 pages).

[20] Appendix 1. Protocol and Protocol Amendments; Appendix 2. Sample Case Report Form (CRF); Appendix 3. Ethics Committee Members; Appendix 4. List of Key Study Personnel and Curricula Vitae; Appendix 5. Signature of Principal Investigator. Appendix 6. List of Subjects Receiving Test Drugs from Specific Batches; Appendix 7. Randomization Scheme; Appendix 8. Audit Certificate; Appendix 9. Statistical Methods; Appendix 10. Quality Assurance Procedures; Appendix 11. Subject Data Listings (Subject Disposition, Eligibility, Demographics, Medical History, Physical Examination, etc); Appendix 12. Pharmacokinetic Report; Appendix 13. Bioanalytic Report; Appendix 15. Pharmacodynamic Report; Appendix 16. Case Report Forms.

[21] The source of this list of appendices was a Clinical Study Report. This Clinical Study Report, together with its appendices, is about 1250 pages long.

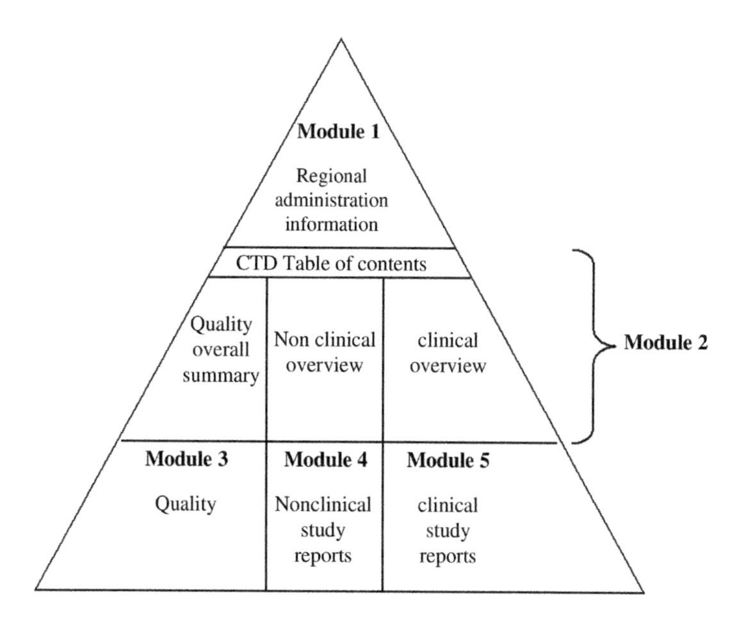

FIGURE 1.1 CTD. Pyramid diagram showing the CTD and its five modules. Module 1 is different from country to country and Modules 2–5 are the same for all countries.

3. Click "Search Drugs@FDA"
4. Type name of drug (chemical name or trade name)
5. Click on "Approval Date(s) and History, Letters, Labels, Reviews for NDA

Package labels can be found in the USPI. In response to a query from this author, an FDA official explained the meaning of USPI and the various sources of package labels:

> The term USPI" refers to "US Prescribing Information (USPI)." USPI is the printed labeling that is approved by FDA and accompanies each FDA approved drug. The prescribing information is sometimes called the package insert or labeling ... [m]ost labels can be found in our Drugs@FDA catalog, FDA Label Repository, and DailyMed (a National Institutes of Health managed site). Please note, Drugs@FDA only contains FDA-approved drugs, the FDA Label Repository and DailyMed searches may result in over the counter (OTC) and also unapproved drug labels.[22]

d. Raw Data Used for Drafting the Clinical Study Report

Raw data used for drafting Clinical Study Reports and other components of the eCTD take the form of CRFs, patient narratives, and TFLs. CRFs and listings are described in this book in Chapter 13, Coding. Patient narratives are described below, as well as in this book in the Chapter 12, Adjudication of Clinical Data, in the account of FDA's review of alvimopan NDA 021-775.[23]

[22] MM. Drug Information Specialist, Division of Drug Information, Center for Drug Evaluation and Research Food and Drug Administration (e-mail of March 28, 2017).

[23] Alvimopan (GI recovery following bowel resection surgery) NDA 021-775. The 383-page Medical Review contains three reviews. These are Special Safety Review (55 pages) on Pages 13–67 of Medical Review, Clinical Review (128 pages) on Pages 91–218 of Medical Review, and Clinical Review (124 pages) on Pages 257–380 of Medical Review.

According to MedEffect Canada, "The objective of the **narrative** is to summarize all relevant clinical and related information, including patient characteristics, therapy dates, medical history, clinical course of the event(s), diagnosis, and AR(s) including the outcome, laboratory evidence (including normal ranges), and any other information that supports or refutes an AR (e.g., rechallenge information). The **narrative** should serve as a comprehensive stand-alone medical story.[24]"

International Conference on Harmonization (ICH) E3 provides an account of patient narratives.[25] According to ICH E3, **patient narratives** on deaths and serious adverse events include, "the nature and intensity of event, the clinical course leading up to event, with an indication of timing relevant to test drug/investigational product administration; relevant laboratory measurements, whether the drug was stopped, and when; countermeasures; postmortem findings; investigator's opinion on causality, and sponsor's opinion on causality, if appropriate. In addition, the following information should be included: Patient identifier – Age and sex of patient ... Disease being treated ... with duration ... of illness – Relevant concomitant/previous illnesses with details of occurrence/duration – Relevant concomitant/previous medication with details of dosage – Test drug/investigational product administered, drug dose, if this varied among patients, and length of time administered.[26]"

III. LEGAL BASIS FOR REGULATED CLINICAL TRIALS

a. CFR, USC, and Courtroom Opinions

The source of law most frequently encountered by medical writers and other personnel in pharmaceutical companies is Title 21 of the CFR. The CFR, which provides "rules," is a form of administrative law, because it applies to one of the administrative branches of the federal government, namely, the FDA. Another source of law applying to drugs is the United States Code (USC). The USC consists of statutes, not rules. The relevant part is Title 21 of the USC. Title 21 of the USC, which concerns the Food and Drug Administration, is a codification of an act of congress. This act is the Federal Food, Drug, and Cosmetic Act.[27]

[24] Canada Vigilance Adverse Reaction Monitoring Program and Database, a program of MedEffect™ Canada. Guidance Document for Industry - Reporting Adverse Reactions to Marketed Health Products; 2011 (43 pages).

[25] International Conference on Harmonisation of Technical Requirements for Registration of Pharmaceuticals for Human Use. ICH Harmonised Tripartite Guideline. Structure and Content of Clinical Study Reports E3; 1995 (41 pages).

[26] International Conference on Harmonisation of Technical Requirements for Registration of Pharmaceuticals for Human Use. ICH Harmonised Tripartite Guideline. Structure and Content of Clinical Study Reports E3; 1995 (41 pages).

[27] Kleinfeld VA. Legislative history of the federal Food, Drug, and Cosmetic Act. Food Drug Law J. 1995;50:65.

Yet another source of law is published opinions from various courts. A large number of published cases provide guidance on the content the package label.[28] The issue of the adequacy of safety warnings on the package label is illustrated by the following excerpt from a courtroom opinion:

> Plaintiff asserts that the warnings on Zometa's label was inadequate to properly warn ... physicians about the risks of osteonecrosis of the jaw (ONJ) when using Zometa in its intended manner ... [the defendent] on the other hand, asserts that the warning was so comprehensive, no other warning would have caused Plaintiff's doctors to make a different prescribing decision ... the adequacy of warnings can become a question ... where the warning is accurate, clear, and unambiguous.[29]

b. Basis for the Package Label in the CFR

Regarding the package label, 21 CFR §314.50(c)(2)(ii) and 21 CFR §601.14(b) each sets forth the same requirements for labeling:

> **21 CFR §314.50(c)(2)(ii) Labeling.** The content of labeling required under §201.100(d)(3) of this chapter (commonly referred to as the package insert or professional labeling), including all text, tables, and figures, must be submitted to the agency in electronic format.
> **21 CFR §601.14 (b) Labeling.** The content of labeling required under §201.100(d)(3) of this chapter (commonly referred to as the package insert or professional labeling), including all text, tables, and figures, must be submitted to the agency in electronic format.

Section 201.100(d)(3), in turn refers to Sections 201.56, 201.57, and 201.80. Section 201.57, which is reproduced in part below, is the part of the CFR that most closely resembles the material in this book. In other words, the topics covered in this book track the requirements set forth by Title 21 CFR §201.57. Shown below are excerpts from the CFR that define these sections of the label. The topics are indicated by the bulletpoints, and the excerpts are indicated by the quotation marks. One unusual instruction, for example, applies to the Contraindications section. The instruction is that, if there are not any contraindications, then the Contraindications section of the package label must read, "none."

- Indications and usage
- Dosage and Administration
- Contraindications
- Warnings and Precautions
- Drug Interactions
- Use in Specific Populations
- Clinical Pharmacology

"**Indications and Usage.** A concise statement of each of the product's indications ... [m]ajor limitations of use (e.g., lack of effect in particular subsets of the population, or second line therapy status) must be briefly noted. If the product is a member of an

[28] Brody T. Package labels, in: Brody T. CLINICAL TRIALS Study Design, Endpoints and Biomarkers, Drug Safety, and FDA and ICH Guidelines, 2nd ed., Elsevier, New York; 2016, Pages 689–718.

[29] Franklin Cubbage, Plaintiff v. Novartis Pharmaceuticals Corp., Defendent. US District Court, M.D. Florida, July 5, 2016. Case No. 5:16-cv-129-0c-30PRL.

established pharmacologic class, the concise statement under this heading in Highlights must identify the class in the following manner: (Drug) is a (name of class) indicated for (indication(s))."

"**Dosage and Administration**. (i) This section must state the recommended dose and, as appropriate: (A) The dosage range, (B) An upper limit beyond which safety and effectiveness have not been established, or beyond which increasing the dose does not result in increasing effectiveness, (C) Dosages for each indication and subpopulation, (D) The intervals recommended between doses, (E) The optimal method of titrating dosage, (F) The usual duration of treatment when treatment duration should be limited, (G) Dosing recommendations based on clinical pharmacologic data (e.g., clinically significant food effects), (H) Modification of dosage needed because of drug interactions or in special patient populations (e.g., in children, in geriatric age groups, in groups defined by genetic characteristics, or in patients with renal or hepatic disease)."

"**Contraindications**. This section must describe any situations in which the drug should not be used because the risk of use (e.g., certain potentially fatal adverse reactions) clearly outweighs any possible therapeutic benefit. Those situations include use of the drug in patients who, because of their particular age, sex, concomitant therapy, disease state, or other condition, have a substantial risk of being harmed by the drug and for whom no potential benefit makes the risk acceptable. Known hazards and not theoretical possibilities must be listed (e.g., if severe hypersensitivity to the drug has not been demonstrated, it should not be listed as a contraindication). If no contraindications are known, this section must state *None*."

"**Warnings and Precautions** ... This section must describe clinically significant adverse reactions (including any that are potentially fatal, are serious even if infrequent, or can be prevented or mitigated through appropriate use of the drug) ... limitations in use imposed by them (e.g., avoiding certain concomitant therapy), and steps that should be taken if they occur (e.g., dosage modification) ... the labeling must be revised to include a warning about a clinically significant hazard as soon as there is reasonable evidence of a causal association with a drug; a causal relationship need not have been definitely established."

"**Drug Interactions**. (i) This section must contain a description of clinically significant interactions, either observed or predicted, with other prescription or over-the-counter drugs, classes of drugs, or foods (e.g., dietary supplements, grapefruit juice), and specific practical instructions for preventing or managing them. The mechanism(s) of the interaction, if known, must be briefly described. Interactions that are described in the "Contraindications" or "Warnings and Precautions" sections must be discussed in more detail under this section."

"**Use in Specific Populations**. This section must contain the following subsections ... **Pregnancy**. This subsection of the labeling must contain the following information in the following order under the subheadings "Pregnancy Exposure Registry," "Risk Summary," "Clinical Considerations," and "Data"... **Risk statement based on animal data**. When animal data are available, the Risk Summary must summarize the findings in animals and based on these findings, describe, for the drug, the potential risk of any adverse developmental outcome(s) in humans ... **Presence of drug in human milk**. The Risk Summary must state whether the drug and/or its active metabolite(s) are present in human milk ... **Geriatric use**. (A) A specific geriatric indication, if any, that is supported by adequate and well-controlled studies in the geriatric population must be described under the "Indications and Usage" section, and appropriate geriatric dosage must be stated under the "Dosage and Administration" section. The "Geriatric use" subsection must cite any limitations on the geriatric indication, need for specific monitoring ... associated with the geriatric indication."

"**Clinical Pharmacology**. (i) This section must contain information relating to the human clinical pharmacology and actions of the drug in humans. Pharmacologic information based on in vitro data using human biomaterials or pharmacologic animal models, or relevant details about in vivo study designs or results (e.g., drug interaction studies), may be included in this section if essential to understand dosing or drug interaction information presented in other sections of the labeling. This section must include the following subsections: (A) 12.1 Mechanism of action ... Pharmacokinetics ... Pharmacodynamics . . ."

c. FDA's Guidance for Industry

FDA's Guidance for Industry documents, as well as the other cited published commentaries, provide instructions for the following package label sections:

- Black Box Warning[30,31,32]
- Adverse Events section[33]
- Clinical Pharmacology section[34]
- Clinical Studies section[35]

[30] US Department of Health and Human Services; Food and Drug Administration; Center for Drug Evaluation and Research (CDER); Center for Biologics Evaluation and Research (CBER). Guidance for Industry. Warnings and Precautions, Contraindications, and Boxed Warning Sections of Labeling for Human Prescription Drug and Biological Products—Content and Format; 2011 (13 pages).

[31] O'Connor NR. FDA boxed warnings: how to prescribe drugs safety. Am. Fam. Physician. 2010;81:298–303.

[32] Panagiotou OA, et al. Different black box warning labeling for same-class drugs. J. Gen. Intern. Med. 2011;26:603–10.

[33] US Department of Health and Human Services; Food and Drug Administration; Center for Drug Evaluation and Research (CDER); Center for Biologics Evaluation and Research (CBER). Guidance for Industry. Adverse Reactions Section of Labeling for Human Prescription Drug and Biological Products—Content and Format; 2006 (13 pages).

[34] US Department of Health and Human Services; Food and Drug Administration; Center for Drug Evaluation and Research (CDER); Center for Biologics Evaluation and Research (CBER). Guidance for Industry. Clinical Pharmacology Section of Labeling for Human Prescription Drug and Biological Products—Content and Format; 2016 (16 pages).

[35] US Department of Health and Human Services; Food and Drug Administration; Center for Drug Evaluation and Research (CDER); Center for Biologics Evaluation and Research (CBER). Guidance for Industry. Clinical Studies Section of Labeling for Human Prescription Drug and Biological Products—Content and Format; 2006 (22 pages).

- Dosage and Administration section[36]
- Drug Class section[37]

d. Clinical Study Protocol, IND, NDA, and BLA

Title 21 of the CFR sets forth various requirements for conducting clinical trials, such as requirements for the Clinical Study Protocol. Regarding the need for a Clinical Study Protocol, the CFR states:

> (iii) A protocol is required to contain ... a statement of the objectives and purpose of the study. (b) The name and address and a statement of the qualifications (curriculum vitae or other statement of qualifications) of each investigator ... criteria for patient selection and for exclusion of patients and an estimate of the number of patients to be studied ... description of the design of the study, including the kind of control group to be used, if any.[38]

Before conducting a clinical trial, the Sponsor needs to submit an IND to the FDA. This requirement is set forth in 21 CFR §312.20, which states that:

> A sponsor shall submit an IND to FDA if the sponsor intends to conduct a clinical investigation with an investigational new drug ... [a] sponsor shall not begin a clinical investigation ... until the investigation is subject to an IND.

Phase 1, Phase 2, and Phase 3 clinical trials are described in 21 CFR §312.22. The content of the IND, which is set forth in 21 CFR §312.23, includes FDA's Form-1571, name of the Sponsor, names of Contract Research Organizations (CROs) (if any), data from animal studies such as data on efficacy and toxicity, and data from prior human studies, if any.

Moving ahead in the FDA submission timeline, the CFR also details procedures that are followed after the Phase 1, Phase 2, and Phase 3 clinical trials are completed, namely, the submission of an NDA or BLA. The data gathered during the animal studies and human clinical trials of an IND become part of the NDA or BLA. The NDA and BLA find a basis in 21 CFR §315.50, which sets forth the requirements that these applications include animal

[36] US Department of Health and Human Services; Food and Drug Administration; Center for Drug Evaluation and Research (CDER); Center for Biologics Evaluation and Research (CBER). Guidance for Industry. Dosage and Administration Section of Labeling for Human Prescription Drug and Biological Products—Content and Format; 2010 (10 pages).

[37] US Department of Health and Human Services; Food and Drug Administration; Center for Drug Evaluation and Research (CDER); Center for Biologics Evaluation and Research (CBER). Guidance for Industry and Review Staff. Labeling for Human Prescription Drug and Biological Products—Determining Established Pharmacologic Class for Use in the Highlights of Prescribing Information; 2009 (9 pages).

[38] 21 CFR 312.23. IND content and format. (6) Protocols.

pharmacology and toxicology data, human pharmacokinetics data, and clinical data on efficacy and safety.

The transition between completing the clinical trials and the deciding to draft and submit an NDA or BLA has been described as, "When the sponsor of a new drug/biologic believes that enough evidence on the drug's safety and effectiveness has been obtained to meet the FDA's requirements for marketing approval, the sponsor submits to FDA a new drug application ... [a] new drug application/biological licensure application (NDA/BLA; for the FDA) or a marketing authorization application (MAA; for the EU) is a comprehensive submission of all relevant information required for a drug to be approved for marketing.[39]"

The length of time between the date of submission of the NDA or BLA to the FDA to the date of FDA's approval, that is, to the date of FDA's Approval Letter is about 7 years.[40] Regarding the request to market a drug in the European Union (EU), background on the Marketing Authorization Application is provided in References.[41,42,43]

The drug development and approval process includes designing and testing candidate drugs, such as small molecule drugs and biologicals. Testing includes nonclinical studies, such as those using cultured cells or animals, and clinical studies (human subjects). In regulated clinical trials, the term "subject" or "study subject" is used (the term "patient" is not used).[44]

Activities coordinated with drug development and FDA submissions include the preparation and submission of publications for various journals as well as preparing and filing patent applications. Funding for a company can be facilitated by publications, by having filed a patent application, and by succeeding in winning allowance patent claims and the

[39] Senderowicz AM, Pfaff O. Similarities and differences in the oncology drug approval process between FDA and European Union with emphasis on *in vitro* companion diagnostics. Clin. Cancer Res. 2014;20:1445–52.

[40] Lanthier ML, et al. Accelerated approval and oncology drug development timelines. J. Clin. Oncol. 2010;28:e226–7.

[41] European Medicines Agency (EMA). European Medicines Agency pre-authorisation procedural advice for users of the centralised procedure; 2016 (349 pages).

[42] Hartman M, et al. Approval probabilities and regulatory review patterns for anticancer drugs in the European Union. Crit. Rev. Oncol. Hematol. 201387:112–21.

[43] Kingham, et al. Key regulatory guidelines for the development of biologics in the United States and Europe: Development and Strategies (ed. W. Wang and M. Singh) John Wiley and Sons, Inc., New York; 2014.

[44] US Department of Health and Human Services; Food and Drug Administration; Center for Drug Evaluation and Research (CDER); Center for Biologics Evaluation and Research (CBER); Center for Devices and Radiological Health (CDRH). Guidance for Industry Investigator Responsibilities—Protecting the Rights, Safety, and Welfare of Study Subjects; 2009 (15 pages).

issue of one or more patents. The cited articles provide an introduction to patenting.[45,46,47,48,49,50,51,52]

Pharmaceutical companies acquire guidance for planning and conducting clinical trials from FDA's Guidance for Industry and from ICH Guidelines. FDA's Guidance for Industry and ICH Guidelines take the form of over 100 documents, each of a length of about 10-50 pages. ICH is not a regulatory agency, but instead it is called an initiative.[53,54] ICH uses the regulatory authorities of the United States, Europe, and Japan together with experts from the pharmaceutical industry. ICH Guidelines are available for drug safety, drug efficacy, and for quality control and manufacturing.

e. Briefing Document

The Briefing Document is also called a Briefing Package and a Briefing Book. The Sponsor drafts the Briefing Document and submits it to FDA for use during meetings with the Sponsor and FDA. The Briefing Document finds a basis in FDA's Guidance for Industry.[55] The Sponsor prepares the Briefing Document for use by FDA's panelists during the meeting with FDA's Advisory Committee. Information in the Briefing Document

[45] Brody T. Patents, in: CLINICAL TRIALS Study Design; Endpoints & Biomarkers; Drug Safety; FDA & ICH Guidelines, 2nd ed., Elsevier, Inc., New York, NY; 2016, Pages 829–46.

[46] Brody T. Rebutting obviousness rejections by way of anti-obviousness case law. J. Pat. Trademark Off. Soc. 2017;99:1–57.

[47] Brody T. Enabling claims under 35 USC §112 to methods of medical treatment or diagnosis, based on *in vitro* cell culture models and animal models. J. Pat. Trademark Off. Soc. 2015;97:328–411.

[48] Brody T. Claims to ranges, the result-effective variable, and *In re Applied Materials*. J. Pat. Trademark Off. Soc. 2016;98:618–726.

[49] Brody T. Rebutting obviousness rejections based on impermissible hindsight. J. Pat. Trademark Off. Soc. 2014;96:366–424. For publishing this article, the author received the 2016 Rossman Memorial Award. The award was presented to the author at a ceremony held Feb. 10, 2016 at the United States Patent and Trademark Office in Alexandria, VA.

[50] Brody T. Functional elements in patent claims, as construed by the Board of Patent Appeals and Interferences. John Marshall Rev. Intel. Prop. Law 2014;13:251–320.

[51] Brody T. Negative claim limitations in patent claims. Am. Intel. Prop. Law Assoc. Q. Rev. 2013;41:29-72.

[52] Brody T. Obviousness in patents following the US Supreme Court's decision of KSR International Co. v. Teleflex, Inc. J. Pat. Trademark Off. Soc. 2010;92:26–70.

[53] Pugsley MK, et al. Principles of safety pharmacology. Br. J. Pharmacol. 2008;154:1382–99.

[54] ICH. Understanding MedDRA. The Medical Dictionary for Regulatory Authorities; 2013 (20 pages).

[55] US Department of Health and Human Services; Food and Drug Administration; Center for Drug Evaluation and Research (CDER). Guidance for Industry and FDA Staff Qualification Process for Drug Development Tools; 2014 (32 pages).

supports the messages in the sponsor's presentation during the meeting and supports the desired labeling language.[56] The Briefing Document must be submitted 22 business days prior to the meeting.[57]

Briefing Documents are quoted at later points in this book, as part of this book's accounts of the FDA submissions for miltefosine,[58,59] ticagrelor,[60,61] dapagliflozin,[62] and two separate Briefing Documents for denosumab.[63,64]

f. The Sponsor and the Investigator

This distinguishes the sponsor and the investigator.[65] According to the CFR, the sponsor is, "A person who takes responsibility for and initiates a clinical investigation. The sponsor may be an individual or pharmaceutical company, governmental agency, academic institutions, private organization, or other organization. The sponsor does not actually conduct the investigation.[66]"

Also, according to the CFR, an investigator is, "an individual who actually conducts a clinical investigation under whose immediate direction the drug is administered or dispensed to a subject.[67]" FDA's Guidance for Industry distinguishes between the sponsor and investigator, "A sponsor takes responsibility for and initiates a clinical investigation. A sponsor can be an individual or pharmaceutical company ... academic institution ... or other organization. An investigator is the individual who actually

[56] Cox V, Scott MC. FDA Advisory Committee Meetings: what they are, why they happen, and what they mean for regulatory professionals. Regulatory Rapporteur. 2014;11:5–8.

[57] US Department of Health and Human Services; Food and Drug Administration. Guidance for Industry Advisory Committee Meetings—Preparation and Public Availability of Information Given to Advisory Committee Members; 2008 (20 pages).

[58] NDA 204-684 Miltefosine (Impavido®) for the Treatment of Visceral, Mucosal and Cutaneous Leishmaniasis. FDA Briefing Document for the AntiInfective Drugs Advisory Committee Meeting October 18, 2013. Page 8 of 43-page Briefing Document.

[59] Miltefosine (visceral leishmaniasis) NDA 204–684.

[60] AstraZeneca (June 2010) Ticagrelor NDA 22-433 Briefing Document for Cardiovascular and Renal Drugs Advisory Committee Meeting. Page 102 of 344-page Briefing Document.

[61] Ticagrelor (thrombic events with acute coronary syndromes) NDA 022-433.

[62] Dapagliflozin (type-2 diabetes mellitus) NDA 202-293. Pages 132 and 133 of 494-page Medical Review.

[63] Briefing Document. Background Document for Meeting of Advisory Committee for Reproductive Health Drugs (August 13, 2009) Denosumab (Proposed trade name: PROLIA) Amgen, Inc. (Page 22 of 87 pages).

[64] Briefing Document. Background Information for the Meeting of the Advisory Committee for Reproductive Health Drugs. August 13, 2009 (Page 32 of 147 pages).

[65] American Academy of Pediatrics. Off-label use of drugs in children. Pediatrics. 2014;133:563–7.

[66] Code of Federal Regulations. 21 CFR §312.3.

[67] Code of Federal Regulations. 21 CFR §312.3.

conducts the investigation.[68]" Also, the Sponsor may train the investigators in the conduct of the study.[69]

g. Legal Basis for Good Clinical Practices

FDA's Guidance for Industry publications are only recommendations and do not have the force of the law. On the other hand, if a pharmaceutical company does not comply with the routines and activities set forth under FDA's Guidance—Good Clinical Practice, FDA can take legal action against the pharmaceutical company by issuing a Warning Letter and by forcing the company to shut down all operations.[70]

FDA's Guidance for Industry—Good Clinical Practices[71] sets forth requirements for activities such as adverse event reporting and drug accountability. Regarding adverse event reporting, Good Clinical Practice requires that, "The sponsor should promptly notify ... the regulatory authority of findings that could affect adversely the safety of subjects."

Regarding drug accountability, Good Clinical Practices requires that, "The sponsor ... [e]nsure timely delivery of investigational product(s) to the investigator ... [m]aintain records that document shipment, receipt, disposition, return, and destruction of the investigational product.[72]"

An accurate and practice view of Good Clinical Practice is provided by the many Warning Letters that are issued by the FDA to various Sponsors. A comprehensive account of FDA's Warning Letters, as they apply to Good Clinical Practices, has been published.[73]

The European Medicines Agency (EMA) has published a report of Good Clinical Practices violations by region, for years 2000-12.[74] The regions covered in this report

[68] US Department of Health and Human Services; Food and Drug Administration; Center for Drug Evaluation and Research (CDER); Center for Biologics Evaluation and Research (CBER). Guidance for Industry. Investigational New Drug Applications Prepared and Submitted by Sponsor-Investigators; May 2015 (28 pages).

[69] US Department of Health and Human Services; Food and Drug Administration; Center for Drug Evaluation and Research (CDER); Center for Biologics Evaluation and Research (CBER); Center for Devices and Radiological Health (CDRH). Guidance for Industry Investigator Responsibilities—Protecting the Rights, Safety, and Welfare of Study Subjects; October 2009 (18 pages).

[70] Brody T. Warning Letters, in: Clinical Trials (Study Design, Endpoints and Biomarkers, Drug Safety, and FDA and ICH Guidelines), 2nd ed., Elsevier, New York, NY; 2016, Pages 719–80.

[71] US Department of Health and Human Services; Food and Drug Administration; Center for Drug Evaluation and Research (CDER); Center for Biologics Evaluation and Research (CBER). Guidance for Industry E6 Good Clinical Practice: Consolidated Guidance; 1996 (63 pages).

[72] US Department of Health and Human Services; Food and Drug Administration; Center for Drug Evaluation and Research (CDER); Center for Biologics Evaluation and Research (CBER). Guidance for Industry E6 Good Clinical Practice: Consolidated Guidance; 1996 (63 pages).

[73] Brody T. Warning Letters, in: Clinical Trials (Study Design, Endpoints and Biomarkers, Drug Safety, and FDA and ICH Guidelines), 2nd ed., Elsevier, New York, NY; 2016, Pages 719–80.

[74] European Medicines Agency. Classification and analysis of the GCP inspection findings of GCP inspections conducted at the request of the CHMP (Inspection reports to EMA 2000-2012). European Medicines Agency, 30 Churchill Place, London, UK; 2014 (50 pages).

include the United States, EU, Canada, Africa, Eastern Europe, and South America. For example, regarding Eastern Europe, the report stated that, "A total of 9 inspections have been carried out in this area, reporting 146 findings ... [a] total of 18 critical findings were reported in this area in 4 of the inspected investigational sites in this area and hence 5 inspections did not record any critical finding." The violations in Good Clinical Practice ("findings") in Eastern Europe concerned, for example, supplying and storage of the study drug, informed consent process, and compliance with the Clinical Study Protocol. EMA's report also provided an account as to whether the violations were those of the sponsor, the investigator, the CRO, or the analytical laboratory.

IV. DRAFTING FDA SUBMISSIONS

This book addresses the need for the Director of Regulatory Affairs, the Chief Medical Officer, medical writers, and other company personnel, to obtain a working knowledge of the following topics:

- **Package labels.** FDA's decision-making processes used in arriving at the final draft of the package label.
- **FDA's critiques of the Sponsor's clinical trial design.** FDA's decision-making processes, where it criticizes the Sponsor's clinical trial design or where it instructs the Sponsor to conduct additional studies.
- **Interpreting ambiguous data.** FDA's decision-making processes in navigating through ambiguous areas, on the topics of safety, efficacy, and clinical trial design.
- **Topics frequent in FDA-submissions but rare in medical journals.** This book reveals topics frequently encountered in FDA submissions but not routinely encountered in medical journals. The term "rest of world (ROW)" is an accepted term used in FDA submissions to distinguish study sites in the United States from study sites in Europe and Asia.[75,76,77] But the term "rest of world" is not often used in medical journals.[78] The terms "System Organ Class (SOC)" and "Preferred Term" are routinely used in FDA submissions when disclosing adverse events and for dividing adverse events into

[75] Bloomgren G, et al. Risk of natalizumab-associated progressive multifocal leukoencephalopathy. N. Engl. J. Med. 2012;366:1870–80.

[76] Smith RE, et al. Darbepoetin alpha for the treatment of anemia in patients with active cancer not receiving chemotherapy or radiotherapy: results of a phase III, multicenter, randomized, double-blind, placebo-controlled study. J. Clin. Oncol. 2008;26:1040–50.

[77] Hack SP, et al. HGF/MET-directed therapeutics in gastroesophageal cancer: a review of clinical and biomarker development. Oncotarget 2014;5:2866–80.

[78] The terms "rest of world" and "rest-of-world" occurs in only one article in *New England Journal of Medicine*, from the past 50 years (search conducted December 4, 2016).

various categories.[79,80] But the "System Organ Class"[81] and "Preferred Term"[82] are not often used in medical journals. Similarly, the term "treatment emergent adverse events (TEAEs)" is commonly used in FDA submissions as a format for disclosing adverse events. Yet another example is[83] the subject of Intrinsic Factors and Extrinsic Factors.[84] However, these terms are rarely used in medical journals.[85]

V. DISTINGUISHING PACKAGE LABELING OF PHARMACEUTICALS FROM OTHER LABELING ISSUES

a. Off-Label Uses

Once a given drug receives FDA approval, it is common for physicians to use their own clinical judgment and prescribe the drug for diseases or for subgroups of patients that deviate from the disease and patient subgroup required by the package label.[86] Off-label uses also encompass the use via routes of administration or dose amounts different from that required by the package label.[87] The most common off-label uses for anticancer drugs

[79] US Department of Health and Human Services; Food and Drug Administration; Center for Drug Evaluation and Research (CDER); Center for Biologics Evaluation and Research (CBER). Guidance for Industry. E9 Statistical Principles for Clinical Trials; September 1998 (46 pages).

[80] US Department of Health and Human Services; Food and Drug Administration; Center for Drug Evaluation and Research (CDER); Center for Biologics Evaluation and Research (CBER). Guidance for Industry. E2F Development Safety Update Report; August 2011 (39 pages).

[81] "System organ class" or "system organ classes" occurs in only nineteen articles in *New England Journal of Medicine* from the past 50 years (search conducted December 4, 2016).

[82] "Preferred term" and "preferred terms" occur in only 39 articles in *New England Journal of Medicine* from the past 50 years (search conducted December 4, 2016).

[83] US Department of Health and Human Services; Food and Drug Administration; Center for Drug Evaluation and Research (CDER). Reviewer Guidance Conducting a Clinical Safety Review of a New Product Application and Preparing a Report on the Review; February 2005 (84 pages).

[84] A search of *New England Journal of Medicine* for the past 20 years revealed only six articles with the term Intrinsic Factor or Intrinsic Factors. Similarly, a search of *Journal of Clinical Oncology* for the past 20 years revealed only seven articles with the term Intrinsic Factors. The searches were conducted January 11, 2017.

[85] A search in *New England Journal of Medicine* using the query terms "treatment emergent," "treatment-emergent," "TEAE," and "TEAEs," revealed that only 28 articles included these terms, in the past 50 years (search conducted December 4, 2016).

[86] Hamel S, et al. Off-label use of cancer therapies in women diagnosed with breast cancer in the United States. SpringerPlus 2015;4:209 (9 pages).

[87] Pfister DG. Off-label use of oncology drugs: the need for more data and then some. J. Clin. Oncol. 2012;30:584–6.

are for rituximab, gemcitabine, and bevacizumab.[88] Other common off-label uses are for anticonvulsants, antipsychotic drugs, and antibiotics.[89]

This concerns the issue of drug expense as a source of motivation for an off-label use. To provide the example of angiogenesis blocking drugs, ranibizumab is an antibody with a structure based on that of the antibody bevacizumab. Ranibuzumab and bevacizumab each bind to vascular endothelial growth factor (VEGF), where the outcome is blocking the formation of new blood vessels (angiogenesis). Angiogenesis is part of the pathology of solid tumors and also part of the pathology of age-related macular degeneration (blindness). But bevacizumab is much less expensive than ranibizumab and, as a consequence, off-label use of bevacizumab for age-related macular degeneration has become common. According to one commentator, the lack of FDA approval of bevacizumab for treating age-related macular degeneration creates the need for "heightened surveillance for systemic adverse effects" when using bevacizumab for this non-FDA-approved indication.[90]

This concerns the issue where existing drugs are not effective and where the physician explores an off-label use of another drug. This provides the example of drugs for systemic lupus erythematosus (SLE), an autoimmune disease. SLE results from overactivity of B cells. B cells are a type of lymphocyte that secretes antibodies into the bloodstream. Most therapies for SLE are off-label, where the relevant drugs are various immunosuppressants. Rituximab is an antibody that binds to CD20 of B cells, resulting in depletion of B cells. But rituximab has not been FDA-approved for the indication of SLE. Rituximab is used for off-label treatment of about 1% of SLE patients, and specifically, where other drugs have not worked for the patient being treated.[91]

Yet another source of motivation for off-label uses is where the package label does not encompass a particular patient subgroup, such as the pediatric population. Off-label uses are common in pediatric patients, where the most common deviation from the package label instructions is the dose amount.[92,93] The relative lack of FDA-approved drugs where the label's indication is for children is due to small number of available study subjects, low financial incentive, and ethical concerns.[94] The drugs used for off-label indications for

[88] Pfister DG. Off-label use of oncology drugs: the need for more data and then some. J. Clin. Oncol. 2012;30:584–6.

[89] Stafford RS. Regulating off-label drug use—rethinking the role of the FDA. N. Engl. J. Med. 2008;358:1427–9.

[90] Schmucker C, et al. A safety review and meta-analyses of bevacizumab and ranibizumab: off-label versus gold standard. PLoS One 2012;7:e42701. DOI: 10.13701.

[91] Ryden-Aulin M, et al. Off-label use of rituximab for systemic lupus erythematosus in Europe. Lupus Sci. Med. 2015;3:e000163. DOI. 10.1136.

[92] Balan S, et al. Awareness, knowledge and views of off-label prescribing in children: a systematic review. Br. J. Clin. Pharmacol. 2015;80:1269–80.

[93] Luedtke KE, Buck ML. Evaluation of off-label prescribing at a children's rehabilitation center. J. Pediatr. Pharmacol. Ther. 2014;19:296–301.

[94] Czaja AS, et al. Patterns of off-label prescribing in the pediatric intensive care unit and prioritizing future research. J. Pediatr. Pharmacol. Ther. 2015;20:186–96.

children include dexmedetomidine, dopamine, hydromorphone, ketamine, lorazepam, methadone, milrinone, and oxycodone.[95]

Pharmaceutical companies are able to recoup their investment, and sometimes earn profit, as long as they own an issued patent. However, once the term of the patent expires, companies will have lost their incentive to perform additional clinical studies for proving efficacy and safety for a new indication of the old drug. The result is that physicians are forced to use their own judgment (rather than rely on judgment of FDA reviewers) in deciding to employ an off-label use.[96]

Off-label uses can best be justified where a medical journal published data showing efficacy and safety for the off-label use. This statement refers to publications reporting clinical data that had not been submitted to the FDA, that is, by way of an NDA or BLA.[97]

Investigators affiliated with clinics, hospitals, or pharmaceutical companies are free to devise and conduct **clinical studies for off-label uses**. Mullins et al.[98] provides a guideline for clinical studies for off-label uses. The guideline for this type of study design is quite similar to the typical study design used for acquiring data on efficacy and safety for submitting to the FDA, but with a notable exception. The exception is the recommendation that clinical studies for off-label indications make use of **community-based trial sites** (unlike clinical studies for FDA-approval, which are generally based in academic centers).

In a communication to this author, Prof. Mullins explained that this difference is to facilitate insurance coverage of the patients who eventually need compensation for the off-label use. The explanation was that, "The issue that we were trying to address is the evidence the FDA requires v. what insurers want for their coverage decisions. Some of this is about internal v external validity. **Insurers want to know about the value of drugs in broader populations**, which includes those treated in community settings. However, **for regulatory purposes, it is easier for drug manufacturers to focus on narrower populations** and study sites.[99]"

An issue with off-label uses occurs where pharmaceutical companies issue advertisements or distribute publications on those uses. This practice has been called, "substituting marketing for science." This issue can arise after a company wins FDA approval of a drug for a given indication and then promotes use of the same drug but for a different indication.[100]

[95] Czaja AS, et al. Patterns of off-label prescribing in the pediatric intensive care unit and prioritizing future research. J. Pediatr. Pharmacol. Ther. 2015;20:186–96.

[96] Pfister DG. Off-label use of oncology drugs: the need for more data and then some. J. Clin. Oncol. 2012;30:584–6.

[97] Pfister DG. Off-label use of oncology drugs: the need for more data and then some. J. Clin. Oncol. 2012;30:584–6.

[98] Mullins CD, et al. Recommendations for clinical trials of off-label drugs used to treat advanced cancer. J. Clin. Oncol. 2012;30:661–6.

[99] E-mail of April 29, 2017 from Dr. C. Daniel Mullins, University of Maryland School of Pharmacy, Baltimore, MD.

[100] Robertson C, Kesselheim AS. Regulating off-label promotion—a critical test. N. Engl. J. Med. 2016;375:2313–5.

TABLE 1.1 NCCN Drugs and Biologics Compendium® Multiple Myeloma Version 1.22016

FDA-Approved Indication (On-Label Uses)	NCCN Recommended Use (Off-Label Uses)
BENDAMUSTINE	
Chronic Lymphocytic Leukemia (CLL): Bendamustine ... for injection is indicated for the treatment of patients with chronic lymphocytic leukemia ... **Non-Hodgkin's Lymphoma:** Bendamustine ... for injection is indicated for the treatment of patients with indolent B-cell non-Hodgkin's lymphoma that has progressed during or within six months of treatment with rituximab.	Therapy for previously treated **myeloma** ... for disease relapse or for progressive or refractory disease as single agent [or] in combination with lenalidomide and dexamethasone.
BORTEZOMIB FOR TREATING MULTIPLE MYELOMA	
Bortezomib for injection is indicated for the treatment of patients with **multiple myeloma**. Bortezomib for injection is indicated for the treatment of patients with **mantle cell lymphoma**.	Primary chemotherapy for progressive solitary plasmacytoma or smoldering myeloma (asymptomatic) that has progressed to active (symptomatic) **myeloma** in combination with dexamethasone with or without cyclosporine, doxorubicin, lenalidomide, or thalidomide for transplant candidates ... [or] combination with dexamethasone with or without cyclophosphamide or lenalidomide or in MPB (melphalan, prednisone, and bortezomib) regimen for nontransplant candidates.

From E-mail of May 3, 2017 from Dr. Marian Birkeland, PhD, National Comprehensive Cancer Network® (NCCN®) Fort Washington, PA 19034.

National Comprehensive Cancer Network (NCCN) provides lists of off-label uses for anticancer drugs. In response to an inquiry from this author, NCCN responded that:

> The NCCN Clinical Practice Guidelines contain recommendations for ... therapies for a particular disease indication without reference to the indications found on the FDA label ... a comparison of on- and off- label usage ... [is provided by] the NCCN Drugs and Biologics Compendium. This is a resource that lists all agents recommended within NCCN guidelines, along with information about the NCCN recommended usage as well as the indication from the FDA label.[101]

Table 1.1 provides two excerpts from NCCN, where these are for the anticancer drugs—bendamustine, for multiple myeloma, and bortezomib, for multiple myeloma. Table 1.1 compares the FDA-approved package label indication with the NCCN recommended use, which encompasses both on- and off-label uses.

b. Food and Dietary Supplement Labeling

To provide additional perspective and contrast to pharmaceutical labeling, packaged foods and dietary supplements also contain FDA-regulated package labels, where these are printed on the package (there is no package insert). FDA's Guidance for Industry provides information for food package labeling. However, food labels are never submitted to

[101] E-mail of May 3, 2017 from Dr. Marian Birkeland, PhD, National Comprehensive Cancer Network® (NCCN®) Fort Washington, PA.

FDA for approval, and there is no requirement or avenue for their submission for approval.[102] Once the food is on the market, and if FDA becomes aware of a defect in the labeling format or if the label makes an inappropriate claim that the food can prevent any disease, FDA will issue a Warning Letter to the food company and threaten to shut down the company's operations.[103]

A common problem is that food or dietary supplement package labels state that the product can be used to treat cancer. When this type of package label comes to FDA's attention, FDA always responds by stating that the product had not been submitted to FDA by way of an NDA, and it is always the case that FDA responds by threatening to shut down the company. This provides the example of a cherry juice product and FDA's Warning Letter to *Traverse Bay Farms* in Bellaire, Michigan.[104]

A product called "Cherry Juice Concentrate" had a label reading "[o]ngoing research shows that tart cherries are a rich source of antioxidants ... which may help to relieve the pain of arthritis, gout and possibly fibromyalgia ... [t]art cherries ... can offer joint pain-relief and ... fight cancer and heart disease, and prevent sleep disorders." FDA responded with a warning letter stating that: "These claims cause your products to be drugs ... [b] ecause these products are not generally recognized as safe and effective when used as labeled, they are also new drugs ... a new drug may not be legally marketed in the United States without an approved New Drug Application (NDA).[105,106]"

FDA's warning concluded with, "Failure to promptly correct these violations may result in enforcement action without further notice. Enforcement action may include seizure of violative products, injunction against the manufacturers and distributors of violative products, and criminal sanctions against persons responsible for causing violations of the Act. Please advise this office in writing, within 15 working days of receipt of this letter, as to the specific steps you have taken or will be taking to correct these violations.[107]"

FDA's Warning Letter against the *Traverse Bay Farms* cherry juice company is typical of all Warning Letters to manufacturers of foods and dietary supplements, where the package label includes a disease claim. What is typical is that the Warning Letter refers to the appropriate avenue for seeking FDA approval of drugs, that is, the avenue that begins by submitting an NDA.

[102] US Department Health and Human Services; Food and Drug Administration. Guidance for Industry: A Food Labeling Guide; 2013 (130 pages).

[103] Brody T. Food and Dietary Supplement Package Labeling—Guidance from FDA's Warning Letters and Title 21 of the Code of Federal Regulations. Comp. Rev. Food Sci. Food Safety 2015;15:92–129.

[104] Warning Letter Ref. No. DT-06-18. Traverse Bay Farms, Bellaire, MI.

[105] Brody T. Food and Dietary Supplement Package Labeling—Guidance from FDA's Warning Letters and Title 21 of the Code of Federal Regulations. Comp. Rev. Food Sci. Food Safety 2015;15:92–129.

[106] Warning Letter Ref. No. DT-06-18. Traverse Bay Farms, Bellaire, MI.

[107] Warning Letter Ref. No. DT-06-18. Traverse Bay Farms, Bellaire, MI.

CHAPTER

2

FDA's Decision-Making Process When Assessing Ambiguous Data

This chapter provides an advance glance of FDA's decision-making processes, as further detailed in all of the subsequent chapters in this book. What is described below are problems that were roadblocks to FDA-approval, communications between the Sponsor and FDA on overcoming these roadblocks, and the strategies taken to win approval of the study drug.

FDA's Medical Reviews and other reviews disclose approaches for overcoming roadblocks taking the form of data on efficacy or safety that are ambiguous, inconclusive, or fragmentary, and roadblocks taking the form of clinical trial design problems. The reason for overcoming these roadblocks is to justify using the Sponsor's data on efficacy and safety as a basis for warnings and for instructions on the package label.

In describing safety data, the term "adverse event" (AE) is used for data from human subjects and from patients, whereas in contrast, the term "toxicity" is usually used for data from animal studies. The term "adverse events" has a built-in intentional ambiguity because, as required by the Code of Federal Regulations (CFR), AEs encompass disorders caused by the study drug, as well as disorders not caused by the study drug such as disorders caused by the disease being treated or by a concomitant medication.

For example, when a Sponsor has safety data that are inconclusive, but still of concern, the Sponsor can place it in the Adverse Reactions section or in the Clinical Trials Experience section of the package label instead of in the Warnings and Precautions section. Safety data that are more alarming and that have a greater need to be communicated to the physician are placed in the Warnings and Precautions section. But safety data of less urgency, and safety data of unsure relation to the study drug, can be placed in the Clinical Trials Experience section or in the Adverse Reactions section, as documented below. Also, when a Sponsor has inconclusive safety data, the Sponsor can agree to perform additional clinical studies in the timeframe after FDA approval to acquire better quality safety data.

FDA's Drug Review Process and the Package Label
DOI: https://doi.org/10.1016/B978-0-12-814647-7.00002-6

The best examples from FDA's Medical Reviews and from FDA's Clinical Pharmacology Reviews on approaching ambiguous data are from the situations outlined below. Each of these situations is detailed in a later chapter in this book:

1. Meetings of Sponsor with FDA prior to submitting an NDA or BLA
2. Where drug interaction data are available from animals or cultured cells but not yet available from clinical studies
3. Postmarketing requirement (PMR)
4. Requirement for a Risk Evaluation and Mitigation Strategy (REMS)
5. Accounting for an AE of concern, but where data are ambiguous, by requiring on the package label that only patients failing prior to treatment be given the drug
6. Negotiating with FDA to modify the clinical study design to reduce the extent of AEs, for example, by including a screening step
7. Delaying warnings until further data are collected
8. Instructions on package label for taking drug with food versus without food
9. Ambiguity in relatedness between study drug and an AE resulted in disclosure of the AE on the package label, not in the Warnings and Precautions section, but instead in the Adverse Reactions section
10. Drug class analysis is a tool for arriving at package label safety warnings, where the Sponsor's clinical studies provide insufficient data on drug safety
11. Drug class analysis may play a dominant role in arriving at AEs on the package label, where clinical AE data are ambiguous
12. Admission of ambiguous or inconclusive data on the package label
13. Clinical Pharmacology section and Use in Specific Populations section of package label
14. Animal toxicity data as a basis for labeling despite the absence of any human AE data
15. Splitting and lumping
16. Outliers
17. Burden of proof
18. Clinical Hold

1. **Meetings with FDA prior to submitting an NDA or BLA**. The CFR (21 CFR §312.47) reveals the use of Pre-NDA meetings and Pre-BLA meetings to resolve the Sponsor's questions prior to submitting an NDA or BLA. Section 312.47 states that, "FDA has found that delays associated with the initial review of a marketing application may be reduced by exchanges of information about a proposed marketing application. The primary purpose of this kind of exchange is to uncover any major unresolved problems." Use of a Pre-NDA meeting for advice on the Sponsor's proposed data analysis strategy is revealed in FDA's Medical Review for fingolimod.[1]
2. **Where drug interaction data are available from animals or cultured cells but not yet available from clinical studies**. The package label for ponatinib (Iclusig®) provides guidance on how to draft the Drug Interactions section of the package label, where data from enzyme assays, cultured cells, or animals are available, but where corresponding clinical studies have not been performed. The example is from drug–drug interaction studies on a drug transporter called, P-glycoprotein (P-gp).

[1] Fingolimod (multiple sclerosis) NDA 022-527. Page 151 of 519-page Medical Review.

One way to draft the package label is to disclose the results from the studies from enzymes, cultured cells, or animals, and to admit that data from humans are not available, as shown here. What is shown here is the Drug Interactions section from the package label for ponatinib. The admission of lack of human data reads, "exposure of these substrates has not been evaluated in clinical studies," appears at the end of the package label excerpt. The term "exposure" refers to concentration of the drug in the bloodstream:

> Drug Interactions: Drugs that are substrates of the P-gp or ABCG2 Transporter Systems. In vitro studies demonstrate that Iclusig inhibits the P-gp and ABCG2 [also known as BCRP] transporter systems. The effect of coadministration of Iclusig with sensitive substrates of the P-gp (e.g., aliskiren, ambrisentan, colchicine, dabigatran etexilate, digoxin, everolimus, fexofenadine, imatinib, lapatinib, maraviroc, nilotinib, posaconazole, ranolazine, saxagliptin, sirolimus, sitagliptin, tolvaptan, and topotecan) and ABCG2 [also known as BCRP] (e.g., methotrexate, mitoxantrone, imatinib, irinotecan, lapatinib, rosuvastatin, sulfasalazine, and topotecan) transporter systems on exposure of these substrates has not been evaluated in clinical studies.[2]

3. **Postmarketing requirement (PMR).** FDA can grant approval to the study drug but, as a condition for drug approval, require postmarketing studies. FDA's Guidance for Industry describes the PMR.[3] As detailed in this textbook, FDA's approvals for bosutinib[4] and lorcaserin[5] illustrate FDA's requirements for various postmarketing studies as part of FDA's drug approval process.

   Where FDA imposes a postmarketing requirement (PMR) to conduct additional experiments, in some cases any interested person can determine if the Sponsor had followed through with the experiments. This was the case ustekinumab, where FDA's Clinical Review required that the Sponsor, "conduct an in vitro study or studies to determine whether IL-12 ... modulate CYP enzyme expression."[6] The package label published on the date of FDA-approval did not contain the results of the experiments,[7] but a supplemental package label dated several years later, did describe the results, "The effects of IL-12 ... on the regulation of CYP450 enzymes were evaluated in an in vitro study using human hepatocytes, which showed that IL-12 ... did not alter human CYP450 enzyme activities."[8] To access package labels and FDA's reviews, go to FDA's website (www.fda.gov), click on the "drugs" tab, click on "Search Drugs@FDA," then type the drug's name, and finally click on "Approval Dates and HIstory."

[2] Package label. ICLUSIG® (ponatinib) tablets for oral use. November 2016 (23 pages). See, Ponatinib (chronic myeloid leukemia) NDA 203-469.

[3] US Department of Health and Human Services; Food and Drug Administration; Center for Drug Evaluation and Research (CDER); Center for Biologics Evaluation and Research (CBER). Guidance for Industry. Postmarketing Studies and Clinical Trials—Implementation of Section 505(o)(3) of the Federal Food, Drug, and Cosmetic Act; 2011 (18 pages).

[4] Bosutininib (chronic myelogenous leukemia; CML) NDA 203-341. Page 36 of 89-page Clinical Pharmacology Review.

[5] Lorcaserin (obesity) NDA 022-529.

[6] Ustekinumab (plaque psoriasis) BLA 125-261. Page 8 of 194-page Clinical Pharmacology Review.

[7] Package label. STELARA™ (ustekinumab) Injection, for subcutaneous use. September 2009 (12 pages).

[8] Package label. STELARA™ (ustekinumab) Injection, for subcutaneous use. March 2014 (20 pages).

4. **Requirement for a Risk Evaluation and Mitigation Strategy (REMS)**. FDA can require that the sponsor create and administer a REMS. Also, FDA can impose a postmarketing requirement (PMA). These are separate requirements and one is not a subset of the other. Risk Evaluation and Mitigation Strategies (REMS) are frequently published on FDA's website, at the time FDA publishes the Approval Letter. In FDA's review of ustekinumab, FDA recommended that an REMS be required because of ambiguity in whether the study drug increases risk for infections and cancer.[9,10]
5. **Accounting for an AE of concern, but where data are ambiguous, by arriving at the solution where the package label required that only patients failing prior to treatment be given the drug**. In FDA's review for pegloticase, the issue was cardiac AEs, such as deaths and arrhythmias. But the number of study subjects was low, and the number of cardiac AEs was much lower. FDA reviewers wondered whether the drug should be approved or not. As a solution to the ethical issues of approving a new drug associated with cardiac AEs, FDA arrived at the solution of requiring that only patients failing conventional therapy should be treated with pegloticase.[11,12]
6. **Negotiating with FDA to modify the clinical study design to reduce the extent of AEs, for example, by including a screening step**. Where AEs are so frequent or so severe as to raise the issue of halting the clinical study, FDA's Medical Reviews reveal the option of modifying the clinical study design to include a screening step that eliminates subjects at increased risk for the AE in question. This strategy is illustrated for the study drug, adalimumab.[13,14]
7. **Delaying warnings until further data are collected**. When faced with a gray area, FDA may recommend refraining from including warnings on the package label until additional information on given type of AE is acquired. This approach by FDA is illustrated for FDA's approval of certolizumab (Cimzia®), where the FDA reviewer wrote, "Until a clearer picture emerges as to the generalizability of this rare adverse reaction to the whole class of TNF blockers, Cimzia should not carry the warning either."[15]
8. **Instructions on the package label for taking drug with or without food**. In drafting the package label, the question of whether to require that the drug be taken with or without food can be answered with the assistance of an algorithm: (1) Does food impair or increase absorption of the drug; (2) If the food increases absorption, does this increase the drug's toxicity; (3) If the food increases absorption, does this increase the

[9] Ustekinumab (psoriasis) BLA 125-261. Pages 216–217 of 246-page Medical Review.

[10] Food and Drug Administration (2011) Guidance for Industry. Postmarketing Studies and Clinical Trials. Implementation of Section 505(o)(3) of the Federal Food, Drug, and Cosmetic Act (18 pages).

[11] Pegloticase (hyperuricemia and gout) BLA 125-293. Page 31 of 238-page Medical Review.

[12] Package label. KRYSTEXXA™ (pegloticase) for injection, for intravenous infusion; September 2010 (14 pages).

[13] Adalimumab (rheumatoid arthritis) BLA 125-057. 25-page Medical Review.

[14] Package label. HUMURA (adalimumab) injection, for subcutaneous use. June 2016 (51 pages).

[15] Certolizumab (Crohn's disease) BLA 125-160. FDA divided the Medical Review into three parts: 1st part: 80 pages, 2nd part: 80 pages, 3rd part: 56 pages.

drug's efficacy; (4) Is compliance by patients for taking the drug enhanced more by a package label that requires fasting versus by a package label that requires food. FDA's Clinical Pharmacology Reviews, together with answers from FDA reviewers enabled the author to devise these four questions and to provide corresponding solutions.

9. **Ambiguity in relatedness between study drug and an AE resulted in disclosure of the AE on the package label, not in the Warnings and Precautions section, but instead in the Adverse Reactions section.**
   i. **Fingolimod**. Disclosing a type of AE on the package label, not as a warning, but simply as one of the AEs that materialized during one of the Sponsor's clinical trials. This situation is illustrated for FDA's review of fingolimod (Gilenya®). In FDA's review of fingolimod, FDA stated that, "It is unclear whether the cases of lymphoma are related to study drug."[16] The disclosure in the package label, which took the form of mere information (and not any warning) in the Adverse Reactions section stated that, "the relationship of lymphoma to GILENYA remains uncertain."[17]
   ii. **Brivaracetam.** Another example of AEs that were not clearly related with the study drug and FDA's recommendation for corresponding information on the package label is from FDA's review of brivaracetam. FDA's requirement was that, "Therefore, these treatment emergent AEs should be included in the Adverse Reactions section of brivaracetam labeling instead of Warnings and Precautions."[18]
   iii. **Sofosbuvir/velpatasvir combination**. A combination drug for treating hepatitis C virus also reveals the situation where ambiguity in relatedness of the drug to AEs resulted in disclosure in the Adverse Reactions section.[19] The FDA reviewer admitted that "no clear indication for an increased risk" for the AE in question[20] and, as a result, the AE was disclosed in the Adverse Reactions section which admitted that there was only a "potential causal relationship" between the drug and the risk for the AE.[21]
   iv. **Regorafenib**. Yet another example of disclosing toxicity data in a section other than the Warnings and Precautions section is found in FDA's review of regorafenib.[22] The Sponsor had toxicity data on dentin and epiphyseal growth plate only from animal studies, and hence it was apparent that Sponsor doubted it to be appropriate to disclose it in the Warnings and Precautions section. Instead,

[16] Fingolimod (multiple sclerosis) NDA 022-527. Page 271 of 519-page Medical Review.

[17] Package label. GILENYA (fingolimod) capsules, for oral use. February 2016 (19 pages).

[18] Brivaracetam (epilepsy) NDA 205-836, NDA 205-837, NDA 205-838. Page 274 of 382-page Medical Review.

[19] Sofosbuvir/velpatasvir combination (hepatitis C virus) NDA 208-341. Page 145 of 171-page Medical Review.

[20] Sofosbuvir/velpatasvir combination (hepatitis C virus) NDA 208-341. Page 145 of 171-page Medical Review.

[21] Package label. EPCLUSA® (sofosbuvir and velpatasvir) tablets, for oral use; February 2017 (32 pages).

[22] Regorafenib (metastatic colorectal cancer after prior treatment with various drugs) NDA 203-085. Pages 10–16 of 69-page Clinical Review.

these toxicities were disclosed in a less prominent part of the package label, namely, in the Use in Specific Populations section of the package label.[23] Another reason for placing the information in the Use in Specific Populations section was that it related to juveniles and not to adults.

v. **Certolizumab pegol**. In still another example of the best way to handle ambiguous or incomplete safety data, the package label for a TNF-α blocker disclosed the safety information in the Adverse Reactions section instead of in the Warnings and Precautions section.[24,25]

vi. **Vorapaxar**. For vorapaxar, a subgroup (low body weight) that was considered for placing in the Warnings and Precautions section for vorapaxar was eventually held to pose minimal or no risk, and as a result, the information about this subgroup found a place in a less prominent part of the package label, namely the Pharmacokinetics section of the package label.[26,27]

vii. **Abatacept**. Abatacept is an immunosuppressant drug for treating rheumatoid arthritis. One of the AEs associated with this drug is the materialization of cancer in patients taking the drug. Ambiguity in relatedness resulted in this AE appearing in the Adverse Reactions section of the label and not in the Warnings and Precautions section.[28,29]

10. **Drug class analysis is a tool for arriving at package label safety warnings, where the Sponsor's clinical studies provide insufficient data on drug safety**. Drug class analysis is used to predict AEs for the study drug, where the goal is to draft package label warnings. The importance for drug class analysis increases, where data on AEs from the Sponsor's own clinical studies are in short supply. FDA's reviews for lorcaserin,[30] nebivolol,[31,32,33] and vemurafenib[34] reveal a technique for improving

[23] Package label. STIVARGA (regorafenib) tablets, oral; September 2012 (14 pages).

[24] Package label. CIMZIA (certolizumab pegol); April 2008 (19 pages).

[25] Certolizumab (Crohn's disease) BLA 125-160.

[26] Package label. ZONTIVITY™ (vorapaxar) Tablets 2.08 mg, for oral use; May 2014 (18 pages).

[27] Baker NC, et al. Overview of the 2014 Food and Drug Administration Cardiovascular and Renal Drugs Advisory Committee meeting about vorapaxar. Circulation 2014;130:1287–94.

[28] Abatacept (rheumatoid arthritis) BLA 125-118. Page 64 of 127-page Pharmacology Review.

[29] ORENCIA (abatacept) for injection, for intravenous use; June 2016 (23 pages).

[30] Lorcaserin (obesity) NDA 022-529. Pages 23, 204, and 273 of 425-page Medical Review.

[31] Nebivolol (hypertension) NDA 021-742. Page 12 of the eighth pdf file (80 pages) of 12 pdf files in Medical Review.

[32] Nebivolol (hypertension) NDA 021-742. Pages 27 and 28 of the first pdf file (80 pages) of seven pdf files for the Pharmacology Review, and pages 20 and 21 of third pdf file (80 pages) in the seven Pharmacology Reviews.

[33] Nebivolol (hypertension) NDA 021-742. Page 21 of the third pdf file (80 pages) of the seven Pharmacology Reviews.

[34] Vemurafenib (BRAF V600E mutation positive unresectable or metastatic melanoma) NDA 202-429. Pages 12 and 13 of 105-page pdf file Medical Review.

accuracy of drug class analysis, where the technique is to screen, dissect, and narrow the field of other relevant drugs in the drug class, and to exclude analyses that would give misleading AEs.

In other words, what is first contemplated is the AEs in the drug class, as initially defined by a broad definition, and what is next contemplated, is the smaller array of AEs detected when using a narrower definition carved out from the broadly defined drug class.

Analysis for vemurafenib started with the drug class of "protein kinases" and then narrowed it to the drug class, "protein kinases excluding VEGFR." Analysis for lorcaserin started with the drug class of 5HT1B, 5HT2B, and 5HT2C inhibitors, then contemplated narrowing it down to only 5HT2C inhibitors, and then expanded it slightly to 5HT2B inhibitors plus 5HT2C inhibitors. Analysis for nebivolol started with the broad class of all β-adrenergic receptor inhibitors but then decided that the only relevant drug was nebivolol, because of its low safety margin with regard to the AE of interest (reproductive toxicity). Other examples of drug class analysis come from FDA's reviews of vilazodone,[35,36] and osimertinib.[37,38]

FDA's reviews of macitentan illustrate FDA's reaction to data that is insufficient, : (1) Resulting from poor experimental design or (2) Resulting from a well-designed experiment that was not optimally suited to provide information on drug toxicity.

FDA's reaction to the poorly designed experiment was to require the package label's warnings to be based on drug class analysis, that is, based on package labels from similar drugs that had already been FDA-approved. Regarding a poorly designed study intended to capture data on testicular toxicity, FDA stated that, "The testicular safety study ... conducted in healthy male subjects was poorly conducted and, therefore, uninformative. This agent will get a statement in the label that reflects the bosentan experience."[39]

FDA's reaction to the well-designed study that could not adequately detect toxicity was to require pharmacovigilence of the marketed drug. Regarding this well-designed study, FDA stated that, "However, there is still need for vigilance because ... the majority of macetentan doses studied has been 10 mg or less, so the safety of higher doses is unknown, and ... the total number of patients who have taken the drug is small."[40]

[35] Vilazodone (antidepressant) NDA 022-567. Page 8 of 126-page Clinical Review. Pages 2–127 of 259-page pdf file Medical Review.

[36] Package label. VIIBRYD™ (vilazodone hydrochloride) tablets. January 2010 (13 pages).

[37] Osimertinib (epidermal growth factor receptor T790M mutation positive NSCLC) NDA 208-065. 137-page pdf file Medical Review.

[38] Package label. TAGRISSON™ (osimertinib) tablet, for oral use; November 2015 (16 pages).

[39] Macitentan (pulmonary arterial hypertension) NDA 204-410. Page 75 of 163-page pdf file Medical Review.

[40] Macitentan (pulmonary arterial hypertension) NDA 204-410. Liver Toxicity Review by Maryann Gordon, MD (3 pages) from 163-page Medical Review.

11. **Drug class analysis may play a dominant role in arriving at AEs on the package label, where clinical AE data are ambiguous or inconclusive**. The best example of using drug class analysis for arriving at package labeling may be from FDA's review of vilazodone, where the number of AEs was small, where relatedness was unclear, and where the AEs were only mild. FDA's Medical Review characterized these AEs as, "bleeding events ... appear to be minor and not necessarily attributable to the drug," "two subjects were identified with probable related," "there was no evidence of hyponatremia in the database."[41] Even though the clinical AE data were ambiguous, mild, or non-existent, drug class analysis took a dominant role and all of these AEs were printed as warnings on the package label.

    A similar scenario occurred for ustekinumab, a drug that blocks one of the cytokines (interleukin-12; IL-12). Ustekinumab is for treating psoriasis. The Sponsor's data from human subjects showed that the incidence of an AE (infections) in the ustekinumab treatment arm was not any greater than in the placebo arm. Despite this lack of evidence that ustekinumab-treated human subjects were at increased risk for infections, FDA used drug class analysis (mechanism of action) as a basis for requiring a package label warning, writing, "labeling should reflect that these **risks are theoretical in nature**, as they have not been evidenced in the database to date."[42]
12. **Admission of ambiguous or inconclusive data on the package label**. Where AE data are ambiguous, the Sponsor might want to consider including a statement to that effect on the package label. An example is found on the package label for tofacitinib. The statement of ambiguous data for white blood cells was, "The clinical significance of these changes is unknown" and the statement of ambiguous data for antibodies was, "changes were small and not dose-dependent."[43]
13. **Clinical Pharmacology section and Use in Specific Populations section of the package label**. Where information on AEs or pharmacokinetics is characterized by ambiguity, the information sometimes finds a place on the Clinical Pharmacology section of the package label, as when the package label states that the information is only strong enough to show "potential"[44] or only strong enough to show that an event is "not expected,"[45] or that the AE is "not been established" (in comments on pediatric patients),[46] or where data are only strong enough to show that the event is "unlikely to substantially alter."[47] Ambiguous information of this type is generally not placed in the Black Box Warning, Warnings and Precautions, Dose and Administration, or Drug Interactions section.

[41] Vilazodone (antidepressant) NDA 022-567. Page 202 of 259-page pdf file Medical Review.

[42] Ustekinumab (psoriasis) BLA 125-261. Page 84 of 246-page Medical Review.

[43] Package label. XELJANZ® (tofacitinib) tablets for oral administration; November 2012 (27 pages) NDA 203-214.

[44] Regorafenib (metastatic colorectal cancer) NDA 203-085. Pages 9, 30 and 31 of 64-page Clinical Pharmacology Review.

[45] Package label. INLYTA® (axitinib) tablets for oral administration; January 2012 (17 pages) NDA 202-324.

[46] Package label. STIVARGA (regorafenib) tablets, oral; September 2012 (14 pages).

[47] Package label. XELJANZ® (tofacitinib); November 2012 (27 pages).

14. **Animal toxicity data as a basis for labeling in the situation where AE data in humans are not available**. This is about animal data revealing an interesting type of toxicity, but where human data on the corresponding toxicity are not available. In the absence of any toxicity data from humans, the Sponsor can disclose the animal toxicity data in the Nonclinical Toxicology section, rather than in any of the other section, of the package label. This type of decision is revealed with FDA's approval of regorafenib, where the toxicity concerned dentin.[48,49] Animal toxicity data are also relied upon for package label warnings relating to pregnancy, embryo–fetal toxicity, appearance of the drug in breast milk, and other toxicity issues that cannot ethically be conducted with human subjects.
15. **Splitting and lumping**. Ambiguity may be encountered in deciding whether to disclose a given AE in the package label, for example, where the number of study subjects experience that AE is small. The process of coding can result in artificially enhanced (or artificially minimized) numbers of AEs. The activity leading to these artificial changes in AE numbers is called, splitting and lumping. FDA reviewers scrutinize the Sponsor's decisions in splitting and lumping, as demonstrated by FDA's Medical Reviews for apremilast,[50] brivaracetam,[51] dimethyl fumarate,[52] droxidopa,[53] and linaclotide,[54] as shown in Chapter 13, Coding.
16. **Outliers**. This is about the situation where there is an "outlier," that is, a data point that has a value that is unusually high or low, when compared to the rest of the data, and where the extreme value cannot be explained. FDA's review for lenalidomide illustrates that an outlier can be handled, by conducting one analysis that includes the outlier and another analysis that excludes the outlier, and additionally by recommending that the Sponsor conduct a clinical study addressing that particular outlier.[55]
17. **Burden of proof**. When faced with the situation of inconclusive data, either the Sponsor or FDA may take the approach that the inconclusive nature of the data in question is not of the type that precludes FDA approval. This type of burden shifting argument is found in FDA's approval for **lenvatinib**, where FDA wrote, "There were also **no specific trends noted in demographic subgroup analysis, that would preclude** lenvatinib's use to the proposed population of RAI refractory thyroid cancer patients."[56]

[48] Regorafenib (metastatic colorectal cancer after prior treatment with various drugs) NDA 203-085. Pages 10-16 of 69-page Clinical Review.

[49] Package label. STIVARGA (regorafenib) tablets, oral; September 2012 (14 pages).

[50] Apremilast (psoriatic arthritis) NDA 206-088. Page 119 of 132-page Medical Review.

[51] Brivaracetam (epilepsy) NDA 205-836, NDA 205-837, NDA 205-838. Pages 192, 193, 239, 240, 253 and 254 of 382-page pdf file Medical Review.

[52] Dimethyl fumarate (multiple sclerosis) Page 61 of 288-page Medical Review.

[53] Droxidopa (neurogenic orthostatic hypotension) NDA 203-202. Pages 38 and 66 of 436-page Medical Review.

[54] Linaclotide (irritable bowel syndrome) NDA 202-811. Pages 42, 76, 185, and 418 of 489-page Medical Review.

[55] Lenalidomide (myelodysplastic syndromes) NDA 021-880. Pages 24 and 25 of 46-page Clinical Pharmacology Review.

[56] Lenvatinib (radioiodine-refractory differentiated thyroid cancer) NDA 206-947. Pages 67–70 and 74 of 167-page Clinical Review; 185-page pdf file Medical Review.

FDA's review of **lorcaserin** provides another example, where the burden of proof was invoked as part of drug safety analysis. In short, FDA observed mechanism of action data predicting that lorcaserin would not cause serious AEs, but then moved forward, and required additional clinical safety data, writing that, "Despite its relative 5HT2C specificity as compared to 5HT2B . . . [t]he "**clinical data as collected up to this point do not exonerate the drug**" for valvular heart disease."[57,58]

Yet another conclusion invoking this same burden of proof was used in FDA's Medical Review for eslicarbazepine. FDAs' conclusion regarding association of the study drug with Stevens Johnson Syndrome was, "Although confounded by lamotrigine, the role of eslicarbazepine cannot be ruled out."[59]

In other words, where it is not clear if additional safety studies should be conducted, or if it is not clear how to draft safety warnings on the package insert, the best way to make a concluding statement that invokes the burden of proof style, such as, "the data do not exonerate the drug."

18. **Clinical Hold**. FDA can impose a Clinical Hold for various reasons, and a number of approaches are available to the Sponsor for overcoming a Clinical Hold. During the course of a clinical study, an excessive number of severe AEs, excessive deviations from the Clinical Study Protocol, and other problems can provoke FDA to impose a Clinical Hold.

A Clinical Hold is a notification issued by FDA to the Sponsor to delay a proposed clinical trial or to suspend an ongoing clinical trial (21 CFR §312.42). The grounds for a Clinical Hold include FDA's assessment that human subjects are exposed to an unreasonable risk of injury, that the clinical investigators named in the IND are not qualified to conduct the clinical trial, and that the Investigator's Brochure is erroneous (21 CFR §312.42).

Where FDA imposes the Clinical Hold because of misconduct by the Sponsor, this misconduct can take the form, for example, of failure to report serious adverse events (SAEs), enrolling study subjects having conditions that put them at increased risk when they are administered the study drug, repeated failure to administer informed consent forms, and failure of Institutional Review Board to review and approve significant changes in the Clinical Study Protocol and falsification of safety data.[60]

FDA may also issue a Partial Clinical Hold, which is, "[a] delay or suspension of only part of the clinical work requested under the IND (e.g., a specific protocol or part of a protocol is not allowed to proceed; however, other protocols or parts of the protocol are allowed to proceed under the IND)."[61]

[57] Lorcaserin (obesity) NDA 022-529. Page 66 of 425-page Medical Review.

[58] Lorcaserin (obesity) NDA 022-529. Page 227 of 425-page Medical Review.

[59] Eslicarbazepine (partial onset seizures) NDA 022-416. Pages 114 and 115 of 620-page Medical Review.

[60] US Department of Health and Human Services; Food and Drug Administration.Guidance for Industry and Clinical Investigators. The use of clinical holds following clinical investigator misconduct; September 2004 (8 pages).

[61] US Department of Health and Human Services; Food and Drug Administration. Guidance for Industry. Submitting and Reviewing Complete Responses to Clinical Holds; April 1998 (3 pages).

Upon receipt of a Clinical Hold Letter, the Sponsor can respond by preparing an amendment to the IND that addresses the issues set forth in the Letter. FDA then evaluates the amendment and determines if the Sponsor's response is satisfactory.[62]

This textbook provides examples from FDA's reviews for Clinical Hold for submissions for the drugs, ivacaftor,[63] evoculumab,[64] and vigabatrin.[65]

Press releases frequently announce a Clinical Hold, and examples are as follows. Vadastuximab for acute myeloid leukemia was put on Clinical Hold because six patients experienced hepatotoxicity, including cases of veno-occlusive disease, with four deaths.[66] A clinical study on an inhaled drug was suspended because the drug was carcinogenic, as determined in a chronic study with rats. In the Clinical Hold, FDA requested additional information from the rat study.[67] A Clinical Hold against study on an osteoarthritis drug was imposed because an occurrence of an infection in the injected knee joint of a study subject.[68] A Clinical Hold against an anticoagulant drug was imposed because of allergic reaction SAEs.[69]

[62] Poole K. The Sponsor's Guide to Regulatory Submissions for an Investigational New Drug. Biological Resources Branch, DCTD, NCI-Frederick. SAIC-Frederick, Inc.; 2005 (104 pages).

[63] Ivacaftor (cystic fibrosis) NDA 203-188. 30-page Cross Discipline Team Leader Review.

[64] FDA Briefing Document. Endocrinologic and Metabolic Drugs Advisory Committee (EMDAC); June 10, 2015, Page 50 of 405 pages.

[65] Vigabatrin (complex partial seizures in adults, infantile spasms in children) NDA 022-006. Page 27 of 219-page Medical Review.

[66] Press release. Seattle Genetics announces clinical hold on several Phase 1 trials of vadastuximab talirine; December 27, 2016.

[67] Press release. Insmed announces clinical hold on ARIKACE® Phase 3 clinical trials; August 1, 2011.

[68] Press release. Flexion Therapeutics announced clinical hold of FX006 Phase 2b clinical trial in osteoarthritis of the knee; September 17, 2014.

[69] Press release. Regado Biosciences announces clinical hold of REGULATE-PCI trial following voluntary halt of trial by Regado; July 9, 2014.

CHAPTER

3

Food Effect Studies

I. INTRODUCTION

a. Pharmacokinetics

The influence of food on drug absorption and pharmacokinetics (PK) is presented early in this book for only one reason. The reason is that all readers will already be familiar with some aspects of food composition and gastrointestinal (GI) physiology.

Drugs can be absorbed in the small intestines by passive diffusion through membranes of gut cells or by transport proteins residing in gut cell membranes. Lipophilic drugs, such as most tyrosine kinase inhibitor (TKI) drugs, tend to be absorbed by passive diffusion. Where transport proteins are used, the transporter may be one of the organic anion transporting proteins. Some transporters act on drugs that have already entered the enterocyte, where they can direct the drug back out (efflux) to the intestinal lumen. These efflux transporters include P-glycoprotein (P-gp), breast cancer resistance protein, and multidrug resistance-associated proteins.[1]

Food can influence the rate of absorption and PK of oral drugs by delaying gastric emptying time, changing stomach pH, stimulating secretion of bile salt detergents, and by interacting directly with the drug.[2] Drugs that are TKIs are especially susceptible to food–drug interactions.[3] Because food can either increase or decrease bioavailability, Sponsors may need to test food–drug interactions in human subjects early in drug development to provide rational dosing instructions for clinical trials to be used as a basis for FDA-approval.[4]

[1] Herbrink M, et al. Variability in bioavailability of small molecular tyrosine kinase inhibitors. Cancer Treat. Rev. 2015;41:412–22.

[2] Mathias N, et al. Food effect in humans: predicting the risk through in vitro dissolution and in vivo pharmacokinetic models. AAPS J. 2015;17:988–98.

[3] Parsad S, Ratain MJ. Food effect studies for oncology drug products. Clin. Pharmacol. Ther. 2017. doi: 10.1002/cpt.610.

[4] Kang SP, Ratain MJ. Inconsistent labeling of food effect for oral agents across therapeutic areas: differences between oncology and non-oncology products. Clin. Cancer Res. 2010;16:4446–51.

FDA's Drug Review Process and the Package Label
DOI: https://doi.org/10.1016/B978-0-12-814647-7.00003-8

The severity of any given food effect can calculated by comparing the bloodstream concentrations of the study drug, when consumed in the presence or absence of food. It is conventional, in PK studies, to express bloodstream concentrations in terms of "area under the curve" (AUC) or maximal concentration (Cmax). Thus, the severity for the food effect for any given drug can be calculated by one or both of[5]:

$$\text{Food effect} = \text{AUC}_{\text{fed}}/\text{AUC}_{\text{fasted}}$$
$$\text{Food effect} = \text{Cmax}_{\text{fed}}/\text{Cmax}_{\text{fasted}}$$

FDA's Guidance for Industry provides for food effect data, as part of the information in the Dosage and Administration section of the package label:

> Information under the Dosage and Administration heading must contain a ... summary of the recommended dosage regimen (e.g., starting dose, dose range, titration regimens, route of administration) ... differences among population subsets, monitoring recommendations, if any, and ... e.g., dosing adjustments recommended for concomitant therapy, specific populations with coexisting conditions, clinically relevant food effects.[6]

This chapter provides FDA's basis for instructions relating to foods, as typically disclosed in the Dosage and Administration section of the package label. The topics in this chapter include:

- need for a concomitant meal with an oral drug,
- need for fasting state with an oral drug,
- need to take an oral drug with or shortly before a meal, because the goal of the drug is to influence processing of nutrients that are in the meal, as is the case with the drugs, ferric citrate, canagliflozin, and lixisenatide,
- need to avoid certain foods, such as grapefruit juice or excessive alcohol, and
- need to avoid drugs that influence gastric physiology, such as omeprazole or antacids, when taking an oral drug.

b. Food Composition

Where an investigator needs to conduct a food effect study, the need typically arises for preparing test meals that are low-fat meals or high-fat meals. Guidance for preparing these meals can take the form of standard recipes, or alternatively, by calculating the amount of fat in each component of the test meal. For most food effect studies, the information needed is likely to be the number of grams of fat, carbohydrate, and protein in a given meal. Data on fat content of eggs, for example, can be acquired from primary

[5] Mathias N, et al. Food effect in humans: predicting the risk through in vitro dissolution and in vivo pharmacokinetic models. AAPS J. 2015;17:988–98.

[6] U.S. Department of Health and Human Services. Food and Drug Administration. Center for Drug Evaluation and Research (CDER). Center for Biologics Evaluation and Research (CBER). Guidance for industry. Labeling for human prescription drug and biological products – implementing the PLR content and format requirements; 2013 (30 pp.).

sources.[7,8,9,10] A convenient source of food composition with data on thousands of common foods is "Handbook 8," a publication of the US Department of Agriculture (USDA).[11] In addition, package labels on dried foods and canned foods disclose the amount of protein, carbohydrate, and fat.[12]

Parsad and Ratain[13] recommend that investigators define the overall caloric content of the meal as well as the calories or grams of fat in the meal. Investigators need to be aware of the composition of any test meals that are used, as revealed by the FDA reviewer's complaints in the NDA for eltrombopag. The FDA reviewer complained that, "Specific calorie breakdown **was not provided by the sponsor** but it is safe to assume it is within FDA's Guidance, Food Effect Bioavailability and Fed Bioequivalance Studies, that is, 800 to 1000 calories with 150, 250, and 500-600 calories derived from protein, carbohydrate, and fat."[14,15]

c. FDA's Guidance for Industry on Food Effect Studies

Regarding the influence of food on the bioavailability of small molecule oral drugs, FDA's Guidance for Industry states that[16]:

Food can alter bioavailability by various means, including:

- Delay gastric emptying
- Stimulate bile flow
- Change GI pH

[7] Grobas S, et al. Influence of source and percentage of fat added to diet on performance and fatty acid composition of egg yolks of two strains of laying hens. Poult. Sci. 2001;80:1171–9.

[8] Edwards HM, et al. Studies on the cholesterol content of eggs from various breeds and/or strains of chickens. Poult. Sci. 1960;39:487–9.

[9] Leskanich CO, Noble RC. Manipulation of the n − 3 polyunsaturated fatty acid composition of avian eggs and meat. Worlds Poult. Sci. 1997;53:155–83.

[10] Hargis PS, Van Elswyk ME. Manipulating the fatty acid composition of poultry meat and eggs for the health conscious consumer. Worlds Poult. Sci. J. 1993;49:251–64.

[11] Agricultural Research Service. United States Department of Agriculture. Composition of foods, raw, processed and prepared (Agriculture Handbook No. 8); 1975.

[12] Brody T. Food and dietary supplement package labeling—guidance from FDA's warning letters and title 21 of the code of federal regulations. Compr. Rev. Food Sci. Food Saf. 2015;15:92–129.

[13] Parsad S, Ratain MJ. Food effect studies for oncology drug products. Clin. Pharmacol. Ther. 2017. doi: 10.1002/cpt.610.

[14] Eltrombopag (chronic immune thrombocytopenic purpura) NDA 022-291. Page 10 of 96-page Clinical Pharmacology Review (3rd of three pdf files).

[15] U.S. Department of Health and Human Services. Food and Drug Administration. Center for Drug Evaluation and Research (CDER). Guidance for industry. Food-effect bioavailability and fed bioequivalence studies; 2002 (12 pp.).

[16] U.S. Department of Health and Human Services. Food and Drug Administration. Center for Drug Evaluation and Research (CDER). Guidance for industry. Food-effect bioavailability and fed bioequivalence studies; 2002 (12 pp.).

- Increase splanchnic blood flow
- Change luminal metabolism of a drug substance
- Physically or chemically interact with a dosage form or a drug substance

Regarding timing, FDA's Guidance provides that, "Food effects on bioavailability are generally greatest when the drug product is administered shortly after a meal is ingested."[17]

FDA's Guidance for Industry expressly recommends using a high-fat meal for bioavailability studies, but without regard to whether the study drug is hydrophilic (water-soluble) or lipophilic (water-insoluble). To this end, FDA's Guidance states, "We recommend that food-effect bioavailability ... studies be conducted using meal conditions that are expected to provide the greatest effects on gastrointestinal physiology so that systemic drug availability is maximally affected. A high-fat (50 percent of total caloric content of the meal) ... meal is recommended as a test meal for food-effect bioavailability ... studies."[18]

FDA's Guidance for Industry also recommends that test meals have specific amounts of protein, carbohydrate, and fat. Thus, investigators will need information on food composition when drafting any Clinical Study Protocol that provides instructions for carrying out a food effect study.

FDA's Guidance recommends, "A high-fat ... 50 percent of total caloric content of the meal ... and high-calorie ... 800 to 1000 calories ... meal is recommended as a test meal for food-effect bioavailability ... [t]his test meal should derive approximately 150, 250, and 500-600 calories from protein, carbohydrate, and fat, respectively."[19]

The term "food effect," without more, does not imply that bioavailability of any given drug is increased or decreased by taking food with the oral drug. FDA's Guidance for Industry is careful to disclose that food can increase or decrease bioavailability. Regarding enhanced bioavailability, FDA's Guidance teaches that, "In general, meals that are high in total calories and fat content are more likely to affect the gastrointestinal physiology and thereby result in a larger effect on the bioavailability of a drug substance or drug product. We recommend use of high-calorie and high-fat meals during food-effect bioavailability... studies."[20]

[17] U.S. Department of Health and Human Services. Food and Drug Administration. Center for Drug Evaluation and Research (CDER). Guidance for industry. Food-effect bioavailability and fed bioequivalence studies; 2002 (12 pp.).

[18] U.S. Department of Health and Human Services. Food and Drug Administration. Center for Drug Evaluation and Research (CDER). Guidance for industry. Food-effect bioavailability and fed bioequivalence studies; 2002 (12 pp.).

[19] U.S. Department of Health and Human Services. Food and Drug Administration. Center for Drug Evaluation and Research (CDER). Guidance for industry. Food-effect bioavailability and fed bioequivalence studies; 2002 (12 pp.).

[20] U.S. Department of Health and Human Services. Food and Drug Administration. Center for Drug Evaluation and Research (CDER). Guidance for industry. Food-effect bioavailability and fed bioequivalence studies; 2002 (12 pp.).

Regarding reduced bioavailability, FDA's Guidance describes water-soluble ("rapidly dissolving") drugs, writing that, "Important food effects on bioavailability are least likely to occur with ... rapidly dissolving, immediate release drug products ... [h]owever, for some drugs in this class, food can influence bioavailability when there is a high first-pass effect, extensive adsorption, complexation, or instability of the drug substance in the GI tract ... [f]or rapidly dissolving formulations ... food can affect Cmax and the time at which this occurs (tmax) by delaying gastric emptying and prolonging intestinal transit time."[21]

d. Published Accounts of Dietary Fat's Influence on Bioavailability

An account of the influence of food on drug bioavailability is available for lumefantrine, an antimalarial drug. The account focused on the ability of dietary fat to increase absorption based on the fact that the drug was lipophilic, where the added fat enhanced absorption by stimulating the body's fat-absorption machinery.

The researchers commented on two antimalarial drugs, lumefantrine and artemether, writing, "Lumefantrine is more lipophilic than artemether, so oral absorption is expected to be affected much more by variation in food fat content than artemether." The researchers measured lumefantrine oral bioavailability when taken with a high-fat diet (porridge with sunflower oil) or with a low-fat diet (porridge only). This particular high-fat diet increased AUC by sixfold, when compared to the porridge-only diet.[22]

Similarly, other investigators commented on the lipophilic nature of lumefantrine and discovered that dietary fat increased absorption of this drug. Lumefantrine absorption was increased fivefold, where the source of fat was soya milk. In the study, subjects were given lumefantrine with various amounts of soya milk (0, 10, 40, 150, and 500 mL), supplying 0, 0.32, 1.28, 4.8, and 16 g of fat, respectively.[23] These soya milk meals are low-fat meals, not high-fat meals.

Referring to other studies on the influence of fat on lumefantrine and artemether absorption, Ashley et al.[24] observed that, "Lumefantrine is a hydrophobic lipophilic compound, which is absorbed slowly ... [i]n plasma, it is bound ... to high-density lipoproteins (HDLs) and partitions to circulating fats ... [a]rtemether bioavailability was increased two-fold and lumefantrine bioavailability was increased 16-fold in those who took the drug with food."

[21] U.S. Department of Health and Human Services. Food and Drug Administration. Center for Drug Evaluation and Research (CDER). Guidance for industry. Food-effect bioavailability and fed bioequivalence studies; 2002 (12 pp.).

[22] Mwebaza N, et al. Comparable lumefantrine oral bioavailability when co-administered with oil-fortified maize porridge or milk in healthy volunteers. Basic Clin. Pharmacol. Toxicol. 2013;113:66–72.

[23] Ashley EA, et al. How much fat is necessary to optimize lumefantrine oral bioavailability? Trop. Med. Int. Health. 2007;12:195–200.

[24] Ashley EA, et al. How much fat is necessary to optimize lumefantrine oral bioavailability? Trop. Med. Int. Health. 2007;12:195–200.

As a general proposition, the situation where a type of meal increases absorption or increases plasma concentration of the drug raised the following issues:

- Where a meal increases plasma drug concentration, it is possible that taking meals will enhance the drug's efficacy.
- Where a meal increases plasma drug concentration and where the result is enhancement of the drug's efficacy, it might be reasonable to draft the package label to require that food be taken with the drug.
- By increasing plasma concentration, it is possible that taking meals will increase the drug's toxicity.
- Where food increases plasma drug concentrations and where the result is increased toxicity, it might be reasonable to draft the package label to require that the drug be taken on an empty stomach.

e. Influence of Food on GI Physiology

The presence of fat in the small intestine slows gastric emptying, stimulates the release of several GI hormones, and suppresses appetite and energy intake. Slowing of gastric emptying by fat is dependent on enzymatic hydrolysis of triglycerides releasing free fatty acids, as catalyzed by gastric and intestinal lipases, where the effect of free fatty acids depends on their chain length.[25,26] Regarding chain length, studies on humans where fatty acids of various chain lengths were infused into the stomach demonstrated that those with a chain length of ≥12 carbon atoms inhibited gastric emptying, whereas those that are only 10 carbons long have relatively little effect.[27,28]

The amount of fat in a meal influences secretion of various gut hormones, notably, ghrelin, cholecystokinin (CCK), glucagon-like peptide 1 (GLP-1), and peptide YY. Studies on human subjects reveal that where fat is infused into the small intestines, the result is slower gastric emptying, and stimulated release of CCK, peptide YY, and GLP-1, and suppressed release of ghrelin. GLP-1 is secreted from L cells in the distal small intestinal mucosa, in response to fat and carbohydrate. Peptide YY is secreted from L cells of the ileum and large intestine in response to long-chain fatty acids. Ghrelin is synthesized by oxyntic cells of the fundic mucosa.[29]

[25] Feinle C, et al. Effects of fat digestion on appetite, APD motility, and gut hormones in response to duodenal fat infusion in humans. Am. J. Phyiol. Gastrointest. Liver Physiol. 2003;284:G798–807.

[26] Little TJ, et al. Modulation by high-fat diets of gastrointestinal function and hormones associated with the regulation of energy intake: implications for the pathophysiology of obesity. Am. J. Clin. Nutr. 2007;86:531–41.

[27] McLaughlin J, et al. Fatty acid chain length determines cholecystokinin secretion and effect on human gastric motility. Gastroenterology. 1999;116:46–53.

[28] Hunt JN, Knox MT. A relation between the chain length of fatty acids and the slowing of gastric emptying. J. Physiol. 1968;194:327–36.

[29] Little TJ, et al. Modulation by high-fat diets of gastrointestinal function and hormones associated with the regulation of energy intake: implications for the pathophysiology of obesity. Am. J. Clin. Nutr. 2007;86:531–41.

At an early time after consuming a meal, the proximal stomach relaxes to accommodate the meal and later, the antrum grinds solids to a small particle size, and then pulsating movements by stomach muscles force the chyme out through the pylorus.[30] These processes are regulated by interactions of nutrients passing through the pylorus, followed by interaction of certain nutrients with receptors in the small intestines. The consequent influence on stomach muscles is mediated by the vagus nerve as well as by secreted gut hormones mentioned earlier, GLP-1, CCK, and peptide YY.[31]

f. Example of Medium-Chain Triglyceride Diet

This concerns the medium-chain triglyceride diet, a diet that is used for diet therapy of epilepsy.[32,33,34] According to Sills et al., "The medium chain triglyceride (MCT) diet is a form of neutral lipid triglyceride containing fatty acid molecules with a **carbon chain length of 6 to 12**, the major constituents being octanoic acid 81% and decanoic acid 15%."[35]

In contrast, note that typical dietary oils and fats consist mainly of longer chain-length fatty acids. Pig fat consists mainly of triglycerides with **16-carbon and 18-carbon fatty acids**.[36] Similarly, the triglycerides of soybeans consist mainly of those with **16-carbon and 18-carbon fatty acids**.[37]

Thus, in the unlikely event where an investigator uses medium-chain triglycerides as a test meal, rather than a more conventional long-chain triglyceride containing meal; the result would be a drastic change in the influence of the meal on GI absorption and on consequent exposure to the study drug. As stated above, fatty acids with a chain length of ≥12 carbon atoms inhibit gastric emptying, whereas those that are only ≤10 carbons long have no effect.[38,39]

[30] Marathe CS, et al. Relationships between gastric emptying, postprandial glycemia, and incretin hormones. Diabetes Care. 2013;36:1396–405.

[31] Marathe CS, et al. Relationships between gastric emptying, postprandial glycemia, and incretin hormones. Diabetes Care. 2013;36:1396–405.

[32] Brody T. Nutritional Biochemistry. New York: Elsevier; 1999, p. 236–243.

[33] Neal EG, et al. A randomized trial of classical and medium-chain triglyceride ketogenic diets in the treatment of childhood epilepsy. Epilepsia. 2009;50:1109–17.

[34] Sills MA, et al. The medium chain triglyceride diet and intractable epilepsy. Arch. Dis. Child. 1986;61:1168–72.

[35] Sills MA, et al. Role of octanoic and decanoic acids in the control of seizures. Arch. Dis. Child. 1986;61:1173–7.

[36] Lu T, et al. Supplementing antioxidants to pigs fed diets high in oxidants: II. effects on carcass characteristics, meat quality, and fatty acid profile. J. Anim. Sci. 2014;92:5464–75.

[37] Scholfield CR. Composition of soybean lecithin. J. Am. Oil Chem. Soc. 1981;58:889–92.

[38] McLaughlin J, et al. Fatty acid chain length determines cholecystokinin secretion and effect on human gastric motility. Gastroenterology. 1999;116:46–53.

[39] Hunt JN, Knox MT. A relation between the chain length of fatty acids and the slowing of gastric emptying. J. Physiol. 1968;194:327–36.

g. Examples of Antacids and Proton Pump Inhibitors

Agents that reduce gastric acid, such as antacids and proton pump inhibitors, can reduce bioavailability of certain drugs. Antacids include magnesium trisilicate and aluminum hydroxide.[40] Proton pump inhibitors include omeprazole, rabeprazole, lansoprazole, and esomeprazole.[41] An introduction to gastric physiology provides a diagram of the proton pump, its location in the parietal cells of the stomach, and the mechanism of action of omeprazole.[42] Concurrent use of proton pump inhibitors is a routine issue in NDA submissions for oral drugs.

van Leeuwen et al.[43] details the influence of a proton pump inhibitor on the bioavailability of an anticancer drug, erlotinib. In short, **omeprazole reduced erlotinib's exposure** by about half. Exposure was determined by the parameters AUC and Cmax. The term "exposure" is conventionally used in FDA-submissions to refer to one or more parameters that measure the concentration of a drug in the bloodstream. As a consequence, the package label for erlotinib instructs that proton pump inhibitors not be taken with erlotinib. In fact, proton pump inhibitors have been found to **decrease erlotinib's efficacy** in treating cancer, in addition to merely decreasing exposure.

van Leeuwen et al.[44] described a clever and inexpensive solution for patients who need to take proton pump inhibitors with erlotinib or other drugs where bioavailability is sensitive to gastric pH. The clever and inexpensive solution is to drink Coca-Cola® when taking erlotinib. Coca-Cola has a pH of 2.5, and the desired result of the Coca-Cola is to lower the pH in the stomach and thus avoid interference of the omeprazole with erlotinib's bioavailability. But other beverages, such as 7-Up® (pH 3.5) or orange juice (pH 4), have pH values that are considerably higher and thus would not be expected to be as effective as Coca-Cola.

The issue with erlotinib is that it is protonated in the environment of a typical gastric pH, but it is less ionized and less protonated at more alkaline pHs. As stated in FDA's ClinPharm Review, "Erlotinib ... is very slightly soluble in water ... [a]queous solubility of erlotinib ... is dependent on pH with **increased solubility at a pH of less than 5** due to protonation of the ... amine. Over the pH range of 1.4 to 9.6, maximal solubility of ... 0.4 mg/mL occurs at a **pH of approximately 2**"[45]

van Leeuwen et al.[46] generalized their technique for overcoming the bioavailability-inhibiting properties of omeprazole, as being potentially applicable to all drugs with a low

[40] Richter JE. Review article: the management of heartburn in pregnancy. Aliment Pharmacol. Ther. 2005;22:749–57.

[41] Shin JM, Sachs G. Pharmacology of proton pump inhibitors. Curr. Gastroenterol. Rep. 2008;10:528–34.

[42] Brody T. Nutritional Biochemistry. New York: Elsevier; 1999, p. 70–87.

[43] van Leeuwen RW, et al. Influence of the acidic beverage cola on the absorption of erlotinib in patients with non-small-cell lung cancer. J. Clin. Oncol. 2016;34:1309–14.

[44] van Leeuwen RW, et al. Influence of the acidic beverage cola on the absorption of erlotinib in patients with non-small-cell lung cancer. J. Clin. Oncol. 2016;34:1309–14.

[45] Erlotinib (non-small cell lung cancer) NDA 021-743. Page 14 of 87-page Clinical Pharmacology Review.

[46] van Leeuwen RW, et al. Influence of the acidic beverage cola on the absorption of erlotinib in patients with non-small-cell lung cancer. J. Clin. Oncol. 2016;34:1309–14.

pKa value. To this end, the authors stated that, "In theory, in patients with elevated intragastric pH and subsequent impaired absorption ... the use of cola may increase bioavailability of erlotinib or other TKIs with a relatively low pKa value."

To summarize, food effect studies encompass the influence of meals on gastric pH and the consequent influence of changed gastric pH on the bioavailability and exposure of the study drug. Although the present account of antacids and proton pump inhibitors and their influence on drug exposure is usually classified as a drug–drug interaction (DDI), this topic also reasonably fits into this food effect chapter.

Regarding antacids, the relevant DDIs are not limited to those relating to reduced gastric acid. Where the antacid is one that comprises a polyvalent cation, and where the antacid is taken concomitantly with a drug that chelates with polyvalent cations, such as aluminum, the following may result.[47] The result is that the drug can be inactivated by complexing with the aluminum.

This is the situation with deferasirox, which is an iron-chelating drug, where the package label warned against concomitant administration of deferasirox and aluminum-containing antacids.[48,49] Eltrombopag, a drug used for treating a platelet disorder[50] also binds to antacids that comprise polyvalent cations. FDA's ClinPharm Review observed that, "Plasma eltrombopag exposure was ... reduced by ... 70% when coadministered with polyvalent cations, e.g., antacids."[51] This finding found a place in the Dosage and Administration section of eltrombopag's package label, which cautions against antacids and polyvalent cations such as aluminum in its warning:

> **DOSAGE AND ADMINISTRATION** ... Allow at least a 4-hour interval between ... eltrombopag and other medications, e.g., **antacids**, calcium-rich foods ... or supplements containing polyvalent cations such as iron, calcium, **aluminum**, magnesium, selenium, and zinc.[52]

A problem can arise from antacids that contain sodium alginate, where the problem is "a raft of alginic acid gel." FDA's ClinPharm Review on eltrombopag, revealed a problem with sodium alginate antacids. FDA complained about this particular antacid, because, "on ingestion ... it reacts rapidly with gastric acid to form **a raft of alginic acid gel** ...

[47] Ogawa R, Echizen H. Clinically significant drug interactions with antacids: an update. Drugs. 2011;71:1839–64.

[48] Package label. EXJADE® (deferasirox) Tablets for oral suspension; November 2005 (15 pp.).

[49] Deferasirox (chronic iron overload due to blood transfusions) NDA 021-882.

[50] Jenkins JM, et al. Phase 1 clinical study of eltrombopag, an oral, nonpeptide thrombopoietin receptor agonist. Blood. 2007;109:4739–41.

[51] Eltrombopag (chronic immune thrombocytopenic purpura) NDA 022-291. Pages 5 and 24 of 85-page Clinical Pharmacology Review (1st of three pdf files).

[52] Package label. PROMACTA® (eltrombopag) tablets; October 2008 (21 pp.).

which floats on the stomach contents ... [t]he effect of this raft on the bioavailability of ... [the study drug] ... could be a confounding factor."[53] These alginate rafts have been described.[54,55]

h. Conundrum on Whether to Use Food to Enhance Drug Absorption or to Delay Drug Absorption

This describes an apparent contradiction regarding instructions to avoid food. Among the examples shown in this chapter, some illustrate the situation where food increases bioavailability and increases exposure (plasma levels) to a given study drug. Upon first impression, it might seem that it would always be desired for a patient to take an oral drug under conditions (with food) that increase bioavailability and exposure. But for some drugs where foods increase bioavailability and exposure, the package label instructs to avoid food. This seems like a contradiction. In response to a question from this author, an FDA official explained the contradiction, writing, "We understand your thinking, that getting more drug into your body seems like it would be beneficial, but this is not always the case. Increased bioavailability of a drug may sometimes be dangerous and lead to increased adverse reactions."[56]

Consistent with this explanation, another FDA official replied that, "For this drug, the recommendation was ... to taken in the fasted state, as increases in exposure greater than 2-fold are anticipated to substantially increase the risk of adverse reactions based on the observed safety data, PK/PD relationships, and PK data."[57]

An alternate explanation why increasing exposure by taking food is not necessarily beneficial was provided by yet another FDA official. The FDA official explained that greater drug exposure does not necessarily mean greater clinical efficacy, "And you will find cases where the effect of food negatively impacts a drug because it increases bioavailability and leads to increased adverse events **without improving efficacy**."[58]

[53] Eltrombopag (chronic immune thrombocytopenic purpura) NDA 022-291. Page 33 of 96-page Clinical Pharmacology Review (3rd of three pdf files).

[54] Hampson FC, et al. Alginate rafts and their characterisation. Int. J. Pharmaceuticals. 2005;294:137–47.

[55] Sweis R, et al. Post-prandial reflux suppression by a raft-forming alginate (Gaviscon Advance) compared to a simple antacid documented by magnetic resonance imaging and pH-impedance monitoring: mechanistic assessment in healthy volunteers and randomised, controlled, double-blind study in reflux patients. Aliment Pharmacol. Ther. 2013;37:1093–102.

[56] MM. Drug Information Specialist, Division of Drug Information, Center for Drug Evaluation and Research Food and Drug Administration (e-mail of March 28, 2017).

[57] KDe, Drug Information Specialist, CDER Small Business and Industry Assistance, Division of Drug Information, Center for Drug Evaluation and Research. Food and Drug Administration (e-mail of March 31, 2017).

[58] KDe, Drug Information Specialist, CDER Small Business and Industry Assistance, Division of Drug Information, Center for Drug Evaluation and Research. Food and Drug Administration (e-mail of April 26, 2017).

But this leads to another apparent contradiction. If adverse events (AEs) are an increased concern and create an enhanced danger to the patient, why not just require that the drug dose be smaller and that the drug be taken with a meal. An answer to this conundrum is provided by FDA's remarks for abiraterone. FDA's review for abiraterone explains why the package label requires avoiding food, "it is impractical to standardize meals in clinical practice a large variability in exposure may be seen if abiraterone acetate is dosed with a meal, because a high fat meal increases the AUC 10-fold and a low-fat meal increases the AUC 4-fold when compared to fasting conditions."[59]

Parsad and Ratain[60] also explain advantages of requiring taking a drug in the fasted state (empty stomach), rather than with food, even though food can enhance bioavailability, writing that, "Labeling should be easy to understand and follow, and aligned with food or fasting states that will enhance adherence. Overnight fasting could be recommended for once-daily agents required to be taken on an empty stomach, eliminating further restrictions on what type of food can be ingested before or after the fast."[61]

Another published comment provides a rationale that it is advantageous to require food with the drug, stating that, "a dosing schedule tied to routine meals will be easier for patients (particularly elderly cancer patients taking multiple oral medications) and can be a great way to improve adherence, which is recognized as a serious challenge in cancer treatments with oral agents."[62]

In response to a query from the author, an FDA official disclosed yet another advantage for taking food with a drug, "You will find cases in which the drug is administered with food. This may be done to alleviate local irritation in the GI tract."[63]

i. Biopharmaceutics Classification System

FDA's Guidance for Industry recognizes the classification of drugs according to Biopharmaceutics Classification System (BCS) class.[64] The BCS classes are:

- Class 1: High Solubility–High Permeability
- Class 2: Low Solubility–High Permeability

[59] Abiraterone (prostate cancer) NDA 202-379. Page 31 of 86-page Clinical Pharmacology Review.

[60] Parsad S, Ratain MJ. Food effect studies for oncology drug products. Clin. Pharmacol. Ther. 2017. doi: 10.1002/cpt.610.

[61] Parsad S, Ratain MJ. Food effect studies for oncology drug products. Clin. Pharmacol. Ther. 2017. doi: 10.1002/cpt.610.

[62] Kang SP, Ratain MJ. Inconsistent labeling of food effect for oral agents across therapeutic areas: differences between oncology and non-oncology products. Clin. Cancer Res. 2010;16:4446–51.

[63] KDe, Drug Information Specialist, CDER Small Business and Industry Assistance, Division of Drug Information, Center for Drug Evaluation and Research. Food and Drug Administration (e-mail of April 26, 2017).

[64] U.S. Department of Health and Human Services. Food and Drug Administration. Center for Drug Evaluation and Research (CDER). Guidance for industry. Waiver of in vivo bioavailability and bioequivalence studies for immediate-release solid oral dosage forms based on a biopharmaceutics classification system; 2015 (14 pp.).

- Class 3: High Solubility–Low Permeability
- Class 4: Low Solubility–Low Permeability

FDA's Guidance for Industry, as well as authors from FDA and various drug companies,[65] provide a definition of "solubility" that tracks the pH conditions found in the stomach. The pH conditions in the stomach reside in the range, pH 1–6.8. The following definition does not involve solubility testing under alkaline conditions, "A drug substance is considered **highly soluble when the highest strength is soluble** in 250 mL or less of aqueous media over the pH range of 1-6.8. The volume estimate of 250 mL is derived from ... study protocols that prescribe administration of a drug product to fasting human volunteers with a glass (about 8 ounces) of water."[66]

FDA's Guidance distinguishes between "solubility" and "dissolution," as is evident from the excerpt reading, "in vivo **dissolution** of an immediate release solid oral dosage form is rapid or very rapid in relation to gastric emptying and the drug has high **solubility**, the rate and extent of drug absorption is unlikely to be dependent on drug **dissolution**."[67]

As stated in a publication by FDA officials, dissolution testing can involve a pH solubility profile of the drug substance, dissolution profiles at different rotational speeds and dissolution media, as separately determined in media at various pHs, for example, in buffers with a pH 1.2, 4.5, and 6.8.[68]

Drug availability can decrease when exposed to an environment where the drug's overall charge is neutral (isoelectric point). For example, if a drug has one carboxylic acid group and one amino group, a solution that is at a pH that permits the carboxylic acid group to be charged (minus charge) and that also permits the amino group to be charged (protonated; plus charge), the result will be that the drug will not have an overall charge. Thus, any given drug can show reduced solubility when exposed to a specific acidic pH, neutral pH, or to an alkaline pH.[69] Regarding drug solubility and bioavailability, Lin et al.[70] stated that, "Solubility is ... an important determinant in drug absorption; a drug must be reasonably soluble in the aqueous environment to be absorbed properly ... poor aqueous solubility limited these inhibitors for oral delivery."

[65] Parr A, et al. The effect of excipients on the permeability of BCS class III compounds and implications for biowaivers. Pharm Res. 2016;33:167–76.

[66] U.S. Department of Health and Human Services. Food and Drug Administration. Center for Drug Evaluation and Research (CDER). Waiver of in vivo bioavailability and bioequivalence studies for immediate-release solid oral dosage forms based on a biopharmaceutics classification system guidance for industry; 2015 (14 pp.).

[67] U.S. Department of Health and Human Services. Food and Drug Administration. Center for Drug Evaluation and Research (CDER). Waiver of in vivo bioavailability and bioequivalence studies for immediate-release solid oral dosage forms based on a biopharmaceutics classification system guidance for industry; 2015 (14 pp.).

[68] Anand O, et al. Dissolution testing for generic drugs: an FDA perspective. AAPS J. 2011;13:328

[69] Lentz KA. Current methods for predicting human food effect. AAPS J. 2008;10:282–8.

[70] Lin JH, Lu AY. Role of pharmacokinetics and metabolism in drug discovery and development. Pharmacol. Rev. 1997;49:403–49.

Mathias et al.[71] detailed the use of in vitro dissolution studies and defined the concept of "in vitro–in vivo relationship." Dissolution studies can be conducted with the active drug ingredient or with a tablet or capsule that contains the active drug ingredient. The compound is added to a liquid medium and mixed at 37°C for a few hours with a magnetic stirring bar, where samples are taken every 5 minutes and used to determine the concentration of solubilized drug. The rate of dissolution is calculated in terms of micrograms per milliliter per minute.

Two different dissolution media are used: the first simulating fasted-state intestinal fluid and the second simulating fed-state intestinal fluid. This type of study assumes that the rate of dissolution of the active drug ingredient is the rate-limiting step (the controlling mechanism) that imposes any food effects that are observed with human subjects.

An example is provided with a lipophilic drug that dissolves very slowly in water. A 3-hour dissolution study with a stirring bar revealed that in fasted state intestinal fluid, dissolution was slow and resulted in solubilization to give a solution of 6 μg/mL after 3 hours. But in fed-state intestinal fluid, solubilization was greater, where after 3 hours of stirring the lipophilic drug reached a concentration of 15 μg/mL.

Dissolution media can take the form of fasted-state simulated gastric fluid, fed-state simulated gastric fluid, fasted-state intestinal fluid, and fed-state intestinal fluid.[72,73] For example, fasted-state simulated gastric fluid can take the form of diluted hydrochloric acid at pH 1.2 with pepsin (3.2 mg/mL).[74]

Consistent with the above example of Mathias et al.,[75] and regarding drugs that are not water-soluble, Lenz[76] teaches that absorption of lipophilic drugs can be increased by a high-fat meal and that lipophilic drugs taken with a high-fat meal may preferably be absorbed into the lymphatics instead of via the blood vessels, where the lymphatics deliver the lipophilic drugs via the thoracic duct to the bloodstream. In other words, intestinal lymphatics drain via the thoracic lymph directly into the systemic circulation. This feature of lymphatic absorption of drugs circumvents the first pass metabolic events that occur with absorption via the portal blood.[77] When absorption is via blood vessels, absorption of nutrients is by way of the portal vein, which leads from the gut to the liver.

[71] Mathias N, et al. Food effect in humans: predicting the risk through in vitro dissolution and in vivo pharmacokinetic models. AAPS J. 2015;17:988–98.

[72] Fotaki N, Vertoni M. Biorelevant methods and their applications in in vitro-in vivo correlations for oral formulations. The Open Drug Delivery Journal. 2010;4:2–13.

[73] Mathias N, et al. (2015) Food effect in humans: predicting the risk through in vitro dissolution and in vivo pharmacokinetic models. The AAPS Journal. 17:988–98.

[74] Fotaki N, Vertoni M. Biorelevant methods and their applications in in vitro-in vivo correlations for oral formulations. Open Drug Deliv. J. 2010;4:2–13.

[75] Mathias N, et al. Food effect in humans: predicting the risk through in vitro dissolution and in vivo pharmacokinetic models. AAPS J. 2015;17:988–98.

[76] Lentz KA. Current methods for predicting human food effect. AAPS J. 2008;10:282–8.

[77] Feeney OM, et al. 50 years of oral lipid-based formulations: provenance, progress and future perspectives. Adv. Drug Deliv. Rev. 2016;101:167–94.

Further details on dissolution testing methods and equipment are provided in the cited papers.[78,79] Dissolution of oral drugs, which can be improved by including a detergent such as sodium dodecyl sulfate (SDS) in the capsule formulation, was done with alectinib capsules.[80] On the other hand, where SDS is present in the capsule, the SDS can irritate the gastric mucosa and cause nausea, vomiting, diarrhea, and abdominal pain.

j. Influence of Food on Gastric pH and Stomach Emptying

Gastric pH and rate of stomach emptying can influence the availability of all oral drugs. For orientation, a time-course study is available, showing changes in gastric pH following meals of protein or of carbohydrate.[81] In detail, gastric pH and rate of emptying can influence exposure. Exposure is conventionally measured in terms of AUC, Cmax, Cmin, and tmax. FDA's Guidance for Industry recommends that, "a food-effect bioavailability study be conducted for all new chemical entities (NCEs) during the IND period."[82]

FDA's Guidance for Industry describes a "standard breakfast," as applicable to food effect studies.[83] The standard breakfast takes the form of, "Thirty minutes before drug administration, each subject should consume a standardized, high fat content meal consisting of one buttered English muffin, one fried egg, one slice of American cheese, one slice of Canadian bacon, one serving of hash brown potatoes, eight fluid oz. (240 mL) of whole milk, six fluid oz. (180 mL) of orange juice."[84]

Koziolek et al.[85] describe the use of an "FDA standard breakfast," the influence of this meal on gastric pH and emptying, and some of the implications on exposure. This account of the standard breakfast and gastric pH values read, "The intraluminal conditions of the fed stomach are critical for drug release from solid oral dosage forms and thus, often

[78] Fotaki N, Vertoni M. Biorelevant methods and their applications in in vitro-in vivo correlations for oral formulations. Open Drug Deliv. J. 2010;4:2–13.

[79] Technical Brief 2010 Volume 5. In vitro dissolution testing for solid oral dosage forms. Bethlehem, PA: Particle Sciences, Inc.

[80] Larkins E, et al. FDA approval: alectinib for the treatment of metastatic, ALK-positive non-small cell lung cancer following crizotinib. Clin. Cancer Res. 2016;22:5171–6.

[81] Lennard-Jones JE, et al. Effect of different foods on the acidity of the gastric contents in patients with duodenal ulcer. Gut. 1968;9:177–82.

[82] U.S. Department of Health and Human Services. Food and Drug Administration. Center for Drug Evaluation and Research (CDER). Guidance for industry. Food-effect bioavailability and fed bioequivalence studies; 2002 (12 pp.).

[83] U.S. Department of Health and Human Services. Food and Drug Administration. Center for Drug Evaluation and Research (CDER). Guidance for industry. Buspirone hydrochloride tablets in vivo bioequivalence and in vitro dissolution testing; 1998 (10 pp.).

[84] U.S. Department of Health and Human Services. Food and Drug Administration. Center for Drug Evaluation and Research (CDER). Guidance for industry. Buspirone hydrochloride tablets in vivo bioequivalence and in vitro dissolution testing; 1998 (10 pp.).

[85] Koziolek M, et al. Intragastric pH and pressure profiles after intake of the high-caloric, high-fat meal as used for food effect studies. J. Control Release. 2015;220(Part A):71–8.

associated with the occurrence of food effects on oral bioavailability ... intragastric pH and pressure profiles present after the ingestion of the high-caloric, high-fat ... FDA **standard breakfast** were investigated ... the **standard breakfast** impeded gastric emptying before lunch in 18 out of 19 subjects ... [t]he median pH value ... [initially] ranged between pH 3.3 and 5.3. Subsequently, the pH decreased ... and reached minimum values of pH 0-1 after approximately 4 h."[86]

II. FOOD EFFECTS WHERE THE DRUG IS INTENDED TO INFLUENCE GI PHYSIOLOGY OR TO INFLUENCE NUTRIENT PROCESSING

a. Canagliflozin (Type-2 Diabetes Mellitus) NDA 204-042

Canagliflozin is a small molecule for treating diabetes. The goal of the drug is to lower plasma glucose concentrations. The main target of the drug is the sodium/glucose co-transporter-2 (SGLT-2), more specifically, SGLT-2 located in the renal proximal tubule. The normal function of renal SGLT-2 is to reabsorb glucose that has been filtered in the glomerulus and that has become part of the nascent urine. Canagliflozin acts at renal SGLT-2 and prevents resorption of the glucose, thereby lowering plasma glucose.

An introduction to transporter physiology provides diagrams of cells of the renal tubule and of the gut, showing the locations of various transporters in the apical and basolateral membrane, including glucose transporters and ion transporters, as well as points of action of various drugs on the transporters.[87]

Two different sodium/glucose co-transporters exist, and these are SGLT-2, which is in the renal tubule, and SGLT-1, which is in the renal tubule and also in the gut (enterocytes). At a given concentration, inhibition of SGLT-2 (Ki = 4.0 nM) is much greater than inhibition of SGLT-1 (Ki = 770 nm). To provide scientific background, where the value for the Ki is relatively high, you need a higher concentration of inhibitor to achieve inhibition of the target. The target may be, for example, an enzyme or a transporter. Canagliflozin acts at the extracellular side (lumenal side) of both of these transporters.[88,89]

Thus, after an oral dose of canagliflozin, its concentration is relatively high in the gut lumen, where it is thought to inhibit gut SGLT-1. But after the drug is absorbed, circulated throughout the bloodstream, and then filtered into the nascent urine, it is much too dilute to cause significant inhibition of SGLT-1 in the renal tubule.

To reiterate, renal tubule SGLT-2 is the main site of action of canagliflozin, but in the time frame shortly after dosing, when canagliflozin is in the gut lumen and thus has direct access to the site of inhibition of SGLT-1. This site of action (SGLT-1) may be responsible

[86] Koziolek M, et al. Intragastric pH and pressure profiles after intake of the high-caloric, high-fat meal as used for food effect studies. J. Control Release. 2015;220(Part A):71–8.

[87] Brody T. Nutritional Biochemistry, 2nd ed. New York: Elsevier;1999, p. 113–115 and 705–721.

[88] Ohgaki R, et al. Interaction of the sodium/glucose cotransporter (SGLT) 2 inhibitor canagliflozin with SGLT1 and SGLT2. J. Pharmacol. Exp. Ther. 2016;358:94–102.

[89] Mori K, et al. Physiologically based pharmacokinetic-pharmacodynamic modeling to predict concentrations and actions of sodium-dependent glucose transporter 2 inhibitor canagliflozin in human intestines and renal tubules. Biopharm. Drug Dispos. 2016;37:491–506.

for the observed suppression of postprandial increase in plasma glucose. Canagliflozin's package label recommends taking **before the first meal** of the day. The above-disclosed scenario regarding gut SGLT-1 was used to justify this recommendation. Regarding the observed suppression by canagliflozin of postprandial glucose, FDA's Medical Review stated that, "Canagliflozen 300 mg given before a meal reduced the postprandial glucose excursion, which was not seen with 150 dose. This may be due to increased SGLT-1 inhibition in the intestinal lumen with the higher dose, before the drug gets absorbed."[90]

Commenting further on canagliflozin's inhibition of gut SGLT-1, FDA's ClinPharm Review stated, "It is speculated that after dosing, and during drug absorption, canagliflozin levels within the lumen ... could transiently be high enough to inhibit gastrointestinal SGLT-1-mediated glucose absorption and thereby reduce prandial plasma glucose excursions. Sponsor conducted a study ... to investigate the effect of canagliflozin (300 mg) on gastrointestinal glucose absorption ... in healthy subjects using a dual-tracer method ... [w]ith canagliflozin, the rate of systemic appearance of orally ingested radioactive-glucose (a measure of intestinal glucose absorption) was lower for the first 90 minutes compared to placebo."[91]

Package label. Consistent with FDA's comments, the Dosage and Administration section of the package label for canagliflozin (Invokana®) instructed:

> **DOSAGE AND ADMINISTRATION**. The recommended starting dose is 100 mg once daily, taken before the first meal of the day.[92]

b. Canagliflozin/Metformin Combination (Type-2 Diabetes Mellitus) NDA 204-353

This concerns a tablet that contains two drugs, canagliflozin and metformin, for treating Type-2 diabetes mellitus. Each of these two drugs has a different mechanism of action:

> **First mechanism of action:** Canagliflozin acts on the kidney to prevent resorption of glucose in the glomerular filtrate, thereby promoting the loss of glucose in the urine
> **Second mechanism of action:** Metformin decreases glucose production by the liver and improves the body's response to insulin[93]

This summarizes the issues and decision trees revealed by FDA's analysis of the canagliflozin/metformin combination:

1. **Twice as many experiments needed for fixed drug combination (FDC) drugs.** Where the drug takes the form of a FDC, what is needed is bioavailability data and exposure data for both of the drugs, and the influence of food on bioavailability and on exposure. This means, separate experiments are needed for each drug by itself (not in a combination form).

[90] Canagliflozin (type-2 diabetes mellitus) NDA 204-042. Page 28 of 255-page Medical Review.

[91] Canagliflozin (type-2 diabetes mellitus) NDA 204-042. Pages 36–37 of Clinical Pharmacology Review.

[92] Package label. Canagliflozin (INVOKANA) tablets, for oral use; February 2017.

[93] Fala L. Invokamet (canagliflozin plus metformin HCl): first fixed-dose combination with an SGLT2 inhibitor approved for the treatment of patients with type 2 diabetes. Am. Health Drug Benefits. 2015;8:70–4.

2. **Food effects can be significant or not significant.** Food effect studies need to measure the PK parameters of AUC, Cmax, and tmax. If the influence of food is not significant for any of these, then package label instructions for taking food may not be needed.
3. **Time for drug to reach targets.** Where the study drug has targets that are reached immediately after eating, such as receptors in the gut mucosa, and also has targets that are reached sometime after eating, such as receptors in the renal tubule lumen, then food effect studies will need to take into account the time factor.
4. **Dosage and administration instructions for drugs with food effects and for drugs that influence food processing.** Even if food effect studies show that the study drug's bioavailability is not influenced by food, such as food that is a high-fat meal or food that is excessive alcohol, the package label instructions might still need instructions on taking food where the goal of the study drug is to influence the body's processing of food or to influence the body's processing of one particular nutrient in the food.

In food effect studies for the canagliflozin/metformin combination, the Sponsor administered the tablet containing the drug combination to human subjects and determined PK values for canagliflozin and PK values for metformin. This study was conducted under two dietary conditions, a high-fat meal and with fasting. FDA's ClinPharm Review commented separately on canagliflozin PK and on metformin PK.

Regarding the canagliflozin component, FDA stated that, "The effect of **high fat meal** on the single-dose pharmacokinetics of the proposed ... tablet of CANA/MET IR at highest dose strength (150 mg/1000 mg) was examined in a randomized, open-label, single-dose, 2-period crossover study (DIA 1037) in 24 healthy adult subjects. The Cmax, AUC_{last}, and AUC_{inf} ratios of geometric means ... for **canagliflozin** between **fed and fasting conditions** were contained within the bioequivalence limits of 80 to 125% indicating no effect of food on the pharmacokinetics of the canagliflozin."[94]

About the metformin component, FDA observed the influence of the test meal on metformin PK, writing, "Regarding the **metformin** component of the canagliflozin/metformin ... tablet, this study demonstrated a decrease in Cmax of about 16% and no change in AUC_{last} or AUC_{inf}."[95]

Regarding the PK parameter of tmax for each drug component, FDA wrote, "Median tmax increased by approximately 2 hours for **canagliflozin** and approximately 1 hour for **metformin** FDC was administered under **fed conditions** compared with administration under fasted conditions."[96]

Regarding the influence of food on the PK parameter of Cmax, the influence on metformin exposure was a decrease in Cmax of only 16%.

FDA's conclusion was, "a food effect study ... evaluating the to-be-marketed CANA/MET IR FDC that showed that **food did not affect canagliflozin bioavailability** following

[94] Canagliflozin and metformin combination (type-2 diabetes mellitus) NDA 204-353. Pages 48–49 of 107-page Clinical Pharmacology Review.

[95] Canagliflozin and metformin combination (type-2 diabetes mellitus) NDA 204-353. Pages 48–49 of 107-page Clinical Pharmacology Review.

[96] Canagliflozin and metformin combination (type-2 diabetes mellitus) NDA 204-353. Pages 48–49 of 107-page Clinical Pharmacology Review.

single-dose administration of the 150/1,000 mg CANA/MET IR ... tablet."[97] Regarding metformin, it should be self-evident to the reader that a decrease of Cmax by only 16% represents essentially no food effect on metformin exposure.

Regarding alcohol, FDA's Medical Review merely stated that, "Labeled safety concerns with metformin include the following: Lactic acidosis: the risk increases with sepsis, dehydration, **excessive alcohol intake**, hepatic insufficiency, renal impairment, and acute congestive heart failure."[98]

The risk for lactic acidosis with metformin, especially where the patient has high alcohol intake, renal impairment, hepatic impairment, and so on, had already been established in the medical literature. For example, Hulisz et al.[99] stated that, "In addition to concurrent medications, it might also be wise to consider a patient's use of alcoholic beverages before prescribing metformin. Ethanol might reduce both the conversion of lactate to glucose and hepatic extraction of lactate. Consequently, heavy alcohol consumption (eg, binge drinking) sufficient to cause hepatic impairment could represent a relative contraindication to metformin use."

This provides food effect information from the DeFronzo et al. and Stein et al. publications. DeFronzo et al.[100] establishes that metformin is a low-risk drug and, at the same time, is a high-risk drug. It is low risk because of the rarity of lactic acidosis, but it is at the same time high risk because the lactic acidosis (when it occurs) is often fatal. DeFronzo et al. teach that, "Though metformin-associated lactic acidosis ... is an extremely rare condition (most estimates are ≤10 events per 100,000 patient-years of exposure), cases continue to be reported and are associated with mortality rates of 30 to 50%."[101]

Taken together, the above information does not suggest or imply that the study drug should be taken with a meal. On the other hand, the goal of the study drug is to improve the patient's metabolic responses to meals. In other words, the study drug is more logically taken with a meal than at other times. As revealed by the title of an article by Stein et al.,[102] "canagliflozin ... reduces post-meal glucose excursion." But this article title, without more, does not suggest or imply that the drug should be taken just before a meal or if the drug should be taken the day before a meal.

The publication from Stein et al.[103] went a step further and explored the time-dependence on the drug's efficacy. Canagliflozin's influence on its renal target and in

[97] Canagliflozin and metformin combination (type-2 diabetes mellitus) NDA 204-353. Pages 98–99 of 107-page Clinical Pharmacology Review.

[98] Canagliflozin and metformin combination (type-2 diabetes mellitus) NDA 204-353. Pages 12 and 70 of 119-page Medical Review.

[99] Hulisz DT, et al. Metformin-associated lactic acidosis. J. Am. Board Fam. Pract. 1998;11:233–6.

[100] DeFronzo R, et al. Metformin-associated lactic acidosis: current perspectives on causes and risk. Metabolism. 2016;65:20–9.

[101] DeFronzo R, et al. Metformin-associated lactic acidosis: current perspectives on causes and risk. Metabolism. 2016;65:20–9.

[102] Stein P, et al. Canagliflozin, a sodium glucose co-transporter 2 inhibitor, reduces post-meal glucose excursion in patients with type 2 diabetes by a non-renal mechanism: results of a randomized trial. Metabolism. 2014;63:1296–303.

[103] Stein P, et al. Canagliflozin, a sodium glucose co-transporter 2 inhibitor, reduces post-meal glucose excursion in patients with type 2 diabetes by a non-renal mechanism: results of a randomized trial. Metabolism. 2014;63:1296–303.

reducing urinary glucose excretion was an effect that was maximally retained for 24 hours (this statement concerns glucose excretion from renal tubules and does not concern glucose absorption by the gut). On the other hand, if canagliflozin was **administered just prior to breakfast**, there were greater reductions in plasma glucose, despite similar urinary glucose excretion to that found with **dosing 24 hours before breakfast**. The greater reduction found with dosing just before breakfast was due to the drug's influence on the gut.

The drug's influence on the gut took the form of, "transient inhibition of gut glucose transport by blockade of SGLT1 by high intraluminal concentrations of canagliflozin in the upper gastrointestinal tract after drug ingestion but prior to drug absorption."[104]

Thus, the package label's Dosage and Administration instructions for "Take twice daily with meals" was directed to the goal of ensuring canagliflozin would effectively block two different targets:

- Block SGLT1 in gut enterocytes
- Block SGLT1 in the renal tubule

Package label. The package label contained information about two types of food: (1) Excess alcohol and (2) Any kind of meal. The information about alcohol resided in the Black Box Warning:

> **BOXED WARNING: LACTIC ACIDOSIS.** Lactic acidosis can occur due to metformin accumulation. The risk increases with conditions such as renal impairment, sepsis, dehydration, excess **alcohol intake**, hepatic impairment, and acute congestive heart failure.[105]

The Dosage and Administration section of the package label contained instructions on meals:

> **DOSAGE AND ADMINISTRATION** ... Take twice daily with meals, with gradual dose escalation to reduce the gastrointestinal side effects due to metformin[106]

c. Ferric Citrate (Control Serum Phosphorus in Chronic Kidney Disease) NDA 205-874

Ferric citrate is a drug for controlling serum phosphorus in patients with chronic kidney disease. The mechanism of action of ferric citrate is to bind to dietary phosphate in the GI tract, preventing absorption and promoting loss in feces.

[104] Stein P, et al. Canagliflozin, a sodium glucose co-transporter 2 inhibitor, reduces post-meal glucose excursion in patients with type 2 diabetes by a non-renal mechanism: results of a randomized trial. Metabolism. 2014;63:1296–303.

[105] Package label. INVOKAMET (canagliflozin and metformin hydrochloride) tablets, for oral use; August 2014 (45 pp.).

[106] Package label. INVOKAMET (canagliflozin and metformin hydrochloride) tablets, for oral use; August 2014 (45 pp.).

Regarding the term "phosphorus," as used in describing phosphate metabolism, please note that free atoms of phosphorus never exist in biology. But it is convenient to use the term "phosphorus" when referring to dietary phosphorus or serum phosphorus. The term "phosphorus" avoids confusion from calculations based on amounts on the various types of phosphate compounds in the body, such as phosphoric acid, sodium phosphate, potassium phosphate, and phosphate esters.

Hyperphosphatemia occurs in patients with end-stage renal disease (ESRD). Therapy with dialysis insufficiently removes dietary phosphorus even with phosphorus-restricted diets. Phosphate binders can control hyperphosphatemia when dietary restriction and dialysis fail.[107] In addition to ferric citrate, other drugs for binding phosphate include aluminum-based binders, calcium-based binders, and sevelamer.

In FDA's Medical Review for ferric citrate, FDA commented that, "As with the approved iron and non-iron containing phosphate binders, adverse events were primarily limited to the gastrointestinal tract."[108] Although the location of the drug's effect is in the gut lumen, the consequent desired effect is to reduce plasma phosphate. Commenting on the goal of reducing serum phosphorus, the FDA reviewer remarked on the use of ferric citrate for treating patients with ESRD, writing that, "Serum phosphorus is an accepted surrogate endpoint for drug approval in the dialysis population (ESRD)."[109]

FDA stated that ferric citrate is to be taken with meals, writing that ferric citrate, "which contains ferric iron as the active ingredient, is a tablet proposed to be **taken with meals that binds phosphate in the food** content thereby reducing the intake of phosphate. As proposed by the applicant, the iron ... reacts with dietary phosphate in the GI tract and precipitates phosphate as ferric phosphate. This compound is insoluble and is excreted in the stool. By binding phosphate in the GI tract and decreasing absorption, ... [the study drug] lowers the serum phosphate concentration."[110]

Package label. FDA's comments that ferric citrate that should be taken with meals found a place on the package label. Ferric citrate's desired effect is not on any receptors and not on any enzymes, but instead, directly on the phosphate component of foods. Also, note that about **200 mg phosphorus** is excreted per day with the fluids of the GI tract, and in this case, the phosphate does not come directly from food (it comes from a person's own body and it is released into the gut lumen).[111] This is in contrast to the main source

[107] Van Buren PN, et al. The phosphate binder ferric citrate and mineral metabolism and inflammatory markers in maintenance dialysis patients: results from prespecified analyses of a randomized clinical trial. Am. J. Kidney Dis. 2015;66:479–88.

[108] Ferric citrate (control serum phosphorus in chronic kidney disease) NDA 205-874. Page 12 of 95-page Medical Review.

[109] Ferric citrate (control serum phosphorus in chronic kidney disease) NDA 205-874. Page 14 of 95-page Medical Review.

[110] Ferric citrate (control serum phosphorus in chronic kidney disease) NDA 205-874. Page 66 of 95-page Clinical Pharmacology Review.

[111] Brody T. Nutritional Biochemistry. New York: Elsevier; 1999, p. 773.

of phosphorus for the intestinal lumen (the diet), where the Recommended Daily Allowance (RDI) is **1250 mg phosphorus.**[112] The Dosage and Administration section required taking with meals:

> **DOSAGE AND ADMINISTRATION**. Starting dose is 2 tablets orally 3 times per day with meals.[113]

d. Lixisenatide (Type-2 Diabetes) NDA 208-471

Lixisenatide is for improving glycemic control in Type-2 diabetes mellitus. The drug is an agonist of GLP-1 receptor, where the drug activates the receptor for the endogenous incretin GLP-1. (The term "agonist" means "stimulant.") Lixisenatide and other drugs of the same class lower glucose levels by inhibiting the secretion of glucagon, promoting the release of insulin in response to hyperglycemia and slowing gastric emptying.[114] Slowing gastric emptying prolongs absorption of meal-derived glucose and dampens increases in postprandial glucose levels in the bloodstream.[115]

Lorenz et al.[116] provide a measure of the duration of efficacy of a single injection of lixisenatide, in comparison with placebo injections. With drug injection prior to breakfast, the postprandial glucose levels are dramatically flattened after a test meal, whereas in the time frame after lunch, postprandial glucose is not so much inhibited, and in the time frame after dinner (10 hours after breakfast), the morning's lixisenatide injection has little or no effect on dampening postprandial glucose levels.[117] Note that the drug influences blood glucose levels by slowing gastric emptying, as well as by stimulating insulin and suppress glucagon secretion.

The Sponsor measured gastric emptying time by measuring a stable isotope exhaled in the breath, "Gastric emptying time (from a ^{13}C-octanoic acid breath test) for the test breakfast meal on the 4th day of the 10 microgram dose level and on Day 28."[118]

[112] Code of Federal Regulations. 21 CFR §101.9 states that RDI for adults and children ≥4 years is 1.250 mg phosphorus.

[113] Package label. AURYXIA (ferric citrate) tablets, for oral use; July 2015 (9 pp.).

[114] Pfeffer MA, et al. Lixisenatide in patients with type 2 diabetes and acute coronary syndrome. New Engl. J. Med. 2015;373:2247–57.

[115] Lorenz M, et al. Effects of lixisenatide once daily on gastric emptying in type 2 diabetes—relationship to postprandial glycemia. Regul. Pept. 2013;185:1–8.

[116] Lorenz M, et al. Effects of lixisenatide once daily on gastric emptying in type 2 diabetes—relationship to postprandial glycemia. Regul. Pept. 2013;185:1–8.

[117] Lorenz M, et al. Effects of lixisenatide once daily on gastric emptying in type 2 diabetes—relationship to postprandial glycemia. Regul. Pept. 2013;185:1–8.

[118] Lixisenatide (diabetes) NDA 208-471. Page 35 of 198-page Clinical Pharmacology Review.

Carbon-13 is a stable isotope and is thus innocuous to human health. The ^{13}C-octanoic acid breath test for measuring gastric emptying time has been detailed.[119,120] The ^{13}C-octanoic acid is mixed with the food. The ^{13}C-labeled octanoic acid reaches the duodenum, where it is rapidly absorbed and transported to the liver where it is oxidized, so that the $^{13}CO_2/^{12}CO_2$ ratio over time provides a measure of gastric emptying.[121] The major sources of the $^{12}CO_2$ (in contrast to the $^{13}CO_2$) are the fats and carbohydrates stored in the body, as well as the fats and carbohydrates in the food.

Hence, gastric emptying can be measured with and without lixisenatide. Also, the ^{13}C-octanoic acid test can measure the amount of time needed, after lixisenatide injection, for the drug to take effect.

Lixisenatide's target, namely GLP-12 receptor, occurs in the vagus nerve and in the brain.[122] A person with training in pharmacology or nutrition can readily understand that for subcutaneous injections of lixisenatide to be effective in slowing gastric emptying, it should be injected **somewhat before** a meal. That said, it might be added that FDA's ClinPharm Review failed to disclose the data that were used to arrive at the package label's instruction regarding, "within one hour."

Meier et al.[123] describe the test meal used in the ^{13}C-octanoic test, in a narrative for a clinical study on lixisenatide's influence on gastric emptying time, "a standardized ^{13}C-labeled breakfast ... consisting of 281 kcal (16% protein, 62% fat, and 24% carbohydrate) and incorporating 91 mg ^{13}C-octanoic acid ... mixed with egg, was given to patients, and ^{13}C-octanoic acid breath tests were performed for evaluation of gastric emptying."

Meier et al.[124] also provide hour-by-hour data on the rate of appearance of carbon-13 in the breath, in the hours following consumption of the test meal, in the presence and absence of lixisenatide.

Fig. 3.1 is from the Sponsor's Briefing Document used at FDA's EMDAC Advisory Committee Meeting.[125] This Briefing Document was used at a meeting held May 25, 2016, where the Sponsor's NDA referred to this same meeting and date.[126] In Fig. 3.1, the upper

[119] Maes BD, et al. [13C]octanoic acid breath test to measure gastric emptying rate of solids. Dig. Dis. Sci. 1994;39(12 Suppl.):104S–106S.

[120] Perri F, et al. 13C-octanoic acid breath test for measuring gastric emptying of solids. Eur. Rev. Med. Pharmacol. Sci. 2005;9(5 Supp. 1):3–8.

[121] Lorenz M, et al. Effects of lixisenatide once daily on gastric emptying in type 2 diabetes—relationship to postprandial glycemia. Regul. Pept. 2013;185:1–8.

[122] Baggio LL, Drucker DJ. Glucagon-like peptide-1 receptors in the brain: controlling food intake and body weight. J. Clin. Invest. 2014;124:4223–6.

[123] Meier JJ, et al. Contrasting effects of lixisenatide and liraglutide on postprandial glycemic control, gastric emptying, and safety parameters in patients with type 2 diabetes on optimized insulin glargine with or without metformin: a randomized, open-label trial. Diabetes Care. 2015;38:1263–73.

[124] Meier JJ, et al. Contrasting effects of lixisenatide and liraglutide on postprandial glycemic control, gastric emptying, and safety parameters in patients with type 2 diabetes on optimized insulin glargine with or without metformin: a randomized, open-label trial. Diabetes Care. 2015;38:1263–73.

[125] Sanofi (May 9, 2016) Lixisenatide and iGlarLixi (insulin glargine/lixisenatide fixed-ratio combination) for the treatment of type 2 diabetes mellitus. Briefing Document (217 pp.).

[126] Lixisenatide (diabetes) NDA 208-471. 20-page Summary Review.

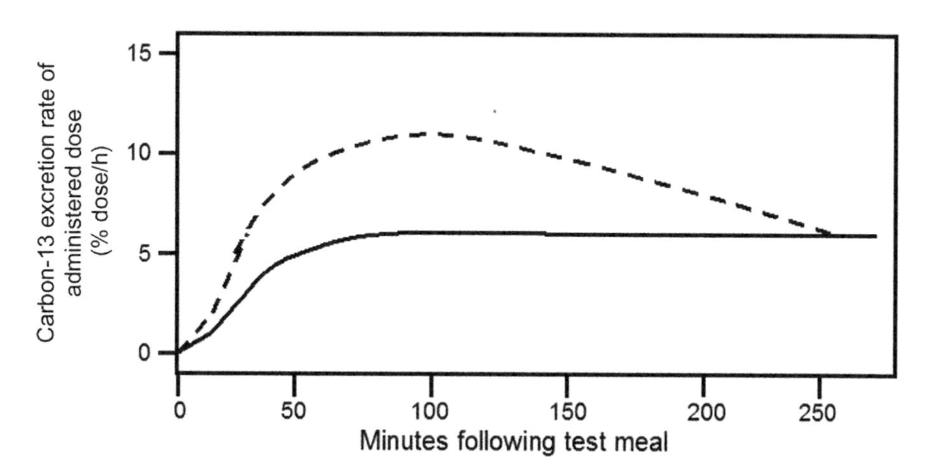

FIGURE 3.1 Rate of oxidation of stable isotope-labeled test meal. The dashed line is from subjects at baseline and the solid line is from subjects after 8 weeks of lixisenatide treatment.

curve (dashed line) shows the hour-by-hour carbon-13 in breath on the first day of the 8-week clinical study, with 20 μg lixisenatide injection, and the lower curve (solid line) shows the hour-by-hour carbon-13 in breath, with 20 μg lixisenatide injection on the last day of the 8-week study, showing greatly reduced rate of carbon-13 release.

A goal of the clinical study was to compare the drug's efficacy at baseline with the drug's efficacy after 8 weeks of treatment, that is, to measure, "change from baseline to week 8 in incremental area under the postprandial plasma glucose curve for 4 h after a standardized solid breakfast." Lixisenatide was administered every day.[127]

Regarding the greater effect found after the 8-week treatment, as compared to the effect on the first day (baseline), the Sponsor explained that, "the profound impact of lixisenatide ... after repeated administration over nearly 8 weeks demonstrates that this effect is durable and provides evidence of a lack of tachyphylaxis (attenuation of the pharmacodynamic effect over time)."[128]

FDA's ClinPharm Review recommended that lixisenatide be given before the first meal of the day. Lixisenatide is an injected drug, not an oral drug. FDA recommended that, "The proposed dosing regimen is that lixisenatide is administered once daily **within one hour prior to the first meal of the day** and can be injected subcutaneously in the abdomen, in the thigh or in the upper arm."[129]

[127] Meier JJ, et al. Contrasting effects of lixisenatide and liraglutide on postprandial glycemic control, gastric emptying, and safety parameters in patients with type 2 diabetes on optimized insulin glargine with or without metformin: a randomized, open-label trial. Diabetes Care. 2015;38:1263–73.

[128] Sanofi (May 9, 2016) Lixisenatide and iGlarLixi (insulin glargine/lixisenatide fixed-ratio combination) for the treatment of type 2 diabetes mellitus. Briefing Document (217 pp.).

[129] Lixisenatide (diabetes) NDA 208-471. Page 10 of 198-page Clinical Pharmacology Review.

Package label. The package label for lixisenatide (Adlyxin®) reflected FDA's recommendation in its instructions:

> **DOSAGE AND ADMINISTRATION**. Administer once daily within one hour before the first **meal** of the day ... Instruct patients to administer an injection of ADLYXIN within one hour before the first **meal** of the day preferably the same **meal** each day. If a dose is missed, administer ADLYXIN within one hour prior to the next **meal**.[130]

e. Pancrelipase (Exocrine Pancreatic Lipase Deficiency) NDA 022-210

Pancrelipase is an enzyme purified from pig pancreas. The enzyme replaces enzymes genetically missing in patients, and it is taken orally with a meal. Pancrelipase is used for patients with cystic fibrosis (pediatric and adult patients) and chronic pancreatitis (adult patients) who are not capable of digesting fats. The goal of pancrelipase is to improve the body's processing of food.

FDA's Guidance for Industry suggests a study design where a Sponsor can demonstrate clinical efficacy of a new pancrelipase product. This is to demonstrate that the product decreases stool fat in a 72-hour stool collection, where the decrease is greater than in a comparison group. For example, stool fat originally >14 g/day is reduced, with the pancrelipase, to under 7 g/day.[131]

In exocrine pancreatic lipase deficiency, fat malabsorption occurs when levels of lipase in the duodenum fall below 5%–10% of normal range. Also, in pancreatic lipase deficiency, there is a decreased output of bicarbonate, where the result is a more acidic pH in the duodenum, a consequent precipitation of bile salts, resulting in the inability of the bile salts to emulsify dietary fats.[132]

The package label included instructions for dose titration, and the basis for these instructions were as follows. Dose titration was directed to the goal of improving fat digestion, where fat digestion (or lack thereof) was detected by measuring fecal fat. The basis for the package label instructions rested on these sources:

1. **Cystic Fibrosis Foundation recommendation.** The Sponsor proposed that titration instructions be based on recommendations from the Cystic Fibrosis Foundation,[133] citing Dodge et al.[134] These recommendations include, "Children >4 years old: begin with 500 lipase units/kg meal. Doses in excess of 2,500 USP lipase units/kg body weight/meal should be used with caution and only when accompanied by documented three-day fecal fat measurements."

[130] Package label. ADLYXIN (lixisenatide) injection, for subcutaneous use; July 2016 (33 pp.).

[131] U.S. Department of Health and Human Services. Food and Drug Administration. Center for Drug Evaluation and Research (CDER). Guidance for industry. Exocrine pancreatic insufficiency drug products – submitting NDAs; 2006 (15 pp.).

[132] Struyvenberg MR, et al. Practical guide to exocrine pancreatic insufficiency - Breaking the myths. BMC Med. 2017;15:29.

[133] Pancrealipase (exocrine pancreatic lipase deficiency) NDA 022-210. Page 62 of 132-page Medical Review.

[134] Dodge JA, et al. Cystic fibrosis: nutritional consequences and management. Best Pract. Res. Clin. Gastroenterol. 2006;20:531–46.

2. **Clinical endpoint.** All Clinical Study Protocols used in clinical trials intended to assess efficacy require one or more efficacy endpoints. The Sponsor's clinical endpoint was, "the percentage of responders after one and two weeks of treatment. Responders were defined as patients without steatorrhea ($<30\%$ fecal fat content) and without signs/symptoms of malabsorption."[135]
3. **Titration clinical study.** FDA's Medical Review revealed the study design, where "The starting dose ... was **titrated** ... to control clinical symptoms of exocrine pancreas insufficiency ... [t]he dose was **titrated** by increases ... rounding to the nearest 5,000 lipase units/capsule ... [t]he total dose was not to exceed 10,000 lipase units/kg body weight/day ... [t]he actual dose ... was **titrated** based on the patient's malabsorption symptoms."[136]
4. **FDA's recommendation.** FDA's Summary Review concluded that the Sponsor's results supported the package labeling, writing, "The clinical efficacy studies ... to support product approval were reviewed ... [t]he clinical reviewers concluded that the efficacy findings support approval and labeling."[137]

Package label. The titration study design used by the Sponsor was mirrored in the package label instructions. The package label for pancrelipase (Zenpep®) provided titration instructions and referred to taking with a meal:

> **DOSAGE AND ADMINISTRATION** ... Children Older than 12 Months and Younger than 4 Years. Enzyme dosing should begin with 1,000 lipase units/kg of body weight per **meal** to a maximum of 2,500 lipase units/kg of body weight per **meal.**[138]

The fact that pancrelipase should be taken with a meal was further emphasized by the instructions:

> **DOSAGE AND ADMINISTRATION** ... Administration. ZENPEP should be swallowed whole. For infants or patients unable to swallow intact capsules, the contents may be **sprinkled on soft acidic food**, e.g., applesauce.[139]

f. Summary of the Above Food Effect Studies

This provides the take-home lessons from FDA's analyses of (1) Canagliflozin/metformin, (2) Lixisenatide, (3) Ferric citrate, and (4) Pancrelipase:

- If the goal of the drug is to influence the body's processing of food (without regard to individual nutrients), then the package label should reasonably state that the drug needs to be administered with or shortly before the meal.

[135] Pancrelipase (exocrine pancreatic lipase deficiency) NDA 022-210. Pages 75, 96, and 99 of 132-page Medical Review.

[136] Pancrealipase (exocrine pancreatic lipase deficiency) NDA 022-210. Pages 75, 96, and 99 of 132-page Medical Review.

[137] Abiraterone (prostate cancer) NDA 202-379. Page 8 of 16-page Summary Review.

[138] Package label. ZENPEP (pancrelipase) delayed release capsules; August 2009 (11 pp.).

[139] Package label. ZENPEP (pancrelipase) delayed release capsules; August 2009 (11 pp.).

- If the goal of the drug is to influence the body's processing of an individual nutrient in food such as glucose, then the package label should reasonably state that the drug needs to be administered with or shortly before the meal.
- If the goal of the drug is to alter events that are subjected to the drug's influence for only a **short duration** after taking the drug, such as an influence on drug targets in the enterocyte, then the package label should likely require taking the drug slightly before a meal or with a meal.
- But if the goal is to alter events that are subjected to the drug's influence for a **longer duration** of time, such as drug influence that is maintained over a 24-hour interval, then it might be adequate for the package label merely to require taking the drug once a day, without regard to timing with a meal.
- Guidance for drafting instructions for taking food, and for drafting dose–titration instructions, can come from medical journal publications, from position statements by organizations devoted to the disease in question, and from the Sponsor's own Clinical Study Protocol.

III. FOOD EFFECT STUDIES WITH ANTICANCER DRUGS

a. Abiraterone (Prostate Cancer) NDA 202-379

Abiraterone is a drug for prostate cancer. Regarding food effects on abiraterone, food increased exposure of abiraterone, where this increased exposure posed a risk for the adverse event of increased QTc interval. Increased QTc interval is a cardiac AE.

The food effect study was designed so that each subject served as his own control, and where a "wash-out" period of 1 week separated the different test meals. FDA described the methodology as, "study to determine the effect of food on the pharmacokinetics of abiraterone ... in healthy male subjects. The three periods tested the effect of **fasting** (overnight fast and no food for 4 hours post dose), a **low fat meal** (298.7 total calories of calories with 7.3% from fat) or a **high fat meal** (total calories of 826.3 with 56.5% from fat) on the PK of abiraterone. PK samples were collected ... and a 7 day wash-out was used between treatments."[140]

Note the definitions of low- and high-fat meals in terms of total calories from fat. Then, FDA described the results of food on drug exposure, where exposure was in terms of the parameters, Cmax and AUC, "At 1000 mg, systemic exposure of abiraterone, increased with the administration of food compared to the **fasted state** ... [c]ompared with the **fasted state**, the ... mean for abiraterone Cmax and $AUC_{0\text{-infinity}}$ increased by approximately 7- and 5-fold, respectively, when administered with a **low-fat meal** and by approximately 17- and 10-fold, respectively, when administered with a high-fat meal."[141]

As part of FDA's observations on the food effect of increasing abiraterone exposure, FDA observed that low- and high-fat food increased risk for AEs. The AE of concern was increases from baseline in the QTc interval, where this increase was >30 milliseconds.

[140] Abiraterone (prostate cancer) NDA 202-379. Page 45 of 86-page Clinical Pharmacology Review.

[141] Abiraterone (prostate cancer) NDA 202-379. Page 45 of 86-page Clinical Pharmacology Review.

With the low- and high-fat meals, the increases were in the range of 31–39 milliseconds.[142] Details on the influence of abiraterone on the QTc interval have been published.[143,144]

Thus, at this point in the train of logic, one might propose the following two approaches to reducing risk for an increased QTc interval that was >30 milliseconds:

- An approach where lower drug doses are used, where patients eat food
- Alternatively, an approach to take a higher drug dose, with fasting

For a variety of drugs, including abiraterone, where food increases exposure, the package label requires that **food not be taken**. In other words, instead of requiring a lower dose where patients are required to take food, the package label requires that food be not taken. FDA's ClinPharm Review for abiraterone provides a rationale for avoiding food, namely, that it would be impractical to require that patients standardize their meals. FDA's comment, which should be contemplated with drafting package labels to any kind of oral drug, was, "However, since it is **impractical to standardize meals** in clinical practice a large variability in exposure may be seen if abiraterone acetate is dosed with a meal, because a high fat meal increases the AUC 10-fold and a low fat meal increases the AUC 4-fold when compared to fasting conditions (overnight fast and fast for 4 hours post-dose)."[145]

Package label. FDA's comment on the fact that it is "impractical to standardize meals" found a place on the package label of abiraterone (Zytiga®). The package label required that food not be taken:

> DOSAGE AND ADMINISTRATION ... ZYTIGA must be taken on an empty stomach. No food should be consumed for at least two hours before the dose of ZYTIGA is taken and for at least one hour after the dose of ZYTIGA is taken.[146]

b. Afatinib (Melanoma With Exon 19 Deletion or Exon 21 L858R Mutation) NDA 201-292

Afatinib is a small molecule for treating melanoma, where the melanoma tumors were characterized by specific genetic mutations. Afatinib inhibits the tyrosine kinase activity of EGFR receptor. Food decreased exposure of afatinib. Exposure was in terms of tmax, AUC, and Cmax.

The food effect study revealed that food induced a delay in tmax and that food reduced the AUC and Cmax. These results were the basis for the package label's instruction to avoid food. FDA's ClinPharm Review provides a basis for the dose (number of milligrams)

[142] Abiraterone (prostate cancer) NDA 202-379. Page 45 of 86-page Clinical Pharmacology Review.

[143] Khan A. Kneale B. Life threatening torsades de pointes due to abiraterone-induced hypokaelemia in a patient with metastatic prostate cancer. New Zealand Med. J. 2016;129:124–7.

[144] Tolcher AW, et al. Effect of abiraterone acetate plus prednisone on the QT interval in patients with metastatic castration-resistant prostate cancer. Cancer Chemother. Pharmacol. 2012;70:305–13.

[145] Abiraterone (prostate cancer) NDA 202-379. Page 31 of 86-page Clinical Pharmacology Review.

[146] Package label. ZYTIGA™ (abiraterone acetate) tablets; April 2011 (23 pp.).

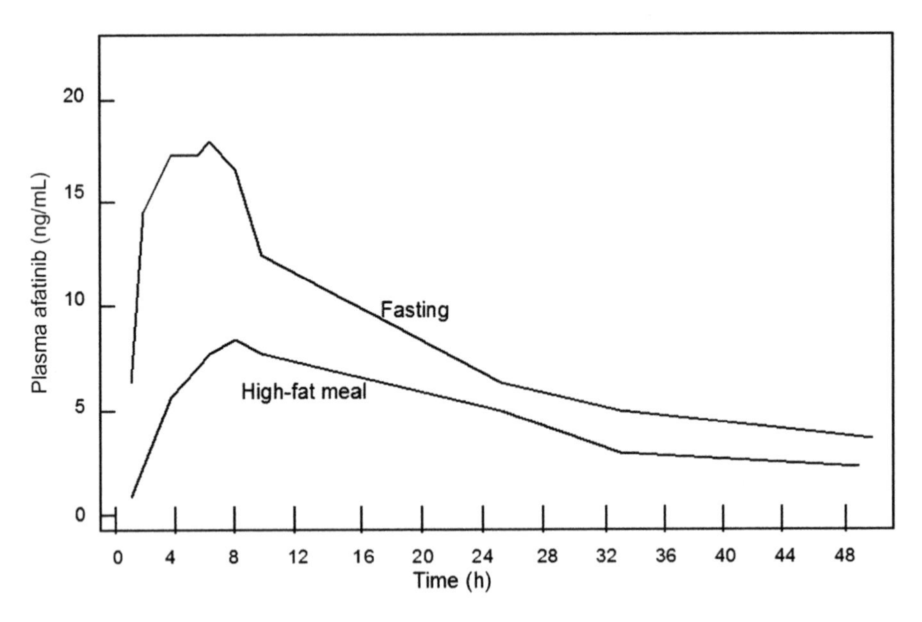

FIGURE 3.2 Afatinib plasma levels. Blood plasma concentrations of the anticancer drug afatinib, over time, are shown after oral tablets are taken with fasting (upper curve) or with a high-fat meal (lower curve).

recited on the package label, in addition to a basis for instructions to avoid food. FDA's dose recommendation for 40 mg was based on adverse events, that is, on the fact that subjects given the higher dose of 50 mg required dose reductions to avoid AEs.[147] FDA's recommendation for 40 mg found a place on the package label, as shown below.

FDA referred to a figure from the Sponsor's food effect study. This figure is reproduced below (Fig. 3.2). FDA's reasoning and recommendations were, "Based on the ... decrease in afatinib exposure (39% in $AUC_{0\text{-}inf}$ and 50% in Cmax) after a high-fat meal, as compared to that under the fasted condition, **FDA recommends afatinib to be taken at least one hour before or two hours after a meal.**"[148] To this point, FDA further stated, "The effect of a high fat/high caloric meal taken 30 minutes before afatinib on the PK of a ... dose of 40 mg afatinib ... was evaluated ... [t]he median tmax was delayed from 3.0 hour to 6.9 hour after a high fat meal."[149] FDA recommended that food be avoided when taking afatinib.

[147] Afatinib (melanoma with exon 19 deletion or exon 21 L858R mutation) NDA 201-292. Page 14 of 126-page Clinical Pharmacology Review.

[148] Afatinib (melanoma with exon 19 deletion or exon 21 L858R mutation) NDA 201-292. Page 14 of 126-page Clinical Pharmacology Review.

[149] Afatinib (melanoma with exon 19 deletion or exon 21 L858R mutation) NDA 201-292. Pages 49–50 of 126-page Clinical Pharmacology Review.

Package label. FDA's recommendations for the dose (40 mg) and for avoiding food found a place on the Dosage and Administration section of the package label for afatinib (Gilotrif®):

> **DOSAGE AND ADMINISTRATION**. Recommended dose: 40 mg orally, once daily ... [i]nstruct patients to take GILOTRIF at least 1 hour before or 2 hours after a meal.[150]

c. Cabozantinib (Metastatic Medullary Thyroid Cancer) NDA 203-756

Food increased exposure of cabozantinib, where this increased exposure resulted in increased risk for cardiac AEs. The package label of cabozantinib instructed patients not to eat food and to take the drug either well before or well after any meal. The package label added that the goal of these instructions was to avoid increases in exposure that could provoke QT prolongation. FDA's ClinPharm Review provided comments on a food effect study that used a high-fat diet. A food effect was detected, and this was an increase in exposure by about 50%. FDA commented:

> What is the effect of food on the bioavailability ... of the drug from the dosage form? What dosing recommendation should be made, if any, regarding administration of the product in relation to meals or meal types? The effect of food on the pharmacokinetics of a single dose of cabozantinib ... was evaluated in a ... study of 47 evaluable healthy subjects ... [t]he Cmax and AUC values ($AUC_{0\text{-}t}$ and $AUC_{0\text{-}inf}$) were ... increased by 41% and 57%, respectively, when cabozantinib was administered with a high-fat, high calorie meal (Figure 13).[151]

The figure referred to by FDA's comments is reproduced in Fig. 3.3. The fast-rising lower curve (dashed line) is fasted subjects. The slow-rising upper curve (solid line) is fed subjects. The *X*-axis is time in hours. The *Y*-axis is exposure, where the scale is logarithmic. Fig. 3.3 shows that the high-fat meal increased exposure by about 50%.

The Sponsor proposed that, "The proposed dosage of cabozantinib is 140 mg administered orally once daily (QD), **taken at least 1 hour before or 2 hours after a meal**."[152] FDA agreed with this proposal, writing, "FDA recommended label ... [t]he recommended daily dose of COMETRIQ is 140 mg (one 80-mg and three 20-mg capsules) **taken at least 1 hour before or 2 hours after a meal**."[153]

FDA's Other Reviews expressly stated that the danger of increased cabozantinib was increased risk for the cardiac adverse event of QT prolongation. Regarding cabozantinib, FDA's Other Reviews stated, "Administration of cabozantinib, under fasted conditions, is appropriate. Based on a dedicated food-effect study, **Cmax and AUC were moderately increased by 41% and 57%, respectively, when cabozantinib was administered with a**

[150] Package label. GILOTRIF® (afatinib) tablets, for oral use; October 2016 (19 pp.).

[151] Cabozantinib (metastatic medullary thyroid cancer) NDA 203-756. Page 29 of 106-page Clinical Pharmacology Review.

[152] Cabozantinib (metastatic medullary thyroid cancer) NDA 203-756. Page 44 of 106-page Clinical Pharmacology Review.

[153] Cabozantinib (metastatic medullary thyroid cancer) NDA 203-756. Page 34 of 106-page Clinical Pharmacology Review.

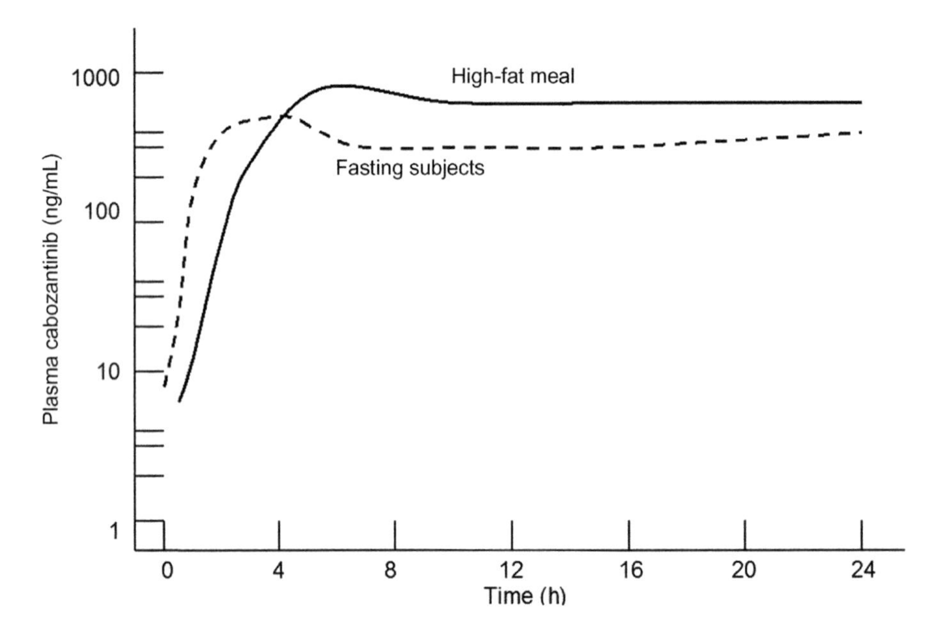

FIGURE 3.3 Cabozantinib plasma concentrations. Blood plasma cabozantinib levels are shown when the oral drug is taken with fasting (dashed curve) or with a high-fat meal (solid curve).

high-fat, high calorie meal ... [t]he proposed label stipulates cabozantinib should be taken under fasting conditions. Patients who take cabozantinib with food may be at greater risk of QT prolongation."[154]

Package label. The Dosage and Administration section of the package label for cabozantinib (Cometriq®) reflected FDA's recommendation:

> **DOSAGE AND ADMINISTRATION** ... Instruct patients **not to eat for at least 2 hours before and at least 1 hour** after taking COMETRIQ.[155]

d. Dabrafenib (Melanoma With BRAF V600E Mutation) NDA 202-806

Dabrafenib is a small-molecule drug for treating melanoma, where the indication required that the melanoma cells be characterized by a specific genetic mutation. The mutation is V600E (valine changed to glutamate, at amino acid 600 in the polypeptide chain). Food decreased exposure of dabrafenib. This provides the example of a package label instructing patients that they must not eat food when taking the drug. The Sponsor conducted a food effect study that measured bioavailability by way of the exposure parameters of Cmax and $AUC_{0-\infty}$.[156] The food effect study compared exposure with fasting subjects versus subjects fed a high-fat meal. Table 3.1 provides the exposure data in fasting versus fed subjects.

[154] Cabozantinib (metastatic medullary thyroid cancer) NDA 203-756. Page 39 of 111-page Other Reviews.

[155] Package label. COMETRIQ™ (cabozantinib) capsules, for oral use; November 2012 (19 pp.).

[156] Dabrafenib (melanoma with BRAF V600E mutation) NDA 202-806. Page 33 of 83-page Clinical Pharmacology Review.

TABLE 3.1 Influence of Food on Dabrafenib Exposure

	Fasting	Fed
Cmax	2160 ng/mL	1006 ng/mL
$AUC_{0-\infty}$	12126 (ng)(hour)/mL	8415 (ng)(hour)/mL

From: Dabrafenib (melanoma with BRAF V600E mutation) NDA 202-806. Page 33 of 83, Clinical Pharmacology Review.

The Sponsor proposed that the package label require administering dabrafenib under fasting conditions, either 1 hour before or 2 hours after a meal, in view of the advantageous increase in drug exposure with fasting. FDA agreed with the Sponsor's proposal, writing, "Since ... dabrafenib taking with a high fat meal resulted in a 30% decrease in AUC and 50% decrease in Cmax and the clinical efficacy was established based on dabrafenib administration under fasted conditions, the review team recommends that do not administer dabrafenib with a high fat meal."[157]

Please note the critical concept in FDA's narrative, namely that "the clinical efficacy was established based on dabrafenib administration under fasted conditions." What this means, is that if the Sponsor wants to convince FDA that the drug will be effective when administered to patients in the postmarketing situation, then the package label will need to require that administration be under the same conditions (fasting) as was used during the Sponsor's clinical studies.

Package label. The Sponsor's proposal found a place in the package label for dabrafenib (Tifinlar®), which provided the instruction:

> **DOSAGE AND ADMINISTRATION** ... The recommended dose of TAFINLAR is 150 mg orally twice daily taken at least 1 hour before or at least 2 hours after a meal.[158]

e. Osimertinib (Metastatic, Epidermal Growth Factor Receptor T790M Mutation-Positive, Non-Small-Cell Lung Cancer) NDA 208-065

Osimertinib (Tagrisso®) is a small-molecule drug for treating lung cancer. Food had almost no influence on osimertinib exposure. FDA's ClinPharm Review made recommendations on the need for food effect studies for the subgroup of patients with renal impairment and for the subgroup of patients with hepatic impairment.

FDA's ClinPharm Review stated, "The recommended dose of Tagrisso is 80 mg once daily taken ***without regard to food***. Administration of 20 mg osimertinib Phase 1 tablet with **a high-fat breakfast showed a minimal increase in Cmax (14%) and AUC (19%)**. Pre-dosing of 40 mg omeprazole tablets for 5 days had no clinically meaningful impact on the exposure of osimertinib at a single 80 mg dose. The appropriate dose of Tagrisso has

[157] Dabrafenib (melanoma with BRAF V600E mutation) NDA 202-806. Page 33 of 83-page Clinical Pharmacology Review.

[158] Package label. TIFINLAR® (dabrafenib) capsules, for oral use; June 2017 (31 pp.).

not been established in patients with **severe renal impairment** or **end-stage-renal disease**. The appropriate dose of Tagrisso has not been determined in patients with moderate or severe **hepatic impairment**."[159]

Regarding patients with hepatic impairment, FDA imposed a postmarketing requirement (PMR), requiring that the Sponsor "conduct a pharmacokinetic trial to determine the appropriate dose of Tagrisso in patients with hepatic impairment."[160] Regarding patients with renal impairment, FDA recommended, "However, further clinical studies should be conducted to evaluate the effects of hepatic/renal impairment on PK of osimertinib."[161] A view of the earliest available package label (November 2015) reveals, on the Use in Specific Populations Section, that there were not any clinical studies on subjects with renal or hepatic impairment. A view of the most recent package label (October 2017) reveals that these studies had still not been carried out ("The pharmacokinetics of osimertinib in patients with end-stage renal disease . . . or with severe hepatic impairment . . . are unknown.").

FDA's recommendation about food was based on clinical studies, where subjects were given a high-fat breakfast or where subjects were given omeprazole. FDA's Pharmacology Review provided data from the omeprazole experiment, where the data showed that omeprazole had little influence on exposure of osimertinib, "Gastric pH modifiers: osimertinib Cmax and AUC were 2% and 7% higher when osimertinib ... tablet was administered to healthy volunteers with prior dosing of 40 mg omeprazole tablets for 5 days to elevate the gastric pH."[162]

Package label. The package label reflected FDA's recommendation that osimertinib should be "taken without regard to food." The package label for osimertinib (Tagrisso®) instructed:

> **DOSAGE AND ADMINISTATION** ... 80 mg orally once daily, with or without food.[163]

f. Venetoclax (Chronic Lymphocytic Leukemia With 17p Deletion) NDA 208-573

Venetoclax is a small-molecule drug for the indication of chronic lymphocytic leukemia.[164] The drug inhibits the activity of a protein called, "B-cell lymphoma-2."[165]

Food increased exposure of venetoclax. The Sponsor and FDA both agreed that this increase was desirable. FDA's ClinPharm Review focused on the design of the food effect

[159] Osimertinib (metastatic, epidermal growth factor receptor (EGFR) T790M mutation-positive, non-small cell lung cancer (NSCLC)) NDA 208-065. 90-page Clinical Pharmacology Review.

[160] Osimertinib (metastatic, epidermal growth factor receptor (EGFR) T790M mutation-positive, non-small cell lung cancer (NSCLC)) NDA 208-065. Page 5 of 90-page Clinical Pharmacology Review.

[161] Osimertinib (metastatic, epidermal growth factor receptor (EGFR) T790M mutation-positive, non-small cell lung cancer (NSCLC)) NDA 208-065. Page 49 of 90-page Clinical Pharmacology Review.

[162] Osimertinib (metastatic, epidermal growth factor receptor (EGFR) T790M mutation-positive, non-small cell lung cancer (NSCLC)) NDA 208-065. Page 23 of 90-page Clinical Pharmacology Review.

[163] Package label. TAGRISSO™ (osimertinib) tablet, for oral use; November 2015 (14 pp.).

[164] Roberts AW, et al. Targeting BCL2 with venetoclax in relapsed chronic lymphocytic leukemia. New Engl. J. Med. 2016;374:311–22.

[165] Rogers KA, Byrd JC. Venetoclax adds a new arrow targeting relapsed CLL to the quiver. Cancer Cell. 2016;29:3–4.

study. In general, the following food effect study designs are relevant where a drug can be used for single dosing or, alternatively, for multiple dosing (repeated dosing for many days). The unique feature of multiple dosing is that the concept of steady-state plasma concentrations becomes relevant. But the concept of steady-state plasma concentrations is not relevant to a single-dose drug.

The following types of study designs are available:

1. Comparing drug exposure with or without food where subjects taking food get multiple drug dosing. The control subjects (no food) get multiple drug dosing.
2. Comparing drug exposure with or without food where subjects taking food get multiple drug dosing. The control subjects (no food) get only single drug dosing.
3. Comparing drug exposure with or without food where subjects taking food get single drug dosing. The control subjects (no food) get single drug dosing.
4. Comparing drug exposure with or without food where subjects taking food get single drug dosing. The control subjects (no food) get multiple drug dosing.

As might be self-evident to the reader, study designs (1) and (3) are appropriate, whereas, in contrast, study designs (2) and (4) make no sense.

FDA's ClinPharm Review described the basis for requiring that the drug be taken with a meal. The food effect was that food enhanced exposure. FDA's analysis dwelled on the methodology for conducting food effect studies, where the issue was use of exposure data acquired after a **single drug dose** versus exposure data acquired after **multiple doses** (with one dose per day).

The FDA reviewer contemplated the influence of food, either high-fat meals or low-fat meals, on plasma venetoclax levels, and observed that food increased drug exposure, writing, "What are the characteristics of drug absorption? In patients ... venetoclax plasma concentrations peaked at 5 to 8 hours after single and multiple doses administration with food. Administration with low- and high-fat meals increased venetoclax PK exposure by 3 to 5 fold[166] ... effect of food on venetoclax PK was evaluated ... following a high-fat meal (Week 1 Day -7), and a low-fat meal (Week 6 Day 1) relative to the same patients dosed under fasting condition (Week 1 Day 1)."[167]

But the FDA reviewer went a step further by complaining about a mistake in the study design. The test with the high-fat meal was paired with an appropriate control, but the test with the low-fat meal was paired with the wrong type of control. Regarding high-fat meals, the FDA reviewer wrote, "The results showed that **single dose** of venetoclax with **high-fat meals** increased venetoclax Cmax and AUC_{inf} approximately 4-fold compared to **single dose** under **fasting condition**."[168]

Unfortunately, the test with the low-fat meal was paired with the wrong type of control. The goal of the experiment was to compare exposure when the drug was taken with a low-fat meal with exposure when the drug was taken with fasting. The Sponsor's mistake

[166] Venetoclax (chronic lymphocytic leukemiawith 17p deletion) NDA 208-573. Page 13 of 75-page Clinical Pharmacology Review.

[167] Venetoclax (chronic lymphocytic leukemiawith 17p deletion) NDA 208-573. Pages 44–45 of 75-page Clinical Pharmacology Review.

[168] Venetoclax (chronic lymphocytic leukemia with 17p deletion) NDA 208-573. Pages 44–45 of 75-page Clinical Pharmacology Review.

was that the food effect experiment with the low-fat meal was with steady-state levels of the drug in the bloodstream, whereas, in contrast, fasting subjects had received only a single dose of drug (because of the single dose regimen, the plasma drug levels could not have achieved any form of steady state).

Regarding this mistake, FDA wrote, "[t]he dose-normalized AUC after multiple doses (steady state) of venetoclax with low-fat meal was approximately 4-fold of that from single dose under fasting condition. However, this comparison might overestimate the food effect from low-fat meal because **PK data for comparison were those after single dose under fasting condition, not at steady state**."[169]

The mistake did not interfere with FDA's ability to arrive at conclusion, and the Sponsor and FDA both agreed that, "The applicant proposed that venetoclax should be **taken with a meal to ensure adequate oral bioavailability**. The applicant also stated that the specification of the fat content of the meals taken with venetoclax is not necessary because there was only a 1.5-fold increase of venetoclax exposure from low-fat to high-fat meal and the exposure-response relationship at steady state dose of 400 mg is relatively flat, which is acceptable."[170]

Package label. The Sponsor's finding that exposure was increased with meals, whether high fat or low fat, found a place in the Dosage and Administration section of the package label. The package label for venetoclax (Venclexta®) instructed:

> **DOSAGE AND ADMINISTRATION** ... VENCLEXTA tablets should be taken orally once daily with a meal and water. Do not chew, crush, or break tablets.[171]

g. Regorafenib (Colorectal Cancer) NDA 203-085

The NDA for regorafenib (Stivarga®) included a food effect study intended to answer the questions: What is the effect of food on the bioavailability of the drug from the dosage form? What dosing recommendation should be made regarding administration of the product in relation to meals or meal types?[172].

Subjects took a single 160 mg dose of regorafenib with 8 ounces of water shortly after the test meal.[173] Blood samples were collected for up to 336 hours and used to calculate PK data. The Sponsor's low-fat breakfast took the following form:

> Two slices of white toast with 1 tablespoon of low-fat margarine and 1 tablespoon of jelly and 8 ounces of skim milk. (319 calories and 8.2 grams of fat). One cup of cereal (i.e., Special K), 8 ounces of skimmed milk, one piece of toast with jam (no butter or marmalade), apple juice, and one cup of coffee or tea (2 g fat, 17 g protein, 93 g of carbohydrate, 520 calories).[174]

[169] Venetoclax (chronic lymphocytic leukemia with 17p deletion) NDA 208-573. Pages 44–45 of 75-page Clinical Pharmacology Review.

[170] Venetoclax (chronic lymphocytic leukemia with 17p deletion) NDA 208-573. Pages 44–45 of 75-page Clinical Pharmacology Review.

[171] VENCLEXTA™ (venetoclax) tablets, for oral use; April 2016 (22 pp.).

[172] Regorafenib (colorectal cancer) NDA 203-085. Page 34 of 64-page Clinical Pharmacology Review.

[173] Regorafenib (colorectal cancer) NDA 203-085. Page 24 of 76-page Medical Review.

[174] Regorafenib (colorectal cancer) NDA 203-085. Page 34 of 64-page Clinical Pharmacology Review.

The Sponsor's high-fat breakfast took this form:

> A high-fat breakfast was defined as two eggs fried in butter, two slices of white toast with two pats of butter, two strips of bacon, four ounces of hash brown potatoes, and eight ounces of whole milk (approximately 945 calories and 54.6 grams of fat).[175]

As stated above, the content of thousands of foods, in terms of weight, fat, carbohydrate, protein, and energy, are disclosed in a book known as "Handbook 8," enabling the calculation of these parameters present in any meal.[176]

The food effect study took into account two of regorafenib's metabolites, namely, "M2" and "M5." M2 is the same as regorafenib, but with a gain of a hydroxyl group on the heterocyclic nitrogen. M5 is the same as regorafenib, but with a gain of a hydroxyl group on the heterocyclic nitrogen, and with loss of the methyl group on the amide nitrogen.[177] Strumberg et al.[178] and Sunakawa et al. [179] published accounts of plasma concentrations of regorafenib, M2, and M5, over the time course of 96 hours after administering regorafenib. Regorafenib has the following structure:

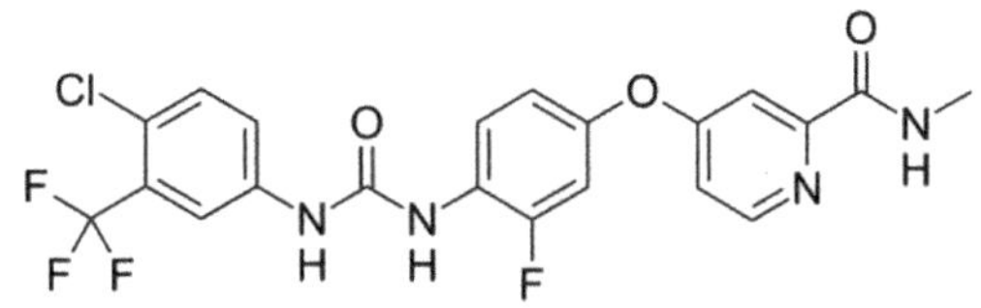

Regorafenib

The structures of M2 and M5 are shown below:

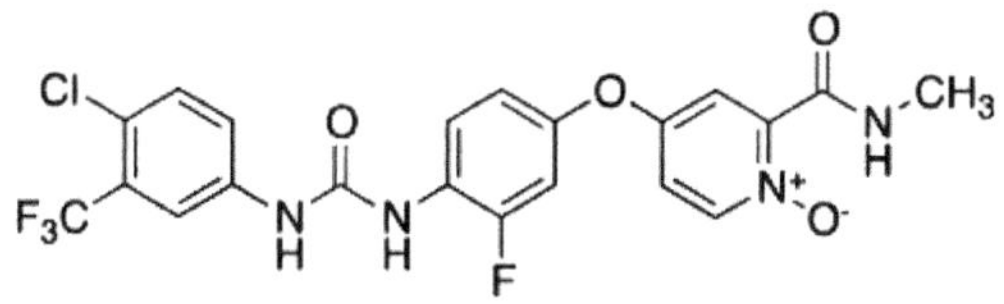

M2

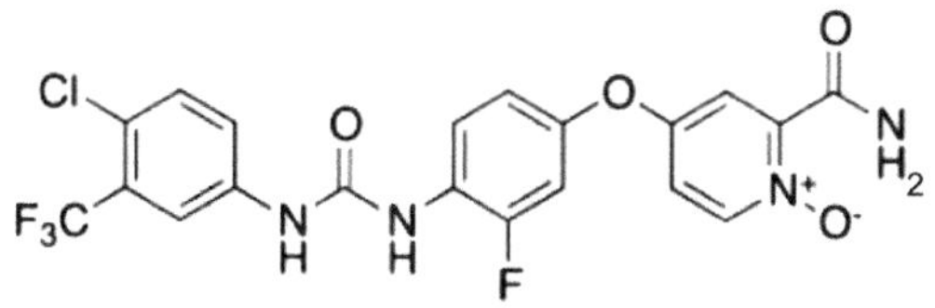

M5

[175] Regorafenib (colorectal cancer) NDA 203-085. Page 34 of 64-page Clinical Pharmacology Review.

[176] Agricultural Research Service. United States Department of Agriculture. Composition of Foods, Raw, Processed and Prepared (Agriculture Handbook No. 8); 1975.

[177] Regorafenib (metastatic colorectal cancer) NDA 203-085. Page 18 of 64-page Clinical Pharmacology Review.

[178] Strumberg D, et al. Regorafenib (BAY 73-4506) in advanced colorectal cancer: a phase I study. Br. J. Cancer. 2012;106:1722–7.

[179] Sunakawa Y, et al. Regorafenib in Japanese patients with solid tumors: phase I study of safety, efficacy, and pharmacokinetics. Invest. New Drugs. 2014;32:104–12.

The need to take into account plasma concentrations of these metabolites, and the need to take into account the influence of food on exposure of these metabolites, arises from the fact that M2 and M5 are pharmaceutically active. Regarding efficacy of regorafenib and M2 and M5, FDA wrote, "M2 and M5 were ... major circulating metabolites at steady-state and exhibited similar anticancer activity compared to regorafenib in tumor models of colorectal cancer. M2 and M5 inhibited the same protein kinases as regorafenib with IC50 values."[180]

Consistently, Zopf et al.[181] demonstrated that M2 and M5 have activities similar to that of regorafenib, in tests of kinase inhibition, and also in tests of antitumor activity using xenograft models. According to FDA's Medical Review, the PK of regorafenib and its metabolites are, "M-2 and M-5 were measured in the pharmacokinetic studies along with regorafenib ... [a]t steady-state, regorafenib reached mean Cmax of 3.9 micrograms/mL and the mean AUC of 58.3 (micrograms)(h)/mL ... [t]he metabolites M-2 and M-5 reached steady-state concentrations that were similar to regorafenib."[182]

Now, regarding the influence of food on PK, a view of FDA's ClinPharm Review reveals that the Sponsor was careful to test exposure values under the conditions of (1) Fasting, (2) After a low-fat breakfast, and (3) After a high-fat breakfast:

> Food Effect: Regorafenib is recommended to be administered with a low-fat meal. As compared to the fasted state, a **low-fat breakfast increased** the mean AUC of regorafenib, M2 and M5 by 36%, 40% and 23%, respectively, whereas a **high-fat meal increased** the mean AUC of regorafenib by 48%, but **decreased** the mean AUC of M2 and M5 by 20% and 51%, respectively.[183]

FDA's ClinPharm Review provided a more detailed account of the exposure results:

> After a high-fat meal, the mean AUC of regorafenib was increased by 48% and the mean AUC of M2 and M5 was decreased by 20% and 51%, respectively, resulting in an **overall exposure approximately 8% lower as compared to the fasted state**. A low-fat breakfast, defined as the first example above, increased the mean AUC of regorafenib by 36% and the mean AUC of M2 and M5 by 40% and 23%, respectively, resulting in **overall exposure approximately 33% higher as compared to the fasted state.**[184]

FDA's review for regorafenib is unique and dramatic, in that the Sponsor tested drug exposure with three types of food conditions (high fat, low fat, and fasting) instead of just two different food conditions. The result was that the low-fat meal was preferred. Even more unique and striking is the fact that the food effect study took into account the influence of food on the drug's metabolites.

[180] Regorafenib (metastatic colorectal cancer) NDA 203-085. Page 37 of 64-page Clinical Pharmacology Review.

[181] Zopf D, et al. Pharmacologic activity and pharmacokinetics of metabolites of regorafenib in preclinical models. Cancer Med. 2016;5:3176–85.

[182] Regorafenib (colorectal cancer) NDA 203-085. Page 20 of 76-page Medical Review.

[183] Regorafenib (colorectal cancer) NDA 203-085. Page 19 of 64-page Clinical Pharmacology Review.

[184] Regorafenib (colorectal cancer) NDA 203-085. Pages 34–35 of 64-page Clinical Pharmacology Review.

Package label. Consistent with the Sponsor's findings on the influence of test meals on PK of regorafenib (Stivarga®), the package label required:

DOSAGE AND ADMINISTRATION ... Take Stivarga with food (a low-fat breakfast).[185]

IV. STUDY DESIGN METHODOLOGY—STEADY-STATE CONCENTRATIONS

FDA's ClinPharm Review for venetoclax, described earlier, identified a mistake in study design where the issue was drug dosing resulting in steady-state drug concentrations versus non-steady-state drug concentrations.

A clinical study by Stone et al. [186] further describes the concept of "steady state" in a clinical study showing how plasma levels of a drug gradually reached steady state over the course of about 3 weeks. This is illustrated by a clinical study on caspofungin (i.v.) for treating aspergillosis infections.

The caspofungin clinical studies by Stone et al. [187] concerned daily dosing with a constant dose level for 3 weeks. In one study, the dose was 15 mg per day, in another it was 35 mg per day, and in the other it was 70 mg per day. The results demonstrated that "trough concentration" was something that "climbed slightly," and that "trough concentration" was something that eventually was "approaching steady state."

In the words of Stone et al.,[188] "Trough concentrations reached a stable plateau by 3 to 4 days at 15 mg daily. At 35 mg daily, trough concentrations climbed slightly after 3 to 4 days, while at 50 or 70 mg daily, they climbed more substantially throughout the 14-day studies ... [t]he individual profiles at 35 mg daily suggest that most subjects were approaching steady state by day 14 ... [t]he 3-week study of individuals at 70 mg daily was conducted in part to further characterize the approach to steady state for this regimen ... steady state was not achieved by day 14 ... [h]owever, the accumulation occurring during the third week was slight and suggests that the subjects were approaching steady state during the third week."

Figure 5 in Stone et al.[189] (not reproduced here) illustrates the trough concentrations for a study where daily drug doses (50 mg/day) were administered over the course of 14 days. The trough concentrations were those from analysis of blood sample taken just before the next day's drug dose.

[185] Package label. STIVARGA (regorafenib) tablets, oral; September 2012 (14 pp.).

[186] Stone JA, et al. Single- and multiple-dose pharmacokinetics of caspofungin in healthy men. Antimicrob. Agents Chemother. 2002;46:739–45.

[187] Stone JA, et al. Single- and multiple-dose pharmacokinetics of caspofungin in healthy men. Antimicrob. Agents Chemother. 2002;46:739–45.

[188] Stone JA, et al. Single- and multiple-dose pharmacokinetics of caspofungin in healthy men. Antimicrob. Agents Chemother. 2002;46:739–45.

[189] Stone JA, et al. Single- and multiple-dose pharmacokinetics of caspofungin in healthy men. Antimicrob. Agents Chemother. 2002;46:739–45.

As can be seen from figure 5 in Stone et al.,[190] the trough concentration after only one dose (Day 1) was relatively low, whereas the trough concentration from the next day (Day 2) was a bit higher, where eventually the rise in trough concentrations was found to plateau (Days 8–15). Each dose was administered by way of a 1-hour infusion.

This defines trough concentration. FDA's Guidance for Industry provides an account of "trough concentrations" in a narrative relating to oral drugs, but the narrative is likely also to be relevant to intravenous (i.v.) drugs. FDA's Guidance for Industry states that trough plasma concentration is abbreviated as Cmin (minimal plasma concentration attained in a given time frame). Trough concentration refers to the lowest blood concentration that is detected in between consecutive doses of a drug administered with a multiple dosing regimen.

FDA's Guidance states that, "collection of multiple plasma samples over a dosing interval is often not practical. As a substitute, a trough plasma sample can be collected just before administration of the next dose … **[t]rough concentrations are often proportional to AUC**, because they do not reflect drug absorption processes, as peak concentrations do in most cases. For many of the drugs that act slowly relative to the rates of their absorption, distribution, and elimination, **trough concentration and AUC can often be equally well correlated** with drug effects."[191]

V. FOOD EFFECT STUDIES ON DRUGS FOR OTHER INDICATIONS

a. Efavirenz (HIV-1) NDA 021-360

Efavirenz inhibits reverse transcriptase of HIV-1 and is used for treating AIDS.[192] Food increased exposure of efavirenz and increased risk for central nervous system (CNS) AEs.

FDA's ClinPharm Review provided an account of the design of food effect studies for efavirenz, "The Applicant evaluated the effect of a high-fat meal on efavirenz exposure … subjects received a … 600 mg dose of efavirenz … under fasting conditions in one study period and a … 600 mg dose of efavirenz … under fed conditions (high-fat/high calorie breakfast meal, 1000 kcal with 60 grams of fat) in the other study period."[193]

FDA further revealed the results of the food effect studies, where the high-fat meal increased exposure, "administration of efavirenz with **a high fat meal increases Cmax and AUC by 79% and 28%**, respectively … [t]he increase seen in efavirenz Cmax

[190] Stone JA, et al. Single- and multiple-dose pharmacokinetics of caspofungin in healthy men. Antimicrob. Agents Chemother. 2002;46:739–45.

[191] U.S. Department of Health and Human Services. Food and Drug Administration. Center for Drug Evaluation and Research (CDER). Center for Biologics Evaluation and Research (CBER). Guidance for industry. Exposure-response relationships — study design, data analysis, and regulatory applications; 2003 (25 pp.).

[192] Bastos MM, et al. Efavirenz a nonnucleoside reverse transcriptase inhibitor of first-generation: approaches based on its medicinal chemistry. Eur. J. Med. Chem. 2016;108:455–65.

[193] Efavirenz (HIV-1) NDA 021-360. 30-page Clinical Pharmacology Review.

following efavirenz tablet in a fed state may be of clinical significance ... there may be a relationship between high efavirenz concentrations and CNS adverse drug experiences (ADEs)."[194]

FDA's Medical Review revealed that the AEs took the form of "nervous system symptoms" and that, "A higher incidence of adverse events was reported when efavirenz was taken with food compared to the fasted state."[195] Finally, FDA recommended that the package label require taking on an empty stomach, to avoid AEs affecting the CNS, "The efavirenz tablet formulation ... is to be taken at bedtime on an **empty stomach**. This efavirenz dosing recommendation is an attempt **to avoid CNS adverse events** ... typically associated with efavirenz."[196]

Package label. Consistent with the recommendations of the FDA reviewer, the package label for efavirenz (Sustiva®) recommended that the drug should be taken on an empty stomach:

> **DOSAGE AND ADMINISTRATION**. Adults. The recommended dosage of SUSTIVA (efavirenz) is 600 mg orally, once daily, in combination with a protease inhibitor and/or nucleoside analogue reverse transcriptase inhibitors (NRTIs). It is recommended that SUSTIVA be taken on an **empty stomach**, preferably at bedtime. The increased efavirenz concentrations observed following administration of SUSTIVA with **food may lead to an increase in frequency of adverse reactions**.[197]

The patient guide that was part of the package label provided further guidance for patients who cannot swallow pills or capsules, for example, infants:

> Preparing a dose of SUSTIVA mixed with food using the capsule sprinkle method. Before you prepare a dose of SUSTIVA mixed with food using the capsule sprinkle method, gather the following supplies: • paper towels • teaspoon for measuring • small spoon for stirring and feeding • small clean container (such as a small cup or bowl) • soft food such as applesauce, grape jelly, or yogurt.[198]

b. Miltefosine (Visceral Leishmaniasis) NDA 204-684

Miltefosine is a small molecule for treating parasitic infections. This provides examples of food effects, where the food effects were desired because the food prevented the AEs of abdominal pain and nausea.

The Sponsor's Briefing Document revealed that miltefosine produced various GI AEs and that taking food ameliorated these AEs. As a consequence, the package label

[194] Efavirenz (HIV-1) NDA 021-360. 30-page Clinical Pharmacology Review.

[195] Efavirenz (HIV-1) NDA 021-360. 14-page Medical Review.

[196] Efavirenz (HIV-1) NDA 021-360. 30-page Clinical Pharmacology Review.

[197] Package label. SUSTIVA® (efavirenz) capsules and tablets. January 2017 (38 pp.).

[198] Package label. SUSTIVA® (efavirenz) capsules and tablets. January 2017 (38 pp.).

instructed that the drug be taken with food. The Sponsor reported that GI AEs associated with the drug were abdominal pain (9.1%), diarrhea (12.0%), dyspepsia (5.7%), nausea (39.2%), and vomiting (17.7%).[199]

The Briefing Document proposed that food be given to reduce these AEs, "Proposed dose and justification ... [a]dministration with food reduces **gastrointestinal adverse reactions** ... [t]he recommended dosing regimen ... is ... 30–44 kg, one 50 mg capsule twice a day with food (breakfast and dinner), ≥45 kg, one 50 mg capsule three times daily with food (breakfast, lunch, and dinner)."[200]

Package label. The Sponsor's proposal for dosing with food for miltefosine (Impadvido®), as well as the reason for including food, found a place on the package label:

> DOSAGE AND ADMINISTRATION. **Administer with food to ameliorate gastrointestinal adverse reactions.** 30 to 44 kg: one 50 mg capsule twice daily for 28 consecutive days. 45 kg or greater: one 50 mg capsule three times daily for 28 consecutive days.[201]

Food instructions to reduce GI AEs, example of pirfenidone. Reducing GI AEs by taking food was also found for another drug, pirfenidone. For this drug, food reduced the Cmax to about 50% and also reduced the GI AE of nausea.[202] The package label for pirfenidone instructed patients:

> DOSAGE AND ADMINISTRATION. Recommended dosage: 801 mg (three capsules) three times daily **taken with food.**[203]

c. Vilazodone (Major Depressive Disorder) NDA 022-567

Vilazodone is used for treating major depressive disorder. Food increased vilazodone exposure and increased vilazodone's efficacy.

The diagnosis of major depressive disorder, as well as assessment of the drug's efficacy, can be made using a depression scale such as the Montgomery–Åsberg Depression

[199] NDA 204-684 Miltefosine (Impavido®) for the Treatment of Visceral, Mucosal and Cutaneous Leishmaniasis. FDA Briefing Document for the AntiInfective Drugs Advisory Committee Meeting October 18, 2013. Page 28 of 43-page Briefing Document.

[200] NDA 204-684 Miltefosine (Impavido®) for the Treatment of Visceral, Mucosal and Cutaneous Leishmaniasis. FDA Briefing Document for the AntiInfective Drugs Advisory Committee Meeting October 18, 2013. Page 8 of 43-page Briefing Document.

[201] Package label. IMPAVIDO (miltefosine) capsules, for oral use; March 2014 (18 pp.).

[202] Pirfenidone (idiopathic pulmonary fibrosis) NDA 022-535. Pages 5, 42, 44, 49, 72, and 119 of 204-page Clinical Pharmacology Review.

[203] Package label. ESBRIET® (pirfenidone) capsules, for oral use; October 2014 (18 pp.).

Rating Scale or Hamilton Depression Rating Scale.[204] Vilazodone is a selective serotonin reuptake inhibitor (SSRI). Other drugs in the SSRI class are escitalopram, paroxetine, and sertraline.[205]

The Sponsor tested the influences of both a high- and low-fat meal on vilazadone's PK. The exposure parameters are AUC to infinite time (AUC_{inf}) and Cmax. FDA's ClinPharm Review described the influence of food on exposure:

> Food Effect. The effect of **high fat (high calorie) and light meal (low calorie)** on single-dose PK of orally administered vilazodone 20 mg tablet was evaluated. The type of food (e.g., high fat or light meal) did not appear to be influential. Meal significantly increased systemic exposure (AUC_{inf}) of vilazodone by 64-85%, and increased Cmax of vilazodone by 147-160%. **The sponsor proposed that vilazodone should be taken with food in the label based on the higher systemic exposure under fed conditions.**[206]

Taking into account the Sponsor's recommendation, FDA concluded that the package label should have the following statement about food. Note that FDA's recommendation included remarks about drug titration, "What are the proposed dosages and routes of administration? Vilazodone is administered orally. The **recommended dose is 40 mg QD with food** after titration, starting with an initial dose of 10 mg once daily for 7 days followed by 20 mg once daily for an additional 7 days."[207]

Regarding the use of titration for patients taking vilazodone, the reason for the Sponsor's use of the stepped-dose (titration) method of administering vilazodone was to reduce AEs, in particular, GI AEs. Regarding this goal, FDA stated, "However, titration did not completely **eliminate the adverse gastrointestinal effects associated with vilazodone**. More than half of the patients exposed to vilazodone experienced a gastrointestinal adverse event. Sexual dysfunction was captured in the spontaneous adverse event reporting at approximately 10% and is addressed later in the safety review."[208]

Package label. FDA's recommendation regarding taking food and regarding titration found a place on the package label for vilazodone (Viibryd®). The package label read:

> **DOSAGE AND ADMINISTRATION**. VIIBRYD should be titrated to the 40 mg dose, starting with an initial dose of 10 mg once daily for 7 days, followed by 20 mg once daily for an additional 7 days, and then increased to 40 mg once daily ... VIIBRYD should be **taken with food**. **Administration without food** can result in inadequate drug concentrations and may diminish effectiveness.[209]

[204] Durgam S, et al. Categorical improvements in disease severity in patients with major depressive disorder treated with vilazodone: post hoc analysis of four randomized, placebo-controlled trials. Neuropsychiatr. Dis. Treat. 2016;12:3073–81.

[205] Khan A, et al. Post hoc analyses of anxiety measures in adult patients with generalized anxiety disorder treated with vilazodone. Prim. Care Companion CNS Disord. 2016;18. doi: 10.4088/PCC.15m01904. eCollection 2016.

[206] Vilazodone (major depressive disorder) NDA 022-567. Page 22 of 112-page Clinical Pharmacology Review.

[207] Vilazodone (major depressive disorder) NDA 022-567. Page 26 of 112-page Clinical Pharmacology Review.

[208] Vilazodone (major depressive disorder) NDA 022-567. Page 225 of 259-page Medical Review.

[209] Package label. VIIBRYD (vilazodone HCl) tablets for oral administration; January 2010 (13 pp.).

d. Ivacaftor (Cystic Fibrosis for CFTR Gene G551D Mutation) NDA 203-188

Ivacaftor is a small molecule for treating cystic fibrosis in patients where the cystic fibrosis transmembrane conductance regulator gene has a G551D mutation. Food increased exposure of ivacaftor, where this increase was especially desired because of the very low water-solubility of ivacaftor.

Ivacaftor's structure is shown below. The aromatic rings and the many methyl groups account for the drug's low solubility in water.

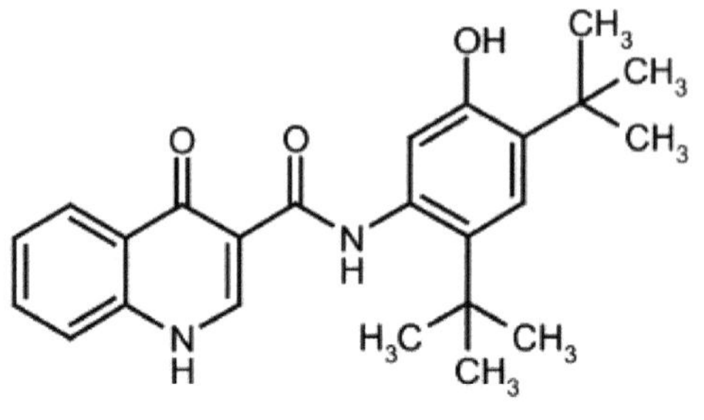

Ivacaftor

FDA's ClinPharm Review observed the low solubility in water, writing that, "The absolute bioavailability of ivacaftor in humans has not been determined because it is very insoluble (<0.001 mg/mL in water) and no intravenous formulation was available."[210]

Absolute bioavailability can be determined from the ratio between the amount of drug reaching the systemic circulation and the total administered dose. The ratio can be calculated from: [AUC after intravenous injection]/[AUC after extravascular administration].[211,212] Note that "extravascular administration" can be by the way of oral administration or subcutaneous injection. FDA then described the influence of food, "Administration with food increases the bioavailability ... by 2- to 4-fold ... therefore ivacaftor should be administered with food. The ... time to reach maximum plasma concentration (tmax) is ... 4.0 hours in the fed state."[213]

In addition to using the exposure parameter of tmax, FDA also referred to the influence of food on AUC and Cmax for ivacaftor, "Coadministration with food increased AUC by ... 2–3 fold and Cmax by 2–4 fold."[214]

[210] Ivacaftor (cystic fibrosis for CFTR gene G551D mutation) NDA 203-188. Pages 33–34 of 102-page Clinical Pharmacology Review.

[211] Jiang W, et al. Modifications of the method for calculating absolute drug bioavailability. J. Pharm. Pharm. Sci. 2016;19:181–7.

[212] Cannady EA, et al. Absolute bioavailability of evacetrapib in healthy subjects determined by simultaneous administration of oral evacetrapib and intravenous [(13) C8]-evacetrapib as a tracer. J. Labelled Comp. Radiopharm. 2016;59:238–44.

[213] Ivacaftor (cystic fibrosis for CFTR gene G551D mutation) NDA 203-188. Pages 33–34 of 102-page Clinical Pharmacology Review.

[214] Ivacaftor (cystic fibrosis for CFTR gene G551D mutation) NDA 203-188. Pages 62–63 of 102-page Clinical Pharmacology Review.

Package label. FDA's observations found a place on the package label for ivacaftor (Kaldyco®). The label required that the drug be taken with a meal that contains fat:

> **DOSAGE AND ADMINISTRATION.** Adults and pediatric patients age 6 years and older: one 150 mg tablet taken orally every 12 hours with fat-containing food.[215]

VI. FDC DRUGS AND INSTRUCTIONS FOR TAKING FOOD

a. Elvitegravir, Cobicistat, Emtricitabine, Tenofovir Alafenamide Combination (HIV-1) NDA 207-561

This concerns a combination drug for HIV-1. FDA's Guidance for Industry provides recommendations for combination of drugs for treating HIV-1.[216] E/C/F/TAF is a combination of four drugs, where the combination is an FDC consisting of:

1. Elvitegravir
2. Cobicistat
3. Emtricitabine
4. Tenofovir alafenamide

One of the four drugs is tenofovir alafenamide (TAF), which is a prodrug of tenofovir.[217] Once taken up by lymphocytes, TAF is hydrolyzed to produce tenofovir.[218] The alafenamide component of TAF is not a drug. A published food effect study on administering the E/C/F/TAF combination, with or without food, followed the PK of one of the drugs (elvitegravir) in this combination and determined that food improved exposure. The published study, which was by Yamada et al. [219] concluded that, "Administration under fasted conditions ... resulted in decreases in the mean AUCinf and Cmax of elvitegravir by 50% and 57%, respectively, relative to the administration with a standard breakfast."

FDA's ClinPharm review for the E/C/F/TAF combination reiterated the Sponsor's data from a food effect study on one of the components, namely TAF. In FDA's review the

[215] Package label. KALYDECO™ (ivacaftor) tablets; January 2012 (9 pp.).

[216] U.S. Department of Health and Human Services. Food and Drug Administration. Center for Drug Evaluation and Research (CDER). Guidance for industry. Fixed dose combinations, co-packaged drug products, and single-entityversions of previously approved antiretrovirals for the treatment of HIV; 2006 (36 pp.).

[217] Ray AS, et al. Tenofovir alafenamide: a novel prodrug of tenofovir for the treatment of human immunodeficiency virus. Antiviral Res. 2016;125:63–70.

[218] Callebaut C, et al. In vitro virology profile of tenofovir alafenamide, a novel oral prodrug of tenofovir with improved antiviral activity compared to that of tenofovir disoproxil fumarate. Antimicrob. Agents Chemother. 2015;59:5909–16.

[219] Yamada H, et al. Effects of a nutritional protein-rich drink on the pharmacokinetics of elvitegravir, cobicistat, emtricitabine, tenofovir alafenamide, and tenofovir compared with a standard meal in healthy Japanese male subjects. Clin. Pharmacol. Drug Dev. 2017. doi: 10.1002/cpdd.365.

terms "light meal" and "low fat meal" are both used to refer to "low fat meal."[220] FDA's comments referred to the tmax with a low-fat meal or a high-fat meal, "When E/C/F/TAF is administered **with food** to HIV-infected subjects, TAF tmax is 1 hour. Relative to fasting conditions, administration of E/C/F/TAF with a **low fat meal or high fat meal** results in TAF AUC increased 15% and 18%, respectively."[221] Thus, food resulted in a very slight increase in TAF exposure. Hence, at this point, it can reasonably be concluded that the package label for the E/C/F/TAF combination should state that the drug can be taken with or without food.

FDA moved a step further in food effect analysis and observed that the package labels for two of the other drugs, which are components of the E/C/F/TAF combination, already required that food be taken with that particular drug. These two other drugs are elvitegravir and cobicistat. FDA concluded that package labeling for the E/C/F/TAF combination should require food in view of the fact that the individual drugs elvitegravir and cobicistat are recommended to be administered with food.[222]

In general, the technique of basing package label writing for a new drug on package labels for similar drugs that are already FDA-approved and on the market is called "drug class" analysis. As can be seen from the package label for a drug formulation that included only one of these four drugs (cobicistat), the label required food. The Dosage and Administration section for cobicistat (Tybost®) required:

> **DOSAGE AND ADMINISTRATION**. Cobicistat (Tybost®) must be co-administered with atazanavir or darunavir at the same time, **with food**, and in combination with other HIV-1 antiretroviral agents.[223]

The above excerpt reproduces the Dosage and Administration section from a drug cobicistat (Tybost®), which contains only one molecule (cobicistat). Cobicistat's package label states that cobicistat must be taken with food. On the other hand, the package label for Tybost® requires that two other drugs be taken with Tybost®, namely, atazanavir and darunavir. FDA's ClinPharm Review for cobicistat (Tybost®) states that the package label's requirement for food when taking Tybost® was not at all based on any food effect study with cobicistat but instead was based on food effect studies with the separately administered drugs (atazanavir and darunavir) that must be taken with Tybost®.

Thus, any person needing information on food effects on the E/C/F/TAF combination drug needs to hunt backwards in time to find food effect information on cobicistat and then needs to hunt even further back in time to find food effect information on atazanavir and darunavir. But at this point in the hunt, it appears that the available information does not have any relevance to the E/C/F/TAF combination (the reason why the available information will not likely be relevant to the E/C/F/TAF combination is that this

[220] Elvitegravir, cobicistat, emtricitabine, tenofovir, alafenamide (E/C/F/TAF) combination (HIV-1) NDA 207-561. Pages 147–150 of 218-page Clinical Pharmacology Review.

[221] Elvitegravir, cobicistat, emtricitabine, tenofovir, alafenamide (E/C/F/TAF) combination (HIV-1) NDA 207-561. Page 12 of 218-page Clinical Pharmacology Review.

[222] Elvitegravir, cobicistat, emtricitabine, tenofovir, alafenamide (E/C/F/TAF) combination (HIV-1) NDA 207-561. Page 151 of 218-page Clinical Pharmacology Review.

[223] Package label. Tybost® (cobicistat) tablets, for oral use; September 2014 (30 pp.). NDA 203-094.

combination does not include atazanavir or darunavir). Please also note that the Dosage and Administration section of the package label for the E/C/F/TAF combination does not mention any option or any requirement for co-administering atazanavir or darunavir. Thus, it might reasonably be suggested that FDA's reasoning for the food requirement for the E/C/T/TAF combination, which was based on studies of totally different drugs (atazanavir and darunavir) was flawed.

The following is from FDA's ClinPharm Review for cobicistat, which states that the food recommendations were not based on any food effect studies for cobicistat but were instead based on food effect studies on drug that were required to be coadministered with cobicistat. FDA's ClinPharm Review for cobicistat (one of the components of the E/C/F/TAF combination) stated:

> There were no trials that were submitted in the current NDA submission that evaluated the effect of food on cobicistat exposure. The recommendation in the proposed cobicistat U.S. prescribing information to administer cobicistat when coadministered with atazanavir or darunavir with food is consistent with the dosing recommendations for atazanavir or darunavir when co-administered with ritonavir.[224]

To reiterate, it might reasonably be suggested that FDA's reasoning for the food requirement for the E/C/T/TAF combination was flawed. Eventually, food effect studies on the E/C/T/TAF combination were published.[225,226] These published studies provided direct support for the package label's recommendation that the E/C/T/TAF combination (Genvoya®) be taken with food.

Package label. FDA's recommendation for the package label of the E/C/T/TAF combination (Genvoya®) found a place on the package label instructions:

> **DOSAGE AND ADMINSTRATION** ... Recommended dosage: One tablet taken orally once daily **with food**[227]

b. Methodologies of General Applicability to Food Effect Studies

FDA's ClinPharm Review for the E/C/T/TAF combination provided methodological details relevant to all food effect studies.[228] One detail shows the variety of pharmacokinetic parameters that can be followed and calculated with a single food effect study. These

[224] Cobicistat (HIV-1) NDA 203-094. Page 41 of 114-page Clinical Pharmacology Review.

[225] Yamada H, et al. Effects of a nutritional protein-rich drink on the pharmacokinetics of elvitegravir, cobicistat, emtricitabine, tenofovir alafenamide, and tenofovir compared with a standard meal in healthy Japanese male subjects. Clin. Pharmacol. Drug Dev. 2017. doi: 10.1002/cpdd.365.

[226] Shiomi M, et al. Effects of a protein-rich drink or a standard meal on the pharmacokinetics of elvitegravir, cobicistat, emtricitabine and tenofovir in healthy Japanese male subjects: a randomized, three-way crossover study. J. Clin. Pharmacol. 2014;54:640–8.

[227] Package label. GENVOYA® (elvitegravir, cobicistat, emtricitabine, and tenofovir alafenamide) tablets for oral use; November 2005 (44 pp.).

[228] Elvitegravir, cobicistat, emtricitabine, tenofovir, alafenamide (E/C/F/TAF) combination (HIV-1) NDA 207-561. Pages 148–150 of 218-page Clinical Pharmacology Review.

parameters are AUC_{inf}, AUC_{last}, Cmax, tmax, and $T_{1/2}$. The values for these parameters were determined where the Sponsor administered the E/C/T/TAF combination, and then measured plasma concentrations of TAF, and also measured plasma concentrations of a metabolite of TAF. The metabolite was TFV (tenofovir).

Another methodological detail takes the form of a clever and careful clinical study, where study subjects were administered with the E/C/T/TAF combination in its pharmaceutical formulation, or alternatively, where the drugs in this combination were coadministered as separate drugs (and not as the pharmaceutical formulation). The FDA reviewer wrote, "In subjects administered E/C/F/TAF or its **components as single agents** with food ... PK was comparable for the components of E/C/F/TAF versus **single agents**, suggesting no formulation-related difference in food effect for E/C/F/TAF versus single agents."[229]

VII. DRUG–PH INTERACTIONS AND DRUG–CATION INTERACTIONS

Regarding pH, FDA's ClinPharm Reviews sometimes assess the influence of gastric pH on the study drug's solubility, bioavailability, and exposure. For example, where neutral stomach pH reduces drug solubility, with consequent impairment in bioavailability, the issue arises that the package label should have instructions to avoid proton pump inhibitors, to avoid antacids, or to avoid foods with an alkaline pH.

a. Axitinib (Renal Cell Carcinoma) NDA 202-324

FDA observed that axitinib's solubility is higher in acid and lower at neutral conditions and contemplated the problem that proton pump inhibitors or antacids could reduce axitinib exposure. The Sponsor's clinical studies of drug exposure involved measuring AUC, Cmax, and tmax after an oral dose of axitinib.

The Sponsor tested the influence of gastric pH on axitinib's bioavailability and exposure by coadministering rabeprazole. The proton pump inhibitor (rabeprazole) reduced the Cmax and AUC of axitinib.

FDA's ClinPharm Review commented on the Sponsor's discovery that a proton pump inhibitor reduced axitinib's bioavailability, at least when measured by the parameter of Cmax. FDA's comment was, "The aqueous solubility of axitinib is low over a wide range of pH values, and lower pH values result in higher solubility ... which raises the question about the potential for gastric pH elevating agents (such as proton pump inhibitors, H2 blockers, or antacids) to alter the solubility of axitinib. The sponsor conducted a PK study ... to evaluate the potential for drug-drug interactions with rabeprazole. In the presence of rabeprazole, **a 42% decrease in axitinib Cmax (i.e., reduction in absorption rate) was observed.**"[230]

[229] Elvitegravir, cobicistat, emtricitabine, tenofovir, alafenamide (E/C/F/TAF) combination (HIV-1) NDA 207-561. Pages 148–150 of 218-page Clinical Pharmacology Review.

[230] Axitinib (renal cell carcinoma) NDA 202-324. Page 23 of 118-page Clinical Pharmacology Review.

TABLE 3.2 Solubility of Bosutinib vs pH

pH 1.0	11.03 mg/mL
pH 2.0	9.4 mg/mL
pH 4.5	6.1 mg/mL
pH 5.0	2.7 mg/mL
pH 6.8	0.02 mg/mL

From: Bosutinib (chronic myelogenous leukemia; CML) NDA 203-341. Page 37 of 89, Clinical Pharmacology Review.

But FDA then turned to the exposure parameter of AUC, found the reduction in AUC to be insignificant, and recommended that the package label refrain from increasing axitinib dosing where the patient is taking a proton pump inhibitor. Regarding AUC, FDA wrote, "However, only a 15% decrease in AUC was observed, which is not considered clinically significant. There were no differences in tmax or T½ with or without rabeprazole. Therefore, no axitinib dose adjustment is recommended."[231]

b. Bosutinib (Chronic, Accelerated or Blast Phase Ph + Chronic Myelogenous Leukemia) NDA 203-341

FDA's ClinPharm Review reveals a chemical analysis of bosutinib's solubility at various pH conditions, in conjunction with clinical bioavailability studies with and without a proton pump inhibitor (lansoprazole).

FDA's analysis included an account of the BCS class, writing, "Bosutinib ... can likely be classified as a BCS Class IV drug, due to low solubility and low permeability."[232] The Sponsor provided an analysis of bosutinib solubility versus pH, showing that bosutinib is more soluble in acid pH and about 500-fold less soluble at neutral pH. Table 3.2 reveals bosutinib solubilities in the range of pH 1.0–6.8.

Then, FDA commented on bosutinib's solubility characteristics versus pH, on the Sponsor's study of proton pump inhibitor on drug exposure, and reiterated the Sponsor's recommendation for patients taking proton pump inhibitors or antacids. FDA's comment was, "Bosutinib has pH dependant solubility in vitro. Bosutinib is highly soluble at or below pH 5 and the solubility reduces above pH 5. When a single 400 mg oral dose of bosutinib was coadministered with ... oral doses of lansoprazole 60 mg, exposures to bosutinib decreased by 46% for Cmax and by 26% for $AUC_{0\text{-}inf}$ compared to when bosutinib was administered alone ... **[t]he applicant recommended antacids should be considered as an alternative to proton pump inhibitors ... and administration times of bosutinib and antacids should be separated whenever possible.**"[233]

[231] Axitinib (renal cell carcinoma) NDA 202-324. Page 23 of 118-page Clinical Pharmacology Review.

[232] Bosutinib (chronic, accelerated or blast phase Ph + chronic myelogenous leukemia (CML)) 203-341. Pages 36–37 of 89-page Clinical Pharmacology Review.

[233] Bosutinib (chronic, accelerated or blast phase Ph + chronic myelogenous leukemia (CML)) 203-341. Pages 36–37 of 89-page Clinical Pharmacology Review.

To summarize, where a proton pump inhibitor reduces the study drug exposure (AUC, Cmax, tmax), the Sponsor should consider including these instructions on the package label:

- Use antacids instead of proton pump inhibitors for treating excess gastric acid
- Instruction for separated administration times for the study drug and proton pump inhibitor, where the study drug is administered first

Package label. FDA's recommendations regarding proton pump inhibitors and antacids and bosutinib's bioavailability found a place a place on the package label:

> **DRUG INTERACTIONS** ... Proton pump inhibitors may decrease bosutnib drug levels. Consider short-acting antacids in place of proton pump inhibitors.[234]

This concerns the Clinical Pharmacology section of the bosutinib package label and its teachings of pH effects. The information can be summarized by these steps of reasoning:

Step 1. Bosutinib is insoluble above pH 5.
Step 2. Proton pump inhibitors, such as lansoprazole, elevate stomach pH.
Step 3. The proton pump inhibitor, lansoprazole, decreases drug exposure.
Step 4. Conclusion that proton pump inhibitors should be avoided when taking bosutinib, and that instead, short acting antacids should be used.

The Clinical Pharmacology section of the package label for bosutinib provided the information:

> **CLINICAL PHARMACOLOGY** ... Bosutinib ... has a pH dependent solubility ... [a]t or below pH 5, bosutinib ... [is] a highly soluble compound. Above pH 5, the solubility ... reduces rapidly ... [i]n a ... trial in 24 healthy volunteers ...bosutinib was either administered alone or in combination with ... lansoprazole under fasting conditions. Lansopraxole decreased bosutinib Cmax and AUC by 46% and 26%, respectively.[235]

c. Osimertinib (Metastatic EGFR T790M Mutation Positive Non-Small-Cell Lung Cancer) NDA 208-065

This concerns FDA's review of osimertinib (Tagrisso®) for treating non-small-cell lung cancer. FDA's ClinPharm Review referred to FDA's recommendation for the package label and observed that food did not influence osimertinib's exposure. FDA observed that, "The recommended dose of Tagrisso is 80 mg once daily taken without regard to food. Administration of 20 mg osimertinib Phase 1 tablet with a high-fat breakfast showed a **minimal increase in Cmax (14%) and AUC (19%)**. Pre-dosing of 40 mg omeprazole tablets for 5 days had **no clinically meaningful impact on the exposure** of osimertinib at a single 80 mg dose."[236]

[234] Package label. OSULIF® (bosutinib) tablets, for oral use; September 2012 (13 pp.).

[235] Package label. BOSULIF® (bosutinib) tablets, for oral use; September 2012 (13 pp.).

[236] Osimertinib (metastatic EGFR T790M mutation positive non-small cell lung cancer (NSCLC) NDA 208-065. Page 3 of 90-page Clinical Pharmacology Review.

Also, FDA observed that osimertinib's exposure was not much influenced by a concomitant proton pump inhibitor (omeprazole). Here, FDA observed that, "Gastric Acid Reducing Agents: The **exposure of osimertinib was not affected** by concurrent administration of a single 80 mg Tagrisso dose following 40 mg **omeprazole** administration for 5 days ... osimertinib Cmax and AUC were 2% and 7% higher when osimertinib 80 mg film-coated tablet was administered to healthy volunteers with prior dosing of 40 mg omeprazole tablets for 5 days to elevate the gastric pH."[237]

Package label. FDA's observations of lack of food effect found a place on the package label for osimertinib (Tagrisso®):

> **DOSAGE AND ADMINISTRATION** ... 80 mg orally once daily, with or without food.[238]

Also, FDA's comment that osimertinib exposure was not much influenced by coadministered omeprazole found a place in the Clinical Pharmacology section of the package label:

> **CLINICAL PHARMACOLOGY** ... Pharmacokinetics. Gastric Acid Reducing Agents: The exposure of osimertinib was not affected by concurrent administration of a single 80 mg TAGRISSO tablet following 40 mg omeprazole administration for 5 days.[239]

d. Eltrombopag (Chronic Immune Thrombocytopenic Purpura) NDA 022-291

Eltrombopag is a small molecule for treating a platelet disorder, chronic immune thrombocytopenic purpura. The structure of eltrombopag is shown below. The package label states that, "Eltrombopag **chelates polyvalent cations** (such as iron, calcium, aluminum, magnesium, selenium, and zinc) in foods, mineral supplements, and antacids[240]" The oxygen and nitrogen atoms are likely involved in complexing polyvalent cations.

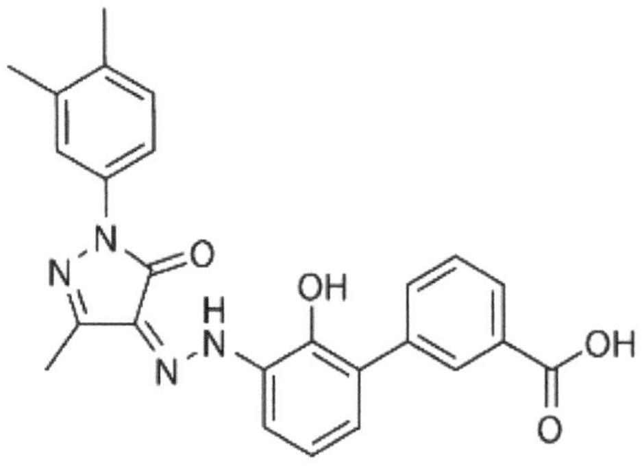

[237] Osimertinib (metastatic EGFR T790M mutation positive non-small cell lung cancer (NSCLC) NDA 208-065. Pages 9 and 23 of 90-page Clinical Pharmacology Review.

[238] Package label. TAGRISSO™ (osimertinib) tablet, for oral use; March 2017 (20 pp.).

[239] Package label. TAGRISSO™ (osimertinib) tablet, for oral use; March 2017 (20 pp.).

[240] Package label. PROMACTA® (eltrombopag) tablets; October 2008 (21 pp.).

Eltrombopag's bioavailability is dramatically reduced by polyvalent cations, as determined by studies with human subjects using a meal high in calcium or with human subjects taking an antacid that contains aluminum plus magnesium.[241] Calcium, aluminum, and magnesium are each polycations.

FDA's ClinPharm Review observed that a high-fat breakfast reduced exposure, in terms of the parameters AUC, Cmax, and tmax. FDA attributed the calcium in the meal (the "dairy content") as contributing to this reduced exposure. FDA went a step further by recommending that the drug be taken on an empty stomach to avoid the food effect.

The FDA reviewer stated that, "Administration of eltrombopag with a standard high-fat breakfast ... decreased plasma eltrombopag exposure ... $AUC_{0\text{-}inf}$ by 59% and Cmax by 65% and delayed tmax by one hour ... [t]hese results were considered to be clinically significant ... [i]t is possible that the **dairy content** of the standard high-fat breakfast is responsible for ... the reduced eltrombopag exposure ... low-calcium meals ... had a lower impact on plasma eltrombopag exposure ... regardless of fat content."[242]

In addition to deciding that the results were clinically significant, the FDA reviewer stated that, "The reviewer recommends the ... product labeling include the statement, 'PROMACTA should be taken only on an empty stomach (1 hour before or 2 hours after a meal)'."[243] Moreover, the reviewer referred to FDA's Guidance for Industry[244] as providing a cutoff point for deciding if, or if not, a food effect was clinically significant. FDA's recommendations found a place on the package label, as shown below.

Package label. The package label for eltrombopag (Promacta®) provided the instructions:

> **DOSAGE AND ADMINISTRATION** ... Give on an empty stomach (1 hour before or 2 hours after a meal.[245]

[241] Williams DD, et al. Effects of food and antacids on the pharmacokinetics of eltrombopag in healthy adult subjects: two single-dose, open-label, randomized-sequence, crossover studies. Clin. Ther. 2009;31:764–76.

[242] Eltrombopag (chronic immune thrombocytopenic purpura) NDA 022-291. Pages 5, 24, and 61 of 85-page Clinical Pharmacology Review (1st of three pdf files).

[243] Eltrombopag (chronic immune thrombocytopenic purpura) NDA 022-291. Pages 5, 24, and 61 of 85-page Clinical Pharmacology Review (1st of three pdf files).

[244] U.S. Department of Health and Human Services. Food and Drug Administration. Center for Drug Evaluation and Research (CDER). Guidance for industry. Food-effect bioavailability and fed bioequivalence studies; 2002 (12 pp.).

[245] Package label. PROMACTA® (eltrombopag) tablets; October 2008 (21 pp.).

VIII. DRUG–PH INTERACTIONS ANALYSIS AND ASSESSMENT BY THE BCS

a. Introduction

This provides additional examples of drug–pH interaction analysis where FDA's review focused on the solubility and permeability of the study drug. FDA's Guidance for Industry on BCS defines the concept, the solubility class:

> The solubility class should be determined by calculating the volume of an aqueous medium sufficient to dissolve the highest strength in the pH range of 1-6.8. A drug substance should be classified as highly soluble when the highest strength is soluble in < 250 mL of aqueous media over the pH range of 1-6.8.[246]

FDA's Guidance for Industry on BCS provides information on permeability class:

> A drug substance is considered to be highly **permeable** when the extent of absorption in humans is determined to be 85 percent or more of an administered dose ... [d]epending on study variability, a sufficient number of subjects, animals, excised tissue samples, or cell monolayers should be used in a study to provide a reliable estimate of drug **permeability**. This relationship should allow precise differentiation between drug substances of low and high intestinal **permeability** attributes.[247]

b. Cabozantinib (Metastatic Medullary Thyroid Cancer) NDA 203-756

FDA's ClinPharm Review for cabozantinib focused on pH and solubility. FDA's ClinPharm Review revealed that cabozantinib is insoluble at pH 4 and above, and that that a dedicated study was needed to test the influence of proton pump inhibitors on cabozantinib exposure. To this end, FDA stated that, "The solubility of cabozantinib is pH-dependent with the solubility at normal gastric pH the highest and practically **insoluble when pH is greater than 4.** The effect of gastric pH modifying drugs (proton pump inhibitors, H2 blockers, antacids) **on pharmacokinetics of cabozantinib** ... **was inconclusive**. A post-marketing requirement (PMR) for conducting a dedicated pH effect study is recommended."[248]

Although the Sponsor had information on solubility, the Sponsor's information on PK with human subjects was not conclusive. Because the Sponsor's information, to date, was

[246] U.S. Department of Health and Human Services. Food and Drug Administration. Center for Drug Evaluation and Research (CDER). Guidance for industry. Waiver of in vivo bioavailability and bioequivalence studies for immediate-release solid oral dosage forms based on a biopharmaceutics classification system; 2015 (14 pp.).

[247] U.S. Department of Health and Human Services. Food and Drug Administration. Center for Drug Evaluation and Research (CDER). Guidance for industry. Waiver of in vivo bioavailability and bioequivalence studies for immediate-release solid oral dosage forms based on a biopharmaceutics classification system; 2015 (14 pp.).

[248] Cabozantinib (metastatic medullary thyroid cancer) NDA 203-756. Page 3 of 106-page Clinical Pharmacology Review.

inconclusive, FDA explained that a PMR was needed. A PMR was needed because of the potential for antacids taken by patients, to severely reduce cabozantinib's bioavailability:

> A PMR will be requested based on the following reasons ... [t]he solubility of cabozantinib is pH-dependent ... and practically insoluble when pH is greater than 4 ... gastric pH modifying drugs (proton pump inhibitors, H2 blockers, antacids) can elevate the stomach pH at levels close to 6 or 7, therefore, co-medication may greatly decrease the solubility of cabozantinib.[249]

PMRs are described by FDA's Guidance for Industry.[250] The Sponsor responded to this PMR by determining the influence of gastric pH modifying agents on cabozantinib exposure, where the results were published on a supplemental package label (May 2016), which replaced the original package label published on the date of FDA approval (November 2012). The second package label reported that there was **not any influence on exposure** of the study drug. This information was in the Clinical Pharmacology section:

> **CLINICAL PHARMACOLOGY** ... Pharmacokinetics ... Gastric pH modifying agents on Cabozantinib. **No clinically-significant effect on plasma cabozantinib exposure** (AUC) was observed following co-administration of the proton pump inhibitor (PPI) esomeprazole (40 mg daily for 6 days) with a single dose of 100 mg cabozantinib to healthy volunteers.[251]

c. Regorafenib (Metastatic Colorectal Cancer) NDA 203-085

BCS class assessment was part of FDA's ClinPharm Review for regorafenib, where the determined class was BCS Class 2 (low solubility, high permeability). FDA's ClinPharm Review commented on solubility and permeability:

> The applicant claims that regorafenib is BCS Class 2, since it demonstrated high permeability in Caco-2 cells ... and poor solubility independent of medium or pH ... [t]he applicant classifies regorafenib as BCS class 2 based on data that showed regorafenib has low solubility and high permeability across Caco-2 cells with a relative bioavailability of 69% to 83%.[252]

This information did not result in any corresponding instructions on the package label, for dosing or regarding concomitant proton pump inhibitors. Instead, the information found a place in the relatively neutral Description section of the package label.

[249] Cabozantinib (metastatic medullary thyroid cancer) NDA 203-756. Page 25 of 106-page Clinical Pharmacology Review.

[250] U.S. Department of Health and Human Services. Food and Drug Administration. Center for Drug Evaluation and Research (CDER). Center for Biologics Evaluation and Research (CBER). Guidance for industry. Postmarketing studies and clinical trials — Implementation of Section 505(o)(3) of the Federal Food, Drug, and Cosmetic Act; 2011 (18 pp.).

[251] Package label. COMETRIQ (cabozantinib) capsules, for oral use; May 2016 (22 pp.). NDA 203-756

[252] Regorafenib (metastatic colorectal cancer) NDA 203-085. Pages 17, 33, and 59 of 64-page Clinical Pharmacology Review.

Package label. The package label for regorafenib (Stivarga®) provided information on drug solubility in the Description section of the package label. The label read:

> **DESCRIPTION** . . . Regorafenib is **practically insoluble in water**, slightly soluble in acetonitrile, methanol, ethanol, and ethyl acetate and sparingly soluble in acetone.[253]

d. Tofacitinib (Rheumatoid Arthritis) NDA 203-214

FDA's ClinPharm Review of tofacitinib provided an analysis of BCS class, where the analysis was directed to solubility and permeability. FDA first contemplated the solubility characteristics of tofacitinib and then determined the BCS class of the drug, where this was BCS Class 3 (high solubility/low permeability).

FDA's observations on tofacitinib's solubility and permeability were, "Based on the biopharmaceutic classification system principles, in what class is this drug and formulation? What solubility, permeability and dissolution data support this classification? **Tofacitinib can be considered a BCS class 3 drug because of high aqueous solubility and moderate permeability**. The aqueous pH solubility of tofacitinib (the citrate salt) was determined to be >0.04 mg/mL, which is the concentration obtained from dissolving the highest dose strength of 10 mg tablet in 250 mL solution. Thus the tofacitinib solubility profile meets the high solubility criteria set forth based on the BCS principles."[254]

Also, FDA provided a rationale for excluding tofacitinib from being a Class I drug, writing that, "The human oral bioavailability study showed . . . **oral bioavailability of the commercial tofacitinib tablet was 74%**, which is **less than the 90% criterion described in the BCS guidance for a Class I agent**. In the human mass balance study, the mean total percentage of administered radioactive dose recovered was 94%, with 80% in the urine and 13.8% in the feces, which did not conclusively show that the fraction of dose absorbed was greater than 90%."[255]

FDA's BCS class analysis for tofacitinib included permeability analysis, according to the BCS classification scheme. Regarding permeability, FDA wrote, "In vitro permeability assessments also showed that . . . permeability . . . values of tofacitinib at concentrations 1×, 0.1×, and 0.001× of clinical dose (10 mg in 250 mL) were lower than that of metoprolol, which is a highly permeable compound and was used as the reference. Based on available data, tofacitinib appears to have low permeability based on BCS principles."[256]

Package label. FDA's observations that tofacitinib (Xeljanz®) has "high aqueous solubility" found a place in the Description section of the package label. However, the label did not provide information on permeability or on BCS class:

> **DESCRIPTION**. XELJANZ is the citrate salt of tofacitinib, a JAK inhibitor. Tofacitinib citrate is a white to off-white powder . . . [i]t is freely soluble in water.[257]

[253] Package label. STIVARGA® (regorafenib) tablets, for oral use; April 2017 (25 pp.).

[254] Tofacitinib (rheumatoid arthritis) NDA 203-214. Page 39 of 181-page Clinical Pharmacology Review.

[255] Tofacitinib (rheumatoid arthritis) NDA 203-214. Page 39 of 181-page Clinical Pharmacology Review.

[256] Tofacitinib (rheumatoid arthritis) NDA 203-214. Page 39 of 181-page Clinical Pharmacology Review.

[257] Package label. XELJANZ® (tofacitinib) tablets for oral administration; November 2012 (27 pp.).

e. Vemurafenib (BRAF Mutation-Positive Melanoma) NDA 202-429

FDA's ClinPharm Review for vemurafenib made use of the BCS class, in comments on solubility and permeability:

> In vitro studies showed that vemurafenib has low permeability, and **low aqueous solubility** ... [b]ased on these data, vemurafenib has limited oral absorption, and it is classified as a **BCS Class IV compound** with low solubility and low permeability ... [t]he solubility studies is aqueous media were performed at 37°C, and the results indicate that vemurafenib is practically insoluble in aqueous media across a pH range of 1 to 7.5.[258]

The comments from the FDA reviewer found a place in the Description section of the package label.

Package label. The Description section for vemurafenib (Zelboraf®) revealed that the drug is insoluble in water and provided additional interesting information on the inactive ingredients:

> **DESCRIPTION.** Vemurafenib is a white to off-white crystalline solid. It is practically insoluble in aqueous media ... [t]he inactive ingredients of ZELBORAF are: Tablet core: hypromellose acetate succinate, croscarmellose sodium, colloidal silicon dioxide, magnesium stearate, and hydroxypropyl cellulose. Coating: pinkish white: poly (vinyl alcohol), titanium dioxide, polyethylene glycol 3350, talc, and iron oxide red.[259]

IX. VITAMINS AND MINERALS

The package labels for pharmaceuticals may include instructions to administer the drug with a vitamin or mineral. The vitamin or mineral can be characterized as a drug when it is bought via a prescription. But the same vitamin or mineral can be characterized as a dietary supplement, where a prescription is not relevant, that is, in the situation where labeling conforms to FDA's Guidance for Industry: A Food Labeling Guide.[260,261]

a. Denosumab (Osteoporosis) BLA 125-320

FDA's Medical Review describes the rationale for requiring denosumab to be coadministered with calcium and vitamin D. Denosumab is an antibody for treating osteoporosis.

[258] Vemurafenib (BRAF mutation-positive melanoma) NDA 202-429. Pages 31 and 45 of 84-page Clinical Pharmacology Review.

[259] Package label. ZELBORAF® (vemurafenib) tablet for oral use; April 2017 (18 pp.).

[260] U.S. Dept. Health and Human Services. Food and Drug Administration. Guidance for industry. A food labeling guide; 2013 (130 pp.).

[261] Brody T. Food and dietary supplement package labeling—guidance from FDA's warning letters and title 21 of the Code of Federal Regulations. Compr. Rev. Food Sci. Food Saf. 2015;15:92–129.

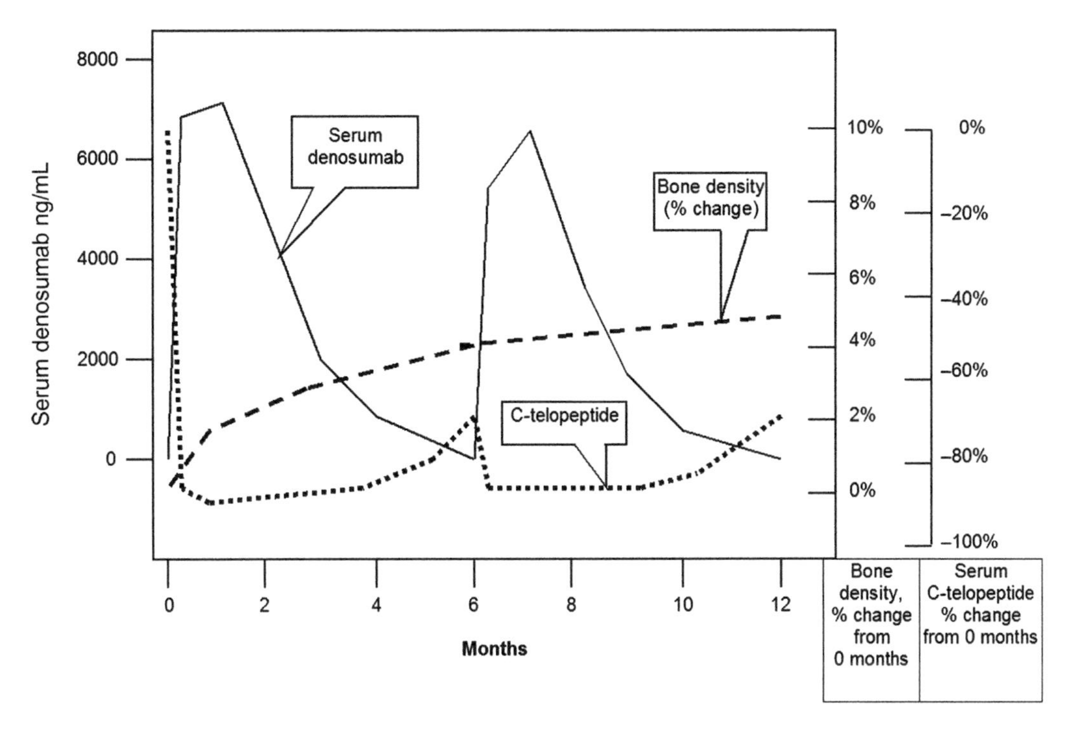

FIGURE 3.4 Denosumab, C-telepeptide, and bone density. Blood levels of denosumab, a drug for treating osteoporosis, are shown over the course of a year. Also shown are two biomarkers for denosumab's efficacy, serum C-telopeptide and bone density.

FDA provided a time course showing PK values and efficacy. Efficacy was measured by serum c-telopeptide and lumbar spine bone marrow density, over a 12-month period.[262] Fig. 3.4 illustrates these values:

- Denosumab concentration in blood
- Serum c-telopeptide (measure of denosumab's efficacy)
- Lumbar spine bone marrow density (measure of denosumab's efficacy)

C-telopeptide is a marker that reflects bone resorption. Bone resorption means that the bone is being dissolved away (bone loss). Fig. 3.4 reveals that the Sponsor's data showed that denosumab reduced serum c-telopeptide, where this reduced c-telopeptide was only found when serum denosumab was at high concentrations. The Sponsor's data also showed a steady increase in bone density over the course of 12 months.

This reveals FDA's recommendations for calcium and vitamin D. FDA's Cross Discipline Team Leader Review revealed the basis for the package label's requirement for taking calcium and vitamin D.[263] FDA observed that, "Hypocalcemia remains a safety

[262] Denosumab (osteoporosis) BLA 125-320. Pages 288–289 of 710-page Medical Review.

[263] Denosumab (osteoporosis) BLA 125-320. 33-page Cross Discipline Team Leader Review.

concern for all antiresorptive agents. Subjects in ... the clinical ... studies received calcium and vitamin D supplementation. The risk of hypocalcemia will be the highest in patients who are not adequately replete or supplemented with calcium and vitamin D. **Product labeling** will include a contraindication for patients with low calcium levels and a warning and precaution outlining hypocalcemia risk because of underlying disturbances of mineral metabolism."[264]

FDA's recommendations regarding labeling found a place in the package label, as shown below.

Package label. The package label for denosumab (Prolia®) required that calcium and vitamin D be taken with the drug, where information on calcium and occurred in three different sections of the package label:

> DOSAGE AND ADMINISTRATION ... Administer 60 mg every 6 months as a subcutaneous injection in the upper arm, upper thigh, or abdomen. Instruct patients to **take calcium 1000 mg daily and at least 400 IU vitamin D daily.**[265]

> CONTRAINDICATIONS ... Pre-existing hypocalcemia must be corrected prior to initiating therapy with Prolia.[266]

> WARNINGS AND PRECAUTIONS ... Hypocalcemia may be exacerbated by the use of Prolia. Pre-existing hypocalcemia must be corrected prior to initiating therapy with Prolia. In patients predisposed to hypocalcemia and disturbances of mineral metabolism ... clinical monitoring of calcium and mineral levels (phosphorus and magnesium) is highly recommended.[267]

b. Lomitapide (Homozygous Familial Hypercholesterolemia) NDA 203-858

1. Introduction

Lomitapide lowers plasma low-density lipoprotein (LDL)–cholesterol in patients with familial hypercholesterolemia. Familial hypercholesterolemia, which affects 1 in 250 persons, results in elevated LDL cholesterol levels throughout life. When familial hypercholesterolemia is not treated, the result is premature atherosclerotic cardiovascular disease.[268]

Lomitapide acts by inhibiting an enzyme that processes dietary fat. The enzyme, which is located in enterocytes and hepatocytes, is microsomal triglyceride transfer protein.

[264] Denosumab (osteoporosis) BLA 125-320. 33-page Cross Discipline Team Leader Review.

[265] Package label. Prolia™ (denosumab) Injection, for subcutaneous use; June 2010 (17 pp.).

[266] Package label. Prolia™ (denosumab) Injection, for subcutaneous use; June 2010 (17 pp.).

[267] Package label. Prolia™ (denosumab) Injection, for subcutaneous use; June 2010 (17 pp.).

[268] McCrindle BW, Gidding SS. What should be the screening strategy for familial hypercholesterolemia? New Engl. J. Med. 2016;375:1685–6.

Where the enzyme in **enterocytes** is inhibited, the result is reduced production of chylomicrons. Where the enzyme in **hepatocytes** is inhibited, the result is reduced production of VLDLs.[269]

As a consequence of lowering VLDLs (which are normally catabolized to generate LDLs), lomitapide lowers LDLs. In normal lipoprotein metabolism, VLDLs secreted by the liver are processed to generate IDLs which, in turn, are processed to generate LDLs, which circulate in the bloodstream for a few days, with eventual uptake by the liver.[270] Lomitapride's lowering of VLDLs does not directly prevent atherosclerosis but instead lomitapride's prevention of atherosclerosis is indirect because its downstream influence is to reduce plasma LDLs.

Because the drug lowers plasma levels of LDLs, the drug lowers plasma levels of all nutrients that are carried by LDLs, including cholesterol, fatty acids, and fat-soluble vitamins. FDA's Medical Review acknowledged this mechanism of action, referring to data where subjects were treated with lomitapide, "the ratio of serum vitamin E:lipid ... was relatively stable, suggesting that the observed decrease in vitamin E was not the result of malabsorption but rather the result of lomitapide's effect on serum lipoproteins."[271]

2. *Clinical Studies Show that Lomitapride Causes Vitamin E Deficiency*

The Sponsor found that human subjects given lomitapide became vitamin E deficient. On this point, FDA wrote that, "vitamin E decreased during lomitapide treatment (-49.5% placebo subtracted)."[272] Further demonstrating that lomitapride was the cause of the deficiency, FDA wrote that, "increases observed in fatty acid levels **after cessation of treatment** supports a causal relationship between lomitapide treatment and the observed decreases."[273]

In another clinical study, the Sponsor found that lomitapride lowered vitamin E but not vitamin A, and in the words of the FDA reviewer, "Consistent with other trials in the development program, lomitapide induced a dose-related decrease in total vitamin E, but not vitamin A, in this 8-week study."[274]

[269] Hooper AJ, et al. Contemporary aspects of the biology and therapeutic regulation of the microsomal triglyceride transfer protein. Circ. Res. 2015;116:193–205.

[270] Brody T. Nutritional Biochemistry, 2nd ed. New York: Elsevier; 1999, p. 332–357.

[271] Lomitapide (homozygous familial hypercholesterolemia) NDA 203-858. Page 154 of 304-page Medical Review.

[272] Lomitapide (homozygous familial hypercholesterolemia) NDA 203-858. Page 63 of 304-page Medical Review.

[273] Lomitapide (homozygous familial hypercholesterolemia) NDA 203-858. Page 203 of 304-page Medical Review.

[274] Lomitapide (homozygous familial hypercholesterolemia) NDA 203-858. Page 80 of 304-page Medical Review.

3. *Study design for dietary supplements*

To ensure that patients in the postmarketing situation would not become deficient in vitamin E and essential fatty acids, the Sponsor conducted a study where dietary supplements were administered with lomitapide. The study included a dietician and, as stated by the FDA reviewer:

> dietician met with subjects at visit 1 . . . the dietician was to review current eating habits . . . and review detailed diet instructions[275] . . . [b]eginning at visit 2, approximately two weeks before starting study drug, subjects were instructed to begin taking daily . . . dietary supplements to supply . . . 400 IU vitamin E, 200 mg linoleic acid (LA), 110 mg eicosapentaenoic acid (EPA), 220 mg alpha-linolenic acid (ALA), and 80 mg docosahexaenoic acid (DHA). They were told to continue this supplement throughout the study.[276]

As part of the study design, laboratory tests were used to measure the subject's levels of fat-soluble vitamins and essential fatty acids. The laboratory tests were for [277]:

- serum concentrations of vitamins A, D, and E, with vitamin E,
- direct assessment of vitamin K via carboxylation of serum osteocalcin,
- beta-carotene levels, and
- fatty acid profile: LA, ALA, EPA, DHA, arachidonic acid (AA), and eicosatrienoic (Mead) acid.

4. *Sponsor's Discovery that Lomitapide Results in Nutrient Deficiencies*

The Sponsor conducted a clinical study demonstrating that lomitapide causes deficiencies in various fat-soluble nutrients. FDA's Medical Review revealed that lomitapide resulted in lower levels of vitamin E and in lower levels of the essential fatty acids. Regarding the lower levels of vitamin E and essential fatty acids, FDA made the prediction that, "Given its mechanism of action, lomitapide could potentially lead to deficiencies in fat soluble nutrients by inducing intestinal fat malabsorption. In the phase 2 . . . study . . . decreases from baseline to the end of treatment were observed in . . . alpha-linolenic acid, gamma-linolenic acid, linoleic acid, AA, eicosapentaenoic acid, docosahexaenoic acid, and docosapentaenoic acid."[278]

5. *Sponsor's Decision to Include Dietary Supplements in Subsequent Clinical Studies with Lomitapide*

After the Sponsor's initial discovery that the study drug provoked the deficiencies, the Sponsor decided to include nutrient supplements in subsequent clinical studies. As stated by FDA, human subjects took vitamin E plus essential fatty acids but did not take supplements of vitamin A or vitamin K. Despite lack of vitamin A and vitamin K supplements, there was no evidence of deficiencies in these two vitamins.

[275] Lomitapide (homozygous familial hypercholesterolemia) NDA 203-858. Page 101 of 304-page Medical Review.

[276] Lomitapide (homozygous familial hypercholesterolemia) NDA 203-858. Page 103 of 304-page Medical Review.

[277] Lomitapide (homozygous familial hypercholesterolemia) NDA 203-858. Page 106 of 304-page Medical Review.

[278] Lomitapide (homozygous familial hypercholesterolemia) NDA 203-858. Page 154 of 304-page Medical Review.

Even though the subjects took vitamin E and fatty acids, there were still signs of deficiencies in these two nutrients. In FDA's own words, "Thus, in the pivotal trial, subjects were provided dietary supplements containing vitamin E, linoleic acid, alpha-linolenic acid, EPA, and DHA ... levels of vitamins A and D tended to increase over time, and there was no ... change in vitamin K ... as assessed by proportion of osteocalcin that was uncarboxylated. Total vitamin E decreased ... [r]egarding fatty acids, the trends over time suggest reductions in all fatty acids during approximately the first 26 weeks of therapy with subsequent stabilization and/or trends toward baseline."[279,280]

Package label. The package label for lomitapide (Juxtapid®) required that the drug be taken with vitamin E plus essential fatty acids:

> **DOSAGE AND ADMINISTRATION** ... Due to reduced absorption of fat-soluble vitamins/fatty acids: Take daily vitamin E, linoleic acid, alpha-linolenic acid (ALA), eicosapentaenoic acid (EPA), and docosahexaenoic acid (DHA) supplements.[281]

X. GRAPEFRUIT JUICE

a. Introduction

Grapefruit juice inhibits one of the cytochrome P450 isozymes (CYP3A4), where the result of this inhibition is increased exposure and increased risk for AEs from any drug that is normally catabolized by CYP3A4. Bergamottin and dihydroxybergamottin are the chemicals in grapefruit juice that inhibit CYP3A4, thereby resulting in increased plasma levels of any drug that is normally catabolized by CYP3A4.[282,283] These chemicals are furanocoumarin compounds. The structure of bergamottin is shown below:

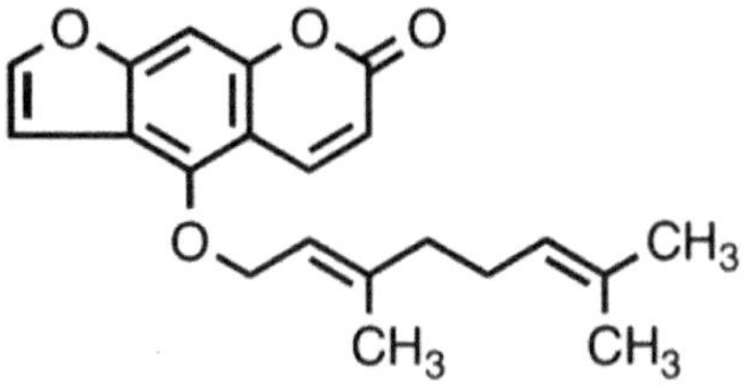

[279] Lomitapide (homozygous familial hypercholesterolemia) NDA 203-858. Page 154 of 304-page Medical Review.

[280] Lomitapide (homozygous familial hypercholesterolemia) NDA 203-858. Page 194 of 304-page Medical Review.

[281] Package label. JUXTAPID® (lomitapide) capsules, for oral use; May 2016 (30 pp.).

[282] Lin HL, et al. Identification of the residue in human CYP3A4 that is covalently modified by bergamottin and the reactive intermediate that contributes to the grapefruit juice effect. Drug Metab. Dispos. 2012;40:998–1006.

[283] He K, et al. Inactivation of cytochrome P450 3A4 by bergamottin, a component of grapefruit juice. Chem. Res. Toxicol. 1998;11:252–9.

Package label instructions on grapefruit juice from various drugs. Grapefruit juice can increase the plasma concentration of drugs, such as cyclosporine,[284] atorvastatin,[285] cabozantinib,[286] and calcium channel antagonists.[287] The package label of **cyclosporine** has the following warning:

> **DRUG INTERACTIONS** ... **Grapefruit juice**: Grapefruit and grapefruit juice affect metabolism, increasing blood concentrations of cyclosporine, thus should be avoided.[288]

Similarly, the package label for **atorvastatin** has the warning:

> **DRUG INTERACTIONS** ... **Grapefruit Juice**: Contains one or more components that inhibit CYP3A4 and can increase plasma concentrations of atorvastatin, especially with excessive grapefruit juice consumption (> 1.2 liters per day).[289,290]

Moreover, the package label for **felodipine** (Plendil®) warns:

> **CLINICAL PHARMACOLOGY** ... The bioavailability of PLENDIL is influenced by the presence of food ... [t]he **bioavailability of felodipine was increased approximately two-fold when taken with grapefruit juice**. Orange juice does not appear to modify the kinetics of PLENDIL. A similar finding has been seen with other dihydropyridine calcium antagonists, but to a lesser extent than that seen with felodipine.[291]

A single glass of grapefruit juice can increase the oral bioavailability of a concomitantly taken drug and can enhance the beneficial or toxic effects of that drug.[292] The cytochrome P450 enzymes that catalyze metabolism of various drugs and other xenobiotics are expressed in hepatocytes and enterocytes. The particular type of cytochrome P450 enzyme that is influenced by grapefruit juice consumption is CYP3A4, where the influence is that CYP3A4 is inhibited. Inhibiting the activity of CYP3A4 prevents this cytochrome P450 from destroying the drug in question, where the consequence is increased availability of that drug to the patients.

[284] Package label. Sandimmune® Soft Gelatin Capsules (cyclosporine capsules, USP); March 2015 (21 pp.).

[285] Package label. LIPITOR® (atorvastatin calcium) tablets for oral administration; March 2015 (22 pp.).

[286] Package label. COMETRIQ™ (cabozantinib) capsules, for oral use; November 2012 (19 pp.).

[287] Ohnishi A, et al. (2006) Major determinant factors of the extent of interaction between grapefruit juice and calcium channel antagonists. Br. J. Clin. Pharmacol. 62:196–9.

[288] Package label. Sandimmune® Soft Gelatin Capsules (cyclosporine capsules, USP); March 2015 (21 pp.). NDA 050-625.

[289] Package label. LIPITOR® (atorvastatin calcium) Tablets for oral administration; June 2017 (22 pp.).

[290] Atorvastatin (hypercholesterolemia) Supplemental NDA (sNDA) 020-702.

[291] Package label. PLENDIL® (felodipine) extended-release tablets; October 2012 (14 pp.). Package label from Supplemental NDA (sNDA) 019-834.

[292] Kim H, et al. Inhibitory effects of fruit juices on CYP3A activity. Drug Metab. Disposit. 2006; 34:512–23.

b. Cabozantinib (Metastatic Medullary Thyroid Cancer) NDA 203-756 (Effect Of Grapefruit Juice)

Cabozantinib provides an example of Dosage and Administration instructions concerning the influence of grapefruit juice on cytochrome P450 enzymes, with the consequent changed metabolism of cabozantinib. DDIs that involve cytochrome P450-mediated metabolism of various drugs are detailed in Chapters 7 and 8, Drug–Drug Interactions—Part One (Small Molecule Drugs) and Drug–Drug Interactions—Part Two (Therapeutic Proteins).

FDA's ClinPharm Review warns against grapefruit juice, in comments relating to the influence of grapefruit juice and of various drugs on one of the cytochrome P450 enzymes, namely, CYP3A4. The warning from the FDA reviewer was, "CYP3A4 Inhibitors. Avoid the use of concomitant strong CYP3A4 inhibitors ... e.g., ketoconazole, itraconazole, clarithromycin, atazanavir ... in patients receiving COMETRIQ ... [d]o not ingest foods ... e.g., **grapefruit, grapefruit juice** ...that are known to inhibit cytochrome P450.[293]

Package label. FDA's concerns for grapefruit juice and cabozantinib (Cometriq®) found a place in the Dosage and Administration section of the package label:

> **DOSAGE AND ADMINISTRATION** ... Do not administer COMETRIQ with food. Instruct patients not to eat for at least 2 hours before and at least 1 hour after taking COMETRIQ ... [d]o not ingest foods (e.g., **grapefruit, grapefruit juice**) or nutritional supplements that are **known to inhibit cytochrome P450** during COMETRIQ.[294]

The Patient Information Guide, which FDA's website provides as an attachment to the package label, also warned against grapefruit juice. The Patient Information Guide warned:

> **PATIENT INFORMATION GUIDE** ... What should I avoid while taking COMETRIQ? **You should not drink grapefruit juice, eat grapefruit** or any foods or supplements that contain these products, during treatment with COMETRIQ. They may increase the amount of COMETRIQ in your blood.[295]

XI. ALCOHOL-INDUCED DOSE DUMPING

a. Introduction

Dietary alcohol can increase the dissolution rate of tablets and capsules. The phenomenon known as "alcohol-induced dose dumping" refers to the rapid release of most or all

[293] Cabozantinib (metastatic medullary thyroid cancer) NDA 203-756. Pages 34–35 of 106-page Clinical Pharmacology Review.

[294] Package label. COMETRIQ™ (cabozantinib) capsules, for oral use; November 2012 (19 pp.).

[295] Package label. COMETRIQ™ (cabozantinib) capsules, for oral use; November 2012 (19 pp.), and Patient Information Guide (5 pp.).

of the dose from a tablet or capsule within a short period of time.[296] Alcohol-induced "dose dumping" is an overly rapid release of the active ingredient of any drug formulation with consequent increased risk for toxicity to the patient.[297]

Dose dumping is an increased concern with drugs manufactured as **modified-release (MR) formulations**, that is, where drug release is controlled by a polymer matrix or a polymer coating.[298] The concern is that dose dumping will result in increased exposure to the drug, with consequent increased risk for toxicity.

Whitaker et al.[299] describe one such MR formulation, called Chronocort®, where the matrix has an enteric coat, a sustained release layer, an inner-most layer of hydrocortisone drug, and an inert core. Karunasena et al.[300] provide data on a dissolution study of the Chronocort® formulation, where this formulation was exposed to solutions with 0% ethanol, 5% ethanol, 20% ethanol, or 40% ethanol. Release of hydrocortisone was increased at 40% alcohol, but the increase was not great enough to be called "dose dumping." In the words of the researchers, "At a high alcohol concentration of 40% ... there was a change in the Chronocort® dissolution ... with a moderate increase in drug release over the 2 h but no significant alcohol-induced dose dumping."[301]

Darwish et al.[302] measured PK of hydrocodone in a study with human subjects. The subjects were given an "extended release" formulation of hydrocodone with increasing amounts of alcohol: hydrocodone (ER) 15 mg with 240 mL water and 240 mL orange juice containing 4,% 20%, and 40% alcohol in a fasted state. The alcohol had almost no influence on PK, and the authors concluded that dose dumping did not occur.

FDA and EMA both use the term "dose dumping" to refer to this phenomenon. FDA recommends that dissolution testing be conducted in 0.1N HCl at 0%, 5%, 20%, and 40% ethanol. These three concentrations were chosen to replicate the ethanol concentrations in beer, mixed drinks, and hard liquor, respectively.[303] EMA also provides recommendations on dose dumping, and recommends studies with human subjects on an empty stomach, where subjects drink alcohol or no alcohol.[304]

[296] Anand O, et al. Dissolution testing for generic drugs: an FDA perspective. AAPS J. 2011;13:328–35.

[297] Jedinger N, et al. Alcohol dose dumping: the influence of ethanol on hot-melt extruded pellets comprising solid lipids. Eur. J. Pharm. Biopharm. 2015;92:83–95.

[298] Friebe TP, et al. Regulatory considerations for alcohol-induced dose dumping of oral modified-release formulations. Pharm. Technol. 2017;39:40–6.

[299] Whitaker MJ, et al. An oral multiparticulate, modified-release, hydrocortisone replacement therapy that provides physiological cortisol exposure. Clin. Endocrinol. 2014;80:554–61.

[300] Karunasena N, et al. Impact of food, alcohol and pH on modified-release hydrocortisone developed to treat congenital adrenal hyperplasia. Eur. J. Endocrinol. 2017;176:405–11.

[301] Karunasena N, et al. Impact of food, alcohol and pH on modified-release hydrocortisone developed to treat congenital adrenal hyperplasia. Eur. J. Endocrinol. 2017;176:405–11.

[302] Darwish M, et al. Assessment of alcohol-induced dose dumping with a hydrocodone bitartrate extended-release tablet formulated with CIMA(®) abuse deterrence technology. Clin. Drug Investig. 2015;53:645–52.

[303] Anand O, et al. Dissolution testing for generic drugs: an FDA perspective. AAPS J. 2011;13:328–35.

[304] European Medicines Agency (EMA). Guideline on the pharmacokinetic and clinical evaluation of modified release dosage forms; 2014 (46 pp.).

Shown below is an account of two different formulations for the same drug, where the account reveals data too moderate to be considered to be "dose dumping" and data severe enough to be characterized as "dose dumping." The moderate data are from hydromorphone (Exalgo®), and the severe data are from hydromorphone (Palladone®).

b. Dose Dumping for Hydromorphone (Exalgo®)

FDA's ClinPharm Review for hydromorphone (Exalgo®) took note of PK data from human subjects and concluded that the influence of alcohol in increasing Cmax was not of a magnitude sufficient to characterize the Cmax increase as indicative of dose dumping. FDA's comment that "the release characteristics were maintained" means that alcohol did not cause a severe deterioration of the extended release properties of the extended-release tablets.

FDA's account took the form, "Alcohol Effect: The controlled-release property of Exalgo tablet is maintained in the presence of alcohol and that **there is no 'dose dumping' of hydromorphone**. With various concentrations of alcohol (4%, 20% and 40%) the median tmax is 12 to 16 hours with the range of 4 to 27 hours, which are similar to 0% alcohol. The Cmax values in the ... alcohol treatments in the fasted state are higher than that in the 0% alcohol treatment, with mean geometric ratios of 117%, 131%, and 128% for the 4%, 20%, and 40% alcohol treatments, respectively ... [a]lthough the in vitro release rate is slightly increased in the presence of 40% alcohol, the release characteristics were maintained."[305]

Package label. A view of the package label for hydromorphone (Exalgo®) reveals data consistent with data in FDA's ClinPharm Review:

> **CLINICAL PHARMACOLOGY** ... Drug Interaction Studies. Alcohol Interaction. An in vivo study examined the effect of alcohol (40%, 20%, 4% and 0%) on the bioavailability of a single dose of 16 mg of EXALGO in healthy, fasted or fed volunteers ... [f]ollowing concomitant administration of 240 mL of 40% alcohol while fasting ... **Cmax increased by 37% and up to 151%** in an individual subject.[306]

The label for hydromorphone (Exalgo®) also had Black Box Warning, which warned against alcohol. The Black Box Warning disclosed the additive influence of hydromorphone and alcohol in depressing the CNS and increased risk for death. But the Black Box Warning apparently was not influenced by the effect of alcohol in enhancing dissolution and increasing Cmax. The influences of alcohol in increasing Cmax were only modest, where these increases were merely "up to 151%."[307]

[305] Hydromorphone (acute and chronic pain) NDA 021-217. Page 4 of 59-page Clinical Pharmacology Review.

[306] Package label. EXALGO (hydromorphone HCl) extended-release tablets, for oral use, CII; December 2016 (26 pp.). NDA 021-217

[307] Package label. EXALGO (hydromorphone HCl) extended-release tablets, for oral use, CII; December 2016 (26 pp.). This label was published as part of NDA 021-217 submission.

c. Dose Dumping for Hydromorphone (Pallodone®)

The following involves two different effects of alcohol. One of these effects involves dose dumping, but the other effect involves additive influence on the CNS. To reiterate this point, the two effects are (1) alcohol-induced dose dumping, as measurable by alcohol-induced concentration in plasma drug levels and (2) additive effects of alcohol and hydromorphone, in depressing the CNS.

The hydromorphone formulation (Pallodone®), approved several years earlier than Exalgo®, warned against alcohol in the Warnings and Precautions section. Pallodone's approval date was September 2004. The warning read:

> **WARNINGS AND PRECAUTIONS.** Interactions with Alcohol and Drugs of Abuse. Hydromorphone may be expected to have additive effects, when used in conjunction with alcohol, other opioids, or drugs, whether legal or illicit, which cause central nervous system depression ... **[p]atients should NOT combine Palladone Capsules with alcohol** or other pain medications, sleep aids, or tranquilizers except by the orders of the prescribing physician, because dangerous additive effects may occur, resulting in serious injury or death.[308]

About a year later (July 2005), FDA published an "FDA Alert" stating that the Sponsor, "has agreed to FDA's request that they voluntarily suspend sales and marketing of Palladone in the United States. At this time, the Agency has concluded that the **overall risk versus benefit profile of Palladone is unfavorable due to a potentially fatal interaction with alcohol**."[309]

The FDA ALERT referred to the fact that alcohol increases Cmax of hydromorphone. The FDA ALERT read, "A pharmacokinetic study ... showed that co-ingestion of a 12-mg Palladone capsule with 240 mL (8 ounces) of **40% (80 proof) alcohol** resulted in an average peak hydromorphone concentration approximately **six times greater than when taken with water**. One subject in this study experienced a **16-fold increase when the drug was ingested with 40% alcohol compared with water**. In certain subjects, 8 ounces of 4% alcohol (equivalent to 2/3 of a typical serving of beer) resulted in almost twice the peak plasma hydromorphone concentration than when the drug was ingested with water."[310]

These particular increases in Cmax for Palladone® ("six times greater" and "16-fold increase") are much greater than the modest increases found with Exalgo®, where alcohol resulted in an increase of Cmax by only 37% and up to 151%.[311]

[308] Package label. PALLADONE™ (Hydromorphone Hydrochloride Extended-Release Capsules. 12 mg, 16 mg, 24 mg, 32 mg; September 2004 (24 pp.). This label was published as part of NDA 021-044 submission.

[309] FDA ALERT. Alcohol-palladone interaction information for healthcare professionals: hydromorphone hydrochloride extended-release capsules (marketed as Palladone); July 2005.

[310] FDA ALERT. Alcohol-palladone interaction information for healthcare professionals: hydromorphone hydrochloride extended-release capsules (marketed as Palladone); July 2005.

[311] Package label. EXALGO (hydromorphone HCl) extended-release tablets, for oral use, CII; December 2016 (26 pp.). This label was published as part of NDA 021-217 submission.

d. Dose Dumping For Topiramate

An example of dose dumping is found in FDA's Medical Review for topiramate, where FDA wrote that, "The Division previously recommended that the Sponsor provide human subject data on this issue. In vitro dissolution studies suggest the **possibility of dose dumping**. Because of ... the absence of a human study, a contraindication for alcohol use will be recommended 6 hours prior to or after alcohol use."[312]

The Dosage and Administration section of the package label for topiramate (Trokendi®) gave the instruction:

> **DOSAGE AND ADMINISTRATION** ... Administration with Alcohol. Alcohol use should be completely avoided within 6 hours prior to and 6 hours after TROKENDI XR® administration.[313]

Also, the Contraindications section of the label instructed physicians not to give the drug where the patient had recent alcohol use:

> **CONTRAINDICATIONS**. With recent alcohol use, that is, within 6 hours prior to and 6 hours after TROKENDI XR® use.[314,315]

XII. CONCLUDING REMARKS

A recurring food effect theme is the situation where food increases the drug's exposure and where this increased exposure results in increased **efficacy**. An equally common theme is the converse situation, where food increases the drug's exposure, and where the consequence is increased **adverse events**. The term "exposure" refers to one or more PK parameters, such as AUC, Cmax, Cmin, Ctrough, and tmax.

The bulleted points given below illustrate the main issues in this chapter. Other issues from this chapter include grapefruit juice and alcohol-induced dose dumping.

- Where food increases exposure, and where increased exposure increases risk for adverse events, the package label can instruct that drug be taken in a fasting state (examples of abiraterone, afatinib, cabozantinib, and efavirenz).
- Where taking a meal with the drug results in decreased exposure, and where decreased exposure results in lower efficacy, the package label may require fasting (dabrafenib and venetoclax).

[312] Topiramate (partial onset seizures, tonic-clonic seizures) NDA 201-635. Page 39 of 65-page Medical Review.

[313] Package label. TROKENDI XR (topiramate) extended-release capsules for oral use; April 2017 (57 pp.).

[314] Package label. TROKENDI XR (topiramate) extended-release capsules for oral use; April 2017 (57 pp.).

[315] This author points out the bad grammar on the package label. The package label has a contraindication that forbids taking the drug, "within 6 hours prior to and 6 hours after TROKENDI XR® use." What is bad grammar, is that it makes no sense to forbid taking the drug within 6 hours prior to and 6 hours after taking the drug. A correct version of this contraindication would be forbidding taking the drug within 6 hours prior to and 6 hours after taking alcohol.

- Where the goal of the drug is to influence how the stomach processes food, for example, gastric emptying time, then the package label needs to instruct taking the drug with food (lixisenatide).
- Where the drug acts directly on nutrients in a meal (and where this action is desired), then the package label needs to instruct that the drug be taken with meals (examples of the drugs, ferric citrate and pancrelipase).
- Where the drug acts directly on nutrients in a meal (and where this action must be avoided or prevented), then the package label needs to instruct taking the drug in a fasting state (example of eltrombopag).
- Need to maintain acidic gastric conditions and thus need to avoid proton pump inhibitors (example of bosutinib).
- Using food to decrease GI AE, thus resulting in instructions on package label to take drug with a meal (example of miltefosine).
- Where the Sponsor's Clinical Study Protocol required fasting, or fed state, or a specific type of meal (high-fat meal), and where the FDA-approval was based on efficacy and safety from the completed clinical study, then it might be preferred for the package label to require the same dietary state as required in the Clinical Study Protocol (example of dabrafenib).
- Where patients are not able to swallow large capsules, the package label may instruct patients how to open the capsule and sprinkle the drug on food. This technique is shown above for the efavirenz package label.

CHAPTER

4

Dose Modification and Dose Titration

I. INTRODUCTION

This chapter reveals FDA's basis for its recommendations for the Dosage and Administration section of the package label. The topics include:

- dose modifications for avoiding drug toxicity,
- dose modifications for increasing drug efficacy,
- impaired renal function and the need to reduce drug dose, and
- impaired hepatic function and the need to reduce drug dose.

Instructions for dose modifications for study subjects can be included in the Clinical Study Protocol, where the need to reduce or increase dosing is anticipated. These instructions may be for dose reduction, interruption, or permanent discontinuation, where a given study subject is experiencing adverse events (AEs). The instructions can also be for dose increases in the situation where greater efficacy is needed for a given study subject or where a gradual increase in dosing is needed to avoid AEs when ramping up to an effective dose.

Although permanent discontinuation of the drug might seem to be the same thing as the subject withdrawing from the clinical study, they are not the same thing. The fact that these are not the same thing is evident from this excerpt from FDA's Medical Review of denosumab:

> **Adverse Events Leading to Study Withdrawal** ... 174 subjects ... discontinued the trial due to an adverse event ... [t]he most commonly reported AEs leading to discontinuation were breast cancer ... back pain ... and constipation.[1]

> **Adverse Events Leading to Discontinuation of Study Drug** ... Subjects had the option of discontinuing the study drug and **remaining in the study for collection of further data** ... 394 subjects ... discontinued study drug ... [t]he most common reasons for study drug discontinuation was cancer ... and fracture ... [b]reast cancer was the most common malignancy reported followed by colon cancer, gastric cancer, ovarian cancer, and pancreatic cancer.[2]

[1] Denosumab (post-menopausal osteoporosis) BLA 125-320. Page 163 of 710-page Medical Review.

[2] Denosumab (post-menopausal osteoporosis) BLA 125-320. Page 163 of 710-page Medical Review.

FDA's Drug Review Process and the Package Label
DOI: https://doi.org/10.1016/B978-0-12-814647-7.00004-X

As might be inferred, withdrawal means that the subject was not interested to remain in the study for collection of further data, whereas, in contrast, drug discontinuation means that the subject was willing to provide further data in the months or years following permanent discontinuation of taking the study drug.

This chapter, as well as most of the chapters in this book, was written, so that the paragraphs reiterate FDA's train of logic used to arrive at FDA's recommendations for the package label. In embarking on each chapter in this book and in reading the accounts of the various study drugs, the reader should first glance at the excerpts from the package label for that study drug. These excerpts from the package label constitute the destination of FDA's train of logic. In other words, the goal of each narrative corresponding to a given drug is to demonstrate the basis for warnings and other information eventually printed on the package label.

a. Clinical Study Protocol Dose Modification Instructions

The precursor for any dose modification instructions in a package label is likely to be corresponding instructions in the Clinical Study Protocol. At an early stage in the drug development and drug-approval process, the Sponsor drafts a Clinical Study Protocol and submits it to the FDA. The following excerpt from the Code of Federal Regulations (CFR) establishes the need for a dosing plan and a dose modification plan.

The CFR teaches that, "In general, protocols for Phase 1 studies may be less detailed and more flexible than protocols for Phase 2 and 3 studies. Phase 1 protocols should be directed primarily at providing an outline of the investigation— an estimate of the number of patients to be involved, a description of safety exclusions, and a description of the **dosing plan** including duration, dose, or method to be used in determining dose ... [a] protocol for a Phase 2 or 3 investigation should be designed in such a way that, if the sponsor anticipates that some **deviation from the study design** may become necessary as the investigation progresses, alternatives or contingencies to provide for such deviation are built into the protocols at the outset."[3]

The best source of Clinical Study Protocols is *New England Journal of Medicine*. This journal publishes Clinical Study Protocols as a supplement to its articles. Shown below are instructions for two different aspects of dosing:

- Dose delay or reduction (example of ponatinib)
- Dose escalation and stopping dose escalation (example of venetoclax)

Regarding the ponatininb study, the Clinical Study Protocol includes the instruction, "**Dose delay and/or reduction** for adverse events (AEs) attributable to the study drug ... [t]here will be no dose modification for grade 1 or 2 non-hematologic toxicities (except for pancreatitis) attributable to the study drug that are manageable with supportive care ... [i]n the event of a persistent grade 1 or 2 non-hematologic adverse drug reaction that is ... intolerable ... the patient may be managed by **dose delay or reduction** as described in Table 14-1... [i]n the event of grade 3 or 4 adverse event attributed to study drug, the patient may be managed by **dose reduction or delay** as well. Guidelines are described in

[3] 21 CFR §312.23. IND content and format.

Table 14-1. Note that grade 3 or 4 myelosuppression might be attributable to disease rather than to the study drug."[4,5]

A corresponding instruction in the Dose and Administration section of the package label for ponatinib reads, "Modify or interrupt dosage for hematologic and non-hematologic toxicity."[6]

Regarding venetoclax and dose-titration, the Clinical Study Protocol included the dose-titration instructions, "Dosing with ... venetoclax will begin at 200 mg/day and escalate to the maximum tolerable dose (MTD) with a minimum of 3 subjects at each dose level. **The study drug dose will be escalated in 100 mg increments until a dose level of 500 mg is reached.** After which dose escalation will continue in standard 25% to 40% increments ... [e]scalation ... to the next dose level will proceed if all assigned subjects at a dose level complete Week 3 without experiencing a dose limiting toxicity (DLT). If one subject within any dose level experiences a DLT, a total of 6 subjects will be enrolled at that dose level. If 1 out of 6 subjects experiences a DLT, then dose escalation will continue. If ≥2 out of 6 subjects at a dose level experience a DLT, a dose escalation will stop because the DLT will have been exceeded, and the previous dose will be considered for the MTD ... [t]he MTD will be defined as the highest dose level at which less than 2 of 6 or n < 33% subjects ... experience a DLT."[7,8]

Corresponding instructions on the package label for venetoclax are, "Initiate therapy with VENCLEXTA at 20 mg once daily for 7 days, followed by a weekly ramp-up dosing schedule to the recommended daily dose of 400 mg."[9]

b. Introduction to the Therapeutic Window

Therapeutic window and the closely related concept, therapeutic index, are sometimes taken into account when determining dosing. The therapeutic window refers to the drug dose needed, where there is a pressing need to maintain "exposure" within a range that is effective and yet avoids undue AEs. Therapeutic window is taken into account when assessing the need for dose modification under circumstances such as patient variability, the influence of food on exposure, and the influence of a coadministered drug on exposure. The term "exposure" usually refers to a parameter of drug concentration in blood, such as AUC, Cmax, tmax, or Cmin.

[4] Clinical Study Protocol. A Pivotal Phase 2 Trial of Ponatinib (AP24534) in Patients with Refractory Chronic Myeloid Leukemia and Ph + Acute Lymphoblastic Leukemia. ARIAD protocol number: AP24534-10-201 (Pages 63–65 of 96 pages).

[5] Cortes JE, et al. A phase 2 trial of ponatinib in Philadelphia chromosome-positive leukemias. N. Engl. J. Med. 2013;369:1783–96.

[6] Package label. ICLUSIG® (ponatinib) tablets for oral use; December 2012 (17 pages).

[7] Clinical Study Protocol M12-175. A Phase 1 Study Evaluating the Safety, Pharmacokinetics and Preliminary Efficacy of ABT-199 in Subjects with Relapsed or Refractory Chronic Lymphocytic Leukemia.

[8] Roberts AW, et al. Targeting BCL2 with venetoclax in relapsed chronic lymphocytic leukemia. N. Engl. J. Med. 2016;374:311–22.

[9] Package label. VENCLEXTA™ (venetoclax) tablets, for oral use; April 2016 (22 pages).

FDA's Guidance for Industry on bioequivalence defines, "**narrow therapeutic range** drug products as those containing ... drug substances that are subject to therapeutic drug concentration or pharmacodynamic monitoring ... or where product labeling indicates a narrow therapeutic range designation. Examples include digoxin, lithium, phenytoin, theophylline, and warfarin."[10] FDA's Guidance for Industry on drug interactions provides further information on drugs with a narrow therapeutic range or window.[11] The cited references describe drugs with a narrow therapeutic window, where these are amiodarone (cardiac antiarrhythmic drug),[12] sunitinib (anticancer drug),[13] tacrolimus (immunosuppressant),[14,15] and warfarin (anticoagulant).[16]

A dramatic account of the narrow therapeutic window for **warfarin** is provided by an account of the Yup'ik native Americans in Alaska, where their warfarin requirement is changed because of a naturally occurring variant in one of their cytochrome P450 enzymes (CYP4F2*3).[17,18]

Cytochrome P450 is abbreviated as "CYP" and the cytochrome P450 enzymes may be called CYP enzymes or CYP isozymes. The CYP enzymes occur as various isozymes, such as CYP1A2, CYP2B6, CYP2D6, and CYP4F2. CYP4F2 catalyzes the hydroxylation of the phytyl side chain of vitamin K, thus leading to the degradation of vitamin K and consequent impaired blood clotting. But a genetic variant of CYP4F2, namely, the variant known as CYP4F2*3, does not much catalyze the degradation of vitamin K. Thus, where a person's genome encodes CYP4F2*3 instead of the wild-type CYP4F2, the consequence is elevated levels of vitamin K in the body. When a patient with CYP4F2*3 needs to be

[10] US Department of Health and Human Services; Food and Drug Administration; Center for Drug Evaluation and Research (CDER). Guidance for Industry. Waiver of In Vivo Bioavailability and Bioequivalence Studies for Immediate-Release Solid Oral Dosage Forms Based on a Biopharmaceutics Classification System; 2015 (14 pages).

[11] US Department of Health and Human Services; Food and Drug Administration; Center for Drug Evaluation and Research (CDER). Guidance for Industry. Drug Interaction Studies—Study Design, Data Analysis, Implications for Dosing, and Labeling Recommendations; 2012 (75 pages).

[12] Naccarato M, et al. Amiodarone and current antiretroviral therapy: a case report and review of the literature. Antiviral Ther. 2014;19:329–39.

[13] Haas B, et al. Is sunitinib a narrow therapeutic index drug? A systematic review and in vitro toxicology analysis of sunitinib. Regulatory Toxicol. Pharmacol. 2016;77:25–34.

[14] Borra LCP, et al. High within-patient variability in the clearance of tacrolimus is a risk factor for poor long-term outcome after kidney transplantation. Nephrol. Dial. Transplant. 2010;25:2757–63.

[15] Iwasaki K. Metabolism of tacrolimus (FK506) and recent topics in clinical pharmacokinetics. Drug Metab. Pharmacokinet. 2007;22:328–35.

[16] Kuruvilla M, Gurk-Turner C. A review of warfarin dosing and monitoring. BUMC Proceedings. 2001;14:305–6.

[17] Fohner AE, et al. Variation in genes controlling warfarin disposition and response in American Indian and Alaska Native people: CYP2C9, VKORC1, CYP4F2, CYP4F11, GGCX. Pharmacogenet. Genomics. 2015;25:343–53.

[18] Au NT, et al. Dietary and genetic influences on hemostasis in a Yup'ik Alaska native population. PLoS One 2017;1:e0173616 (16 pages).

treated with the "blood thinner" warfarin, the warfarin dose must be increased to counteract the greater stores of vitamin K in the body. In short, because of the narrow therapeutic window of warfarin, any increase in the amount of warfarin's target (vitamin K) in the body, will need a corresponding change in the amount of warfarin administered to the patient. To provide scientific background, warfarin inhibits epoxide reductase and prevents the recycling of vitamin K in the body.

Klein[19] provides an example of a drug with a narrow therapeutic window and of the goal of maintaining plasma concentration to achieve efficacy and to avoid AEs. The drug is **theophylline**, a drug for treating airway obstruction. Optimal therapeutic serum levels are 8–15 μg/mL. AEs, such as nausea, vomiting, abdominal cramps, cardiac arrhythmias, and convulsions, can occur where theophylline in serum is >25 μg/mL.

To ensure that the blood concentration is maintained within the narrow concentration range, theophylline is provided in an extended release form and not as an immediate release form.

Although some drugs have a broad therapeutic window, the need to be vigilant arises where the drug shows high variability between patients or where the drug shows high variability with the same patient from dose-to-dose. The term "critical dose drug" refers to this situation, that is, the situation where the drug shows high inter- and intraindividual variability.[20,21]

FDA provided a thorough and dedicated account of "narrow therapeutic window" in its review of **dofetilide**. Information on assessing "narrow therapeutic window" and the implications for package labeling are found in FDA's reviews for NDA 020-091.[22] Dofetilide is for treating cardiac arrhythmia. The following commentary includes the topics of drug–drug interactions and renal impairment (topics detailed later on in this book). The concepts of QT interval and QT prolongation, which are measures of cardiac physiology, are detailed in a chapter in the cited reference.[23] The present account is intended only to illustrate the concept of therapeutic window.

c. Therapeutic Window Illustrated by FDA's Review of Dofetilide

FDA's Medical Review revealed that the narrow therapeutic window for **dofetilide** caused this drug to be very susceptible to undesirable exposure changes, where the exposure changes are caused by a coadministered drug. As a general proposition, plasma concentrations of any given drug can be influenced by a coadministered second drug, where

[19] Klein S. Predicting food effects on drug release from extended-release oral dosage forms containing a narrow therapeutic index drug. Dissolution Technol. 2009;28–40.

[20] Editorial. Finding the optimal therapeutic for tacrolimus. Pediatr. Transplantation. 2014;18:783–5.

[21] Filler G. Calcineurin inhibitor in pediatric renal transplant recipients. Paediatr. Drugs. 2007;9:165–74.

[22] FDA's review for dofetilide NDA 020-931 included these reviews: Medical Review (eight pdf files), Clinical Pharmacology Review (five pdf files), and Administrative Documents (two pdf files).

[23] Brody T. QT Interval Prolongation in Clinical Trials in CLINICAL TRIALS Study Design, Endpoints and Biomarkers, Drug Safety, and FDA and ICH Guidelines, 2nd ed., Elsevier, Inc., New York, NY; 2016, pp. 502–14.

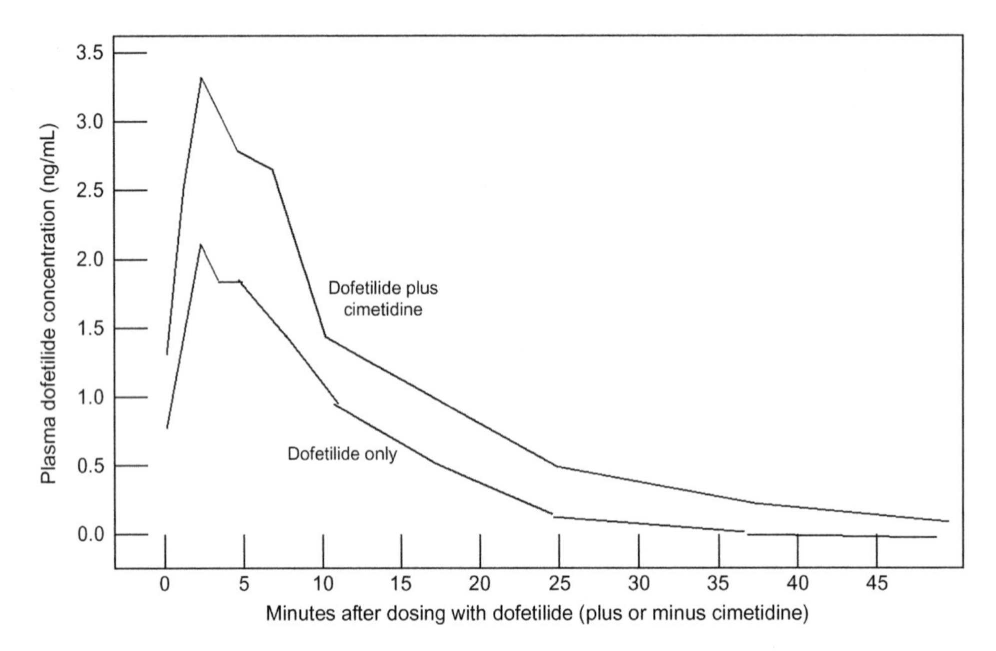

FIGURE 4.1 Blood plasma concentrations of dofetilide with and without cimetidine. Dofetilide has a narrow therapeutic window, meaning that excessive blood levels can lead to AEs, in this case, to cardiac toxicity. Coadministered cimetidine increases dofetilide's concentration and thus increases the risk for dofetilide's cardiac toxicity.

the second drug can provoke undesired increases (or decreases) in the first drug. Among the drug–drug interaction studies conducted by the Sponsor, the Sponsor tested the influence of coadministered cimetidine on plasma concentrations of dofetilide. Fig. 4.1 shows that the coadministered cimetidine provoked an increase in dofetilide concentrations.

FDA expressed concern about this particular drug–drug interaction, writing, "This is of great concern because dofetilide has a **very narrow therapeutic window.**"[24] To provide a word on the biochemical mechanism of the drug–drug interaction, the effect where cimetidine causes an increase in dofetilide plasma concentration occurs when cimetidine inhibits cytochrome P450 3A4 (CYP3A4). The "very narrow therapeutic window" had a corresponding warning dofetilide's package label, in the Dosage and Administration section, the Drug Interactions section, and the Clinical Pharmacology section.

Turning to FDA's recommendations for the package label, FDA considered instructions on the label for tailoring dofetilide doses according to the patient's renal function (creatinine values) and cardiac function (QT values). FDA referred to the drug's narrow window, writing, "Because of the **narrow window between effective doses** ... **and doses with excessive risk** [cardiac adverse events] ... clinical trial protocols have been designed ... to start at 500 micrograms per day, and allow downward dose adjustment for QTc

[24] Dofetilide (cardiac arrhythmia) NDA 020-091. Page 25 of eighth pdf file (45-page pdf file) Medical Review. The Medical Review has eight pdf files in all.

TABLE 4.1 Algorithm for Dofetilide Dosing, Based on a Patient's Renal Function (Creatinine Clearance) and Cardiac Function (QTc Prolongation)

Creatinine Clearance	Without Any QTc Prolongation	With QTc Prolongation
>60 mL/minute	0.5 mg bid[a]	0.25 mg bid
40–60 mL/minute	0.25 mg bid	0.125 mg bid
20–40 mL/minute	0.125 mg bid	0.125 qd (once a day)
<20 mL/minute	individualized	discontinue

[a] "bid" means two doses per day. The usual initial dose is 0.5 mg bid.
From: Dofetilide (cardiac arrhythmia) NDA 020-091. Pages 24 and 25 of second pdf file (36 page pdf file) of Administrative Documents.

prolongation and low creatinine clearance."[25] QTc prolongation is a measure of cardiac pathology. Low creatinine is an indicator of renal pathology.

Then, FDA's review turned its attention to an algorithm for dosing. The Sponsor had recommended using this algorithm, and the fact that FDA approved the algorithm is evident from the fact that it appears on the package label. On this point, FDA wrote, "Accordingly in the package label, the sponsor recommends a usual initial dose of 500 micrograms twice a day (BID) for patients ... which may be reduced as follows." Table 4.1 reproduces data from FDA's review, which provides a scheme for dose modification depending on a patient's renal function and further depending on the patient's cardiac function (QTc prolongation).

At first, FDA complained about the complexity of the dosing algorithm but then realized that the complex algorithm was needed because of dofetilide's narrow therapeutic window, that is, because of the fact that overly **high plasma dofetilide posed a risk for cardiac toxicity**. FDA's complaint and rationalization was, "Dosage recommendation and adjustment to minimize proarrhythmic risk are therefore no trivial matters for dofetilide, although one may argue that the proposed scheme is justified by the complexity of the disease to be treated."[26]

FDA's Medical Review referred to dofetilide's "narrow therapeutic window" and provided a figure, reproduced here in Fig. 4.1, showing plasma concentrations of dofetilide over time (plus or minus coadministered cimetidine), and then warned about the danger of cardiac toxicity due to the fact that cimetidine provoked huge increases in plasma dofetilide.

On these points, FDA's Medical Review wrote, "**Narrow Therapeutic Window**. The effective dose being proposed by the sponsor is 500 micrograms bid, with adjustments based on creatinine clearance and QT prolongation. Doses of 750 micrograms and above were dropped from the development program because of excessive proarrhythmia. One patient ... erroneously received 2 doses of 500 micrograms in 1 hour instead of 12 hours apart ... and developed ... cardiac arrest about 2 hours after the second dose."[27]

[25] Dofetilide (cardiac arrhythmia) NDA 020-091. Pages 24 and 25 of second pdf file (36-page pdf file) of Administrative Documents.

[26] Dofetilide (cardiac arrhythmia) NDA 020-091. Pages 24 and 25 of second pdf file (36-page pdf file) of Administrative Documents.

[27] Dofetilide (cardiac arrhythmia) NDA 020-091. Page 6 of sixth pdf file (45-page pdf file) for Medical Review.

FDA's attention then turned to drug–drug interactions, writing, "Drug Interactions. To avoid higher dofetilide concentrations ... dofetilide should not be co-administered with cimetidine."[28] The Drug Interactions section of the package label for dofetilide (Tikosyn®) warned about taking the antistomach acid drug, cimetidine. Dofetilide's package label instructed the physician to not give cimetidine with dofetilide and, instead, to consider other drugs against stomach acid:

> **DRUG INTERACTIONS**. Cimetidine: **Concomitant use of cimetidine is contraindicated**. Cimetidine at 400 mg BID (the usual prescription dose) co-administered with TIKOSYN (500 mcg BID) for 7 days has been shown to increase dofetilide plasma levels by 58% ... [i]f a patient requires TIKOSYN and anti-ulcer therapy, it is suggested that omeprazole, ranitidine, or antacids (aluminum and magnesium hydroxides) be used as alternatives to cimetidine, as these agents have no effect on the pharmacokinetic profile of TOKOSYN.[29]

II. FDA'S REVIEWS REVEAL THE ORIGINS OF DOSE MODIFICATION OR DOSE TITRATION INSTRUCTIONS IN THE PACKAGE LABEL

Excerpts from FDA's Medical Reviews and Clinical Pharmacology Reviews illustrate FDA's decision-making processes for arriving at recommendations on the package label, as applied to dose modifications and dose titrations. The first several excerpts are from FDA's reviews for:

- Alirocumab
- Canagliflozin
- Ferric citrate
- Lenalidomide
- Nivolumab
- Sebelipase alfa

a. Alirocumab (Hyperlipidemia) BLA 125-559

A publication described the alirocumab study design for clinical study called "OPTIONS." The publication described the starting dose, the technique of using blood low-density lipoprotein (LDL) cholesterol (LDL-C) levels as a basis for increasing (or for not increasing) alirocumab dose, and the fact that the study design ensured that study subjects and investigator were blinded as to which subject received an increased (up-titrated) dose.

The publication stated, "**At week 12, based on their week 8 LDL-C level** ... patients randomized to alirocumab may, in a blinded manner, **have had their dose increased from 75 mg Q2W to 150 mg Q2W**. To maintain blinding, lipid values obtained at week 8 for the

[28] Dofetilide (cardiac arrhythmia) NDA 020-091. Page 6 of sixth pdf file (45-page pdf file) for Medical Review.

[29] TIKOSYN® (dofetilide) capsules. March 2016 (26 pages).

purpose of up-titration occurred in an automated process and were not communicated to investigators or patients. Patients with baseline LDL-C levels greater or equal to 70 mg/dL ... continued to receive alirocumab 75 mg Q2W if their week 8 LDL-C level was less than 70 mg/dL ... otherwise, they were dose up-titrated to receive alirocumab 150 mg Q2W ... if their week eight LDL-C level was greater or equal to 70 mg/dL."[30]

Turning to FDA's description of the same clinical study, FDA's ClinPharm Review described the dose titration scheme that eventually found a place on the package label. FDA's recommendation on a reasonable dosing scheme included these elements:

- **Exposure.** Increasing alirocumab resulted in increasing exposure, as revealed by FDA's observation, "Alirocumab exposure increased in a dose-dependent manner in patients and LDL-C reduction reached apparent nadir after 150 mg Q2W."[31]
- **Efficacy.** Efficacy took the form of reduced LDL cholesterol. Efficacy was greater with greater dose amounts.
- **Dosing scheme matching that in Clinical Study Protocol.** A basis for the highest dose in FDA's recommendation was that it was the highest dose used in the Sponsor's studies. In the Sponsor's studies, the highest dose (150 mg Q2W) was characterized by the advantage that "LDL-C reduction reached apparent maximum after 150 mg Q2W with mean reduction in LDL-C of 67.26%."[32]

FDA's ClinPharm Review observed that upwards titration of alirocumab resulted in a corresponding reduction in LDL cholesterol:

> **Alirocumab exposure increased** in a dose-dependent manner in patients and **LDL-cholesterol reduction reached apparent nadir** after 150 mg Q2W ... additional LDL-cholesterol reduction was noted among 6 of 8 trials with the **titration scheme**, which ranged from 1.5 to 23.1%, in patients who were titrated in the pivotal trials up to 150 mg Q2W, and baseline LDL-cholesterol values in the titrated patients were higher than those of 75 mg Q2W. Further, both 75 and 150 mg Q2W had superior efficacy compared to placebo. Therefore, it seems reasonable to consider 75 mg Q2W as the starting dose and **alirocumab can be titrated up to 150 mg Q2W** in patients needing additional LDL-cholesterol reduction.[33]

Q2W means one dose every 2 weeks. Clinical data showed that, with 150 mg alirocumab Q2W, LDL cholesterol levels were reduced to the lowest levels possible with the drug. These data are revealed in Fig. 4.2. The *Y*-axis shows efficacy in terms of percent reduction in LDL cholesterol, as compared to baseline levels. The *X*-axis shows exposure of alirocumab, in terms of plasma drug levels (in micrograms/milliliter). Fig. 4.2 is a composite of data from three different clinical studies, having similar experimental designs, where in the first design subjects received 50 mg alirocumab, in the second design they

[30] Robinson JG, et al. Efficacy and safety of alirocumab as add-on therapy in high-cardiovascular-risk patients with hypercholesterolemia not adequately controlled with atorvastatin (20 or 40 mg) or rosuvastatin (10 or 20 mg): design and rationale of the ODYSSEY OPTIONS Studies. Clin. Cardiol. 2014;37:597–694.

[31] Alirocumab (hyperlipidemia) BLA 125-559. Page 10 of 91-page, Clinical Pharmacology Review.

[32] Alirocumab (hyperlipidemia) BLA 125-559. Page 14 of 91-page, Clinical Pharmacology Review.

[33] Alirocumab (hyperlipidemia) BLA 125-559. Page 10 of 91-page, Clinical Pharmacology Review.

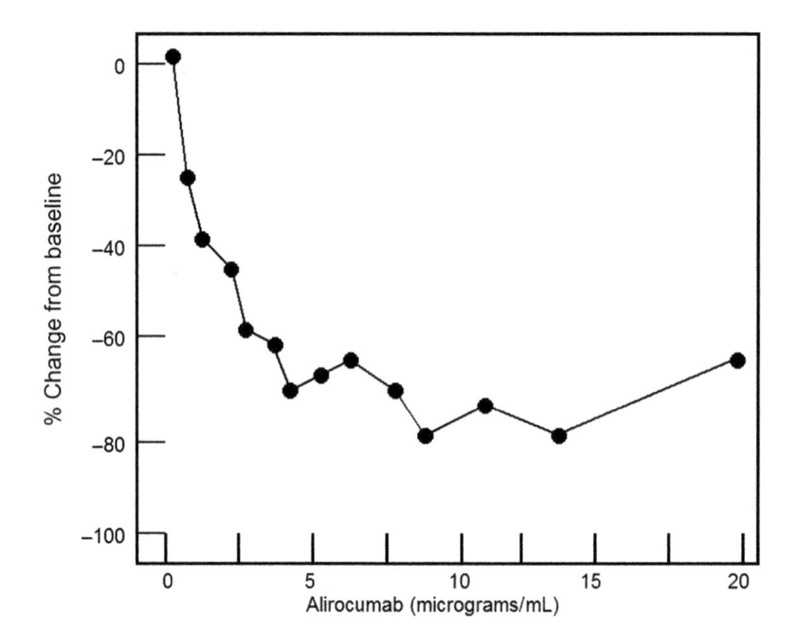

FIGURE 4.2 Plasma LDL cholesterol reduction vs alirocumab in bloodstream. The changes in blood alirocumab concentration were the result of giving study subjects doses of 50, 100, or 150 mg alirocumab. Alirocumab treatment was for 12 weeks.

received 100 mg, and in the third design they received 150 mg.[34] FDA's ClinPharm Review referred to this type of clinical study design as a "dose-finding trial" because the goal of the study was to find the optimal dose.[35]

Data from all three studies showed a drop from baseline LDL cholesterol, and that is what is plotted in the figure. FDA emphasized the benefit of the highest dose, writing, "Is there a benefit of titration of alirocumab dose from 75 mg Q2W to 150 mg Q2W? Yes, there is a benefit of titration of alirocumab dose from 75 mg Q2W to 150 mg Q2W."[36]

Package label. The package label for alirocumab (Praluent®) provided instructions for dose titration:

> **DOSAGE AND ADMINISTRATION**. The recommended starting dose for PRALUENT is 75 mg administered subcutaneously once every 2 weeks, since the majority of patients achieve sufficient LDL-cholesterol reduction with this dosage. If the LDL-cholesterol response is inadequate, the **dosage may be increased** to the maximum dosage of 150 mg administered every 2 weeks ... [m]easure LDL-cholesterol levels within 4 to 8 weeks of initiating or **titrating** PRALUENT, to assess response and adjust the dose, if needed.[37]

[34] Alirocumab (hyperlipidemia) BLA 125-559. Page 14 of 91-page, Clinical Pharmacology Review.

[35] Alirocumab (hyperlipidemia) BLA 125-559. Pages 14–23 of 91-page, Clinical Pharmacology Review.

[36] Alirocumab (hyperlipidemia) BLA 125-559. Page 25 of 91-page, Clinical Pharmacology Review.

[37] Package label. PRALUENT™ (alirocumab) injection, for subcutaneous use. July 2015 (16 pages). BLA 125-559.

b. Canagliflozin (Type-2 Diabetes Mellitus) NDA 204-042

This concerns FDA's review of canagliflozin (Invokana®) for treating Type-2 diabetes mellitus. Canagliflozin inhibits a transporter in the renal tubule. The drug prevents resorption of glucose that exists in the forming urine, thereby enhancing removal of glucose in the urine, with the desired consequence of reducing plasma glucose.

1. **Origin of dose-titration instructions in the Clinical Study Protocol.** A published account of the clinical study, named "CANVAS," revealed that the clinical study design involved, study subjects in a, "1:1 ratio to receive canagliflozin, administered at an initial dose of 100 mg daily with an optional increase to 300 mg starting from week 13, or matching placebo."[38] As shown below, the study design as set forth in the Clinical Study Protocol and as published, found a place in the package label.
2. **Dose-titration based on efficacy, as determined by reducing glycosylated Hb.** FDA's ClinPharm Review recommended that the package label include dose-titration instructions. This recommendation was based on the fact that the Sponsor determined that greater efficacy was found at greater doses. Efficacy was measured by HbA1c. HbA1c, which is a type of glycosylated hemoglobin, is routinely used to assess the influence of drugs in treating diabetics.[39,40]

   Regarding basing proposed canagliflozin dosing on corresponding reductions in glycosylated HB, FDA wrote, "the dose-response relationship, evident among 100 mg and 300 mg QD doses for **reduction in HbA1c**, supports the proposed doses of canagliflozin as 100 mg once daily (QD) and 300 mg QD."[41]
3. **Dose-titration based on efficacy, as determined by increasing urinary glucose.** In addition to the measure of glycosylated hemoglobin (HbA1c), the Sponsor also did a dose titration to assess influence on urinary glucose excretion. FDA's account of this dose titration read, "Following single dose administration of canagliflozin in healthy subjects, the **increase in urinary glucose excretion (UGE)** was dose-dependent up to 400 mg dose of canagliflozin given QD. When the dose was increased from 400 mg to 800 mg, no further increase in 24-h UGE was observed, suggesting saturation of UGE response."[42]

   FDA's ClinPharm review recommended, "Given a modest increase in benefit with an increased risk of adverse events for 300 mg QD dose, compared to the 100 mg QD dose, this reviewer recommends **a titration based dosing strategy** ... recommends the

[38] Neal B et al. Canagliflozin and cardiovascular and renal events in type 2 diabetes. N. Engl. J. Med. 2017;377:644–57. DOI: 10.1056/NEJMoa1611925.

[39] Larsen ML, et al. Effect of long-term monitoring of glycosylated hemoglobin levels in insulin-dependent diabetes mellitus. N. Engl. J. Med. 1990;323:1021–5.

[40] Norcliffe D, Turner EM. Comparison of five commercial kits for the determination of glycosylated haemoglobin. J. Clin. Pathol. 1984;37:1177–81.

[41] Canagliflozin (type-2 diabetes mellitus) NDA 204-042. Pages 12 and 114 of 165-page Clinical Pharmacology Review.

[42] Canagliflozin (type-2 diabetes mellitus) NDA 204-042. Pages 53 and 54 of 165-page-Clinical Pharmacology Review.

following ... labeling actions ... [i]n the package label ... [s]tarting dose of 100 mg QD in all patients. **Titrate to 300 mg** based on individual patient's tolerability and need of further glycemic control."[43] In other words, the FDA reviewer wanted physicians to customize the dose titration according to the responses by individual patients.

4. **Need to establish a cutoff for excluding patients at high risk for kidney injury.** FDA's Medical Review focused on the subset of study subjects with renal impairment and, in particular, on subjects with low values for estimated glomerular filtration rate (eGFR). FDA's concern focused on the cutoff point of eGFR under 45 mL/min/1.73 m^2. FDA's concern was increased risk for safety issues, where the drug was given to patients with renal disease, old age, and concomitant drugs. FDA's Medical Review stated:

> The amount of safety data in subjects with diabetes and moderate renal impairment is limited (particularly in subjects with an eGFR <45 mL/min/1.73 m^2) and what data exist suggest a high absolute risk of ... acute kidney injury ... [t]his risk may be related to the presence of underlying renal disease, age, concomitant therapies ... [i]n addition, this risk may be magnified in the postmarketing setting when canagliflozin is used outside the carefully monitored setting of a clinical trial and in a less selected population. Given these issues, we think considerable uncertainty remains regarding renal safety in patients with diabetes and moderate renal impairment.[44]

Various FDA reviewers expressed concern about patients with this same cutoff point. For example, one reviewer wrote, "Dr. Lewis was specifically concerned about using canagliflozin in patients with low GFR (<45 mL/min/1.73 m^2)."[45] Another reviewer wrote, "Several panel members who voted yes and no discussed discomfort of using canagliflozin in patients with moderate renal impairment, where the benefit risk is unfavorable and safety issues are concerning. Dr. Palevsky again expressed concerns about patients with eGFR <45 mL/min/1.73 m^2."[46]

As revealed by comments in FDA's Medical Review, FDA reviewers focused on the cutoff point of eGFR below 45 mL/min/1.73 m^2, where this value found a place in the Dosage and Administration section. But an even lower cutoff point found a place in the Contraindications section, as shown below. FDA's Medical Review provides a basis for this even lower cutoff point, in its comments that, "Due to its mechanism of action, canagliflozin has not been studied and is not expected to provide any benefit in patients with severe renal insufficiency (<30 mL/min/1.73 m^2)."[47]

[43] Canagliflozin (type-2 diabetes mellitus) NDA 204-042. Pages 12 and 114 of 165-page, Clinical Pharmacology Review.

[44] Canagliflozin (type-2 diabetes mellitus) NDA 204-042. Page 163 of 255-page Medical Review.

[45] Canagliflozin (type-2 diabetes mellitus) NDA 204-042. Page 241 of 255-page Medical Review.

[46] Canagliflozin (type-2 diabetes mellitus) NDA 204-042. Page 244 of 255-page Medical Review.

[47] Canagliflozin (type-2 diabetes mellitus) NDA 204-042. Page 95 of 255-page Medical Review.

5. **Summary of steps taken to arrive at the Dosage and Administration section.** FDA's review for canagliflozin reveals a number of steps and options that may reasonably be used to arrive at the Dosage and Administration section for a great variety of drugs. These steps and options include:
   - **Clinical Study Protocol.** Consider using the dosing instructions in the Clinical Study Protocol as part of the basis for instructions in the package label.
   - **Efficacy measure.** Consider which measure of efficacy, for example, glycosylated hemoglobin, urinary glucose, or fasting glucose, is best for assessing dose–response relationships.
   - **Package label excludes specific subgroups (Dosage and Administration).** Where a subgroup in the expected patient population is of increased risk for AEs from the study drug, consider drafting exclusion criteria for this subgroup in the Dosage and Administration section as a subgroup that may receive the lowest dose (but that may not receive higher doses).
   - **Package label excludes specific subgroups (Contraindications).** Where a subgroup in the expected patient population is of increased risk for AEs from the study drug, consider drafting exclusionary criteria in a Contraindications section. Also, where the study drug cannot benefit a particular subgroup in the patient population, consider drafting exclusionary criteria in a Contraindications section.

Package label. FDA's recommendations for dose titration found a place on the package label, where the dosing range is 100 mg/day–300 mg/day. The Dosage and Administration section provides instructions for dose titration and also instructions to refrain from giving the drug to patients with impaired renal function.

> **DOSAGE AND ADMINISTRATION**. The recommended starting dose of INVOKANA (canagliflozin) is 100 mg once daily, taken before the first meal of the day. In patients ... who have an eGFR of 60 mL/min/1.73 m^2 or greater and require additional glycemic control, the dose can be increased to 300 mg once daily.[48]

The phrase, "patients ... who have an eGFR of 60 mL/min/1.73 m^2 or greater," refers to an exclusion criterion, which instructs the physician not to administer the drug to patients with impaired renal function as evidenced by glomerular filtration rate that is substantially $<$60 mL/min/1.73 m^2. These exclusion criteria took the form:

> **DOSAGE AND ADMINISTRATION** ... Assess renal function before initiating and periodically thereafter ... [l]imit the dose of INVOKANA to 100 mg once daily in patients who have an eGFR of 45 to less than 60 mL/min/1.73 m^2 ... [i]nitiation or use of INVOKANA is not recommended if eGFR is below 45 mL/min/1.73 m^2.[49]

[48] Package label. INVOKANA™ (canagliflozin) tablets, for oral use. February 2017 (42 pages). NDA 204-042.

[49] Package label. INVOKANA™ (canagliflozin) tablets, for oral use. February 2017 (42 pages). NDA 204-042.

The label's Contraindications section read:

> **CONTRAINDICATIONS** ... Severe renal impairment (eGFR less than 30 mL/min/1.73 m^2), end stage renal disease (ESRD), or patients on dialysis.[50]

The unit "estimated glomerular filtration rate" (eGFR) is explained.[51] EGFR can be calculated using the Cockcroft–Gault equation or the MDRD Study equation.[52] FDA's Medical Review revealed that the Sponsor used the MDRD Study equation, as is evident from FDA's description of one of the Sponsor's studies, "Placebo-controlled trial in ... subjects with moderate renal impairment (estimated glomerular filtration rate of ≥ 30 to <50 mL/min/1.73 m^2 based on MDRD equation)."[53]

National Kidney Foundation explains the role of body surface area as, "Inulin clearance is widely regarded as the gold standard for measuring glomerular filtration rate. Inulin clearance measurements in healthy, hydrated young adults (adjusted to a standard body surface area of 1.73 m^2) have mean values of 127 mL/min/1.73 m^2 in men and 118 mL/min/1.73 m^2 in women."[54]

c. Ferric Citrate (Serum Phosphorus in Chronic Kidney Disease) NDA 205-874

Chronic kidney disease can result in hyperphosphatemia. Where kidney disease patients are treated with dialysis, additional therapy is needed for hyperphosphatemia. In health, phosphate is mainly eliminated via the kidneys to the urine. In kidney disease, oral phosphate binding agents, such as ferric chloride, aluminum, and calcium salts, have been used to bind dietary phosphate in the gut lumen, thereby preventing absorption and, at the same time, promoting excretion in the feces.[55,56]

This reveals the use of dose titration of ferric citrate, a drug for reducing serum phosphorous levels in patients with chronic kidney disease. FDA's ClinPharm Review for ferric citrate revealed that the Sponsor's clinical studies included a dose-titration study design.

[50] Package label. INVOKANA™ (canagliflozin) tablets, for oral use. February 2017 (42 pages). NDA 204-042.

[51] Yale JF, et al. Efficacy and safety of canagliflozin in subjects with type 2 diabetes and chronic kidney disease. Diabetes Obes. Metab. 2013;15:463–73.

[52] National Kidney Foundation DOQI Kidney Disease Outcomes Quality Initiative. Clinical Practice Guidelines for Chronic Kidney Disease: Evaluation, Classification, and Stratification. National Kidney Foundation, Inc. New York, NY; 2002, p. 86–96. ISBN 1-931472-10-6.

[53] Canagliflozin (type-2 diabetes mellitus) NDA 204-042. Page 37 of 255-page Medical Review.

[54] National Kidney Foundation DOQI Kidney Disease Outcomes Quality Initiative. Clinical Practice Guidelines for Chronic Kidney Disease: Evaluation, Classification, and Stratification. National Kidney Foundation, Inc. New York, NY; 2002, p. 86–96. ISBN 1-931472-10-6.

[55] Floege J, et al. A phase III study of the efficacy and safety of a novel iron-based phosphate binder in dialysis patients. Kidney Int. 2014;86:638–47.

[56] Hsu CH, et al. New phosphate binding agents: ferric compounds. J. Am. Soc. Nephrol. 1999;10:1274–80.

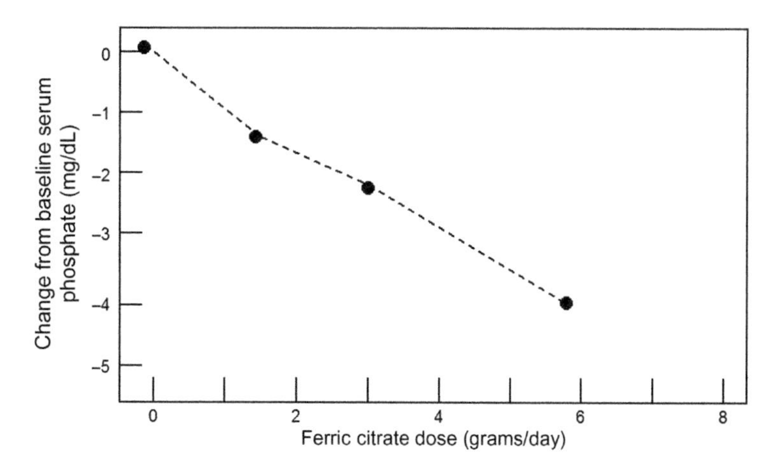

FIGURE 4.3 Serum phosphate versus oral ferric citrate. Greater oral doses of ferric citrate result in lower levels of serum phosphate, showing efficacy for treating hyperphosphatemia.

The Sponsor's dose titration was described as, "All subjects initiated on ferric citrate were started with a fixed dose ... of 6 caplets per day (1,260 mg of ferric iron as ferric citrate). Subjects were titrated at Weeks 1, 2, and 3. The maximum number of ... caplets was 12 g/day. Subjects were to take ferric citrate orally with meals or snacks or within one hour after their meals or snacks."[57]

FDA's ClinPharm Review referred to the proposed titration scheme for the package label and to its timing, and to the proposed starting dose and maximum dose, "What are the proposed dosages ... [t]he sponsor's proposed starting dose is 2 tablets orally three times per day with meals. The dose can be increased or decreased by 1 to 2 tablets per day at 2- to 4-week interval. The maximum dose is 12 tablets daily."[58]

FDA's ClinPharm Review revealed the dose–response relationship that was used as a basis for the dose-titration scheme on the package label. Fig. 4.3 reveals a dose–response relationship. The data are from four separate groups of subjects, each receiving a fixed dose of the study drug.[59]

This describes the washout period as a feature of study design. In addition to the dose–response study shown in Fig. 4.3, which discloses *reduction in serum phosphate*, the Sponsor conducted a related study showing how halting drug treatment resulted in an

[57] Ferric citrate (serum phosphorus in chronic kidney disease) NDA 205-874. Pages 23 and 67 of 95-page Clinical Pharmacology Review.

[58] Ferric citrate (serum phosphorus in chronic kidney disease) NDA 205-874. Page 66 of 95-page Clinical Pharmacology Review.

[59] Ferric citrate (serum phosphorus in chronic kidney disease) NDA 205-874. Pages 67 and 68 of 95-page Clinical Pharmacology Review.

increase in serum phosphate, where the increase was back to the baseline level. This component of study design is called a "washout" period. "Washout" refers to the situation where subjects are administered the study drug, followed by a period of a few days or weeks to allow substantial elimination of the drug from the body (or to allow recovery from the physiological effects of the drug), and then by resuming treatment with a different drug. FDA's ClinPharm Review reveals that it took about 2 weeks during the washout period for serum phosphate to increase to the original baseline level.[60]

Washout periods are used where study design requires that the subject serves as his own control, where following the washout period the same subject is treated with the study drug at a different dosage than that used prior to the washout period. In addition to being used between treatments with different doses of the same study drug, washout periods are used to eliminate earlier nonstudy drugs that were being taken immediately prior to initiating any clinical trial. Also, washout periods are used in crossover studies, where a given subject is first given one drug and then, a few weeks later, switched to a comparator drug. FDA's review for bosutinib reveals a crossover study, where an 8-day washout period was used between treatment with bosutinib plus moxifloxacin and treatment with bosutinib plus ketoconazole.[61] Furthermore, washout periods are used to determine the duration of the therapeutic effects of the drug, in the absence of continued dosing. Other examples of washout periods are shown in clinical trials with ticagrelor,[62] etanercept,[63] amantadine,[64] and soluble TNF receptor.[65]

Package label. The proposals for dose titration in FDA's ClinPharm review found a place in Dosage and Administration section of the package label:

> **DOSAGE AND ADMINISTRATION.** Starting dose is 2 tablets orally 3 times per day with meals ... [a]djust dose by 1 to 2 tablets as needed to maintain serum phosphorus at target levels, up to a maximum of 12 tablets daily. **Dose can be titrated** at 1-week or longer intervals.[66]

[60] Ferric citrate (serum phosphorus in chronic kidney disease) NDA 205-874. Pages 69 and 72 of 95-Clinical Pharmacology Review.

[61] Bosutininib (chronic myelogenous leukemia; CML) NDA 203-341. Page 17 of 89-page Clinical Pharmacology Review.

[62] Pelletier-Galarneau M, et al. Randomized trial comparing the effects of ticagrelor versus clopidogrel on myocardial perfusion in patients with coronary artery disease. J. Am. Heart Assoc. 2017;6:pii: e005894. DOI: 10.1161/JAHA.117.005894.

[63] Combe B. Update on the use of etanercept across a spectrum of rheumatoid disorders. Biol. Target Ther. 2008;2:165–73.

[64] Hader W, et al. A randomized controlled trial of amantadine in fatigue associated with multiple sclerosis. Can. J. Neurol. Sci. 1987;14:273–8.

[65] Moreland LW, et al. Treatment of rheumatoid arthritis with a recombinant human tumor necrosis factor receptor (p75)-Fc fusion protein. N. Engl. J. Med. 1997;337:141–7.

[66] Package label. AURYXIA (ferric citrate) tablets, for oral use. July 2015 (9 pages).

d. Lenalidomide (Myelodysplastic Syndromes) NDA 021-880

Lenalidomide is a small-molecule drug used for a hematological cancer, myelodysplastic syndromes (MDS). FDA's ClinPharm Review for lenalidomide reveals the situation where the FDA was concerned about AEs and took note of dose reductions required for study subjects during the course of the Sponsor's clinical studies. The same type of dose reduction found a place in instructions on the package label.

This also illustrates the need for dose reduction instructions where patients have renal impairment. Renal impairment is a common complication in patients with hematological cancers. The development of renal impairment, taking the form of acute kidney injury, can increase the toxicity of chemotherapy and limit further cancer treatment. Where there is renal impairment, patients may need to receive hemodialysis.[67] In the Dose Modification and Administration section of the package label, the instructions referred to patients with various degrees of renal impairment and instructed the physician to give lenalidomide after (not before) hemodialysis to avoid removing the drug from the patient's body.[68]

1. **Need to reduce dose with neutropenia and thrombocytopenia.** In the initial clinical study, the Clinical Study Protocol required a relatively high dose of study drug. But the Sponsor encountered drug-related toxicity and then submitted an amended Clinical Study Protocol to the FDA, where this amended Protocol required lower doses. This decision-making process is memorialized in FDA's ClinPharm Review. The toxicities in question were detected by blood tests and these were thrombocytopenia and neutropenia. The following concerns doses of 25 and 10 mg/day:

> Based on the prior determination that 25 mg/day was the maximum tolerated dose ... the initial starting dose of lenalidomide in this study was 25 mg daily ... [a]lthough erythroid responses [this refers to efficacy of the drug] were achieved within 16 weeks, a high incidence of **neutropenia and thrombocytopenia was observed within the first 4 to 8 weeks of treatment** ... [a]s a result of these findings, the protocol was amended to study 2 lower-dose regimens ... 10 mg lenalidomide was administered daily ... and, although erythroid responses [drug efficacy] were observed ... dose-limiting **neutropenia or thromobocytopenia was found to be 13 weeks.**[69]

FDA also contemplated doses lower than 10 mg, where the FDA made recommendations for exploring the efficacy and safety of the 10 mg/day dose. Regarding the need to explore the efficacy and safety of low doses, FDA wrote, "Approximately 80% of patients ... had dose reductions during the study. As the **activity of doses lower than 10 mg is unknown, it is possible that a lower dose could provide less toxicity while retaining efficacy**. A study is currently under way ... that includes an arm where patients are dosed at 5 mg/day ... [l]ack of data on the ability

[67] Lahoti A, et al. Predictors and outcome of acute kidney injury in patients with acute myelogenous leukemia or high-risk myelodysplastic syndrome. Cancer. 2010;116:4063–8.

[68] Package label. REVLIMID (lenalidomide) capsules for oral use. February 2017 (36 pages).

[69] Lenalidomide (myelodysplastic syndromes) NDA 021-880. Page 8 of 46-page Clinical Pharmacology Review.

of less toxic regiments to produce efficacy is an unresolved significant omission. This omission will be ... remedied by the study of a 5 mg dose."[70]

2. **Need to reduce dose with renal impairment.** FDA's Medical Review referred to the fact that lenalidomide is excreted mainly by the kidneys and to the need to exclude renal impaired subjects from the clinical trial.[71] FDA wrote, "Because lenalidomide is mainly excreted by the kidney, renal function should be carefully monitored to avoid excess toxicity." Regarding excluding renal impaired subjects, FDA referred to the cutoff point of serum creatinine of 2.5 mg/dL, writing, "The study excluded patients with serum creatinine >2.5 mg/dL."[72]

   FDA's ClinPharm Review included an account of renal function, as assessed by CL/F and half-life of lenalidomide in the bloodstream ($t_{1/2}$), writing that, "the Applicant attributes the differences between the populations as due to differences in renal function ... between the groups."

Table 4.2, which is from FDA's Review, compares renal function (CL/F) and half-life ($t_{1/2}$) in the healthy controls and in cancer patients.[73] As can be seen, clearance was less in the cancer patients.

FDA's recommendations for the label included, "The following language ... is reproduced from the Precautions section of the Applicant's Proposed Package Insert: 'This drug is known to be substantially excreted by the kidney, and the risk of toxic reactions to this drug may be greater in patients with impaired renal function ... care should be taken in dose selection, and ... to monitor renal function.' We recommend that a similar statement regarding monitoring of patients with renal impairment be added to the package insert."[74] Please note FDA's recommendation for monitoring. Monitoring is requirement that is set forth on the package labels for many drugs.

Package label. The Dosage and Administration section referred to tables in the Full Prescribing Information section of the package label, as indicated by numbers, 2.1, 2.2, and 2.3, where the tables provided instructions on dose modification in response to AEs. Also, this section also referred to the text at "2.4" which provided instructions on dose modification in response to renal impairment:

> **DOSAGE AND ADMINISTRATION** ... Continue or modify dosing based on clinical and laboratory findings (2.1, 2.2, 2.3). Renal impairment: Adjust starting dose based on the creatinine clearance value (2.4).[75]

[70] Lenalidomide (myelodysplastic syndromes) NDA 021-880. Page 28 of 46-page Clinical Pharmacology Review.

[71] Lenalidomide (myelodysplastic syndromes) NDA 021-880. Pages 43 and 167 of 184-page Medical Review.

[72] Lenalidomide (myelodysplastic syndromes) NDA 021-880. Pages 43 and 167 of 184-page Medical Review.

[73] Lenalidomide (myelodysplastic syndromes) NDA 021-880. Page 14 of 46-page Clinical Pharmacology Review.

[74] Lenalidomide (myelodysplastic syndromes) NDA 021-880. Page 24 of 46-page Clinical Pharmacology Review.

[75] Package label. REVLIMID (lenalidomide) capsules for oral use. February 2017 (36 pages).

TABLE 4.2 Comparison of PK Parameters in Healthy Subjects and in MDS Subjects

	Healthy Subjects	Subjects With MDS
CL/F (mL/min)	287 mL/min	178 mL/min
$t_{1/2}$ (h)	3.4 h	3.6 h

From: Lenalidomide (myelodysplastic syndromes) NDA 021-880. Page 14 of 46, Clinical Pharmacology Review.

TABLE 4.3 Platelet Counts

When Platelets	Recommended Course
IF THROMBOCYTOPENIA DEVELOPS WITHIN 4 WEEKS OF STARTING TREATMENT AT 10 MG DAILY	
Fall to <50,000/μL	Interrupt REVLIMID treatment
Return to ≥50,000/μL	Resume REVLIMID at 5 mg daily
IF THROMBOCYTOPENIA DEVELOPS AFTER 4 WEEKS OF STARTING TREATMENT AT 10 MG DAILY	
<30,000/μL	Interrupt REVLIMID treatment
Return to ≥30,000/μL	Resume REVLIMID at 5 mg daily
IF THROMBOCYTOPENIA DEVELOPS DURING TREATMENT AT 5 MG DAILY	
<30,000/μL	Interrupt REVLIMID treatment
Return to ≥30,000/μL	Resume REVLIMID at 2.5 mg daily

From: Package label. REVLIMID (lenalidomide) capsules for oral use. February 2017 (36 pages).

The instructions in one of the tables, reproduced here as Table 4.3, refer to the need to modify the dose where the patient has thrombocytopenia and neutropenia. The starting treatment is 10 mg per day, and this dose was stepped down to 5 mg daily and further stepped down to 2.5 mg daily (Table 4.3):

To summarize, Table 4.3 shows dose modification instructions in response to platelet counts. Table 4.4, also from the package label, provides instructions for dose modification in patients with renal impairment.

e. Nivolumab (Metastatic Melanoma) BLA 125-554

Nivolumab is an antibody that binds to PD-1, where the consequence is the blocking of signaling from the ligand (PD-L1) to the receptor (PD-1). Nivolumab is used for treating melanoma and a variety of other cancers.[76] PD-1 is expressed on the surface of T cells,

[76] Larkin J, et al. Combined nivolumab and ipilimumab or monotherapy in untreated melanoma. N. Engl. J. Med. 2015;373:23–34.

TABLE 4.4 Starting Dose Adjustments for Patients With Renal Impairment

Renal Function (Cockcroft–Gault)	Dose in REVLIMID Combination Therapy	Dose in REVLIMID Maintenance Therapy
CLcr 30 or 60 mL/min	10 mg once daily	5 mg once daily
CLcr < 30 mL/min (not requiring dialysis)	15 mg every other day	2.5 mg once daily
CLcr < 30 mL/min (requiring dialysis)	5 mg once daily. On dialysis days, administer the dose following dialysis	2.5 mg once daily. On dialysis days, administer the dose following dialysis

CLcr, creatinine clearance.
From: Package label. REVLIMID (lenalidomide) capsules for oral use. February 2017 (36 pages).

where binding of PD-1 with one of its ligands (PD-L1; PD-L2) on a tumor cell inhibits the T cell and prevents it from killing the tumor cells.[77] Nivolumab blocks this type of inhibition with consequent activation of T cells and increased antitumor immunity.

FDA's ClinPharm Review suggested that renal and hepatic impairment are not potential problems and that there is no need for dose modification instructions for patients with renal or hepatic impairment. But FDA observed that not enough information existed on organ impairment to make any corresponding recommendation on the package label, writing:

> **Renal Impairment.** No dedicated clinical studies were conducted to evaluate the effect of renal impairment on the PK of nivolumab. Based on ... analysis which included patients with mild, ... moderate, ... and severe ... renal impairment, the effect of mild and moderate renal impairment on clearance of nivolumab was minor. **Data is not sufficient for drawing a conclusion on severely renal impaired patients.**[78]

> **Hepatic Impairment. No dedicated clinical studies were conducted to evaluate the effect of hepatic impairment** on the PK of nivolumab. Based on ... analysis which included patients with mild hepatic impairment ... and normal hepatic function ... there was no clinically important differences in the clearance of nivolumab between patients with mild hepatic impairment and patients with normal hepatic function.[79]

In contrast to the situation with FDA's ClinPharm Review, FDA's Medical Review provided a more practical analysis of the renal AEs and liver AEs, where the result was dose modification information to be included on the package label. FDA's analysis was for a variety of AEs, including the AEs of pneumonitis, renal AEs, and liver AEs.

[77] Nivolumab (metastatic melanoma) BLA 125-554. Clinical Pharmacology Review (58 pages).

[78] Nivolumab (metastatic melanoma) BLA 125-554. Clinical Pharmacology Review. Pages 19 and 20 of 58 pages.

[79] Nivolumab (metastatic melanoma) BLA 125-554. Clinical Pharmacology Review. Pages 19 and 20 of 58 pages.

(i) **Liver AEs.** FDA's Medical Review observed the potential for drug-induced liver injury (DILI), and recommended, "Due to the potential for severe DILI, the reviewer **recommends inclusion of routine monitoring of liver functions tests** in the label for AST, ALT, total bilirubin, and alkaline phosphatase every two weeks in patients receiving nivolumab, and post-marketing surveillance. Drugs with a predisposition to hepatotoxicity **should be used with caution** in patients treated with nivolumab."[80]

FDA's recommendation for "routine monitoring of liver function tests" and for "caution" was based on the observations that elevations of AST or ALT to greater than 5 × upper limit of normal (ULN) in the nivolumab group (7 patients, 2.6%) and the observation that six (2.2%) patients experienced a shift of baseline bilirubin to greater than 2 × ULN.

(ii) **Renal AEs.** FDA's review began by contemplating the results from the Sponsor's analysis of the safety database. The Sponsor had counted the number of AEs that were detected and recorded by these terms (the terms are listed here), "blood creatinine increased, blood urea increased, creatinine renal clearance decreased, hypercreatinemia, nephritis, nephritis allergic, nephritis autoimmune, renal failure, acute renal failure, renal tubular necrosis, tubulointerstitial nephritis, and urine output decreased, selected to encompass those most likely to be reported in a patient with nephritis."[81]

FDA's narrative continued, observing that, "Adverse event terms belonging to the renal select AE category occurred in 18 (6.7%) patients in the nivolumab group ... [t]he most frequently occurring AE, regardless of causality, were blood creatinine increased in the nivolumab group."[82]

(iii) **Pneumonitis.** Turning to the AE of pneumonitis, FDA observed that, "Pneumonitis was the most frequently reported term in the nivolumab group (8 patients, 3.0%) ... [t]hree patients had Grade 1 pneumonitis, five patients had Grade 2 pneumonitis, and one patient had Grade 3 pneumonitis.[83]

The FDA reviewer went as far as to recommend a Black Box Warning for pneumonitis, writing, "Despite the initiation of a pneumonitis management ... for early recognition and management of pneumonitis, a fatal AE occurred in another study ... [t]he reviewer **recommends a boxed warning in the label** for pneumonitis."[84]

FDA's observations and recommendations for dose modification found a counterpart on the package label, as shown below.

Package label. The Dosage and Administration section for nivolumab (Opdivo®) was unusual, as compared to package labels of most other drugs, in that it consisted of instructions for responding to about a dozen separately identified AEs. The comments in FDA's ClinPharm Review regarding (1) pneumonitis, (2) liver injury (AST, ALT, bilirubin), and

[80] Nivolumab (metastatic melanoma) BLA 125-554. Page 98 of 156-page Medical Review.

[81] Nivolumab (metastatic melanoma) BLA 125-554. Page 98 of 156-page Medical Review.

[82] Nivolumab (metastatic melanoma) BLA 125-554. Page 98 of 156-page Medical Review.

[83] Nivolumab (metastatic melanoma) BLA 125-554. Page 102 of 156-page Medical Review.

[84] Nivolumab (metastatic melanoma) BLA 125-554. Page 104 of 156-page Medical Review.

(3) renal function (creatinine), each found a place in the Dosage and Administration section. The information took the form of instructions to withhold or to permanently discontinue nivolumab[85]:

> **DOSAGE AND ADMINISTRATION**. Dose Modifications ... Withhold OPDIVO for any of the following:
>
> - Grade 2 pneumonitis
> - Aspartate aminotransferase (AST) or alanine aminotransferase (ALT) greater than 3 and up to 5 times upper limit of normal (ULN) or total bilirubin greater than 1.5 and up to 3 times ULN
> - Creatinine greater than 1.5 and up to 6 times ULN or greater than 1.5 times baseline ...
> - Resume OPDIVO in patients whose adverse reactions recover to Grade 0-1.
> - Permanently discontinue OPDIVO for any of the following:
> - Grade 3 or 4 pneumonitis
> - AST or ALT greater than 5 times ULN or total bilirubin greater than 3 times ULN
> - Creatinine greater than 6 times ULN.[86]

f. Sebelipase Alfa (Lysosomal Acid Lipase Deficiency) BLA 125-561

Sebelipase is recombinant human lysosomal acid lipase (LAL). The genetic disease LAL deficiency results from mutations in LAL/cholesteryl esterase, with consequent increases in LDL cholesterol, as well as increases in cholesteryl esters and triglycerides in the lysosomes. Untreated infants usually die by the age of 6 months.

FDA's ClinPharm Review for sebelipase alfa provides a step-by-step basis for justifying a dose-escalation scheme on the package label. FDA's review for sebelipase used the following steps of analysis as a basis for the dose escalation scheme in the package label:

Step 1: According to the mechanism of action considerations, an increase in study drug concentration in the body would reasonably result in greater clinical efficacy.[87]
Step 2: Clinical data demonstrate that higher dose levels result in greater exposure.[88]
Step 3: Greater exposure results in greater clinical efficacy.[89]
Step 4: The result is that FDA recommended dose escalation instructions on the package label.[90]

[85] Package label. OPDIVO (nivolumab) injections, for intravenous use; December 2014 (16 pages).

[86] Package label. OPDIVO (nivolumab) injections, for intravenous use; December 2014 (16 pages).

[87] Sebelipase alfa (lysosomal acid lipase (LAL) deficiency) BLA 125-561. Page 34 of 87-page Clinical Pharmacology Review.

[88] Sebelipase alfa (lysosomal acid lipase (LAL) deficiency) BLA 125-561. Pages 17 and 62 of 87-page Clinical Pharmacology Review.

[89] Sebelipase alfa (lysosomal acid lipase (LAL) deficiency) BLA 125-561. Page 37 of 87-page Clinical Pharmacology Review.

[90] Sebelipase alfa (lysosomal acid lipase (LAL) deficiency) BLA 125-561. Page 11 of 229-page Medical Review, and Page 80 of 87-page Clinical Pharmacology Review.

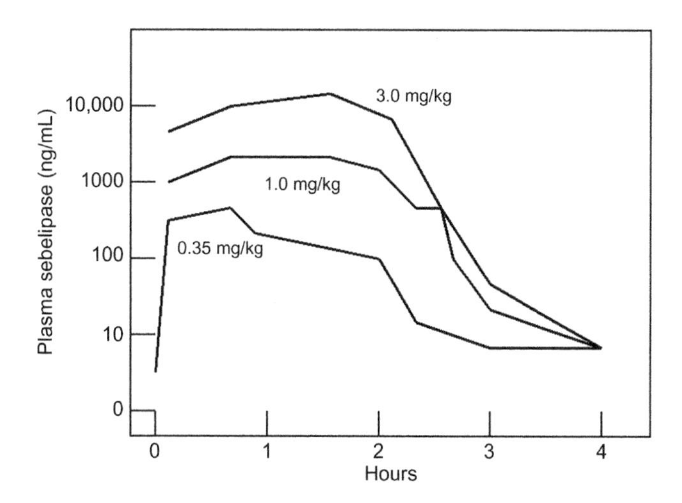

FIGURE 4.4 Plasma sebelipase alfa vs amount of sebelipase alfa injected.

This illustrates the FDA's accounts of Steps 1–4:

Step 1. Regarding mechanism of action, FDA observed, "The biological activity is primarily driven by enzyme concentrations and duration of exposure in the lysosomes."[91] The importance of maintaining concentration of the study drug is especially vital, in view of the rapid clearance of the study drug.

Regarding rapid clearance, FDA observed that, "Sebelipase alfa has a very short plasma half-life (6 minutes) and lysosomes are the site of action."[92] Clinical data also supported mechanism-of-action–based arguments for a dose-escalation scheme. The FDA reviewer observed that, "the decision to dose escalate was based on suboptimal response in one or more clinical measures."[93] In other words, the need for dose escalation was based on the fact that lower doses did not always work.

Step 2. Pharmacokinetic data revealed that higher doses resulted in greater exposure, as shown in Fig. 4.4. It reveals study drug concentrations in the bloodstream over the course of 4 hours, at different dosages.

Regarding the greater exposure with higher doses, FDA observed, "There was a reasonably dose-proportional increase in exposure between the 0.35 and 1 mg/kg doses, and there was a greater than dose-proportional increase in exposure between the 1 and 3 mg/kg doses."[94]

[91] Sebelipase alfa (lysosomal acid lipase (LAL) deficiency) BLA 125-561. Page 34 of 87-page Clinical Pharmacology Review.

[92] Sebelipase alfa (lysosomal acid lipase (LAL) deficiency. BLA 125-561. Page 34 of 87-page Clinical Pharmacology Review.

[93] Sebelipase alfa (lysosomal acid lipase (LAL) deficiency) BLA 125-561. Page 9 of 87-page Clinical Pharmacology Review.

[94] Sebilipase alfa (lysosomal acid lipase (LAL) deficiency) BLA 125-561. Pages 17 and 62 of 87-page Clinical Pharmacology Review.

Step 3. FDA acknowledged that higher exposure results in greater efficacy, "The trend of exposure-response—higher exposures appear to be associated with higher efficacy—with the ... efficacy endpoint of low density lipoprotein cholesterol (LDLs) provides supportive evidence of effectiveness."[95]

Step 4. In view of the above, FDA recommended approval of the dose escalation scheme on the package label, writing, "this reviewer recommends approval ... this reviewer recommends an initial dose of 1 mg/kg with a dose escalation to 3 mg/kg once weekly."[96]

Reiterating details of the proposed dose escalation scheme, FDA wrote, "Proposed dosing regiments ... 1 mg/kg administered as an intravenous ... infusion once every other week. In patients presenting with rapidly progressive disease in infancy, the recommended starting dose is 1 mg/kg weekly. In clinical studies, these patients were dose escalated to 3 mg/kg once weekly."[97]

Taking a step forward to make recommendations for the package label, FDA's ClinPharm Review commented, "Key changes to the labeling are ... [f]or patients with rapidly progressing LAL deficiency presenting within the first 6 months of life, added a **dosing instruction to increase to 3 mg/kg once weekly if an optimal clinical response is not achieved**."[98]

Package label. FDA's dosing recommendations for sebelipase alfa (Kanuma®) found a place on the package label, which provided the instructions:

> **DOSAGE AND ADMINISTRATION**. Patients with ... LAL deficiency presenting within the first 6 months of life: The recommended starting dosage is 1 mg/kg ... [f]or patients who do not achieve an optimal clinical response, increase to 3 mg/kg once weekly.[99]

III. PACKAGE LABEL INSTRUCTIONS FOR DOSE MODIFICATION INFLUENCED BY PHENOMENON OF TARGET-MEDIATED DRUG DISPOSITION

a. Background on Target-Mediated Drug Disposition

Target-Mediated Drug Disposition (TMDD) is described. Various types of therapeutic proteins exhibit TMDD. These include antibodies, cytokines, and growth factors.[100] Where

[95] Sebilipase alfa (lysosomal acid lipase (LAL) deficiency) BLA 125-561. Page 37 of 87-page Clinical Pharmacology Review. This statement on exposure/efficacy referred to a 20-week study with a constant dose of 1 mg/kg without any dose-escalation.

[96] Sebelipase alfa (lysosomal acid lipase (LAL) deficiency) BLA 125-561. Page 11 of 229-page Medical Review, and Page 80 of 87-page Clinical Pharmacology Review.

[97] Sebelipase alfa (lysosomal acid lipase (LAL) deficiency) BLA 125-561. Page 52 of 87-page Clinical Pharmacology Review.

[98] Sebelipase alfa (lysosomal acid lipase (LAL) deficiency. BLA 125-561. 30-page Summary Review.

[99] Package label. KANUMA (sebelipase alfa) injection, for intravenous use. December 2015 (13 pages).

[100] Dua P, et al. A tutorial on target-mediated drug disposition (TMDD). Pharmacomet. Syst. Pharmacol. 2015;4:324–37.

TMDD occurs, the therapeutic protein binds with high affinity to its target, typically the extracellular domain of a membrane-bound protein. Removal of a therapeutic antibody from the bloodstream occurs when the antibody binds to its target and where the antibody/target complex is internalized into the cell and then degraded.[101]

For antibodies, the main route of elimination can be via uptake and degradation by cells, which has been found for antibodies of the IgG class and excretion in the bile as has been found for the IgA class antibodies.[102] Antibody uptake and degradation has been demonstrated, for example, for anti-CD22 antibodies that bind to and are taken up by B cells. This uptake/degradation scenario requires that the membrane-bound receptor be capable of internalization. Regarding CD22, Carnahan et al.[103] found that adding an anti-CD22 antibody (epratuzumab) to cultured cells caused internalization of CD22 and where prolonged incubation of anti-CD22 antibody with the cells resulted in 80% of this membrane-bound protein (CD22) being internalized.

Similarly Pagel et al.[104] observed that most anti-CD22 antibodies are rapidly internalized into B cells after binding to CD22. Please note that just because an antibody binds to CD22, and is followed by the antibody/CD22 complex being internalized, does not mean that the antibody is degraded. In other words, there is always the possibility that the antibody is released from the cell or that CD22 is recycled and inserted back into the plasma membrane.

Press et al.[105] addressed this scenario and found that once the antibody/CD22 complex is internalized, the antibody moves to the lysosomal compartment of the cell where it is degraded. Press et al. provided side-by-side comparisons of degradation rates of various antibodies that are specific for various membrane-bound proteins of B cells. The antibodies tested were anti-CD19, anti-CD20, anti-CD22, anti-CD37, anti-HLA class II, and anti-CD40. The degradation of antibodies can be measured using radioactively labeled antibodies, where release of the radioactive label from the antibody has been digested into peptides and amino acids.

Where the goal is to determine dosage and administration of any therapeutic protein and where TMDD is one of the pathways of drug clearance, this degradative pathway needs to be explored on an individual basis. Urinary excretion is the main elimination

[101] Peletier LA, Gabrielsson J. Dynamics of target-mediated drug disposition: characteristic profiles and parameter identification. J. Pharmacokinet. Pharmacodyn. 2012;39:429–51.

[102] Wang W, et al. Monoclonal antibody pharmacokinetics and pharmacodynamics. Clin. Pharmacol. Ther. 2008;84:S48–58.

[103] Carnahan J, et al. Epratuzumab, a humanized monoclonal antibody targeting CD22: characterization of in vitro properties. Clin. Cancer Res. 2003;9:3982S–90S.

[104] Pagel JM, et al. Evaluation of CD20, CD22, and HLA-DR targeting for radioimmunotherapy of B-cell lymphomas. Cancer Res. 2007;67:5921–8.

[105] Press OW, et al. Endocytosis and degradation of monoclonal antibodies targeting human B-cell malignancies. Cancer Res. 1989;49:4906–12.

pathway for small molecules because they can easily pass through pores in the glomerulus. TMDD is much less common for small molecules, but examples have been found.[106,107,108]

b. Alemtuzumab (B-Cell Chronic Lymphocytic Leukemia) BLA 103-948

Alemtuzumab is an antibody specific for CD52, a membrane-bound peptide of B cells. CD52 takes the form of a 12-amino acid peptide where the peptide is anchored to the cell membrane with a glycosylphosphatidylinositol (GPI) anchor. The 12-amino acid peptide is covalently bound to the GPI anchor. The sequence of the 12-amino acid peptide is Gly-Gln-Asn-Asp-Thr-Ser-Gln-Thr-Ser-Ser-Pro-Ser.[109] Alemtuzumab binds to an epitope consisting of part of the peptide and part of the GPI anchor.[110] In other words, the antibody simultaneously binds to an amino acid moiety and to a phospholipid moiety. TMDD with alemtuzumab involved these steps:

(1) Alemtuzumab binds to cancer cells, with the consequent eradication of the cancer cells.
(2) The elimination of the cancer cells results in concomitant increase in exposure of the antibody, even though the antibody dosage is maintained at a constant rate.
(3) Increased alemtuzumab exposure resulted in increased AEs.

As mentioned earlier, the term "exposure" refers to the concentration of the drug in the bloodstream in terms of a unit such as AUC, Cmax, tmax, Cmin, and Ctrough. Ctrough refers to the lowest blood concentration (the trough) that occurs in the time frame between two drug doses, for example, where the drug doses are separated by a day. FDA's Medical Review warned that physicians might need to adjust antibody dose to account for variability of the tumor burden, "The tumor burden and the density of CD52 antigen on the cell surface contribute to the marked inter-patient variability noted in the pharmacokinetic studies."[111]

What could account for the eventual increase in antibody exposure, despite the constant antibody dose? Referring to the concomitant decrease in number of tumor cells and increase in exposure, FDA observed, "**As the malignant lymphocytosis decline** -- with the nadir CLL count usually attained about week 4 -- **the peak and trough levels of ... alemtuzumab began to rise** and after about two weeks reached a steady state ... [o]ver eight

[106] An G. Small-molecule compounds exhibiting target-mediated drug disposition (TMDD): a minireview. J. Clin. Pharmacol. 2017;57:137–50.

[107] An G, et al. Small-molecule compounds exhibiting target-mediated drug disposition—a case example of ABT-384. J. Clin. Pharmacol. 2015;55:1079–85.

[108] Yamazaki S, et al. Application of target-mediated drug disposition model to small molecule heat shock protein 90 inhibitors. Drug Metab. Dispos. 2013;41:1285–94.

[109] Treumann A, et al. Primary structure of CD52. J. Biol. Chem. 1995;270:6088–99.

[110] Holgate RG, et al. Characterization of a novel anti-CD52 antibody with improved efficacy and reduced immunogenicity. PLoS One. 2015;10:e0138123.

[111] Alemtuzumab (B-cell chronic lymphocytic leukemia (B-cell CLL)) BLA 103-948. Page 7 of 133-page Medical Review.

weeks of therapy CLL patients were noted to have ... an increase in ... peak serum level [of alemtuzumab] from 0.59 micrograms/mL to a ... peak level of 8.82 micrograms/mL, and an increase in the ... trough level from 0.09 micrograms/mL to a ... trough level of 6.12 micrograms/mL."[112]

As a consequence of the fact that cancer cell elimination results in increases in alemtuzumab exposure, FDA warned that the increased blood levels of alemtuzumab can provoke an increase in AEs, "A proportional increase in AUC and Cmax is noted with ... doses up to 80 mg ... [i]ncreased toxicity, especially hematologic, is observed with doses greater than 80 mg. The **increase in ... serum drug concentrations may contribute to the prolonged hematologic toxicity observed in some patients**."[113]

The Sponsor responded to the alemtuzumab-induced AEs by using a type of dose modification (dose delay), when administering alemtuzumab to study subjects, "In the twenty-four instances where **hematological toxicity** was reported as a reason for dose delay -- including the two instances of febrile neutropenia and two instances of infection with neutropenia -- the average **delay between doses was 16.8 days**."[114]

FDA's ClinPharm Review developed the topic of TMDD, stating that greater numbers of cancer cells in the patient's body resulted in greater clearance of alemtuzumab from the bloodstream (lesser exposure), whereas, in contrast, lower numbers of tumor cells resulted in greater alemtuzumab exposure.

> Since malignant cells participate in the clearance of ... alemtuzumab, higher numbers of [malignant] cells reduces both peak and trough values [of alemtuzumab]. An inverse relationship was determined between the duration of dosing and high levels of circulating lymphocytes ... **[a]n inverse relationship exists between disease burden and pharmacokinetic parameters such as Cmax and AUC**, as malignant cells participate in the removal of ... alemtuzumab ... [s]teady-state levels are achieved at approximately 6 weeks of repeated dosing and are a consequence of the dosing interval and disease burden.[115]

Package label. The Dosage and Administration section mirrored FDA's reviews in its instructions to escalate the dose and to eventually reduce the dose. The May 2004 package label read as follows. This label referred to the need for an escalation scheme to minimize toxicities. But the May 2004 label did not mention any eventual need to reduce the dose or to delay the dose:

> **DOSAGE AND ADMINISTRATION**. Campeth [alemtuzumab] therapy should be initiated at a dose of 3 mg administered as a 2 hour IV infusion daily ... [w]hen the Campath 3 mg daily dose is tolerated

[112] Alemtuzumab (B-cell chronic lymphocytic leukemia (B-cell CLL)) BLA 103-948. Page 7 of 133-page Medical Review.

[113] Alemtuzumab (B-cell chronic lymphocytic leukemia (B-cell CLL)) BLA 103-948. Page 7 of 133-page Medical Review.

[114] Alemtuzumab (B-cell chronic lymphocytic leukemia (B-cell CLL)) BLA 103-948. Page 26 of 133-page Medical Review.

[115] Alemtuzumab (B-cell chronic lymphocytic leukemia (B-cell CLL)) BLA 103-948. 4-page Clinical Pharmacology Review.

(e.g., infusion-related toxicities are ≤ Grade 2), the daily dose should be escalated to 10 mg and continued until tolerated. When the 10 mg dose is tolerated, the maintenance dose of Campath 30 mg may be initiated. The maintenance dose of Campath is 30 mg/day administered three times per week on alternate days (i.e., Monday, Wednesday, and Friday) for up to 12 weeks.[116]

The package label for the same drug, but 10 years later (November 2014), instructed physicians to start with a higher dose (treatment course) and then at a later time, use a lesser dose (treatment course), possibly to account for the expected lower number of tumor cells in the patient's body:

DOSAGE AND ADMINISTRATION. Administer LEMTRADA [alemtuzumab] by intravenous infusion over 4 hours for 2 treatment courses: First course 12 mg/day on 5 consecutive days. Second course: 12 mg/day on 3 consecutive days 12 months after first treatment course.[117]

c. Obinutuzumab (Chronic Lymphocytic Leukemia) BLA 125-486

This describes the dose escalation and dose reduction schemes for obinutuzumab. Obinutuzumab is an antibody that binds CD20, a membrane-bound protein of B cells. The antibody binds to CD20 of normal B cells and to CD20 of malignant B cells. Plasma concentrations of obinutuzumab can be influenced by:

- Amount of obinutuzumab injected (the more that is injected, the greater will be the plasma concentration).
- Removal of obinutuzumab from the bloodstream by endocytosis of the obinutuzumab/CD20 complex. Where obinutuzumab binds to CD20 expressed on the surface of B cells, this binding together with endocytosis of the antibody/CD20 complex reduces the bloodstream concentration of free antibody.
- Loss of cancer cell burden in the patient during successful cancer treatment with obinutuzumab. Where more cancer cells are killed and thus unable to bind the administered obinutuzumab, the greater will be plasma exposure of obinutuzumab where drug doses remain at the same amount.

FDA's reviews of obinutuzumab[118] and alemtuzumab[119] provide examples of these topics, as they apply to dosage and administration. Details on the topics of endocytosis of antibody/CD20 complexes and of similar complexes, the various fates of the

[116] Package label. Campath® (Alemtuzumab); April 2004 (14 pages).

[117] Package label. LEMTRADA™ (alemtuzumab) injection, for intravenous use; November 2014 (29 pages).

[118] Obinutuzumab (chronic lymphocytic leukemia (CLL)) BLA 125-486. 83-page Clinical Pharmacology Review.

[119] Alemtuzumab (B-cell chronic lymphocytic leukemia (B-cell CLL)) BLA 103-948. 133-page Medical Review and 4-page Clinical Pharmacology Review.

endocytosed antibody, and of clearance of antibodies from the bloodstream by a high tumor burden where the tumor acts as a sink for the antibody are provided by the cited papers.[120,121,122]

FDA's ClinPharm Review revealed the need for initial dosing to start with a low dose, to reduce infusion-related AEs, and then to escalate the dose. FDA's recommend for a dose-escalation scheme was, "The sponsor's proposed dosing regimen as reviewed by Drs. Grillo and Florian, achieves its purpose of reducing infusion related adverse events with the first dose while the dose intensification over cycle 1 results in obinutuzumab exposures closer to steady state by cycle 2."[123] In other words, the initial low dose was to reduce infusion-related AEs, and the subsequent higher dose was to achieve a steady-state concentration in the bloodstream.

FDA contemplated the need for determining a dose–response relationship, where the response was **efficacy**. Efficacy endpoints for most cancer clinical trials are overall survival, progression-free survival (PFS), and objective response.[124] Also, FDA considered the need to determine a dose–response relationship, where the response was **AEs**:

> While the information currently in hand is not sufficient to determine if 1000 mg is the optimal obinutuzumab dose for various patient subgroups (low vs. high body weight, males vs. females, patients with low vs. high tumor burden) the safety and efficacy results support that 1000 mg administered over 6 cycles (1000 mg weekly for 3 weeks in cycle 1) **improved progression free survival (PFS)** ... [a]n exposure-response relationship was identified between obinutuzumab and progression-free survival (PFS); however, no exposure-response relationships were identified between obinutuzumab and **adverse event rate.**[125]

Reiterating the fact that no dose–response relationship was detected for AEs, the FDA reviewer contemplated, "Is there evidence of exposure-safety relationships for obinutuzumab? No, the available data did not identify an exposure-response relationship between obinutuzumab exposure and **adverse event rate**. There is evidence of increased likelihood of certain **adverse events** with obinutuzumab treatment compared to control, including ... higher cardiac events ... neutropenia, and thrombocytopenia."[126]

[120] Golay J, et al. Lessons for the clinic from rituximab pharmacokinetics and pharmacodynamics. MAbs. 2013;5:826–37.

[121] O'Reilly MK, et al. CD22 is a recycling receptor that can shuttle cargo between the cell surface and endosomal compartments of B cells.J. Immunol. 2011;186:1554–63.

[122] Beers SA, et al. Antigenic modulation limits the efficacy of anti-CD20 antibodies: implications for antibody selection. Blood. 2010;115:5191–201.

[123] Obinutuzumab (chronic lymphocytic leukemia (CLL)) BLA 125-486. Page 2 of 83-page Clinical Pharmacology Review.

[124] Brody T. CLINICAL TRIALS Study Design, Endpoints and Biomarkers, Drug Safety, and FDA and ICH Guidelines), 2nd ed., Elsevier, New York; 2016, pp. 247–376.

[125] Obinutuzumab (chronic lymphocytic leukemia (CLL)) BLA 125-486. Page 2 of 83-page Clinical Pharmacology Review.

[126] Obinutuzumab (chronic lymphocytic leukemia (CLL)) BLA 125-486. Page 41 of 83-page Clinical Pharmacology Review.

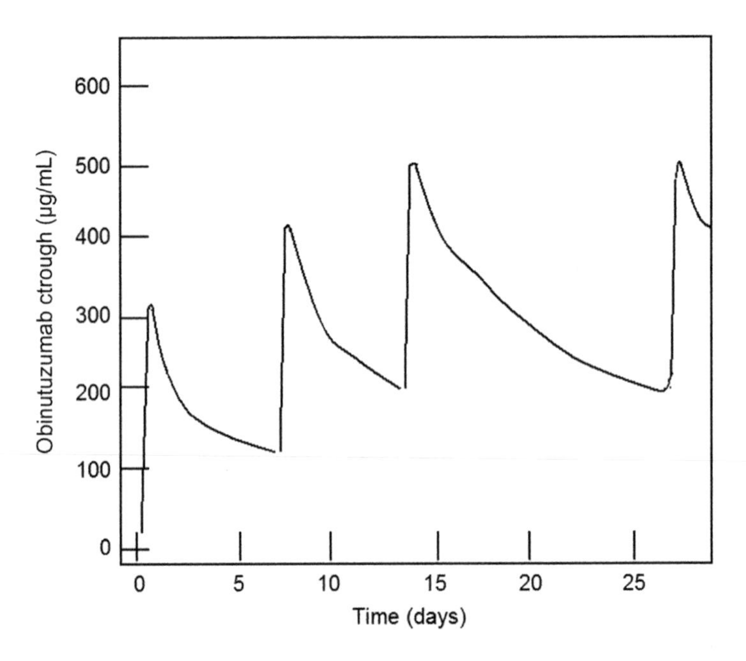

FIGURE 4.5 Obinutuzumab concentration vs treatment day.

The overall dosing scheme involved:

- an initial gradual ramping-up to minimize AEs,
- followed by maintenance at a high dose (Cycle 1) to overcome the disappearance of the antibody via TMDD, and
- then followed finally by a lower dose to account for the lower tumor burden (with lowered level of TMDD).

Referring to these dose-titration steps, FDA concluded that the dosing scheme was acceptable, writing, "The applicant's proposed dosing regimen is based on a 28 day cycle and includes three components: **(i)** splitting of the initial dose over two days to reduce infusion related reaction risk; **(ii)** administration of 3000 mg in cycle 1 to overcome target-mediated drug disposition; and **(iii)** selection of a 1000 mg dose for cycles 2-6. This dose and dosing strategy is acceptable given the PFS response observed across exposure quartiles."[127]

FDA also referred to the fact that obinutuzumab eventually reached a steady-state concentration, "Also, the dose intensification over cycle 1 results in obinutuzumab **exposures closer to steady state by cycle 2**."[128] Fig. 4.5 shows this steady state. Please note that the

[127] Obinutuzumab (chronic lymphocytic leukemia (CLL)) BLA 125-486. Pages 7 and 8 of 83-page Clinical Pharmacology Review.

[128] Obinutuzumab (chronic lymphocytic leukemia (CLL)) BLA 125-486. Page 8 of 83-page Clinical Pharmacology Review.

last two peaks in Fig. 4.5 have the same height, indicating that a steady state was reached (all of the data points in Fig. 4.5 are Ctrough levels, that is, the lowest blood concentration for the indicated day).

The commentary below concerns these goals, where each of these could compel the need for dose modification instructions in the package label:

(1) Does increased drug exposure (AUC, tmax, Cmax) correlate with increased AEs?
(2) Does renal impairment increase drug exposure?
(3) What is the best dosing scheme for attaining steady state drug concentrations?

FDA's ClinPharm Review set forth the goal of determining a correlation of plasma drug concentrations with AEs, the goal determining if renal impairment increases plasma drug concentrations, and the goal of using PK studies was to arrive at the dose amount and number of days needed to reach steady state.

To the first goal, FDA observed that there was not any particular relationship between exposure and AEs, "There is evidence of increased likelihood of certain adverse events with obinutuzumab treatment, including the percentage of subjects with ... cardiac events, tumor lysis syndrome, neutropenia, and thrombocytopenia ... [h]owever, the available data did not support an exposure-response relationship between obinutuzumab exposure and adverse event rate."[129]

Regarding the second goal, FDA concluded, "No difference in obinutuzumab PK was observed across categories for normal, mild, or moderate renal function ... [h]owever, renal impairment is not expected to be a major factor affecting exposure as monoclonal antibodies are generally catabolized by ubiquitous proteolytic enzymes."[130]

Regarding the third goal, FDA stated, "Also, the dose intensification over cycle 1 results in obinutuzumab exposures closer to steady state by cycle 2."[131]

FDA's comments on this figure refer to a period of dosing that encompasses Cycle 1 and to the start of Cycle 2. Apparently referring to this same dosing scheme, FDA observed that a steady state had been reached by Cycle 2, "while the dose intensification over cycle 1 results in obinutuzumab exposures closer to steady state by cycle 2." See, pages 2 and 42 and 43 of 83, ClinPharm Review.

TMDD was also a topic of FDA's concern. FDA commented on the ability of cancer cells to clear antibody drugs from the bloodstream:

> The sponsor identified increased elimination of obinutuzumab following initial dosing, which was believed to be associated with **target-mediated drug disposition.** Patients with higher initial tumor burden and a **higher number of CD20-positive tumor cells clear the drug faster** than patients with a lower initial

[129] Obinutuzumab (chronic lymphocytic leukemia (CLL)) BLA 125-486. Page 8 of 83-page Clinical Pharmacology Review.

[130] Obinutuzumab (chronic lymphocytic leukemia (CLL)) BLA 125-486. Page 23 of 83-page Clinical Pharmacology Review.

[131] Obinutuzumab (chronic lymphocytic leukemia (CLL)) BLA 125-486. Pages 2, 8, and 16, of 83-page Clinical Pharmacology Review.

tumor burden. This was supported by observations from the early phase trials, the population PK modeling analysis, and is in agreement with the pharmacokinetics observed for other monoclonal antibodies (i.e., rituximab).[132]

In addition to merely describing the phenomenon of TMDD, FDA commented on using an initial higher obinutuzumab dosing for overcoming TMDD, "To overcome this increased initial clearance and **saturate the target early in treatment**, the sponsor proposed weekly dosing of 1000 mg obinutuzumab for three weeks over the first cycle of treatment. Based on the ... predicted exposures for a typical subject, this regimen is predicted to result in obinutuzumab Ctrough exposures that are 6.6-fold higher at the beginning of cycle 2 compared to the scenario where no dose-intensification was used during cycle 1."[133]

Package label. Obinutuzumab's package label provided instructions for the initial dose escalation (starting from a very low dose to minimize AEs) (Days 1 and 2), followed by further dose escalation (two closely situated days (Days 8 and 15) in Cycle one to achieve saturation and then reduced dosing (only one dose per cycle) for the subsequent cycles. The Dosage and Administration section, which reflected FDA's comments, read[134]:

DOSAGE AND ADMINISTRATION ... Recommended dose for 6 cycles (28 day cycles)

- 100 mg on day 1 Cycle 1
- 900 mg on day 2 Cycle 1
- 1000 mg on day 8 and 15 of Cycle 1
- 1000 mg on day 1 of Cycles 2–6

IV. INTRODUCTION TO RENAL IMPAIRMENT AND HEPATIC IMPAIRMENT

Renal impairment and hepatic impairment are issues separate from AEs due to the disease being treated and separate from drug-induced AEs. Where drug-induced AEs is a risk, there is sometimes a need to adjust drug dosing where the patient has underlying renal or hepatic impairment.

a. Renal Impairment

Renal impairment encompasses conditions such as chronic kidney disease and end-stage renal failure. Uremia is a disorder caused by renal failure. If a drug is eliminated in

[132] Obinutuzumab (chronic lymphocytic leukemia (CLL)) BLA 125-486. Page 43 of 83-page Clinical Pharmacology Review.

[133] Obinutuzumab (chronic lymphocytic leukemia (CLL)) BLA 125-486. Page 43 of 83-page Clinical Pharmacology Review.

[134] Package label. GAZYVA (obinutuzumab) injection, for intravenous infusion. November 2013 (15 pages).

the urine, impaired renal function can increase drug exposure to the extent that the dose amount or frequency needs to be reduced. FDA's Guidance on Pharmacokinetics in Patients with Impaired Renal Function provides a cutoff point of 30%, in its recommendation that:

> A PK study should be conducted in patients with impaired renal function when the drug is ... to be used in such patients and when renal impairment is likely to mechanistically alter the PK of the drug and/or its active metabolites. This would ... be the case ... if the fraction of dose excreted unchanged in the urine is at least 30%.[135]

FDA's Guidance distinguishes between single-dose PK studies for assessing renal function from multiple-dose PK studies for assessing renal function. FDA's Guidance on Exposure-Response Relationships teaches when to use single-dose versus multiple-dose PK studies.[136] Regarding renal function, FDA's Guidance states that:

> A single-dose study is satisfactory for cases where ... single-dose studies accurately describe the PK for the pertinent drug and potentially active metabolites. This will be true when the drug and active metabolites exhibit **linear and time-independent PK** at the concentrations anticipated in the patients to be studied. A multiple-dose study is usually recommended when the drug or an active metabolite exhibits **nonlinear or time-dependent PK.**[137]

b. Methods for Assessing Renal Function

Renal function can be assessed by eGFR and measured glomerular filtration rate (mGFR). eGFR is sufficient for adjusting drug dosage to account for any reduced ability of the kidneys to excrete drugs. Dose adjustment is especially critical where the drug has a narrow therapeutic window. But mGFR is preferred where the drug dosing is prolonged and potentially toxic, as with cancer chemotherapy.[138]

Glomerular filtration rate can be measured directly, where it is called "measured GFR" (mGFR)[139] or it can be estimated by a number of available formulas, where it is called

[135] US Department of Health and Human Services; Food and Drug Administration; Center for Drug Evaluation and Research (CDER). Guidance for Industry. Pharmacokinetics in Patients with Impaired Renal Function—Study Design, Data Analysis, and Impact on Dosing and Labeling; 2010 (18 pages).

[136] US Department of Health and Human Services; Food and Drug Administration; Center for Drug Evaluation and Research (CDER). Center for Biologics Evaluation and Research (CBER). Guidance for Industry. Exposure-Response Relationships—Study Design, Data Analysis, and Regulatory Applications; 2003 (25 pages).

[137] US Department of Health and Human Services; Food and Drug Administration; Center for Drug Evaluation and Research (CDER). Guidance for Industry. Pharmacokinetics in Patients with Impaired Renal Function—Study Design, Data Analysis, and Impact on Dosing and Labeling (18 pages).

[138] Stevens LA, Levey AS. Measured GFR as a confirmatory test for estimated GFR. J. Am. Soc. Nephrol. 2009;20:2305–13.

[139] Deng F, et al. Applicability of estimating glomerular filtration rate equations in pediatric patients: comparison with a measured glomerular filtration rate by iohexol clearance. Transl. Res. 2015;165:437–45.

"estimated GFR" (eGFR). Determining the mGFR is by subcutaneous injection of a tracer, iodine-125 labeled iothalamate, followed by measuring the appearance of the labeled iothalamate in the serum and in the urine at various time points over several hours.[140]

FDA's Guidance for Industry recognizes various ways to assess renal function by eGFR and to determine if a given study subject has impaired renal function. FDA's Guidance discloses the Cockcroft–Gault equation and the MDRD equation. FDA's Guidance states that either way of assessing renal function is acceptable in its comment that, "Either the Cockcroft-Gault or MDRD equation can be used to assign subjects to a renal impairment group."[141]

Use of the MDRD equation is illustrated by clinical studies on empagliflozin,[142] fenofibrate,[143] and antihypertensive drugs.[144] Reliable guidance on using iothalamate to determine mGFR are cited.[145,146,147,148] Iothalamate has the following structure:

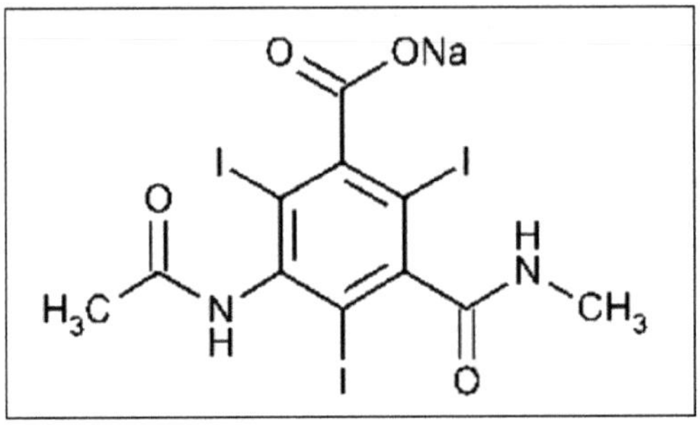

[140] Lin YC, et al. Determinants of the creatinine clearance to glomerular filtration rate ratio in patients with chronic kidney disease: a cross-sectional study. BMC Nephrol. 2013;134:43–9.

[141] US Department of Health and Human Services; Food and Drug Administration; Center for Drug Evaluation and Research (CDER). Guidance for Industry. Pharmacokinetics in Patients with Impaired Renal Function—Study Design, Data Analysis, and Impact on Dosing and Labeling; 2010 (18 pages).

[142] Wanner C, et al. Empagliflozin and progression of kidney disease in type 2 diabetes. N. Engl. J. Med. 2016;375:323–34.

[143] ACCORD Study Group, et al. Effects of combination lipid therapy in type 2 diabetes mellitus. N. Engl. J. Med. 2010;362:1563–74.

[144] Appel LJ, et al. Intensive blood-pressure control in hypertensive chronic kidney disease. N. Engl. J. Med. 2010;363:918–29.

[145] Appel LJ, et al. Intensive blood-pressure control in hypertensive chronic kidney disease. N. Engl. J. Med. 2010;363:918–29.

[146] Parsa A, et al. APOL1 risk variants, race, and progression of chronic kidney disease. N. Engl. J. Med. 2013;369:2183–96.

[147] DCCT/EDIC Research Group, et al. Intensive diabetes therapy and glomerular filtration rate in type 1 diabetes. N. Engl. J. Med. 2011;365:2366–76.

[148] Inker LA, et al. Estimating glomerular filtration rate from serum creatinine and cystatin C. N. Engl. J. Med. 2012;367:20–9.

c. Hepatic Impairment

FDA's Guidance teaches that hepatic impairment can be caused by alcoholic liver disease, hepatitis C virus infections, primary biliary cirrhosis, and liver other diseases.[149]

FDA's Guidance contrasts the creatinine clearance test, which is widely used for estimating a given drug's PK characteristics in patients with renal impairment, with corresponding tests for liver function. In contrast to the creatinine clearance test, there does not exist any widespread test for liver function that can predict any given drug's PK characteristics. The available tests for liver function include[150]:

- serum bilirubin;
- serum albumin;
- prothrombin time;
- liver catabolism of antipyrine,[151] indocyanine green, or galactose;
- liver fibrosis assessed by liver biopsy or by non-invasive techniques[152,153,154]; and
- Child–Pugh score. FDA's Guidance for Industry and other sources teach how to assess hepatic impairment by the Child-Pugh score (or classification).[155,156,157,158]

[149] US Department of Health and Human Services; Food and Drug Administration; Center for Drug Evaluation and Research (CDER); Center for Biologics Evaluation and Research (CBER). Guidance for Industry. Pharmacokinetics in Patients with Impaired Hepatic Function: Study Design, Data Analysis, and Impact on Dosing and Labeling; 2003 (16 pages).

[150] US Department of Health and Human Services; Food and Drug Administration; Center for Drug Evaluation and Research (CDER); Center for Biologics Evaluation and Research (CBER). Guidance for Industry. Pharmacokinetics in Patients with Impaired Hepatic Function: Study Design, Data Analysis, and Impact on Dosing and Labeling; 2003 (16 pages).

[151] Everson GT, et al. Quantitative liver function tests improve the prediction of clinical outcomes in chronic hepatitis C: results from the Hepatitis C Antiviral Long-term Treatment Against Cirrhosis Trial. Hepatology. 2012;55:1019–29.

[152] Ovchinsky N, et al. Liver biopsy in modern clinical practice: a pediatric point-of-view. Adv. Anat. Pathol. 2012;19:250–62.

[153] Zhang Z, et al. The diagnostic accuracy and clinical utility of three noninvasive models for predicting liver fibrosis in patients with HBV infection. PLoS One. 2016;11:e0152757.

[154] Muga R, et al. Unhealthy alcohol use, HIV infection and risk of liver fibrosis in drug users with hepatitis C. PLoS One. 2012;7:e46810.

[155] US Department of Health and Human Services; Food and Drug Administration; Center for Drug Evaluation and Research (CDER); Center for Biologics Evaluation and Research (CBER). Guidance for Industry. Pharmacokinetics in Patients with Impaired Hepatic Function: Study Design, Data Analysis, and Impact on Dosing and Labeling; 2003 (16 pages).

[156] Albarmawi A, et al. CYP3A activity in severe liver cirrhosis correlates with Child-Pugh and model for end-stage liver disease (MELD) scores. Br. J. Clin. Pharmacol. 2014;77:160–9.

[157] Hollebecque A, et al. Safety and efficacy of sorafenib in hepatocellular carcinoma: the impact of the Child-Pugh score. Aliment Pharmacol. Ther. 2011;34:1193–201.

[158] Rossle M, et al. The transjugular intrahepatic porosystemic stent-shunt procedure for variceal bleeding. N. Engl. J. Med. 1994;330:165–71.

d. Influence of Renal Impairment on Liver Function

Dose adjustment may be needed in the special situation where renal impairment causes hepatic impairment. This special situation is recognized by FDA's Guidance for Industry by its statement, "Renal impairment can adversely affect some pathways of hepatic/gut drug metabolism .... [t]hese changes may be particularly prominent in patients with severely impaired renal function and have been observed even when the renal route is not the primary route of elimination of a drug."[159]

If a given drug is excreted mainly in the bile, with very little excretion in the urine, then renal impairment would not be expected to result in increased exposure, and thus any renal impairment would not be a reason to reduce dosage of the drug. Or so it seems. A well-documented phenomenon is where renal impairment prevents the liver from clearing a given drug. This is where renal impairment in chronic kidney disease prevents hepatocytes from uptaking the drug from the bloodstream, with a consequent reduction in transport out of the hepatocyte into the bile canaliculi with excretion in the bile.

Also, renal impairment in chronic kidney disease can prevent hepatocytes from uptaking the drug from the bloodstream, with the consequent reduction in catabolism by cytochrome P450 enzymes in the hepatocyte.[160]

Chronic kidney disease results in increased concentrations of a toxin (CMPF). CMPF is 3-carboxy-4-methyl-5-propyl-2-furanpropionic acid. Regarding the origin of CMPF, CMPF is a furan fatty acid metabolite.[161] Increased bloodstream CMPF is associated with conditions identified as chronic kidney disease,[162] chronic renal failure,[163] and end-stage renal disease.[164] CMPF inhibits drug transport mediated by the OATP transporter, where this transporter mediates transport from the bloodstream into hepatocytes. Sun et al.[165] studied the pharmacokinetics of erythromycin in normal human subjects and in end-stage renal disease human subjects, where the results suggest the scenario shown in Fig. 4.6. Erythromycin was used as a probe substrate in the human subject studies. Fig. 4.6 shows erythromycin (filled ovals) and its uptake by the hepatocyte (mediated by OATP) and efflux into the bile duct (mediated by P-glycoprotein).

[159] US Department of Health and Human Services; Food and Drug Administration; Center for Drug Evaluation and Research (CDER); Center for Biologics Evaluation and Research (CBER). Guidance for Industry. Pharmacokinetics in Patients with Impaired Renal Function—Study Design, Data Analysis, and Impact on Dosing and Labeling; 2010 (21 pages).

[160] Momper JD, et al. Nonrenal drug clearance in CKD: searching for the path less traveled. Adv. Chronic Kidney Dis. 2010;17:384–91.

[161] Prentice KJ, et al. The furan fatty acid metabolite CMPF is elevated in diabetes and induces β cell dysfunction. Cell Metab. 2014;19:653–66.

[162] Boelaert J, et al. A novel UPLC-MS-MS method for simultaneous determination of seven uremic retention toxins with cardiovascular relevance in chronic kidney disease patients. Anal. Bioanal. Chem. 2013;405:1937–47.

[163] Sun H, et al. Effects of renal failure on drug transport and metabolism. Pharmacol. Ther. 2006;109:1–11.

[164] Sun H, et al. Hepatic clearance, but not gut availability, of erythromycin is altered in patients with end-stage renal disease. Clin. Pharmacol Ther. 2010;87:465–72.

[165] Sun H, et al. Hepatic clearance, but not gut availability, of erythromycin is altered in patients with end-stage renal disease. Clin. Pharmacol Ther. 2010;87:465–72.

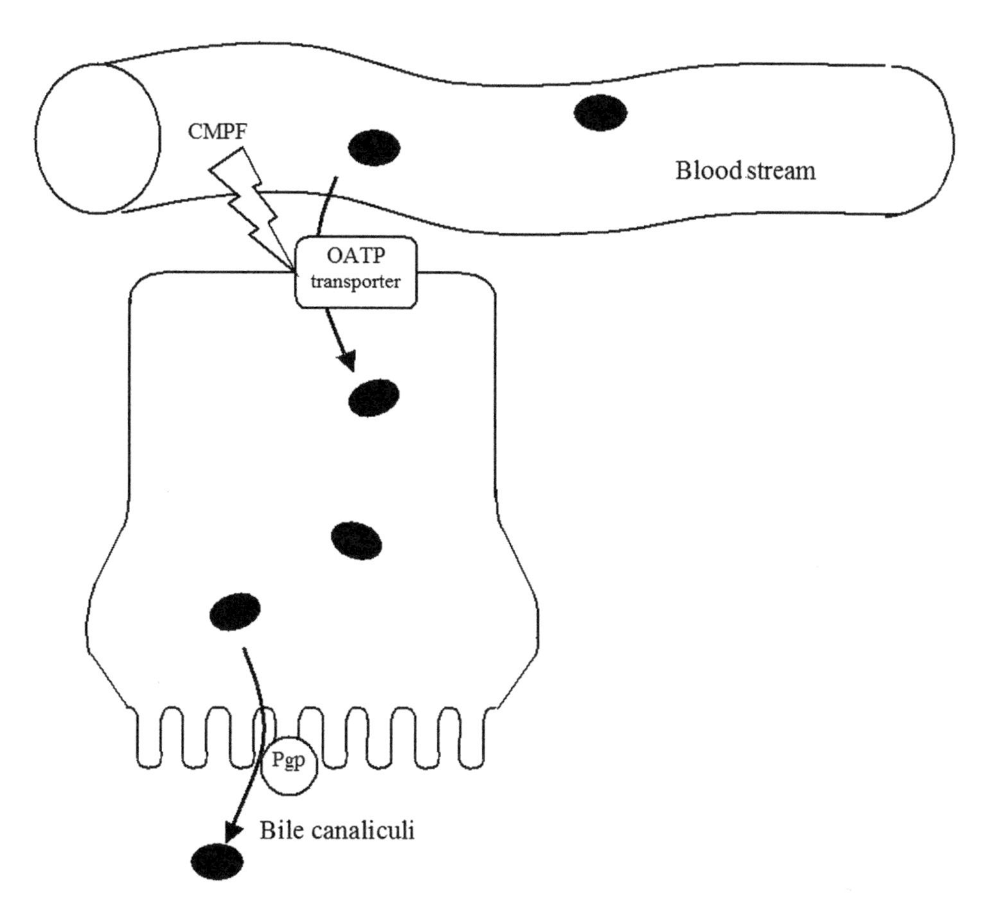

FIGURE 4.6 Drug transport through the hepatocyte. OATP transporter mediates transport of the drug into the hepatocyte, where this transport can be inhibited by the renal toxin, CMPF, with consequent rise in plasma toxin concentrations.

Fig. 4.6 shows the basolateral membrane and apical membrane of the hepatocyte. Also, it illustrates the point (lightning bolt) where CMPF inhibits OATP, thus reducing transport of erythromycin from the bloodstream into the hepatocyte, with consequent reduction in clearance to the bile, where the end result is an increase in bloodstream erythromycin concentration.

To provide a context for the above CMPF data, the "uremic toxins" are solutes that are excreted by the kidneys in health, but are retained in patients with chronic renal failure, and that are harmful to the patient.[166] Vanholder et al.[167] provides a huge list of uremic toxins,

[166] Duranton F, et al. Normal and pathologic concentrations of uremic toxins. J. Am. Soc. Nephrol. 2012;23:1258–70.

[167] Vanholder R, et al. Review on uremic toxins: classification, concentration, and interindividual variability. Kidney Int. 2003;63:1934–43.

including CMPF, hippuric acid, and indoxyl sulfate, together with their concentrations in blood plasma or serum, in health vs chronic renal failure. The pathological effects caused by uremic toxins include toxicity to neutrophils and macrophages and toxic influence on endothelial cells resulting in hypertension and disturbances in the blood clotting cascade.[168]

In the CMPF experiments by Sun et al.,[169] the researchers introduced the uremic toxins CMPF, indoxyl sulfate, hippuric acid, indole-3-acetic acid, guanidino succinic acid, and indoxyl-β-glucuronide into medium containing hepatocytes. Each hepatocyte incubation mixture was supplemented with a different uremic toxin, where the uremic toxin was used at the highest concentration observed in uremic patients. **CMPF (uremic toxin) inhibited transport of erythromycin (probe substrate)** into the hepatocytes more than any of the other uremic toxins. One of the authors (L.F.) pointed out that one must not discount the influence of the renal toxin **indoxyl sulfate** on impairing erythromycin clearance from the body.[170] Sun et al.[171] also measured cytochrome P450-mediated catabolism of erythromycin in hepatocytes and discovered that **indoxyl sulfate** directly inhibits the CYP3A-catalyzed demethylation of erythromycin. "CYP3A" refers to the cytochrome P450-3A isozyme. Erythromycin is partially metabolized by CYP3A4 to its N-demethylated metabolite but is primarily excreted unchanged in the bile by P-glycoprotein (P-gp).[172]

Rat hepatocyte studies revealed that chronic renal failure can reduce OATP expression, as measured by the amount of OATP protein expressed. The amount of OATP protein expressed was determined by Western blots. Reduced OATP expression was measured by a variety of clever experiments, for example, where OATP expression was measured in hepatocytes from normal rats and in hepatocytes from partial nephrectomized rats, or where rat hepatocytes were exposed to normal human serum versus serum from chronic renal failure patients.[173,174,175] Nolin et al.[176] provide evidence from human metabolic studies that end-stage renal disease results in decreased activity of hepatic OATP expression, a finding consistent with the above-cited rat hepatocyte studies.

[168] Vanholder R, et al. A bench to bedside view of uremic toxins. J. Am. Soc. Nephrol. 2008;19:863–70.

[169] Sun H, et al. Effects of uremic toxins on hepatic uptake and metabolism of erythromycin. Drug Metab. Dispos. 2004;32:1239–46.

[170] E-mail of July 11, 2017, from Lynda Frassetto, MD, Division of Nephrology, UCSF School of Medicine, San Francisco, CA.

[171] Sun H, et al. Effects of uremic toxins on hepatic uptake and metabolism of erythromycin. Drug Metab. Dispos. 2004;32:1239–46.

[172] Sun H, et al. Hepatic clearance, but not gut availability, of erythromycin is altered in patients with end-stage renal disease. Clin. Pharmacol Ther. 2010;87:465–72.

[173] Nolin TD, et al. ESRD impairs nonrenal clearance of fexofenadine but not midazolam.J. Am. Soc. Nephrol. 2009;20:2269–76.

[174] Naud J, et al. Effects of chronic renal failure on liver drug transporters. Drug Metab. Dispos. 2008;36:124–8.

[175] Dreisbach AW. The influence of chronic renal failure on drug metabolism and transport. Clin. Pharmacol. Ther. 2009;86:553–6.

[176] Nolin TD, et al. The pharmacokinetics of fexofenadine but not midazolam are altered in end-stage renal disease. Clin. Pharmacol. Ther. 2008;83:S58.

Chronic renal failure may also reduce clearance of drugs by hepatocytes, by way of inhibiting activity of one or more of the cytochrome P450 isozymes. Each of these isozymes is designated by "CYP" as in the designations, CYP3A, CYP2C9, and CYP2B6. Nolin et al.[177] provide an overview of the various CYP isozymes, and the issue of CYP isozyme inhibition in chronic renal disease.

e. Summary

For drafting the Dosage and Administration section, it might be valuable to assess the influence of chronic renal failure on the exposure (AUC, Cmax, tmax) of the study drug. An appropriate clinical study is to compare the PK of the study drug with healthy control subjects vs chronic renal failure subjects. **Where renal failure increases AUC or Cmax**, the investigator might want to draft **dose reduction instructions**, as applied to renal failure patients. Clinical studies can be conducted with standard drug probes to determine if renal impairment-induced hepatic impairment takes the form of reduced CYP isoenzyme activity or if it takes the form of reduced clearance to the bile duct.

As a general proposition, when drafting the Clinical Study Protocol, renal impairment and hepatic impairment can be used for defining the various subsets of study subjects enrolling in any given clinical study.[178]

f. Renal Impairment and Hepatic Impairment Fit Into the Concept of Intrinsic Factors

Analysis of efficacy or safety data for any given drug can focus on any subgroup in the population of study subjects. FDA uses the terms "intrinsic factors" and "extrinsic factors" for defining subgroups.[179] The terms "intrinsic factors" and "extrinsic factors" are not related to the intrinsic factor used in cobalamin absorption or to the extrinsic factor of the blood clotting pathway. In subgroup analysis, the term "intrinsic factors" refers to age group, gender, race or ethnicity, genetic markers, and degree of renal impairment and hepatic impairment.[180] "Extrinsic factors" refers to smoking habits, concomitant medications, diet, and so on.

[177] Nolin TD, et al. Hepatic drug metabolism and transport in patients with kidney disease. Am. J. Kidney Dis. 2003;42:906–25.

[178] Editorial, 2010, Mechanism matters. Nature Medicine. 16:347.

[179] US Department of Health and Human Services; Food and Drug Administration; Center for Drug Evaluation and Research (CDER); Center for Biologics Evaluation and Research (CBER). Guidance for Industry M4E: The CTD—Efficacy; 2001 (58 pages).

[180] Small DS, et al. Effect of intrinsic and extrinsic factors on the clinical pharmacokinetics and pharmacodynamics of prasugrel. Clin. Pharmacokinet. 2010;49:777–98.

V. FDA'S ANALYSIS OF RENAL IMPAIRMENT AND HEPATIC IMPAIRMENT, AND THE PACKAGE LABEL

FDA's analysis of renal impairment and hepatic impairment, as it applies to arriving at package label instructions, is illustrated by FDA's ClinPharm Reviews for these drugs:

- Bosutinib
- Canagliflozin/metformin combination
- E/C/F/TAF combination
- Ivacaftor
- Lenalidomide

a. Bosutinib (Chronic Myeloid Leukemia) NDA 203-341

FDA's ClinPharm Review for bosutinib reveals FDA's decision-making process for arriving at the dose modification instructions. The Sponsor recommended that no dose modification instructions were needed for renal impairment but that, in contrast, dose modification instructions were needed for hepatic impairment. Regarding renal impairment, FDA wrote:

> Based upon what is known about exposure-response relationships and their variability and the groups studied, healthy volunteers vs. patients vs. specific populations, what dose adjustments, if any, are recommended for each of these groups? ... Renal impairment. Based on the lack of exposure-response relationships for efficacy and safety ... within the studied dose range, dose adjustment for the 30% reduction in bosutinib clearance for patients with moderate renal impairment ... are ... **not necessary.**[181]

Concerning hepatic impairment and dose adjustment, FDA stated:

> Hepatic impairment. The applicant evaluated the relationship between hepatic impairment and exposure ... [b]osutinib Cmax and AUC increased by approximately 2-fold in those with ... hepatic impairment as compared to the healthy volunteers ... [t]he applicant proposes a dose adjustment to 200 mg in patients with hepatic impairment. This dose adjustment **appears reasonable** based on bosutinib exposure.[182]

Package label. The Dosage and Administration section for bosutinib (Bosulif®) is shown. Note the instructions for hepatic impairment, but the absence of any instructions for patients with renal impairment, which is consistent with FDA's review. The following instruction is in the package label, as initially approved in 2012:

> **DOSAGE AND ADMINISTRATION** ... Hepatic impairment (at baseline): reduce BOSULIF dose to 200 mg daily.[183]

[181] Bosutinib (chronic myeloid leukemia) NDA 203-341. Pages 29 and 31 of 89-page Clinical Pharmacology Review.

[182] Bosutinib (chronic myeloid leukemia) NDA 203-341. Pages 29 and 31 of 89-page Clinical Pharmacology Review.

[183] Package label. BOSULIF® (bosutinib) tablets, for oral use; September 2012 (13 pages).

TABLE 4.5 Recommended Starting Dosage With Hepatic Impairment or Renal Impairment

Organ Function Status	Recommended Starting Dosage
Normal hepatic and renal function	500 mg once daily
HEPATIC IMPAIRMENT	
Mild (Child–Pugh A), moderate (Child–Pugh B), or severe (Child–Pugh C)	200 mg daily
RENAL IMPAIRMENT	
Creatinine clearance 30–50 mL/min	400 mg daily
Creatinine clearance <30 mL/min	300 mg daily

From: Package label. BOSULIF® (bosutinib) tablets, for oral use. April 2017 (15 pages).

On the other hand, for the revised package label issued several years later (2017), the Dosage and Administration section had separate instructions for patients with hepatic impairment and for patients with renal impairment. Apparently, the revision was the result of PK data acquired following FDA's initial approval in 2012. The more recent label had the following instructions and table, as shown in Table 4.5:

> **DOSAGE AND ADMINISTRATION** ... Adjust dosage for ... organ impairment.[184]

b. Canagliflozin/Metformin Combination (Type-2 Diabetes Mellitus) NDA 204-353

This concerns the canagliflozin/metformin combination for treating Type-2 diabetes mellitus. Canagliflozin promote the loss of glucose in the urine, whereas metformin improves the body's response to insulin and reduces the liver's production of glucose.[185]

In evaluating renal impairment and hepatic impairment for the study drug, FDA turned to previously published warnings for the two components of the study drug, namely canagliflozin and metformin. Regarding canagliflozin, FDA used the concept drug–disease interactions, writing:

> The drug-disease interactions ... are expected to be similar to the individual components. As noted in my review for NDA 204-042, the **efficacy of canagliflozin depends on renal function** and its efficacy is modest in those with moderate renal function. As noted in canagliflozin labeling, canagliflozin has not been studied and **should not be used in those with severe renal impairment** (e.g., ≤ 30 mL/min/1.73 m^2), end-stage renal disease (ESRD), or on dialysis. In addition, subjects with **moderate renal impairment** experienced more significant changes in renal function and had **more renal-related events**, and canagliflozin is to be discontinued in patients with eGFR <45 mL/min/1.73 m^2.[186]

[184] Package label. BOSULIF® (bosutinib) tablets, for oral use; April 2017 (15 pages).

[185] Fala L. Invokamet (canagliflozin plus metformin HCl): first fixed-dose combination with an SGLT2 inhibitor approved for the treatment of patients with type 2 diabetes. Am. Health Drug Benefits. 2015;8:70–4.

[186] Canagliflozin and metformin combination (type-2 diabetes mellitus) NDA 204-353. Pages 12, 70, and 108 of 119-page Medical Review.

Note the FDA Reviewer's mention of an earlier submission, that is, NDA 204-042, which concerned canagliflozin. Also, note that FDA's recommendation on dose modification was based on two different patient characteristics:

- eGFR with a value <45 mL/min/1.73 m^2
- AEs relating to the kidneys

Regarding metformin, FDA referred to both renal impairment and hepatic impairment but disclosed that renal impairment resulted in AEs severe enough to warrant a contraindication in a previously issued package label. On this point FDA stated that, "Labeled safety concerns with metformin include the following: Lactic acidosis: **the risk increases** with sepsis, dehydration, excessive alcohol intake, **hepatic insufficiency, renal impairment**, and acute congestive heart failure ... [c]ontraindicated in patients with **renal impairment** (e.g., serum creatinine ≥ 1.5 mg/dL for males, ≥ 1.4 mg/dL for females, or abnormal creatinine clearance)."[187]

Thus, in reiterating warnings about hepatic impairment and renal impairment from the individual components of the study drug,[188] FDA stated that the biggest safety risks were renal impairment in patients taking metformin and renal impairment in patients taking canagliflozin.

Package label. The Dosage and Administration section of the package label for the canagliflozin/metformin combination (Invokamet®) set forth warnings only about renal impairment (not hepatic impairment):

> **DOSAGE AND ADMINISTRATION** ... Assess renal function before initiating INVOKAMET. Do not initiate or continue INVOKAMET if creatinine levels are greater than or equal to 1.5 mg/dL for males or 1.4 mg/dL for females, or if eGFR is below 45 mL/min/1.73 m^2.[189]

The value 1.73 m^2 is for adjusting for body surface area.[190,191] Although information on hepatic impairment was not in the Dosage and Administration section, it found a place in the Black Box Warning. In other words, the Black Box Warning included instructions for dose modification:

> **BOXED WARNING**: LACTIC ACIDOSIS ... Lactic acidosis can occur due to metformin accumulation ... risk increases with conditions such as **renal impairment**, sepsis, dehydration, excess alcohol intake, **hepatic impairment**, and acute congestive heart failure ... [i]f acidosis is suspected, discontinue INVOKAMET and hospitalize the patient immediately.[192]

[187] Canagliflozin and metformin combination (type-2 diabetes mellitus) NDA 204-353. Pages 12, 70, and 108 of 119-page Medical Review.

[188] See, comments in "Important Safety Issues With Consideration to Related Drugs" on Page 12 of 119-page Medical Review for NDA 204-353.

[189] Package label. INVOKAMET (canagliflozin and metformin hydrochloride) tablets, for oral use; August 2014 (45 pages).

[190] Stevens LA, et al. Comparison of drug dosing recommendations based on measured GFR and kidney function estimating equations. Am. J. Kidney Dis. 2009;54:33–42.

[191] Warnock DG. Estimated glomerular filtration rate: fit for what purpose? Nephron. 2016;134:43–9.

[192] Package label. INVOKAMET (canagliflozin and metformin hydrochloride) tablets, for oral use; August 2014 (45 pages).

Hepatic and renal impairment also found a place in the Warnings and Precautions section, which read:

> **WARNINGS AND PRECAUTIONS** ... Lactic acidosis ... INVOKAMET is not recommended in hepatic impairment ... [e]nsure normal renal function before initiating and at least annually thereafter.[193]

FDA's analysis was based on previous FDA submissions for canagliflozin alone and for metformin alone. This approach for assessing renal and hepatic impairment, and corresponding package label information is evident from FDA's comment, "The drug-disease interactions ... are expected to be similar to the individual components."[194]

c. Elvitegravir, Cobicistat, Emtricitabine, Tenofovir Alafenamide Combination (HIV-1) NDA 207–561

Elvitegravir, Cobicistat, Emtricitabine, Tenofovir Alafenamide (E/C/F/TAF) combination takes the form of a combination of four drugs, where the combination was a fixed-dose combination (FDC). Tenofovir alafenamide (TAF) is a prodrug, where enzymatic hydrolysis in lymphocytes converts it to the active drug, tenofovir.[195,196]. FDA's Guidance for Industry provides advice for submissions on FDC drugs, for example, that the Sponsor should determine if bioavailability of each component drug in the combination is the same as the bioavailability of the individually administered drug.[197]

A published clinical study comparing healthy control subjects vs renally impaired subjects administered TAF to subjects and then followed the plasma concentrations of TAF (prodrug) and tenofovir (active drug). The results were as follows: The exposure parameters of TAF (AUC, Cmax) were 80%–90% higher in the subjects with severe renal impairment than the healthy controls. Regarding the active drug (tenofovir), Cmax was 180% higher and the tenofovir AUC was 470% higher in subjects with severe renal impairment relative to the values for the healthy controls.[198]

[193] Package label. INVOKAMET (canagliflozin and metformin hydrochloride) tablets, for oral use; August 2014 (45 pages).

[194] Canagliflozin and metformin combination (type-2 diabetes mellitus) NDA 204-353. Pages 12, 70, and 108 of 119-page Medical Review.

[195] Ray AS, et al. Tenofovir alafenamide: a novel prodrug of tenofovir for the treatment of human immunodeficiency virus. Antiviral Res. 2016;125:63–70.

[196] Callebaut C, et al. In vitro virology profile of tenofovir alafenamide, a novel oral prodrug of tenofovir with improved antiviral activity compared to that of tenofovir disoproxil fumarate. Antimicrob. Agents Chemother. 2015;59:5909–16.

[197] US Department of Health and Human Services; Food and Drug Administration; Center for Drug Evaluation and Research (CDER). Guidance for Industry. Fixed Dose Combinations, Co-Packaged Drug Products, and Single-EntityVersions of Previously Approved Antiretrovirals for the Treatment of HIV; 2006 (36 pages).

[198] Custodio JM, et al. Pharmacokinetics and safety of tenofovir alafenamide in HIV-uninfected subjects with severe renal impairment. Antimicrob. Agents Chemother. 2016;60:5135–40.

TAF is related to an earlier-developed anti-HIV-1 drug, tenofovir disoproxil fumarate (TDF). TDF has an established nephrotoxicity, whereas TAF shows much reduced nephrotoxicity.[199,200] The present narrative on renal impairment only concerns how renal impairment influences TAF exposure or influences TAF toxicity (but the present narrative is not about the issue of TAF-induced renal toxicity).

1. **FDA's analysis of renal impairment.** FDA requested that the Sponsor conduct a clinical study comparing exposure in normal subjects versus exposure in renal-impaired subjects to support package labeling. About the requested study, FDA wrote, "[it] was a study in HIV-infected subjects with **mild to moderate renal impairment** ... there was no cohort with normal renal function ... [i]n order to inform the impact of renal impact on E/C/F/TAF exposures, we asked the sponsor to repeat the analysis by comparing PK data from this renal impairment study ... relative to PK data from HIV-infected subjects with **normal renal function**."[201] In short, FDA was complaining about poor study design.

Apparently in response to FDA's request for a side-by-side study with healthy and renal-impaired subjects, the Sponsor conducted this type of clinical study, which FDA described as, "Relative to healthy controls, subjects with severe renal impairment had TAF AUC increased 92% and Cmax increased 80% ... renal clearance was reduced 88%."[202] The side-by-side study showed that severe renal impairment resulted in a near-doubling of exposure. From the same side-by-side study, Table 4.6 shows that renal impairment reduces renal clearance rate, reduces the percent of dose appearing in urine, and reduces micrograms of drug recovered in the urine. Renal function was measured in terms of renal clearance CLrenal (mL/min), percentage of dose recovered in urine (%), and Ae (ng) (Table 4.6). "Ae" refers to the amount of drug excreted unchanged in the urine.[203]

The Sponsor had determined exposure (area under the curve; AUC) for each of the individual components of the E/C/F/TAF (Genvoya®) combination, following administration of the combination drug. Table 4.7 shows the AUC results for each of the four drug components. Study subjects were divided into those with **greater** renal impairment (eGFR <50 mL/min) and into those with **lesser** renal impairment (eGFR ≥ 50 mL/min). Table 4.7 shows that for each of the component drugs, exposure was moderately greater in subjects with the lesser glomerular filtration rate (eGFR).

[199] Post FA, et al. Brief report: switching to tenofovir alafenamide, coformulated with clvitegravir, cobicistat, and emtricitabine, in HIV-infected adults with renal impairment: 96-week results from a single-arm, multicenter, open-label phase 3 study. J. Acquired Immune Defic. Syndr. 2017;74:180–4.

[200] Wang H, et al. The efficacy and safety of tenofovir alafenamide versus tenofovir disoproxil fumarate in antiretroviral regimens for HIV-1 therapy: meta-analysis. Medicine (Baltimore) 2016;95:e5146.

[201] Elvitegravier, cobicistat, emtricitabine, tenofovir alafenamide (E/C/F/TAF) combination (HIV-1) NDA 207-561. Page 6 of 218-page Clinical Pharmacology Review.

[202] Elvitegravier, cobicistat, emtricitabine, tenofovir alafenamide (E/C/F/TAF) combination (HIV-1) NDA 207-561. Page 115 of 218-page Clinical Pharmacology Review.

[203] Gad SC (ed.). Drug clearance, in: Panuganti SD, Svensson CK. Preclinical Development Handbook: ADME and Biopharmaceutical Properties. Wiley and Sons, Inc., Hoboken, NJ; 2008 (pp. 724 and 725).

TABLE 4.6 TAF Urine Pharmacokinetics

	Normal Renal Function ($n = 13$ subjects)	Severe Renal Impairment ($n = 14$ subjects)
CLrenal (mL/min)	35.8 mL/min	4.2 mL/min
Percentage of dose recovered in urine (%)	2.00%	0.47%
Ae (micrograms of drug excreted in urine)	500 μg	117 μg

From: Elvitegravier, cobicistat, emtricitabine, tenofovir alafenamide (E/C/F/TAF) combination (HIV-1) NDA 207–561. Page 115 of 218 page Clinical Pharmacology Review.

TABLE 4.7 Exposure for Each Component of the Study Drug Combination, Following Administration of the Study Drug Combination

	Exposure (AUC Unit (ng)(h)/mL)	
	Degree of Renal Impairment of Study Subjects	
	eGFR ≥ 50 mL/min	eGFR < 50 mL/min
Elvitegravir	26,468	28,705
Cobicistat	9393	11,316
Emtricitabine	19,380	25,140
TAF	227	340

From: Elvitegravier, cobicistat, emtricitabine, tenofovir alafenamide (E/C/F/TAF) combination (HIV-1) NDA 207–561. Pages 136 and 137 of 218, Clinical Pharmacology Review.

In view of the Sponsor's data (Table 4.7), FDA found alarm in the moderate exposure increases for each of the drugs and recommended that the E/C/F/TAF *not be given* to renally impaired patients. FDA wrote, "The sponsor's label proposes that E/C/F/TAF fixed drug combination (FDC) can be administered without dose adjustment to subjects with CrCl ≥ 30 mL/min. Unless safety is deemed adequately established in subjects with CrCL of 30-49 mL/min ... we disagree and propose ***not to recommend administration of E/C/F/TAF to subjects with CrCL < 50 mL/min.***"[204]

2. **FDA's analysis of hepatic impairment.** FDA commented on the lack of much influence of mild or moderate hepatic impairment on drug exposure and concluded that the exposure changes were not clinically relevant. But for patients with severe hepatic impairment, FDA recommended that E/C/F/TAF combination drug should not be given to these patients.

FDA made a separate recommendation for patients with mild (Child–Pugh Class A), moderate (Child–Pugh Class B), and severe hepatic impairment (Child–Pugh Class C).

[204] Elvitegravier, cobicistat, emtricitabine, tenofovir alafenamide (E/C/F/TAF) combination (HIV-1) NDA 207-561. Pages 139 of 218-page Clinical Pharmacology Review.

FDA recommended that the E/C/F/TAF combination *not be given* to severe hepatic impairment patients. FDA wrote, "In subjects with mild to moderate (Child-Pugh Class A and B) hepatic impairment ... exposures in subjects with hepatic impairment were altered by ≤13%. Based on exposure-response relationships, these exposure changes are not clinically relevant. We agree with proposed E/C/F/TAF labeling wherein there is no dose adjustment for Child-Pugh A-B, and E/C/F/TAF ***is not recommended for Child-Pugh C.***"[205]

FDA's recommendation for dose adjustment in hepatic-impaired patients was based on exposure studies where only TAF (and not the drug combination) was administered.[206]

Package label. Following FDA's recommendations, the package label for the E/C/F/TAF combination (Genvoya®) recommended not to administer the E/C/F/TAF combination to patients with renal impairment:

> **DOSAGE AND ADMINISTRATION** ... **Renal impairment:** GENVOYA is not recommended in patients with estimated creatinine clearance below 30 mL per minute.[207]

Also following FDA's recommendations, the label recommended against administering the drug to patients with severe hepatic impairment:

> **DOSAGE AND ADMINISTRATION** ... **Hepatic impairment:** GENVOYA is not recommended in patients with severe hepatic impairment.[208]

d. Ivacaftor (Cystic Fibrosis for CFTR Gene G551D Mutation) NDA 203-188

Ivacaftor is for treating cystic fibrosis in patients where the cystic fibrosis transmembrane conductance regulator gene has a G551D mutation. FDA's review of ivacaftor provides the example of the disease–drug interaction where renal impairment inhibits hepatic clearance, resulting in increased drug exposure.[209]

1. **Renal impairment.** FDA's ClinPharm Review observed that the ivacaftor was not much excreted in the urine, raising the issue that any degree of renal impairment would not likely increase exposure (and thus not result in any AEs). However, the FDA reviewer

[205] Elvitegravier, cobicistat, emtricitabine, tenofovir alafenamide (E/C/F/TAF) combination (HIV-1) NDA 207-561. Pages 13 and 40 of 218-page Clinical Pharmacology Review.

[206] Elvitegravier, cobicistat, emtricitabine, tenofovir alafenamide (E/C/F/TAF) combination (HIV-1) NDA 207-561. Pages 13 and 40 of 218-page Clinical Pharmacology Review.

[207] Elvitegravier, cobicistat, emtricitabine, tenofovir alafenamide (E/C/F/TAF) combination (HIV-1) NDA 207-561. Pages 13 and 40 of 218-page Clinical Pharmacology Review.

[208] Package label. GENVOYA® (elvitegravir, cobicistat, emtricitabine, and tenofovir alafenamide) tablets for oral use; November 2005 (44 pages).

[209] Ivacaftor (cystic fibrosis for CFTR gene G551D mutation) NDA 203-188. Pages 43 and 44 of 102-page Clinical Pharmacology Review.

referred to FDA's Guidance for Industry[210] and was careful to consider the situation where renal impairment does result in impaired hepatic impairment. FDA's word of caution referred to the situation where renal toxins that accumulate in chronic renal disease inhibit hepatic clearance of drugs:

> Only a small fraction of administered ivacaftor dose is eliminated by renal route ... [t]herefore, and effect of renal impairment on ivacaftor clearance is unlikely, and so no dose adjustment is necessary for patients with mild or moderate renal impairment. However, **renal impairment may also affect some pathways of hepatic and gut drug metabolism** ... therefore ... caution is recommended while using ivacaftor in these patients.[211]

The scenario where renal impairment blocks pathways of hepatic drug metabolism is described in the cited references.[212,213,214,215,216,217]

Package label (renal impairment). FDA's comments on renal impairment found a place in the Use in Specific Populations section of the package label for ivacaftor (Kalydeco®):

> **USE IN SPECIFIC POPULATIONS** ... Renal Impairment. KALYDECO has not been studied in patients with mild, moderate, or severe renal impairment or in patients with end stage renal disease. No dose adjustment is necessary for patients with mild to moderate renal impairment; however, **caution is recommended while using KALYDECO in patients with severe renal impairment (creatinine clearance less than or equal to 30 mL/min)** or end stage renal disease.[218]

2. **Hepatic impairment.** Regarding the influence of hepatic impairment in increasing ivacaftor exposure, FDA observed that, "After single dose, the Cmax for ... ivacaftor ...

[210] US Department of Health and Human Services; Food and Drug Administration; Center for Drug Evaluation and Research (CDER); Center for Biologics Evaluation and Research (CBER). Guidance for Industry Pharmacokinetics in Patients with Impaired Renal Function—Study Design, Data Analysis, and Impact on Dosing and Labeling; 1998 (14 pages).

[211] Ivacaftor (cystic fibrosis for CFTR gene G551D mutation) NDA 203-188. Pages 43 and 44 of 102-page Clinical Pharmacology Review.

[212] Prentice KJ, et al. The furan fatty acid metabolite CMPF is elevated in diabetes and induces β cell dysfunction. Cell Metab. 2014;19:653–66.

[213] Boelaert J, et al. A novel UPLC-MS-MS method for simultaneous determination of seven uremic retention toxins with cardiovascular relevance in chronic kidney disease patients. Anal. Bioanal. Chem. 2013;405:1937–47.

[214] Sun H, et al. Effects of renal failure on drug transport and metabolism. Pharmacol. Ther. 2006;109:1–11.

[215] Sun H, et al. Hepatic clearance, but not gut availability, of erythromycin is altered in patients with end-stage renal disease. Clin. Pharmacol Ther. 2010;87:465–72.

[216] Duranton F, et al. Normal and pathologic concentrations of uremic toxins. J. Am. Soc. Nephrol. 2010;23:1258–70.

[217] Vanholder R, et al. Review on uremic toxins: classification, concentration, and interindividual variability. Kidney Int. 2003;63:1934–43.

[218] Package label. KALYDECO™ (ivacaftor) Tablets; January 2012 (9 pages).

was similar in subjects with moderate hepatic impairment and ... healthy subjects ... [h]owever, the $AUC_{0\text{-infinity}}$ for ... ivacaftor was approximately 2-fold higher in subjects with moderate hepatic impairment than in ... healthy subjects."[219]

FDA's analysis of exposure distinguished between Cmax and AUC. The value for Cmax was not influenced by hepatic impairment, while, in contrast, the value for AUC was doubled with hepatic impairment. FDA's explanation was that Cmax is primarily a function of the drug's absorption by the gut (and less a function of liver metabolism). FDA explained, "These results show that moderate hepatic impairment has no ... effect on absorption (Cmax) of ivacaftor ... but slows down the clearance ... by approximately two-fold of that in healthy subjects."[220]

FDA recommended that the dose be cut in half for patients with moderate hepatic impairment, writing that, "the recommended dose in patients with moderate hepatic impairment was 150 mg once daily."[221] FDA's comments found a place on the package label, which instructed that the dose was 150 mg every 12 hours, and that with severe hepatic impairment, the dose should be half of this or even lower.[222]

Package label (hepatic impairment). FDA's recommendations for ivacaftor (Kalydeco®) found a place on the package label in the Dosage and Administration section, Use in Specific Populations section, and in the Warnings and Precautions section:

> **DOSAGE AND ADMINISTRATION** ... Reduce dose in patients with moderate and severe hepatic impairment ... The dose of KALYDECO should be reduced to 150 mg once daily for patients with moderate hepatic impairment (Child-Pugh Class B). KALYDECO should be used with caution in patients with severe hepatic impairment (Child-Pugh Class C) at a dose of 150 mg once daily or less frequently.[223]

> **USE IN SPECIFIC POPULATIONS** ... Hepatic Impairment. No dose adjustment is necessary for patients with **mild hepatic impairment** ... [a] reduced dose of 150 mg once daily is recommended in patients with **moderate hepatic impairment** ... [s]tudies have not been conducted in patients with severe hepatic impairment ... but exposure is expected to be higher than in patients with moderate hepatic impairment. Therefore, use with caution at a dose of 150 mg once daily or less frequently in patients with **severe hepatic impairment** after weighing the risks and benefit of treatment.[224]

> **WARNINGS AND PRECAUTIONS** ... Elevated transaminases (ALT or AST). Transaminases ... should be assessed prior to initiating KALYDECO ... [p]atients who develop increased transaminase levels should be closely monitored until abnormalities resolve. **Dosing should be interrupted** in patients with ALT or AST or greater than 5 times the upper limit of normal (ULN).[225]

[219] Ivacaftor (cystic fibrosis for CFTR gene G551D mutation) NDA 203-188. Pages 43 and 44 of 102-page Clinical Pharmacology Review.

[220] Ivacaftor (cystic fibrosis for CFTR gene G551D mutation) NDA 203-188. Pages 43 and 44 of 102-page Clinical Pharmacology Review.

[221] Ivacaftor (cystic fibrosis for CFTR gene G551D mutation) NDA 203-188. Pages 43 and 44 of 102-page Clinical Pharmacology Review.

[222] Package label. KALYDECO™ (ivacaftor) Tablets; January 2012 (9 pages).

[223] Package label. KALYDECO™ (ivacaftor) Tablets; January 2012 (9 pages).

[224] Package label. KALYDECO™ (ivacaftor) Tablets; January 2012 (9 pages).

[225] Package label. KALYDECO™ (ivacaftor) Tablets; January 2012 (9 pages).

e. Lenalidomide (MDS) NDA 021-880

FDA's ClinPharm Review of lenalidomide revealed FDA's recommendation for instructions for patients with renal impairment and to recommend that there not be any instructions for patients with hepatic impairment.[226] In FDA's analysis of exposure (AUC) data, FDA's attention and care is shown by the fact that FDA twice analyzed the exposure data—with and without taking into account an "outlier" data point. FDA recommended that the Sponsor perform an additional clinical study to resolve ambiguity on the PK data from subjects with renal impairment:

> We recommend that a ... **statement regarding monitoring of patients with renal impairment be added to the package label** ... four patients with mild renal impairment were sampled for pharmacokinetics ... their AUCs are compared to those of patients ... with normal renal function ... ***if the apparent outlier in the "Normal" renal function group is excluded,*** the mild renal impairment group has an AUC 85% greater than the unimpaired group. If the ***"outlier" is included,*** the difference is 56% ... [b]ased on the data showing that 2/3 of lenalidomide is excreted renally ... and consistent with the ... effect of renal impairment ... we recommend **a Phase 4 commitment to perform a pharmacokinetic study in subjects with renal impairment.**[227]

Thus, FDA's decision to recommend another clinical study was based first on the fact that the study drug's pathway of excretion is mainly by the renal route and secondly on the fact that data indicated that renal impairment resulted in a significant increase in exposure. In addition to this, FDA added a brief remark on patients with hepatic impairment, "Hepatic impairment. No dosage regimen adjustments are recommended."[228] Accordingly, the package label did not mention anything about dose adjustment in hepatic impairment.

Actually, the package label did instruct physicians to halt lenalidomide (Revlimid®) in the situation of lenalidomide-induced hepatic toxicity ("Hepatotoxicity: Hepatic failure including fatalities; monitor liver function. Stop REVLIMID and evaluate if hepatotoxicity is suspected"[229]), but this is not exactly the same thing as dose modification for hepatic impaired subjects, because the term hepatic impairment usually implies a preexisting condition.

Package label. The Dosage and Administration section read:

> **DOSAGE AND AMINISTRATION** ... Renal impairment: Adjust starting dose in patients with moderate or severe renal impairment and on dialysis (CLcr < 60 mL/min).[230]

[226] Lenalidomide (myelodysplastic syndromes) NDA 021-880. Pages 24 and 25 of 46-page Clinical Pharmacology Review.

[227] Lenalidomide (myelodysplastic syndromes) NDA 021-880. Pages 24 and 25 of 46-page Clinical Pharmacology Review.

[228] Lenalidomide (myelodysplastic syndromes) NDA 021-880. Pages 24 and 25 of 46-page Clinical Pharmacology Review.

[229] Package label. REVLIMID (lenalidomide) capsules for oral use; February 2015 (36 pages).

[230] Package label. REVLIMID (lenalidomide) capsules for oral use; February 2015 (36 pages).

TABLE 4.8 Starting Dose Adjustments for Patients With Renal Impairment

Category	Renal Function (Cockcroft–Gault)	Dose in MDS
Moderate renal impairment	CLcr 30–60 mL/min	5 mg, every 24 h
Severe renal impairment	CLcr < 30 mL/min (not requiring dialysis)	2.5 mg, every 24 h
End-stage renal impairment	CLcr < 30 mL/min (requiring dialysis)	2.5 mg, once daily

CLcr, creatinine clearance.
From: Package label. REVLIMID (lenalidomide) capsules for oral use. February 2015 (36 pages).

The package label included a table that broke down the instructions in terms of moderate, severe, and end-stage renal impairment, which is reproduced in Table 4.8.

The Use in Specific Population section of the package label referred to doses for renal impaired and hepatic impaired patients:

> **USE IN SPECIFIC POPULATIONS** ... **Renal Impairment.** Since lenalidomide is primarily excreted unchanged by the kidney, adjustments to the starting dose of REVLIMID are recommended to provide appropriate drug exposure in patients with moderate (CLcr 30–60 mL/min) or severe renal impairment (CLcr < 30 mL/min) and in patients on dialysis ... **Hepatic Impairment.** No dedicated study has been conducted in patients with hepatic impairment. The elimination of unchanged lenalidomide is predominantly by the renal route.[231]

VI. CONCLUDING REMARKS

Dose modification instructions on the package label can encompass:

- Dose escalations, for example, in the form of a step-by-step titration
- Dose reductions
- Dose interruptions
- Permanent discontinuations
- Requirement for periodic monitoring of laboratory values or monitoring AEs.

FDA's ClinPharm Pharm reviews illustrate FDA's decision-making processes for recommending dose-escalation instructions for alirocumab, canagliflozin, ferric citrate, and sebelipase. The alirocumab instructions read, for example, "If the LDL-cholesterol response is inadequate, the dosage may be increased."[232,233]

FDA's recommended dose-reduction instructions are shown for lenalidomide and nivolumab, where the goal was to reduce risk for AEs. Dose-reduction instructions are also given for the preexisting conditions of renal impairment and hepatic impairment.

[231] Package label. REVLIMID (lenalidomide) capsules for oral use; February 2015 (36 pages).

[232] Package label. PRALUENT™ (alirocumab) injection, for subcutaneous use; July 2015 (16 pages).

[233] Alirocumab (hyperlipidemia) BLA 123-339. Page 10 of 91-page Clinical Pharmacology Review.

Renal impairment can reduce excretion of the drug, resulting in increases in exposure, with a consequent increased risk for AEs. Hepatic impairment can reduce catabolism of the drug by liver enzymes, increasing exposure, and increasing risk for AEs. FDA's reviews for bosutinib, canagliflozin, the E/C/F/TAF combination drug, lenalidomide, and nivolumab, all made observations on organ impairment.

This chapter reveals the situation, where, "Renal impairment can adversely affect some pathways of hepatic/gut drug metabolism."[234] This refers to the mechanism where, in chronic renal disease, uremic toxins accumulate in the blood, where the uremic toxins can block drug excretion by the liver into the bile. In this way, renal failure can impair the excretion of drug that is primarily eliminated by the liver (and not by the kidneys).

The phenomenon of TMDD can result in the need for dose modification instructions in the package label. The phenomenon of TMDD is illustrated in this chapter for:

- Alemtuzumab
- Obinutuzumab

For any recommendation of drug dose, including recommendations for dose escalation, dose reduction, and the specific route of administration (oral, subcutaneous, and intravenous), the medical writer should take into account the therapeutic window for the drug. Where the therapeutic window is narrow, more attention needs to be given to arriving at dosages on the package label.

[234] US Department of Health and Human Services; Food and Drug Administration; Center for Drug Evaluation and Research (CDER); Center for Biologics Evaluation and Research (CBER). Guidance for Industry. Pharmacokinetics in Patients with Impaired Renal Function—Study Design, Data Analysis, and Impact on Dosing and Labeling; 2010 (21 pages).

CHAPTER

5

Contraindications

I. INTRODUCTION

a. Code of Federal Regulations

The package label must include an Indications and Usage section (21 CFR §201.57(c)(2)). Section 201.57 requires:

> **Indications and Usage**. This section must state that the drug is indicated for the treatment, prevention, mitigation, cure, or diagnosis of a recognized disease or condition, or of a manifestation of a recognized disease or condition, or for the relief of symptoms associated with a recognized disease or condition.

The Contraindications section carves out exceptions from the Indications and Usage section and identifies subgroups of patients that must not receive the drug. These subgroups include, for example, patients with low neutrophil counts, pregnancy, or patients with renal impairment. Section 201.57(c)(5) requires:

> **Contraindications**. This section must describe any situations in which the drug should not be used because the risk of use ... clearly outweighs any possible therapeutic benefit. Those situations include use of the drug in patients who, because of their particular age, sex, concomitant therapy, disease state, or other condition, have a substantial risk of being harmed by the drug and for whom no potential benefit makes the risk acceptable. Known hazards and not theoretical possibilities must be listed ... [i]f no contra-indications are known, this section must state None.

In contrast, the Black Box Warning and the Warnings and Precautions section of the package label may refer to all patients or to one particular subgroup, but where drug administration is still permitted. For example, the Warnings and Precautions section may instruct physicians to monitor patients for drug toxicity in the event that the physician decides to administer the drug.

FDA's Drug Review Process and the Package Label
DOI: https://doi.org/10.1016/B978-0-12-814647-7.00005-1

b. Contraindications Using the Example of Drug–Drug Interactions

Where a package label discloses the potential for drug–drug interactions (DDIs), this interaction is almost always not desirable for the patient. Typically, DDIs are not desired, because the first drug stimulates the body to inactivate the second drug (thereby reducing its plasma concentration and reducing its efficacy). Another type of DDI is where the first drug inhibits the body's ability to inactivate the second drug (thereby increasing its plasma concentration and increasing risk for toxicity from that drug). Yet another type of DDI is where two drugs have the same type of toxicity and where coadministration of both drugs results in additive toxicity.

In comments on DDIs, Hatton et al.[1] provided this account on when the interaction should find a place in the Contraindications section, "Contraindicated DDIs are those for which no situations have been identified where the benefit of the combination outweighs the risk. Using this definition, there are no circumstances where an override is an acceptable action for contraindicated DDIs."

On the other hand, Hatton et al.[2] suggested that the term "contraindication" does not always have this absolute meaning and that more flexible meanings can be used, and if the more flexible meaning applies the information can be placed instead in the Warnings and Precautions section. The suggestion was that "most contraindicated drug pairs were **not absolute contraindications and could be downgraded** ... [f]or example, according to the sildenafil ... labeling, sildenafil is contraindicated in patients ... using organic nitrates ... [h]owever, evidence indicates that sildenafil and nitrates can be used intermittently with adequate separation of dosing ... nitroglycerine may be considered 24 hours after sildenafil dosing ... or with appropriate blood pressure monitoring."[3] Please note the situation of flexibility in the contraindication against a dangerous drug combination, which Hatton et al.[4] described as a dosing that is "used intermittently with adequate separation of dosing."

The Contraindications section for sildenafil (Viagra®) warned against coadministering sildenafil with organic nitrates, but it refrained from mentioning anything about dosing that is "intermittently with adequate separation of dosing":

> **CONTRAINDICATIONS**. Administration of VIAGRA to patients using nitric oxide donors, such as **organic nitrates or organic nitrites in any form**. VIAGRA was shown to potentiate the hypotensive effect of nitrates ... Known hypersensitivity to sildenafil or any component of tablet ... Administration with guanylate cyclase (GC) stimulators, such as riociguat.[5]

Another example of a contraindication against coadministering two drugs because of additive toxicity is provided by the package label for 4-hydroxybutyrate (Xyrem®), a

[1] Hatton RC, et al. Evaluation of contraindicated drug-drug interaction alerts in a hospital setting. Ann. Pharmacother. 2011;45:297–308.

[2] Hatton RC, et al. Evaluation of contraindicated drug-drug interaction alerts in a hospital setting. Ann. Pharmacother. 2011;45:297–308.

[3] Hatton RC, et al. Evaluation of contraindicated drug-drug interaction alerts in a hospital setting. Ann. Pharmacother. 2011;45:297–308.

[4] Hatton RC, et al. Evaluation of contraindicated drug-drug interaction alerts in a hospital setting. Ann. Pharmacother. 2011;45:297–308.

[5] Package label. VIAGRA® (sildenafil citrate) tablets, for oral use. September 2015 (24 pp.).

drug for narcolepsy. The Contraindications section instructed the physician not to coadminister 4-hydroxybutyrate with sedative hypnotics or with alcohol:

> **CONTRAINDICATIONS** ... In combination with sedative hypnotics or alcohol.[6,7]

The Warnings and Precautions section of the same package label stated that the combination of 4-hydroxybutryate (Xyrem®) and sedative hypnotics can increase risk for profound sedation, syncope, and death:

> **WARNINGS AND PRECAUTIONS**. The concurrent use of Xyrem with other CNS depressants, including but not limited to opioid analgesics, benzodiazepines, sedating antidepressants or antipsychotics ... may increase the risk of ... profound sedation, syncope, and death.[8]

II. FDA'S DECISION-MAKING PROCESS FOR ARRIVING AT RECOMMENDATION FOR THE CONTRAINDICATIONS SECTION

a. Introduction

For each drug described below, each paragraph is correlated with one or more sections of the package label. Thus, the reader might want to glance at the package label excerpt prior to reading the narratives corresponding to each of the drugs. This discloses FDA's observations and analysis for arriving at recommendations for the package label, for the following drugs:

- Cabazitaxel
- Canagliflozin
- Dolutegravir
- Lomitapide
- Macitentan
- Pegloticase
- Tapentadol
- Topiramate
- Vorapaxar
- 4-Hydroxybutyrate

b. Subgroup Analysis

Subgroup analysis is a tool for drafting the Contraindications section. This analysis involves contemplating which subgroups in the patient population are expected to be at unacceptable risk for AEs, and then identifying the AEs relevant to each subgroup. After

[6] Package label. XYREM® (sodium oxybate) oral solution, CIII. January 2017 (25 pp.).

[7] 4-Hydroxybutyrate (cataplexy in narcolepsy; excessive daytime sleepiness in narcolepsy) NDA 021-196.

[8] Package label. XYREM® (sodium oxybate) oral solution, CIII. January 2017 (25 pp.).

establishing the array of AEs related to the study drug, FDA breaks the study subjects into subgroups to determine which subgroup should be excluded from receiving the marketed drug, and thus be identified in the Contraindications section. Subgroups include age group, pregnancy, degree of renal impairment, degree of hepatic impairment, amount of study drug dose (e.g., 100, 200, 500 mg), known hypersensitivity to the study drug, and concomitant drugs.

For **lomitapide**, the subgroups on the Contraindications section were patients with pregnancy, patients taking CYP3A4 inhibiting drugs, and patients with hepatic impairment.[9]

For **macitentan**, the only subgroup of interest was pregnancy, where FDA's recommendation for the Contraindications section was based on drug class analysis and on animal toxicity data.[10,11] In addition to the contraindication of pregnancy, the label had a Black Box Warning against administering the drug during pregnancy. This is an example of a contraindication communicated via the Black Box Warning.

For **tapentadol**, the list of AEs was derived from drug class analysis, from the Sponsor's own clinical studies, and from animal toxicity data. After assembling the list of AEs, the Sponsor subjected these AEs to subgroup analysis, where the subgroups were concomitant disorders (respiratory depression, asthma, paralytic ileus), hypersensitivity to the study drug, and concurrent drugs (monoamine oxidase (MAO) inhibitors).[12,13]

For **topiramate**, an AE of concern was central nervous system (CNS) depression. CNS depression was a concern only for one particular subgroup of patients, namely, patients taking concomitant alcohol. The contraindication on the package label was directed against patients taking alcohol.[14]

For **vorapaxar**, the AE of concern was intracranial hemorrhage. This AE was derived from drug class analysis (platelet inhibitor drugs). The Sponsor subjected this adverse event to subgroup analysis, where the result was that the Contraindications section instructed physicians not to give vorapaxar to patients with a history of stroke or with a history of prior transischemic attack.[15]

c. Cabazitaxel (Hormone-Refractory Prostate Cancer) NDA 201-023

This provides FDA's pathways of reasoning for arriving at the Contraindications section for cabazitaxel (Jevtana®). The AE of most concern was neutropenia, and this AE found a place, not only in the Contraindications section, but also in other sections, as shown here.[16] In the package label sections indicated by the bulletpoints, the only

[9] Package label. JUXTAPID™ (lomitapide) capsules, for oral use. December 2012 (30 pp.).

[10] Macitentan (pulmonary arterial hypertension) NDA 204-410. Page 63 of 163-page Medical Review.

[11] Package label. OPSUMIT® (macitentan) tablets, for oral use. October 2013 (16 pp.).

[12] Tapentadol (severe pain, neuropathic pain) NDA 200-533. Page 48 of 236-page Medical Review.

[13] Package label. NUCYNTA® ER (tapentadol) extended-release tablets for oral use. December 2016 (31 pp.).

[14] Package label. TROKENDI XR (topiramate) extended-release capsules for oral use. April 2017 (57 pp.).

[15] Package label. ZONTINVITY™ (vorapaxar) Tablets 2.08 mg, for oral use. May2014 (18 pp.).

[16] Package label. JEVTAN (cabazitaxel) injection, for intravenous use. September 2016 (24 pp.).

writing taking the form of a "contraindication" was that in the Contraindications section. In contrast, the writing in the other sections took the form of warnings, monitoring instructions, or information not requiring any particular action by the physician.

- **Black Box Warning.** "Neutropenic deaths have been reported."
- **Contraindications**. "Neutrophil counts of $\leq 1{,}500/mm^3$."
- **Warnings and Precautions.** "Bone marrow suppression (particularly neutropenia) and its clinical consequences."
- **Dosage and Administration.** "Delay treatment until neutrophil count is $>1{,}500$ cells/mm^3, then reduce dosage of JEVTANA to 20 mg/m^2." The unit "mg/m^2" refers to body surface area.
- **Adverse Reactions.** "Most common all grades adverse reactions ($\geq 10\%$) are neutropenia, anemia, leukopenia, thrombocytopenia, diarrhea ..."
- **Use in Specific Populations.** "Geriatric use ... The incidence of neutropenia, fatigue, asthenia, pyrexia, dizziness, urinary tract infection and dehydration occurred at rates $\geq 5\%$ higher in patients who were 65 years of age or greater compared to younger patients."

An early publication on cabazitaxel (formerly known as XRP6258) demonstrates that neutropenia was the AE of greatest concern. This publication compared cabazitaxel with other taxane drugs, "Neutropenia was the principal dose limiting toxicity ... observed in this study. Although neutropenia was common at the higher dose levels, the duration of severe neutropenia was typically brief and rarely associated with fever. At the recommended dose level, 20 mg/m^2, grade 4 neutropenia was observed in only 4% of courses, which compares favorably to other taxanes."[17]

The Clinical Study Protocol can provide definitions for any AE and instructions for detecting any given AE. Also, in the Clinical Study Protocol, any AE can be one of the inclusion criteria or exclusion criteria. What this means, is that the criteria used for recruiting study subjects into the clinical study can require that potential subjects have already experienced a particular adverse event. Also, what this means is that the critiria used for recruting study subjects can require that potential subject have never experienced a particular adverse event. To provide background information, AEs can include conditions such as a rash, vomiting, fainting, as well as abnormal values in laboratory chemistry, such as low neutrophil counts. Of use for drafting the package label is that the exact wording in the definition (or in the inclusion/exclusion criteria) can find a place in the Contraindications section. This was the case for cabazitaxel, where one of the exclusion criteria in the Clinical Study Protocol required:

"Neutrophils $\leq 1.5 \times 10^9/L$"

and where the instructions on the Contraindications section contained exactly the same value for the neutrophil count (but using different units):

"Neutrophil counts of $\leq 1{,}500/mm^3$"

[17] Mita AC, et al. Phase I and pharmacokinetic study of XRP6258 (RPR 116258A), a novel taxane, administered as a 1-hour infusion every 3 weeks in patients with advanced solid tumors. Clin. Cancer Res. 2009;15:723–30.

As a general proposition, Clinical Study Protocols include definitions for adverse events, such as neutropenia. Where a clinical study is published in *New England Journal of Medicine*, the journal also publishes a supplement that includes the Clinical Study Protocol. A publication from this journal on abiraterone for prostate cancer[18] provides the Clinical Study Protocol, where the protocol sets forth a definition for neutropenia:

> **Neutropaenia and its complications. Adverse event.** Grade 4 neutropenia for 7 days or more. Grade 3–4 neutropenia with oral fever >38.5°C. Infection (i.e. documented infection with grade 3–4 neutropenia).[19]

The implication of this definition is as follows. This provides a generic account of the flow of information for clinical studies for any kind of drug: Step one: For every AE that matched this definition, the same AE is captured on the Case Report Form during the course of the clinical study. Step two: With analysis of the Case Report Forms, the eventual result is that the Sponsor is able to identify neutropenia as an AE of especial concern (a "safety signal"). Step three: The end result is that neutropenia is used in drafting one of more sections in the package label.

1. *Inclusion and Exclusion Criteria in Clinical Study Protocol*

FDA's Medical Review for cabazitaxel reiterated the inclusion and exclusion criteria from the Clinical Study Protocol for cabazitaxel. The goal of these particular inclusion criteria was to ensure that the enrolled subjects all had the disease to be treated by the study drug and that the subjects were healthy enough to survive at least 2 months of treatment.[20] The **inclusion criteria** recited[21]:

- Diagnosis of histologically or cytologically proven prostate adenocarcinoma that was refractory to hormone therapy and previously treated with a Taxotere (or docetaxel)-containing regimen. Patients had documented progression of disease during or within 6 months after prior hormone therapy and disease progression during or after Taxotere (or docetaxel)-containing therapy.
- Life expectancy >2 months.
- Eastern Cooperative Oncology Group (ECOG) performance status 0 to 2 (i.e., patient was to be ambulatory, capable of all self-care, and up and about more than 50% of waking hours).

[18] James ND, et al. Abiraterone for prostate cancer not previously treated with hormone therapy. New Engl. J. Med. 2017; 377:338–51. DOI: 10.1056/NEJMoa1702900.

[19] Page 78 of 110-page Clinical Study Protocol STAMPEDE Systemic Therapy in Advancing or Metastatic Prostate Cancer: Evaluation of Drug Efficacy. MRC PR08. ISRCTN number: ISRCTN78818544. EUDRACT number: 2004-000193-31. A 5-stage multi-arm randomised controlled trial. Ver. 1.0, May 2004. The protocol was published as a supplement to James ND. et al. Abiraterone for prostate cancer not previously treated with hormone therapy. New Engl. J. Med. 2017;377:338–51. DOI: 10.1056/NEJMoa1702900.

[20] Information provided by Israel Rios, MD, Chief Medical Officer, SciClone Pharmaceuticals, Foster City, CA. Personal communication in March 2008.

[21] Cabazitaxel (hormone-refractory prostate cancer) NDA 201-023. Pages 26–27 of 90-page Medical Review.

A goal of the **exclusion criteria** was to ensure that the subjects enrolling in the clinical study were healthy enough to reduce risk for drug-induced AEs. The exclusion criteria read[22]:

- **Neutrophils $\leq 1.5 \times 10^9$/L**
- Hemoglobin ≤ 10 g/dL
- Platelets $\leq 100 \times 10^9$/L
- Total bilirubin $\geq$ upper limit of normal (ULN)
- Aspartate aminotransferase/serum glutamic oxaloacetic transaminase (AST/SGOT) $\geq 1.5 \times$ ULN
- Alanine aminotransferase/serum glutamate pyruvate transaminase (ALT/SGPT) $\geq 1.5 \times$ ULN
- Creatinine $\geq 1.5 \times$ ULN

2. *FDA's Medical Review Expressed Concern for Neutropenia and Where Low Neutrophil Counts Permitted Fatal Infections*

FDA observed that the cabazitaxel-treated study subjects experienced death due to infections occurring during neutropenia. FDA's comments rationalized approval of the drug, despite the deaths due to infection, based on the fact that neutropenia in patients could likely be prevented by treating with GCSF (a growth factor) and because of the lack of any other effective drugs.

FDA's articulation of the AE of neutropenia was "there were deaths due to toxicity on the cabazitaxel arm ... some of the **deaths were due to infectious complications during a period of neutropenia**, infection-related deaths may be better prevented in the postmarketing setting with the use of prophylactic GCSF in patients at high risk of neutropenic complications. The proposed patient population currently has no treatment options which offer a survival benefit, and the robust results in overall survival demonstrated by cabazitaxel would provide a new treatment option for these patients."[23]

FDA focused further on neutropenia and observed that this AE was responsible for dose modifications occurring during the course of the clinical study. Dose modifications took the form of dose delay, dose interruptions, and dose reductions. FDA's Medical Review observed, "Grade 3–4 neutropenia accounted for the majority of dose modifications on both arms, but was a more frequent cause on the cabazitaxel arm. Diarrhea, dehydration, abdominal pain, nausea, dysuria, flushing, and hypersensitivity each accounted for dose modification in at least three patients on the cabazitaxel arm but none on the comparator arm."[24]

Converging more on the issue of neutropenia, FDA's attention turned to the fact that neutropenia was such a serious AE that it led to discontinuations (permanently stopping the study drug). FDA's concern is shown by the data in Table 5.1 and by FDA's comments about discontinuations, "Twenty-two (6%) patients discontinued JEVTANA treatment due

[22] Cabazitaxel (hormone-refractory prostate cancer) NDA 201-023. Pages 26–7 of 90-page Medical Review.

[23] Cabazitaxel (hormone-refractory prostate cancer) NDA 201-023. Page 9 of 90-page Medical Review.

[24] Cabazitaxel (hormone-refractory prostate cancer) NDA 201-023. Pages 49 and 51 of 90-page Medical Review.

TABLE 5.1 Discontinuations due to Adverse Events

	Cabazitaxel (Study Drug Arm) $N = 371$ Subjects	Mitoxantrone (Control Treatment Arm) $N = 371$ Subjects
Neutropenia	**9%**	**0%**
Renal failure	7%	0%
Infection	6%	1%
Hematuria	5%	1%
Sepsis	5%	1%
Fatigue	4%	1%
Diarrhea	4%	1%
Abdominal pain	3%	0%
Febrile neutropenia	3%	0%

Cabazitaxel (hormone-refractory prostate cancer) NDA 201-023. Page 60 of 90-page Medical Review.

to neutropenia, febrile neutropenia, sepsis, or infection ... **[n]eutropenia led to more treatment discontinuations** than any other adverse event on the cabazitaxel arm. Note that no discontinuations due to neutropenia occurred on the comparator arm."[25]

3. *FDA's Recommendations for the Package Label*

FDA recommended that neutropenia be included in the Contraindications section of the package label, as well as in the Black Box Warning and in the Warnings and Precautions section. One might ask why the Black Box Warning was not sufficient and why a Contraindication was also needed. The answer is that the Code of Federal Regulations requires that if there is a contradiction, then it must be disclosed in the Contraindications section.

FDA's Cross Discipline Team Leader Review dwelled extensively on the Contraindications section, the AE of neutropenia, and the need to instruct physicians to give GCSF (a growth factor) to patients at risk for neutropenia:

> Much labeling discussion focused on ... content of the Boxed Warning, **Contraindications**, Warnings and Precautions sections. Emphasis was on neutropenia, febrile neutropenia, infection, diarrhea, hypersensitivity reactions and renal failure. Neutropenia and related complications were the main safety issue in the randomized controlled trial ... [m]ost neutropenic deaths occurred on the first cycle. Accordingly the **FDA review team revised the package insert** to indicate that, "Primary prophylaxis with G-CSF should be considered in patients with high-risk clinical features (age $>$ 65 years, poor performance status, previous episodes of febrile neutropenia, extensive prior radiation ports, poor nutritional status, or other serious comorbidities) that predispose them to increased complications from prolonged neutropenia." ... **[t]he package insert already indicated that "JEVTANA should not be administered to patients with neutrophils $\leq$ 1,500/mm^3."**[26]

[25] Cabazitaxel (hormone-refractory prostate cancer) NDA 201-023. Page 60 of 90-page Medical Review.

[26] Cabazitaxel (hormone-refractory prostate cancer) NDA 201-023. Page 35 of 43-page Cross Discipline Team Leader Review. See also, page 14 of 16-page Summary Review.

Package label. FDA's recommendations regarding neutropenia found a place on the Contraindications section for cabazitaxel (Jevtana®), the Black Box Warning, and the Warnings and Precautions section:

CONTRAINDICATIONS. Neutrophil counts of $\leq$1,500/mm^{3}.[27]

The Black Box Warning also included a contraindication against giving the drug to patients with neutropenia:

BOXED WARNING. NEUTROPENIA ... Neutropenic deaths have been reported. Obtain frequent blood counts to monitor for neutropenia. Do not give JEVTANA if neutrophil counts are $\leq$1,500 cells/mm^{3}.[28]

The Warnings and Precautions section mentioned the need for treating with the growth factor, GCSF:

WARNINGS AND PRECAUTIONS ... Bone marrow suppression (particularly neutropenia) and its clinical consequences (febrile neutropenia, neutropenic infections): Monitor blood counts frequently to determine if dosage modification or initiation of GCSF is needed.[29]

d. Canagliflozin (Type-2 Diabetes Mellitus) NDA 204-042

Canagliflozin inhibits the glucose transporter (SGLT2) in the renal tubule, thus preventing glucose resorption, with the desired consequence of reducing blood glucose levels. To review renal physiology, with normal operation of the circulatory system, a small proportion of blood during each pass through the kidneys is filtered through the glomerulus, resulting in newly formed urine. In the renal tubule, many nutrients in the newly forming urine are transported back into the circulation, thus minimizing loss of nutrients in the urine. Glucose reabsorption takes place in the proximal kidney tubules, where this resorption is mediated by SGLT2 (responsible for 90% of reabsorption) and SGLT1 (responsible for 10% of resorption).[30]

FDA's Medical Review commented on AEs found during the Sponsor's clinical trial, where FDA scrutinized renal AEs and hypersensitivity AEs.

1. Renal Adverse Events

FDA's Medical Review revealed safety data from study subjects with renal impairment, that is, where the subjects had renal impairment in addition to diabetes. FDA's comments referred to these subgroups in the population of study subjects:

- Elderly
- Concomitant drugs (loop diuretics)
- Subjects with renal impairment (low baseline eGFR)

[27] Package label. JEVTANA® (cabazitaxel) injection, for intravenous use. May 2017 (22 pp.).

[28] Package label. JEVTANA® (cabazitaxel) injection, for intravenous use. May 2017 (22 pp.).

[29] Package label. JEVTANA® (cabazitaxel) injection, for intravenous use. May 2017 (22 pp.).

[30] Seufert J. SGLT2 inhibitors – an insulin-independent therapeutic approach for treatment of type 2 diabetes: focus on canagliflozin. Diabetes Metab. Syndr. Obes. 2015;8:543–54.

FDA's Medical Review referred to these subgroups and to the AEs associated with these subgroups, writing, "Because canagliflozin increases urinary glucose excretion, it acts as an osmotic diuretic with an increase in urine output, and there was an increased incidence of adverse events related to intravascular volume depletion with canagliflozin (e.g., postural dizziness and hypotension) ... canagliflozin was also associated with ... renal failure. The incidence of volume depletion and renal-related events were dose-dependent and notably increased in subjects with moderate renal impairment. Subgroup analyses also showed that **elderly** (≥75 years of age) and **concomitant use of loop diuretics** were at an increased, dose-related risk for **volume depletion events**. I do not recommend the higher dose (300 mg) of canagliflozin in **subjects with low baseline eGFR**."[31]

Speaking on the need for package label warnings in patients with renal impairment, an FDA reviewer wrote, "I recommend approval of canagliflozin for the proposed indication ... [h]owever, given the available information related to efficacy and safety submitted in this NDA, **I do not recommend: Canagliflozin for use in patients with moderate renal impairment** who are Chronic Kidney Disease (CKD) Stage 3b and below (e.g., eGFR $<45\ mL/min/1.73\ m^2$)."[32]

Continuing with the recommendation for the package label, FDA was careful to distinguish the subgroups of patients receiving high dose (300 mg) versus low dose (100 mg) canagliflozin. The FDA reviewer recommended:

> Since **I will not be recommending** the use of 300 mg dose of canagliflozin in patients with **moderate renal impairment and elderly** where the incidences of ... renal-related events were notably elevated, I believe prescribers can monitor for and manage ... renal-related changes ... in patients with normal to mild renal impairment[33] ... [a]lso, subjects with **moderate renal impairment had a higher risk for renal-related adverse events (i.e., acute renal failure)**: 8.9% and 9.3% of canagliflozin 100 mg and 300 mg groups compared to 3.7% of placebo group had renal-related adverse events in the Moderate Renal Impairment Dataset ... [t]his data suggest that the adverse consequences of renal-related changes were elevated with both doses of canagliflozin in subjects with moderate renal impairment.[34]

2. *Hypersensitivity*

FDA's observations compelled the decision to include the adverse event of hypersensitivity in the Contraindications section. The relevant and compelling facts are shown by the bullet points:

- The fact that hypersensitivity reactions were serious enough to provoke study subjects to discontinue or to drop out
- The fact that hypersensitivity reactions required medical treatment for one of the study subjects
- The fact that hypersensitivity reactions occurred to a greater extent in the treatment arm than in the placebo arm

[31] Canagliflozin (type-2 diabetes mellitus) NDA 204-042. Page 15 of 255-page Medical Review.

[32] Canagliflozin (type-2 diabetes mellitus) NDA 204-042. Page 12 of 255-page Medical Review.

[33] Canagliflozin (type-2 diabetes mellitus) NDA 204-042. Page 99 of 255-page Medical Review.

[34] Canagliflozin (type-2 diabetes mellitus) NDA 204-042. Page 17 of 255-page Medical Review.

- The fact that there was a temporal association with dosing and the hypersensitivity reactions

Regarding hypersensitivity, FDA made these observations and recommended that the AEs can be mitigated by appropriate labeling:

> There was an increased incidence in serious adverse skin reactions and in discontinuations/dropouts due to skin events (rash, pruritus, urticaria) indicative of **hypersensitivity reactions** with canagliflozin compared to placebo. This included a case of angioedema of upper lip with canagliflozin after 22 days of canagliflozin treatment, which **required treatment** ... and the review of cases that led to discontinuation of treatment showed positive temporal relationship suggesting drug causality; in one case, there was a positive dechallenge and rechallenge[35] ... skin and hypersensitivity reactions ... were not dose-dependent and can be **mitigated by appropriate labeling.**[36]

Package label. FDA's recommendations for canagliflozin (Invokana®) found a place in the Contraindications section and the Dosage and Administration section. The Contraindications section instructed the physician not to administer the drug to patients with serious hypersensitivity reaction to the drug or to patients with severe renal impairment:

> **CONTRAINDICATIONS**. History of serious hypersensitivity reaction to INVOKANA. Severe renal impairment, end stage renal disease (ESRD), or on dialysis.[37]

The Dosage and Administration section concerned dose escalation, and it warned about renal impairment. This section included information taking the form of a contraindication (do not initiate INVOKANA if eGFR is below 45 mL/min/1.73 m^2):

> **DOSAGE AND ADMINISTRATION**. The recommended starting dose is 100 mg once daily, taken before the first meal of the day ... Dose can be increased to 300 mg once daily in patients tolerating INVOKANA 100 mg once daily who have an eGFR of 60 mL/min/1.73 m^2 or greater and require additional glycemic control ... INVOKANA is limited to 100 mg once daily in patients who have an eGFR of 45 to less than 60 mL/min/1.73 m^2 ... Assess renal function before initiating INVOKANA. **Do not initiate INVOKANA if eGFR is below 45 mL/min/1.73 m^2** ... Discontinue INVOKANA if eGFR falls below 45 mL/min/1.73 m^2.[38]

e. Dolutegravir (HIV-1 in Combination with Other HIV-1 Drugs) NDA 204-790

Dolutegravir (DTG) is a small molecule for treating HIV-1 viral infections. The package label contraindicates coadministration with dofetilide (anticardiac arrhythmia drug). Shown below is FDA's reasoning and logic for recommending this contraindication.

[35] Canagliflozin (type-2 diabetes mellitus) NDA 204-042. Page 16 of 255-page Medical Review.

[36] Canagliflozin (type-2 diabetes mellitus) NDA 204-042. Page 16 of 255-page Medical Review.

[37] Package label. INVOKANA (canagliflozin) tablets, for oral use. March 2013 (35 pp.).

[38] Package label. INVOKANA (canagliflozin) tablets, for oral use. March 2013 (35 pp.).

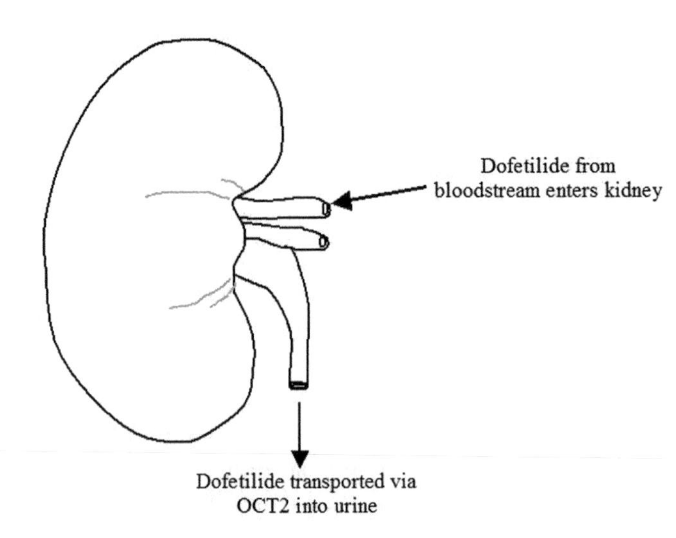

FIGURE 5.1 Clearance of dofetilide from the body mediated by OCT2. After transport via OCT2, which resides in the basolateral membrane of proximal tubule cells, and then transport across the apical membrane, drugs and metabolites enter the forming urine.

1. ***Coadministering Dolutegravir with Dofetilide***

This concerns the DDI between the DTG (anti-HIV-1) and dofetilide (anticardiac arrhythmia drug). FDA's analysis of DDIs was based on biochemical data showing that the study drug inhibited the drug transporter called OCT2. In humans, **OCT2** is expressed by renal tubule cells, where it resides in the basolateral membrane.[39] In contrast, **OCT1** is expressed in humans in the basolateral membrane of hepatocytes.

FDA referred to the fact that dofetilide has a narrow therapeutic window, thus increasing the need for vigilance regarding DDIs that can increase (or decrease) plasma dofetilide levels. FDA's recommendation for the label's contraindication took this form:

> Several drug–drug interaction studies were conducted to evaluate effects of dolutegravir (DTG) coadministration with other agents ... [i]n vitro, DTG inhibits OCT2 and therefore can ... increase **exposures** of drugs excreted by OCT2. Agents eliminated by this **renal cation transporter** include **dofetilide**, an antiarrhythmic agent used to treat atrial fibrillation ... [n]o DTG in vivo drug interaction studies were conducted ... [d]ofetilide has a narrow therapeutic window. Based on the DTG effects on OCT2, increases in dofetilide **exposures** and resultant toxicity including **life-threatening events can be expected**, hence coadministration of dofetilide with DTG is **contraindicated.**[40]

Figs. 5.1 and 5.2 show the sequence of events, starting with the administration of DTG and ending with excretion by the kidney into the urine. Fig. 5.1 shows this sequence where

[39] Higgins JW, et al. Ablation of both organic cation transporter (OCT)1 and OCT2 alters metformin pharmacokinetics but has no effect on tissue drug exposure and pharmacodynamics. Drug Metab. Dispos. 2012;40:1170–7.

[40] Dolutegravir (HIV-1 in combination with other HIV-1 drugs) NDA 204-790. Page 169 of 198-page Medical Review.

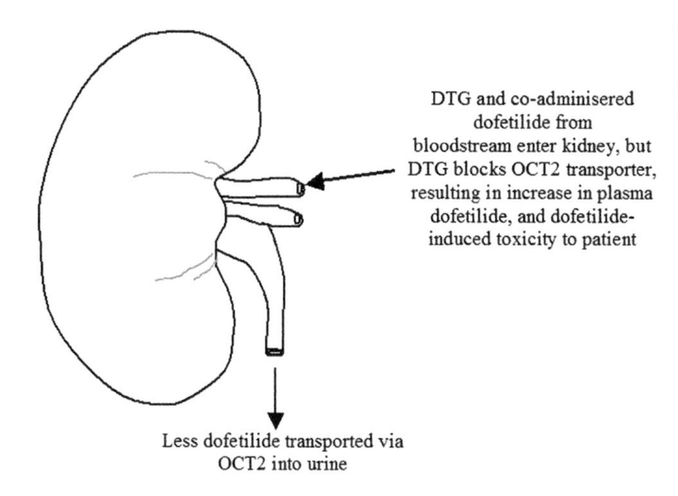

FIGURE 5.2 Inhibition by dolutegravir (DTG), an anti-HIV-1 drug, of OCT2, resulting in less excretion of dofetilide (anticardiac arrhythmia drug) into the urine.

dofetilide (cardiac drug) is administered alone, while Fig. 5.2 shows this sequence where dofetilide (cardiac drug) is coadministered with DTG (anti-HIV-1 drug). OCT2 is a transporter located in the membrane of renal tubule cell membrane. OCT2 mediates dofetilide excretion into the urine. OCT2 transporter is inhibited by DTG.

2. *Hypersensitivity to Dolutegravir*

The package label's instruction not to give to patients previously experiencing hypersensitivity originated from the Sponsor's own clinical study. The terms "challenge" and "rechallenge" conventionally refer to administration of an infective agent or some other antigen, followed some time later by a second administration of the same agent. FDA described the hypersensitivity reactions from the Sponsor's clinical studies:

> Subject 6929 ... self-stopped medication and self re-challenged resulting in similar, but **more severe hypersensitivity** ... [b]ased on these hypersensitivity events we agree with the Applicant that data support DTG **product labeling** for drug hypersensitivity, characterized by rash, constitutional findings and sometimes organ dysfunction ... **language indicating DTG should not be used in patients who have experienced a previous hypersensitivity reaction to DTG** is warranted based on subject 6929 who worsened after re-challenge ... in light of the **life-threatening** potential of **hypersensitivity reactions**, the clinical review team agrees with the Applicant's proposal for ... language indicating that DTG should not be used in patients who have experienced a previous hypersensitivity reaction to DTG is acceptable.[41]

[41] Dolutegravir (HIV-1 in combination with other HIV-1 drugs) NDA 204-790. Pages 92, 95, and 96 of 198-page Medical Review.

Package label. FDA's recommendations found a place in the Contraindications section of the DTG (Tivicay®) package label. There were two contraindications, one relating to history of hypersensitivity to the study drug (DTG) and the other relating to coadministration of a second drug (dofetilide):

> **CONTRAINDICATIONS**. Previous hypersensitivity reaction to dolutegravir ... Co-administration with dofetilide.[42]

f. Lomitapide (Homozygous Familial Hypercholesterolemia) NDA 203-858

Lomitapide is a small molecule for treating familial hypercholesterolemia. This disease involves very high serum cholesterol, present from birth, due to mutations affecting the activity of low-density lipoprotein receptor.[43]

This introduces the topic of steatohepatitis, which was a safety concern expressed in FDA's Medical Review for lomitapide. Steatohepatitis is characterized by liver biopsy findings that are indistinguishable to those seen in alcoholic hepatitis. About 50% of nonalcoholic steatohepatitis patients eventually develop liver fibrosis, 15% develop cirrhosis, and 3% may progress further to terminal liver failure.[44]

1. *Hepatic Impairment*

FDA asked if elevated liver transaminases, as appearing in the serum, were (or were not) a sign of steatohepatitis caused by lomitapide. In view of this ambiguity, an FDA reviewer commented, "I have the greatest concern for the potential risk for hepatotoxicity. Although it is straightforward to measure transaminases and other liver-related laboratories, it is currently not known what these elevated levels ... are signaling. **Are these patients developing steatohepatitis?** If so, what degree of transaminase elevation should raise alarm? ... Or, is the hepatic steatosis caused by lomitapide relatively benign? We do not have answers to these questions, especially given the absence of liver histology from lomitapide treated patients."[45]

FDA expanded on the hepatic AEs for lomitapide and observed that these hepatic AEs were due to a mechanism of action unique to this drug, "Lomitapide induces hepatic steatosis, by limiting the liver's ability to export triglycerides via VLDL particles; the **long-term consequences** of the accumulation of hepatic fat are unknown. Whether lomitapide-induced hepatic steatosis will behave similarly to simple steatosis with a

[42] Package label. TIVICAY® (dolutegravier) tablets, for oral use. March 2017 (36 pp.).

[43] France M, et al. HEART UK statement on the management of homozygous familial hypercholesterolaemia in the United Kingdom. Atherosclerosis. 2016;255:128–39.

[44] Amacher DE, Chalasani N. Drug-induced hepatic steatosis. Semin. Liver Dis. 2014;34:205–14.

[45] Lomitapide (homozygous familial hypercholesterolemia) NDA 203-858. Page 250 of 304-page Medical Review.

benign clinical course, or whether it will induce inflammation (steatohepatitis) and **progress to cirrhosis over many years**, is unknown."[46]

FDA's comments about lomitapide's potential to "induce inflammation ... and progress to cirrhosis over many years" found a place in the Contraindications section of the package label, which contraindicated the drug in patients with unexplained persistent abnormal liver function tests.[47]

2. *Pregnancy*

The Sponsor conducted toxicity studies to explore toxicities on the embryo and fetus of rats, ferrets, and rabbits. The Sponsor was careful to use a dose to animals that was equivalent to the human dose, as shown by the comment that included the phrase, "based on 60 mg dose." The Sponsor's findings on teratogenicity were as follows: "Nonclinical data also suggest the **lomitapide is a teratogen**. Lomitapide was associated with significant fetal malformations in rats and ferrets at less than clinical exposure (based on 60 mg dose) when administered during organogenesis. In rabbits, adverse effects were not observed at $3\times$ clinical dose, but embryo-fetal lethality was observed $\geq 6\times$ clinical dose. There have been no reports of pregnancy among women taking lomitapide."[48]

FDA recommended that the label needed a warning about teratogenicity, "the teratogenic potential of lomitapide should be able to be mitigated through labeling and a medication guide."[49] This recommendation found a place in the Contraindications section of the package label.

3. *Concomitant CYP3A4 Inhibitors*

The Sponsor conducted PK studies with human subjects, where the subjects were given lomitapide with a standard CYP3A4 inhibitor. FDA's Medical Review described this study as follows. As stated earlier, the term "exposure" is conventionally used to mean blood plasma concentration, where concentration is expressed in terms of parameters such as AUC, Cmax, and tmax. FDA described DDIs occurring at the point of CYP3A4, referring to a "study to evaluate the effects of the strong CYP3A4 inhibitor ketoconazole on the single-dose PK of lomitapide ... in 30 healthy volunteers ... Ketoconazole markedly increased lomitapide exposure. The ... Cmax and AUCinf of lomitapide increased approximately 15-fold and 27-fold, respectively, with ketoconazole co-administration."[50]

The DDI involved lomitapide coadministered with a CYP3A4 inhibitor. This DDI found a place in the Contraindications section, which instructed the physician to not give lomitapide in patients taking CYP3A4 inhibitor.

[46] Lomitapide (homozygous familial hypercholesterolemia) NDA 203-858. Page 20 of 304-page Medical Review.

[47] Package label. JUXTAPID™ (lomitapide) capsules, for oral use. December 2012 (30 pp.).

[48] Lomitapide (homozygous familial hypercholesterolemia) NDA 203-858. Pages 18 and 40 of 304-page Medical Review.

[49] Lomitapide (homozygous familial hypercholesterolemia) NDA 203-858. Page 19 of 304-page Medical Review.

[50] Lomitapide (homozygous familial hypercholesterolemia) NDA 203-858. Page 31 of 304-page Medical Review.

FDA made recommendations for the Contraindications section, "I suggest revising the contraindications to include: Pregnancy, Concomitant administration with moderate or strong CYP3A inhibitors, and Moderate or severe hepatic impairment (i.e., Child–Pugh B and C) and patients with active liver disease."[51] All of these recommendations found a place on the label, as shown below.

Package label. FDA's recommendations for lomitapide (Juxtapid®) found a place in the Contraindications section, which referred to: (1) Pregnancy, (2) Concomitant cytochrome P450 inhibitors, and (3) Hepatic impairment:

> **CONTRAINDICATIONS.** Pregnancy ... Concomitant use with strong or moderate CYP3A4 inhibitors ... Moderate or severe hepatic impairment or active liver disease including unexplained persistent abnormal liver function tests.[52]

g. Macitentan (Pulmonary Arterial Hypertension) NDA 204-410

Macitentan is a small molecule drug used for treating hypertension. Macitentan inhibits endothelin type-1 (ET-1) receptor. ET-1 (the ligand) binds to ET-1 receptor (the receptor).

Patients with hypertension have enhanced vascular expression of ET-1. Antagonists of endothelin receptors lower blood pressure in hypertensive patients.[53] The endothelins are oligopeptides that are 21 amino acids long. There are three types of endothelins (ET-1, ET-2, and ET-3), and these transmit signals that cause vasoconstriction. Vasoconstriction, which is constriction of smooth muscles in blood vessels, increases the blood pressure. In health, ET-1 maintains vascular tone and favors vasodilation. But in kidney disease, renal vasoconstriction caused by ET-1 to endothelin receptor-mediated signaling leads to salt and water retention with consequent hypertension.[54]

FDA's Medical Review listed the AEs of greatest concern. The AEs of fetal toxicity and decreased sperm count were characterized as drug class effects. The FDA reviewer made this list of AEs[55]:

Primary safety concerns include

1. Fetal toxicity (class effect)
2. Decrease in hemoglobin/anemia
3. Pulmonary veno-occlusive disease with pulmonary edema (class effect)
4. Liver function abnormalities
5. Decreased sperm count (class effect)

[51] Lomitapide (homozygous familial hypercholesterolemia) NDA 203-858. Page 357 of 304-page Medical Review.

[52] Package label. JUXTAPID™ (lomitapide) capsules, for oral use. December 2012 (30 pp.).

[53] Schiffrin EL. Vascular endothelin in hypertension. Vascul. Pharmacol. 2005;43:19–29.

[54] Dhaun N, et al. Role of endothelin-1 in clinical hypertension: 20 years on. Hypertension. 2008;52:452–9.

[55] Macitentan (pulmonary arterial hypertension) NDA 204-410. Page 63 of 163-page Medical Review.

As shown in FDA's list, the FDA reviewer was aware of the issue of fetal toxicity because other drugs of the same class posed a risk for fetal toxicity. In general, predicting AEs of any study drug can be based entirely on known AEs for other drugs of the same drug class (drug class effect), on the Sponsor's own clinical studies, on the Sponsor's animal toxicity studies, or on any combination of the above.

Macitentan belongs to the same drug class as ambrisentan and bosentan, where the drug class was defined by the drug's mechanism of action. This shared mechanism of action was inhibiting endothelin-1 receptor.[56]

The package label included contraindications against use in pregnancy. But this contraindication was not solely the result of drug class analysis. The Sponsor's animal toxicity studies also revealed fetal toxicity. FDA's Summary Review referred to these animal studies, "there was an increased incidence of early intrauterine deaths and post-implantation loss on dames mated to exposed male rats. Fetal developmental effects were evaluated in rats and rabbits. Serious malformations were observed at all doses tested and comprised craniofacial abnormalities and cardiovascular abnormalities."[57]

FDA's recommended contraindication for pregnancy was based, at least in part, on drug class analysis. FDA's drug class analysis is revealed by the fact that FDA mentioned two other drugs in the same drug class (ambrisentan and bosentan). FDA's recommendation for the label's contraindication read:

> Reviewer comment: These risks will be described in labeling. **Macitentan should be contraindicated in pregnancy** (Boxed Warning) and should be approved with a Risk Evaluation and Mitigation Strategy (REMS) that links drug dispensing to mandatory monthly pregnancy testing and use of adequate contraception in females of reproductive potential—similar to the REMS programs for **ambrisentan** and **bosentan.**[58]

This author points out that FDA's prediction of AEs by drug class analysis went into greater depth than the form of drug class analysis used for most other study drugs, in that FDA's drug class analysis referred to the REMS strategies used with other drugs of the same class. The package labels for ambrisentan and bosentan each had CONTRAINDICATIONS against use in pregnancy, as shown below for ambrisentan (Letairis®) and bosentan (Tracleer®). The package label for ambrisentan read:

> **CONTRAINDICATIONS**. Pregnancy ... Idiopathic Pulmonary Fibrosis.[59]

The package label for bosentan read:

> **CONTRAINDICATIONS**. Pregnancy ... Use with Cyclosporine ... Use with Glyburide.[60]

[56] Angus JA, et al. Functional estimation of endothelin-1 receptor antagonism by bosentan, macitentan and ambrisentan in human pulmonary and radial arteries in vitro. Eur. J. Pharmacol. 2017;804:111–6. DOI: 10.1016.

[57] Macitentan (pulmonary arterial hypertension) NDA 204-410. Page 5 of 14 page Summary Review.

[58] Macitentan (pulmonary arterial hypertension) NDA 204-410. Page 10 of 14 page Summary Review.

[59] Package label. LETAIRIS (ambrisentan) tablets for oral use. October 2015 (25 pp.).

[60] Package label. TRACLEER® (bosentan) tablets, for oral use. October 2016 (21 pp.). NDA 021-290.

Package label. FDA's recommendations for macitentan (Opsumit®) found a place on the Black Box Warning and in the Contraindications section. The information in the Black Box Warning was not merely a warning, but also included a contraindication—the instruction never to give the drug to pregnant patients:

> **BOXED WARNING**: EMBRYO FETAL TOXICITY ... Do not administer OPSUMIT to pregnant female because it may cause fetal harm.[61]

> **CONTRAINDICATIONS**. Pregnancy. OPSUMIT may cause fetal harm when administered to a pregnant woman. OPSUMIT is contraindicated in females who are pregnant. OPSUMIT was consistently shown to have teratogenic effects when administered to animals. If OPSUMIT is used during pregnancy, apprise the patient of the potential hazard to a fetus.[62]

h. Pegloticase (Hyperuricemia and Gout) BLA 125-293

This illustrates the unique and rarely occurring situation where FDA assesses the indication of the study drug and engages in a dialogue with the Sponsor for changing the proposed indication. This situation is disclosed here in the Contraindications chapter, because the Indications and Usage section included a contraindication.

Pegloticase (Krystexxa™) is for treating gout. Pegloticase takes the form of an enzyme (urate oxidase) modified by covalent attachment of nine groups of methoxy polyethylene glycol (PEG).[63,64] For therapeutic proteins, it is conventional for the manufacturer to connect a PEG moiety in order to increase the lifetime of the therapeutic protein in the patient's bloodstream.

As shown by the Indications and Usage section of the package label, the drug had been FDA-approved for the indication of "chronic gout in adult patients **refractory to conventional therapy**." This instructs the physician to refrain from administering pegloticase unless the patient had failed to respond to conventional therapy. In other words, this instruction is a contraindication telling the physician not to administer the drug if the patient falls into a specific subgroup. Here, the contraindication was that pegloticase should not be administered to patients where conventional therapy had already been shown to be effective. The contraindication took this form:

> **INDICATIONS AND USAGE**. KRYSTEXXA™ (pegloticase) is a PEGylated uric acid specific enzyme indicated for the treatment of chronic gout in adult patients refractory to conventional therapy.[65]

[61] Package label. OPSUMIT® (macitentan) tablets, for oral use. October 2013 (16 pp.).

[62] Package label. OPSUMIT® (macitentan) tablets, for oral use. October 2013 (16 pp.).

[63] Pegloticase (hyperuricemia and gout) BLA 125-293. Page 4 of 28-page Chemistry Review.

[64] Lipsky PE, et al. Pegloticase immunogenicity: the relationship between efficacy and antibody development in patients treated for refractory chronic gout. Arthritis Res. Ther. 2014;16:R60. DOI: 10.1186/ar4497 (8 pp.).

[65] Package label. KRYSTEXXA™ (pegloticase) for injection, for intravenous infusion. September 2010 (14 pp.).

The commentary in the next few paragraphs provides the basis for the requirement in the Indications and Usage section, that the drug is to be used only in patients who failed to respond to conventional therapy. FDA expressed concern for cardiovascular AEs. FDA's Medical Review noted that the placebo treatment arm (43 subjects) did not experience any cardiovascular AEs while, in contrast, the pegloticase treatment arm (169 subjects) experienced cardiovascular AEs such as cardiovascular deaths, congestive heart failure, cardiac arrhythmias, and angina.[66]

FDA observed that many patients are refractory to treatment with drugs for lowering plasma uric acid and treating gout, writing, "A significant minority of these patients are **refractory** to, or intolerant of, these uric acid lowering drugs and are subject to prolonged episodes of recurrent arthritis and large tophi deposition."[67]

Regarding cardiovascular AEs and patients refractory to conventional therapy for gout, FDA's Medical Review stated that a meeting was convened "to discuss the risks and benefits associated with pegloticase treatment ... refractory to conventional therapy. The committee could not reach agreement as to whether there was a true **cardiovascular safety signal**, given the small number of cases and the multiple comorbidities of the subjects ... [w]hile the committee voted 14 to 1 in favor of approval, several members stated that it would be important to limit the use of pegloticase to patients who were truly **refractory to conventional treatment.**"[68]

To reiterate, the goal of this group of paragraphs is to demonstrate the origin of the requirement, in the Indications and Usage section, that the drug be given only to patients not responding to conventional therapy. Further commenting on cardiovascular AEs, an FDA reviewer stated, "I concur with the clinical review team that this is a reasonably sized database ... [t]here were three deaths in the pegloticase every two weeks group and one in the every four weeks group. Two of the deaths in the very two weeks group were due to **cardiac events** in subjects with prior cardiovascular histories."[69]

FDA continued to express concern for cardiovascular AEs, but added that the data were not conclusive, writing:

> Both Dr. Grant and the division clinical review team concluded that, while the numbers of events were too low to reach a definite conclusion regarding the cardiotoxicity of pegloticase, this signal does raise a concern that will require further evaluation[70] ... [a]n internal cardiology consultant, Dr. Stephen Grant ... reviewed the individual cases ... [t]his analysis again showed events in several different categories with no clear pattern to suggest a particular mechanism. Overall, the events fit into categories of ischemic cardiovascular disease, heart failure, and cardiac arrhythmias ... [t]he Applicant conducted their own analysis of the cardiovascular serious adverse events using a blinded adjudication committee ... [t]heir results are similar to those of Dr. Grant ... [h]owever, there remains a degree of uncertainty about cardiac safety of pegloticase because so few events were observed.[71]

[66] Pegloticase (hyperuricemia and gout) BLA 125-293. Pages 16 and 43–4 of 238-page Medical Review.

[67] Pegloticase (hyperuricemia and gout) BLA 125-293. Page 4 of 238-page Medical Review.

[68] Pegloticase (hyperuricemia and gout) BLA 125-293. Page 3 of 238-page Medical Review.

[69] Pegloticase (hyperuricemia and gout) BLA 125-293. Page 42 of 238-page Medical Review.

[70] Pegloticase (hyperuricemia and gout) BLA 125-293. Page 44 of 238-page Medical Review.

[71] Pegloticase (hyperuricemia and gout) BLA 125-293. Page 70 of 238-page Medical Review.

FDA continued to express concern about the ambiguity of the cardiovascular AE data, referring to "uncertainty" and the "fragility of conclusions," but managed to arrive at a recommendation for the package label. FDA proposed that the package label requires limiting pegloticase to patients that had failed other drugs, writing, "In approaching how to handle this issue, I am very cognizant that the population at risk **should be limited to those that have failed other therapies,** and this drug did demonstrate substantial efficacy for a population that otherwise would be suffering. As such, I believe it would be proper to not let this issue, given the fragility of conclusions, by itself disallow patient access. **Labeling should reflect this finding** and our uncertainty."[72]

Package label. Consistent with FDA's comments, the package label for pegloticase requires limiting patients to those that had failed other therapies:

> **INDICATIONS AND USAGE**. KRYSTEXXA™ (pegloticase) is a PEGylated uric acid specific enzyme indicated for the treatment of chronic gout in adult patients refractory to conventional therapy.[73]

i. Tapentadol (Severe Pain, Neuropathic Pain) NDA 200-533

Tapentadol is a small molecule for treating pain. The drug acts has a mechanism similar to that of opioids, where it stimulates mu-opioid receptor. The drug also has a nonopioid mechanism, where it inhibits reuptake of norepinephrine.[74] Tapentadol is effective in treating chronic low back pain, osteoarthritis knee pain, cancer-related pain, and neuropathic pain.[75]

AEs revealed by drug class analysis. FDA's analysis of contraindications focused on the pre-existing condition of respiratory depression. The FDA reviewer expressly required that the contraindications be consistent with information on package labels for other drugs of the same class. The relevant drug class was long-acting opioids: "The tapentadol ER labeling should be **consistent with current labeling of approved long-acting opioids** and contain contraindications in unmonitored patients with severely impaired pulmonary function and in patients receiving MAO inhibitors."[76]

[72] Pegloticase (hyperuricemia and gout) BLA 125-293. Page 31 of 238-page Medical Review.

[73] Package label KRYSTEXXA™ (pegloticase) for injection, for intravenous infusion. September 2010 (14 pp.).

[74] Langford RM, et al. Is tapentadol different from classical opioids? A review of the evidence. Br. J. Pain. 2016;10:217–21.

[75] Sanchez Del Aguila MJ, et al. Practical considerations for the use of tapentadol prolonged release for the management of severe chronic pain. Clin. Ther. 2015;37:94–113.

[76] Tapentadol (severe pain, neuropathic pain) NDA 200-533. Pages 7 and 108 of 236-page Medical Review.

As part of FDA's drug class analysis, FDA listed the contraindications on package labels for other drugs of the same class. FDA added a qualifying statement regarding the drug class analysis, "Not all opioids are labeled for all of these contraindications." FDA consulted package labels for these other long-acting opioid drugs used in the treatment of chronic pain[77]:

- Oxycodone CR (Oxycontin®)
- Morphine sulfate and naltrexone (Embeda®)
- Tramadol (Ultram ER, Ryzolt®)
- Hydromorphone (Exalgo®)

FDA's list of contraindications, as acquired from these other package labels, took this form[78]:

1. Not to be used in patients with ... **respiratory depression** ... in unmonitored settings or in the absence of resuscitative equipment. Opioids can worsen hypoventilation in patients with baseline hypoventilation.
2. Not to be used in patients with acute or severe **asthma** in unmonitored settings or in the absence of resuscitative equipment. Opioids can stimulate histamine release which may cause an asthma exacerbation.
3. Not to be used in patients with **paralytic ileus**.

FDA's Medical Review focused on AEs caused by tramadol. FDA was careful to define the drug class comprising tapentadol (the study drug) and tramadol as "a centrally-acting analgesic with opioid and non-opioid activity." That said, FDA turned to AEs caused by a concomitant drug (MAO inhibitors). In other words, the issue was whether the study drug (tapentadol) had the same problem as tramadol, namely, the dangerous DDI with coadministered MAO inhibitors. On this point FDA warned that "Tapentadol is a centrally-acting synthetic analgesic combining opioid and non-opioid activity, similar to Tramadol ... **[a]dministration of Tramadol may enhance the seizure risk in patients taking MAO inhibitors** ... [c]oncomitant use of Tramadol with MAO inhibitors ... may increase the risk of serotonin syndrome."[79]

The Warnings and Precautions section of tapentadol's package label, which was based on drug class analysis, revealed the form of toxicity resulting from coadministering tapentadol with MAO inhibitors. The toxicity is "serotonin syndrome," where the symptoms include agitation, hallucinations, coma, tachycardia, and lack of muscle coordination.[80]

[77] Tapentadol (severe pain, neuropathic pain) NDA 200-533. Page 185 of 236-page Medical Review.

[78] Tapentadol (severe pain, neuropathic pain) NDA 200-533. Page 114 of 236-page Medical Review.

[79] Tapentadol (severe pain, neuropathic pain) NDA 200-533. Page 14 of 236-page Medical Review.

[80] Package label. NUCYNTA® ER (tapentadol) extended-release tablets for oral use. December 2016 (31 pp.)

1. *AEs Revealed by Sponsor's Own Clinical Study*

FDA's review also demonstrated that the Sponsor's own human subjects experienced one of same AEs as was brought to light by the drug class analysis (respiratory depression). One of the Sponsor's study subjects was described as follows: "This subject (Subject 22-05) in Study JPN-C01 had ... **respiratory depression** ... [t]he subject was subsequently discontinued from the study ... [a]pproved labeling reflects the types of vital sign events noted."[81]

2. *AEs Predicted by Sponsor's Animal Toxicity Studies*

The Sponsor's animal toxicity studies also revealed a toxic effect that found a place in the Contraindications section, respiratory depression. FDA's Pharmacology Review observed that in animals, "At high doses of tapentadol ... fearfulness, sedation or excited behavior, recumbency and hunched posture, **impaired respiratory function** ... were observed."[82]

Package label. FDA's Medical Reviews brought to light various preexisting conditions (asthma, paralytic ileus, respiratory depression) and concomitant drugs (MAO inhibitors) that defined subgroups of patients that should not receive tapentadol (Nucynta®). These conditions found a place in the Contraindications section:

> **CONTRAINDICATIONS**. Significant respiratory depression ... Acute or severe bronchial asthma ... Known or suspected paralytic ileus ... Hypersensitivity to tapentadol or to any other ingredients of the product ... Concurrent use of monoamine oxidase inhibitors (MAOIs) or use of MAOIs within the last 14 days.[83]

j. Topiramate (Partial Onset Seizures, Tonic–Clonic Seizures) NDA 201-635

Topiramate (Trokendi XR®) is a small molecule drug for treating seizures. The drug has several targets in the nervous system. It blocks voltage-gated sodium channels, increases gamma-aminobutyric acid (GABA), and inhibits AMPA receptors and kainate glutamate ionotropic receptors.[84]

FDA's Medical Review and FDA's ClinPharm Review focused on alcohol and alcohol's ability to increase absorption of topiramate, thus increasing risk for drug toxicity. The Sponsor had conducted PK studies on the influence of an alcohol drink with dogs, but FDA determined that the study design was defective. Also, the Sponsor had not conducted any PK studies with human subjects. As a consequence of the suspected safety issue (but

[81] Tapentadol (severe pain, neuropathic pain) NDA 200-533. Page 48 of 236- page Medical Review.

[82] Tapentadol (severe pain, neuropathic pain) NDA 200-533. Page 12 of 48-page Pharmacology Review.

[83] Package label. NUCYNTA® ER (tapentadol) extended-release tablets for oral use. December 2016 (31 pp.).

[84] Clark AM, et al. Intravenous topiramate: comparison of pharmacokinetics and safety with the oral formulation in healthy volunteers. Epilepsia. 2013;54:1099–105.

availability of only in vitro data), FDA required a contraindication against alcohol on the package label. Another risk with alcohol is that as a consequence of the ***initial overly high plasma values***, what ***followed was a period of ineffectively low plasma values***.

As disclosed in Chapter 3, Food Effect Studies and in Chapter 7, Drug–Drug Interactions—Part One (Small Molecule Drugs) and Chapter 8, Drug–Drug Interactions—Part Two (Therapeutic Proteins), Sponsors and FDA routinely assess situations where drug exposure is too-high and also the situation where drug exposure is too low. But this example of topiramate may be the only example from FDA's reviews, where one type of food resulted in both drug levels that were both too-high and too-low.

1. *Altered Drug Pharmacokinetics After Alcohol Consumption*

FDA's ClinPharm Review summarized the problem of drinking alcohol when taking topiramate: "Interaction with Alcohol. In vitro data show that, in the presence of alcohol, the pattern of topiramate release from Trokendi XR capsules is markedly altered. As a result, plasma levels of topiramate with Trokendi XR may be **dangerously high soon after dosing and subtherapeutic later in the day**. Therefore, alcohol use should be completely avoided within 6 hours prior to and 6 hours after Trokendi XR administration."[85]

FDA warned about toxicity taking the form of "CNS depression" and "breakthrough seizures." Apparently, "breakthrough seizures" refer to seizures that materialize when blood topiramate levels fall below an effective concentration: "Based on the in vitro data, in the presence of alcohol, plasma levels of topiramate with Trokendi XR™ may be **markedly higher soon after dosing and subtherapeutic later in the day**. Topiramate is a CNS depressant. Concomitant administration of topiramate with alcohol can result in significant CNS depression. And the altered pattern of topiramate release from Trokendi XR™ capsules with alcohol may increase the likelihood of breakthrough seizures."[86]

2. *FDA Complained About Study Design with Dogs*

The FDA reviewer characterized the initial overly high plasma drug values as resulting from dose dumping:

> In vitro data show that, in the presence of alcohol, the pattern of topiramate release from Trokendi XR capsules is significantly altered. Although at the pre-NDA meeting, a clinical **alcohol dose dumping study** Trokendi XR was requested from the Agency, the Applicant has not responded. Instead, an in vivo study in dogs was conducted to evaluate the potential dose-dumping with alcohol (0%, 10%, and 40%). However, the dog was never shown to be a suitable in vivo predictive model for potential dosage form and alcohol interaction in humans.[87]

[85] Topiramate (partial onset seizures, tonic-clonic seizures) NDA 201-635. Page 57 of 146-page Clinical Pharmacology Review.

[86] Topiramate (partial onset seizures, tonic-clonic seizures) NDA 201-635. Page 28 of 65-page Medical Review.

[87] Topiramate (partial onset seizures, tonic-clonic seizures) NDA 201-635. Pages 28 and 29 of 65-page Medical Review.

3. *Difference Between Dog and Human Physiology*

FDA complained that the dog studies were irrelevant to humans, based on dramatic differences in stomach pH values in the dog and human, pointing out that the gastric pH in humans is pH 1–3 while that in dogs was pH 7–9. FDA complained that, "The physiological difference ... between humans (more acidic gastric pH 1~3) and dogs (gastric pH 7~9) is noted. However, potential impact in gastric pH difference on topiramate release from the ER formulation in the presence of alcohol is unclear[88] ... [a]lthough the division repeatedly informed the sponsor that the alcohol interaction study performed in **dogs was not an acceptable substitute** for the required study in people, the sponsor did not perform a human alcohol interaction study."[89]

4. *Pentagastrin for Use in Canine Food Effect Studies*

In this author's opinion, the Sponsor and FDA were unaware of the following advice for canine food effect studies (advice to use pentagastrin). Mathias et al.[90] acknowledged that food effect studies with dogs can predict human food effects on drugs, but provided the following words of caution and advice. A problem is that the fasted-state gastric pH can vary between pH 1.6 (very acidic) and pH 8.5 (mild alkaline) between laboratory colonies of dogs. To maintain consistency of study design, and to mimic the human gastric pH, dogs should be treated with pentagastrin in the fasted state and in the fed state. Pentagastrin stimulates gastric acid secretion where the result is that the pH of the dog's stomach is caused to be the acidic range more typical of the human stomach.

5. *Lack of Any Human PK Data and Consequent Need for Package Label Contraindications*

FDA observed that the Sponsor had not conducted any PK studies with humans and concluded that, as a consequence, the package label must require no drinking of alcohol. FDA's recommendation for the Contraindication labeling for topiramate (Trokendi®) took this form:

> As there are no in-vivo results regarding the effect of alcohol on Trokendi ... in humans, per the current available information on the kinetics of alcohol, variability of gastrointestinal emptying time and other confounding factors (e.g., meal etc.), **restriction for the alcohol consumption 6 hours prior and after administration of Trokendi XR™** dosing is recommended by the Agency for the labeling (Contraindication and Warnings and Precautions sections).[91]

[88] Topiramate (partial onset seizures, tonic-clonic seizures) NDA 201-635. Page 103 of 146-page Clinical Pharmacology Review.

[89] Topiramate (partial onset seizures, tonic-clonic seizures) NDA 201-635. Pages 37–8 of 65-page Medical Review.

[90] Mathias N, et al. Food effect in humans: predicting the risk through in vitro dissolution and in vivo pharmacokinetic models. AAPS J. 2015;17:988–98.

[91] Topiramate (partial onset seizures, tonic-clonic seizures) NDA 201-635. Pages 22, 37, and 38 of 65-page Medical Review.

FDA explained the basis for recommendations on the package label regarding timing of the drug dose and any alcohol drinking:

> From a clinical pharmacology standpoint, we do not anticipate that a significant PK interaction with alcohol consumption is likely to occur at approximately 2 to 3 hours prior to or after the Trokendi XR™ dosing. The rationale for this ... is based on ... information on the gastrointestinal (GI) absorption and gastric emptying of ethanol, that is, near complete disappearing from stomach by 30 min and 126 min under fasted and fed conditions, respectively) (Levitt MD et al (1997) Am J Physiol Gastrointest Liver Physiol. 273:G951-G957; Lennernäs H (2009) Mol Pharm. 6:1429-40).[92]

Package label. FDA's recommendations for topiramate (Trokendi®), regarding alcohol, found a place on the package label:

> **CONTRAINDICATIONS**. With recent alcohol use, that is, within 6 hours prior to and 6 hours after TROKENDI XR® use.[93,94]

k. Vorapaxar (Thrombotic Cardiovascular Events) NDA 204-886

The package label for vorapaxar is unique, in that FDA's Medical Review explained that the variety of different contraindications on the package label resulted from the fact that physicians would not easily be able to distinguish between two different subgroups of patients. The problem was that the drug should be contraindicated in patients with **history of stroke**, but physicians might not be able to distinguish history of stroke patients from patients with **prior transient ischemic attack** (TIA).

The idea for this contraindication was suggested by the Data Safety and Monitoring Board (DSMB) during the course of one of the Sponsor's clinical studies. The cited reference provides an account of the composition and duties of the DSMB.[95] As recollected by the FDA reviewer, "Notably, during the trial the ... Data Safety and Monitoring Board (DSMB) recommended discontinuation of study treatment in all patients with a **prior history of stroke** because of an increased rate of ... hemorrhagic stroke, in **patients with a stroke history**, along with continuation of the trial in other study subjects. This recommendation was made after enrollment had been closed. The trial leadership and the Applicant accepted this recommendation and promptly implemented it."[96]

[92] Topiramate (partial onset seizures, tonic-clonic seizures) NDA 201-635. Page 57 of 146-page Clinical Pharmacology Review.

[93] Package label. TROKENDI XR (topiramate) extended-release capsules for oral use. April 2017 (57 pp.).

[94] This author points out the bad grammar on the package label. The package label has a contraindication against taking Trokendi XR, where the contraindication forbids taking the drug "within 6 hours prior to and 6 hours after TROKENDI XR® use." However, any physician can easily discern the meaning.

[95] Brody T. Clinical trials: study design, endpoints and biomarkers, drug safety, and FDA and ICH Guidelines. 2nd ed. New York: Elsevier; 2016. pp. 23, 500–1, 561–2, 720, 727, 746–9, 881.

[96] Vorapaxar (thrombotic cardiovascular events) NDA 204-886. Pages 11, 98, and 99 of 234-page Medical Review.

FDA used drug class analysis in arriving at the contraindication against giving the drug to patients with a history of stroke. This drug class analysis involved predicting AEs of the study drug (vorapaxar) by the AEs on the package label for another drug of the same class, referring to antiplatelet drugs in general, and specifically to prasugrel. FDA based its recommendation for vorapaxar's contraindication labeling on the contraindication on the label for another drug (prasugrel), writing:

> The findings in ... [the clinical study] related to a history of prior stroke and TIA are analogous to the prasugrel experience in ... acute coronary syndromes (ACS) subjects. If vorapaxar is approved, it merits a contraindication in patients with a history of prior stroke or TIA, **similar to prasugrel**[97] ... [t]he most important safety risk of other antiplatelet drugs is the risk of bleeding ... prasugrel, an approved ... P2Y12 antagonist, was associated with an increased rate of intracranial bleeding in patients with a prior history of ischemic stroke ... this risk resulted in a boxed warning and **contraindication.**[98]

This is about concomitant disorders that cannot easily be distinguished from each other. This provides useful and practical guidance for drafting the Contraindications section, where the goal is to exclude certain patients from treatment and where the exclusion is based on a concomitant disorder that is similar to another concomitant disorder. The Sponsor's way out of this problem was to include both of the concomitant disorders in the Contraindications section. This refers to concomitant disorders that are concomitant with thrombotic cardiovascular events (vorapaxar's indication). A component of this analysis was the extreme danger of the AE at risk (intracranial hemorrhage). The Sponsor's reasoning was as follows:

> One additional population used to evaluate efficacy and safety represents an important post-hoc subset of subjects: The Proposed Label Population ... is the Applicant's proposed target population. The Applicant's rationale for limiting use to this population is that **it might be difficult for a physician to distinguish between a prior stroke and a prior TIA**, and **it would more prudent to simply exclude patients with either** from treatment with vorapaxar because of the risk of intracranial hemorrhage.[99]

FDA's additional remarks on the need to exclude patients with prior stroke and the need to exclude patients with prior transient ischemic attack (TIA), took the form, "However, a warning against its use in anyone who has had a TIA may be justified if the drug is approved."[100]

As shown below, the label contraindicated the use of vorapaxar in patients with a history of stroke and also in patients with history of TIA.

[97] Vorapaxar (thrombotic cardiovascular events) NDA 204-886. Page 41 of 234-page Medical Review.

[98] Vorapaxar (thrombotic cardiovascular events) NDA 204-886. Page 43 of 234-page Medical Review.

[99] Vorapaxar (thrombotic cardiovascular events) NDA 204-886. Pages 111 and 127 of 234-page Medical Review.

[100] Vorapaxar (thrombotic cardiovascular events) NDA 204-886. Page 130 of 234-page Medical Review.

Package label. The package label for vorapaxar (Zontivity®) included a Black Box Warning and a Contraindications section, where both contraindicated use of vorapaxar in patients with history of stroke or history of TIA:

> **WARNING:** BLEEDING RISK ... Do not use ZONTIVITY in patients with a history of stroke, transient ischemic attack (TIA), or intracranial hemorrhage (ICH); or active pathological bleeding ... Antiplatelet agents, including ZONTIVITY, increase the risk of bleeding, including intracranial hemorrhage (ICH) and fatal bleeding.[101]

> **CONTRAINDICATIONS**. History of stroke, TIA, or intracranial hemorrhage (ICH). Active pathological bleeding.[102]

I. 4-Hydroxybutyrate (Cataplexy in Narcolepsy; Excessive Daytime Sleepiness in Narcolepsy) NDA 021-196

4-Hydroxybutyrate is for treating sleep disorders. 4-Hydroxybutryate, also known as oxybate and as gamma-hydroxybutyrate, has a structure similar to that of succinate semialdehyde. Both of these molecules are shown below:

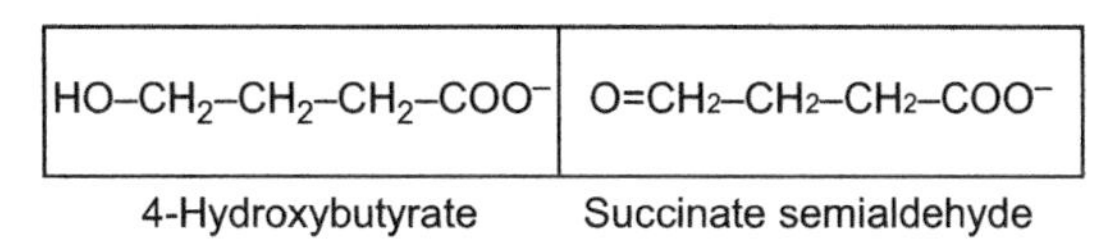

4-Hydroxybutyrate   Succinate semialdehyde

The FDA reviewers commented on the catabolic pathways used to metabolize 4-hydroxybutyrate. The pathways were determined using animals. The main pathway was oxidation of 4-hydroxybutyrate to succinate semialdehyde, followed by further oxidation to produce succinic acid. The chemical "4-hydroxybutyrate" is the same as "gamma-hydroxybutyrate (GHB)" and is the same as "oxybate." FDA's Medical Review revealed that

> Animal studies indicated that metabolism is the major elimination pathway for sodium oxybate, producing carbon dioxide and water via the ... Krebs cycle, and secondarily by beta-oxidation ... [t]he primary pathway involves ... GHB dehydrogenase, that catalyses the conversion of ... oxybate to succinic semialdehyde, which is then biotransformed to succinic acid by the enzyme succinic semialdehyde dehydrogenase.[103]

FDA's ClinPharm Review described the alternate catabolic pathway, "An alternate pathway of biotransformation involves beta-oxidation via 3,4-dihydroxybutyrate to carbon dioxide and water."[104]

[101] Package label. ZONTINVITY™ (vorapaxar) tablets 2.08 mg, for oral use. May 2014 (18 pp.).

[102] Package label. ZONTINVITY™ (vorapaxar) tablets 2.08 mg, for oral use. May 2014 (18 pp.).

[103] 4-Hydroxybutyrate (cataplexy in narcolepsy; excessive daytime sleepiness in narcolepsy) NDA 021-196. Page 32 of 41-page pdf file (sixth pdf file of 12 pdf files for Medical Review).

[104] 4-Hydroxybutyrate (cataplexy in narcolepsy; excessive daytime sleepiness in narcolepsy) NDA 021-196. Pages 5–6 of 47-page Clinical Pharmacology Review.

FDA described a biochemical pathway that provided part of the basis for including a contraindication. Published medical articles provided additional information which, when combined with the biochemical pathway information, provided a solid basis for the contraindication. The medical articles described mutations on human succinic semialdehyde dehydrogenase deficiency and psychiatric disorders resulting from succinic semialdehyde dehydrogenase deficiency.

These psychiatric disorders include **hypotonia, ataxia, dysfunctional or absent speech, seizures, hallucinations, and aggressive behaviors.**[105,106] Victims of succinic semialdehyde dehydrogenase deficiency can be identified by increased levels of 4-hydroxybutyrate in the urine[107] as well as by identifying their genetic mutations.[108,109]

FDA's commentary, together with the medical literature's account of hypotonia and ataxia in the genetic disease, found a place on the package label.

Package label. The package label for 4-hydroxybutryate (Xyrem®) included a contraindication instructing physicians not to administer 4-hydroxybutyrate to patients with a genetic disorder causing deficiency in the enzyme succinic semialdehyde dehydrogenase:

> **CONTRAINDICATIONS**. Succinic semialdehyde dehydrogenase deficiency. Xyrem is contraindicated in patients with succinic semialdehyde dehydrogenase deficiency. This is a rare disorder of inborn error of metabolism variably characterized by **mental retardation, hypotonia, and ataxia.**[110]

III. GENETIC MARKERS USED IN PACKAGE LABELING

The above commentary on succinic semialdehyde dehydrogenase and the package label for 4-hydroxybutyrate (Xyrem®) is an example of a genetic disease and its corresponding biomarker that resulted in instructions in the Contraindications section. Genetic markers for a variety of drugs have found a place on the package label, where the information on genetics can reside in one or more of the Indications and Usage section, Dosage and Administrations section, Warnings and Precautions section, and Clinical Studies section, as shown below.

[105] Malaspina P, et al. Comparative genomics of aldehyde dehydrogenase 5a1 (succinate semialdehyde dehydrogenase) and accumulation of gamma-hydroxybutyrate associated with its deficiency. Hum. Genomics. 2009;3:106–20.

[106] Liechti ME, et al. Pharmacokinetics and pharmacodynamics of γ-hydroxybutyrate in healthy subjects. Br. J. Clin. Pharmacol. 2016;81:980–8.

[107] Bekri S, et al. The molecular basis of succinic semialdehyde dehydrogenase deficiency in one family. Mol. Genet. Metab. 2004;81:347–51.

[108] Malaspina P, et al. Comparative genomics of aldehyde dehydrogenase 5a1 (succinate semialdehyde dehydrogenase) and accumulation of gamma-hydroxybutyrate associated with its deficiency. Hum. Genomics. 2009;3:106–20.

[109] Akaboshi S, et al. Mutational spectrum of the succinate semialdehyde dehydrogenase (ALDH5A1) gene and functional analysis of 27 novel disease-causing mutations in patients with SSADH deficiency. Hum. Mutat. 2003;22:442–50.

[110] Package label. XYREM® (sodium oxybate) oral solution, CIII. January 2017 (19 pp.).

Genetic markers can be used to detect various kinds of mutations, for example, mutations that change a single amino acid, mutations that delete part of a polypeptide, single nucleotide polymorphisms, chromosomal rearrangements, and gene duplications. Laboratory tests for detecting genetic biomarkers include DNA sequencing, PCR-dependent methods such as the TaqMan® technique, and chromosomal probes such as those used for fluorescence in situ hybridization (FISH technique).

The bullet points illustrate biological situations where genetic markers can be used in package labeling:

- Genetic markers that are essential for identifying the particular cancer that is to be treated by the drug
- Genetic markers that are essential for identifying the particular strain of virus, bacterium, or parasite that is to be targeted by the drug
- Genetic markers that correspond to the patient's genome and that identify patients for which the drug was FDA-approved
- Genetic markers that identify patients where the drug should be contraindicated, or where a Black Box Warning is needed, or where adverse DDIs are expected

These bullet point concepts are illustrated as follows:

a. Genetic Markers for Cancer

Osimertinib (Tagrisso®) is for treating **non-small cell lung cancer**. The package label refers to a mutation in the EGFR gene, namely, the T790M mutation. The test for the genetic mutation is conducted on a tumor biopsy or on free tumor cells circulating in the bloodstream. The Indications and Usage section refers to this mutation in the narrative, "TAGRISSO is a kinase inhibitor indicated for the treatment of patients with metastatic epidermal growth factor receptor (EGFR) T790M mutation positive non-small cell lung cancer (NSCLC), as detected by an FDA approved test." The Dosage and Administration section also refers to this mutation in the instructions, "Confirm the presence of T790M mutation in tumor or ... plasma specimens prior to initiation of treatment with TAGRISSO."[111]

b. Genetic Markers for Viral Infections

Simeprevir (Olysio®) is used for treating **hepatitis C virus** (HCV) infections. The package label refers to a mutation in the viral genome which takes the form of a Q80K mutation. This means that the genetic mutation causes wild-type glutamine residue to be converted to the mutant lysine residue. The Indications and Usage section warns that, "Efficacy of OLYSIO ... is substantially reduced in patients infected with HCV ... with an NS3 Q80K polymorphism." "NS3" is one of the genes in the viral genome.[112] In other words, this means that the drug does not work well where the virus has the Q80K mutation.

[111] Package label. TAGRISSO® (osimertinib) tablets, for oral use. March 2017 (20 pp.).

[112] Brody T. Clinical trials study design, endpoints and biomarkers, drug safety, and FDA and ICH Guidelines. 2nd ed. New York: Elsevier; 2016. pp. 439–42 and 637–9.

Also, the Dosage and Administration section recommends that patients be screened for the presence of this mutation (Q80K mutation in the viral NS3 gene) and use alternative drugs, instead of simeprevir.[113] In this way, the Q80K genetic marker identifies patients that should be treated with other drugs, in preference over simeprevir.

c. Genetic Markers That Identify Patients That Can Be Treated with the Drug (Lomitapide)

This concerns a genetic disease resulting in high cholesterol. The drug is lomitapide (Juxtabid®). Lomitapide inhibits **microsomal triglyceride transfer protein** (MTP). The package label refers to the genetic defect and describes a genetic test for diagnosing patients where the test result was condition that must be satisfied before administering the drug. The Indications and Usage section disclosed that the genetic defect was a homozygous mutation (affecting both copies of the same gene in the patient):

> **INDICATIONS AND USAGE**. JUXTAPID is a microsomal triglyceride transfer protein inhibitor indicated as an adjunct to a low-fat diet and other lipid-lowering treatments ... to reduce low-density lipoprotein cholesterol (LDL-C), total cholesterol (TC), apolipoprotein B (apo B), and non-high density lipoprotein cholesterol (non-HDL-C) in patients with homozygous familial hypercholesterolemia.[114]

The Clinical Studies section of the package label for lomitapide described three different tests for diagnosing homozygous familial hypercholesterolemia, where one of the tests is a genetic test. The test requires detecting mutations in both LDL receptor alleles:

> **CLINICAL STUDIES**. A diagnosis ... was defined by the presence of at least one of the following clinical criteria: (1) Documented functional mutation(s) in both LDL receptor alleles or alleles known to affect LDL receptor functionality, or (2) Skin fibroblast LDL receptor activity $<20\%$ normal, or (3) Untreated TC >500 mg/dL and TG <300 mg/dL and both parents with documented untreated TC >250 mg/dL.[115]

Regarding the genetics of homozygous familial hypercholesterolemia, Cuchel et al.[116] provides a diagram of chromosome 19 showing the location of the LDL receptor gene and explains that, in a mating between heterozygous parents who each carry one copy of a gene that is mutated, 25% of children will carry two copies of **wild-type alleles** (homozygous normal), 50% will be **heterozygotes**, and 25% will carry two copies of familial hypercholesterolemia-mutation-bearing genes where this 25% of the children will have **homozygous** familial hypercholesterolemia. In another paper, Cuchel et al.[117] disclose some of the inactivating mutations of LDL receptor in patients with the homozygous

[113] Package label. OLYSIO (simeprevir) capsules, for oral use. May 2017 (48 pp.).

[114] Package label. JUXTAPID® (lomitapide) capsules for oral use. May 2016 (31 pp.).

[115] Package label. JUXTAPID® (lomitapide) capsules for oral use. May 2016 (31 pp.).

[116] Cuchel M, et al. Homozygous familial hypercholesterolaemia: new insights and guidance for clinicians to improve detection and clinical management. A position paper from the Consensus Panel on Familial Hypercholesterolaemia of the European Atherosclerosis Society. Eur. Heart J. 2014;35:2146–57.

[117] Cuchel M, et al. Inhibition of microsomal triglyceride transfer protein in familial hypercholesterolemia. New Engl. J. Med. 2007;356:148–56.

mutation, where some of the mutations were deletion mutations and others were mutations changing one amino acid to another.

Further regarding lomitapide, another type of genetic test has been proposed. This test distinguishes patients expected to respond best to lomitapide from those that are expected not to respond much. The test is to determine (using the patient's DNA) the DNA sequence for the gene encoding lomitapide's target (MTP), to compare the patient's DNA sequence with those previously determined to **predict good response to the drug**, and to compare the patient's DNA sequence with those previously determined to **predict lack of response to the drug.**[118]

d. Genetic Markers That Identify Patients That Can Be Treated with the Drug (Butalbital/Codeine Combination)

This concerns a combination drug (Fiorinal®) for treating tension headache. The drug is butalbital/codeine combination. The package label warned that the drug must never be given to patients with a certain cytochrome P450 genotype, that is, a genotype of cytochrome P450 2D6 (CYP2D6). The relevant genetic abnormality took the form of duplications of this gene in the patient's genome. The gene duplication caused a biochemical abnormality resulting in ultra-rapid metabolism of the drug to morphine, and the consequent toxicity resulting in death. The Black Box Warning read:

> **BOXED WARNING** ... Death Related to Ultra-Rapid Metabolism of Codeine to Morphine. Respiratory depression and death have occurred in children who received codeine following tonsillectomy and/or adenoidectomy and had evidence of being ultra-rapid metabolizers of codeine due to a CYP2D6 polymorphism.[119]

The Warnings and Precautions section explains that the genetic change took the form of extra copies of the CYP2D6 gene and that the result is increased risk for death following codeine administration. The names of the genetic variants are "CYP2D6 gene duplication *1/*1xN" and "CYP2D6 gene duplication *1/*2xN." As revealed below, this type of genetic variant occurs in up to 10% of Caucasians and about 3% of African Americans. The package label warned:

> **WARNINGS AND PRECAUTIONS.** Death Related to Ultra-Rapid Metabolism of Codeine to Morphine ... Respiratory depression and death have occurred in children who received codeine in the post-operative period following tonsillectomy and/or adenoidectomy and had evidence of being ultra-rapid metabolizers of codeine (i.e., multiple copies of the gene for cytochrome P450 isoenzyme 2D6 or high morphine concentrations). Deaths have also occurred in nursing infants who were exposed to high levels of morphine in breast milk because their mothers were ultra-rapid metabolizers of codeine. Some individuals may be ultra-rapid metabolizers because of a specific CYP2D6 genotype (gene duplications denoted as *1/*1xN or *1/*2xN). The prevalence of this CYP2D6 phenotype varies widely and has been estimated at 0.5 to 1% in Chinese and Japanese, 0.5 to 1% in Hispanics, 1 to 10% in Caucasians, 3% in African Americans, and 16 to 28% in North Africans, Ethiopians, and Arabs ... [t]hese individuals convert codeine into its active metabolite, morphine, more rapidly and completely than other people. This rapid conversion results in higher than expected serum morphine levels. Even at labeled dosage regimens, individuals who

[118] Kolovou GD, et al. MTP gene variants and response to lomitapide in patients with homozygous familial hypercholesterolemia. J. Atheroscler. Thromb. 2016;23:878–83.

[119] Package label. FIORINAL® with CODEINE (butalbital, aspirin, caffeine, and codeine phosphate, USP) capsules, for oral use, CIII. December 2016 (35 pp.).

are ultra-rapid metabolizers may have life-threatening or fatal respiratory depression or experience signs of overdose (such as extreme sleepiness, confusion, or shallow breathing).[120]

Turning to the published literature, Crews et al.,[121] Zahari et al.,[122] and Racoosin et al.[123] described genetic variants in the CYP2D6 gene, as it occurs in various ethnic and racial groups, how these CYP2D6 variants influence efficacy and safety of various drugs, and how they influence DDIs. To provide another example of genetic variation of one of the cytochrome P450 enzymes, Fohner et al.[124] describe the situation of the Yup'ik indigenous people in Alaska that have a naturally occurring variant in one of the cytochrome P450 enzymes (CYP4F2*3), where the variant influences the dose amount of warfarin for Yup'ik people needing this anticoagulant.

The cytochrome P450 enzymes mediate the catabolism of many drugs, where information on package labels typically accounts for this sort of catabolism where the drug is administered alone or where the drug is coadministered with a second drug. The literature on the CYP enzymes demonstrates that the polymorphic nature of their genes influences the exposure, efficacy, safety profile, and DDIs for a variety of drugs. The naturally occurring wild-type CYP enzymes include CYP3A5,[125] CYP2B6,[126] CYP2C8,[127] CYP2C19,[128] and CYP2D6.[129,130] The cited references describe polymorphisms (genetic variations) for each of the genes encoding each of these CYP enzymes.

[120] Package label. FIORINAL® with CODEINE (butalbital, aspirin, caffeine, and codeine phosphate, USP) capsules, for oral use, CIII. December 2016 (35 pp.).

[121] Crews KR, et al. Clinical Pharmacogenetics Implementation Consortium guidelines for cytochrome P450 2D6 genotype and codeine therapy: 2014 update. Clin. Pharmacol. Ther. 2014;95:376–82.

[122] Zahari Z, Ismail R. Influence of cytochrome P450, family 2, subfamily D, polypeptide 6 (CYP2D6) polymorphisms on pain sensitivity and clinical response to weak opioid analgesics. Drug Metab. Pharmacokinet. 2014;29:29–43.

[123] Racoosin JA, et al. New evidence about an old drug--risk with codeine after adenotonsillectomy. New Engl. J. Med. 2013;368:2155–7.

[124] Fohner AE, et al. Variation in genes controlling warfarin disposition and response in American Indian and Alaska Native people: CYP2C9, VKORC1, CYP4F2, CYP4F11, GGCX. Pharmacogenet. Genomics. 2015;25:343–53.

[125] Belkhir L, et al. Interaction between darunavir and etravirine is partly mediated by CYP3A5 polymorphism. PLoS One. 2016;11:e0165631. DOI: 10.1371/journal.pone.0165631.

[126] Ilic K, et al. The influence of sex, ethnicity, and CYP2B6 genotype on bupropion metabolism as an index of hepatic CYP2B6 activity in humans. Drug Metab. Dispos. 2013;41:575–81.

[127] Aquilante CL, et al. Impact of the CYP2C8 *3 polymorphism on the drug-drug interaction between gemfibrozil and pioglitazone. Br. J. Clin. Pharmacol. 2013;75:217–26.

[128] Michaud V, et al. Efavirenz-mediated induction of omeprazole metabolism is CYP2C19 genotype dependent. Pharmacogenomics. 2014;14:151–9.

[129] Teh LK, Bertilsson L. Pharmacogenomics of CYP2D6: molecular genetics, interethnic differences and clinical importance. Drug Metab. Pharmacokinet. 2012;27:55–67.

[130] Tracy TS, et al. Interindividual variability in cytochrome P450-mediated drug metabolism. Drug Metab. Dispos. 2016;44:343–51.

IV. CONCLUDING REMARKS

The Contraindications section provides instructions to the physician that the drug must not be administered to one or more subgroups of patients. The subgroup may be one that is a permanent feature of the patient and that cannot be reversed, such as having had prior stroke or being over the age of 70 years or being refractory to another drug for the same disease. The subgroup may be one that is chronic but that is reversible, such as hepatic impairment or pregnancy. Also, the subgroup may be one that is temporary and that can be quickly dissipated at will, such as drinking alcohol on the day the drug is to be taken or taking a concomitant drug on the day the drug is to be taken.

FDA's analysis for arriving at the Contraindications section involves the following steps:

Step one. Contemplate the various adverse events associated with the drug, for example, AEs occurring during the Sponsor's own clinical studies or AEs predicted by drug class analysis.

Step two. Divide the patient population into various subgroups. This refers to patients expected to be prescribed the drug, in the postmarketing situation. The subgroup can be those with a preexisting condition or disease (in addition to the disease to be treated), patients with a defined genetic makeup, patients taking a specific concomitant drug such as MAO inhibitors, and so on.

Step three. The third step is to recommend that the Contraindications section be drafted so that it instructs physicians not to administer the drug to the subgroups of greatest risk for a given AE.

CHAPTER

6

Animal Studies

I. INTRODUCTION

a. Introduction to Toxicology and Animal Efficacy Studies

Animal toxicology data are used to predict adverse events (AEs) for human patients, where the predicted AEs can find a place in the Black Box Warning, Warnings and Precautions section, and Adverse Reactions section of the package label. Animal toxicity studies have enhanced value in situations where it would not be ethical to perform corresponding studies on human subjects, such as studies assessing embryofetal toxicity, teratology, or defective bone growth in infancy.

This chapter also illustrates the use of animal models for assessing drug efficacy. After an animal model has been identified and developed, one can establish reliability and reproducibility of that model by validation procedures. Validating an animal model includes developing procedures or tests that use a specific type of animal, followed by employing the laboratory procedure, independently conducted in blind trials, by at least three different laboratories.[1]

A stand-alone topic regarding animal studies is the "Animal Rule." The Animal Rule provides the extreme situation where animal studies completely take the place of clinical studies. The Animal Rule finds a basis in Section 314.600 (small molecule drugs) and Section 601.90 (biologicals) of the Code of Federal Regulations. According to FDA's Guidance for Industry,[2] "The Animal Rule states that for drugs developed to ameliorate or prevent serious or life threatening conditions caused by exposure to lethal or permanently disabling toxic substances, when human challenge studies would not be ethical to perform ... FDA may grant marketing approval based on ... animal efficacy studies when the results of those studies establish that the drug is reasonably likely to produce clinical benefit in humans."

[1] Varga OE, et al. Validating animal models for preclinical research: a scientific and ethical discussion. Altern. Lab. Anim. 2010;38:245–8.

[2] U.S. Dept. of Health and Human Services. Food and Drug Administration. Guidance for Industry. Product development under the animal rule. May 2014 (53 pp.).

FDA's Drug Review Process and the Package Label
DOI: https://doi.org/10.1016/B978-0-12-814647-7.00006-3

b. FDA's Guidance for Industry and Animal Studies

Animal toxicity data can serve as a basis for AE warnings on package labels. To use animal toxicity data for predicting AEs in human subjects one can use the no observed adverse effect level (NOAEL) and the safety margin. NOAEL just means the highest dose level that does not produce a significant increase in toxicity to animals. The safety margin predicts if a given form of toxicity in animals is likely to pose a risk for a corresponding AE in patients.

FDA's Guidance for Industry on Exploratory Investigational New Drug (IND) Studies provides for animal studies, for supporting efficacy of the study drug, and for including in the New Drug Application (NDA) or BLA. FDA's Guidance provides that:

> animal studies ... incorporate endpoints that are mechanistically based on the pharmacology of the new chemical entity and thought to be important to clinical effectiveness. For example, if the degree of saturation of a receptor or the inhibition of an enzyme were considered ... related to effectiveness, this parameter would be characterized and determined in the animal study.[3]

FDA's Guidance for Industry on Nonclinical Safety Studies provides for:

> 7-day repeated-dose toxicity study in one species, usually rodent, by intended route of administration with toxicokinetic data, or via the i.v. route. Hematology, clinical chemistry, necropsy, and histopathology data should be included.[4]

The Code of Federal Regulations (21 CFR §201.57(14)) requires that the package label include a section devoted to nonclinical toxicology. Nonclinical toxicology encompasses animal toxicology experiments, the Ames mutagenicity test,[5] and tests on human cardiac monocytes.[6] Section 201.57(14) requires:

> Nonclinical toxicology. This section must contain ... as appropriate ... Carcinogenesis, mutagenesis, impairment of fertility. This subsection must state whether long term studies in animals have been performed to evaluate carcinogenic potential and, if so, the species and results. If results from reproduction

[3] U.S. Department of Health and Human Services. Food and Drug Administration. Center for Drug Evaluation and Research (CDER). Guidance for Industry, Investigators, and Reviewers. Exploratory IND studies. 2006 (13 pp.).

[4] U.S. Department of Health and Human Services. Food and Drug Administration. Center for Drug Evaluation and Research (CDER). Center for Biologics Evaluation and Research (CBER). Guidance for Industry M3(R2). Nonclinical safety studies for the conduct of human clinical trials and marketing authorization for pharmaceuticals. 2000 (25 pp.).

[5] U.S. Department of Health and Human Services. Food and Drug Administration. Center for Drug Evaluation and Research (CDER). Center for Biologics Evaluation and Research (CBER). Guidance for Industry S2(R1). Genotoxicity testing and data interpretation for pharmaceuticals intended for human use. 2012 (31 pp.).

[6] U.S. Department of Health and Human Services. Food and Drug Administration. Center for Drug Evaluation and Research (CDER). Center for Biologics Evaluation and Research (CBER). Guidance for Industry S7B. Nonclinical evaluation of the potential for delayed ventricular repolarization (QT interval prolongation) by human pharmaceuticals. 2005 (10 pp.).

studies or other data in animals raise concern about mutagenesis or impairment of fertility in either males or females, this must be described ... Animal toxicology and/or pharmacology. Significant animal data necessary for safe and effective use of the drug in humans.[7]

c. Developing and Validating Animal Models

An animal model for showing efficacy is shown here, using the example of the disease enterocolitis. Lu et al.[8] commented on the relevance of animal models to human diseases, stating that the animal disease should replicate features of the human disease and that the same drug (or an equivalent) as proposed for humans should be effective in the animal model:

> "animal models should **replicate most of the** ... **features** that are seen in human ... [enterocolitis]." Moreover, "[i]n addition, as a validation strategy, it should be possible to attenuate the severity of ... [enterocolitis] in a particular animal model through the use of ... strategies that have been shown to attenuate ... [enterocolitis] in humans ... the **mechanisms by which the models are induced should parallel the clinical condition** of human ... [enterocolitis] to the extent that is possible."[9]

Brody[10] describes a number of appropriate animal models for various diseases, as well as animal models and cell culture models suitable for discovering drug mechanisms, but not likely to be suitable or appropriate for predicting efficacy or safety in humans.

Rashid et al.[11] describes animal model validation as "when the appropriate mouse model is identified, it is important to validate the degree to which it reliably produces the data and endpoints desired before evaluating the efficacy of any novel therapeutics or testing novel hypotheses of breast cancer biology in vivo." Zuluaga et al.[12] describes animal model validation as "the translation of the results to humans ... which is the statistical comparison of pharmacodynamic parameters to demonstrate reliability and relevance. Reliability requires repeatability (intralaboratory) and reproducibility (interlaboratory), and relevance relates to the accuracy (sensitivity and specificity) of the model to predict

[7] 21 CFR §201.57(14).

[8] Lu P, et al. Animal models of gastrointestinal and liver diseases. Animal models of necrotizing enterocolitis: pathophysiology, translational relevance, and challenges. Am. J. Physiol. Gastrointest. Liver Physiol. 2014;2306:G917–28.

[9] Lu P, et al. Animal models of gastrointestinal and liver diseases. Animal models of necrotizing enterocolitis: pathophysiology, translational relevance, and challenges. Am J. Physiol. Gastrointest. Liver Physiol. 2014;2306:G917–28.

[10] Brody T. Enabling claims under 35 USC §112 to methods of medical treatment or diagnosis, based on in vitro cell culture models and animal models. J. Pat. Trademark Off. Soc. 2015;97:328–411.

[11] Rashid OM, Takabe K. Animal models for exploring the pharmacokinetics of breast cancer therapies. Expert Opin. Drug Metab. Toxicol. 2015;11:221–30.

[12] Zuluaga AF, et al. About the validation of animal models to study the pharmacodynamics of generic microbiologicals. Clin. Infect. Dis. 2014;59:459–61.

the biological response."[13] According to Varga et al.,[14] validating an animal model involves developing tests and procedures that use a specific type of animal, followed by employing the laboratory procedure, independently conducted in a blind trial, by at least three different laboratories.

d. Animal Models for Arriving at Starting Dose in Humans

Animals are used to arrive at an appropriate starting dose for humans. Starting dose refers to an appropriate dose, where a proposed clinical trial is the first to administer a given drug to humans. The starting dose can be at the upper limit, for example, at a limit that results in an acceptable risk for adverse events, or the starting dose can be lower.[15,16]

FDA's Guidance for Industry recommends that, "The results from the pre-clinical program can be used to select starting and maximum doses for the clinical trials. The starting dose is anticipated to be no greater than 1/50 of the NOAEL from the 2-week toxicology study in the sensitive species on a mg/m^2 basis."[17] The unit "mg/m^2" refers to milligrams of drug per square meter of body surface area. Body surface area can be calculated from the cited formulas.[18,19]

For small molecule drugs and some biologics, FDA uses the concept maximum recommended starting dose (MRSD).[20] FDA provides an algorithm for deriving the MRSD for human clinical trials. FDA's algorithm includes determining the NOAEL in animals, conversion of NOAEL to human equivalent dose (HED), and application of a safety factor.

[13] Zuluaga AF, et al. About the validation of animal models to study the pharmacodynamics of generic microbiologicals. Clin. Infect. Dis. 2014;59:459–61.

[14] Varga OE, et al. Validating animal models for preclinical research: a scientific and ethical discussion. Altern. Lab. Anim. 2010;38:245–8.

[15] U.S. Department of Health and Human Services. Food and Drug Administration. Guidance for Industry. Estimating the maximum safe starting dose in initial clinical trials for therapeutics in adult healthy volunteers. 2005 (27 pp.).

[16] Reigner BG, Blesch KS. Estimating the starting dose for entry into humans: principles and practice. Eur. J. Clin. Pharmacol. 2002;57:835–45.

[17] U.S. Department of Health and Human Services. Food and Drug Administration. Center for Drug Evaluation and Research (CDER). Guidance for Industry, Investigators, and Reviewers. Exploratory IND studies. 2006 (13 pp.).

[18] Sawyer M, Ratain MJ. Body surface area as a determinant of pharmacokinetics and drug dosing. Invest. New Drugs. 2001;19:171–7.

[19] Kouno T, et al. Standardization of the body surface area (BSA) formula to calculate the dose of anticancer agents in Japan. Jpn. J. Clin. Oncol. 2003;33:309–13.

[20] U.S. Department of Health and Human Services. Food and Drug Administration. Guidance for Industry. Estimating the maximum safe starting dose in initial clinical trials for therapeutics in adult healthy volunteers. 2005 (27 pp.).

The animal species that generates the lowest human equivalent dose (HED) is called the most sensitive species.[21] The algorithm for converting the NOAEL value from animal studies to the HED can be based on body surface area[22] or, alternatively, it can be based on milligrams of drug per kilograms body weight.

This concerns adjusting drug dose to take into account differences in the drug's affinity for the target in animals versus the target in humans. In comments on antibody drugs, Lynch et al.[23] recommended that "the dose in the animal model should be adjusted to reflect the difference in affinity between the animal and humans in order to ensure adequate exposure in the toxicity study." For example, an antibody against interleukin-1β (IL-1β) (gevokizumab) has an affinity for **human IL-1β** that is about 1000 times greater than for **mouse IL-1β**. Despite the relatively low affinity for the mouse target, the affinity was found to be adequate for testing this antibody in mice, for mouse models for the human disease.[24]

e. Animal Models for Safety Testing

For testing safety, animal models have been developed for testing drug toxicity on a variety of organ systems, including kidneys, liver, and heart.

1. *Renal Toxicity*

Publications from FDA provide guidance for using various animals for testing renal toxicity, for example, mouse, rat, guinea pig, hamster, rabbit, dog, mini-pig, rhesus monkey, and cynomolgus monkeys.[25,26] Biomarkers used in animal studies for renal toxicity include changes in creatinine, blood urea nitrogen (BUN), potassium, phosphorus, and calcium.[27]

[21] U.S. Department of Health and Human Services. Food and Drug Administration. Guidance for Industry. Estimating the maximum safe starting dose in initial clinical trials for therapeutics in adult healthy volunteers. 2005 (27 pp.).

[22] Blanchard OL, Smoliga JM. Translating dosages from animal models to human clinical trials--revisiting body surface area scaling. FASEB J. 2015;29:1629–34.

[23] Lynch C, et al. Practical considerations for nonclinical safety evaluation of therapeutic monoclonal antibodies. MAbs. 2009;1:2–11.

[24] Owyang AM, et al. XOMA 052, a potent, high-affinity monoclonal antibody for the treatment of IL-1beta-mediated diseases. MAbs. 2011;3:49–60.

[25] Blank B, et al. Review of qualification data for biomarkers of nephrotoxicity. Submitted by the Predictive Safety Testing Consortium. 2009 (79 pp.).

[26] U.S. Department of Health and Human Services. Food and Drug Administration. Center for Drug Evaluation and Research (CDER). Guidance for Industry. Immunotoxicology evaluation of investigational new drugs. 2002 (35 pp.).

[27] Blank B, et al. Review of qualification data for biomarkers of nephrotoxicity. Submitted by the Predictive Safety Testing Consortium. Center for Drug Evaluation and Research. U.S. Food and Drug Administration. January 16, 2009 (79 pp.).

An animal model for testing drug-induced renal damage was developed by Pavokvic et al.[28] This renal damage test involves measuring biomarkers in the urine, such as glutathione *S*-transferase alpha, clusterin (CLU), and kidney injury molecule-1.[29,30]

2. *Hepatic Toxicity*

Rats have been used as an animal model for acetaminophen toxicity, where the biomarker was alanine transaminase (ALT).[31] Further details on ALT and other liver enzymes for assessing liver damage by acetaminophen, as well as by other drugs, are described.[32] Liver damage can also be assessed by the ratio of [aspartate transaminase]/[alanine transaminase] (AST/ALT).[33,34] An elevated AST/ALT ratio predicts fibrosis and cirrhosis.

3. *Cardiac Toxicity*

Animal models are available for testing cardiac toxicity. For example, Duran et al.[35] describes animal cardiac tests for the drug class known as VEGF-signaling pathway inhibitors, where the member drugs include pazopanib, sorafenib, and sunitinib. These animal tests included those for myocyte survival. Another battery of safety tests for cardiac toxicity involves plasma biomarkers such as tropinin I and microRNAs (MiR1, miR133a, and miR133b).[36]

f. Animal Models for Assessing Efficacy

Many human diseases have corresponding animal models for testing efficacy. The following outlines animal models for testing drug efficacy for diseases of the kidney, liver, and heart.

[28] Pavokvic M, et al. Comparison of the MesoScale discovery and Luminex multiplex platforms for measurement of urinary biomarkers in a cisplatin rat kidney injury model. J. Pharmacol. Toxicol. Methods. 2014;69:196–204.

[29] Pavokvic M, et al. Comparison of the MesoScale discovery and Luminex multiplex platforms for measurement of urinary biomarkers in a cisplatin rat kidney injury model. J. Pharmacol. Toxicol. Methods. 2014;69:196–204.

[30] Blank B, et al. Review of qualification data for biomarkers of nephrotoxicity. Submitted by the Predictive Safety Testing Consortium. Center for Drug Evaluation and Research. U.S. Food and Drug Administration. January 16, 2009 (79 pp.).

[31] Kondo K, et al. Enhancement of acetaminophen-induced chronic hepatotoxicity in restricted fed rats: a nonclinical approach to acetaminophen-induced chronic hepatotoxicity in susceptible patients. J. Toxicol. Sci. 2012;37:911–29.

[32] Chalasani P, et al. ACG Clinical Guideline: The diagnosis and management of idiosyncratic drug-induced liver injury. Am. J. Gastroenterol. 2014;109:950–66. DOI: 10.1038 (17 pp.).

[33] Botrus M, Sikaris KA. The De Ritis ratio: the test of time. Clin. Biochem. Rev. 2013;34:117–30.

[34] Brody T. Nutritional biochemistry. 2nd ed. New York, NY: Elsevier; 1999. pp. 426–7, 542–8.

[35] Duran JM, et al. Sorafenib cardiotoxicity increases mortality after myocardial infarction. Circ. Res. 2014;114:1700–12.

[36] D'Alessandra Y, et al. Circulating microRNAs are new and sensitive biomarkers of myocardial infarction. Eur. Heart J. 2010;31:2765–73.

1. *Renal Diseases*

For testing efficacy of drugs for renal diseases, animal models include mice implanted with Renca cells to produce renal cell carcinoma.[37,38] Renal fibrosis can be studied with an animal model taking the form of the COL4A3 knockout mouse.[39] Efficacy of treatment with paricalcitol and ramipril was tested with this mouse model.[40] The COL4A3 gene encodes collagen type IV alpha 3 chain.

2. *Hepatic Diseases*

For testing efficacy of drugs for hepatic diseases, available animal models include the following. Regarding research on hepatitis B virus (HBV), the chimpanzee and woodchuck are accepted animal models for HBV infections.[41,42] In contrast to the situation with HBV, the chimpanzee is the only recognized animal model for the study of hepatitis C virus.[43] Regarding another hepatic disease, animal models for studying jaundice include the Gunn rat.[44]

3. *Cardiac Diseases*

For testing drug efficacy against atherosclerosis, mouse models include mice genetically deleted in the LDL receptor.[45] The arterial lesions in human familial hypercholesterolemia (lack of LDL receptor) have some of the characteristics of the mouse lesion, including

[37] Hughes S, et al. Interleukin 21 efficacy in a mouse model of metastatic renal cell carcinoma. J. Clin. Oncol. 2004;22(14 Suppl.):2598.

[38] Chen L, et al. Modification of antitumor immunity and tumor microenvironment by resveratrol in mouse renal tumor model. Cell Biochem. Biophys. 2015;72:617–25.

[39] Rubel D, et al. Antifibrotic, nephroprotective effects of paricalcitol versus calcitriol on top of ACE-inhibitor therapy in the COL4A3 knockout mouse model for progressive renal fibrosis. Nephrol. Dial. Transplant. 2014;29:1012–9.

[40] Rubel D, et al. Antifibrotic, nephroprotective effects of paricalcitol versus calcitriol on top of ACE-inhibitor therapy in the COL4A3 knockout mouse model for progressive renal fibrosis. Nephrol. Dial. Transplant. 2014:29:1012–9.

[41] Wieland SF. The chimpanzee model for hepatitis B virus infection. Cold Spring Harbor Perspect. Med. 2015;5:a021469.

[42] Korolowicz KE, et al. Antiviral efficacy and host innate immunity associated with SB 9200 treatment in the woodchuck model of chronic hepatitis B. PLoS One. 2016;11(8):e0161313. DOI: 10.1371 (21 pp.).

[43] Bukh J. A critical role for the chimpanzee model in the study of hepatitis C. Hepatology. 2004;39:1469–75.

[44] Stanford JA, et al. Hyperactivity in the Gunn rat model of neonatal jaundice: age-related attenuation and emergence of gait deficits. Pediatr. Res. 2015;77:434–9.

[45] Getz G, Reardon CA. Animal models of atherosclerosis. Arterioscler. Thromb. Vasc. Biol. 2012;32:1104–15.

lesions in the aortic valves and in the aortic root. For testing drug efficacy for cardiac diseases, animal models have been developed for heart diseases such as myocardial infarction[46] and congestive heart failure.[47]

g. Other Animal Models for Testing Efficacy

Regarding other diseases, the most common animal model for multiple sclerosis is experimental autoimmune encephalitis (EAE),[48] the most common model for rheumatoid arthritis is collagen-induced arthritis,[49] a model for Alzheimer's disease is mice deleted in neprilysin (neprilysin degrades beta-amyloid),[50] and a model for prostate cancer is LNAI cells implanted in athymic mice.[51] LNAI cells are androgen-independent prostate cancer cells.

II. NO OBSERVED ADVERSE EFFECT LEVEL (NOAEL)

a. Introduction to NOAEL

NOAEL is defined by FDA's Guidance for Industry as "Several definitions of NOAEL exist, but for selecting a starting dose, the following is used: the highest dose level that does not produce a significant increase in adverse effects in comparison to the control group. In this context, adverse effects that are biologically significant ... should be considered in the determination of the NOAEL."[52] Animal toxicology studies of three types are used to determine the NOAEL. These are studies that detect:

1. Overt toxicity, such as lesions;
2. Biomarkers such as levels of liver enzymes in the blood serum; and
3. Exaggerated pharmacodynamic effects.

Knowledge of NOAEL, as determined in animal toxicity studies, can be used to arrive at a safe starting dose in human subjects. FDA has pointed out that data taking the form

[46] Wu Y, et al. Acute myocardial infarction in rats. J. Vis. Exp. 2011;(48):2464. DOI: 10.3791/2464 (4 pp.).

[47] Chen J, et al. A new model of congestive heart failure in rats. Am. J. Physiol. Heart Circ. Physiol. 2011;310:H994–1003.

[48] Rossi B, Constantin G. Live imaging of immune responses in experimental models of multiple sclerosis. Front. Immunol. 2016;7:Article 506.

[49] Shimizu K, et al. IL-1 receptor type 2 suppresses collagen-induced arthritis by inhibiting IL-1 signal on macrophages. J. Immunol. 2015;194:3156–68.

[50] Yuede CM, et al. Rapid in vivo measurement of β-amyloid reveals biphasic clearance kinetics in an Alzheimer's mouse model. J. Exp. Med. 2016;213:677–85.

[51] Benitez A, et al. Targeting hyaluronidase for cancer therapy: antitumor activity of sulfated hyaluronic acid in prostate cancer cells. Cancer Res. 2011;71:4085–95.

[52] U.S. Department of Health and Human Services. Food and Drug Administration. Center for Drug Evaluation and Research (CDER). Guidance for Industry. Estimating the maximum safe starting dose in initial clinical trials for therapeutics in adult healthy volunteers. 2005 (27 pp.).

of a pharmacokinetic value in humans, such as AUC or Cmax, without more, are generally not sufficient to arrive at a safe starting dose.[53]

This chapter provides examples of NOAEL studies, mainly from FDA's Pharmacological Reviews for the lumacaftor/ivacaftor combination.[54] Exemplary accounts of the determination and interpretation of NOAEL values are also available from FDA's Pharmacological Reviews for lorcaserin (obesity) NDA 022-529, tofacitinib (rheumatoid arthritis) NDA 203-214, apremilast (psoriatic arthritis) NDA 205-437, and macitentan (hypertension) NDA 204-410.

A study of the lumacaftor/ivacaftor combination drug provides examples for NOAEL values, as calculated from various animal toxicology studies.[55] FDA's Pharmacology Review for the lumacaftor/ivacaftor combination drug provides these concepts relating to NOAEL values:

- **Choice of test animals**. NOAEL values can be calculated using a variety of animals, for example, the rat or dog.
- **Male and female animals**. NOAEL values from a given animal can be calculated using both male and female.
- **Choice of toxic effect**. For any given NOAEL study, the investigator can use one or more types of toxic effects as the read-out for determining NOAEL.
- **Pregnant animals**. Where NOAEL study is conducted with a pregnant animal, toxicities can be simultaneously detected in the mother and also in the embryo, with calculation of NOAEL values that apply to the mother's toxicity and to the embryo's toxicity.
- **Combination drugs**. For a combination drug, NOAEL values can be experimentally determined for only the first drug in the combination. Also, NOAEL can be determined for only the second drug in the combination. In addition, NOAEL can be determined for the combination itself, by simultaneously treating animals with both drugs.

The following excerpts demonstrate how the term NOAEL fits into the scheme of animal toxicology studies and the types of study design and the types of AE data that are collected for arriving at a value for NOAEL.

b. Types of Toxicities Detected and Arrival at Value for NOAEL

The following dog study was designed to enable calculating the NOAEL. The Sponsor monitored the animals for dose-dependent toxicities relating to vomiting, hematology, liver weight, and histology. Because none of the toxicities were significant, the Sponsor concluded that the highest dose was the NOAEL. FDA's Pharmacology Review described the dog study, which involved a series of progressively higher doses, the fact that the highest dose was not toxic to organs, and the conclusion that the highest dose establishes the NOAEL.

[53] U.S. Department of Health and Human Services. Food and Drug Administration. Center for Drug Evaluation and Research (CDER). Guidance for Industry. Estimating the maximum safe starting dose in initial clinical trials for therapeutics in adult healthy volunteers. 2005 (27 pp.).

[54] Lumacaftor/ivacaftor (cystic fibrosis) NDA 206-038. 502-page pdf file for Pharmacology Review.

[55] Lumacaftor/ivacaftor (cystic fibrosis) NDA 206-038. 502-page pdf file for Pharmacology Review.

FDA's account of the dog NOAEL study focused on various forms of toxicity (vomiting, RBC counts, platelet counts, reticulocyte counts, histopathology) with increasing drug doses. The study used a range of relatively low doses.

Regarding vomiting, RBC counts, and platelet counts in dogs, FDA wrote:

> In a 12-month oral ... toxicology study ... dogs received lumacaftor at doses of 0, 125, 250, or 500 mg/kg/day ... [a] slight increase in the incidence/occurrence of vomiting (food) was observed in both sexes at 500 mg/kg/day. Red blood parameters (RBC counts, hemoglobin, and hematocrit) were slightly decreased ... at 250 and 500 mg/kg/day during months 3, 6, 9, and 12. Platelet counts were slightly increased ... at 250 and 500 mg/kg/day during months 3, 6, 9, and 12. Slight decreases of absolute reticulocyte counts were observed ... at 500 mg/kg/day during months 9 and 12 and also ... at 250 mg/kg/day during month 12.[56]

Regarding histology of organs in the dogs, FDA wrote, "Liver weights were increased for ... at 6 and 12 months; however, there were no corresponding histopathological findings. Judging the totality of the histopathological findings from the 6-month interim sacrifice and 12-month terminal sacrifice ... no treatment-related target organs of toxicity were identified with doses up to 500 mg/kg/day. The NOAEL was identified as the high dose of 500 mg/kg/day."[57]

In another dog study intended to determine NOAEL, the dosing range went to a greater dose range, that is, up to 1000 mg/day. The following 3-month dog study used doses from 0 to 1000 mg/kg/day. The result was that the NOAEL was still 500 mg/kg/day (just as in the above-described low dose range study). Thus, the results from both of the dog studies gave the NOAEL value of 500 mg/kg/day. FDA wrote:

> In a 3-month oral ... toxicology study, dogs received lumacaftor at doses of 0, 125, 250, 500, or 1000 mg/kg/day.... Two dogs developed irregular gait, jerky movements and/or muscle rigidity prior to sacrifice. Irregular gait, trembling, jerky movements, and/or muscle rigidity were observed sporadically in individual animals at 1000 mg/kg/day ... [h]istopathological findings were evident for dogs at 1000 mg/kg/day and included the thymus (increased severity of lymphocyte depletion), male reproductive organs (delays of maturation in the testes and prostate and sloughed germ cells in the epididymides), and liver (extramedullary hematopoiesis) ... **The NOAEL was identified as 500 mg/kg/day based upon deaths, neurological clinical signs, and histopathological findings in the male reproductive organs at 1000 mg/kg/day**.[58]

c. NOAEL Determined Simultaneously for Mother and Fetus

This is about simultaneous determination of toxicity to a pregnant mother and to the embryo or fetus. In a study of pregnant rabbits, NOAEL values were determined with regard to maternal and embryo toxicity. Rabbits received lumacaftor at 50, 100, or 200 mg/kg/day, during gestation. At the middle dose (100 mg), **maternal rabbit toxicities** took the form of reduction of body weight and death, thus compelling the conclusion that

[56] Lumacaftor/ivacaftor (cystic fibrosis) NDA 206-038. Pages 356–7 of 502-page Pharmacology Review.

[57] Lumacaftor/ivacaftor (cystic fibrosis) NDA 206-038. Pages 356–7 of 502-page Pharmacology Review.

[58] Lumacaftor/ivacaftor (cystic fibrosis) NDA 206-038. Pages 356–7 of 502-page Pharmacology Review.

the NOAEL was the next dose down (*50 mg/kg/day*). Regarding **fetal toxicity**, the tests did not reveal any toxic effects and thus the NOAEL as it applied to the fetus was determined to be the highest dose tested (*200 mg/kg/day*).[59]

III. CALCULATING THE SAFETY MARGIN

a. Background on the Safety Margin

The safety margin is a function of this ratio:

[NOAEL in the most sensitive species]/[expected therapeutic dose in humans]

To account for differences between humans and laboratory species, a safety margin is established based on the NOAEL in the most sensitive of the tested species.[60] The safety margin has been defined as the ratio between the exposure achieved at the highest dose, at which no dose-limiting toxicity is discovered in the most sensitive animal species, as compared to the exposure expected at the highest dose targeted for achieving efficacy in humans.[61]

The safety margin is calculated using animal exposure data after the NOAEL dose, and human exposure data. The formula using the exposure unit of AUC is[62]

$[AUC_{animal}$ at NOAEL$]/[AUC_{human,\ when\ tested\ at\ max.\ recommended\ human\ dose}]$

According to this AUC-based formula, safety margin is the fold-difference between the AUC at the drug concentration that causes toxicity in animals, when measured at NOAEL, and the AUC in humans, at the maximum recommended human dose.[63] Alternatively, the

[59] Lumacaftor/ivacaftor (cystic fibrosis) NDA 206-038. Page 374 of 502-page pdf file for Pharmacology Review.

[60] Steinmetz KL, Spack EG. The basics of preclinical drug development for neurodegenerative disease indications. BMC Neurol. 2009;9(Suppl. 1):S2. DOI 10.1186 (13 pp.).

[61] Bloom J, Dean RA. Biomarkers in clinical drug development. Boca Raton, FL: CRC Press; 2003. p. 99.

[62] Vogel H, Maas J, Gebauer A. Drug discovery and evaluation: methods in clinical pharmacology. Berlin: Springer-Verlag; 2011 (p. 247).

[63] Vogel H, Maas J, Gebauer A. Drug discovery and evaluation: methods in clinical pharmacology. Berlin: Springer-Verlag; 2011 (p. 247).

safety margin can be calculated with animal exposure data when measured at the NOAEL, and human exposure data, using the exposure unit of Cmax. The formula using the Cmax unit is[64,65]

$$[Cmax_{animal} \text{ at NOAEL}] / [Cmax_{human,\ when\ tested\ at\ maximum\ recommended\ human\ dose}]$$

According to this Cmax-based formula, safety margin is the fold-difference between the drug concentration that causes toxicity in animals, when measured at NOAEL, and the Cmax in humans, at the maximum recommended human dose. Comparison between animal exposure and human exposure is generally based on AUC, but sometimes it may be more appropriate to use Cmax.[66]

Regarding the experimental setup for safety margin studies, "The route of administration in these studies must be the same as the proposed clinical route. If the proposed route is oral, drug is administered by gavage to rats and by gavage or capsule to dogs. The duration of administration ... must ... conform to the proposed clinical protocol. For example, if 14 days of continuous drug administration is proposed for the phase 1 clinical trial, then animal toxicity studies of at least 14 to 28 days are typically required to support a clinical study of this length."[67]

b. Example of Safety Margin Calculation for Antibacterial drug (ETX0914)

Figs. 6.1 and 6.2 show data from human pharmacokinetic studies, with arrows indicating parameters acquired from animal studies.[68] The study was conducted by AstraZeneca and other companies. Values for NOAEL as determined from animal studies are notated in the figures (but these values were not used for plotting the curves). Fig. 6.1 shows the data relevant to **AUC**, while Fig. 6.2 shows data relevant to **Cmax**. Both figures show a plot of exposure versus dose from human subjects.

The administered drug was an antibacterial drug (ETX0914). ETX0914 inhibits topoisomerase II and is used to treat *Neisseria gonorrhoea* infections.[69] The researchers based

[64] Vogel H, Maas J, Gebauer A. Drug discovery and evaluation: methods in clinical pharmacology. Berlin: Springer-Verlag; 2011 (p. 247).

[65] Ollerstam A. Safety assessment in the discovery and development of a new drug. Danmark: Leo Pharma A/S; November 2014 (50 pp.).

[66] Vogel H, Maas J, Gebauer A. Drug discovery and evaluation: methods in clinical pharmacology. Berlin: Springer-Verlag; 2011 (p. 247).

[67] Steinmetz KL, Spack EG. The basics of preclinical drug development for neurodegenerative disease indications. BMC Neurol. 2009;9(Suppl. 1):S2. DOI 10.1186 (13 pp.).

[68] Basarab GS, et al. Responding to the challenge of untreatable gonorrhea: ETX0914, a first-in-class agent with a distinct mechanism-of-action against bacterial type II topoisomerases. Sci. Rep. 2015;5:11827. DOI: 10.1038 (13 pp.).

[69] Foerster S, et al. Genetic resistance determinants, in vitro time-kill curve analysis and pharmacodynamic functions for the novel topoisomerase II Inhibitor ETX0914 (AZD0914) in *Neisseria gonorrhoeae*. Front. Microbiol. 2015;6:1377 (14 pp.).

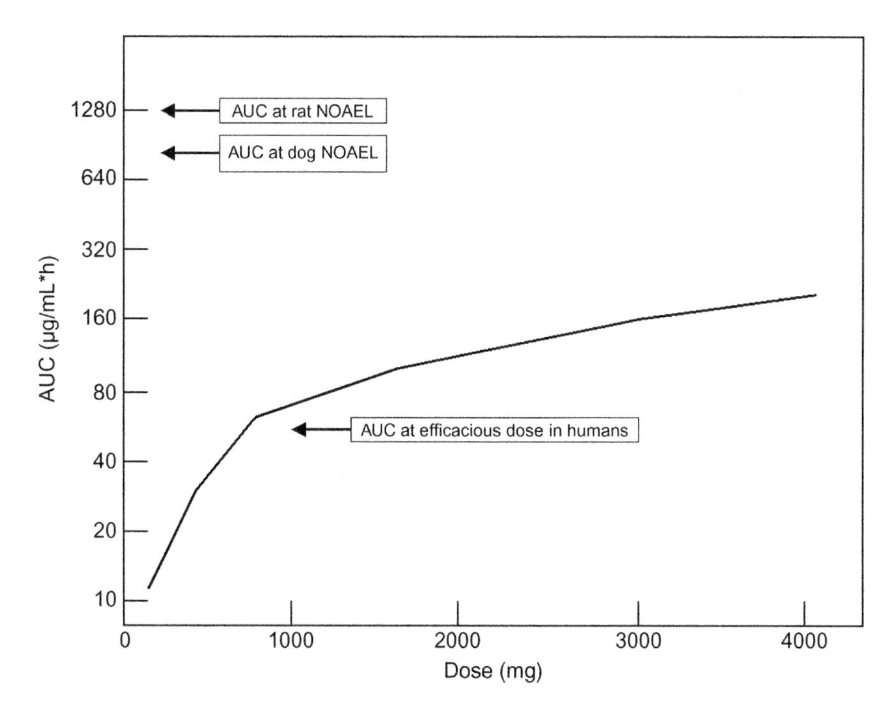

FIGURE 6.1 Exposure (AUC) versus ETX0914 dose in human subjects. The efficacious AUC occurred at 800 mg and was exceeded at doses greater than 800 mg. The margins to the NOAEL in rat and dog were 6.4- and 4.8-fold, respectively, at the 4000 mg dose. Permission to reproduce figure granted on July 2, 2017 by Scientific Reports.

their calculations on exposure data only from the highest dose (4000 mg) that was used with humans. The figures show doses ranging from 250 mg up to 4000 mg. For both figures, arrows are used to indicate rat exposure (AUC; Cmax) values at the rat's NOAEL and the dog exposure (AUC; Cmax) values at the dog's NOAEL.

After plugging the animal exposure data into the numerator, and the human exposure data into the denominator, and calculating the ratio, Basarab et al.[70,71] arrived at safety margins with a value of about 5. The AUC-based safety margins was **6.4, based on rat exposure data and human exposure data**. The AUC-based safety margin was **4.8, based on dog exposure data and human exposure data.**

[70] Basarab GS, et al. Responding to the challenge of untreatable gonorrhea: ETX0914, a first-in-class agent with a distinct mechanism-of-action against bacterial type II topoisomerases. Sci. Rep. 2015;5:11827. DOI: 10.1038 (13 pp.).

[71] "The article for which you have requested permission has been distributed under a Creative Commons CC-BY license (please see the article itself for the license version number). You may reuse this material without obtaining permission from Nature Publishing Group, providing that the author and the original source of publication are fully acknowledged, as per the terms of the license." (Website of Scientific Reports accessed July 2, 2017).

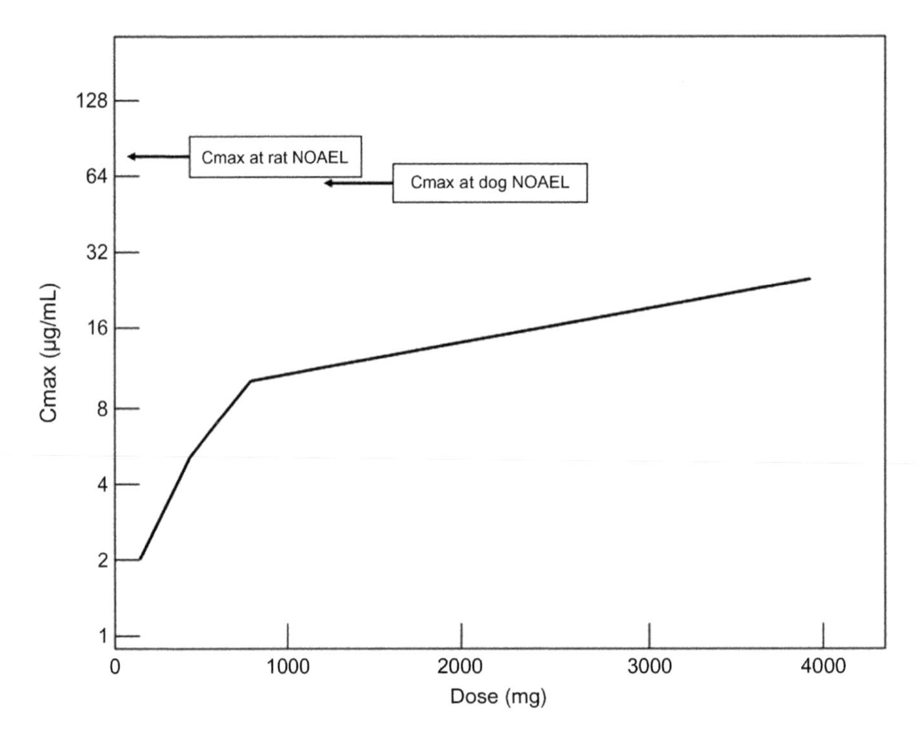

FIGURE 6.2 Exposure (Cmax) versus ETX0914 dose in human subjects. The margins to the NOAEL in rat and dog were 4.9- and 3.7-fold, respectively, at the 4000 mg dose. Permission to reproduce figure granted on July 2, 2017 by Scientific Reports.

c. Example of Safety Margins for Cardiac Drugs and for an Antifungal Agent

Redfern et al.[72] provides an account of how safety margin is used, with respect to cardiac drugs, and for arriving at a recommended drug dose that minimizes risk for AEs. The AE in question was QT interval prolongation (a type of arrhythmia). Redfern et al. began as follows: "Our dataset suggests that a margin of 30-fold ... would be adequate to ensure an acceptable degree of safety from arrhythmogenesis ... [h]owever, for the future, one should aim for higher margins where possible, as then concerns over drug interactions, variable pharmacokinetics, and all the other risk factors would recede."

But then, Redfern et al.[73] recommended that smaller safety margins might be preferred, where the patient's disease poses a high risk for death. In other words, if the disease is extremely dangerous to the patient, the Sponsor should be more willing to continue

[72] Redfern WS, et al. Relationships between preclinical cardiac electrophysiology, clinical QT interval prolongation and torsade de pointes for a broad range of drugs: evidence for a provisional safety margin in drug development. Cardiovasc. Res. 2003;58:32–45.

[73] Redfern WS, et al. Relationships between preclinical cardiac electrophysiology, clinical QT interval prolongation and torsade de pointes for a broad range of drugs: evidence for a provisional safety margin in drug development. Cardiovasc. Res. 2003;58:32–45.

developing the drug, even where the safety margin is low. To this end, Redfern et al.[74] recommended that "margins should reflect disease severity and medical need. For example, one could envisage that **a 10-fold margin might be acceptable for drugs used in diseases which are lethal if untreated**, e.g., cancer, AIDS."

A study of an antifungal agent provides another example of calculating the safety margin.[75] The antifungal agent used exposure data from rats and monkeys, where calculations of safety margin used the AUC formula and also used the Cmax formula.

IV. USE IN SPECIFIC POPULATIONS SECTION OF THE PACKAGE LABEL

This provides a context for the Use in Specific Populations section of the package label and, more specifically, for information on pregnancy and lactation, including toxicity tests in pregnant animals. The Code of Federal Regulations (21 CFR §201.56) states that drug labeling must contain this information:

Boxed Warning
Indications and Usage
Dosage and Administration
Contraindications
Warnings and Precautions
Adverse Reactions
Drug Interactions
Use in Specific Populations

Section 201.56 breaks down the Use in Specific Populations section into these categories:

8.1 Pregnancy
8.2 Lactation
8.3 Females and Males of Reproductive Potential
8.4 Pediatric use
8.5 Geriatric use

Section 201.56 further breaks down **8.1 Pregnancy** into several topics, including the topic of animals, "When animal data are available, the Risk Summary must summarize the findings in animals and based on these findings, describe, for the drug, the potential risk of any adverse developmental outcome(s) in humans. This statement must include ... species affected ... animal doses expressed in terms of human dose or exposure equivalents,

[74] Redfern WS, et al. Relationships between preclinical cardiac electrophysiology, clinical QT interval prolongation and torsade de pointes for a broad range of drugs: evidence for a provisional safety margin in drug development. Cardiovasc. Res. 2003;58:32–45.

[75] Schmitt-Hoffman A, et al. Single-ascending-dose pharmacokinetics and safety of the novel broad-spectrum antifungal triazole BAL4815 after intravenous infusions (50, 100, and 200 milligrams) and oral administrations (100, 200, and 400 milligrams) of its prodrug, BAL8557, in healthy volunteers. Antimicrob. Agents Chemother. 2006;50:279–85.

and outcomes for pregnant animals and offspring ... When there are no animal data, the Risk Summary must so state."

In the following account of FDA's Pharmacology Reviews for a variety of drugs, the Use in Specific Populations section includes information on toxicity to pregnant animals for these small molecule drugs and biologicals:

- Hydromorphone
- Regorafenib
- Sumatriptan/naproxen combination
- Atezolizumab
- Belimumab
- Evolocumab
- Nivolumab
- Pegloticase

V. ANIMAL DATA FROM FDA'S REVIEWS FIND A CORRESPONDING DISCLOSURE IN THE PACKAGE LABEL (SMALL MOLECULE DRUGS)

a. Introduction

The information below tracks the topics in FDA's Pharmacology Reviews, regarding animal data and their impact on the Sponsor's clinical trial design and on the package label. The small molecule drugs covered are:

- Avibactam/ceftazidime (AVYCAZ) combination
- Hydromorphone
- Lorcaserin
- Lumacaftor/ivacaftor combination
- Nebivolol
- Regorafenib
- Sumatriptan/naproxen combination
- Vigabatrin

FDA's review of **avibactam/ceftazidime (AVYCAZ)** reveals that data from animal model studies can take the place of Phase 3 clinical studies, thus relieving the Sponsor of the burden of conducting an additional study. This strategy was made possible because the individual components of the avibactam/ceftazidime (AVYCAZ) had previously been FDA-approved.

FDA's review of **hydromorphone** reveals that toxicity studies on animal pregnancy can be used as a basis for package label information for pregnancy in humans.

FDA's review of **lorcaserin** and **lumacaftor/ivacaftor combination** reveals the use of safety margin calculations as a technique for predicting risk for AEs in humans. The relevant AEs for these two drugs were cancer and cataracts, respectively.

FDA's review of **nebivolol** discloses that arguments focusing on mechanism of a given AE (Leydig cell tumors) can be used to convince FDA that this nebivolol-related AE as detected in mice is not likely to occur in humans. FDA's review of nebivolol helps establish that these types of mechanism of action arguments are relevant to FDA submissions:

- Whether mechanism of the untreated disease of animal model corresponds to mechanism of the untreated disease in humans;
- Whether mechanism of action of the study drug's efficacy in treating the animal model corresponds to the mechanism of action of the same drug's efficacy in treating the human disease; and
- Whether mechanism of action of the study drug's toxicity in animals corresponds to mechanism of action of the study drug's toxicity in humans.

FDA's review of **regorafenib** dwelled on the mechanism of drug degradation. Differences in drug-degrading pathways in dogs and humans were found to be responsible for differing toxicities in dogs and humans. Regorafenib caused renal toxicity in animals, but did not cause renal toxicity in humans. This difference was traced to the fact that human metabolism resulted in regorafenib's rapid detoxification while, in contrast, regorafenib was not detoxified in animals.

FDA's review for the **sumatriptan/naproxen combination** focused on embryofetal toxicity data from rats and rabbits, where these data found a place on the package label. The package label admitted to the lack of human embryofetal toxicity data, in its statement, that "There are no adequate and well-controlled studies in pregnant women."[76]

FDA's review of **vigabatrin** tells a story where animal and human toxicity studies were integrated with each other, over a period of time, for sorting out visual AEs. As part of the closely integrated animal and human toxicity work, FDA required that "a noninvasive method for clinical monitoring ... of neurotoxicity be developed. In response, the Sponsor ... conducted studies in dogs to develop methods (MRI) ... which were then used to monitor humans."[77]

b. Avibactam/Ceftazidime Combination (AVYCAZ) (Urinary Tract Infections) NDA 206-494

Avibactam/ceftazidime combination (AVYCAZ) is a fixed drug combination for treating infections. Ceftazidime is a cephalosporin drug that kills various bacteria, while avibactam inhibits bacterial enzymes that catalyze ceftazidime degradation.

1. *Justification of Animal Studies Revealed in FDA's Approval Timeline*

The timeline of FDA's evaluation of AVYCAZ reveals the contribution of animal studies leading to FDA approval. Ceftazidime had been earlier approved by a separate NDA (NDA 050-578) for treating various infections, including urinary tract infections.[78] This

[76] Package label. TREXIMET (sumatriptan and naproxen sodium) tablets, for oral use. May 2016 (27 pp.).

[77] Vigabatrin (complex partial seizures in adults, infantile spasms in children) NDA 022-006. Page 5 of 68-page Pharmacology Review

[78] Avibactam/ceftazidime combination (AVYCAZ) (urinary tract infections) NDA 206-494. Page 3 of 33-page Cross Discipline Team Leader Review.

earlier approval was on July 1985, whereas the submission date for AVYCAZ was about 30 years later, June 25, 2014. AVYCAZ was tested in human subjects in Phase 1 clinical trials and Phase 2 clinical trials, but because ceftazidime had already been approved via NDA 050-578, FDA did not require any Phase 3 clinical trial for approval of AVYCAZ.

FDA explained the rationale for using animal studies instead of any Phase 3 clinical trial with human subjects: "The Applicant is relying ... on FDA's previous findings of efficacy and safety of ceftazidime ... [b]ecause confirmatory clinical trials comparing ceftazidime alone to ceftazidime-avibactam [AVYCAZ] would not be feasible, the FDA agreed that the combination of both components ... can be demonstrated by ... **animal models** for infection, where the **addition of avibactam restores the activity of ceftazidime** against ceftazidime-nonsusceptible microorganisms."[79]

FDA further explained how animal studies could take the place of clinical studies, "At a CDER Regulatory Briefing held 29 May 2009, the panel discussed the 'Combination Rule,' that is, demonstrating the contribution of each component in a combination ... as it applies to a proposed ... combination product. When confirmatory clinical trials comparing the β-lactam alone to the combination product are not feasible, the panel concluded that there are other ways to reach the conclusion that both components contribute, such as supportive data from in vitro microbiology, PK/PD models, and **animal studies**."[80]

FDA's explanation referred to the Combination Rule, which is set forth in 21 CFR §300.50 and explained in FDA's Guidance for Industry.[81]

2. *Sponsor's Animal Models*

The Sponsor used several different animal models for testing AVYCAZ against infections. Each of these animal models used infection with bacterial strains expressing beta-lactamase, the enzyme that catalyzes ceftazidime degradation. The bacteria administered to the animals were *Klebsiella pneumoniae*, *Pseudomonas aeruginosa*, and various enterobacteriaceae. The experimental design compared ceftazidime alone with the AVYCAZ combination and also compared other antibacterial drugs with AVYCAZ.

FDA's review described the Sponsor's animal model studies, which substituted for a Phase 3 clinical trial with human subjects: "In this model, separate experimental systemic infections induced by seven enterobacteriaceae isolates were established by intraperitoneal injection to obtain an inoculum between 10-100 times the lethal dose. Mice were treated

[79] Avibactam/ceftazidime combination (AVYCAZ) (urinary tract infections) NDA 206-494. Page 4 of 33-page Cross Discipline Team Leader Review.

[80] Avibactam/ceftazidime combination (AVYCAZ) (urinary tract infections) NDA 206-494. Page 47 of 163-page Medical Review.

[81] U.S. Dept. Health and Human Services. Food and Drug Administration. Center for Drug Evaluation and Research (CBER). Guidance for Industry. Chronic obstructive pulmonary disease: developing drugs for treatment. 2007 (14 pp.).

subcutaneously at 0 and 4 hours post infection with CAZ-AVI ... and comparators ... cefepime, piperacillin-tazobactam ... [t]he activity of ceftazidime was restored when combined with avibactam against all seven isolates."[82,83]

3. *FDA Complained About the Animal Models*

FDA complained about the animal models used by the Sponsor. FDA's complaint was lack of validation: "animal models of infection were not validated ... [and were] conducted without GLP specification ... [f]or example, it was not clear that the methods used for the delivery of challenge agent ... ensured adequate standardization, replication of test conditions, and comparability between treatment arms."[84]

In addition to complaining about lack of validation, FDA complained that the mechanism of action of the infections in the animal models did not resemble that of the corresponding infections in humans: "Ultimately, the natural histories of the infections used in these **models** may not ... have sufficient similarities to extrapolate to humans."[85]

Despite these complaints, the FDA's review of the study drug moved forward, and the AVYCAZ combination received FDA approval. These complaints provide the general lessons that animal models for diseases should be validated and that the mechanisms of the disease in animals should track those in the corresponding disease in humans.

Package label. FDA's analysis and comments on animal models found a place on the package label, where the animal model information occurred in the Clinical Pharmacology section:

> CLINICAL PHARMACOLOGY ... Pharmacodynamics. As with other beta-lactam antimicrobial drugs, the time that ... plasma concentrations of ceftazidime exceeds the ... minimum inhibitory concentration ... against the infecting organism has been shown to best correlate with efficacy in a neutropenic **murine thigh infection model** with *Enterobacteriaceae* and *Pseudomonas aeuroginosa*. The time above a threshold concentration has been determined to be the parameter that best predicts the efficacy in vitro and in vivo nonclinical models.[86]

c. Hydromorphone (Acute and Chronic Pain) NDA 021-217

Opioids are used for treating acute severe pain. However, prolonged use of opioids has only partial analgesic efficacy for management of chronic pain.[87] Hydromorphone is an opioid agonist. The drug stimulates one of the opioid receptors, μ-opioid receptor.

[82] Avibactam/ceftazidime combination (AVYCAZ) (urinary tract infections) NDA 206-494. Page 35 of 163-page Medical Review.

[83] Avibactam/ceftazidime combination (AVYCAZ) (urinary tract infections) NDA 206-494. Pages 11–13 of 181-page Microbiology Review.

[84] Avibactam/ceftazidime combination (AVYCAZ) (urinary tract infections) NDA 206-494. Page 31 of 156-page Clinical Review (page 37 of 163-page pdf Medical Review).

[85] Avibactam/ceftazidime combination (AVYCAZ) (urinary tract infections) NDA 206-494. Page 31 of 156-page Clinical Review (page 37 of 163-page pdf Medical Review).

[86] Package insert. AVYCAZ (ceftazidime–avibactam) for infection, for intravenous use. February 2015 (20 pp.).

[87] Gendron L, et al. Molecular pharmacology of δ-opioid receptors. Pharmacol. Rev. 2016;68:631–700.

There are three opioid receptors, μ-opioid receptor, δ-opioid receptor, and κ-opioid receptor.[88] Agonists of μ-opioid receptor include β-endorphin (the endogenous ligand) and the drugs fentanyl, hydromorphone, and nalbuphine.[89] The endogenous agonists of δ-opioid receptor are the enkephalins and the deltorphins.[90] The endogenous agonists of κ-opioid receptor are the dynorphins.[91]

FDA's Pharmacology Review described testing in pregnant animals and resultant toxicity to the fetal animals. Regarding **infant rats**, "There were no effects on ... overall survival ... at the low dose. The viability and survival ... were decreased at 6.25 mg/kg/day."[92]

Regarding **Syrian hamsters**, the FDA reviewer wrote, "studies reported in the literature showed increased incidence of cranioschisis and exencephaly in hamsters given a single dose of hydromorphone at 125 mg/kg ... on gestation day 8 ... [f]etal brain lesions were also demonstrated in hamsters."[93]

Regarding **mice**, the FDA reviewer wrote, "[m]ice administered hydromorphone ... during organogenesis produced offspring with increased incidence of ... skeletal defects and soft tissue abnormalities ... split supraoccipital, checkerboard and split sternebrae, delayed ossification of paws, ectopic ossification sites, cryptorchidism, cleft palate, and malformed ventricles and retina."[94]

Regarding the package label, FDA concluded, "It is recommended that the results of ... studies on hydromorphone reproductive toxicity ... be included in the product label ... using exposure data for comparison between animals and humans."[95]

Package label. The package label for hydromorphone (Exalgo®) made use of the Sponsor's animal studies, as revealed by the warnings in the Use in Specific Populations section. Apparently as a justification for relying on animal toxicity, the package label stated that "There are no adequate and well-controlled studies in pregnant women."[96] The package label referred to Syrian hamsters and rats and was careful to compare rat dose with human dose (they were equivalent to each other):

> **USE IN SPECIFIC POPULATIONS** ... Pregnancy. Hydromorphone administration to pregnant Syrian hamsters ... produced skull malformations (exencephaly and cranioschisis) ... soft tissue malformations

[88] Sheng WS, et al. Human neural precursor cells express functional kappa-opioid receptors. J. Pharmacol. Exp. Ther. 2007;322:957–63.

[89] Gharagozlou P, et al. Activity of opioid ligands in cells expressing cloned mu opioid receptors. BMC Pharmacol. 2003;3:1 (8 pp.).

[90] Gharagozlou P, et al. Activity of opioid ligands in cells expressing cloned mu opioid receptors. BMC Pharmacol. 2003;3:1 (8 pp.).

[91] Sheng WS, et al. Human neural precursor cells express functional kappa-opioid receptors. J. Pharmacol. Exp. Ther. 2007;322:957–63.

[92] Hydromorphone (acute and chronic pain) NDA 021-217. Page 197 of 207-page Pharmacology Review.

[93] Hydromorphone (acute and chronic pain) NDA 021-217. Page 198 of 207-page Pharmacology Review.

[94] Hydromorphone (acute and chronic pain) NDA 021-217. Page 198 and 202 of 207-page Pharmacology Review.

[95] Hydromorphone (acute and chronic pain) NDA 021-217. Page 203 of 207-page Pharmacology Review.

[96] Package label. EXALGO (hydromorphone hydrochloride) extended release tablets. March 2010 (30 pp.).

(cryptorchidism, cleft palate, malformed ventricles and retina), and skeletal variations ... In the pre- and post-natal effects study in rats, neonatal viability was reduced at 6.25 mg/kg/day (about 1.2 times the human exposure following 32 mg/day).[97]

In referring to "about 1.2 times the human exposure," the warning complied with the recommendation in FDA's Pharmacology Review that the Sponsor should use "exposure data for comparison between animals and humans."[98] Regarding methodology for arriving at safety warnings, the Sponsor's comparison and prediction of human AEs relating to reduced viability was based on a crude estimate (mg/kg/day) instead of on the more accurate estimate provided by safety margin analysis.

d. Lorcaserin (Weight Loss) NDA 022-529

Lorcaserin is a weight loss drug that acts on the brain.[99,100] Lorcaserin is an agonist of 5-HT2C. receptor. The drug decreases food intake and body weight in rodents.[101] "5-HT" refers to 5-hydroxytryptamine, also known as serotonin. Garfield et al.[102] describe lorcaserin's site of action in the brain, namely, 5-HT2C receptors in the hypothalamus of the brain. In detail, 5-HT2C receptors are expressed on arcuate nucleus of POMC neurons in the hypothalamus.

The Sponsor's animal studies on lorcaserin focused on the AEs of various cancers, including astrocytoma (brain cancer), mammary adenocarcinoma (breast cancer), and benign fibroadenoma. FDA's Medical Review described safety margin calculations. Also, the package label includes an account of the safety margin and its implications on the AE of cancer, as was expected in patients.

FDA's Medical Review assessed these concepts relating to toxicity and exposure:

- Need to base safety margin calculations on the male rat (not female rat).
- Need to take into account highest dose where no astrocytoma occurs in male rats and lowest dose where astrocytoma does occur in male rats.
- Need to compare exposure value in male rats, for highest lorcaserin exposure where no astrocytoma occurs, and exposure in humans for the highest recommended dose in humans. Exposure is usually expressed in terms of blood plasma parameters (AUC, Cmax, tmax, Cmin) though drug concentrations in other body compartments can also be used.

[97] Package label. EXALGO (hydromorphone hydrochloride) extended release tablets. March 2010 (30 pp.).

[98] Hydromorphone (acute and chronic pain) NDA 021-217. Page 203 of 207-page Pharmacology Review.

[99] Farr OM, et al. Lorcaserin administration decreases activation of brain centers in response to food cues and these emotion- and salience-related changes correlate with weight loss effects: a 4-week-long randomized, placebo-controlled, double-blind clinical trial. Diabetes. 2016:65:2943–53.

[100] Lorcaserin (weight loss) NDA 022-529. 425-page pdf file for Medical Review.

[101] Martin CK, et al. Lorcaserin, a 5-HT(2 C) receptor agonist, reduces body weight by decreasing energy intake without influencing energy expenditure. J. Clin. Endocrinol. Metab. 2011;96:837–45.

[102] Garfield AS, et al . Increased alternate splicing of Htr2c in a mouse model for Prader-Willi syndrome leads disruption of 5HT2C receptor mediated appetite. Mol. Brain. 2016;9:95 (9 pp.).

FDA focused on the toxicity of astrocytoma (brain cancer). Regarding the appropriate gender of rats, FDA wrote that "these tumors were only seen in one specie and one sex."[103] This specie was rats and the gender was male. Lorcaserin toxicity (astrocytoma) was greater in male rats than in female rats. FDA's review revealed that the Sponsor had conducted studies in a **dose range of 10–30 mg/kg/day for male rats** and up to 100 mg/kg/day for female rats. Regarding female rats, FDA stated that "In the female rat, where astrocytoma was not increased even at the 100 mg/kg/day dose, exposure margin was calculated to be greater than 1000."[104]

In contrast, in male rats, "At the **10 mg/kg/day** (no astrocytoma seen) and **30 mg/kg/day** (astrocytoma seen) doses used in the ... male rat ... study, brain exposure margins relative to human brain at the maximum recommended dose were greater than or equal to 70 and 360, respectfully."[105]

FDA calculated the safety margin from toxicity data using rat astrocytoma tumor data, writing that "A safety margin of 70-fold for astrocytoma in rats, based on estimated brain levels of lorcaserin, presents a negligible clinical risk."[106] FDA reiterated this point, stating, "Our decision to allow the clinical program to proceed ... was based on ... increased safety margin to clinical exposure."[107]

Where the AE occurs in brain (astrocytoma), is it better to calculate safety margin from brain exposure values instead of from blood plasma exposure values? The safety margin provides a go/no-go answer to FDA as to whether the Sponsor's animal data establish that the study drug shows low risk for AEs in humans. The safety margin compares dosing and exposure levels for the test animal (at the NOAEL value) with dosing and exposure levels for the highest recommended dose for patients. A methodological issue that arises is whether the AE occurs in a protected compartment in the body, such as the brain. By "protected compartment," this means an area where flow of metabolites from the bloodstream to the compartment, and from the compartment to the bloodstream is impaired or restricted, when compared to flow from the bloodstream into hepatocytes (liver cells) and of flow from hepatocytes to the bloodstream.

FDA's Pharmacology Review took into account this glitch in safety margin calculations and provided an account of exposure measurements in animals (brain levels, cerebrospinal fluid (CSF) levels) and exposure measurements in humans (CSF levels only). Lorcaserin concentrations in human CSF were used as a surrogate for lorcaserin concentrations in human brain.

FDA found it reasonable that CSF lorcaserin levels be a surrogate for brain lorcaserin levels and observed that rat brain exposure was over 300 times that of human brain exposure, writing, "Brain astrocytomas in male rats were increased in groups treated with the mid and high doses (30 and 100 mg/kg) ... [t]he applicant ... submitted information on cerebrospinal fluid levels of lorcaserin from mice, rats, monkeys and humans. These data

[103] Lorcaserin (weight loss) NDA 022-529. Page 148 of 425-page pdf file for Medical Review.

[104] Lorcaserin (weight loss) NDA 022-529. Page 25 of 425-page pdf file for Medical Review.

[105] Lorcaserin (weight loss) NDA 022-529. Pages 25 and 146 of 425-page pdf file for Medical Review.

[106] Lorcaserin (weight loss) NDA 022-529. Page 23 of 425-page pdf file for Medical Review.

[107] Lorcaserin (weight loss) NDA 022-529. Page 204 of 425-page pdf file for Medical Review.

indicated that cerebrospinal fluid levels could be used to reasonably predict brain levels. Based on these data, the AUC in the rat brain at **30 mg/kg** appears to be over 300 times higher than the likely AUC in human brain at the clinical dose."[108]

Rat exposure data originally submitted by the Sponsor in an original NDA misrepresented the risk for astrocytomas in humans, as shown by some careful control studies conducted by the Sponsor. Following submission of the original NDA, the Sponsor measured plasma/CSF lorcaserin ratios in the rat and plasma/CSF lorcaserin ratios in human subjects.

The results showed that, with passage from ***rat blood to rat CSF***, lorcaserin concentrations increased greatly and that, in contrast, with passage from ***human blood to human CSF***, there was only a slight increase in lorcaserin. On this point, FDA's Pharmacology Review wrote:

> The additional clinical study conducted by the sponsor showed that the level of lorcaserin in the cerebrospinal fluid (CSF) of humans **is much lower than anticipated** based on studies in ... rodents ... [r] easons for the unexpectedly low level of lorcaserin in human CSF were not addressed. Based on a relatively constant relationship of CSF to (total) brain levels of drug as measured in ... rodents, it is estimated that exposure in human brain tissue is 1.7-fold higher than plasma drug levels. This data substantially changed the safety margin for brain neoplasms to 70-fold the clinical dose compared to 5- or 14-fold from the original NDA. Generally, a safety margin in excess of 25-fold the clinical dose for a rodent carcinogen would not be considered likely to reflect a relevant risk to humans.[109]

FDA's Pharmacology Review reiterated and emphasized the fact that in the human, lorcaserin levels in CSF were only slightly higher than in the blood: "Brain distribution studies found preferential distribution of lorcaserin to the brain in mice (26 × the plasma), rats (13–35 × the plasma) and monkeys (10 × the plasma) ... [u]sing human CSF data, the brain lorcaserin exposure in humans was estimated to be 1.7 × the plasma exposure."[110]

To add a word of extra detail, FDA's Medical Review provided data on human exposure, in terms of AUC and Cmax, for both human plasma and human CSF. With a lorcaserin dose of 10 mg twice a day,[111] exposure in humans was plasma Cmax = 61.7 ng/mL.[112] CSF exposure was Cmax = 0.87 ng/mL.[113] The ratio of drug exposure for [CSF]/[blood plasma], in terms of AUC, was 0.017. The ratio of drug exposure for [CSF]/[blood plasma], in terms of Cmax, was 0.014.[114]

Muller et al.[115] teach the generally applicable concept that, "Toxicity often occurs in tissues other than those targeted to achieve drug efficacy. However, as direct measurements of drug exposure in tissues are generally not feasible (particularly in humans), plasma

[108] Lorcaserin (weight loss) NDA 022-529. Page 2 of 539-page pdf file for Pharmacology Review.

[109] Lorcaserin (weight loss) NDA 022-529. Page 6 of 539-page pdf file for Pharmacology Review.

[110] Lorcaserin (weight loss) NDA 022-529. Page 25 of 539-page pdf file for Pharmacology Review.

[111] Lorcaserin (weight loss) NDA 022-529. Pages 17 and 22 of 425-page pdf file for Medical Review.

[112] Lorcaserin (weight loss) NDA 022-529. Page 25 of 425-page pdf file for Medical Review.

[113] Lorcaserin (weight loss) NDA 022-529. Page 25 of 425-page pdf file for Medical Review.

[114] Lorcaserin (weight loss) NDA 022-529. Page 25 of 425-page pdf file for Medical Review.

[115] Muller PY, Milton MN. The determination and interpretation of the therapeutic index in drug development. Nat. Rev. Drug Discov. 2012;11:751–61.

exposure to the drug is usually used as a surrogate for tissue exposure."[116] FDA's Pharmacology Review referred to the published literature, regarding the partitioning of drugs from blood to CSF, in rodents, monkeys, and humans.[117,118]

Package label. FDA's observations on rat astrocytoma for lorcaserin (Belviq®) found a place in the Nonclinical Toxicology section of the package label. The label referred to the fact that astrocytoma did occur in rats, where rat doses resulted in 17 times or greater than the plasma exposure in humans treated with lorcaserin. Also, the label referred to the fact that astrocytoma did not occur in rats, at a rat dose corresponding to 70-fold less than exposure in humans treated with lorcaserin. Both of these statements, that is, the statement regarding "17-times human clinical dose" and the statement regarding "clinical dose is ... 70-fold lower than brain exposure in rats") support the conclusion that the study drug does not pose risk for astrocytoma in humans. The label read:

> **NONCLINICAL TOXICOLOGY**... In male rats, treatment-related neoplastic changes were observed in ... the brain (astrocytoma) at greater than or equal to 30 mg/kg (plasma exposure 17-times human clinical dose) ... [h]uman brain exposure ($AUC_{24h,ss}$) to lorcaserin at the clinical dose is estimated to be 70-fold lower than brain exposure in rats at the dose at which no increased incidence of astrocytoma was observed.[119]

e. Lumacaftor/Ivacaftor (Cystic Fibrosis) NDA 206-038

This concerns the lumacaftor/ivacaftor combination for treating cystic fibrosis. The Sponsor determined the safety margin using PK studies of one of the components of this combination (lumacaftor). Table 6.1 provides all the relevant numbers for calculating the safety margin. Walking through the steps, the rat NOAEL is 1000 mg, and this dose gives a rat exposure of 1300 (value for AUC). The rat exposure is 4.3 × greater than the human value for AUC, given the expected clinical dose of 600 mg. This accounts for the number "4.3" in Table 6.1. According to FDA's Pharmacology Review, "No dose-limiting toxicity or target organs of toxicity were identified in these studies." For this reason, the value for NOAEL was the highest dose given in the study.[120]

In fact, the Sponsor had specifically tested for various types of toxicity, but none were detected, as stated by FDA's review, "There were no biologically significant changes of hematology, coagulation, or clinical chemistry parameters. No target organs of toxicity were identified."[121]

Also, the Sponsor did separate calculation of the safety margin, but based on a 12-month dog study. Here also, safety margin calculations were based on the 600 mg/day dose for humans. Table 6.2 shows data from the 12-month dog study.

[116] Muller PY, Milton MN. The determination and interpretation of the therapeutic index in drug development. Nat. Rev. Drug Discov. 2012;11:751–61.

[117] Lin JH. CSF as a surrogate for assessing CNS exposure: an industrial perspective. Curr. Drug Metab. 2008;9:46–59.

[118] Watson J, et al. Receptor occupancy and brain free fraction. Drug Metab. Dispos. 2009;37:753–60.

[119] BELVIQ (lorcaserin hydrochloride) tablets, for oral use. June 2012 (23 pp.).

[120] Lumacaftor/ivacaftor (cystic fibrosis) NDA 206-038. Page 360 of 502-page Pharmacology Review.

[121] Lumacaftor/ivacaftor (cystic fibrosis) NDA 206-038. Page 356 of 502-page Pharmacology Review.

TABLE 6.1 Six-Month Rat Data Used to Calculate Safety Margin

NOAEL (mg/kg/day)		AUC in Rats (μg hour/mL)	Safety Margin for the Proposed Clinical Dose (Clinical Dose of 600 mg/day Gives AUC = 300 μg hour/mL)
1000	Male rats	1300	4.3
	Female rats	3160	10.5

Lumacaftor/ivacaftor (cystic fibrosis) NDA 206-038. Page 360 of 502-page Pharmacology Review.

TABLE 6.2 Twelve-Month Dog Data Used to Calculate Safety Margin

NOAEL (mg/kg/day)		AUC (μg hour/mL)	Safety Margin for the Proposed Clinical Dose (Clinical Dose of 600 mg/day Gives AUC = 300 μg hour/mL)
500	Male dogs	429	1.4
	Female dogs	515	1.7

Lumacaftor/ivacaftor (cystic fibrosis) NDA 206-038. Page 360 of 502-page Pharmacology Review.

FDA stated that the safety margins as determined in the rat studies and dog studies were greater or equal to 1.0 and concluded that these are adequate safety margins: "margins for clinical exposures to lumacaftor at proposed doses of 600 mg/day and 400 mg q12hr (800 mg/day) relative to NOAELs in the 6-month rat and 12-month dog studies with lumacaftor alone ... are shown ... [s]afety margins were ≥ 1, which were considered adequate."[122]

The same study design was used for ivacaftor alone. The 6-month rat study NOAEL was 50 mg/kg/day and the 12-month dog study NOAEL was 60 mg/kg/day. The male rat AUC was 445 μg hour/mL and the male dog AUC was 351 μg hour/mL. To calculate the exposure margin, the value for human AUC was needed and for the 250 mg human dose, this was 7.6 μg hour/mL. Performing the calculations revealed that the **safety margin was 58.6**, based on comparing male rat AUC with human AUC. Also, the **safety margin was 46.2**, based on comparing male dog AUC with human AUC.[123]

This concerns cataracts. Despite the fact that the safety margins for both lumacaftor and ivacaftor were within the acceptable limit, the Warnings and Precautions section did warn against one type of AE, cataracts. The Sponsor had detected cataracts only in a special study of juvenile rats. (Apparently, the toxicity studies used for calculating NOAEL values were with older rats.) Regarding the juvenile rat studies, FDA's Pharmacology review referred to the fact that the juvenile rats were only 7 to 35 days old. Please note that rats are weaned at the age of 21 days.

FDA's account of package label warnings for cataracts described the basis in animal toxicity studies: "Cataracts were observed in a study in juvenile rats; cataracts were also

[122] Lumacaftor/ivacaftor (cystic fibrosis) NDA 206-038. Page 360 of 502-page Pharmacology Review.

[123] Lumacaftor/ivacaftor (cystic fibrosis) NDA 206-038. Page 364 of 502-page Pharmacology Review.

observed in pediatric patients and **the finding is listed in the 'Warnings and Precautions' section of the product label**[124] ... [c]ataracts were seen in juvenile rats dosed with ivacaftor from postnatal day 7–35 at dose levels of 10 mg/kg/day and higher ... [t]he observation of bilateral posterior subcapsular catacts of the lens in one ... female was attributed to treatment with ivacaftor."[125]

FDA's account of package label warnings for cataracts also referred to the level of risk for children: "there is ongoing clinical concern regarding this risk in children with cystic fibrosis who are treated with ivacaftor. However, the isolated cataract finding in the combination study was judged to represent an acceptable risk for the proposed lumacaftor/ivacaftor combination patient population of CF patients age 12 years and older[126] ... [b]aseline and follow-up ophthalmological examinations are recommended in pediatric patients initiating ivacaftor treatment."[127]

Package label. The Warnings and Precautions section and the Nonclinical Toxicology section for the lumacaftor/ivacaftor combination (Orkambi®) warned against cataracts. The warnings stated that the product was a combination drug and that only the ivacaftor component was responsible for cataracts:

> WARNINGS AND PRECAUTIONS ... Cataracts: Non-congenital lens opacities/cataracts have been reported in pediatric patients treated with ivacaftor, a component of ORKAMBI. Baseline and follow-up examinations are recommended in pediatric patients initiating ORKAMBI.[128]

> NONCLINICAL TOXICOLOGY ... Animal Toxicology ... Cataracts were seen in juvenile rats dosed with ivacaftor from postnatal day 7–35 at oral dose levels of 10 mg/kg/day and higher (approximately 0.3 times the maximum recommended human dose (MRHD) for the ivacaftor component of ORKAMBI based on summed AUCs of ivacaftor and metabolites). This finding has not been observed in older animals.[129]

f. Nebivolol (Hypertension) NDA 021-742

Nebivolol is for treating hypertension. This concerns animal models for safety. The Sponsor discovered that nebivolol resulted in Leydig cell tumors in mice. This discovery was made in a 2-year study with nebivolol administered to mice at 40 mg/kg/day. The goal of the study was to discover any associations of the study drug with tumors.

FDA complained about the observation that nebivolol was associated with Leydig cell tumors and expressed skepticism on the relevance of the mouse data to humans. The outcome of FDA's skepticism is shown by FDA's request, "It will therefore be necessary for

[124] Lumacaftor/ivacaftor (cystic fibrosis) NDA 206-038. Pages 2, 18, 32, and 18 of 502-page Pharmacology Review.

[125] Lumacaftor/ivacaftor (cystic fibrosis) NDA 206-038. Pages 2, 18, 32, and 82 of 502-page Pharmacology Review.

[126] Lumacaftor/ivacaftor (cystic fibrosis) NDA 206-038. Pages 2, 18, 32, and 82 of 502-page Pharmacology Review.

[127] Lumacaftor/ivacaftor (cystic fibrosis) NDA 206-038. Page 60 of 99-page Medical Review.

[128] Package label. ORKAMBI™ (lumacaftor/ivacaftor) tablet, for oral use. July 2015 (12 pp.).

[129] Package label. ORKAMBI™ (lumacaftor/ivacaftor) tablet, for oral use. July 2015 (12 pp.).

you to establish the mechanisms by which nebivolol is responsible for these findings and demonstrate that these findings are not relevant in humans."[130]

The Sponsor argued that the mouse tests did not support the notion that nebivolol is a carcinogen (causes mutations in the genome). To this point, the Sponsor argued that nebivolol's association with cancer was mediated by an endocrinological mechanism (and not by a carcinogenic mechanism) and that the occurrence of Leydig cell tumors in mice was not relevant to humans. The Sponsor's train of thought is shown below:

1. *First Reason Mouse Cancer Tests Not Relevant*

Animal gene-toxicity tests where nebivolol was administered to mice showed that nebivolol did not produce mutations. The lack of mutations suggested that nebivolol is not a carcinogen.

2. *Second Reason Mouse Cancer Tests Not Relevant*

In tests on mice, nebivolol was found to influence hormone levels. Nebivolol increased luteinizing hormone (LH), where these increases were associated with Leydig cell hyperplasia. The Sponsor's goal was to discount the possibility that nebivolol was a carcinogen and to argue that nebivolol's influence on Leydig cells was endocrinological, and not due to genetic mutations. Some cancers have an endocrinological component, such as breast cancer and prostate cancer. Note that breast cancer treatment includes ovarian suppression with tamoxifen,[131,132] while prostate cancer treatment includes androgen deprivation therapy with leuprolide or goserelin.[133]

3. *Third Reason Mouse Cancer Tests Not Relevant*

The Sponsor compared nebivolol dosing in mice and human subjects. The Sponsor argued that the Leydig cell tumors found in mice were not relevant to patients, writing that "Mice tumors occur at doses 424-fold higher than the 10 mg daily maximum recommended human dose."[134]

4. *Fourth Reason Mouse Cancer Tests Not Relevant*

The Sponsor also argued that, unlike the situation with mice, the Sponsor's clinical studies with human subjects did not reveal any endocrinologic tumors and did not reveal

[130] Nebivolol (hypertension) NDA 021-742. Page 3 of 80 pages of first pdf file of 12 pdf files of the Medical Review.

[131] Francis PA, et al. Adjuvant ovarian suppression in premenopausal breast cancer. New Engl. J. Med. 2015;372:436–46.

[132] Griggs JJ, et al. American Society of Clinical Oncology endorsement of the cancer care Ontario practice guideline on adjuvant ovarian ablation in the treatment of premenopausal women with early-stage invasive breast cancer. J. Clin. Oncol. 2011;29:3939–42.

[133] Djavan B, et al. Testosterone in prostate cancer: the Bethesda consensus. BJU Int. 2012;110:344–52.

[134] Nebivolol (hypertension) NDA 021-742. Page 53 of 80-page pdf file, in the second pdf file of 12 pdf files in the Medical Review.

any Leydig cell tumors, writing, "No increase in endocrinologic or Leydig cell tumors have been observed in nebivolol's ... adverse event database."[135]

5. *FDA Insisted That the Sponsor Produce More Data on the Etiology of Leydig Cell Tumors*

FDA focused on mechanism-of-action analysis as a way to establish relevance (or irrelevance) of nebivolol to Leydig cell tumors. FDA observed that there were seven mechanisms in rodents, where five of these were relevant to humans. Thus, mechanism-of-action analysis failed to exclude the possibility that the study drug increased risk for tumors in patients.

FDA's accounting of the mechanisms of the Leydig cell tumors took the form, "As stated in our original consult ... there appear to be 7 mechanisms described in the literature for the development of Leydig cell tumors in rodents. Two of these are considered to be **not relevant to humans** (GnRH and dopamine agonist). Five are potentially **clinically relevant**. These are: (1) Estrogen receptor agonism. (2) Androgen receptor antagonism. (3) Inhibition of testosterone biosynthesis. (4) Inhibition of 5-alpha-reductase activity. (5) Inhibition of aromatase activity."[136]

The medical literature describes the endocrinological basis for Leydig cell tumors, as it applies to mice and humans.[137,138] FDA's condition for approving nebivolol was that "nebivolol was Approvable if the sponsor could establish the **mechanism** by which nebivolol was responsible for ... [tumors] in male mice, prove that the findings were **not relevant in humans**."[139]

FDA's reasoning and logic regarding conditions needed to prove that the Sponsor's mouse data were not relevant to humans are shown by the following arguments.

6. *First Argument. Nebivolol Stimulates Luteinizing Hormone (LH) in Mice*

The Sponsor conducted these tests with male mice:

- Nebivolol alone
- Nebivolol plus hormone injections (dihydrotestosterone, DHT)
- Dihydrotestosterone (DHT) alone

The result was that nebivolol alone increases **Leydig cell tumors**, that DHT alone resulted in **Leydig cell atrophy**, and that the combination of nebivolol with DHT also

135 Nebivolol (hypertension) NDA 021-742. Page 53 of 80-page pdf file, in the second pdf file of 12 pdf files in the Medical Review.

136 Nebivolol (hypertension) NDA 021-742. Page 3 of 80 pages of first pdf file of 12 pdf files of the Medical Review.

137 Mikola M, et al. High levels of luteinizing hormone analog stimulate gonadal and adrenal tumorigenesis in mice transgenic for the mouse inhibin-alpha-subunit promoter/simian virus 40 T-antigen fusion gene. Oncogene. 2003;22:3269–78.

138 Canto P, et al. Mutational analysis of the luteinizing hormone receptor gene in two individuals with Leydig cell tumors. Am. J. Med. Genet. 2002;108:148–52.

139 Nebivolol (hypertension) NDA 021-742. Page 47 of 80 pages of first pdf file of 12 pdf files of the Medical Review.

resulted in **Leydig cell atrophy**. The Sponsor concluded that the real cause of the mouse tumors resulted from nebivolol's stimulation of LH and that DHT's influence in preventing this effect was due to DHT's inhibition of LH.[140]

The Sponsor detected a species difference between mice and rats. Although nebivolol stimulated an increase in LH in mice, the Sponsor determined that nebivolol did not have this influence in rats. Regarding this species difference, FDA wrote, "In a 28-day study of nebivolol ... administered ... to male mice increased serum luteinizing hormone (LH) ... and . Leydig cell hyperplasia were observed at day 28 ... [t]aken together, this data supports Leydig cell tumors in mice may be mediated by LH surges ... [h]owever, the absence of LH elevations in rats despite testicular and sperm effects remains unexplained by this proposed mechanism."[141]

FDA's Pharmacology Review provided the data from mice treated with nebivolol alone and mice treated with nebivolol plus DHT:

- **Nebivolol alone**. "Of the 57 animals given just nebivolol, 44/57 showed Leydig cell hyperplasia."[142]
- **Nebivolol plus DHT**. "Of the 52 animals given nebivolol + DHT, 0 were reported to have Leydig cell hyperplasia."[143]

7. Second Argument. Nebivolol Does Not Stimulate Serum Luteinizing Hormone (LH) in Humans

In humans, LH binds to receptors in the Leydig cells of the testes.[144] Leydig cells from human testes can be stimulated by LH.[145] LH stimulates the Leydig cells of the testes, where this stimulation results in testosterone production. The Sponsor's findings with human subjects demonstrated that nebivolol did not increase serum levels of LH. This fact led to the Sponsor concluding that the mouse tumor data were only relevant to mice and not relevant to humans.

FDA's Medical Review arrived at the conclusion that "healthy male volunteers to determine the effects of nebivolol on ... leuteinizing hormone ... pharmacodynamic endpoints were the ... levels of leuteinizing hormone (LH) measurements ... [t]he study consisted of ... nebivolol or placebo ... [and] demonstrated **nebivolol had no significant effect on ... serum leuteinizing hormone (LH)** ... [t]herefore, after 49 days of daily nebivolol

[140] Nebivolol (hypertension) NDA 021-742. Page 47 of 80 pages of first pdf file of 12 pdf files of the Medical Review.

[141] Nebivolol (hypertension) NDA 021-742. Pages 22, 27, 28, and 29 of 80-page pdf file of Pharmacology Review. First of seven pdf files for Pharmacology Review.

[142] Nebivolol (hypertension) NDA 021-742. Page 17 of 80-page pdf file of Pharmacology Review. First of seven pdf files for Pharmacology Review.

[143] Nebivolol (hypertension) NDA 021-742. Page 17 of 80-page pdf file of Pharmacology Review. First of seven pdf files for Pharmacology Review.

[144] Davies TF, et al. Regulation of primate testicular luteinizing hormone receptors and steroidogenesis. J. Clin. Invest. 1979;64:1070–3.

[145] Vaucher L, et al. Activation of GPER-1 estradiol receptor downregulates production of testosterone in isolated rat Leydig cells and adult human testis. PLoS One. 2014;9:e92425 (7 pp.).

treatment ... nebivolol did not demonstrate any significant changes in ... gonadal function. These findings suggest that the **Leydig cell tumors in male mice are species specific.**"[146]

Despite the fact that FDA agreed with the Sponsor's arguments that nebivolol's association with Leydig cell tumors in mice had no relevance to humans, the Sponsor's data on Leydig cell tumors found a place on the package label. But the disclosure on the package label simply reiterated the Sponsor's argument that nebivolol's association with mouse Leydig cell tumors had no relevance to humans. Apparently because of the neutral nature of these data and because the data were not a cause for alarm, the description of Leydig cell tumors found a place in the Nonclinical Toxicology section and not in the Warnings and Precautions section.

8. Third Argument. Finasteride Experiment Used to Confirm the Proposed Mechanism of Nebivolol

The Sponsor conducted a positive control experiment using a chemical known to stimulate LH production. The positive control experiment was with finasteride, a chemical known to stimulate production of LH. Finasteride was expected to mimic the effects of nebivolol in causing Leydig cell tumors. The result from the finasteride test was that "Chronic administration of finasteride ... produce ... **Leydig cell adenomas** ... in mice after 19 months of treatment ... **Leydig cell hyperplasia** was observed in rats."[147]

The Sponsor's finasteride experiment further tracked the Sponsor's nebivolol study by addressing the effect of the combination of finasteride plus DHT. The result was that "Administration of exogenous DHT to rats treated with ... finasteride ... **prevented** elevation of serum leuteinizing hormone (LH) levels and consequent **Leydig cell hyperplasia.**"[148]

The Sponsor's finasteride experiments further tracked the Sponsor's nebivolol experiments by showing that finasteride effects as observed in rodents did not occur in human subjects. To this end, the Sponsor determined that "Chronic finasteride treatment produces only slight, **nonsignificant increases in leuteinizing hormone (LH) in men**. Therefore, the ... drug-related alterations in hormonal balance is a rodent-specific effect."[149]

Package label. The package label for nebivolol (Bystolic®) referred to the mouse studies showing the association of nebivolol with Leydig cell hyperplasia in mice. The package label took care to explain why the mouse Leydig cell cancer and Leydig cell hyperplasia found in mice were not relevant to humans: (1) The mouse nebivolol dose (mg/kg body

[146] Nebivolol (hypertension) NDA 021-742. Page 47 of 80 pages of first pdf file of 12 pdf files of the Medical Review.

[147] Nebivolol (hypertension) NDA 021-742. Pages 22, 26, and 27 of 80-page pdf file of Pharmacology Review. First of seven pdf files for Pharmacology Review.

[148] Nebivolol (hypertension) NDA 021-742. Pages 22, 26, and 27 of 80-page pdf file of Pharmacology Review. First of seven pdf files for Pharmacology Review.

[149] Nebivolol (hypertension) NDA 021-742. Pages 22, 26, and 27 of 80-page pdf file of Pharmacology Review. First of seven pdf files for Pharmacology Review.

weight) was much greater than the human dose. (2) In humans, nebivolol does not stimulate LH, unlike the situation with mice. The Nonclinical Toxicology section of the package label stated:

> **NONCLINICAL TOXICOLOGY** ... In a two-year study of nebivolol in mice, a ... significant increase in the incidence of testicular Leydig cell hyperplasia ... was observed at 40 mg/kg/day (5 times the maximally recommended human dose of 40 mg on a mg/m^2 basis). Similar findings were not reported in mice administered doses equal to approximately 0.3 or 1.2 times the maximum recommended human dose. No evidence of a tumorigenic effect was observed in a 24-month study in Wistar rats receiving doses of nebivolol 2.5, 10 and 40 mg/kg/day (equivalent to 0.6, 2.4, and 10 times the maximally recommended human dose). Co-administration of dihydrotestosterone (DHT) reduced blood luteinizing hormone (LH) levels and prevented the Leydig cell hyperplasia, consistent with an indirect LH-mediated effect of nebivolol in mice and not thought to be clinically relevant in man.[150]

g. Regorafenib (Colorectal Cancer) NDA 203-085

Regorafenib is a tyrosine kinase inhibitor that targets VEGF receptor, FGF receptor, and PDGF receptor. VEGF, FGF, and PDGF are the natural ligands of these three receptors, respectively, and each transmits a signal to the cell that promotes angiogenesis.[151,152] In treating cancer, regorafenib blocks these signals, thereby preventing the angiogenesis that stimulates new blood vessel formation, where the vessels provide nutrients to growing tumor cells.

FDA's Medical Review of regorafenib provides take-home lessons on these topics:

1. **Animal toxicity data and the package label.** Data on the study drug's animal toxicity (in absence of data on toxicity to humans) found a place in the package label.
2. **Metabolism in animals versus metabolism in humans.** The animal that is chosen for studies on efficacy and safety can be based on the following criteria. One criterion is whether the catabolic pathways that degrade the drug are similar in animals and humans. Another criterion is whether the metabolic pathway targeted by the drug is similar in animals and humans.
3. **Testing drug safety in an isolated organ versus testing in a live animal.** Toxicity of a given drug may differ markedly when tested in a living animal versus with an isolated organ.
4. **Efficacy of administered drug metabolites.** Just as study drugs are subjected to efficacy and safety testing, metabolites of the same study drug can be tested for efficacy and safety. Purified reagent-grade metabolites are used for these studies.

The Sponsor's data and FDA's comments on these topics are revealed below.

[150] Package label. BYSTOLIC® (nebivolol) tablets for oral use. December 2011 (14 pp.).

[151] Wilhelm SM, et al. Regorafenib (BAY 73-4506): a new oral multikinase inhibitor of angiogenic, stromal and oncogenic receptor tyrosine kinases with potent preclinical antitumor activity. Int. J. Cancer. 2011;129:245–255.

[152] Adenis A, et al. Survival, safety, and prognostic factors for outcome with regorafenib in patients with metastatic colorectal cancer refractory to standard therapies: results from a multicenter study (REBACCA) nested within a compassionate use program. BMC Cancer. 2016;16:412 (8 pp.).

1. *Animal Toxicity Data and the Package Label*

Regarding the animal study results, FDA observed toxicity to dentin, epiphyseal growth plates, and thickening of the atrioventricular valve. FDA's Clinical Review and Pharmacology Review disclosed that "changes in dentin and epiphyseal growth plates were present in both species [rats and dogs] ... and may be relevant to the pediatric population[153] ... in rats there were histopathological findings in the heart including perivascular/interstitial edema and pericarditis in the 4-week rat study and thickening of the atrioventricular valve in the 26-week rat study."[154]

Showing further concern on regorafenib's toxicity to bone during growth, FDA's Pharmacology Review observed, "Findings of changes in the epiphyseal growth plate and alterations in dentin suggest increased toxicity in developing organs and have been observed with ***other compounds*** that inhibit VEGFR signaling. These findings may be more relevant to a pediatric patient population." FDA's use of the term ***"other compounds"***[155] invokes the concept of drug class analysis. Drug class analysis is a tool for predicting AEs in humans for the purpose of package labeling.

2. *Metabolism in Animals Versus Metabolism in Humans*

In general, Sponsors conduct metabolic studies that identify metabolites of the study drug, and where these metabolites are typically given the names "M-1," "M-2," "M-3," "M-4," "M-5," and so on. Regorafenib's metabolism was compared using rats, dogs, and human subjects.[156] Regorafenib's renal toxicity is different in humans versus in animals.

Regorafenib is detoxified in humans, where human metabolism converts regorafenib into metabolites that are not toxic to kidneys. This conversion occurs in humans, but is not much found in rats and dogs. FDA's Pharmacology Review stated:

> Renal findings in **rats and dogs** included glomerulopathy, tubular degeneration/regeneration, tubular dilation, and interstitial fibrosis. No renal toxicity was noted in 1-month studies with either the M-2 or the M-5 metabolite which suggests that **differences in metabolism between humans, rats, and dogs** leading to significantly higher human exposures to M-2 and M-5 compared to the species used for toxicological assessment may account for **higher levels of renal toxicity seen in animals** compared to humans in trials.[157]

In view of the lack of renal toxicity in humans, the package label did not mention renal toxicity for humans and refrained from mentioning the metabolic differences between

[153] Regorafenib (metastatic colorectal cancer after prior treatment with various drugs) NDA 203-085. Pages 10–16 of 69-page Clinical Review.

[154] Regorafenib (colorectal cancer) NDA 203-085. Pages 12, 183, and 214 of 219-page Pharmacology Review.

[155] Regorafenib (colorectal cancer) NDA 203-085. Pages 12, 183, and 214 of 219-page Pharmacology Review.

[156] Regorafenib (colorectal cancer) NDA 203-085.

[157] Regorafenib (colorectal cancer) NDA 203-085. Page 12 of 219-page Pharmacology Review. Page 17 of 69-page Clinical Review.

humans and animals.[158] M-2 is the N-oxide of regorafenib and M-5 is the demethylated N-oxide of regorafenib.[159] Consistent with the picture that administering study drug is harmful to the kidney in animals (because of lack of catabolism) but not in humans (because of rapid catabolism), FDA wrote that "the applicant conducted a review of ... **acute renal failure** and did not identify an increased risk of renal failure in regorafenib-treated **patients**."[160]

This observation was part of FDA's analysis of AEs that were categorized by system of the body, that is, by the System Organ Class (SOC) organization. The organization of AEs by SOC is detailed in the cited references.[161,162,163] The take-home lesson is that investigators should make every effort to confirm that any given type of toxicity, as found in animals, can reasonably be used to predict AEs in humans. As stated above, the package label was complete silent on toxicity to kidneys or on renal toxicity. Even the Nonclinical Toxicology section of the package label was silent regarding renal toxicity.[164,165]

The Sponsor's observations regarding regorafenib to M-2 and M-5 metabolism in humans versus in animals was confirmed by Zopf et al.[166] Zopf et al. stated that metabolism in humans is robust: "Clinical trials in patients ... demonstrated that total plasma exposure to each of the metabolites, M-2 and M-5, at steady state was comparable to that of the parent compound after administration of regorafenib."[167]

Continuing with an account of the slower regorafenib metabolism in animals, Zopf et al. stated, "In contrast, when regorafenib was administered to mice ... the majority of exposure attributed to regorafenib (82%), while M-2 and M-5 accounted for 16% and 2% of total exposure, respectively ... [a] Cmax of 4146 μg/L was measured for regorafenib (the administered compound), whereas the Cmax of in vivo formed metabolites M-2 and M-5 reached only 753 μg/L and 63 μg/L."[168]

[158] Package label. STIVARGA (regorafenib) tablets, oral. September 2012 (14 pp.).

[159] Zopf D, et al. Regorafenib (BAY 73-4506): preclinical pharmacology and quantification of its major metabolites. Abstract 1666. 2010. DOI:10.01158.

[160] Regorafenib (colorectal cancer) NDA 203-085. Page 55 of 69-page Clinical Review.

[161] International Federation of Pharmaceutical Manufacturers and Associations. Introductory Guide MedDRA Version 14.0. Chantilly, VA: MedDRA Maintenance and Support Services Organization. 2011 (77 pp.).

[162] U.S. Department of Health and Human Services. Food and Drug Administration. Center for Drug Evaluation and Research (CDER). Center for Biologics Evaluation and Research (CBER). Guidance for Industry. E2C(R2) Periodic Benefit Risk Evaluation Report (PBRER). 2016 (51 pp.).

[163] Babre D. Medical coding in clinical trials. Perspect. Clin. Res. 2010;1:29–32.

[164] Package label. STIVARGA® (regorafenib) tablets, oral. September 2012 (14 pp.).

[165] Package label. STIVARGA® (regorafenib) tablets, oral. April 2017 (23 pp.).

[166] Zopf D, et al. Pharmacologic activity and pharmacokinetics of metabolites of regorafenib in preclinical models. Cancer Med. 2016;5:3176–85.

[167] Zopf D, et al. Pharmacologic activity and pharmacokinetics of metabolites of regorafenib in preclinical models. Cancer Med. 2016;5:3176–85.

[168] Zopf D, et al. Pharmacologic activity and pharmacokinetics of metabolites of regorafenib in preclinical models. Cancer Med. 2016;5:3176–85.

3. *Testing Drug Safety in an Isolated Organ Versus Testing in a Live Animal*

This illustrates the difference in detected cardiac toxicity, when assessed using isolated Purkinje fibers versus in a live animal. The tests involved administering regorafenib, purified M-2, or purified M-5 to animals, or by including in culture media of the Purkinje fibers.

A. TESTS WITH ISOLATED PURKINJE FIBERS.

Using Purkinje fiber assays, the Sponsor tested the toxicity of regorafenib, of pure M-2, and of pure M-5. Regorafenib was not toxic while, in contrast, M-2 and M-5 were toxic using the in vitro Purkinje fiber test. FDA's Pharmacology Review referred to the toxicities of regorafenib, M-2, and M-5: "With in vitro experiments **regorafenib** itself showed low potential for QTc prolongation with no prolongation of action potential duration in a Purkinje fiber assay and an IC50 of 27 μM in the hERG assay; however, **the M-2 and M-5 metabolites** had IC50s of 1.1 and 1.8 μM, respectively, in the hERG assay suggesting a considerably higher potential for QTc prolongation."[169]

The methodologies for Purkinje fiber assays and hERG assays for cardiac toxicity are detailed in the cited references.[170,171,172]

B. TESTS WITH DOGS.

With tests on live animals, cardiac toxicity was not found. FDA's Pharmacology Review reported that "The applicant evaluated hemodynamics, ECG and respiration in dogs following a single dose of regorafenib administered intraduodenally ... and intravenously ... no treatment-related toxicity was observed. These endpoints were evaluated in dogs for the M-2 ... and M-5 ... metabolites administered iv."[173]

Consistently, FDA's Medical Review reported that, "Cardiovascular safety was examined in ... studies in dogs. None of the studies revealed significant changes in ECG parameters ... single dose cardiovascular safety studies in dogs were conducted using each of the metabolites. There were no clearly adverse effects noted for either metabolite in these studies and in ... studies conducted in mice using each of the metabolites, no unique toxicities compared to those observed in animals administered regorafenib were identified."[174]

C. TESTS WITH HUMAN SUBJECTS.

Regarding tests with human subjects, where only regorafenib (and not M-2 or M-5) was administered, FDA's Medical Review reported that "A 12-lead ECG was performed on Day 1 of each cycle for the first 6 cycles. **No clinically relevant changes were observed** for

[169] Regorafenib (colorectal cancer) NDA 203-085. Page 12 of 219-page Pharmacology Review.

[170] Regorafenib (colorectal cancer) NDA 203-085. Page 179 of 219-page Pharmacology Review.

[171] Thomas G, et al. Effect of 4-aminopyridine on action potential parameters in isolated dog Purkinje fibers. Arch. Drug Inf. 2009;3:19–25.

[172] Aubert M, et al. Evaluation of the rabbit Purkinje fibre assay as an in vitro tool for assessing the risk of drug-induced torsades de pointes in humans. Drug Saf. 2006;29:237–54.

[173] Regorafenib (colorectal cancer) NDA 203-085. Page 39 of 219-page Pharmacology Review.

[174] Regorafenib (colorectal cancer) NDA 203-085. Page 18 of 76-page Medical Review.

any of the ECG parameters ... [t]he applicant has completed enrollment in cardiac safety study ... to evaluate QTc prolongation, if any ... The study is performed in approximately 50 patients ... [o]verall, the effect of regorafenib at tmax on the QTc intervals of the ECG, observed in the study were **minimal**, and even the most conservative evaluation, the maximal median change, was **modest and unlikely to be of clinical significance.**"[175]

D. THE PACKAGE LABEL.

A view of the package label reveals that it is completely silent regarding electrocardiogram (ECG) data, QTc interval data, and Purkinje fiber data. This silence is consistent with the Sponsor's data on human subjects administered with the study drug. Although it might have been reasonable for the package insert to state that enhanced metabolism of the study drug to M-2 or M-5 could have led to cardiac toxicity, for example, as might occur with an overdose of regorafenib, the package insert was silent on this possibility.

4. *Efficacy of Administered Drug Metabolites*

Animal studies with regorafenib and the M-2 and M-5 metabolites revealed that each of these three blocked the intended signaling pathway and that each was effective in treating cancer. Regarding the effect on the signaling pathway, FDA wrote, "An in vivo experiment specifically examining anti-VEGF activity was also conducted with regorafenib and the M-2 and M-5 metabolites ... rats were administered iv with control vehicle, 0.1, or 1 mg/kg of regorafenib, 1 mg/kg of M-2, or 1 mg/kg of M-5, followed by 9 μg/kg of recombinant human VEGF 10 minutes later ... VEGF induced an immediate reduction of blood pressure ... in control rats; however, previous administration of ... regorafenib or either metabolite **prevented this decrease in blood pressure**, demonstrating a clear effect of regorafenib on inhibition of VEGF activity."

Regarding efficacy against cancer, FDA wrote, "The activities of regorafenib, M-2, and M-5 were also examined xenograft experiments using ... mice implanted with human tumor cell lines ... treatment of implanted mice with regorafenib, M-2, or M-5 resulted in dose-dependent inhibition of tumor growth compared to negative control treated animals ... and there were no clear differences in anti-tumor activity apparent among regorafenib, M-2, and M-5."

Package label. The Sponsor's animal data on toxicity relating to dentin alteration and epiphyseal growth plate found a place in the Use in Specific Populations section of the package label, where the population in question was pediatric patients. The package label for regorafenib (Stivarga®) warned:

> **USE IN SPECIFIC POPULATIONS**. The safety and efficacy of Stivarga in pediatric patients less than 18 years of age have not been established. In ... studies in rats there were ... findings of **dentin alteration** ... [t]hese findings were observed at regorafenib doses as low as 4 mg/kg (approximately 25% of the AUC in humans at the recommended dose). In ... studies in dogs there were similar findings of **dentin alteration** at doses as low as 20 mg/kg (approximately 43% of the AUC in humans at the recommended dose). Administration of regorafenib in these animals also led to ... growth and thickening of the femoral **epiphyseal growth plate.**[176]

[175] Regorafenib (colorectal cancer) NDA 203-085. Page 62 of 76-page Medical Review.

[176] Package label. STIVARGA (regorafenib) tablets, oral. September 2012 (14 pp.).

Also, the Sponsor's animal data on thickening of the atrioventricular valve found a place in another section of the package label, namely, the Nonclinical Toxicology section, which read:

> **NONCLINICAL TOXICOLOGY** ... In a chronic 26 week repeat dose study in rats there was a dose-dependent increase in the finding of **thickening of the atrioventricular valve**. At a dose that resulted in an exposure of approximately 12% of the human exposure at the recommended dose, this finding was present in half of the examined animals.[177]

h. Sumatriptan/Naproxen Combination (Acute Treatment of Migraine) NDA 021-926

Sumatriptan/naproxen combination (Treximet®) was formerly known as Trexima. FDA's reviews use both of these terms. FDA's analysis for the sumatriptan/naproxen combination drug reveals use of animal toxicity data only (no human data) for warnings on the package label. **Sumatriptan** alone is used for treating acute migraine[178] and chronic cluster headaches.[179] The mechanism of action of sumatriptan, as well as other triptans used as antimigraine drugs, is to bind to three serotonin (5-HT) subtypes, 5-HT1B, 5-HT1D, and 5-HT1F.[180] **Naproxen** inhibits cyclooxygenase (COX) enzymes. These enzymes catalyze the conversion of arachidonic acid to prostaglandin G2. COX inhibitors include naproxen, ibuprofen, mefenamic acid, and lumiracoxib, which are rapid inhibitors. In contrast, the COX inhibitors diclofenac, indomethacin, and flurbiprofen are slow acting, time-dependent COX inhibitors.[181] Naproxen alone is an NSAID for treating acute migraine,[182] osteoarthritis,[183] rheumatoid arthritis,[184] fibromyalgia,[185] and back pain.[186]

[177] Package label. STIVARGA (regorafenib) tablets, oral. September 2012 (14 pp.).

[178] Becker WJ. Acute migraine treatment. Continuum (Minneap. Minn.). 2015;21(4 Headache):953–72.

[179] Leone M, Proiett C. Long-term use of daily sumatriptan injections in severe drug-resistant chronic cluster headache. Neurology. 2016;86:194–5.

[180] Mitsikostas DD, Tfelt-Hansen P. Targeting to 5-HT1F receptor subtype for migraine treatment: lessons from the past, implications for the future. Cent. Nerv. Syst. Agents Med. Chem. 2012;12:241–9.

[181] Blobaum AL, et al. Action at a distance: mutations of peripheral residues transform rapid reversible inhibitors to slow, tight binders of cyclooxygenase-2. J. Biol. Chem. 2015;290:12793–803.

[182] Becker WJ. Acute migraine treatment. Continuum (Minneap. Minn.). 2015;21(4 Headache):953–72.

[183] Sanders D, et al. Pharmacologic modulation of hand pain in osteoarthritis: a double-blind placebo-controlled functional magnetic resonance imaging study using naproxen. Arthritis Rheumatol. 2015;67:741–51.

[184] Nissen SE, et al. Cardiovascular safety of celecoxib, naproxen, or ibuprofen for arthritis. New Engl. J. Med. 2016;375:2519–29.

[185] Bennett RM, et al. An internet survey of 2,596 people with fibromyalgia. BMC Musculoskelet. Disord. 2007;8:27 (11 pp.).

[186] Slawson D. Naproxen alone may be best for acute low back pain. Am. Fam. Physician. 2016;93:316.

FDA's Pharmacology Review for the sumatriptan/naproxen combination disclosed teratogenic effects of the sumatriptan/naproxen combination in rabbits:

> Developmental toxicity study ... in rabbits ... reductions in litter size, and increases in total resorptions per litter ... malformations (interventricular septal defect ... fused caudal vertebrae ... absent intermediate lobe of the lung, irregular ossification of the skull, and incompletely ossified sternal centra)[187] ... [g]ross external malformations were observed with two fetuses ... one with gastroschisis and one with a short tail ... and two late resorptions ... one with acrania, gastroschisis, medial rotation of right hindlimb, short tail, and fused forepaw digits ... downward flexed forepaws, absent tail, no anal opening, and no external urogenital area.[188]

FDA further stated that "a study in rabbits would suffice to evaluate the potential for additive or synergistic effects of the combination of sumatriptan and naproxen on reproduction and development, since the components are currently marketed in the U.S."[189] FDA dissected the toxicities of each of the two drugs in the sumatriptan/naproxen combination, stating that some forms of toxicity were additive while other forms of toxicity were due to only the naproxen:

> Taken together, the ... embryo-fetal toxicity studies demonstrating that naproxen and sumatriptan toxicities resulted in reduction of maternal and fetal body weights **appeared to be additive**. In contrast ... incidences of malformations ... observed in animals treated with **naproxen were not** ... **increased further by co-administration of sumatriptan**[190] ...the toxicities reported for ... the combination of sumatriptan and naproxen is not likely to induce greater reproductive and developmental toxicity than naproxen alone.[191]

The FDA reviewer referred to FDA's earlier approval for one of the components (naproxen) of the study drug. This earlier approval was for Anaprox®.[192,193] Anaprox® contains only one drug (naproxen). In addition to commenting on Anaprox®, FDA invoked drug class analysis, where the relevant drug class was NSAIDs. FDA's invocation of drug class analysis is demonstrated by the term, ***other NSAIDs***.

FDA's concern over the study drug, which was inspired by this drug class analysis, took the form, "While there are no teratogenic effects described in the ... labeling for ANAPROX®, there is evidence in the published literature that interventricular defects are

[187] Sumatriptan/naproxen combination (acute treatment of migraine) NDA 021-926. Page 8 of 34 pages of fourth pdf file for Pharmacology Review.

[188] Sumatriptan/naproxen combination (acute treatment of migraine) NDA 021-926. Page 12 of 34 pages of fourth pdf file for Pharmacology Review.

[189] Sumatriptan/naproxen combination (acute treatment of migraine) NDA 021-926. Page 1 of 34 pages of fourth pdf file for Pharmacology Review.

[190] Sumatriptan/naproxen combination (acute treatment of migraine) NDA 021-926. Page 30 of 34 pages of fourth pdf file for Pharmacology Review.

[191] Sumatriptan/naproxen combination (acute treatment of migraine) NDA 021-926. Page 8 of 34 pages of fourth pdf file for Pharmacology Review.

[192] Naproxen (rheumatoid arthritis, osteoarthritis, ankylosing spondylitis) NDA 018-164.

[193] Package label. ANAPROX (naproxen sodium tablets). July 2008 (31 pp.).

increased after ... administration of ***other NSAIDs***, ibuprofen, ketoralac, meloxican, diflunisal, and aspirin."

FDA's drug class analysis contributed to the warnings on the label for the sumatriptan/naproxen combination. FDA concluded, "The positive findings of embryofetal toxicity and teratogenicity described above warrant inclusion in the labeling for TREXIMA."[194] A further basis for disclosing the teratogenic data from animals on the package label is found on the package label itself, which admits that "There are no adequate and well-controlled studies in pregnant women."[195]

Package label. The package label for the sumatriptan/naproxen combination (Treximet®) contained warnings relating to pregnancy and infancy. The Contraindications section read:

> **CONTRAINDICATIONS** ... Third trimester of pregnancy.[196]

The Use in Specific Populations section referred to animal studies:

> **USE IN SPECIFIC POPULATIONS.** PREGNANCY. There are no adequate and well-controlled studies in pregnant women. TREXIMET (sumatriptan and naproxen) should be used during the first and second trimester of pregnancy only if the potential benefit justifies the potential risk to the fetus. TREXIMET should not be used during the third trimester of pregnancy because inhibitors of prostaglandin synthesis (including naproxen) are known to cause premature closure of the ductus arteriosus in humans. In **animal studies**, administration of sumatriptan and naproxen, alone or in combination, during pregnancy resulted in developmental toxicity (increased incidences of fetal malformations, embryofetal and pup mortality, decreased embryofetal growth) at clinically relevant doses.
>
> Oral administration of sumatriptan combined with naproxen sodium ... or each drug alone ... to **pregnant rabbits** during the period of organogenesis resulted in increased total incidences of fetal ... malformations (cardiac interventricular septal defect in the 50/90 mg/kg/day group, fused caudal vertebrae ... absent intermediate lobe of the lung, irregular ossification of the skull, incompletely ossified sternal centra) at ... doses of sumatriptan and naproxen alone and in combination ...
>
> In previous developmental toxicity studies of sumatriptan, oral administration to **pregnant rats** during the period of organogenesis resulted in an increased incidence of fetal blood vessel abnormalities and decreased pup survival at doses ... Oral administration of sumatriptan to **pregnant rabbits** during the period of organogenesis resulted in increased incidences of vascular and skeletal abnormalities at a dose of 50 mg/kg/day and embryolethality at 100 mg/kg/day.[197]

The Use in Specific Populations section referred to animal studies and devoted a paragraph to labor and delivery. In disclosing risk for one of the components (naproxen) of the drug combination, the package label referred to a drug class analysis for the drug

[194] Sumatriptan/naproxen combination (acute treatment of migraine) NDA 021-926. Page 30 of 34 pages of fourth pdf file for Pharmacology Review.

[195] Package label. TREXIMET (sumatriptan and naproxen sodium) tablets, for oral use. May 2016 (27 pp.).

[196] Package label. TREXIMET (sumatriptan and naproxen sodium) tablets, for oral use. May 2016 (27 pp.).

[197] Package label. TREXIMET (sumatriptan and naproxen sodium) tablets, for oral use. May 2016 (27 pp.).

class of NSAIDs and predicted that naproxen would increase risk for dystocia and other toxicities:

> **USE IN SPECIFIC POPULATIONS** ... LABOR AND DELIVERY. Naproxen-containing products are not recommended in labor and delivery because, through its prostaglandin synthesis inhibitory effect, naproxen may adversely affect fetal circulation and inhibit uterine contractions, thus increasing the risk of uterine hemorrhage. In rat studies with NSAIDs, as with other drugs known to inhibit prostaglandin synthesis, an increased incidence of dystocia, delayed parturition, and decreased pup survival occurred.[198]

The Use in Specific Populations section included an account of nursing mothers and issued a warning about milk:

> **USE IN SPECIFIC POPULATIONS** ... NURSING MOTHERS. Both active components of TREXIMET, sumatriptan and naproxen, have been reported to be secreted in **human milk**. Because of the potential for serious adverse reactions in nursing infants from TREXIMET, a decision should be made whether to discontinue nursing or to discontinue the drug, taking into account the importance of the drug to the mother.[199]

Breast milk. In view of the above warning against the study drug appearing in breast milk, it should be pointed out that this type of warning is frequently found in package labels for other small molecule drugs. The following reproduces the milk warnings in the package labels for:

- Hydromorphone
- Lumacaftor/ivacaftor combination
- Bosutinib

Hydromorphone. "Nursing Mothers. Low concentrations of **hydromorphone have been detected in human milk** in clinical trials. Withdrawal symptoms can occur in breast-feeding infants when maternal administration of an opioid analgesic is stopped. Nursing should not be undertaken while a patient is receiving EXALGO since hydromorphone is excreted in the milk."[200]

Lumacaftor/ivacaftor combination. "Nursing Mothers. Both **lumacaftor and ivacaftor** are excreted into the milk of lactating female rats. Excretion of lumacaftor or ivacaftor into human milk is probable. There are no human studies that have investigated the effects of lumacaftor and ivacaftor on breast-fed infants. Caution should be exercised when ORKAMBI is administered to a nursing woman."[201]

Bosutinib. "Nursing Mothers. It is not known whether **bosutinib** is excreted in human milk. Bosutinib and/or its metabolites were excreted in the milk of lactating rats ... [b] ecause many drugs are excreted in human milk and because of the potential for serious

[198] Package label. TREXIMET (sumatriptan and naproxen sodium) tablets, for oral use. May 2016 (27 pp.).

[199] Package label. TREXIMET (sumatriptan and naproxen sodium) tablets, for oral use. May 2016 (27 pp.).

[200] Package label. EXALGO (hydromorphone hydrochloride) extended release tablets. March 2010 (30 pp.).

[201] Package label. ORKAMBI -- lumacaftor and ivacaftor tablet, film coated. September 2016 (32 pp.).

adverse reactions in nursing infants from BOSULIF, a decision should be made whether to discontinue nursing or to discontinue the drug, taking into account the importance of the drug to the mother."[202]

Milk warnings are also found on package labels for other small molecule drugs, such as axitinib,[203] cabozantinib,[204] dabrafenib,[205] ponatinib,[206] regorafenib,[207] and tofacitinib.[208] The milk warnings on these particular labels were based only on animal milk data.

i. Vigabatrin (Complex Partial Seizures in Adults, Infantile Spasms in Children) NDA 022-006

Vigabatrin is for treating partial onset seizures. About half of patients respond to vigabatrin treatment, where the response takes the form of greater than 50% seizure reduction.[209] Vigabatrin (gamma-vinyl-GABA) irreversibly binds to GABA-transaminase in neurons. The drug has an anticonvulsant effect of several days, even though its half-life in plasma is only 5–8 hours.

1. *Animal Toxicities of Edema and Vision*

FDA's Pharmacology Review revealed the animal toxicity of edema as "Brain lesions, referred to as intramyelinic **edema** (IME), were detected in adult animals (mouse, rat, dog; equivocal in monkey). These lesions were characterized by microvacuoles in white matter, resulting from splitting of the interperiod line of the myelin sheath, and were observed after 3 months or more of dosing."[210]

The animal edema and visual toxicity data had a direct influence on the package label, as revealed by FDA's comments, "In a Supervisory Overview (March 23, 1995), Glenna G. Fitzgerald, Ph.D. concurred on the adequacy of the nonclinical data, and recommended that findings of intramyelinic **edema** and **retinal degeneration** be included in labeling."[211]

Initially, FDA doubted the relevance of animal vision toxicity data to human patients, in view of the fact that only **albino rodents suffered from retinal degeneration** whereas **pigmented rats, dogs, and monkeys did not have vision toxicity**. But at a later point in time, FDA became aware of clinical toxicity data on vision, thereby making the vision toxicity data from albino rodents more relevant and more compelling. Regarding FDA's

[202] Package label. BOSULIF® (bosutinib) tablets, for oral use. September 2012 (13 pp.).

[203] Package label. INLYTA® (axitinib) tablets for oral administration. August 2014 (21 pp.).

[204] Package label. COMETRIQ® (cabozantinib) capsules, for oral use. May 2016 (22 pp.).

[205] Package label. TAFINLAR (dabrafenib) capsules for oral use. May 2013 (19 pp.).

[206] Package label. ICLUSIG® (ponatinib) tablets for oral use. November 2016 (23 pp.).

[207] Package label. STIVARGA® (regorafenib) tablets, for oral use. August 2016 (21 pp.).

[208] Package label. XELJANZ® (tofacitinib). November 2012 (27 pp.).

[209] Krauss GL. Evaluating risks for vigabatrin treatment. Epilepsy Curr. 2009;9:125–9.

[210] Vigabatrin (complex partial seizures in adults, infantile spasms in children) NDA 022-006. Page 7 of 68-page Pharmacology Review.

[211] Vigabatrin (complex partial seizures in adults, infantile spasms in children) NDA 022-006. Page 5 of 68-page Pharmacology Review.

change of heart, FDA wrote, "Vigabatrin-induced retinal degeneration was observed in **albino mouse and rat**, but **not in pigmented rat, dog, or monkey** ... [r]enewed interest in the retinal findings in rodent resulted from a report ... of visual field defects ... sometimes severe, in **40% of patients treated with vigabatrin** versus 0% in non-medicated patients or patients treated with carbamazepine."[212]

FDA's Pharmacology Review and FDA's Medical Review reveal strategies available to the Sponsor, when working together with FDA, to ensure that FDA will grant approval to the drug despite severe and unexpected toxicity issues[213]:

- **Strategy No. 1**. Arguing that the disease is a serious disease with no alternative drugs
- **Strategy No. 2**. Implementing a risk evaluation and mitigation strategy (REMS) to monitor the AE of visual toxicity
- **Strategy No. 3**. In response to a Clinical Hold, devise a new test for early detection of the AE of visual toxicity, for mitigating harm to patients

2. *Strategy No. 1. Arguing That the Disease Is a Serious Disease with No Alternative Drugs*

FDA's Pharmacology Review justified approval of vigabatrin despite the severe animal toxicity data, on the basis that there is "no other approved therapy." FDA wrote, "The ... reviewer ... did not find the nonclinical information adequate to support approval ... based on evidence that juvenile animals were sensitive to neurotoxic effects of vigabatrin. The pharm/tox supervisor recognized this concern but did not object to the approval ... based on the clinical benefit of vigabatrin in infantile spasms, which is a **serious indication with no other approved therapy.**"[214]

An FDA reviewer embellished on this basis for approval, adding that "refractory epilepsy is a serious ... life-threatening condition, and despite the availability of many newer antiepileptic drugs, I believe that ... additional therapies should be made available."[215]

FDA emphasized the fact that, at the time of the Sponsor's clinical study, other available drugs for infantile spasms did not work, thereby adding to the justification for approving the study drug. One FDA reviewer added, "First, refractory epilepsy is a serious, life-altering and life-threatening condition, and despite the availability of many newer anti-epilepsy drugs, I believe that ... additional therapies should be made available. Although patients in these trials did not fail on the 'newer' anti-epilepsy drugs, they were 'refractory' ... to one or several of the standard anti-epilepsy drugs available at the time (e.g., phenytoin, carbamazepine)."[216]

[212] Vigabatrin (complex partial seizures in adults, infantile spasms in children) NDA 022-006. Page 9 of 68-page Pharmacology Review.

[213] Vigabatrin (complex partial seizures in adults, infantile spasms in children) NDA 022-006.

[214] Vigabatrin (complex partial seizures in adults, infantile spasms in children) NDA 022-006. Page 2 of 68-page Pharmacology Review.

[215] Memorandum. From Russell Katz, MD, Director, Division of Neurology Products. NDA 22-006. Vigabatrin (pp. 14–19 of 21 pages).

[216] Vigabatrin (complex partial seizures in adults, infantile spasms in children) NDA 022-006. 21-page Memorandum dated August 14, 2009, on page 18 of 219-page Medical Review.

3. Strategy No. 2. Implementing a Risk Evaluation and Mitigation Strategy (REMS) to Monitor the Adverse Event of Visual Toxicity

In working together with FDA to acquire FDA approval, the Sponsor agreed to implement a REMS. FDA observed that, although AEs of visual toxicity did occur, significant visual loss was not common and that visual toxicity was "tolerated" by patients.

FDA's Medical Review described the elements of the proposed REMS: "Further, despite the occurrence of visual toxicity, it does not appear that there are many patients who have suffered significant visual loss ... [t]his is not to minimize the toxicity, but only to point out that patients have, generally, tolerated whatever pathology the drug has produced (here it should be noted that the drug has been available in many countries since the mid 1980's). In this regard, **the REMS that has been discussed with the sponsor is fairly restrictive, and commits physicians to perform periodic ophthalmologic examinations.**"[217]

Referring to FDA's approval process for vigabatrin and the requirement for the REMS, Cohen et al.[218] remarked that FDA "had become aware of a unique visual field defect associated with ... vigabatrin, and, as a result, the sponsor had proposed that vigabatrin be approved as a last resort treatment under very restrictive conditions."

The Black Box Warning and the Warnings and Precautions section of the package label referred to the REMS. Excerpts from the REMS are[219]: "At product launch ... [the Sponsor] will send a Dear Healthcare Letter ... to all registered ophthalmologists. The Sabril package insert will accompany this letter. Additionally, ... [the Sponsor's] field representatives will call on ... ophthalmologists at key epilepsy centers at product launch to disseminate the Sabril package inserts."

Also, the REMS required the following. Please note the confluence of expertise for two different fields of medicine: (1) ophthalmology and vision and (2) epilepsy and spasms:

> Healthcare providers who prescribe Sabril are specially certified ... [p]rescribers must be enrolled in the REMS program and attest to their understanding of the REMS program requirements and the risks associated with Sabril. Prescribers commit to ... [h]aving experience in treating epilepsy ... [i]f prescribing for infantile spasms, having knowledge of the risk of magnetic resonance imaging (MRI) abnormalities with use of Sabril ... [c]ounseling the patient if the patient is not complying with the required vision assessment, and removing the patient from therapy if the patient still fails to comply with the required vision assessment.[220]

FDA's Approval Letter required a REMS as a condition for FDA approval, writing that "we have also determined that Sabril (vigabatrin) can be approved only if elements ... to

[217] Vigabatrin (complex partial seizures in adults, infantile spasms in children) NDA 022-006. Page 2 of 68-page Pharmacology Review. Page 18 of 219-page Medical Review.

[218] Cohen JA, et al. The potential for Vigabatrin-induced intramyelinic edema in humans. Epilepsia. 2000;41:148–57.

[219] Vigabatrin (complex partial seizures in adults, infantile spasms in children) NDA 022-006. 6-page Risk Evaluation and Mitigation Strategy (REMS) in 36-page pdf file.

[220] Vigabatrin (complex partial seizures in adults, infantile spasms in children) NDA 022-006. 6-page Risk Evaluation and Mitigation Strategy (REMS) in 36-page pdf file.

assure safe use are required as part of a REMS to mitigate these risks ... by ensuring that patients receive ... monitoring of vision ... [y]our proposed REMS, submitted on August 18, 2009 ... is approved."[221] An introductory account of REMS and Dear Healthcare Professional letters has been published.[222]

4. *Strategy No. 3. In Response to a Clinical Hold, Devise a New Test for Early Detection of the Adverse Event of Visual Toxicity, for Mitigating Harm to Patients*

The serious nature of the animal toxicity was such that FDA placed the Sponsor's ongoing clinical studies on a Clinical Hold (21 CFR §312.42). FDA's Medical Review describes the timeline, where the study was put on a Clinical Hold, where the Sponsor responded by using animals (dogs) for devising a method to test the visual toxicity AE when it was still in the reversible stage, and where, as a result, FDA withdrew the clinical hold:

> The IND for vigabatrin was submitted in 1980. In 1983, the Agency became aware of the occurrence of a ... histopathologic finding in animals ... given vigabatrin ... at doses approximating those to be given to humans ... so-called intramyelinic edema (IME) was seen. FDA placed the IND on **clinical hold** until the sponsor was able to develop **a non-invasive method** that could ***detect the occurrence of the lesion in a sufficiently early stage to ensure that it would be reversible*** if the drug was discontinued. After several years, the sponsor was able to validate visual evoked potentials and MRI (in the dog) as a ... test, and clinical testing was permitted to resume in 1989.[223]

This same feature in the timeline of FDA approval was shown by Krauss et al.[224] who described the use of MRI for detecting vigabatrin toxicity, both in animals and in human subjects. Development of vigabatrin was initially halted after animal research showed that it produces IME, where the Sponsor then developed an MRI method for detecting IME in animals. Later, it was shown that adult patients treated with vigabatrin did not develop similar changes and the drug was eventually FDA-approved and marketed.

FDA recommended that the Sponsor develop MRI using animals for later use in humans, writing that "it was the Committee's recommendation that ... a noninvasive method for clinical monitoring of the onset of neurotoxicity be developed. In response, the Sponsor ... conducted studies in dogs to develop methods (MRI) ... which were then used to monitor humans."[225]

MRI was for detecting a lesion called IME: "the sponsor conducted investigative studies in animals that demonstrated that IME could be monitored in animals using MRI and measurement of evoked potentials."

[221] Vigabatrin (complex partial seizures in adults, infantile spasms in children) NDA 022-006. 12-page Approval Letter.

[222] Brody T. Clinical trials: study design, endpoints and biomarkers, drug safety, and FDA and ICH Guidelines. 2nd ed. New York: Elsevier; 2016. pp. 546–58.

[223] Vigabatrin (complex partial seizures in adults, infantile spasms in children) NDA 022-006. Page 27 of 219-page Medical Review.

[224] Krauss GL. Evaluating risks for vigabatrin treatment. Epilepsy Curr. 2009;9:125–9.

[225] Vigabatrin (complex partial seizures in adults, infantile spasms in children) NDA 022-006. Page 5 of 68-page Pharmacology Review.

The MRI technique demonstrated that IME did not occur in adults. But FDA was still worried about IME in pediatric and adult patients, because of the serious nature of the edema lesion, because the edema lesion was correlated with a particular mechanism of action that exists in humans (vigabatrin-induced increases in GABA), and because of the irreversible nature of another type of lesion ("microscopic changes").

As can be seen, animal toxicity studies played a major part in FDA's assessment of IME. FDA's Pharmacology Review teaches that "intramyelinic edema (IME) was not detected in older children and adults using this monitoring strategy. Although the negative MRI findings certainly reduce the safety concern, at least for adults, it is difficult to completely dismiss the concern regarding this potentially serious finding since (1) **IME was detected in multiple species**, i.e., in those species in which vigabatrin increases central GABA levels (mouse, rat, dog), as it does in humans and (2) Although the vacuoles appeared reversible, prolonged administration of vigabatrin in **mouse and rat** resulted in additional microscopic changes reflective of **irreversible injury.**"[226]

FDA turned to the Sponsor's MRI data from human infants (data corresponding to the IME data from animals). FDA observed that "There are also vigabatrin-induced MRI signal changes occurring in about **20% of infants less than age 3 years** that may correspond to the **intramyelinic edema** (IME) observed in the rat and dog model."[227]

5. ***Another Example Where Additional Animal Studies Overcame a Clinical Hold (Example of Evolocumab)***

An FDA submission for another drug, evolocumab, further illustrates the situation where FDA imposed a Clinical Hold and where the Sponsor responded by conducting animal toxicity studies, where the result was that FDA withdrew the Clinical Hold.[228] FDA's request for more animal toxicity studies was as follows:

> On 10 June 2009, IND 105-188 was placed on partial clinical hold ... [t]he applicant was informed that ... multiple dose studies were on clinical hold. They were advised that they needed to obtain ... repeat-dose toxicity data from a **second species** before multiple dose studies would be permitted to proceed. The FDA explained that ... toxicity data from a **second species** were needed to better evaluate the toxicity profile of ... evolocumab before repeat-dose studies were conducted in humans.[229]

Animal toxicity data were provided by the Sponsor in response to the Clinical Hold:

> In order to address the partial clinical hold for repeat-dose clinical studies, a repeat-dose toxicity study was to be conducted in a pharmacologically relevant rodent species (e.g. hamster) ... [o]n March 10, 2010,

[226] Vigabatrin (complex partial seizures in adults, infantile spasms in children) NDA 022-006. Page 2 of 68-page Pharmacology Review. Pages 5 and 8 of 219-page Medical Review.

[227] Vigabatrin (complex partial seizures in adults, infantile spasms in children) NDA 022-006. Page 27 of 219-page Medical Review.

[228] FDA Briefing Document. Endocrinologic and Metabolic Drugs Advisory Committee (EMDAC). June 10, 2015, Page 50 of 405 pages.

[229] FDA Briefing Document. Endocrinologic and Metabolic Drugs Advisory Committee (EMDAC). June 10, 2015, Page 50 of 405 pages.

> the applicant submitted a complete response to the partial clinical hold. Two study reports were submitted and reviewed ... results of the 28-Day toxicity study in the **Golden Syrian Hamster** and ... results of the tissue cross-reactivity study in the **Golden Syrian Hamster.** ... the partial clinical hold was removed and repeat-dose studies ... under this IND were allowed.[230]

6. Yet Another Example Where Additional Animal Studies Overcame a Clinical Hold (Example of Ivacaftor)

FDA's review of ivacaftor for cystic fibrosis provided a timeline for the imposing of the Clinical Hold and FDA's withdrawal of the Clinical Hold in response to additional animal data from the Sponsor.

FDA's Cross Discipline Team Leader Review stated, "In December 2008 ... Sponsor submitted protocols ... for phase 3 clinical trials. However, review of the **non-clinical data** showed that there was **not adequate non-clinical support** for the proposed trials because of a **lack of chronic toxicity data** to support the proposed duration of the trial. On January 9, 2009, the Division placed the clinical studies on **clinical hold** for **lack of non-clinical support.**"[231]

FDA provided an account of the Sponsor's response in providing more animal data and FDA's response by withdrawing the Clinical Hold: "**After adequate nonclinical support was provided**, ... the Sponsor submitted protocols for 3 new clinical studies ... which would provide the basis for demonstrating the safety and efficacy of ivacaftor in the indicated population ... comments were sent to ... the Sponsor on April 13, 2009, which reflected that positive results from ... the study and positive trending efficacy data from ... safety data ... would be adequate for filing an NDA."[232]

Package label. FDA's comments on visual field defects with vigabatrin (Sabril®) found a place in the Black Box Warning:

> **BOXED WARNING:** PERMANENT VISION LOSS ... SABRIL can cause permanent bilateral concentric visual field constriction, including tunnel vision that can result in disability. In some cases, SABRIL may also decrease visual acuity ... [r]isk increases with increasing dose and cumulative exposure, but there is no dose or exposure to SABRIL known to be free of risk of vision loss ... SABRIL is available only through a **restricted program** called the SABRIL REMS Program.[233]

The Warnings and Precautions section of the label referred to animal toxicity studies where the study drug resulted in **edema**. The label referred to edema in young animals and in adult animals:

> **WARNINGS AND PRECAUTIONS** ... Vacuolation, characterized by fluid accumulation and separation of the outer layers of myelin, has been observed in brain white matter tracts in adult and juvenile rats

[230] FDA Briefing Document. Endocrinologic and Metabolic Drugs Advisory Committee (EMDAC). June 10, 2015, Page 50 of 405 pages.

[231] Ivacaftor (cystic fibrosis) NDA 203-188. 30-page Cross Discipline Team Leader Review.

[232] Ivacaftor (cystic fibrosis) NDA 203-188. 30-page Cross Discipline Team Leader Review.

[233] Package label. SABRIL® (vigabatrin) tablets, for oral use. SABRIL® (vigabatrin) powder for oral solution. June 2016 (25 pp.).

and adult mice, dogs, and possibly monkeys following administration of vigabatrin. This lesion, referred to as **intramyelinic edema** (IME), **was seen in animals at doses within the human therapeutic range** ... [v]acuolation in adult animals was correlated with alterations in MRI and changes in visual and somatosensory evoked potentials (EP).[234]

The Warnings and Precautions section for vigabatrin (Sabril®) also referred to the REMS:

WARNINGS AND PRECAUTIONS ... SABRIL is available only through a **restricted distribution program** called the SABRIL **REMS Program**, because of the risk of permanent vision loss ... [p]rescribers must be certified by enrolling in the program, agreeing to counsel patients on the risk of vision loss and the need for periodic monitoring of vision, and reporting any event suggestive of vision loss ... [p]atients must enroll in the program. Pharmacies must be certified and must only dispense to patients authorized to receive SABRIL.[235]

VI. ANIMAL DATA FROM FDA'S REVIEWS FIND A CORRESPONDING DISCLOSURE IN THE PACKAGE LABEL (BIOLOGICS)

Introduction. FDA's reviews for the following biologicals further illustrate the use of animal data as a guide for drafting the package label:

- Atezolizumab
- Belimumab
- Denosumab
- Evolocumab
- Nivolumab
- Pegloticase
- Ustekinumab

Antibody drugs intended for human use are designed to bind specifically to their target in human patients, without regard to the ability to bind to the corresponding target in animals. Also, antibody drugs, as well as other biologicals, are usually designed to minimize undesired responses, by the immune system of the patient, against the drug. Wherever a special topic arises in FDA's review of a particular therapeutic protein, this chapter provides a page of background information on that topic. As described below, these special topics are:

- Antibodies secreted during lactation
- FDA regulations for labeling on pregnancy, nursing, and lactation
- Mechanism of transfer of antibodies from maternal circulatory system to the fetus

[234] Package label. SABRIL® (vigabatrin) tablets, for oral use. SABRIL® (vigabatrin) powder for oral solution. June 2016 (25 pp.).

[235] Package label. SABRIL® (vigabatrin) tablets, for oral use. SABRIL® (vigabatrin) powder for oral solution. June 2016 (25 pp.).

a. Atezolizumab (Urothelial Carcinoma) BLA 761-034

Atezolizumab is a recombinant humanized antibody that binds to PD-L1 and prevents binding of PD-L1 (ligand) to PD-1 (receptor). The PD-L1 to PD-1 signaling pathway is used by cancer cells to prevent $CD8^+$ T-cells from killing cancer cells and to dampen the body's immune response against the cancer. Atezolizumab prevents cancer cells from using this tactic to evade the immune system.

For use in animal studies, the Sponsor prepared a specially engineered version of atezolizumab consisting of the constant region of mouse IgG2a plus the variable region (PD-L1 binding region) of the humanized antibody. The goal of this engineered version of atezolizumab was to prevent the administered antibody from generating, in the mouse, an immune response against the antibody.[236]

FDA described the antibody, "The Applicant generated mouse IgG2a/human chimeric antibodies for murine in vivo studies in order to reduce immunogenicity, which was observed in prior mouse studies ... [a]tezolizumab was immunogenic in mice, resulting in significantly reduced exposures by Week 3."[237]

The Sponsor had conducted a drug class analysis based on the mechanism of action to predict AEs in humans. The relevant drug class is drugs that block PD-L1 to PD-1 interactions ("interference with PD-L1").

FDA's Pharmacology Review stated that "The Applicant submitted a non-product ... literature-based assessment to characterize the ... risk of reproductive and developmental toxicity ... the scientific literature demonstrates that **interference with PD-L1** leads to loss of fetal tolerance and increases risk of immune-mediated abortion. This ... provides evidence that atezolizumab can cause fetal harm ... to a pregnant woman."[238]

FDA referred to articles in the scientific literature, revealing that one of the goals of the PD-L1 to PD-1 signaling pathway in normal physiology is to prevent the mother from mounting an immune response against the fetus and to prevent the mother's immune system from killing the fetus. FDA cited D'Addio et al.[239] and Taglauer et al.[240] who describe the mechanism for preventing harm to the fetus.

D'Addio et al.[241] inform us that "Acceptance of the ... fetus by the mother during pregnancy represents a physiologic model of in vivo immune tolerance ... [o]ur group has ... shown that the **PD1-PDL1 ... pathway plays a key role in inducing and maintaining fetomaternal tolerance in a mouse model** ... expression of PDL1 on the surface of Tregs

[236] Atezolizumab (urothelial carcinoma) BLA 761-034. Page 4 of 58-page Pharmacology Review.

[237] Atezolizumab (urothelial carcinoma) BLA 761-034. Pages 5–6 and 12 of 58-page Pharmacology Review.

[238] Atezolizumab (urothelial carcinoma) BLA 761-034. Page 5 of 58-page Pharmacology Review.

[239] D'Addio F, et al. The link between the PDL1 costimulatory pathway and Th17 in fetomaternal tolerance. J. Immunol. 2011;187:4530–41.

[240] Taglaure ES, et al. Maternal PD-1 regulates accumulation of fetal antigen-specific $CD8^+$ T cells in pregnancy. J. Reprod. Immunol. 2009;80:12–21.

[241] D'Addio F, et al. The link between the PDL1 costimulatory pathway and Th17 in fetomaternal tolerance. J. Immunol. 2011;187:4530–41.

is essential to exert their suppressive effect and to control the maternal immune response. In fact, blocking PDL1 resulted in a loss of regulatory function and reduction in fetal survival rate."

The term "Tregs," which is pronounced "tee-regs,"[242] refers to T-regulatory cells. In normal physiology, Tregs dampen immune responses in the body and prevent the immune system from inflicting damage on the body's own tissues, that is, to prevent autoimmunity.[243]

Further addressing harm to the fetus, FDA observed that atezolizumab can bind to the FcRn receptor. FcRn receptor mediates transfer of antibodies across the placenta to the fetal bloodstream. The constant region (Fc region) of the antibody is what binds to FcRn receptor. Porter et al. warned that "The active transfer of ... antibodies across the placenta by binding of the Fc-region to the neonatal Fc receptor (FcRn) may result in adverse fetal or neonatal effects."[244]

Porter et al.[245] describe this binding to neonatal Fc receptor: "Binding affinities of certolizumab pegol, infliximab, adalimumab and etanercept to human FcRn and FcRn-mediated transcytosis were determined using in vitro assays. Human placentas were perfused ... to measure transfer of [antibodies] ... from the maternal to fetal circulation. FcRn binding affinity (KD) was 132 nM, 225 nM and 1500 nM for infliximab, adalimumab and etanercept, respectively. There was no measurable certolizumab pegol binding affinity."

FDA's Pharmacology Review warned against risk of atezolizumab to the fetus: "Atezolizumab maintains binding to the FcRn receptor, so fetal exposure may occur if a patient is treated during pregnancy ... [i]t is unclear whether fetal exposure to atezolizumab would occur at levels sufficient to cause adverse effects on the developing immune system [of the fetus] ... there is a ... risk of developing immune-mediated disorders ... in the offspring due to the mechanism of action."[246]

FDA then turned its attention to the package label and made recommendations regarding pregnancy as well as for lactation and breast feeding. Drug class analysis was used for arriving at this warning, as is evident from the writing, "This was included to be consistent with labels for FDA-approved products inhibiting PD-1." FDA referred to other FDA-approved products that had the same mechanism of action as atezolizumab (blocking PD-L1 to PD-1 binding). FDA's labeling recommendation took the form, "Labeling ... the

[242] The author learned the pronunciation of various immunology terms, as he was employed at DNAX Research (Schering-Plough) in Palo Alto in the years 2001–2004. Researchers at DNAX Research created the field of molecular immunology, as described in Kornberg A. The golden helix: inside biotech ventures. University Science Books; 2002.

[243] Badhan K, et al. The PD1:PD-L1/2 pathway from discovery to clinical implementation. Front. Immunol. 2016;7:550 (17 pp.).

[244] Porter C, et al. Certolizumab pegol does not bind the neonatal Fc receptor (FcRn): consequences for FcRn-mediated in vitro transcytosis and ex vivo human placental transfer. J. Reprod. Immunol. 2016;116:7–12.

[245] Porter C, et al. Certolizumab pegol does not bind the neonatal Fc receptor (FcRn): consequences for FcRn-mediated in vitro transcytosis and ex vivo human placental transfer. J. Reprod. Immunol. 2016;116:7–12.

[246] Atezolizumab (urothelial carcinoma) BLA 761-034. Page 5 of 58-page Pharmacology Review.

Division recommended that patients should not breastfeed during treatment ... due to the potential for serious adverse reactions in breastfed infants from atezolizumab. This was included to be consistent with labels for FDA-approved products inhibiting PD-1 ... the Division recommends that females ... should use effective contraception during treatment."[247]

FDA's narratives on pregnancy, lactation, and breast-feeding found a place on the package label, as shown below.

Package label. The Warnings and Precautions section of the package label for atezolizumab (Tecentriq®) warned against harm to the fetus. This warning was based on the mechanism of action of the drug and on animal toxicity studies:

> **WARNINGS AND PRECAUTIONS**. Embryo-Fetal Toxicity. Based on its mechanism of action, TECENTRIQ can cause harm when administered to pregnant women. Animal studies have demonstrated that inhibition of the PD-L1/PD-1 pathway can lead to increased risk of immune-related rejection of the developing fetus resulting in fetal death ... [a]dvise females of reproductive potential to use effective contraception during treatment with TECENTRIQ.[248]

The Use in Specific Populations section referred to mechanism of action and to animal toxicity studies:

> **USE IN SPECIFIC POPULATIONS** ... Animal reproduction studies have not been conducted with TECENTRIQ to evaluate its effect on reproduction and fetal development. A literature-based assessment of the effects on reproduction demonstrated that a ... function of the **PD-L1/PD-1 pathway is to preserve pregnancy** by maintaining maternal immune tolerance to a fetus. Blockage of PD-L1 signaling has been shown in **murine models of pregnancy to disrupt tolerance to a fetus and to result in** ... **fetal loss**, therefore ... risks of administering TECENTRIQ during pregnancy include ... abortion or stillbirth.[249]

The Use in Specific Populations section included a separate warning about lactation:

> **USE IN SPECIFIC POPULATIONS** ... Lactation ... There is no information regarding ... atezolizumab in human milk, the effects on the breastfed infant, or the effects on milk production. As human IgG is excreted in human milk, the potential for ... harm to the infant is unknown. Because of the potential for serious adverse reactions in breastfed infants from TECENTRIQ, advise a lactating woman not to breastfeed.[250]

b. Antibodies Secreted During Lactation

Skorpen et al.[251] provides a review of small molecule drugs and antibodies, their levels in milk, and the ratio of levels in plasma/milk. To understand the package label's statement that "human IgG is excreted in human milk," the following should be considered as basic background information.

[247] Atezolizumab (urothelial carcinoma) BLA 761-034. Page 15 of 58-page Pharmacology Review.

[248] TECENTRIQ™ (atezolizumab) injection, for intravenous use. May 2016 (15 pp.).

[249] TECENTRIQ™ (atezolizumab) injection, for intravenous use. May 2016 (15 pp.).

[250] TECENTRIQ™ (atezolizumab) injection, for intravenous use. May 2016 (15 pp.).

[251] Skorpen C, et al. The EULAR points to consider for use of antirheumatic drugs before pregnancy, and during pregnancy and lactation. Ann. Rheum. Dis. 2016;75:795–810.

Immediately after an infant is born, human milk takes the form of colostrum. After that, it takes the form of transitional milk and later takes the form of mature milk. Colostrum provides IgA, IgG, and IgM antibodies to the infant.[252] IgA antibodies are highest in colostrum and lowest in mature milk.[253] Mother's milk IgA antibodies at birth and in the next few days occur at about 275 mg/100 mL and by eight weeks of lactation have dropped to about 50 mg/mL.[254] Also, milk IgG antibodies at birth and in the next few days occur at 30 mg/100 mL and by eight weeks of lactation have dropped to about 5 mg/100 mL. The colostrum IgA and IgG levels on the day of birth are 1,000 mg IgA/100 mL and 90 mg IgG/100 mL, respectively.[255]

Transitional milk represents a period of ramped up milk production to support the nutritional needs of the infant and occurs from 5 days to 2 weeks postpartum, after which the milk is considered mature. Colostrum antibodies protect the infant from infections.[256]

Therapeutic antibodies administered to female human subjects can occur in mother's milk during lactation, as has been documented for natalizumab[257] and infliximab.[258] Thus, Sponsors should consider measuring the levels of therapeutic antibodies in breast milk, in the situation where the antibody is administered by injection (sc, iv, im) and, if detectable, should consider evaluating risk for AEs to the breastfeeding infant.

c. Belimumab (Systemic Lupus Erythematosus; SLE) BLA 125-370

Belimumab is an antibody for treating systemic lupus erythematosus (SLE; lupus). Although there is no animal model for lupus, damage to the body in this disease results from the body's overproduction of antibodies. Because belimumab is able to halt antibody production (without regard to the disease suffered by an animal or patient), it is the case that belimumab can be used to treat lupus.

Belimumab was created, using, as a starting point, bacteriophage libraries which, in turn, provided a lead single-chain fragment chain and ultimately produced a single-chain

[252] Madi A, et al. Tumor-associated and disease-associated autoantibody repertoires in healthy colostrum and maternal and newborn cord sera. J. Immunol. 2015;194:5272–81.

[253] Castellote C, et al. Premature delivery influences the immunological composition of colostrum and transitional and mature human milk. J. Nutr. 2011;141:1181–7.

[254] Miranda R, et al. Effect of maternal nutritional status on immunological substances in human colostrum and milk. Am. J. Clin. Nutr. 1983;37:632–40.

[255] Kulski JK, Hartmann PE. Changes in human milk composition during the initiation of lactation. Aust. J. Exp. Biol. Med. Sci. 1981;59:101–14.

[256] Ballard O, Morrow AL. Human milk composition: nutrients and bioactive factors. Pediatr. Clin. North. Am. 2013;60:49–74.

[257] Baker TE, et al. Transfer of natalizumab into breast milk in a mother with multiple sclerosis. J. Hum. Lact. 2015;31:233–236.

[258] Chaparro M, Gisbert JP. How safe is infliximab therapy during pregnancy and lactation in inflammatory bowel disease? Expert Opin. Drug Saf. 2014;13:1749–62.

antibody called scFV01. The Sponsor then used the single-chain antibody to create an immunoglobulin (IgG), which was called mABA01.[259] The use of bacteriophages for producing humanized antibodies has been described.[260,261]

FDA's Pharmacology Review provided details of the humanization procedure: "In the process of generating a fully humanized anti-BLyS antibody, a lead single-chain fragment chain variant (scFv) ... was isolated from a library of antibody derived from B cells ... displayed in bacteriophage. An affinity matured variant of ... [the lead scFV] ... was then isolated from a randomized library by ... isolating the DNA by polymerase chain reaction (PCR) ... and ligating it into a phagemid vector and ultimately introducing it into E. coli ... and growing the bacteria. The therapeutic candidate was called scFvA01."[262]

In general, the less that any therapeutic protein possesses the same amino acid sequence of a corresponding protein encoded by the human genome, the greater will be the risk that the patient will generate antibodies against the therapeutic protein. Belimumab is notable, in that it has been called a "milestone."

Regarding this milestone, an article published in *Nature* revealed the timeline of FDA's approval for belimumab, stating, "On March 9, 2011, the ... FDA ... did something it had not done in more than 50 years—it approved a drug specifically for the treatment of SLE. The drug, belimumab, is a human monoclonal antibody ... that binds and neutralizes B lymphocyte stimulator (BLyS) ... [t]he **milestone is all the more remarkable** in that as recently as 1998, the target of the approved therapeutic agent (BLyS) was itself an unknown entity to the scientific community."[263]

Belimumab binds to a cytokine that circulates in the bloodstream. The cytokine is BLyS (B-lymphocyte stimulator). By binding to BLyS, the antibody prevents BLyS from binding to BLyS receptor on B cells and thus prevents BLyS from stimulating proliferation of the B cells. BLyS binds to three different receptors that are expressed on the surface of B cells: BR3, TACI, and BCMA.[264,265,266]

[259] Belimumab (systemic lupus erythematosus) BLA 125-370. Pages 37–38 of 171-page Pharmacology Review.

[260] Bostrom J, Fuh G. Design and construction of synthetic phage-displayed Fab libraries. Methods Mol. Biol. 2009;562:17–35.

[261] Beerli RR, et al. Isolation of human monoclonal antibodies by mammalian cell display. Proc. Natl. Acad. Sci. 2008;105:14336–41.

[262] Belimumab (systemic lupus erythematosus) BLA 125-370. Pages 37–38 of 171-page Pharmacology Review.

[263] Stohl W, Hilbert DM. The discovery and development of belimumab: the anti-BLyS–lupus connection. Nature Biotechnol. 2012;30:69–77.

[264] Belimumab (systemic lupus erythematosus) BLA 125-370. Page 23 of 171-page Pharmacology Review.

[265] Miller JP, et al. Space, selection, and surveillance: setting boundaries with BLyS. J. Immunol. 2006;176:6405–10.

[266] BLyS receptor-3, transmembrane activator 1 and calcium-modulator and cyclophilin ligand-interactor (TACI), and B cell maturation Ag (BCMA).

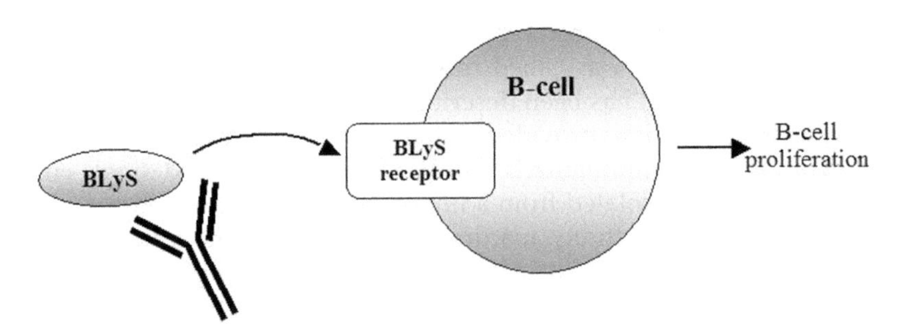

FIGURE 6.3 BLyS action at the BLyS receptor. This shows BLyS, which is a cytokine, binding to BLyS receptor on a B-cell, resulting in proliferation of the cell. Also shown is belimumab, which can bind to BLyS, thus preventing BLyS-mediated proliferation of B-cells, for use in treating lupus (SLE).

Although BLyS binds three different receptors, its greatest influence is mediated by signaling through BR3 (BLyS receptor-3).[267] B-cells require the continuous signaling of BLyS (ligand) to BR3 (receptor). If this signaling pathway is experimentally deleted, then the B-cells die.[268] Bluml et al.[269] provides an account of the various stages of B-cell differentiation, including the stage where antibodies are produced. B-cells express antibodies and release them into the bloodstream. Antibodies are used in immune responses against bacteria, viruses, and cancer cells. However, in lupus (SLE), autoantibodies are generated, which result in damage to the kidneys, renal failure, hypertension, and eventually death.[270]

Fig. 6.3 shows the signaling pathway between BLyS and BLyS receptor, along with the mechanism of action of belimumab. The forked structure in Fig. 6.3 is conventionally used to represent antibodies. The diagram shows BLyS about to bind to BLyS receptor, where the result of this binding is stimulation of B-cell proliferation. Where belimumab is present, the antibody binds to BLyS, blocks its binding to BLyS receptor, and prevents B-cell proliferation.[271]

FDA stated that "There are no animal models available in which to test belimumab's **efficacy** to treat SLE ... [t]here is no animal model of lupus; therefore, efficacy of the

[267] Dosenovic P, et al. BLyS-mediated modulation of naive B cell subsets impacts HIV env-induced antibody responses. J. Immunol. 2012;188:6016–26.

[268] Stadanlick JE, et al). Tonic B cell antigen receptor signals supply an NF-kappaB substrate for prosurvival BLyS signaling. Nature Immunol. 2008;9:1379–87.

[269] Bluml S, et al. B-cell targeted therapeutics in clinical development. Arthritis Res Ther. 2013;15 (Suppl. 1):54 (21 pp.).

[270] Ryan MJ. The pathophysiology of hypertension in systemic lupus erythematosus. Am. J. Physiol. Regul. Integr. Comp. Physiol. 2009;296:R1258–67.

[271] Stohl W, et al. Belimumab reduces autoantibodies, normalizes low complement levels, and reduces select B cell populations in patients with systemic lupus erythematosus. Arthritis Rheum. 2012;64:2328–37.

product could not be tested in an animal model."[272,273] But for testing **safety**, the Sponsor was able to move forward and assess suitability of the best animal model for drug toxicity studies. As a first step, the Sponsor compared binding of belimumab to its target (BLyS), where the target was acquired from human, monkey, and mouse.

1. *Step One. Mouse Is Not a Suitable Model Because of Low Binding to Target*

The Sponsor found that binding of belimumab to human BLyS (0.9 nM) and binding to mouse BLyS (9.48 nM) had the indicated binding constants. As one can see, binding to mouse BLyS is weaker than to human BLyS. FDA wrote, "These results indicate that there is ... 10-fold lower affinity of binding of murine BLyS receptor ... as derived using immobilized human BLyS-receptor and immobilized mouse BLyS-receptor."[274,275] The marked difference in binding affinity indicates that the **mouse is not a suitable animal model** for determining toxicity in humans.

2. *Step Two. Mouse Is Not a Suitable Model Because of Immune Reaction*

The Sponsor discovered that the mouse mounted a vigorous immune response against belimumab, which prevented the detection of other forms of toxicity. In other words, this immune response prevented the detection of toxicity resulting from belimumab's blocking of BLyS/BLyS-receptor-mediated signaling. FDA concluded, "Given the strong neutralizing immunogenic response observed in mice after administration of belimumab, resulting in clearance of belimumab and in some cases death, **the mouse is not an appropriate species** for ... toxicity studies of belimumab."[276]

3. *Step Three. Monkey May Be a Suitable Model*

The Sponsor conducted an in vitro binding experiment, which revealed identical affinity of the study drug for the human target and for the corresponding target in the monkey. FDA wrote, "**Belimumab binds to human and ... monkey BLyS with nearly identical affinity** of 274 picomolar and 264 picomolar, respectively. Belimumab's affinity for binding to mouse BLyS is ... 10-fold lower (9.98 nanomolar)."[277]

4. *Step Four. Monkey Is a Suitable Model*

The Sponsor tested if belimumab's binding actually blocked functioning of the BLyS/BLyS-receptor signaling pathway. Just because an antibody binds to a given cytokine does not necessarily mean that the antibody will block the cytokine's ability to transmit a signal to the cytokine receptor.

[272] Belimumab (systemic lupus erythematosus) BLA 125-370. Page 12 of 171-page Pharmacology Review.

[273] News and Analysis. Landmark lupus approval opens door for next wave of drugs. Nat. Rev. Drug Discov. 2011;10:243–5.

[274] Belimumab (systemic lupus erythematosus) BLA 125-370. Pages 47–8 of 171-page Pharmacology Review.

[275] In these comments the FDA review used the term "BAFF-receptor" instead of "BLyS-receptor." BAFF is a synonym for BLyS. "BAFF" and "BLyS" refer to the same protein.

[276] Belimumab (systemic lupus erythematosus) BLA 125-370. Page 64 of 171-page Pharmacology Review.

[277] Belimumab (systemic lupus erythematosus) BLA 125-370. Pages 12 and 26 of 171-page Pharmacology Review.

Cultured splenocytes (white blood cells in the spleen) were used to test BLyS function and for the ability of the antibody to block this function. The function that was measured was the ability of added BLyS to stimulate splenocyte proliferation. FDA's Pharmacology Review described this influence on splenocytes, and the fact that the monkey is a good toxicity model: "A dose related proliferation of the splenocytes was noted with BLyS derived from human as well as ... monkey and the ... antibody inhibited splenocyte proliferation. The data showed that ... [belimumab] can cross-react with BLyS derived from ... monkey, suggesting that monkey might be an appropriate species for ... toxicity studies."[278]

The above information led FDA to conclude that the **monkey** is an appropriate species and to conclude that the **mouse** is not an appropriate species.

5. *Sponsor's Monkey Toxicity Data*

The monkey toxicity studies involved these parameters:

- Teratogenicity
- Infant and fetal death
- B-cell number
- Serum immunoglobulin levels (IgM)
- Serum levels of IgG, IgA, and IgE immunoglobulins

The Sponsor's monkey toxicology studies revealed that "Belimumab was not teratogenic in monkeys. Some infant and fetal death was observed in belimumab-treated monkeys although the cause is unknown."[279] The monkey toxicity studies revealed that "B cell depletion was observed in peripheral blood and in the spleen and mesenteric lymph node at Day 29[280] ... [t]he changes were not dose related ... [t]he finding is considered as the pharmacological effect of the drug product."[281]

Regarding the ability of belimumab to reduce antibodies in the bloodstream of infant monkeys, FDA wrote, "The serum immunoglobulin levels (IgM) was decreased one year post-partum in the infants. No changes in IgG, IgA, and IgE were noted."[282]

Finally, FDA disclosed that the study drug was secreted by female monkeys into breast milk, which provided a basis for a corresponding warning in the package label. FDA's review stated that "Milk samples were collected ... for the bioanalytical analysis of belimumab ... [o]nly four milk samples could be collected from 2 females, both of these samples had measurable concentrations of belimumab, indicating that the drug product could be secreted into milk."[283]

[278] Belimumab (systemic lupus erythematosus) BLA 125-370. Pages 28, 46, and Figure 25B, of 171-page Pharmacology Review.

[279] Belimumab (systemic lupus erythematosus) BLA 125-370. Pages 2 and 4 of 171-page Pharmacology Review.

[280] Belimumab (systemic lupus erythematosus) BLA 125-370. Page 82 of 171-page Pharmacology Review.

[281] Belimumab (systemic lupus erythematosus) BLA 125-370. Pages 101, 118, and 164 of 171-page Pharmacology Review.

[282] Belimumab (systemic lupus erythematosus) BLA 125-370. Page 134 of 171-page Pharmacology Review.

[283] Belimumab (systemic lupus erythematosus) BLA 125-370. Page 148 of 171-page Pharmacology Review.

Package label. The Sponsor's belimumab (Benlysta®) monkey toxicology findings of fetal deaths, infant deaths, and reduction of IgM, all found a place in the Use in Specific Populations section of the package label:

> **USE IN SPECIFIC POPULATIONS** ... Pregnancy. Nonclinical reproductive studies have been performed in **pregnant cynomolgus monkeys** receiving belimumab at doses of 0, 5 and 150 mg/kg by intravenous infusion (the high dose was approximately 9 times the anticipated maximum human exposure) every 2 weeks from gestation 280 day 20 to 150. Belimumab was shown to cross the placenta. Belimumab was not associated with direct or indirect teratogenicity under the conditions tested. Fetal deaths were observed in 14%, 24% and 15% of pregnant females in the 0, 5 and 150 mg/kg groups, respectively. Infant deaths occurred with an incidence of 0%, 8% and 5%. The cause of fetal and infant deaths is not known. The relevance of these findings to humans is not known. Other treatment-related findings were limited to the expected reversible reduction of B cells in both dams and infants and reversible reduction of IgM in infant monkeys. B cell numbers recovered after the cessation of belimumab treatment by about one year postpartum in adult monkeys and by 3 months of age in infant monkeys. IgM levels in infants exposed to belimumab in utero recovered by 6 months of age.[284]

The Use in Specific Populations section of the package label also warned against belimumab being secreted into mother's milk, based solely on data from female monkeys:

> **USE IN SPECIFIC POPULATIONS** ... Nursing Mothers. It is not known whether BENLYSTA is excreted in human milk or absorbed systemically after ingestion. However, **belimumab was excreted into the milk of cynomolgus monkeys**. Because maternal antibodies are excreted in human breast milk, a decision should be made whether to discontinue breastfeeding or to discontinue the drug, taking into account the importance of breastfeeding to the infant and the importance of the drug to the mother.[285]

6. FDA Regulations for Labeling on Pregnancy, Nursing, and Lactation

When FDA proposes changes to labeling rules, the proposal is published in Federal Register, and FDA invites comments from the public. At a later point, the comments are published, and then later FDA publishes its "Final Rule." In December 2014, FDA's proposed changes for pregnancy, nursing, and lactation labeling were published, as shown in this excerpt from the Federal Register, "FDA proposed the following format and content changes to the 'Pregnancy,' 'Labor and delivery,' and 'Nursing mothers' subsections of prescription drug labeling ... [m]erge the current 'Pregnancy' and 'Labor and delivery' subsections into a single 'Pregnancy' subsection ... under ... 'USE IN SPECIFIC POPULATIONS' ... [r]ename the 'Nursing mothers' subsection as 'Lactation' ... under the section 'USE IN SPECIFIC POPULATIONS.'"[286]

Persons drafting the package label should consult the Federal Register on a periodic basis for revisions to the content and format of all sections of the label.

[284] Package label. BENLYSTA® (belimumab) for injection, for intravenous use only. March 2011 (16 pp.).

[285] BENLYSTA® (belimumab) for injection, for intravenous use only. March 2011 (16 pp.).

[286] Federal Register. December 4, 2014. Vol. 79. No. 233. Dept. Health and Human Services. Food and Drug Administration. 21 CFR Part 201. Content and Format of Labeling for Human Prescription Drug and Biological Products; Requirements for Pregnancy and Lactation Labeling.

d. Denosumab (Postmenopausal Osteoporosis) BLA 125-320

Denosumab is a recombinant antibody for treating osteoporosis.[287] Denosumab is a fully human antibody, which was produced by XenoMouse® technology.[288,289] The antibody was not produced by the process of humanization, that is, by immunizing a wild-type mouse, followed by isolating the mouse gene encoding the antibody, with subsequent humanization.[290,291,292]

Acquiring human antibodies directly from the XenoMouse®, without need for humanization, was described as "Laboratory mice provide a ready source of diverse, ... monoclonal antibodies (mAbs). However, development of rodent antibodies as therapeutic agents has been impaired by the inherent immunogenicity of these molecules. One technology that has been explored to generate low immunogenicity mAbs for in vivo therapy involves the use of transgenic mice expressing repertoires of human antibody gene sequences."[293]

This describes FDA's analysis of the best-suited animal model. FDA observed that the study drug is an antibody that binds to **human** RANK ligand and to **monkey** RANK ligand, but that the antibody does not bind to **rodent** RANK ligand.[294] RANK means "receptor activator of nuclear factor-kappa B." FDA stated that, as a consequence, **wild-type rodents** could not be used. The Sponsor overcame this problem by creating a genetically engineered mouse that expressed human RANK ligand, replacing mouse RANK ligand, thus providing a suitable rodent animal model for use with the antibody.

1. *Mouse Toxicity Data*

The genetically engineered mouse provided toxicity data and efficacy data. The mouse toxicity studies revealed that denosumab "induced a **failure of incisor tooth eruption.**"[295]

[287] Denosumab (post-menopausal osteoporosis) BLA 125-320. 710-page pdf Medical Review.

[288] Foltz IN, et al. Discovery and bio-optimization of human antibody therapeutics using the XenoMouse® transgenic mouse platform. Immunol. Rev. 2016;270:51−64.

[289] Kostenuik PJ, et al. Denosumab, a fully human monoclonal antibody to RANKL, inhibits bone resorption and increases BMD in knock-in mice that express chimeric (murine/human) RANKL. J. Bone Miner. Res.2009;24:182−95.

[290] Presta LG Molecular engineering and design of therapeutic antibodies. Curr. Opin. Immunol. 2008;20:460−70.

[291] Apgar JR, et al. Beyond CDR-grafting: structure-guided humanization of framework and CDR regions of an anti-myostatin antibody. MAbs. 2016;8:1302−18.

[292] Frenzel A, et al. Phage display-derived human antibodies in clinical development and therapy. MAbs. 2016;8:1177−94.

[293] Lonberg N. Human antibodies from transgenic animals. Nat. Biotechnol. 2005;23:1117−25.

[294] Denosumab (post-menopausal osteoporosis) BLA 125-320, BLA 125-331. Deputy Division Director Review. George S. Benson, MD (41 pp. total), October 16, 2009, Pages 151−89 of 710-page pdf file of Medical Review.

[295] Denosumab, BLA 125-320, BLA 125-331. Pharmacology/Toxicology Review and Evaluation (pp. 14, 19, 69−71, and 90 of 211 pages total).

Further regarding toxicity, FDA observed that "fracture healing was morphologically delayed, mechanical strength was not negatively impacted ... [c]allus remodeling was ... noticeably delayed."[296]

2. *Mouse Efficacy Data*

The mouse efficacy data showed that "denosumab treatment induced ... **increases in torsional rigidity** ... relative to vehicle control fractured bones." Note that the purpose of denosumab is to treat osteoporosis.

3. *Monkey Toxicity Data*

The Sponsor tested toxicity in adult monkeys and young monkeys. FDA's Pharmacology Review observed that "adult ... monkeys were dosed weekly with 0, 01, 1, or 10 mg/kg ... [t]here were no significant non-pharmacodynamically related effects." Regarding young monkeys, FDA's review observed that "In animals where the epiphyseal plates had not fully closed prior to treatment, growth plates were markedly enlarged with reduced chondroclasis and expanded growth plates, which were associated with ... **cartilage calcification and cartilage erosion.**"[297]

4. *Monkey Efficacy Data*

Regarding monkey data on monkey efficacy, FDA described a monkey model for osteoporosis and observed that "a 16-month ... study was conducted in ovariectomized ... monkeys, a model which mimics postmenopausal bone loss ... monthly treatment with denosumab ... prevented ovariectomized-induced bone mineral density changes in both cortical and cancellous bone, and **increased bone strength** ... [f]ollowing cessation of treatment, bone mineral density ... returned to original baseline levels."[298]

5. *FDA's Recommendation*

FDA recommended that "denosumab should not be used in patients where the epiphyseal plates were not fully closed." Referring to the monkey toxicity finding of reduced chondroclasis, expanded growth plates, and cartilage erosion, FDA recommended that "[t]his growth plate finding suggests that denosumab should only be used for patients in which the epiphyses are closed (adult populations only)."[299]

Package label. FDA's observations on monkey toxicity found a place in the package label for denosumab (Prolia®):

> NONCLINICAL TOXICOLOGY ... Adolescent primates treated with denosumab at doses >10 times (10 and 50 mg/kg dose) higher than the recommended human dose of 60 mg administered once every 6 months, based on mg/kg, had abnormal growth plates.[300]

[296] Denosumab (post-menopausal osteoporosis) BLA 125-320, BLA 125-331. Page 33 of 91-page Clinical Review.

[297] Denosumab (post-menopausal osteoporosis) BLA 125-320. Page 38 of 254-page Pharmacology Review.

[298] Denosumab (post-menopausal osteoporosis) BLA 125-320. Pages 56, 67, 69, and 71 of 254-page Pharmacology Review.

[299] Denosumab (post-menopausal osteoporosis) BLA 125-320. Page 241 of 254-page Pharmacology Review.

[300] Package label. PROLIA™ (denosumab). June 2010 (17 pp.).

In view of the fact that the indication for denosumab is treating postmenopausal women with osteoporosis at high risk for fracture,[301] it might be asked why FDA's Pharmacology Review included information on young monkeys, that is, in animals where the epiphyseal plates had not fully closed,[302] and it might be asked why the package label stated that adolescent primates treated with denosumab had abnormal growth plates. An answer is available. The published literature suggests that this information was gathered to support an off-label use for denosumab, that is, for osteoporosis in children. The American Academy of Pediatrics has published a policy statement on off-label drugs for children.[303]

Alternatively, it could be that this information was gathered to support a separate FDA submission for the indication of osteoporosis in children. In an article on osteoporosis in children, Saraff et al.[304] report that "New potent antiresorptive drugs such as denosumab, a monoclonal RANKL antibody with a favorable safety profile in adults ... is currently being tested in multicentre trials in children."

e. Evolocumab (Hyperlipidemia and Hypercholesterolemia) BLA 125-522

Evolocumab is an antibody that binds to PCSK9, where the antibody's therapeutic effect is to reduce LDL-cholesterol and where the therapeutic goal is to reduce risk for heart attack and stroke. Evolocumab is a human IgG2 monoclonal antibody[305] that has been described as "fully human."[306] The Sponsor stated that evolocumab, which is an anti-PCSK9 antibody, was developed using the XenoMouse®.[307]

Despite being fully human, evolocumab binds with high affinity to its target in two animals, the hamster and monkey. FDA wrote that "The Applicant identified the hamster and monkey as pharmacologically relevant species for toxicology testing with evolocumab; both species express PCSK9, to which evolocumab binds with high affinity."[308] However, evolocumab binds very poorly to the corresponding target in the mouse.

Regarding the poor binding to the target in mice, the Sponsor found that "Evolocumab binds to **human**, cynomolgus **monkey**, and **hamster** PCSK9 with high affinity (KD = 16, 8,

[301] Package label. PROLIA™ (denosumab). June 2010 (17 pp.).

[302] Denosumab (post-menopausal osteoporosis) BLA 125-320. Page 38 of 254-page Pharmacology Review.

[303] Frattarelli DA, et al. Off-label use of drugs in children. Pediatrics. 2014;133:563–7.

[304] Saraff V, Hogler W. Endocrinology and adolescence: osteoporosis in children: diagnosis and management. Eur. J. Endocrinol. 2015;173:R185–97.

[305] Evolocumab (hyperlipidemia and hypercholesterolemia) BLA 125-522. Page 2 of 163-page Pharmacology Review.

[306] European Medicines Agency (EMA). Assessment report Repatha International non-proprietary name: evolocumab. Procedure No. EMEA/H/C/003766/0000. 2015 (122 pp.).

[307] Chan JC, et al. A proprotein convertase subtilisin/kexin type 9 neutralizing antibody reduces serum cholesterol in mice and nonhuman primates. Proc. Natl. Acad. Sci. 2009:106:9820–5.

[308] Evolocumab (hyperlipidemia and hypercholesterolemia) BLA 125-522. Page 11 of 163-page Pharmacology Review.

and 14 pM, respectively) ... [but] evocolumab has lower affinity for **mouse** PCSK9 (17,000 pM)."[309] The binding affinity to mouse PCSK9 was one thousand times less than the binding affinity to PCSK9 of the human, monkey, and hamster.

1. *Assessing Toxicity on Fertility, Pregnancy, and Toxicity in Neonatal and Infant Monkeys*

FDA observed that evolocumab does not impair fertility, writing that "Effects on fertility were ... assessed in the 6 month chronic monkey toxicity study at exposure multiples of up to 744-, 300- and 134-fold compared to the recommended human doses of 140 mg Q2W, 420 mg QM and 420 mg Q2W, respectively. No effects on fertility endpoints were observed."[310]

Also FDA observed that in pregnant monkeys, neonatal monkeys, and infant monkeys, there was no detectable toxicity: "Evolocumab was tested in pregnant monkeys during the period of embryofetal development to parturition with subcutaneous administration once every two weeks at doses that provide exposure multiples of 30-, 12- and 5.2-fold the recommended human doses of 140 mg Q2W, 420 mg QW and 420 mg Q2W. Offspring were followed to 6 months of infancy ... [n]o clearly drug-related toxicity was observed in maternal or infant monkeys."[311]

Actually, spontaneous abortions were detected in pregnant monkeys, but FDA remarked that the rates of spontaneous abortions were similar to that found with control monkeys. The term "GD" means gestation day. FDA's remarks about the lack of effect on monkey abortions were, "Two fetuses (2/18; 11%) were aborted between GD20 and GD50, which is a slightly higher incidence than observed in the same laboratory for the same species of monkey (mean of 8.1%), but was within the historical control range (6.7 to 39%). Two fetuses were aborted between GD51 and GD99 (11%), which is slightly outside of the historical range (0 to 10%)."[312]

2. *Assessing Toxicity on Fertility in Hamsters*

FDA observed that evolocumab showed no toxic effects on fertility in hamsters. FDA remarked, "Effects of evolocumab on fertility and mating were assessed in hamsters. No effects of evolocumab (subcutaneous dosing once every two weeks) on mating, fertility, estrous cycling, or male reproduction were observed at exposure multiples up to 30-, 12- and 5.3-fold the plasma exposures measured in humans at the 140 mg Q2W, 420 mg QM and 420 mg Q2W evolocumab doses."[313]

[309] Evolocumab (hyperlipidemia and hypercholesterolemia) BLA 125-522. Page 29 of 163-page Pharmacology Review.

[310] Evolocumab (hyperlipidemia and hypercholesterolemia) BLA 125-522. Pages 12–13 of 163-page Pharmacology Review.

[311] Evolocumab (hyperlipidemia and hypercholesterolemia) BLA 125-522. Pages 12–13 of 163-page Pharmacology Review.

[312] Evolocumab (hyperlipidemia and hypercholesterolemia) BLA 125-522. Page 158 of 163-page Pharmacology Review.

[313] Evolocumab (hyperlipidemia and hypercholesterolemia) BLA 125-522. Pages 12–13 of 163-page Pharmacology Review.

Package label. The results from the monkey toxicity studies in pregnancy found a place on the package label of evolocumab (Repatha®):

> **USE IN SPECIFIC POPULATIONS.** Pregnancy. There are no data available on use of REPATHA in pregnant women to inform a drug-associated risk. In animal reproduction studies, there were **no effects on pregnancy or neonatal/infant development** when ... monkeys were ... administered evolocumab from organogenesis through parturition at dose exposures up to 12 times the exposure at the maximum recommended human dose of 420 mg every month.[314]

The Use in Specific Populations section describes the transfer of evolocumab from the mother, across the placenta, and to the circulatory system of the fetus with consequent detection in newborn monkeys:

> **USE IN SPECIFIC POPULATIONS** ... Measurable evolocumab serum concentrations were observed in the infant monkeys at birth at comparable levels to maternal serum, indicating that evolocumab, like other IgG antibodies, crosses the placental barrier. FDA's experience with monoclonal antibodies in humans indicates that they are unlikely to cross the placenta in the first trimester; however, they are likely to cross the placenta in increasing amounts in the second and third trimester.[315]

3. *Mechanism of Transfer of Antibodies from Maternal Circulatory System to the Fetus*

This is background information. In pregnancy, placental villi are submerged in maternal blood. In the third trimester, IgG antibodies are efficiently transported across the placenta from the mother, resulting in fetal blood IgG concentrations approaching that of the mother. A receptor called neonatal Fc receptor (FcRn) is used for transport of IgG and thus provides the fetus and newborn with immune protection until the infant starts producing its own IgG.[316] Antibody transport experiments demonstrated that pregnant rats, pregnant rabbits, and pregnant guinea pigs can transport human antibodies from the maternal blood, across the placenta, to the fetal circulatory system. Experiments with the guinea pig have shown that all human IgG subclasses can pass guinea pig placenta at the last third of pregnancy.[317].

f. Nivolumab (Melanoma) BLA 125-554

Nivolumab is for treating melanoma and other cancers. Nivolumab is an antibody that binds to PD-1, blocks interaction of PD-L1 (ligand) with PD-1 (receptor), with consequent blocking of the PD-L1/PD-1 signaling pathway. The PD-L1/PD-1 signaling pathway can also be blocked by antibody drugs that bind to PD-L1, such as atezolizumab.

[314] Package label. REPATHA (evolocumab) injection, for subcutaneous use. August 2015 (16 pp.).

[315] Package label. REPATHA (evolocumab) injection, for subcutaneous use. August 2015 (16 pp.).

[316] Mathiesen L, et al. Maternofetal transplacental transport of recombinant IgG antibodies lacking effector functions. Blood. 2013;122:1174–81.

[317] Struble EB, et al. Human antibodies can cross guinea pig placenta and bind its neonatal Fc Receptor: implications for studying immune prophylaxis and therapy during pregnancy. Clin. Dev. Immunol. 2012;2012:538701.

1. *Efficacy of Nivolumab for Treating Tumors in Mice*

The Sponsor realized that nivolumab does not bind to mouse PD-1 and hence could not be used in mice. Thus, the Sponsor prepared an antibody that was effective in binding to mouse PD-1. The FDA reviewer articulated the problem as, "Nivolumab does not recognize mouse PD-1. Therefore, to assess the efficacy of PD-1 blocking on ... tumors in mice, a surrogate anti-mouse PD-1 antibody was derived."[318] In order to create a suitable anti-PD-1 antibody, the Sponsor **immunized rats with mouse PD-1** and screened the generated and isolated the generated antibodies for one that binds mouse PD-1."[319] Thus, because the antibody was raised by injecting rats with mouse PD-1, the resulting antibody was able to bind to mouse PD-1.

The antibody prepared in rats had the ability to bind mouse PD-1, and hence was used in mouse efficacy studies. Efficacy studies in mice revealed that the antibody successfully reduced growth of tumors in mice.

As stated by FDA's Pharmacology Review, "the Sponsor assessed efficacy of the antibody in mice bearing tumors, and discovered that in these mice, there was one instance of **delayed tumor growth** and two instances of **tumor regression** ... one of the mice ... had a tumor volume of 514 mm^3 on Day 14 which **regressed** to ... 0 mm^3 on Day 21[320] ... delayed or prevented outgrowth of tumors."[321]

2. *Toxicity of Nivolumab in Monkeys*

In monkeys, "there were dose-related increases in first- and third-trimester pregnancy losses in nivolumab-treated monkeys ... including an increase in incidence of infant loss."[322] In addition, in monkeys there was "one umbilical thrombus in one ... female that aborted on gestation day 47 (GD47) ... [t]hree of the 4 infants lost ... were delivered prematurely (GD 131, 135, and 143) and died within the first two weeks."[323] The Sponsor's findings on monkey fetal deaths and deaths of monkey infants found a place on the package label, as shown below.

Package label. The Sponsor's toxicity studies are exemplary, in that they demonstrate that an antibody (not merely a small molecule) can pass from the mother's bloodstream via the placenta to the bloodstream of the fetus and cause abortions of the fetus and death of the newborn. The Sponsor's animal data occurred in the Use in Specific Populations section:

> **USE IN SPECIFIC POPULATIONS**. Pregnancy ... Animal Data. A central function of the PD-1/PD-L1 pathway is to preserve pregnancy by maintaining maternal immune tolerance to the fetus. Blockade of PD-L1 signaling ... disrupt[s] tolerance to the fetus and ... increase[s] fetal loss. The effects of nivolumab on prenatal and postnatal development were evaluated in monkeys that received nivolumab twice weekly from the onset of organogenesis through delivery, at exposure levels of between 9 and 42 times higher than those observed at the clinical dose of 3 mg/kg of nivolumab (based on AUC). **Nivolumab ... resulted in ... increase in spontaneous abortion and increased neonatal death.**[324]

[318] Nivolumab (melanoma) BLA 125-554. Page 36 of 99-page Pharmacology Review.

[319] Nivolumab (melanoma) BLA 125-554. Page 38 of 99-page Pharmacology Review.

[320] Nivolumab (melanoma) BLA 125-554. Page 38 of 99-page Pharmacology Review.

[321] Nivolumab (melanoma) BLA 125-554. Page 91 of 99-page Pharmacology Review.

[322] Nivolumab (melanoma) BLA 125-554. Pages 5 and 14 of 99-page Pharmacology Review.

[323] Nivolumab (melanoma) BLA 125-554. Page 67 of 99-page Pharmacology Review.

[324] Package label. OPDIVO (nivolumab) injection, for intravenous use. April 2017 (59 pp.).

g. Pegloticase (Pegylated Urate Oxidase) for Chronic Gout. BLA 125-293

FDA's review for pegloticase (Krystexxa®) reveals the situation where, at the time that FDA granted approval to the study drug, FDA recommended that the Sponsor carry out additional toxicity studies in animals. FDA's review also reveals the value in reading the original package label and supplementary package labels that are published in the years following initial FDA approval. Studying the package label issued with FDA's Approval Letter, and studying subsequent supplemental labels, can reveal how the Sponsor responded to FDA's requests that the Sponsor conduct additional testing.[325,326,327]

Pegloticase is a therapeutic protein used for treating gout. Pegloticase consists of an enzyme (uricase) with a covalently bound group of polyethylene glycol. Gout is a type of arthritis that is initiated by sodium urate crystals depositing in the synovial fluid of joints. Gout is associated with high serum urate (hyperuricemia), defined as a serum urate of 6.8 mg/100 mL or more. This number is the limit of solubility of urate, where beyond this level the result is precipitation. The human genome does not express any enzyme that catabolizes urate, but an enzyme from animals (uricase) catabolizes urate to allantoic acid. Hyperuricemia results from overproduction of urate and deficient excretion via the kidneys.[328]

FDA observed that "The sponsor performed a ... reproductive toxicity study in the rat and no maternal toxicity ... or teratogenic effects were seen." After reviewing the Sponsor's data from maternal toxicity and teratogenicity tests, FDA questioned the adequacy of the performed tests and decided that additional animal testing was needed. Regarding this need, FDA recommended, "Based on a prior agreement with [FDA] ... no studies were conducted to assess fertility and early embryonic development or prenatal and postnatal development ... [h]owever ... [FDA] has reevaluated these requirements ... and ... are recommending post-marketing ... studies in rats and ... in rabbits."[329]

Package label. The Sponsor complied with FDA's recommendations for the rabbit toxicity studies. The Sponsor's compliance is evident by comparing the package label that was published along with FDA's Approval Letter (September 2010) with the revised package label dated September 2016. The package label accompanying FDA's Approval Letter did not contain the word "rabbit."[330] In contrast, the revised package label dated September 2016 did have information on rabbit teratogenicity:

> **USE IN SPECIFIC POPULATIONS**. Pregnancy. There are no adequate and well-controlled studies of KRYSTEXXA in pregnant women ... [p]egloticase was not teratogenic in ... **rabbits** at ... 75 times the maximum recommended human dose ... in ... **rabbits** ... [n]o effects on ... fetal body weights were observed ... in **rabbits** ... at maternal doses up to 10 mg/kg twice weekly.[331]

[325] Pegloticase (pegylated urate oxidase) for chronic gout, BLA 125-293. 238-page pdf file for Medical Review.

[326] Package label. KRYSTEXXA™ (pegloticase) injection, for intravenous infusion. September 2010 (14 pp.).

[327] Package label. KRYSTEXXA® (pegloticase injection), for intravenous infusion. September 2016 (16 pp.).

[328] Neogi T. Gout. New Engl. J. Med. 2011;364:443–52.

[329] Summary of Review for Regulatory Action (page 4 of 16 pages total). Located on page 37 of 238-page pdf file containing Medical Review.

[330] Package label. KRYSTEXXA™ (pegloticase) injection, for intravenous infusion. September 2010 (14 pp.).

[331] Package label. KRYSTEXXA® (pegloticase injection), for intravenous infusion. September 2016 (16 pp.).

h. Ustekinumab (Psoriasis) BLA 125-261

This concerns ustekinumab for treating psoriasis. Psoriasis is characterized by thickened epidermis (acanthosis), which results from overproliferation of keratinocytes. Psoriasis also involves increased vascularity and accumulation of T cells, neutrophils, and dendritic cells.[332] Psoriasis is a uniquely human disease, as it appears not to occur spontaneously in animals.[333] The package label warning is exemplary in that it was based, in part, on theoretical considerations from ustekinumab's mechanism of action.

1. *Mouse Models for Psoriasis*

One mouse model for psoriasis is a xenograft model, where normal human skin or psoriatic human skin is transplanted to mouse skin.[334,335,336,337] Where normal human skin is transplanted, the transplant is then further treated to stimulate the formation of psoriatic lesions.

For transplanting skin to mice, immunodeficient SCID mice are used. These mice are not able to reject the human skin because they are not capable of producing T cells or B cells. Seven to 10 days after transplantation of psoriatic plaques originating from human skin, treatment of mice with a drug can be started. Where nonlesional human skin is transplanted to the mouse, the skin needs to be activated to induce the phenotype of psoriasis.[338] A closer picture of xenotransplantation mouse models for psoriasis is provided by these excerpts.

De Oliveira et al.[339] described this mouse model as follows: "In these models, healthy human skin ... is transplanted onto immunodeficient mice, allowed to become vascularized and heal, and in some of these models human immune cells are infused that will reconstitute the recipient with human immune cells to induce skin inflammation."

Similarly, Guerrero-Aspizua et al.[340] describe this type of mouse model as follows: "Nine to 12 weeks after transplantation, when the maturation of the regenerated human

[332] Kulig P, et al. IL-12 protects from psoriasiform skin inflammation. Nature Commun. 2016;7:13466. DOI: 10.1038/ncomms13466.

[333] Peterson TK. In vivo pharmacological disease models for psoriasis and atopic dermatitis in drug discovery. Basic Clin Pharmacol Toxicol. 2006;99:104–15.

[334] Boehncke WH. The SCID-hu xenogeneic transplantation model: complex but telling. Arch. Dermatol. Res. 1999;291:367–73.

[335] Schon MP. Animal models of psoriasis: a critical appraisal. Exp. Dermatol. 2008;17:703–12.

[336] Igney FH, et al. Humanised mouse models in drug discovery for skin inflammation. Expert Opin. Drug Discov. 2006;1:53–68.

[337] Boehncke WH, Schon MP. Animal models of psoriasis. Clin. Dermatol. 2007;25:596–605.

[338] Peterson TK. In vivo pharmacological disease models for psoriasis and atopic dermatitis in drug discovery. Basic Clin. Pharmacol. Toxicol. 2006;99:104–15.

[339] de Oliveira VL, et al. Humanized mouse model of skin inflammation is characterized by disturbed keratinocyte differentiation and influx of IL-17A producing T cells. PLoS One. 2012;7:e45509. DOI:10.1371.

[340] Guerrero-Aspizua S, et al. Development of a bioengineered skin-humanized mouse model for psoriasis: dissecting epidermal-lymphocyte interacting pathways. Am. J. Pathol. 2010;177:3112–24.

TABLE 6.3 Species Cross-Reactivity of Ustekinumab with IL-12

Human	+ Binding
Baboon	+ Binding
Cynomolgus	+ Binding
Mouse	No binding
Rat	No binding
Dog	No binding

Ustekinumab (psoriasis) BLA 125-261. Table 6.1, page 16 of 64-page Pharmacology Review.

skin is complete ... in vitro derived [human] T lymphocyte subpopulations ... were inoculated by intradermal injection into the stable engrafted human skin every other day for two weeks."

Another mouse model for psoriasis, which does not involve any skin transplant, is produced by injecting mice with imiquimod.[341,342]

2. *Biochemistry of Ustekinumab*

Ustekinumab is an antibody that binds to a subunit that is a component of two different cytokines, interleukin-12 (IL-12) and interleukin-23 (IL-23). The common subunit is p40.[343] As such, the antibody can bind to and inactivate both of these cytokines. Ustekinumab was originally produced in mice, but it was produced in mice as a human antibody. According to Benson et al.,[344] "In these mice ... genetic modifications replaced the mouse Ig loci with human antibody transgenes ... [t]hese genetic modifications resulted in a mouse strain capable of producing human antibodies in response to immunizations to any antigen of interest."

The Sponsor's data revealed that ustekinumab binds to IL-12 of the human, baboon, and cynomolgus monkey, but not of mouse, rat, or dog, as shown in Table 6.3.

3. *Efficacy of Ustekinumab in Mice*

Ustekinumab's efficacy against psoriasis was then evaluated in mice. FDA's Pharmacology Review described the animal model: "Ustekinumab was evaluated for efficacy in a humanized mouse model of psoriasis. Non-lesional skin from human psoriasis donors was transplanted onto immunodeficient ... mice and the psoriatic process was

[341] Hawkes JE, et al. The snowballing literature on imiquimod-induced skin inflammation in mice: a critical appraisal. J. Invest. Dermatol. 2017;137:546–9.

[342] Ueyama A, et al. Mechanism of pathogenesis of imiquimod-induced skin inflammation in the mouse: a role for interferon-alpha in dendritic cell activation by imiquimod. J. Dermatol. 2014;41:135–43.

[343] Tang C, et al. Interleukin-23: as a drug target for autoimmune inflammatory diseases. Immunology. 2012;135:112–24.

[344] Benson JM, et al. Discovery and mechanism of ustekinumab: a human monoclonal antibody targeting interleukin-12 and interleukin-23 for treatment of immune-mediated disorders. MAbs. 2011;3:535–45.

triggered by ... injection of autologous activated T cells [human T cells] after acceptance of the grafts ... mice were treated ... with ustekinumab ... [t]he transplanted skin biopsies were then evaluated for psoriasis pathologies."[345]

The result was that ustekinumab was effective, as determined by the fact that it inhibited thickening of the epidermis and inhibited keratinocyte proliferation.

4. *Safety of Ustekinumab in Monkeys (Increased Infections)*

FDA's Pharmacology Review revealed that ustekinumab was associated with increased rates of infections. FDA stated that the nonclinical information was relevant to humans and required that the animal infection data be included in the package label: "Nonclinical safety issues **relevant to clinical use** ... one out of 10 monkeys ... administered 45 mg/kg ustekinumab ... for 26 weeks had a bacterial infection. The dose of 45 mg/kg in monkeys is 45 times (based on mg/kg) the highest intended clinical dose in psoriasis patients ... **[a] dequate labeling on nonclinical information** ... **are necessary**."[346]

FDA's attention turned to infections from the Sponsor's clinical studies, writing that "There were no cases of active tuberculosis or serious fungal infections. The most frequently reported infection requiring antimicrobial treatment was upper respiratory tract infection, which was reported in 1.1% (placebo), 0.8% (ustekinumab 45 mg), and 0.8% (ustekinumab 90 mg) of patients."[347]

As is self-evident, the rate of infections in the ustekinumab treatment arm was not greater than in the placebo treatment arm. Hence one might expect the package label to be silent on any warnings about infections. On the other hand, FDA recommended a warning about infections based on the mechanism of action of ustekinumab (it blocks IL-12). FDA's recommendation was "Product labeling should advise of these potential risks, that is, infections seen in those genetically deficient [in interleukin-12] ... but labeling should reflect that these risks are theoretical in nature, as they have not been evidenced in the database to date."[348]

This recommendation found a place on the package label, as shown below.

Package label. Consistent with data in FDA's reviews, the package label for ustekinumab (Stelara®) warned against infections in patients and stated that, based on theory (biochemical mechanism of blocking IL-12), ustekinumab posed an increased risk for certain bacterial infections:

> **WARNINGS AND PRECAUTIONS** ... Serious infections have occurred. Do not start STELARA® during any clinically important active infection. If a serious infection develops, stop STELARA® until the infection resolves. **Theoretical Risk for Particular Infections:** Serious infections from mycobacteria, salmonella and Bacillus Calmette-Guerin (BCG) vaccinations have been reported in **patients genetically deficient in IL-12/IL-23** ... [e]valuate patients for tuberculosis prior to initiating treatment with STELARA®. Initiate treatment of latent tuberculosis before administering STELARA®.[349]

[345] Ustekinumab (psoriasis) BLA 125-261. Page 19 of 64-page Pharmacology Review.

[346] Ustekinumab (psoriasis) BLA 125-261. Pages 12–13 of 64-page Pharmacology Review.

[347] Ustekinumab (psoriasis) BLA 125-261. Page 54 of 246-page Medical Review.

[348] Ustekinumab (psoriasis) BLA 125-261. Page 84 of 246-page Medical Review.

[349] Package label. STELLARA® (ustekinumab) injection for subcutaneous use. March 2014 (19 pp.).

The package label also disclosed infections in animals treated with ustekinumab, in the Nonclinical Toxicology section:

> **NONCLINICAL TOXICOLOGY** ... In a 26-week toxicology study, one out of 10 monkeys subcutaneously administered 45 mg/kg ustekinumab twice weekly for 26 weeks had a **bacterial infection**.[350]

VII. CONCLUDING REMARKS

Animal studies are used to acquire data on safety and efficacy. Animal data from toxicity studies and pharmacological studies have increased importance in the IND submission because at that stage in the drug-development process, clinical data might not yet be available. FDA's Guidance for Industry states that, "The sponsor-investigator should include a discussion of the rationale for the investigational drug's intended dose, duration, schedule, and route of administration in the proposed trial. This rationale, particularly for phase 1 trials, is best supported by in vitro and available animal data."[351]

According to the CFR, the IND should include "A summary of the pharmacological and toxicological effects of the drug in animals and, to the extent known, in humans."[352] Also, the CFR requires that the IND disclose "Adequate information about pharmacological and toxicological studies of the drug involving laboratory animals or in vitro, on the basis of which the sponsor has concluded that it is ... safe to conduct the proposed clinical investigations. The kind, duration, and scope of animal and other tests required varies with the duration and nature of the proposed clinical investigation."[353]

This chapter's account of animal data includes use for predicting AEs relating to pregnancy and fetal toxicity. Another theme in this chapter is the comparison of mechanisms in animals versus in humans. This encompasses mechanisms of the disease itself, mechanisms of the drug's efficacy, and mechanisms of the drug's toxicity.

Regarding the package label, the most dramatic connections between animal studies and the package label are found in warnings about AEs relating to pregnancy, embryo-fetal toxicity, secretion into mother's milk, and infancy. Use of animal data for package label warnings for these situations is revealed above, in the accounts of:

- Hydromorphone (skull malformations)
- Lumacaftor/ivacaftor (cataracts)
- Nebivolol (dystocia)
- Regorafenib (dentin alteration)
- Sumatriptan/naproxen (fetal malformations)
- Vigabatrin (retinal degeneration)

[350] Package label. STELLARA® (ustekinumab) injection for subcutaneous use. March 2014 (19 pp.).

[351] U.S. Department of Health and Human Services. Food and Drug Administration. Center for Drug Evaluation and Research (CDER). Center for Biologics Evaluation and Research (CBER). Investigational new drug applications. Prepared and Submitted by Sponsor-Investigators. 2015 (25 pp.).

[352] 21 CFR §312.23(a)(5). IND content and format.

[353] 21 CFR §312.23(a)(8). IND content and format.

- Atezolizumab (immune reactions against the fetus and fetal death)
- Belimumab (fetal deaths and infant deaths, secretion of belimumab into mother's milk)
- Denosumab (failure of epiphyseal plates to close)
- Nivolumab (spontaneous abortions, death of neonate)

Additional examples are revealed in **Chapter 10, Drug Class Analysis**, for example, for

- Bocentan (birth defects)[354]
- Macetentan (birth defects)[355]
- Pomalidomide (teratogenicity in rats and rabbits)[356]

A knowledge of mechanism of action of the study drug is useful for drafting the Clinical Study Protocol, in particular, for arriving at appropriate efficacy endpoints and safety endpoints. FDA's Guidance for Industry teaches that "primary and secondary efficacy endpoints should be chosen based on the drug's putative mechanism of action and the proposed indication."[357] Assessing the mechanism of action is more easily obtained with animals, where biopsies can easily be taken and where FDA approval is not needed, as compared to the situation with human subjects. Mechanism of action considerations include the following:

1. **Is the mechanism of the disease the same in animals as in the corresponding disease in humans?** This was an issue for avibactam/ceftazidime (AVYCAZ), where the FDA reviewer complained, "Ultimately, the natural histories of the infections used in these models may not ... have sufficient similarities to extrapolate to humans."[358]
2. **Is the mechanism of action for a given toxicity the same in animals as the corresponding toxicity in humans?** This was an issue for nebivolol. The problem was that nebivolol caused cancer in mice. The Sponsor determined the mechanism of action that connected nebivolol with mouse cancer and determined that this mechanism does not occur in humans. As part of the Sponsor's clever and thorough study on mechanism, the Sponsor tested another drug (finasteride) that has a mechanism that parallels that of nebivolol. Regarding choice of animal models, the Sponsor also determined that nebivolol causes cancer in mice, but not in rats. This demonstrates the need for toxicity testing in a variety of animal species, rather than in only one specie.
3. **Is the mechanism for the study drug's catabolism the same in animals and humans?** This was an issue for regorafenib, where FDA observed that "higher human exposures

[354] Package label. TRACLEER® (bosentan) tablets, for oral use. October 2016 (20 pp.).

[355] Macitentan (pulmonary arterial hypertension) NDA 204-410. Page 19 of 163-page pdf file Medical Review.

[356] Pomalidomide (multiple myeloma) NDA 204-026. Page 14 of 94-page Clinical Review. 101-page pdf file for Medical Review.

[357] U.S. Department of Health and Human Services. Food and Drug Administration. Center for Drug Evaluation and Research (CDER). Chronic obstructive pulmonary disease: developing drugs for treatment. 2016 (17 pp.).

[358] Avibactam/ceftazidime combination (AVYCAZ) (urinary tract infections) NDA 206-494. Page 31 of 156-page Clinical Review (page 37 of 163-page pdf Medical Review).

to M-2 and M-5 compared to the species used for toxicological assessment may account for higher levels of renal toxicity seen in animals compared to humans in trials."[359]

4. **Does the study drug bind adequately to its target in animals, as compared to binding to its target in humans?** This was an issue for belimumab, where belimumab showed a "10-fold lower affinity of binding of murine BLyS receptor" and where, in contrast, "belimumab binds to human and ... monkey BLyS with nearly identical affinity."[360] Binding of the antibody to its target in animals was also an issue for ustekinumab, where the problem was solved by grafting human skin on the animal model. Ustekinumab's target organ was skin, where the disease to be treated was psoriasis.
5. **Is the drug compatible with the immune system of animals?** This was an issue for atezolizumab, where the Sponsor prepared a specially engineered version of atezolizumab to prevent the administered antibody from generating, in the mouse, immune responses against the antibody.[361]

[359] Regorafenib (colorectal cancer) NDA 203-085. Page 12 of 219-page Pharmacology Review. Page 17 of 69-page Clinical Review.

[360] Belimumab (systemic lupus erythematosus) BLA 125-370. Pages 47–8 of 171-page Pharmacology Review.

[361] Atezolizumab (urothelial carcinoma) BLA 761-034. Page 4 of 58-page Pharmacology Review.

CHAPTER

7

Drug–Drug Interactions: Part One (Small Molecule Drugs)

I. INTRODUCTION

Drug–drug interaction (DDI) studies are designed to detect the influence of a second drug on metabolism of the study drug and, conversely, the influence of the study drug on metabolism of a second drug. Also, DDI studies are designed to determine the influence of a second drug on exposure of the study drug and conversely, the influence of the study drug on exposure of a second drug. The term "exposure" refers to parameters of a drug's concentration in the bloodstream, such as AUC, Cmax, Cmin, and tmax. The choice of the second drug is usually based on one of these criteria:

- *Both drugs for treating same disease.* The second drug is coadministered with the study drug, where both drugs are for treating the same disease. This is the situation with the DDI study between regorafenib (study drug) and irinotecan. Irinotecan is commonly coadministered with regorafenib where each of these drugs is for treating colorectal cancer.[1]
- *Second drug for treating a different condition.* The second drug is administered in the same time-frame as the study drug, as is the case where the second drug is a contraceptive, statin, acetaminophen, warfarin, or a monoamine oxidase inhibitor.[2,3] In this situation, the condition to be treated by the second drug is not the same as the condition being treated by the study drug.
- *Second drug conventionally used for PK studies.* This concerns clinical pharmacokinetic studies that are not intended to measure efficacy or safety, but instead are used as a basis for predicting which class of drugs will likely engage in DDIs with the study drug. Here, the second drug is an established inhibitor of one of the cytochrome P450 enzymes and is conventionally used as a model "second drug" in drug–drug inhibition studies.

[1] Regorafenib (metastatic colorectal cancer) NDA 203-085. Page 31 of 64-page Clinical Pharmacology Review.

[2] Package label. ADLYXIN (lixisenatide) injection, for subcutaneous use; July 2016 (33 pp.).

[3] Package label. EXALGO (hydromorphone hydrochloride) extended release tablets; March 2010 (30 pp.).

DOI: https://doi.org/10.1016/B978-0-12-814647-7.00007-5

Ketoconazole is conventionally used as a model "second drug" in DDI studies. When used in drug-drug interaction studies, ketoconazole inhibits cytochrome P450 3A4 (CYP3A4). Although ketoconazole is actually an antifungal drug, the goal of treating fungal infections has no relevance to this type of clinical pharmacokinetic study that uses ketoconazole.[4]

The best introduction to DDIs is examples from clinical studies. Consider the example of the two drugs, *rifampicin* and midazolam. Also, consider the example of the two drugs, *rifampicin* and ibrutinib. In each example, the mechanism of the DDI involves cytochrome P450 3A (CYP3A), an enzyme of hepatocytes. In each example, the result of the DDI is that the first drug (rifampicin) reduced the blood concentration of the second drug. The term "CYP enzyme" means a cytochrome P450 enzyme. Rifampicin is also called rifampin (both of these terms refer to the same drug).

For in vitro DDI studies, the source of CYP enzymes can be purified recombinant CYP enzymes, microsomes, or intact cultured hepatocytes. Guidance is available for isolating the microsomal fraction of the liver and for using the isolated microsomes in enzyme assays.[5] The example of rifampicin and midazolam and the example of rifampicin and ibrutinib are shown below.

a. Influence of Rifampicin on Exposure of Midazolam

This reveals how administering a CYP3A inducer reduces exposure of a second drug. The CYP3A inducer was *rifampicin* and the second drug was *midazolam*.[6] Exposure was expressed in terms of the unit, $AUC_{0-10hours}$. Midazolam "clearance" was defined as [midazolam dose]/[$AUC_{0-10hours}$]. As the study design involved taking blood samples at various intervals over a period of 10 hours, the unit of exposure was $AUC_{0-10hours}$ rather than $AUC_{0-infinity}$.

Human subjects were titrated with various levels of rifampicin, where 10, 20, or 100 mg *rifampicin* was given to each subject, once a day, for 2 weeks in order to induce CYP3A. *Midazolam* was given on only two occasions, once at baseline (day 0) and once at the 2-week time point. Blood plasma concentrations were taken throughout the day and used to measure midazolam; baseline exposure was used to obtain control values; and 2-week exposure values were used to assess the influence of CYP3A induction on midazolam catabolism. The result was that the rifampicin treatment decreases midazolam exposure, where this reduction was most dramatic with the highest level (100 mg/day) of rifampicin dosing. In this type of study design, greater induction of CYP3A resulted from two things: (1) greater amount of rifampicin dose for the daily dose (10, 20, or 100 mg); and (2) greater number of days, during the 2-week period where rifampicin was dosed once per day. The explanation for the reduced midazolam exposure is that CYP3A catalyzes the destruction of midazolam. More precisely stated, CYP3A catalyzes the conversion of midazolam to hydroxymidazolam.[7]

[4] Package label. Ketoconazole tablets USP, 200 mg; April 1998 (4 pp.). NDA 075-273.

[5] Burns K, et al. The nonspecific binding of tyrosine kinase inhibitors to human liver microsomes. Drug Metab. Dispos. 2015;43:1934–7.

[6] Bjorkhem-Bergman L, et al. Comparison of endogenous 4b-hydroxycholesterol with midazolam as markers for CYP3A4 induction by rifampicin. Drug Metab. Dispos. 2013;41:1488–93.

[7] Gorski JC, et al. Regioselective biotransformation of midazolam by members of the human cytochrome P450 3A (CYP3A) subfamily. Biochem. Pharmacol. 1994;47:1643–53.

b. Influence of Rifampicin on Exposure of Ibrutinib

The following provides another example where administering a CYP3A inducer reduces exposure of a second drug. The study was with human subjects, where the study detected the influence of rifampicin on ibrutinib exposure. Ibrutinib is catabolized by CYP3A, where this catabolism can be increased by coadministering rifampin. As shown by de Jong et al.,[8] *rifampin* caused a 10-fold decrease in *ibrutinib* exposure in humans, which is consistent with rifampicin's established property of inducing CYP3A. In the words of de Jong et al., "ibrutinib showed...a decreased exposure in the presence of a strong CYP3A inducer (rifampin)." The reason why ibrutinib's exposure decreased is that CYP3A catalyzes the destruction of ibrutinib. More precisely stated, CYP3A catalyzes the conversion of ibrutinib to a dihydrodiol metabolite of ibrutinib.[9]

c. Pharmacokinetic Boosters and Enhancers (Example of Ritonavir)

The use of a pharmacokinetic booster or enhancer appears to be the only situation where a drug-drug interaction is desirable. This concerns the problem where there is excessive catabolism of a study drug, where catabolism is mediated by one of the CYP enzymes, and where FDA required a *CYP enzyme inhibitor* to be coadministered with the study drug.

The example with elvitegravir is dramatic and unique, in that it is one of the few situations where the package label requires coadministration with a *CYP enzyme inhibitor*. The goal of the CYP enzyme inhibitor was to increase exposure of the study drug. FDA's ClinPharm Review for elvitegravir described the problem of excessive catabolism and also revealed the solution of using a "pharmacokinetic booster." This "pharmacokinetic booster" took the form of an inhibitor of cytochrome P450 3A (CYP3A).

FDA's ClinPharm Review referred to the coadministration of elvitegravir and ritonavir or cobicistat, where the goal was to increase elvitegravir exposure, "Elvitegravir (EVG) primarily undergoes CYP3A-mediated hydroxylation (generating a chlorofluorophenyl group hydroxide of eltegravir [GS-9202]) and glucuronidation via UGT1A1/3 (generating an acyl glucuronide of elvitegravir...[i]n clinical practice, elvitegravir will therefore be co-administered with **a pharmacokinetic 'booster' – a potent CYP3A inhibitor such as ritonavir or cobicistat** – in order to increase EVG exposures."[10]

The package label for elvitegravir (Vitekta®) explains why a second drug (ritonavir) must be coadministered with elvitegravir, "Inform patients that VITEKTA must be taken...with ritonavir in order to achieve adequate drug levels."[11] The Dosage and Administration section for Vitekta® instructs that Vitekta® at a dosage of "85 mg once daily" be coadministered with a dosage of concomittant ritonavir of "100 mg orally once daily."

[8] de Jong J, et al. Effect of CYP3A perpetrators on ibrutinib exposure in healthy participants. Pharmacol. Res. Perspect. 2015;3:e00156 (11 pp.).

[9] Lee C-S, et al. A review of a novel, Bruton's tyrosine kinase inhibitor, ibrutinib. J. Oncol. Pract. 2016;22:92–104.

[10] Elvitegravir (HIV-1) NDA 203-093. Page 5 of 245 Clinical Pharmacology Review.

[11] Package label. VITEKTA® (elvitegravir) tablets, for oral use; July 2015 (26 pp.).

Consistent with FDA's use of the term "pharmacokinetic booster" in its account of ritonavir, the literature also uses the term "pharmacokinetic enhancer" to refer to ritonavir's ability to inhibit CYP3A. Cobicistat is another pharmacokinetic enhancer, which is used to inhibit CYP3A and thus prevent CYP3A-mediated catabolism of various drugs. Marzolina et al.,[12] Deeks,[13] and Gervasoni et al.[14] describe the use of ritonavir or cobicistat as useful "pharmacokinetic enhancers" for preventing CYP3A-mediated drug catabolism.

Greenblatt et al.[15] advocate the use of ritonavir or cobicistat for testing if inhibiting CYP3A changes the exposure of any study drug of interest. Greenblatt et al. suggest that ritonavir or cobicistat can be used, for this purpose, as an alternative to ketoconazole. The problem with ketoconazole, which has been conventionally used as a reagent in DDI studies, is that it can result, on rare occasions, in liver injury.

d. Definitions, Parameters, and Formulas for Drug–Drug Interaction Studies

The term drug-drug interactions refers to the influence of one drug upon another, e.g., where a second drug stimulates catabolism of the first drug, thereby reducing efficacy of the first drug, or where a second drug inhibits catabolism of the first drug, thereby increasing its efficacy (or increasing its toxicity). Two drugs are coadministered in the following situations:

- Where the two drugs are physically combined in one tablet, as in a type of formulation called a fixed drug combination (FDC).
- Where the package label requires that the drug provided by the packaged pill or capsule be taken with a second drug, and where the second drug is provided by a separate pill or capsule.
- Where the physician decides to prescribe two different drugs for the same disease, and where one patient takes both drugs, and where the physician's goal is to get better control over the disease. Physicians are free to prescribe drugs in a manner outside of a package label's indication, in a technique called off label uses.[16,17,18]
- In the situation where the patient is taking a second drug that is irrelevant to the condition treated by the first drug, as where the second drug is for back pain, heartburn, or an infection, or is an oral contraceptive.

[12] Marzolini C, et al. Cobicistat versus ritonavir boosting and differences in the drug–drug interaction profiles with co-medications. J. Antimicrob. Chemother. 2016;71:1755–8.

[13] Deeks ED. Cobicistat: a review of its use as a pharmacokinetic enhancer of atazanavir and darunavir in patients with HIV-1 infection. Drugs 2014;74:195–206.

[14] Gervasoni C, et al. Effects of ritonavir and cobicistat on dolutegravir exposure: when the booster can make the difference. J. Antimicrob. Chemother. 2017;72:1842–4.

[15] Greenblatt DJ, Harmatz JS. Ritonavir is the best alternative to ketoconazole as an index inhibitor of cytochrome P450-3A in drug–drug interaction studies. Br. J. Clin. Pharmacol. 2015;80:342–50.

[16] Kim J, Kapczynski A. Promotion of drugs for off-label uses: the US Food and Drug Administration at a crossroads. JAMA Intern. Med. 2017;177:157–8.

[17] Oliphant CS, et al. Ivabradine: a review of labeled and off-label uses. Am. J. Cardiovasc. Drugs 2016;16:337–47.

[18] Barlas S. FDA to reassess policies on unsolicited Requests for off-label information. Pharm. Ther. 2012;37:318–9.

The Code of Federal Regulations (21 CFR §201.57) requires that the package label disclose DDIs:

> Drug interactions. (i) This section must contain a description of clinically significant interactions, either observed or predicted, with other prescription or over-the-counter drugs, classes of drugs, or foods (e.g., dietary supplements, grapefruit juice)...[i]nteractions that are described in the "Contraindications" or "Warnings and Precautions" sections must be discussed in more detail under this section.

FDA's Guidance for Industry also requires that DDIs be disclosed on the package label. FDA's Guidance refers to drugs that are inhibitors of cytochrome P450 3A (CYP3A inhibitors):

> Information under the Drug Interactions heading must include a...summary of those drugs...or foods that interact...in clinically significant ways with the subject drug, and practical instructions for preventing or managing the interaction...subheadings of summary concepts (e.g., CYP3A inhibitors) can precede specific information.[19]

Drug–drug interaction studies that are in vitro studies, as with microsomes or with cultured cells, and drug–drug interaction studies that are clinical studies are components of FDA-submissions for most drugs. Assessing DDIs involving one or more of the cytochrome P450 enzymes (CYP enzymes), drug-conjugating enzymes, and drug transporters is a routine part of the drug development and submission process.

Mechanisms of DDIs include those where a concomitant drug *inhibits activity* of one or more CYP enzymes, and where the concomitant drug *induces expression* of one or more CYP enzymes.[20] Where a drug inhibits CYP enzyme activity or induces CYP enzyme expression, the drug can influence its own metabolism, and also, the drug can influence metabolism of a concomitantly administered drug.[21] The term "enzyme expression" usually refers to the induction or repression of synthesis of messenger RNA (mRNA) followed by translation of that mRNA into a polypeptide

Regarding CYP enzymes, the most frequently encountered CYP enzyme in DDI studies is CYP3A.[22,23,24] Regarding transport, the most studied cells in DDI studies are enterocytes, hepatocytes, and renal tubule cells, and the most commonly studied transporter is P-glycoprotein (P-gp).

[19] U.S. Department of Health and Human Services. Food and Drug Administration. Center for Drug Evaluation and Research (CDER). Center for Biologics Evaluation and Research (CBER). Guidance for industry. Labeling for human prescription drug and biological products—implementing the PLR content and format requirements; 2013 (30 pp.).

[20] Templeton IE, et al. Quantitative prediction of drug–drug interactions involving inhibitory metabolites in drug development: how can physiologically based pharmacokinetic modeling help? CPT Pharmacometrics Syst. Pharmacol. 2016;5:505–15.

[21] Shin JM, Sachs G. Pharmacology of proton pump inhibitors. Curr. Gastroenterol. Rep. 2008;10:528–34.

[22] Jones BD, et al. Managing the risk of CYP3A induction in drug development: a strategic approach. Drug Metab. Dispos. 2017;45:35–41.

[23] Lee E, et al. Simultaneous evaluation of substrate-dependent CYP3A inhibition using a CYP3A probe substrates cocktail. Biopharm. Drug Dispos. 2016;37:366–72.

[24] Fujita K. Cytochrome P450 and anticancer drugs. Curr. Drug Metab. 2006;7:23–37.

e. Parameters for Measuring Exposure

DDI studies measure how exposure of the study drug is influenced by a second drug. Typical parameters of exposure are AUC, Cmax, tmax Cmin, and Ctrough.[25] Another exposure parameter is the ratio of [AUCi]/[AUC], which measures the ratio of AUC in the presence and absence of a second drug.[26,27] In the words of Galetin et al.[28,29] "the metric for the degree of DDI is the [AUCi]/[AUC] ratio for the plasma concentration-time profiles in the presence and absence of the inhibitor." AUCi refers to area under the curve in the presence of the inhibitor.[30]

Williams et al.[31] provide the example of the study drug terfenadine (metabolized by CYP3A) and ketoconazole (potent CYP3A inhibitor). They describe coadministration of ketoconazole and terfenadine, and where, "the potent CYP3A inhibitor ketoconazole increases the AUC of terfenadine, which is primarily metabolized by CYP3A enzymes." In drug–drug interaction studies, the term "perpetrator" and "victim" are occasionally used to refer to the drugs that are administered to study subjects. The drug that is being followed for its substrate properties is the "victim." The term "substrate properties" means ability to be metabolized. And the drug that is administered, where it is expected to inhibit or induce one of the CYP enzymes, is called the "perpetrator." Thus in drafting FDA-submissions, the medical writer has the option of using the terms "perpetrator" and "victim."

f. Metabolic Consequences of Drug–Drug Interactions

For any given DDI study, these questions are asked:

- Does the *study drug* inhibit metabolism of a second drug?
- Does the *study drug* stimulate metabolism of the second drug?
- Does the second drug inhibit metabolism of the *study drug*?
- Does the second drug stimulate metabolism of the *study drug*?
- Does the *study drug* induce expression of mRNA encoding a CYP enzyme, thus stimulating CYP-mediated catabolism of a second drug?

[25] U.S. Department of Health and Human Services. Food and Drug Administration Center for Drug Evaluation and Research (CDER). Center for Biologics Evaluation and Research (CBER) Guidance for industry. Exposure-response relationships—study design, data analysis, and regulatory applications; 2003 (25 pp.).

[26] Williams JA, et al. Drug–drug interactions for UDP-glucuronosyltransferase substrates: a pharmacokinetic explanation for typically observed low exposure (AUCi/AUC) ratios. Drug Metab. Dispos. 2004;32:1201–8.

[27] Galetin A, et al. CYP3A4 substrate selection and substitution in the prediction of potential drug–drug interactions. J. Pharmacol. Exp. Ther. 2005;314:180–90.

[28] Galetin A, et al. CYP3A4 substrate selection and substitution in the prediction of potential drug–drug interactions. J. Pharmacol. Exp. Ther. 2005;314:180–90.

[29] Williams JA, et al. Drug–drug interactions for UDP-glucuronosyltransferase substrates: a pharmacokinetic explanation for typically observed low exposure (AUCi/AUC) ratios. Drug Metab. Dispos. 2004;32:1201–8.

[30] Galetin A, et al. CYP3A4 substrate selection and substitution in the prediction of potential drug–drug interactions. J. Pharmacol. Exp. Ther. 2005;314:180–90.

[31] Williams JA, et al. Drug–drug interactions for UDP-glucuronosyltransferase substrates: a pharmacokinetic explanation for typically observed low exposure (AUCi/AUC) ratios. Drug Metab. Dispos. 2004;32:1201–8.

- Does the second drug induce expression of mRNA encoding a CYP enzyme, thereby stimulating CYP-mediated catabolism of the *study drug*?

g. The 25% Cutoff Value From In Vitro DDI Studies, as a Trigger to Move Forward with Clinical DDI Studies

Regarding the target of inhibition, e.g., a CYP enzyme or a drug transport system, the European Medicines Agency (EMA) provides a 25% cutoff value, for determining if a particular target of inhibition should be part of a Sponsor's clinical DDI analysis.

According to the EMA, "enzymes involved in metabolic pathways estimated to contribute to ≥25% of drug elimination should be identified…and the in vivo contribution quantified."[32] The ratio of $[Cmax_i]/[Cmax]$ and the ratio of $[tmax_i]/[tmax]$ have unique implications for use of the study drug in the clinical situation, and thus have unique implications for corresponding warnings on the package label.

EMA provides comments on the ratio [I]/Ki, referring to the use of this ratio to predict if a clinical DDI study should be conducted (study drug and probe substrates coadministered to human subjects). This prediction functions as a bridge that connects results from in vitro studies with the decision to design a corresponding in vitro study with human subjects. EMA provides a cutoff value of 10 ([I]/Ki ≥10) and another cutoff value of 0.2 ([I]/Ki ≥0.02). Also, EMA distinguishes between orally administered drugs and intravenously administered drugs:

> an in vivo interaction study with a…probe substrate is recommended, if the conditions below are fulfilled for **orally administered drugs** if the enzyme has marked abundance in the enterocyte, for example, CYP3A, **[I]/Ki ≥ 10** where [I] is the maximum dose taken at one occasion/250 ml. For drugs **regardless of mode of administration** and inhibition of enzymes in the liver, or in organs, exposed to the drug through the systemic circulation **[I]/Ki ≥ 0.02** where [I] is the unbound mean Cmax obtained during treatment with the highest recommended dose.[33]

h. The [I]/K Ratio Cutoff Value (of About 0.1–1.0) From In Vitro DDI Studies, as a Trigger to Move Forward With Clinical DDI Studies

The value of the [I]/Ki ratio is often used by FDA reviewers to determine if the Sponsor needs to conduct a clinical study that directly addresses the potential for DDIs. Ki is the inhibition constant, as determined by in vitro studies. [I] is the blood plasma drug concentration.

The two values acquired can be plugged into the ratio ([I]/Ki), where a value of less than 0.1 means that the DDI is of low risk for patients, and where the value is greater than 1, the DDI is of high risk for patients. Medium risk is where the value is between 0.1 and 1.[34,35] The ratio [I]/Ki is routinely used by FDA reviewers to determine if data from CYP

[32] European Medicines Agency. Guideline on the Investigation of Drug Interactions; 2012. Pages 10 and 13 of 59 pages.

[33] European Medicines Agency. Guideline on the investigation of drug interactions; 2012 (59 pp.).

[34] Ito K, et al. Database analysis for the prediction of in vivo drug–drug interactions from in vitro data. Br. J. Pharmacol. 2004;57:473–86.

[35] Williams JA, et al. Drug–drug interactions for UDP-glucuronosyltransferase substrates: a pharmacokinetic explanation for typically observed low exposure (AUCi/AUC) ratios. Drug Metab. Dispos. 2004;32:1201–8.

enzyme inhibition studies, in combination with a blood plasma value of blood plasma for the study drug, is sufficient to recommend that the Sponsor conduct a clinical DDI study.

Ito et al.[36] advocate the use of various equations in assessing risk for inhibition, writing that, "use of in vitro data to predict inhibition potential of a drug is attractive…interactions are regarded to be with low risk of the…[I]/Ki ratio is less than 0.1 and high risk if it is greater than 1."

i. Example of Plugging Numbers into Formula for [I]/Ki Ratio

The inhibition constant Ki is the concentration of Drug A (study drug) that causes half-maximal inhibition of the metabolism of Drug B (probe drug). The [I] value is the concentration of the study drug detected in the bloodstream of human subjects, when it is administered to human subjects in the maximal amount expected to be needed for effective treatment of patients.

The Sponsor's drug–drug interaction study included in vitro experiments to evaluate the ability of fingolimod to inhibit various human CYP enzymes. As part of this in vitro study, the Sponsor tested the ability of fingolimod to inhibit CYP2C9. The probe substrate that was used in this test was, diclofenac, because diclofenac is specifically recognized as a substrate by CYP2C9. CYP2C9 catalyzes the 4′-hydroxylation of diclofenac. The concentration of fingolimod giving half-maximal inhibition of CYP2C9 was 55 micromolar, and this is the value for Ki. FDA's ClinPharm Review observed that the maximally expected plasma concentration, when used in the postmarketing situation and administered to real patients, was about 0.0325 micromolar fingolimod. This number was derived by the Sponsor from studies with human subjects (not from a drug–drug interaction study, but only from a single oral dose of fingolimod followed by measurements of plasma concentrations).

Regarding the general rule for handling the [I]/Ki formula, and regarding the maximal plasma fingolimod concentration expected in ordinary clinical practice, FDA's ClinPharm Review stated that, "An estimated [I]/Ki ratio of greater than 0.1 is considered positive…[m]aximum plasma concentration at 1 mg fingolimod dose…was approximately 10 ng/mL which corresponds to 0.0325 μmol/L. A drug interaction is remote as [I]/Ki ratio is less than 0.1."[37,38,39]

FDA's ClinPharm Review did not show how to plug the numbers into the [I]/Ki formula, but instead concluded that, "A drug interaction is remote as [I]/Ki ratio is less than 0." FDA also concluded that fingolimod "is unlikely to reduce the in vivo clearance of drugs mainly cleared through metabolism by … CYP2C9." Turning to the formula for the [I]/Ki ratio, and plugging in the above numbers, we find that the result is [0.0325 μmol/L]/55 μmol/L, which equals 0.0006. Thus it can be seen that FDA's conclusion that "a drug interaction is remote as [I]/Ki ratio is less than 0.1" is correct. In other words, it is a fact that 0.0006 is less than 0.1.

The following concerns repeating the DDI experiments separately for each of a variety of probe drugs where each probe drug is specfically recognized by only one type of CYP enzyme.

[36] Ito K, et al. Database analysis for the prediction of in vivo drug–drug interactions from in vitro data. Br. J. Clin. Pharmacol. 2004;57:473–86.

[37] Fingolimod (multiple sclerosis) NDA 022-527. Page 87 of 371-page Clinical Pharmacology Review.

[38] Fingolimod (multiple sclerosis) NDA 022-527. Page 86 of 371-page Clinical Pharmacology Review.

[39] Fingolimod (multiple sclerosis) NDA 022-527. Page 85 of 371-page Clinical Pharmacology Review.

The value for Ki can be separately calculated from results provided by in vitro microsome incubation tests with each of the separate probe drugs. With the in vivo value for [I] in hand, it is easy to plug the various Ki values and the one [I] value into the formula, and to arrive at numbers that predict if there will or will not be DDIs in humans. The degree of enzyme-catalyzed destruction of the probe drug reflects the extent that fingolimod inhibits the corresponding CYP enzyme. The take-home lesson for the physician is that the conclusion made from the DDI experiment with the probe drug applies when the physician needs to coadminister the probe drug, and it also applies when the physician needs to coadminister any other drug that is metabolized by the same CYP enzyme, as metabolized by the probe drug.

j. Correction Factor

This provides a correction factor that interconverts values for IC50 and Ki. The influence of a drug on inhibiting drug metabolism in cultured cells produces an IC50 value, while the influence of a drug on inhibiting drug metabolism by a source of isolated enzyme produces a Ki value. FDA's Guidance for Industry provides an algorithm for converting a IC50 value to a Ki value, "Sometimes inhibitor concentration causing 50% inhibition (IC50) is determined, and Ki can be calculated as IC50/2 by assuming competitive inhibition."[40]

k. Diagram of Drug–Drug Interaction Where the Site of Interaction is CYP3A

Fig. 7.1 illustrates the coadministration of Drug A and Drug B, where Drug B inhibits activity of CYP3A, and where this inhibition blocks constitutive CYP3A-mediated oxidation of Drug A, and where the final result is increased exposure of Drug A. In the situation where increased exposure to Drug A results in adverse events, the package label may warn against coadministering Drug B.

Fig. 7.1 shows CYP3A inside a hepatocyte. The syringe is injecting both drugs into a human subject. The lightning bolt represents inhibition. The reaction catalyzed by CYP3 is disclosed as a hydroxylation reaction. Inhibition of CYP3 causes the concentration of Drug A to increase in the bloodstream. This increase is shown by the greater number of "Drug A" symbols in the figure.

l. FDA's Decision-Making Process for the Daclatasvir Package Label

FDA's ClinPharm Review reveals the decision-making pathway in arriving at recommendations for daclatasvir's package label. FDA used two sources of information, Guidance for Industry[41] and the USPI, as the basis for inputting CYP3A DDI information on the label. The USPI is used in this way:

> The United States Prescribing Information (USPI) is...for communicating the benefit-risk information of a Food and Drug Administration (FDA) approved prescription drug. The USPI is typically the last step of

[40] U.S. Dept of Health and Human Services. Food and Drug Administration. Center for Drug Evaluation and Research (CDER). Guidance for industry. Drug interaction studies—study design, data analysis, implications for dosing and labeling recommendations; 2012 (Page 20 (Figure 2) of 75 pages).

[41] U.S. Dept of Health and Human Services. Food and Drug Administration. Center for Drug Evaluation and Research (CDER). Guidance for industry. Drug interaction studies—study design, data analysis, implications for dosing and labeling recommendations; 2012 (75 pp.).

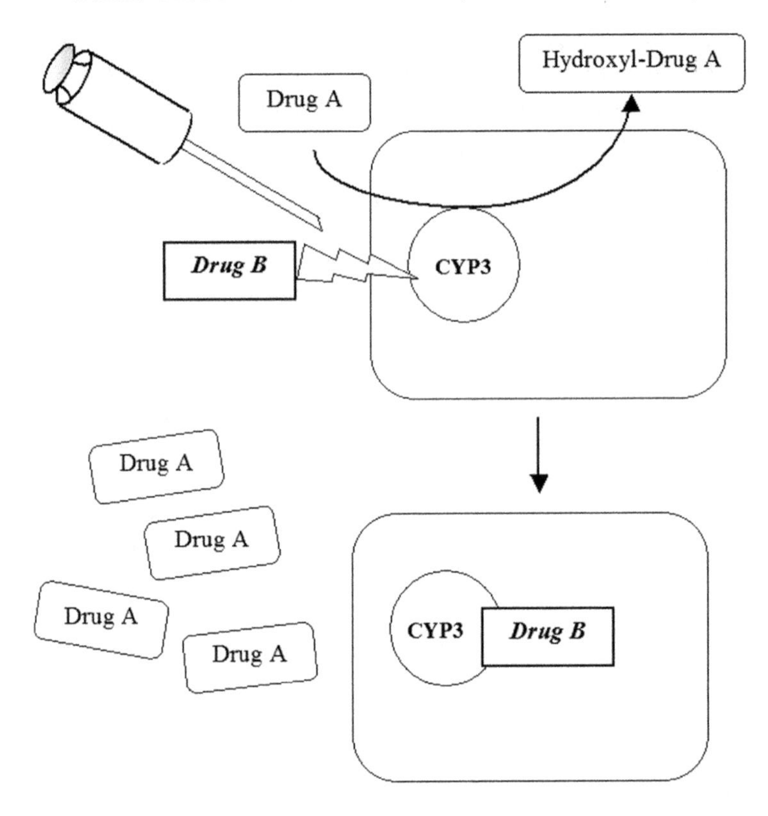

FIGURE 7.1 Drug–drug interaction (CYP inhibition). The top part of the drawing shows CYP3 catalyzing oxidation of Drug A. Drug B enters the hepatocyte and eventually inhibits CYP3. A syringe is shown, injecting Drug A and Drug B into the bloodstream. The lower part of the drawing shows Drug B inhibiting CYP3, where the consequence that any Drug A that enters the hepatocyte is not catabolized by CYP3, where the consequence is that Drug A concentrations rise in the bloodstream.

the drug development process and requires discourse between the FDA and the sponsor for a new drug application. The USPI may also be updated after obtaining FDA approval.[42]

Referring to FDA's Guidance and to the USPI, FDA's analysis of the proposed daclatasvir package label referred to drugs that induce CYP3A, writing, "May lead to loss of virologic response to daclatasvir...[f]or the examples of the strong CYP3A inducers, the...medications were edited to be consistent with the list of strong CYP3A induces that are included in...Guidance for Industry or the available information in the...U.S. prescribing information (USPI)...the dexamethasone USPI states that dexamethasone is a moderate CYP3A inducer."[43]

FDA's ClinPharm Review provided DDI data from a clinical study, where the study drug was administered with a CYP3A inducer. The CYP3A inducer was rifampin. FDA's

[42] Daizadeh I. J. Commer. Biotechnol. 2011;17:218–29.

[43] Daclatasvir (hepatitis C virus) NDA 206-843. Pages 4–5 of 368-page Clinical Pharmacology Review.

recommendation for the label was, "When **rifampin**...once daily was co-administered with a single dose of **daclatasvir**...the daclatasvir Cmax...and AUC...were decreased...when compared with a single dose of daclatasvir...[t]he applicant is proposing to **contraindicate concomitant use of strong CYP3A inducers** with daclatasvir. Based on the changes in daclatasvir exposure that were observed in the...trial, the **contraindication is appropriate**."[44]

FDA's recommendation found a place in the package label, as revealed by the label's contraindication against strong CYP3A inhibitors.[45]

Package label. The package label for daclatasvir (Daklinza®), a drug for hepatitis C virus, provides an example of a DDI warning. The danger of DDIs is evident from the fact that they found a place in the Dosage and Administration section and in the Contraindications section (and not merely in the Drug Interactions section). The label reads:

> **DOSAGE AND ADMINISTRATION.** 60 mg taken orally once daily...[d]ose modification: Reduce dosage to 30 mg once daily with **strong CYP3A inhibitors** and increase dosage to 90 mg once daily with **moderate CYP3A inducers.**[46]

The instruction to reduce dosage of the study drug, in the situation where patients are coadministered "strong CYP3A inhibitors," is based on the need to avoid toxic concentrations of daclatasvir. A similar warning occurs in the Contraindications section, which read:

> "**CONTRAINDICATIONS. Strong inducers of CYP3A**, including phenytoin, carbamazepine, rifampin, and St. John's Wort.[47]

Moreover, the Drug Interactions section of the label provided information on *CYP3A inducers* (these stimulate the catabolism of daclatasvir) and the need to adjust dosage of the daclatasvir:

> **DRUG INTERACTIONS**...Examples: bosentan, dexamethasone, efavirenz, betravirine, modafinil, nafcillin, rifapentine...[i]ncrease DAKLINZA dose to 90 mg once daily when co-administered with moderate inducers of CYP3A.[48]

Also, the Drug Interactions section provided corresponding information on *CYP3A inhibitors* (these reduce catabolism of daclatasvir, thereby undesirably increasing daclatasvir exposure):

> **DRUG INTERACTIONS**...Examples: atazanavir/ritonavir, clarithromycin, indinavir, itraconazole, ketoconazole, nefazodone, nelfinavir, posaconazole, saquinavir, telithromycin, voriconazole...[d]ecrease DAKLINZA dose to 30 mg once daily when coadministered with strong inhibitors of CYP3A.[49]

[44] Daclatasvir (hepatitis C virus) NDA 206-843. Pages 178–179 of 368-page Clinical Pharmacology Review.

[45] Package label. DAKLINZA™ (daclatasvir) tablets, for oral use; July 2015 (21 pp.).

[46] Package label. DAKLINZA™ (daclatasvir) tablets, for oral use; July 2015 (21 pp.).

[47] Package label. DAKLINZA™ (daclatasvir) tablets, for oral use; July 2015 (21 pp.).

[48] Package label. DAKLINZA™ (daclatasvir) tablets, for oral use; July 2015 (21 pp.).

[49] Package label. DAKLINZA™ (daclatasvir) tablets, for oral use; July 2015 (21 pp.).

II. BACKGROUND ON CYTOCHROME P450 ENZYMES AND DRUG TRANSPORTERS

FDA's Guidance for Industry teaches that in vitro DDI studies can be used as a screening tool to assess the potential for DDIs.[50] FDA's Guidance recommends using assays that are sensitive to changes in the activity, or changes in the expression, of various cytochrome P450 enzymes, such as CYP1A2, CYP2B6, CYP2C8, CYP2C9, CYP2C19, CYP2D6, and CYP3A.

These assays determine if the test drug inhibits or induces enzyme expression. In addition to assaying expression of these enzymes, FDA's Guidance recommends measuring the activity of one or more transporters. In FDA-submissions for most small molecule drugs, Sponsors include DDI studies that can detect drug–drug interactions occurring on transporter proteins that can mediate transport of drugs into or out of various cells. These transporters are membrane-bound proteins of the plasma membrane, and they include:

- ATP-Binding Cassette (ABC transporters). P-gp, as well as other members of the ABC transporter family, has been reviewed[51];
- P-glycoprotein (P-gp). P-gp is the most frequently studied member of the ABC transporter family. It is a single polypeptide of 1280 amino acids[52];
- Solute Carrier (SLC transporters);
- Organic Anion Transporters (OAT1, OAT2, and OAT3); and
- Breast Cancer Resistance Protein (BCRP).

a. Probe Substrate Drugs for Detecting CYP Enzyme Activity

Probe substrates that are available are specifically recognized as substrates by only one of the many CYP enzymes. For example, phenacetin (CYP1A2), tolbutamide (CYP2C9), *S*-mephenytoin (CYP2C19), desipramine or dextromethorphan (CYP2D6), and midazolam (CYP3A4) are probe substrates specifically recognized by the indicated cytochrome P450 enzyme. In testing each probe substrate, the researcher needs to identify the product that is formed, in order to conclude that the probe substrate had actually been catabolized by the indicated CYP enzyme.[53,54,55] For probe substrates that are used for in vitro studies, it

[50] U.S. Department of Health and Human Services. Food and Drug Administration. Center for Drug Evaluation and Research (CDER). Guidance for industry. Drug interaction studies—study design, data analysis, implications for dosing, and labeling recommendations; 2012 (75 pp.).

[51] Montari F, Ecker GF. Prediction of drug-ABC-transporter interaction—recent advances and future challenges. Adv. Drug Deliv. Rev. 2015;86:17–26.

[52] Loo TW, Clarke DM. The transmission interfaces contribute asymmetrically to the assembly and activity of human P-glycoprotein. J. Biol. Chem. 2015;290:16954–63.

[53] Foti RS, Wahlstrom JL. CYP2C19 inhibition: the impact of substrate probe selection on in vitro inhibition profiles. Drug Metab. Dispos. 2008;36:523–8.

[54] Ko JW, et al. In vitro inhibition of the cytochrome P450 (CYP450) system by the antiplatelet drug ticlopidine: potent effect on CYP2C19 and CYP2D6. Br. J. Clin. Pharmacol. 2000;49:343–51.

[55] Brown HS, et al. Prediction of in vivo drug–drug interactions from in vitro data: impact of incorporating parallel pathways of drug elimination and inhibitor absorption rate constant. Br. J. Clin. Pharmacol. 2005;60:508–18.

makes little difference if the probe substrate is poisonous to humans. But for probe substrates that are used for clinical pharmacokinetic studies, the probe substrate must not be poisonous to humans.

For studies with human subjects, a cocktail of several probe drugs can be coadministered with a study drug. The use of a cocktail of this sort is a study design typically used with human subjects. Each probe drug is transformed by only one specific type of cytochrome P450 enzyme, and each probe drug is transformed to a unique catabolite. Alternatively, the study drug can be coadministered with only one probe drug. Various drug probes and the corresponding CYP isozyme that is specific for each of these drug probes include[56,57,58,59]

- Caffeine (CYP1A2)
- Losartan (CYP2C9)
- Omeprazole (CYP2C19)
- Dextromethorphan (CYP2D6)
- Midazolam (CYP3A)
- Bupropion (CYP2B6)
- Tolbutamide (CYP2C9)
- Chlorzoxazone (CYP2E1)

b. Mechanism of Action of Drugs in Stimulating Cytochrome P450 Expression

Fig. 7.2 provides a generic account of the action of a drug in stimulating expression of one or more CYP enzymes. The figure includes these structures:

- Hepatocyte
- The study drug
- Transcription factor
- Gene encoding CYP enzyme, including a response element, promoter, and coding sequence

Fig. 7.2 shows a study drug **(1)** entering the nucleus of a hepatocyte, and binding to a transcription factor **(2)** to form a complex **(3)**.[60,61,62] The complex **(3)** then binds to a specific

[56] Grangeon A, et al. Highly sensitive LC-MS/MS methods for the determination of seven human CYP450 activities using small oral doses of probe-drugs in human. J. Chromatogr. B. Analyt. Technol. Biomed. Life Sci. 2017;1040:144–58.

[57] Snyder BD, et al. Evaluation of felodipine as a potential perpetrator of pharmacokinetic drug–drug interactions. Eur. J. Clin. Pharmacol. 2014;70:1115–22.

[58] Rowland A, et al. Optimized cocktail phenotyping study protocol using physiological based pharmacokinetic modeling and in silico assessment of metabolic drug–drug interactions involving modafinil. Front. Pharmacol. 2016;7:517–25.

[59] Tran et al. Therapeutic protein-drug interaction assessment for daclizumab high-yield process in patients with multiple sclerosis using a cocktail approach. Br. J. Clin. Pharmacol. 2016;82:160–7.

[60] Prakash C, et al. Nuclear receptors in drug metabolism, drug response and drug interactions. Nucl. Receptor Res. 2015;2 (35 pp.). doi: 10.11131/2015/101178.

[61] Chen Y, et al. Nuclear receptors in the multidrug resistance through the regulation at drug-metabolizing enzymes and drug transporters. Biochem. Pharmacol. 2012;83:1112–26.

[62] Chen J, Raymond K. Roles of rifampicin in drug–drug interactions: underlying molecular mechanisms involving nuclear receptor PXR. Ann. Clin. Microbiol. Antimicrob. 2006;5:1–11.

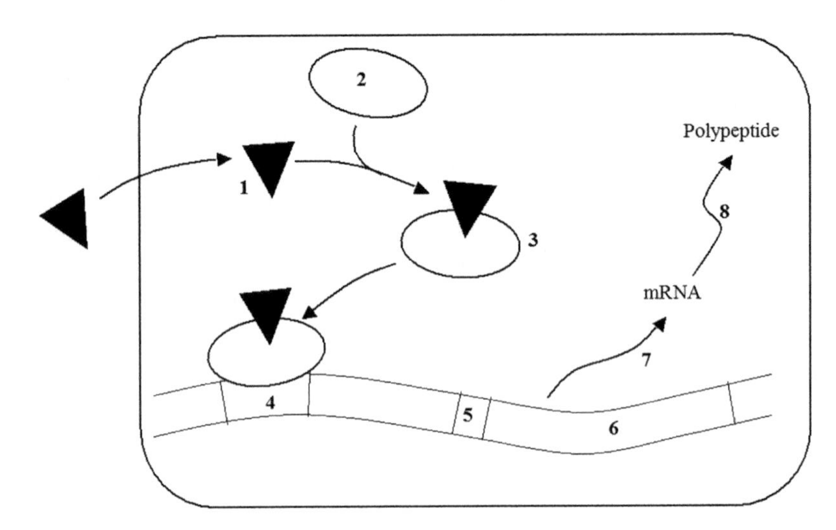

FIGURE 7.2 Drug–drug interaction (CYP induction). The diagram can represent the situation where the study drug induces expression of one or more CYP enzymes, where the result is increased catabolism of a coadministered drug. Also, the diagram can represent the scenario where a coadministered drug induces expression of one or more of the CYP enzymes, where the result is increased catabolism of the study drug. In addition, the diagram illustrates the situation where the study drug can induce its own catabolism.

"response element" that is part of a gene encoding one of the cytochrome P450 enzymes. The term "gene" refers to the sum of the coding sequence with associated regulatory elements, including promoters, enhancers, and response elements. The response element **(4)**, promoter **(5)**, and coding sequence **(6)** are shown. The wavy parallel lines represent both strands of the DNA double helix of the human genome. Transcription **(7)** and translation **(8)** are indicated. DDI studies can measure the amount of mRNA produced by transcription, or the enzymatic activity of the polypeptide produced by translation.[63,64,65]

c. Rifampicin Stimulates Expression of CYP3A4

Rifampicin is used to treat tuberculosis, but is also a standard drug used in pharmacokinetic studies for assessing DDIs. It activates genetic expression of one of the CYP enzymes, with the downstream consequence of increasing destruction of any coadministered drug that happens to be a substrate for that CYP enzyme. It binds to the transcription factor Pregnane X Receptor (PXR) which, in turn, binds to a response element in the

[63] Fahmi OA, et al. Cytochrome P4503A4 mRNA is a more reliable marker than CYP3A4 activity for detecting pregnane X receptor-activated induction of drug-metabolizing enzymes. Drug Metab. Dispos. 2010;38:1605–11.

[64] Prueksaritamont T, et al. Drug–drug interaction studies: regulatory guidance and an industry perspective. AAPS J. 2013;15:629–45.

[65] European Medicines Agency (EMA). Guideline on the investigation of drug interactions. Clin. Pharmacol. Ther. 2015;97:247–62.

genome, and where this binding stimulates expression of the gene encoding CYP3A4.[66,67]

The transcription factor PXR can be bound by several drugs, notably, rifampicin, dexamethasone, indinavir, and paclitaxel, where the result is formation of the drug/PXR complex, and where this complex activates expression of CYP3A4.

The CYP3A4 gene has two response elements, each of which can bind the drug/PXR complex. One of these is close to the translation start site, while the other is about 8000 nucleotides upstream of the proximal response element. Both response elements are located upstream of the gene's promoter.

Most response elements have a nucleic acid sequence that includes AGGTCA or that includes AGTTCA. These particular sequences are called consensus sequences in that the response elements can include these exact sequences or slight variations thereof. The consensus sequence can be aligned, on the strand of genomic DNA, in the forward or reverse direction. The response element for PXP has the polynucleotide sequence: TGAACTCAAAGGAGGTCA.[68]

III. P-GLYCOPROTEIN

Drug-drug interactions studies are routinely used to assess inhibition of drug transport. The transporters may be those mediating passage of drugs from the gut lumen through the enterocyte to the bloodstream, or from the bloodstream through the hepatocyte to the bile duct. Measuring the influence of a first drug on a second drug's transport is most often assessed with the transporter called P-gp. But DDIs are also assessed for the transporters, breast cancer resistance protein (BCRP), organic anion transporters (OAT1, OAT2, and OAT3), and organic cation transporter (OCT).[69]

a. Biology of P-Glycoprotein

Fig. 7.3 shows a generic cell. The figure shows the locations of P-gp in gut cells (enterocytes) as well as in liver cells (hepatocytes). For both types of cells, P-gp is an integral

[66] Chen J, Raymond K. Roles of rifampicin in drug–drug interactions: underlying molecular mechanisms involving nuclear receptor PXR. Ann. Clin. Microbiol. Antimicrob. 2006;5:1–11.

[67] Chen Y, et al. The nuclear receptors constitutive androstane receptor and pregnane X receptor cross-talk with hepatic nuclear factor 4alpha to synergistically activate the human CYP2C9 promoter. J. Pharmacol. Exp. Ther. 2005;314:1125–33.

[68] Song X, et al. The pregnane X receptor binds response elements in a genome context-dependent manner, and PXR activator rifampicin selectively alters the binding among target genes. Drug Metab. Dispos. 2004;32:35-42.

[69] U.S. Department of Health and Human Services. Food and Drug Administration. Center for Drug Evaluation and Research (CDER). Guidance for industry. Drug interaction studies—study design, data analysis, implications for dosing, and labeling recommendations; 2012 (75 pp.).

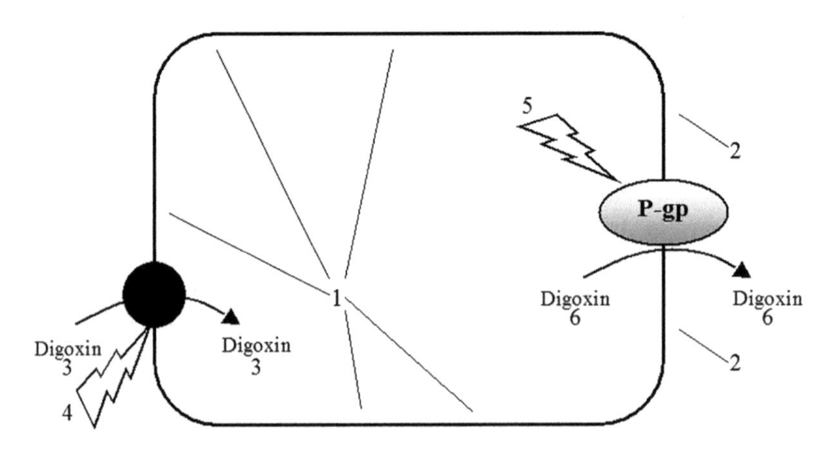

FIGURE 7.3 Drug transporters in the basolateral membrane and apical membrane. The drawing shows a test substrate (digoxin) being transported across the basolateral membrane (transporter shown by filled circle) and across the apical membrane (transporter is P-gp). Each of the two transporters may be inhibited by a coadministered drug, as shown by the lightning bolts.

membrane protein situated in the apical membrane (not in the basolateral membrane).[70,71] The apical membrane of the enterocyte faces the gut lumen and mediates efflux of drugs *out of the cell and into the gut lumen,* while the apical membrane of the hepatocyte faces the bile duct and mediates efflux of drugs *out of the cell and into the bile duct.*[72]

P-gp spans the inner phospholipid layer and the outer phospholipid layer of the phospholipid bilayer of the plasma membrane, as is the case with nearly all membrane-bound proteins. The substrate binding site of P-gp is located within the inner monolayer of the apical membrane, as might be expected, in view of the fact that P-gp is used for removing drugs that reside in the cytosol and for forcing the drugs out of the cell.[73]

In the kidney, P-gp is situated in apical membrane of the proximal tubule epithelial cells, where P-gp mediates efflux of drugs from the tubule cell into the lumen of the nephron where the urine is being formed.[74] To summarize, P-gp in the enterocyte, hepatocyte, and renal tubule directs drugs into the gut lumen, the bile and on to the gut lumen, and the urine, respectively.

[70] Johnson BM, et al. Compartmental modeling to an examination of in vitro intestinal permeability data: assessing the impact of tissue uptake, P-glycoprotein CYP3A. Drug Metab. Dispos. 2003;31:1151–60.

[71] Lin X, et al. Attenuation of intestinal absorption by major efflux transporters: quantitative tools and strategies using a Caco-2 model. Drug Metab. Dispos. 2011;39:265–74.

[72] Mattaloni SM, et al. AKAP350 is involved in the development of apical "canalicular" structures in hepatic cells HepG2. J. Cell Physiol. 2012;227:160–71.

[73] Meng Z, et al. Extrapolation of elementary rate constants of P-glycoprotein-mediated transport from MDCKII-hMDR1-NKI to caco-2 cells. Drug Metab. Dispos. 2017;45:190–7.

[74] Wessler JD, et al. The P-glycoprotein transport system and cardiovascular drugs. J. Am. Coll. Cardiol. 2013;61:2495–502.

For in vitro cell culture studies that measure the influence of a first drug on transport of a second drug, the researcher needs to create a monolayer culture, where the cultured cells tightly adhere to each other, and where one side of the cells in the layer rests on and faces a permeable membrane, and where the other side of the cells resides under a liquid culture medium. For example, the experimental setup can involve transwell plates containing collagen-coated, micropore (0.4 mm pore size) polycarbonate filter membranes.[75] The integrity of the cell monolayer can be checked by way of a standard transepithelial electrical resistance (TEER) assay.[76]

P-gp recognizes a diverse variety of substrates, such as drugs and other xenobiotics. The substrate binds to a large binding site or "pocket" which is open to the cytosol. Once a substrate binds, P-gp utilizes the energy of ATP to drive the substrate out of the cell, and ATP hydrolysis is coupled to substrate extrusion.[77] Molecules that are P-gp substrates do not share any obvious structural characteristics, though many are cationic and hydrophobic.[78] Using histology and staining with an antibody specific for P-gp, Fakhoury et al.[79] provide photographs demonstrating that P-gp is located in the apical membrane of the cell and not in the basolateral membrane.

b. Life History of Orally Administered Drugs and Their Journeys Through Drug Transporters

Initially, a given oral drug encounters enterocytes in the gut lumen, where the drug diffuses through or, alternatively, is transported into the enterocyte. Once inside the enterocyte, the drug can either be destroyed by CYP enzymes in the enterocyte, be driven back out of the enterocyte by P-gp, or pass through the basolateral membrane to the bloodstream.[80]

Hall et al.[81] referred to the uptake of drugs from the gut lumen into the enterocyte as a "passive absorption process." As long as the drug is not catabolized by CYP and does not pass through the enterocyte and out the basolateral membrane to the blood plasma,

[75] Lumen AA, et al. Transport inhibition of digoxin using several common P-gp expressing cell lines is not necessarily reporting only on inhibitor binding to P-gp. PLoS One 2013;8:e69394. doi: 10.1371/journal.pone.0069394.

[76] Li W, et al. Glycine regulates expression and distribution of claudin-7 and ZO-3 proteins in intestinal porcine epithelial cells. J. Nutr. 2016;146:964–9.

[77] Esser L, et al. Structure of the multidrug transporter P-glycoprotein reveal asymmetric ATP binding and the mechanism of polyspecificity. J. Biol. Chem. 2017;292:446–61.

[78] Wessler JD, et al. The P-glycoprotein transport system and cardiovascular drugs. J. Am. Coll. Cardiol. 2013;61:2495–502.

[79] Fakhoury M, et al. Localization and mRNA expression of CYP3A and P-glycoprotein in human duodenum as a function of age. Drug Metab. Dispos. 2005;33:1603–7.

[80] von Richter O, et al. Cytochrome P450 and P-glycoprotein expression in human small intestinal enterocytes and hepatocytes: a comparative analysis in paired tissue speciments. Clin. Pharmacol. Ther. 2004;75:172–83.

[81] Hall SD, et al. Molecular and physical mechanisms of first-pass extraction. Drug Metab. Dispos. 1999;27:161–6.

the drug can repeatedly be passively absorbed and then pumped back out into the gut lumen.[82,83]

In another scenario, P-gp and CYP enzymes function in a manner that is complimentary and that form a coordinated intestinal barrier that prevents drugs from entering the bloodstream. In yet another scenario, an orally administered drug diffuses into the enterocyte, escapes P-gp mediated efflux and escapes catabolism by the enterocyte's CYP enzymes, then enters the bloodstream, and manages to escape metabolism by liver cytochrome P450 enzymes, with eventual elimination from the body via P-gp to the bile, or with eventual elimination via P-gp into the urine.[84]

Where a concomitantly administered second drug inhibits efflux activity of P-gp, this inhibition can involve competition for efflux transport with the first drug.[85] Where two or more drugs are administered orally, the potential for DDIs may be greater at P-gp in the enterocyte than at P-gp in the hepatocyte, because the concentrations of the drugs will always be greater in the gut lumen (and inside the enterocyte) than in the blood plasma (and inside the hepatocyte).[86]

c. Using Verapamil as a Second Drug to Increase Anticancer Effect of a First Drug

P-gp expressed by cancer cells can confer resistance to various anticancer drugs. P-gp can mediate efflux of the anticancer drugs such as vinblastin, doxorubicin, and paclitaxel, thus reducing the efficacy of these drugs against cancer.[87] One strategy for overcoming the problem of efflux of anticancer drugs is to coadminister another drug that impairs P-gp's transport activity. One such drug is verapamil. The proposal to administer combinations of an anticancer drug with verapamil represents a desirable type of DDI.

The nature of the problem and the nature of a solution are summarized by Callaghan et al.,[88] as, "the plethora of investigations that have described the association of P-gp with drug resistance and the positive relationship between expression and poor prognosis. . .[a] strategy to overcome multidrug resistance has been to co-administer **chemical inhibitors of P-gp** with anticancer drugs. Inhibition of P-gp would thereby lead to increased accumulation of anticancer drug within the cell and produce cell cytotoxicity."

[82] Hall SD, et al. Molecular and physical mechanisms of first-pass extraction. Drug Metab. Dispos. 1999;27:161–6.

[83] Christians U, et al. Functional interactions between P-glycoprotein and CYP3A in drug metabolism. Expert Opin. Drug Metab. Toxicol. 2005;1:641–54.

[84] Hall SD, et al. Molecular and physical mechanisms of first-pass extraction. Drug Metab. Dispos. 1999;27:161–6.

[85] Wessler JD, et al. The P-glycoprotein transport system and cardiovascular drugs. J. Am. Coll. Cardiol. 2013;61:2495–502.

[86] Mikkaichi T, et al. P-gp is in canalicular membrane of hepatocytes. Drug Metab. Dispos. 2014;42:520–8.

[87] Callaghan R, et al. Inhibition of the multidrug resistance P-glycoprotein: time for a change in strategy? Drug Metab. Dis. 2014;42:623–31.

[88] Callaghan R, et al. Inhibition of the multidrug resistance P-glycoprotein: time for a change in strategy? Drug Metab. Dis. 2014;42:623–31.

Unfortunately though, attempts to improve anticancer drugs by coadministering verapamil have not worked with patients. Joshi et al.[89] have also described this strategy for increasing efficacy of anticancer drugs, "Over the years, **inhibitors of this pump [P-gp]** have been discovered to administer them in combination with chemotherapeutic agents. The clinical failure of first and second generation P-gp inhibitors (such as verapamil and cyclosporine analogs) has led to the discovery of third generation potent P-gp inhibitors (tariquidar, zosuquidar, laniquidar)."

Verapamil can be used as a chemical reagent for determining if a given drug is subjected to P-gp mediated efflux. Kim et al.[90] used verapamil as a reagent to determine if a cancer cell's resistance to the anticancer drug paclitaxil is mediated by P-gp.

Similarly, Salvatorelli et al.[91] used verapamil to determine that one anticancer drug (doxorubixin) was effluxed by P-gp, but that another anticancer drug (amrubicin) was not effluxed via P-gp. The in vitro study used myocardial strips as the source of cardiac tissue. Verapamil's role was described as, "Amrubicin clearance was not mediated by P glycoprotein. . .as judged from the **lack of effect of verapamil** on the partitioning of amrubicin. . .across myocardial strips and plasma."[92]

d. Digoxin is a Probe for Assessing P-Glycoprotein Mediated Transport

Digoxin is a standard compound (probe substrate) for testing P-gp mediated transport. The transport data are not confounded by digoxin's metabolism by cytochrome P450, because digoxin is not a substrate for any of the CYP enzymes.[93]

Despite the standard use of digoxin as a P-gp substrate, Lumen et al.[94] provided a word of caution regarding usage of cultured cells in a monolayer for testing transport. The problem is the confounding nature that any of these mechanisms can influence transport measurements of the probe drug:

- Study drug inhibits P-gp mediated transport of digoxin across apical membrane
- Study drug inhibits transport of digoxin across basolateral membrane
- Study drug inhibits transport across basolateral membrane and also P-gp mediated transport across apical membrane

Thus the observed value for IC50 could be a function of any one of these three inhibitory mechanisms. Please note the reports from other researchers, described earlier in this

[89] Joshi P, et al. Natural alkaloids as P-gp inhibitors for multidrug resistance reversal in cancer. Eur. J. Med. Chem. 2017;138:273–92.

[90] Kim HJ, et al. P-glycoprotein confers acquired resistance to 17-DMAG in lung cancers with an ALK rearrangement. BMC Cancer 2015;15:553. doi:10.1186/s12885-015-1543-z.

[91] Salvatorelli E, et al. Pharmacokinetic characterization of amrubicin cardiac safety in an ex vivo human myocardial strip model. I. amrubicin accumulates to a lower level than doxorubicin or epirubicin. J. Pharmacol. Exp. Ther. 2012;341:464–73.

[92] Kim HJ, et al. P-glycoprotein confers acquired resistance to 17-DMAG in lung cancers with an ALK rearrangement. BMC Cancer 2015;15:553. doi:10.1186/s12885-015-1543-z.

[93] Fenner KS, et al. Drug–drug interactions mediated through P-glycoprotein: clinical relevance and in vitro–in vivo correlation using digoxin as a probe drug. Clin. Pharmacol. Ther. 2009;85:173-81.

[94] Lumen AA, et al. Transport inhibition of digoxin using several common P-gp expressing cell lines is not necessarily reporting only on inhibitor binding to P-gp. PLoS One 2013;8:e69394.

chapter, that P-gp occurs only in apical membranes. Also, please note that the data on transport protein expression may differ, depending on the type of cell that is studied.

Lumen et al.[95] established that there does exist a transporter in the basolateral membrane that transports digoxin. This was established by the identification and use of a chemical (GF120918)[96] that inhibits the uptake of digoxin into cells. In the words of the researchers, the scenario that is encountered when a probe substrate such as digoxin is used, is, "Since digoxin is a substrate of uptake and efflux transport in…cells and since the uptake transport is inhibitable by GF120918, inhibition of basolateral membrane to apical membrane digoxin transport across these cells is expected to be due to inhibition of P-gp, the basolateral uptake transporter or both."[97]

The scenario considered by Lumen et al.[98] is illustrated in Fig. 7.3. This figure shows the following:

- The basolateral membrane **(1)** and the apical membrane **(2)**
- Digoxin **(3)** entering cell via the GF120918-inhibitable transporter and passing through basolateral membrane
- Test drug inhibiting the GF120918-inhibitable transporter **(4)**
- Test drug inhibiting P-gp **(5)**
- Digoxin efflux via P-gp to extracellular fluid bathing apical membrane **(6)**

The take-home lesson from the study of Lumen et al. is that the value for IC50 that is eventually calculated will be a function of the second drug's inhibition of the basolateral transporter (BT) in the basolateral membrane (if this inhibition exists), and that it will be a function of the second drug's inhibition of the P-gp that is situated in the apical membrane (if this inhibition exists). The exact quotation from Lumen et al.[99] is that, "The value of the IC50 depends upon digoxin and inhibitor binding to both P-gp and basolateral transporter (BT)."

IV. INTEGRATED PICTURE OF P-GLYCOPROTEIN'S INFLUENCE ON DRUG EXPOSURE

P-gp is expressed by cells of the gut (enterocytes), liver (hepatocytes), and kidneys (renal proximal tubule cells), where P-gp mediates efflux of drugs from the cells to the gut lumen, bile duct, and urine, respectively.

[95] Lumen AA, et al. Transport inhibition of digoxin using several common P-gp expressing cell lines is not necessarily reporting only on inhibitor binding to P-gp. PLoS One 2013;8:e69394.

[96] GF120918 is also known as elacridar (see, Kuppens IE, et al. A phase I, randomized, open-label, parallel-cohort, dose-finding study of elacridar (GF120918) and oral topotecan in cancer patients. Clin. Cancer Res. 2007;13:3276–85).

[97] Lumen AA, et al. Transport inhibition of digoxin using several common P-gp expressing cell lines is not necessarily reporting only on inhibitor binding to P-gp. PLoS One 2013;8:e69394.

[98] Lumen AA, et al. Transport inhibition of digoxin using several common P-gp expressing cell lines is not necessarily reporting only on inhibitor binding to P-gp. PLoS One 2013;8:e69394.

[99] Lumen AA, et al. Transport inhibition of digoxin using several common P-gp expressing cell lines is not necessarily reporting only on inhibitor binding to P-gp. PLoS One 2013;8:e69394.

P-gp occurs in the apical membrane (brush border side) of enterocytes,[100] in the apical membrane (canalicular side) of hepatocytes,[101] and in the apical membrane of the renal proximal tubule cells.[102,103] Location of any given transporter may differ, somewhat, in a given type of cell where it is isolated from different animals, and where the cell is isolated from infants versus adults.[104] Also, if the cell is isolated from tissue and then cultured, the transporter's location in the cell may change.[105]

Fromm et al.[106] described the fact that a given first drug, such as quinidine, can influence transport of a probe drug (digoxin) in the gut, liver, and kidney, all at the same time. Thus, where any coadministered drug inhibits P-gp, the coadministered drug may inhibit P-gp activity in enterocytes, hepatocytes, and renal tubule cells.

Fromm et al. stated that, "Each of these sites...is now recognized to express P-glycoprotein in a polarized fashion. Inhibition of P-glycoprotein function would result in decreased **enterocyte efflux into the gut** (thereby accounting for reported increased digoxin absorption) as well as decreased **excretory efflux by the kidney** and **biliary tract**."[107]

Fig. 7.4 reveals that the rate of P-gp mediated efflux from enterocytes and from renal tubule cells can influence exposure (blood concentration) of any given drug. The apical membrane is conventionally represented in histology books by a wiggly line of cells, while the basolateral membrane is shown by a flat line of cells. Adjacent cells adhere to each other by an intermediary region called the tight junction.[108]

Fig. 7.4 shows a drug (filled ovals) in the gut lumen being absorbed at the apical membrane of the enterocyte, and either being effluxed back out via P-gp, or crossing through the cytosol and exiting the enterocyte via the basolateral membrane into the bloodstream. The drug's entry into the enterocyte can be mediated by one or both passive diffusion or by a drug transporter.

After crossing the basolateral membrane and entering the bloodstream, the drug enters various cells in the body. With entry into renal tubule cells, e.g., by passive diffusion or

[100] Fakhoury M, et al. Localization and mRNA expression of CYP3A and P-glycoprotein in human duodenum as a function of age. Drug Metab. Dispos. 2005;33:1603–7.

[101] Bow DA, et al. Localization of P-gp (Abcb1) and Mrp2 (Abcc2) in freshly isolated rat hepatocytes. Drug Metab. Dispos. 2008;36:198–202.

[102] Scotcher D, et al. Delineating the role of various factors in renal disposition of digoxin through application of physiologically-based kidney model to renal impairment populations. J. Pharmacol. Exp. Ther. 2017;116. doi:10.1124/jpet.116.237438.

[103] Mikkaichi T, et al. Isolation and characterization of a digoxin transporter and its rat homologue expressed in the kidney. Proc. Natl. Acad. Sci. 2004;101:3569–74.

[104] Fakhoury M, et al. Localization and mRNA expression of CYP3A and P-glycoprotein in human duodenum as a function of age. Drug Metab. Dispos. 2005;33:1603–7.

[105] Bow DA, et al. Localization of P-gp (Abcb1) and Mrp2 (Abcc2) in freshly isolated rat hepatocytes. Drug Metab. Dispos. 2008;36:198–202.

[106] Fromm MF, et al. Inhibition of P-glycoprotein-mediated drug transport. Circulation 1999;99:552–7.

[107] Fromm MF, et al. Inhibition of P-glycoprotein–mediated drug transport. Circulation 1999;99:552–7.

[108] Hou J. The kidney tight junction (review). Int. J. Mol. Med. 2014;34:1451–7.

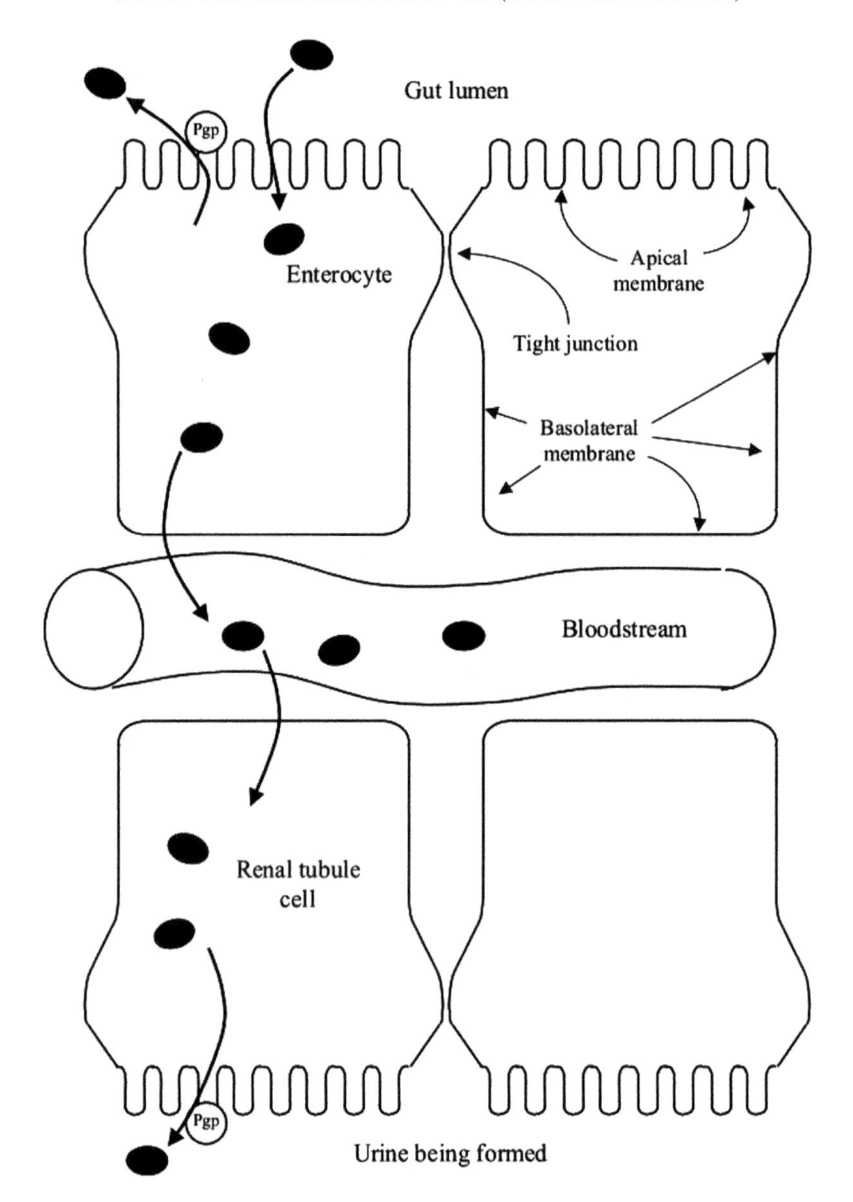

FIGURE 7.4 Enterocyte and renal tubule cell. This shows the passage of an oral drug, from the gut lumen, transport across the enterocyte, then circulating in the bloodstream, and finally transport across a renal tubule cell to the forming urine. The positions of P-gp, a drug efflux transporter, in the enterocyte and the renal tubule cell are shown.

active transport, the drug can then be effluxed via P-gp out the apical membrane into the forming urine (Fig. 7.4).

The net effect of efflux via P-gp in the enterocyte's apical membrane (back into the gut lumen) and efflux via P-gp in the renal tubule cell's apical membrane (into the forming urine) is *reduced exposure*. But if a coadministered drug inhibits P-gp, both of these efflux pathways can be inhibited, resulting in *increased drug exposure*. In pharmacokinetics studies, the term "exposure" usually refers to parameters of blood concentration, such as AUC, Cmax, Cmin, Ctrough, or tmax.

V. CLEARANCE OF DRUGS BY CONJUGATION CATALYZED BY UDP-GLUCURONOSYLTRANSFERASE

UDP-glucuronosyltransferase (UGT) enzymes catalyze the attachment of a glucuronic acid moiety to various drugs and other xenobiotics, as well as to endogenous compounds such as bilirubin. This conjugation promotes their excretion.[109] UGT enzymes can catalyze the attachment of a glucuronic acid moiety to the hydroxyl, carboxyl, amino, or sulfhydryl group of a target compound.[110] The following diagram illustrates the catalytic reaction of UGT, as well as the event of UGT inhibition, as occur where there is a drug-drug interaction involving UGT. The lightning bolt represents inhibition by a coadministered drug. Drugs that *inhibit* UGT include desloratadine,[111] lapatinib, pazopanib, regorafenib, and sorafenib.[112]

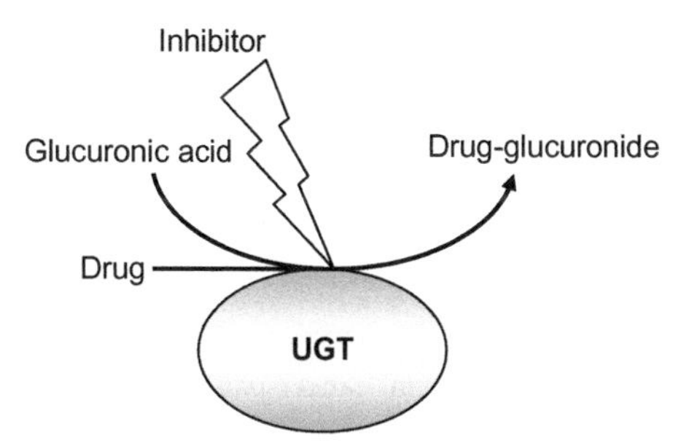

If the activity of UGT is impaired the result can be increased toxicity of a drug that would otherwise be processed by UGT. Impaired UGT activity can result by direct action of a drug that inhibits UGT. Also, impaired UGT activity results from naturally occurring mutations in genes encoding UGT enzymes. Changes in genetic sequence of

[109] Fujiwara R, et al. Structure and protein-protein interactions of human UDP-glucuronosyltransferases. Front. Pharmacol. 2016. eCollection 2016.

[110] Fujiwara R, et al. Structure and protein-protein interactions of human UDP-glucuronosyltransferases. Front. Pharmacol. 2016. eCollection 2016.

[111] Pattanawongsa A, et al. Human UDP-glucuronosyltransferase (UGT) 2B10: validation of cotinine as a selective probe substrate, inhibition by UGT enzyme-selective inhibitors and antidepressant and antipsychotic drugs, and structural determinants of enzyme inhibition. Drug Metab. Dispos. 2016;44:378–88.

[112] Miners JO, et al. Inhibition of human UDP-glucuronosyltransferase enzymes by lapatinib, pazopanib, regorafenib and sorafenib: implications for hyperbilirubinemia. Biochem. Pharmacol. 2017. doi:10.1016/j.bcp.2017.01.002.

the UGT variants that influence expression of UGT1A1 and UGT1A9 have been identified.[113,114,115,116]

Glucuronidation, as it is normally mediated by UGT1A1 and UGT1A9, reduces irinotecan toxicity. Naturally occurring mutations in UGT enzymes result in a dramatic increase in *irinotecan* toxicity. As described by Ceddin et al.,[117] a mutation in **UGT1A1** results in irinotecan causing "severe neutropenia/diarrhea." Also, a mutation in **UGT1A9** was associated with severe toxicity with irinotecan treatment. Regarding the toxicity resulting in patients mutated in **UGT1A9**, Ceddin et al. state that, "Concerning severe toxicity during the entire course of therapy, UGT1A9*22 was associated with severe hematologic toxicity." The asterisk (*) indicates that UGT1A9*22 is a variant of the gene, UGT1A9. DDIs occurring at the location of UGT, and involving the coadministration of *irinotecan* with regorafenib, are detailed at a later point in this chapter.

Another example of UGT-mediated glucuronidation is that of propofol. Intravenous propofol is used for anesthesia. Shortly after infusion is started, the patient experiences a transition period of sleepiness lasting 10 seconds or so. Infusion is continued during the entire medical procedure, e.g., a routine colonoscopy. After the medical procedure, the infusion is stopped, and shortly thereafter the patient regains full-alertness without any transition period of grogginess.[118] This rapid recovery from anesthesia is due to the ongoing glucuronidation and rapid inactivation of propofol.[119,120,121] Propofol is a substrate for UGT1A9, which catalyzes the transfer of a glucuronic acid group to propofol, producing propofol glucuronide. Human studies with radioactive propofol reveal that the inactivated metabolites of propofol are propofol glucuronide, 1-quinol-propofol glucuronide, and

[113] Cecchin E, et al. Predictive role of the UGT1A1, UGT1A7, and UGT1A9 genetic variants and their haplotypes on the outcome of metastatic colorectal cancer patients treated with fluorouracil, leucovorin, and irinotecan. J. Clin. Oncol. 2009;27:2457–65.

[114] Inoue K, et al. Polymorphisms of the UDA-glucuronosyl transferase 1A genes are associated with adverse events in cancer patients receiving irinotecan-based chemotherapy. Tohoku J. Exp. Med. 2013;229:107–14.

[115] Goetz MP, et al. UGT1A1 Genotype-guided Phase I study of irinotecan, oxaliplatin, and capecitabine. Investig. New Drugs 2013;31. doi:10.1007/s10637-013-0034-9 (16 pp.).

[116] Yamanaka H, et al. A novel polymorphism in the promoter region of human UGT1A9 gene (UGT1A9*22) and its effects on the transcriptional activity. Pharmacogenetics 2004;14:329–32.

[117] Cecchin E, et al. Predictive role of the UGT1A1, UGT1A7, and UGT1A9 genetic variants and their haplotypes on the outcome of metastatic colorectal cancer patients treated with fluorouracil, leucovorin, and irinotecan. J. Clin. Oncol. 2009;27:2457–65.

[118] Personnel communication from Dr. Paul Chard, MD of East Bay Center for Digestive Health, Oakland, CA; July 13, 2017.

[119] Jones RD, et al. Pharmacokinetics of propofol in children. Br. J. Anaesth. 1990;65:661–6

[120] Patterson KW, et al. Propofol sedation for outpatient upper gastrointestinal endoscopy: comparison with midazolam. Br. J. Anaesth. 1991;67:108–11.

[121] Vargo JJ, et al. Practice efficiency and economics: the case for rapid recovery sedation agents for colonoscopy in a screening population. J. Clin. Gastroenterol. 2007;41:591–8.

4-quinol-propofol glucuronide.[122,123] What is distinguished about propofol's glucuronidation is that it is a desired feature of the drug's properties, in that it is responsible for the desired property of quick recovery from anesthesia. The package label provides a graph of propofol pharmacokinetics, showing that when infusion is ceased, plasma concentrations drop quickly, where after 5 minutes the propofol concentration drops to half the infusion level, with a continued fall in plasma levels over the next 30 minutes.[124]

Naturally occurring mutations in UGT1A9 in humans include D256N and Y483D. These mutations reduce catalytic efficiency of the enzyme, and are expected to increase exposure of propofol, and possibly increase propofol's toxicity and slow down recovery from anesthesia after the infusion is brought to a halt.[125] The structures of propofol and propofol glucuronide are shown below.[126]

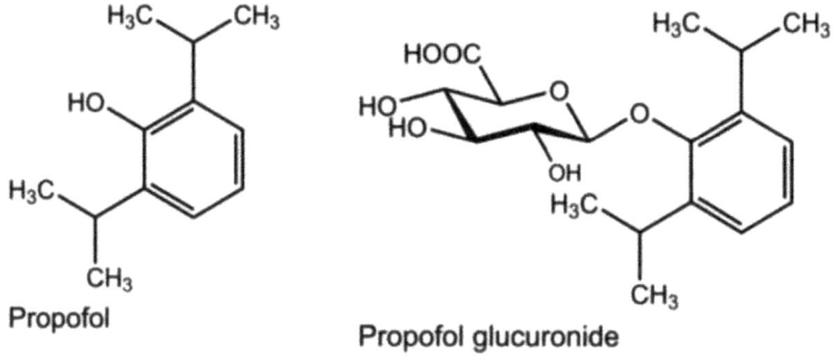

Propofol

Propofol glucuronide

VI. METABOLITES OF STUDY DRUGS

DDI studies can include an analysis of the study drug's metabolites. Where one or more metabolites accumulate in the bloodstream, these issues can arise:

- Does the metabolite induce expression of any of the CYP enzymes?
- Does the metabolite inhibit any of the CYP enzymes?
- Does the metabolite provide efficacy against the disease, with a mechanism of action being similar to that of the study drug?

[122] Kanto J, Gepts E. Pharmacokinetic implications for the clinical use of propofol. Clin. Pharmacokinet. 1989;17:308–26.

[123] Bleeker C, et al. Recovery and long-term renal excretion of propofol, its glucuronide, and two di-isopropylquinol glucuronides after propofol infusion during surgery. Br. J. Anaesth. 2008;101:207–12.

[124] Package label. DIPRIVAN® (propofol) Injectable emulsion, USP; April 2017 (54 pp.).

[125] Takahashi H, et al. Effect of D256N and Y483D on propofol glucuronidation by human uridine 5′-diphosphate glucuronosyltransferase (UGT1A9). Basic Clin. Pharmacol. Toxicol. 2008;103:131–6.

[126] Mukai M, et al. In vitro glucuronidation of propofol in microsomal fractions from human liver, intestine and kidney: tissue distribution and physiological role of UGT1A9. Pharmazie 2014;69:829–32.

- Is the metabolite nontoxic, as compared with the parent study drug?
- Is the metabolite toxic and does it provoke AEs?

Regarding metabolites, FDA's Guidance for Industry recommends, "drug interactions with **metabolites** of investigational drugs . . . metabolites present at greater or equal to 25% of parent drug AUC should be considered.[127]" Examples of DDI studies of metabolites are shown below.

a. Example of Diltiazem Metabolites

FDA's Guidance provides the example of a metabolite that inhibits one of the CYP enzymes, where the inhibition is essentially irreversible. FDA's Guidance states that, "In the case of diltiazem, both parent drug diltiazem and its primary **metabolite, N-desmethyldiltiazem**, are time-dependent CYP3A inhibitors."[128] Burt et al.[129] also describe the inhibition that is caused by diltiazem and its metabolite, using in vitro assays of CYP3A4 activity where quinidine is the probe substrate. Where quinidine is the substrate, CYP3A4 catalyzes its conversion to 3-hydroxy-quinidine.

A big-picture issue is whether the inhibition of a CYP enzyme by *a given drug* is more significant or less significant than the inhibition of the CYP enzyme by *the drug's metabolite*. This issue is illustrated by the following narrative.

Zhang et al.[130] contemplated the plasma concentrations of diltiazem (DTZ) and of its metabolite (desmethyl-diltiazem; desmethyl-DTZ), considered their Ki values with CYP3A4, and reasoned that, "desmethyl-DTZ is a **more potent inhibitor**...of CYP3A4 than the parent drug in vitro...[h]owever, the steady-state plasma concentration of...desmethyl-DTZ is approximately **one-third of that of DTZ**...[t]herefore, given the higher inhibition potency but lower exposure of desmethyl-DTZ compared with DTZ, it is not clear whether desmethyl-DTZ contributes to the overall inhibitory effect observed in vivo after DTZ administration."[131]

The take-home lesson from this is that all assessments of a metabolite's inhibition of cytochrome P450 enzymes need to compare the plasma concentration and inhibition

[127] U.S. Dept of Health and Human Services. Food and Drug Administration. Center for Drug Evaluation and Research (CDER). Guidance for industry. Drug interaction studies—study design, data analysis, implications for dosing and labeling recommendations; 2012 (page 25 of 75 pages).

[128] U.S. Dept of Health and Human Services. Food and Drug Administration. Center for Drug Evaluation and Research (CDER). Guidance for industry. Drug interaction studies—study design, data analysis, implications for dosing and labeling recommendations; 2012 (page 29 of 75 pages).

[129] Burt HJ, et al. Progress curve mechanistic modeling approach for assessing time-dependent inhibition of CYP3A4. Drug Metab. Dispos. 2012;40:1658–67.

[130] Zhang X, et al. Semiphysiologically based pharmacokinetic models for the inhibition of midazolam clearance by diltiazem and its major metabolite. Drug Metab. Dispos. 2009;37:1587–97.

[131] Zhang X, et al. Semiphysiologically based pharmacokinetic models for the inhibition of midazolam clearance by diltiazem and its major metabolite. Drug Metab. Dispos. 2009;37:1587–97.

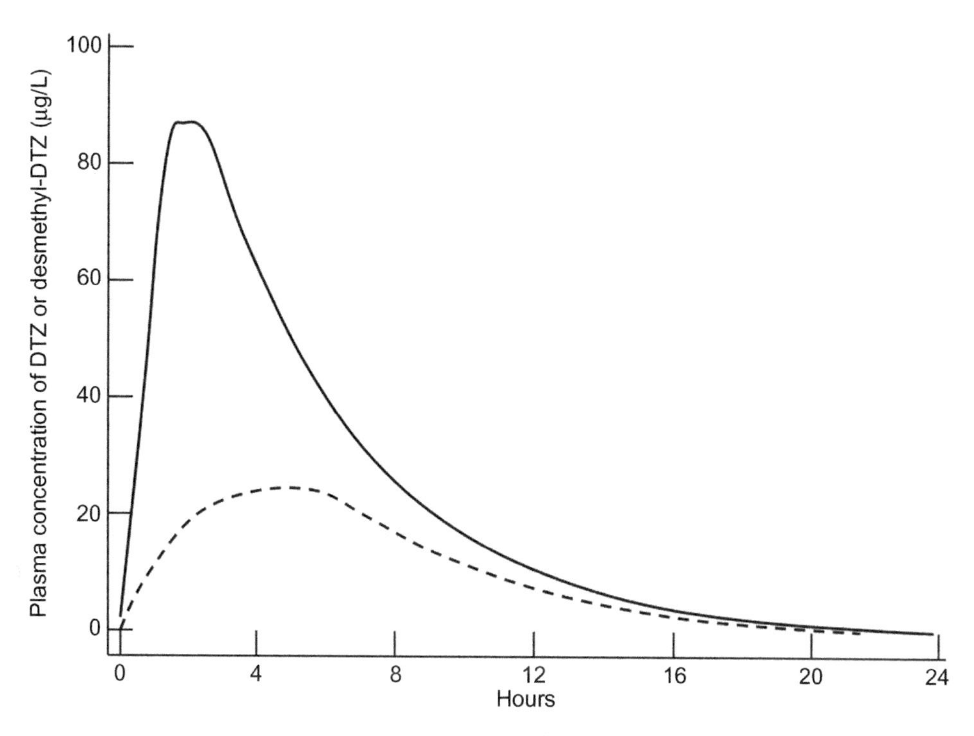

FIGURE 7.5 Plasma levels of diltiazem and its metabolite. Human subjects took a single oral dose of 60 mg DTZ. Plasma DTZ is shown by the upper solid line. The DTZ metabolite is shown by the lower dashed line. *Permission to reproduce figure granted by Richard Dodenhoff, journals director; June 29, 2017. Rockville, MD: ASPET .*

constant of the metabolite, with the plasma concentration and inhibition constant of the parent drug. A practical outcome of this is as follows. If the data show that the study drug is absorbed and reaches a high blood concentration which is then followed by a steady decline to a very low blood concentration, the Sponsor should refrain from assuming the following. What should not be assumed is that at the time after the study drug levels have declined, there will not be any DDIs involving a second coadministered drug. The reason that this should not be assumed is that the study drug could have been converted to a metabolite that does engage in DDIs.

Fig. 7.5 from Zhang et al.[132,133] provides the time course for plasma concentrations of DTZ (solid line) and of desmethyl-DTZ (dashed line), after an oral administration in human subjects.

[132] Zhang X, et al. Semiphysiologically based pharmacokinetic models for the inhibition of midazolam clearance by diltiazem and its major metabolite. Drug Metab. Dispos. 2009;37:1587–97.

[133] Permission to reproduce figure granted by Richard Dodenhoff, journals director; June 29, 2017. Rockville, MD: ASPET.

b. Examples of Dabrafenib Metabolites and of Regorafenib Metabolites

FDA's review of dabrafenib provides an account of metabolites that cause DDIs, while FDA's review of regorafenib provides an account of metabolites that are therapeutically active.[134] These accounts are provided at a later point in this chapter.

VII. DRUG CLEARANCE

The term clearance is used in describing pharmacokinetic properties of drugs, though the meaning of this term can be ambiguous where it is not clear if the narrative refers to clearance by way of CYP enzyme-mediated clearance, to conjugation-mediated clearance, or to transporter-mediated clearance. Also, what can be ambiguous is whether "clearance" refers to a drug's net movement from one compartment of the body to another compartment or, alternatively, if clearance refers to net reduction of the amount of drug from the patient's entire body.

FDA's Guidance for Industry uses the term clearance to mean various different things:

- "A study. . .should be considered when a drug. . .exhibits. . .high hepatic clearance. . .relative to hepatic blood flow."[135]
- "Therapeutic proteins. . .typically do not undergo metabolism or transport as their clearance pathway."[136]
- "renal/biliary clearances. . .can be evaluated together."[137]
- "the need for in vivo drug interaction studies with enzyme inhibitors. . .are based on . . .the measurement of the contribution of the enzyme to the overall systemic clearance of the substrates."[138]

Dreisbach et al.[139] demonstrate that "clearance" can be defined in various ways, in its narrative, "chronic renal failure (CRF) and end-stage renal disease (ESRD) alter drug disposition by affecting protein and tissue binding and reducing **systemic clearance** of

[134] Regorafenib (metastatic colorectal cancer) NDA 203-085. Page 37 of 64-page Clinical Pharmacology Review.

[135] U.S. Department of Health and Human Services. Food and Drug Administration. Center for Drug Evaluation and Research (CDER). Center for Biologics Evaluation and Research (CBER). Guidance for industry. Pharmacokinetics in patients with impaired renal function—study design, data analysis, and impact on dosing and labeling; 1998 (16 pp.).

[136] U.S. Department of Health and Human Services. Food and Drug Administration. Center for Drug Evaluation and Research (CDER). Guidance for industry. Drug Interaction studies—study design, data analysis, implications for dosing, and labeling recommendations; 2012 (75 pp.).

[137] U.S. Department of Health and Human Services. Food and Drug Administration. Center for Drug Evaluation and Research (CDER). Guidance for industry. Drug interaction studies—study design, data analysis, implications for dosing, and labeling recommendations; 2012 (75 pp.).

[138] U.S. Department of Health and Human Services. Food and Drug Administration. Center for Drug Evaluation and Research (CDER) Guidance for industry. Drug interaction studies—study design, data analysis, implications for dosing, and labeling recommendations; 2012 (75 pp.).

[139] Dreisbach AW, Lertora JJ. The effect of chronic renal failure on hepatic drug metabolism and drug disposition. Semin. Dial. 2003;16:45–50.

renally cleared drugs. What is not fully appreciated is that chronic renal failure can significantly reduce **nonrenal clearance**."

FDA's Clinical Pharmacology review for the elvitegravir/cobicistat/emtricitabine/tenofovir alafenamide (E/C/F/TAF) combination drug refers to the parameter of "renal clearance," in its commentary, "Relative to healthy controls, subjects with severe renal impairment had tenofovir alafenamide AUC increased 92% and Cmax increased 80%, renal clearance was reduced 88%.[140]"

Where there is a need to use the term "clearance," this author suggests that researchers define the meaning of "clearance." In other words, does it mean the rate of net removal of a drug from the body, where the net removal is the sum of removal by CYP enzyme-mediated catabolism, plus drug-conjugating enzymes, plus removal in the urine and bile duct? Or does it only mean removal by transport proteins in the kidneys that transport the drug into the urine?

VIII. PUBLISHED CLINICAL STUDIES ON DRUG–DRUG INTERACTIONS

Snyder et al.[141] describe the study drug *felodipine*, and a study measuring the influence of felodipine on the body's processing of an administered cocktail of probes. The study measured the felodipine's effect on the body's processing of each of the individual drug probes in the cocktail.

The readout for the experiment took the form of the ratio of probe exposure plus or minus the study drug. The ratio was

[AUC of drug probe with administered study drug]/[AUC of drug probe without any administered study drug]

To provide yet another example with human subjects, Wennerholm et al.[142] administered the study drug *amodiaquine*, and determined its influence on each of the drugs in the drug probe cocktail. The cocktail contained debrisoquine, omeprazole, losartan, and caffeine. Similarly, Tran et al.[143] described an experiment where the study drug was *daclizumab*, and where the experiment measured the influence of daclizumab on the body's processing of the administered cocktail. As was the case in all of the experiments described here, the blood concentrations (exposure) of each of the drug probes were measured with and without administration of the study drug.

[140] Elvitegravier, cobicistat, emtricitabine, tenofovir alafenamide (E/C/F/TAF) combination (HIV-1) NDA 207-561. Page 115 of 218-page Clinical Pharmacology Review.

[141] Snyder BD, et al. Evaluation of felodipine as a potential perpetrator of pharmacokinetic drug–drug interactions. Eur. J. Clin. Pharmacol. 2014;70:1115–22.

[142] Wennerholm A, et al. Amodiaquine, its desethylated metabolite, or both, inhibit the metabolism of debrisoquine (CYP2D6) and losartan (CYP2C9) in vivo. Eur. J. Clin. Pharmacol. 2006;62:539–46.

[143] Tran et al. Therapeutic protein-drug interaction assessment for daclizumab high-yield process in patients with multiple sclerosis using a cocktail approach. Br. J. Clin. Pharmacol. 2016;82:160–7.

IX. DRUG–DRUG INTERACTIONS IN FDA'S CLINICAL PHARMACOLOGY REVIEWS

a. Introduction

This provides FDA's observations, analyses, and logic and reasoning for arriving at information on the Drug Interactions section on the package label. The following provides FDA's analysis for these drugs:

- Axitinib
- Bosutinib
- Brivaracetam
- Cabozantinib
- Dabrafenib
- Fingolimod
- Lumacaftor/ivacaftor combination
- Ponatinib
- Regorafenib
- Tofacitinib

b. Arriving at Recommendations That the Sponsor Conduct a Clinical Drug–Drug Interactions Study

This reveals FDA's use of the [I]/Ki ratio (CYP inhibition data) or the [I]/IC50 ratio (transporter inhibition data) to determine if the Sponsor needs to conduct a clinical DDI study. This also shows FDA's use of the cutoff value of greater than 0.1, as tool in arriving at this recommendation. The excerpts below also reveal how FDA translates IC50 values to Ki values.

While the concentration of the study drug in human plasma (this involves administering only one drug) does constitute clinical data, it is not the same thing as conducting a clinical DDI study (this requires coadministering the study drug plus a second drug). The following excerpts reveal how the concepts of the [I]/Ki ratio, cutoff point, and Cmax are used. These are excerpts from full descriptions that occur at later points in this chapter:

1. *Excerpt about bosutinib.* "Based on an in vitro study...with Caco-2 cells, bosutinib is a substrate of P-gp with concentration dependant permeability (1, 10 and 100 μM)...[b] osutinib is an inhibitor of P-gp with an IC50 of 2 micromolar, in vitro. **The [I/]Ki** (I based on steady-state Cmax of 270 ng/mL at the 500 mg dose and Ki assumed to be IC50/2) **was approximately 0.5. Based on the I/Ki of >0.1**, it has the potential to inhibit P-gp in humans."[144]
2. *Excerpt about fingolimod.* FDA stated how to use the [I]/Ki ratio algorithm, writing that, "An estimated **[I]/Ki ratio of greater than 0.1** is considered positive and a follow-up in vivo evaluation is recommended."[145]

[144] Bosutininib (chronic myelogenous leukemia; CML) NDA 203-341. Page 22 of 89-page Clinical Pharmacology Review.

[145] Fingolimod (multiple sclerosis) NDA 022-527. Page 87 of 371-page Clinical Pharmacology Review.

3. *Excerpt about tofacitinib.* "Tofacitinib has low potential to inhibit P-gp with an estimated IC50 value of 311 micromolar. At a steady-state unbound Cmax of 310 nanomolar and projected gut concentration of 128 micromolar. . .following a 10 mg BID dose, the systemic [I]/IC50 ratio is 0.001 and the gut [I]/IC50 ratio is 0.4. . .[b]oth of these ratios are significantly below the level where a digoxin interaction study would be warranted, i.e., >0.1 and >10, respectively."[146]

c. Axitinib (Renal Cell Carcinoma) NDA 202-324

Axitinib is a small molecule drug that inhibits three different tyrosine kinases, VEGFR-1, VEGFR-2, and VEGFR-3. The structure of axitinib is shown below. FDA's Medical Review for axitinib reveals the influence of a second coadministered drug on the PK of axitinib, and the influence of axitinib on the PK of a second, coadministered drug.

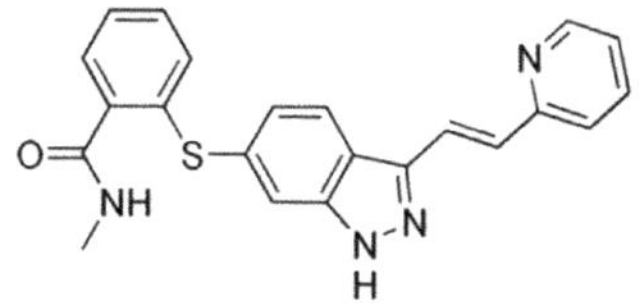

i. Influence of a Second Drug (Ketoconazole) on Exposure of Axitinib—CYP Enzyme Activity

FDA observed that, "Axitinib is a substrate of CYP3A4/5, P-gp, and UGT1A1."[147] CYP3A/5 is a cytochrome P450 enzyme 3A/5. UGT1A1 is UDP-glucuronosyl transferase 1A1, an enzyme that catalyzes glucuronidation of various drugs. This observation prompts the question of whether any CYP3A4/5 inhibitors or CYP3A4/5 inducers change the rate of CYP enzyme-mediated oxidation of axitinib, with consequent changes in axitinib exposure.

As is the situation with all DDIs, the concern is that undue increases in exposure will be toxic, and that significant decreases in exposure will reduce the drug's efficacy.

FDA observed that, "Ketoconazole (a strong CYP3A4/5 inhibitor) **increased axitinib exposure by 106%**, while rifampin (a strong CYP3A4/5 inducer) **decreased axitinib exposure by 80%**. . .drug–drug interaction studies indicate a 106% increase in axitinib exposure (AUC) when administered with ketoconazole and an 80% reduction in axitinib AUC when administered with rifampin."[148]

[146] Tofacitinib (rheumatoid arthritis) NDA 203-214. Page 33 of 181-page Clinical Pharmacology Review.

[147] Axitinib (renal cell carcinoma) NDA 202-324. Page 9 of 118-page Clinical Pharmacology Review.

[148] Axitinib (renal cell carcinoma) NDA 202-324. Pages 9–10 of 118-page Clinical Pharmacology Review.

FDA arrived at this recommendation for the package label, "Therefore, concomitant use of strong inhibitors or inducers of CYP3A4/5 should be avoided. However, if a strong CYP3A4/5 inhibitor must be co-administered, the axitinib dose should be reduced by half."[149]

Although FDA recommended that axitinib's dose be "reduced by half" the package label's recommendation was not so exact, and instead required, "If unavoidable, reduce the INLYTA dose."[150] The reader is invited, at this point in time, to look at the corresponding package label excerpt, which appears below.

ii. Influence of Axitinib on Exposure of a Second Drug (Paclitaxel)—CYP Enzyme Activity

Data on axitinib's inhibition of CYP2C8 were acquired, where enzyme assays measured paclitaxel oxidation. The inhibition constant (Ki), in combination with the blood concentration value for axitinib, [I], allowed calculating the ratio of [I]/Ki = 0.15.

Generally, a value for this ratio of less than 0.1 means that the DDI is of low risk for patients, and a value greater than 1 means that DDI is of high risk for patients. Medium risk is where the value is between 0.1 and 1.[151]

From the results of the in vitro inhibition data [Ki], and the clinical bloodstream concentration of axitinib [I], FDA concluded from the ratio that axitinib might be expected to exert a DDI in vivo. But note FDA use of the term "however," referring to a turning point in FDA's train of thought, "**Axitinib** inhibited CYP2C8 (I/Ki = 0.15) and CYP1A2 (I/Ki = 0.11) in vitro; **however**, in vivo, coadministration of axitinib did not increase **paclitaxel** plasma concentrations, indicating a lack of CYP2C8 inhibition."[152]

In other words, plugging the values into the [I]/Ki ratio equation revealed medium risk for DDIs, but a clinical study with coadministered axitinib and paclitaxel reveals that if there was any DDI, it was not great enough to change exposure.

iii. Clinical Drug–Drug Interaction Study

Turning to the clinical data, Table 7.1 reproduces a table from FDA's review. FDA reiterated its conclusion that, "Axitinib does not appear to alter paclitaxel pharmacokinetics." As one can readily see from Table 7.1, there is not much difference in exposure (AUC) of the second drug (paclitaxel), with or without axitinib.

iv. Influence of Axitinib on Exposure of a Second Drug (Digoxin)—P-glycoprotein Mediated Transport Data

Digoxin is a standard probe substrate for monitoring the influence of inhibitors on P-gp mediated transport of the probe substrate. Digoxin is conventionally used as a probe transport substrate for the following reason. Data on digoxin transport reflect only the influence

[149] Axitinib (renal cell carcinoma) NDA 202-324. Pages 9–10 of 118-page Clinical Pharmacology Review.

[150] Package label. INLYTA® (axitinib) tablets for oral administration; August 2014 (16 pp.).

[151] Ito K, et al. Database analysis for the prediction of in vivo drug–drug interactions from in vitro data. Br. J. Pharmacol. 2004;57:473–86.

[152] Axitinib (renal cell carcinoma) NDA 202-324. Page 10 of 118-page Clinical Pharmacology Review.

TABLE 7.1 Paclitaxel Exposure

Treatment	Cmax (ng/mL)	AUC_{inf} ((ng)(h)/mL)
Paclitaxel alone	3821	5942
Paclitaxel + axitinib	4053	6157

Axitinib (renal cell carcinoma) NDA 202-324. Page 43 and Table 19 of 118-page Clinical Pharmacology Review.

of a coadministered drug and are not confounded by digoxin's metabolism by CYP enzymes, because digoxin is not a substrate for any of the CYP enzymes.[153]

FDA's ClinPharm Review observed that, "axitinib is an inhibitor of P-gp mediated transport in vitro, with an estimated IC50 of 4.5 μM on digoxin efflux. An in vitro study-...was performed to assess the P-gp inhibitory effect of axitinib. This was done by measuring...digoxin...as a probe substrate in Caco-2 cells. Eleven concentrations of axitinib (between 0.1 to 75 micromolar) were used for the study. The data showed that axitinib is a concentration-dependent inhibitor of digoxin efflux."[154]

Persons with a background in enzymology will recognize and appreciate the care that the researchers took, in titrating the Caco-2 cells with the 11 different concentrations of axitinib. To reiterate, FDA observed that axitinib did, in fact, inhibit P-gp's transport of the digoxin substrate.

Package label. The package label mentioned the influence of axitinib (Inlyta®) on CYP3A4/5 in the Drug Interactions section, and instructed the physicians to reduce axitinib's dose in the situation where it was not possible to avoid coadministering the CYP3A4/5 inhibitor:

> **DRUG INTERACTIONS.** Avoid strong CYP3A4/5 inhibitors. If unavoidable, reduce the INLYTA dose. Avoid strong CYP3A4/5 inducers.[155]

Further on in the same package label, in the Full Prescribing Information, the Drug Interactions section identified various strong CYP3A4/5 inducers:

> **DRUG INTERACTIONS...CYP3A4/5 Inducers.** Co-administration of rifampin, a strong inducer of CYP3A4/5, reduced the plasma exposure of axitinib in healthy volunteers. Co-administration of INLYTA with strong CYP3A4/5 inducers (e.g., rifampin, dexamethasone, phenytoin, carbamazepine, rifabutin, rifapentin, phenobarbital, and St. John's Wort) should be avoided...[m]oderate CYP3A4/5 inducers (e.g., bosentan, efavirenz, etravirine, modafinil, and nafcillin) may also reduce the plasma exposure of axitinib and should be avoided if possible.[156]

[153] Fenner KS, et al. Drug–drug interactions mediated through P-glycoprotein: clinical relevance and in vitro–in vivo correlation using digoxin as a probe drug. Clin. Pharmacol. Ther. 2009;85:173-81.

[154] Axitinib (renal cell carcinoma) NDA 202-324. Page 37 of 118-page Clinical Pharmacology Review.

[155] Package label. INLYTA® (axitinib) tablets for oral administration; August 2014 (16 pp.).

[156] Package label. INLYTA® (axitinib) tablets for oral administration; August 2014 (16 pp.).

Additional information was in the Clinical Pharmacology section:

> **CLINICAL PHARMACOLOGY**...In vitro studies demonstrated that axitinib has the potential to inhibit CYP1A2 and CYP2C8. However, co-administration of axitinib with paclitaxel, a CYP2C8 substrate, **did not increase plasma concentrations of paclitaxel in patients**. In vitro studies indicated that axitinib does not inhibit CYP2A6, CYP2C9, CYP2C19, CYP2D6, CYP2E1, CYP3A4/5, or UGT1A1 at therapeutic plasma concentrations. In vitro studies in human hepatocytes indicated that **axitinib does not induce CYP1A1, CYP1A2, or CYP3A4/5**. Axitinib is an inhibitor of the efflux transporter P-glycoprotein (P-gp) in vitro. However, **INLYTA is not expected to inhibit P-gp** at therapeutic plasma concentrations.[157]

d. Bosutinib (Chronic Myelogenous Leukemia; CML) NDA 203-341

Bosutinib is a tyrosine kinase inhibitor used for treating chronic myelogenous leukemia (CML), also known as chronic myeloid leukemia (CML). A view of bosutinib's structure, shown below, reveals that it contains methyl groups and chlorine groups. The bosutinib metabolites M2 (oxydechlorinated bosutinib) and M5 (*N*-desmethyl bosutinib) are detectable in human plasma after dosing with bosutinib.[158]

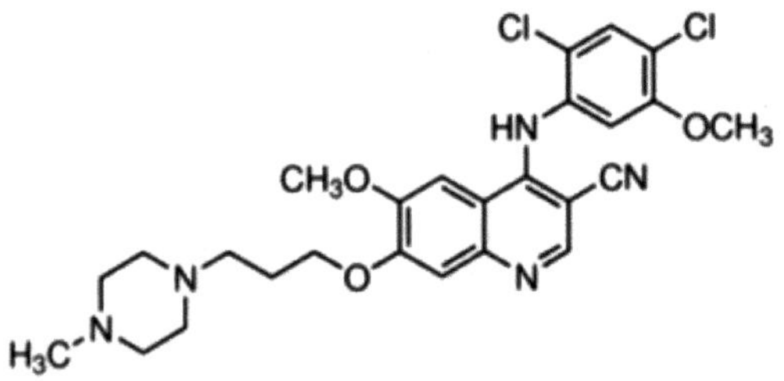

Cl Cl HN OCH_3 CH_3O CN N O N N H_3C

i. Data on Cytochrome P450. Influence of a Second Drug (Aprepitant) on Bosutinib Exposure

Published studies by Hsyu et al.[159] on bosutinib describe DDIs at the level of CYP3A4. The data revealed the influence of a second drug (*aprepitant*) on bosutinib metabolism, as mediated by CYP3A4. Aprepitant is an established inhibitor of CYP3A4. As aprepitant inhibits one of the CYP enzymes, thereby reducing bosutinib catabolism, it might be predicted that bosutinib exposure would be increased, and that was what was found.

In a clinical study comparing bosutinib (no aprepitant) and bosutinib (plus aprepitant), the AUC exposure values were 2268 and 4719 (ng)(h)/mL, respectively. The result was a doubling of bosutinib exposure in the blood of human subjects. Hsyu et al.[160] concluded that, "These results are consistent with a moderate CYP3A4 inhibitor effect of aprepitant on bosutinib."

[157] Package label. INLYTA® (axitinib) tablets for oral administration; August 2014 (16 pp.).

[158] Bosutininib (chronic myelogenous leukemia; CML) NDA 203-341. Page 22 of 89-page Clinical Pharmacology Review.

[159] Hysu PH, et al. Effect of aprepitant, a moderate CYP3A4 inhibitor, on bosutinib exposure in healthy subjects. Eur. J. Clin. Pharmacol. 2017;73:49–56.

[160] Hysu PH, et al. Effect of aprepitant, a moderate CYP3A4 inhibitor, on bosutinib exposure in healthy subjects. Eur. J. Clin. Pharmacol. 2017;73:49–56.

Turning to the Sponsor's own data on a second drug's influence on bosutinib exposure, FDA's ClinPharm Review disclosed that, "Clinical trials showed that the strong CYP3A4 inhibitor **ketoconazole** increased bosutinib AUC 9-fold while the strong CYP3A4 inducer **rifampin** decreased bosutinib AUC by 94%."[161] Because of this influence of strong CYP3A4 inhibitors, FDA became worried about the influence of moderate CYP3A4 inhibitors and, as a result, required that the Sponsor conduct additional studies focusing on moderate CYP3A4 inhibitors, and FDA requested, "Conduct a drug–drug interaction trial to evaluate the effect of a moderate CYP3A4 inhibitor (e.g. erythromycin) on the pharmacokinetics of bosutinib...moderate CYP3A4 inhibitors may increase the exposure of bosutinib 2–4 fold. Therefore, a Post-Marketing Requirement (PMR) will be issued to identify the appropriate dose of bosutinib when used concomitantly with moderate inhibitors."[162]

ii. Data on Cytochrome P450. Influence of a Second Drug (Ketoconazole) on Exposure of Bosutinib (CYP Enzyme Assays)

Turning to the details of the Sponsor's in vitro data and clinical data, the following account illustrates the frequently used algorithm that combines an in vitro-determined value (Ki) with an in vivo-determined value ([I]) for predicting the seriousness of DDI in human subjects. This also illustrates the recurring theme in FDA's ClinPharm Reviews of using the cutoff value of 0.1, for interpreting the value for the [I]/Ki ratio.

FDA's review disclosed three calculated inhibition constants (Ki), from enzyme inhibition assays for CYP3A4, CYP2C19, and CYP2D6, "As a CYP substrate: CYP3A4 is the major isozyme responsible for the metabolism of bosutinib. The effect of a strong CYP3A4 inhibitor (**ketoconazole**) and an inducer (**rifampin**) on the PK **bosutinib** were evaluated in humans...[b]ased on in vitro studies, bosutinib is unlikely an inhibitor or inducer of any of the major CYPs in humans at clinically relevant doses...bosutinib inhibited CYP 3A4, 2C19 and 2D6 with Ki values of 27, 27 and 10 μM, respectively."[163]

Then, FDA moved a step further and commented on the relevance of the in vitro enzyme inhibition studies to the clinical situation. FDA referred to the [I]/Ki ratio. FDA had referred to separate calculations based on plasma concentration of bosutinib in human subjects, and to gastrointestinal (GI) tract concentrations of bosutinib, in human subjects. The GI tract calculations were used to plug into the parameter [I]. Also the bloodstream concentrations were used to plug into the parameter [I]. In this way, the data enabled predicting via the [I]/Ki ratio equation if DDIs could be expected to be mediated by cytochrome P40 enzymes located in gut cells (enterocytes) and by cytochrome P450 enzymes in liver cells (hepatocytes).

[161] Bosutininib (chronic myelogenous leukemia; CML) NDA 203-341. Page 6 of 89-page Clinical Pharmacology Review.

[162] Bosutininib (chronic myelogenous leukemia; CML) NDA 203-341. Pages 6 and 35 of 89-page Clinical Pharmacology Review.

[163] Bosutininib (chronic myelogenous leukemia; CML) NDA 203-341. Page 32 of 89-page Clinical Pharmacology Review.

FDA's calculations regarding the hepatocyte predictions and gut cell predictions took this form:

> The [I]/Ki at the 500 mg dose is approximately 0.05 for CYP2D6 **which is less than 0.1** and suggests that a clinically significant interaction is unlikely. Since CYP3A4 is expressed in the gastrointestinal tract, the ratio value for gut exposure was considered for the 500 mg dose and was approximately 1.1 [$Ratio_{gut} = 1 + ((500/250)/ (548) * 1000/27)$], **which is less than 11**. CYP3A4 inhibition is unlikely in the gut or plasma.[164]

iii. Cardiac Adverse Effects was the Source of Concern for Excessive Bosutinib Exposure

One reason to be concerned about unintentionally high bosutinib was cardiac AEs. Cardiac AEs can be measured by QTc interval and heart rate. A review article on QT interval and QTc interval ("c" means corrected), and their use for assessing AEs in clinical trials, is cited.[165] FDA expressed concern for cardiac AEs where ketoconazole was coadministered with bosutinib:

> Does this drug prolong the QT or QTc interval? Bosutinib does not appear to prolong the QTc interval at clinically relevant exposures. The...review concluded that in a study with a demonstrated ability to detect small effects, no significant changes in placebo adjusted, baseline-corrected QTc...were observed... [a] statistically significant increase in heart-rate was seen with the use of bosutinib and ketoconazole compared to placebo and ketoconazole at time-points after 3 hours.[166]

FDA took a step further and made a recommendation for the package label, "the risk of QTc prolongation greater than 10 msec is mitigated by **labeling language** to avoid the use of strong and moderate CYP3A4 inhibitors."[167] The reader is invited to glance ahead, and view the excerpts from the package label, which warn against CYP3A4 inhibitors.

iv. Influence of Bosutinib on Exposure of a Second Drug

FDA's referred to assays testing the ability of bosutnib to inhibit CYP enzymes. The Sponsor acquired inhibition constants, and plugged the values into the [I]/Ki ratio equation, and determined that the calculated ratio was below the cutoff point. FDA also commented on bosutinib's induction of CYP enzymes. FDA observed

> Based on in vitro studies, bosutinib is unlikely an inhibitor or inducer of any of the major CYPs in humans at clinically relevant doses. Based on in vitro experiments...bosutinib inhibited CYP 3A4, 2C19 and 2D6 with Ki values of 27, 27 and 10 μM, respectively. The **[I/]Ki ratio at the 500 mg dose is**

[164] Bosutininib (chronic myelogenous leukemia; CML) NDA 203-341. Page 32 of 89-page Clinical Pharmacology Review.

[165] Brody T. QT interval prolongation in clinical trials in CLINICAL TRIALS study design, endpoints and biomarkers, drug safety, and FDA and ICH guidelines, 2nd ed. New York, NY: Elsevier, Inc.; 2016. p. 502–514.

[166] Bosutininib (chronic myelogenous leukemia; CML) NDA 203-341. Page 17 of 89-page Clinical Pharmacology Review.

[167] Bosutininib (chronic myelogenous leukemia; CML) NDA 203-341. Page 18 of 89-page Clinical Pharmacology Review.

approximately 0.05 for CYP2D6 which is less than 0.1 and suggests that a **clinically significant interaction is unlikely**...[b]ased on in vitro experiments...bosutinib **did not exhibit**...**inhibition** of CYP 2C9, 2C19, 2D6 or 3A. Based on in vitro experiments...bosutinib **did not induce** CYP 1A2, 2B6, 2C9, 2C19 or 3A4.[168]

v. P-Glycoprotein Efflux Data for Bosutinib

This illustrates cell culture experiments to determine IC50 values, and the often-used calculation of Ki from the IC50 value, with subsequent calculation of the ratio, [I]/Ki. Converting IC50 values to Ki values utilized a fudge factor. The clinical study used for generating the value for blood plasma concentration [I] used bosutinib at a 500 mg dose.

FDA's narrative of cell monolayer efflux studies described how the IC50 value from in vitro efflux data in cell monolayer studies, in combination with the clinical blood concentration value (270 mg/mL bosutinib), was used to predict if bosutnib has the potential to increase exposure of a second coadministered drug (digoxin). The starting information is the [I] value and the IC50 value. The fudge factor is dividing IC50 by two to get the value for Ki.

From FDA's ClinPharm Review we learn, "Based on an in vitro study with Caco-2 cells, bosutinib is a substrate of P-gp with concentration dependant permeability (1, 10 and 100 micromolar)...[b]osutinib is an inhibitor of P-gp with an IC50 of 2 micromolar, in vitro. The [I]/Ki ([I] based on steady-state Cmax of 270 ng/mL at the 500 mg dose and Ki assumed to be IC50/2) was approximately 0.5. Based on the I/Ki of >0.1, it has the potential to inhibit P-gp in humans."[169] Because the calculated [I]/Ki value of 0.5 was greater than the cutoff value of 0.1, FDA concluded that bosutinib has the "potential" to inhibit P-gp in patients.

Another part of FDA's ClinPharm Review plugged the IC50 value directly into an equation, i.e., the equation taking the form, "[Cmax]/IC50." FDA's comments shown below plugged the "IC50" value into the ratio equation. To this end, FDA stated that, "In the study, bosutinib showed a concentration dependent inhibition on P-gp mediated digoxin efflux with an IC50 value of 2 micromolar and resulted in 95% inhibition of P-gp activity at 50 micromolar. Based on the Cmax value of 206 ng/mL (0.4 micromolar), following an oral dose of 600 mg in patients, it would give rise to a ratio of [Cmax]/IC50 of 0.2."[170]

vi. Clinical Drug–Drug Interaction Data

The Sponsor provided clinical data showing that the CYP3A4 inhibitor ketoconazole provoked increases in bosutinib exposure. Fig. 7.6 and Table 7.2 show the clinical results. The figure shows plasma concentration (ng/mL) of bosutinib over the time-frame of 100 hours. The lower curve shows bosutinib alone, and the upper curve shows coadministration of bosutinib and ketoconazole.

[168] Bosutininib (chronic myelogenous leukemia; CML) NDA 203-341. Page 32 of 89-page Clinical Pharmacology Review.

[169] Bosutinib (chronic myelogenous leukemia; CML) NDA 203-341. Pages 22 of 89-page Clinical Pharmacology Review.

[170] Bosutinib (chronic myelogenous leukemia; CML) NDA 203-341. Pages 85 of 199-page Pharmacology Review.

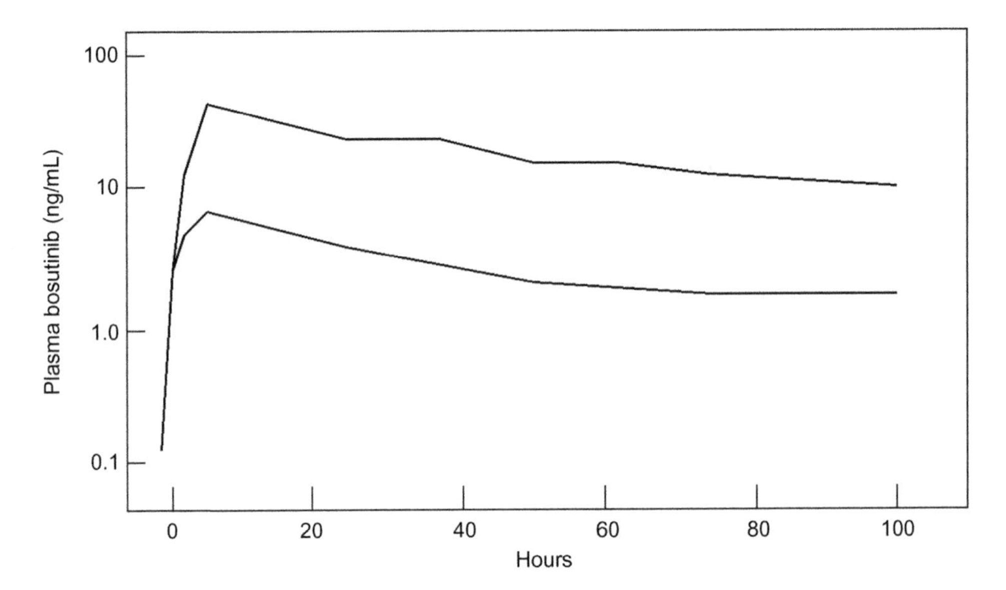

FIGURE 7.6 Bosutinab pharmacokinetics. The lower curve shows plasma bosutinib when administered alone; the upper curve shows plasma bosutinib when coadministered with ketoconazole (CYP3A4 inhibitor). Ketoconazole resulted in a 5× increase in Cmax and a 9× increase in AUC. Note the logarithmic *Y*-axis.

TABLE 7.2 Bosutinib Exposure Parameters After Administering Bosutinib Alone, or Bosutinib in Combination With Ketoconazole

	Bosutinib Only	Bosutinib in Combination With Ketoconazole
Cmax (ng/mL)	7.01	38.36
tmax (h)	6.0	6.0
$t_{1/2}$	46.19	69.04
$AUC_{0-inf.}$	323	2631

Bosutinib (chronic myelogenous leukemia; CML) NDA 203-341. Page 34 of 89-page Clinical Pharmacology Review.

FDA observed that ketoconazole provoked an increase in Cmax and AUC, "A 5-fold increase in Cmax and a 9-fold increase in AUC were observed when **bosutinib** was co-administered with **ketoconazole** as compared with bosutinib administered alone. **Rifampin** decreased the Cmax of bosutinib by 86% and the AUC by 94% compared to that observed when bosutinib was dosed alone."[171]

[171] Bosutininib (chronic myelogenous leukemia; CML) NDA 203-341. Page 32 of 89-page Clinical Pharmacology Review.

The package label. The package label for bosutinib (Bosulif®) contained warnings about CYP3A in the Drug Interactions section. As is evident, FDA's worry about CYP3A inhibitors found a place in the label:

> **DRUG INTERACTIONS.** CYP3A Inhibitors and Inducers: Avoid concurrent use of BOSULIF with strong or moderate CYP3A inhibitors and inducers...Proton Pump Inhibitors: May decrease bosutinib drug levels. Consider short-acting antacids in place of proton pump inhibitors.[172]

The Dosage and Administration section contained instructions to the physician to avoid CYP3A inhibitors as well as to avoid P-gp inhibitors:

> **DOSAGE AND ADMINISTRATION**...**Concomitant Use With CYP3A Inhibitors.** Avoid the concomitant use of strong or moderate CYP3A and/or P-gp inhibitors with BOSULIF as an increase in bosutinib plasma concentration is expected (strong CYP3A inhibitors include ritonavir, indinavir, nelfinavir, saquinavir, ketoconazole, boceprevir, telaprevir, itraconazole, voriconazole, posaconazole, clarithromycin, telithromycin, nefazodone and conivaptan. Moderate CYP3A inhibitors include fluconazole, darunavir, erythromycin, diltiazem, atazanavir, aprepitant, amprenavir, fosamprevir, crizotinib, imatinib, verapamil, grapefruit products and ciprofloxacin).
>
> **Concomitant Use With CYP3A Inducers.** Avoid the concomitant use of strong or moderate CYP3A inducers with BOSULIF as a large reduction in exposure is expected (strong CYP3A inducers include rifampin, phenytoin, carbamazepine, St. John's Wort, rifabutin and phenobarbital. Moderate CYP3A inducers include bosentan, nafcillin, efavirenz, modafinil and etravirine).[173]

vii. Drug–Drug Interactions Involving Bosutinib and the Problem of Drug Insolubility When Stomach pH was Raised.

FDA's review of bosutinib, as well as of a few other drugs, reveal a type of DDI where an increase in gastric pH reduced solubility of the study drug, where the outcome was reduced bioavailability. FDA turned to the Sponsor's clinical data on gastric interactions, and commented on interactions between bosutinib and lansoprazole (proton pump inhibitor).

The problem was that the proton pump inhibitor reduced bosutinib's exposure thus reducing bosutinib's bioavailability and efficacy. FDA described the problem of reduced efficacy caused by proton pump inhibitors, "Bosutinib has pH dependant solubility in vitro....[w]hen...**bosutinib** was co-administered with...**lansoprazole**...exposures to bosutinib decreased by 46% for Cmax and by 26% for $AUC_{0-inf.}$ compared to when bosutinib was administered alone...[t]he applicant recommended antacids should be considered as an alternative to proton pump inhibitors."[174]

The same issue arose during FDA's review of cabozantinib. The issue arose, but FDA was not able to arrive at any conclusion as to any instructions on the package label.

[172] Package label. BOSULIF® (bosutinib) tablets, for oral use; September 2012 (13 pp.).

[173] Package label. BOSULIF® (bosutinib) tablets, for oral use; September 2012 (13 pp.).

[174] Bosutininib (chronic myelogenous leukemia; CML) NDA 203-341. Page 37 of 89-page Clinical Pharmacology Review.

Instead, FDA recommended that further studies be conducted. FDA's review for cabozantinib stated

> The solubility of cabozantinib is pH-dependent with the **solubility at normal gastric pH the highest**, and practically insoluble when pH is greater than 4...[t]he gastric pH modifying drugs...can elevate the stomach pH at levels close to 6 or 7, therefore, co-medication may greatly decrease the solubility of cabozantinib...[t]he effect of gastric pH modifying drugs (proton pump inhibitors, H_2 blockers, antacids) on PK of cabozantinib based on a population PK analysis was inconclusive...[a] post-marketing requirement (PMR) for conducting a dedicated pH effect study is recommended.[175]

Yet another account of gastric DDI analysis is provided, which is from FDA's review of axitinib. FDA's review for axitinib stated, "The aqueous solubility of axitinib is low over a wide range of pH values, and lower pH values result in higher solubility...which raises the question about the potential for gastric pH elevating agents (such as proton pump inhibitors...antacids) to alter the solubility of axitinib. The Sponsor conducted a PK substudy...to evaluate the potential for drug–drug interactions with **rabeprazole**. In the presence of rabeprazole, a **42% decrease in axitinib Cmax**...was observed. However, there were only a **15% decrease in AUC** was observed, which is not considered clinical significant. There were no differences in **tmax** or **$T_{½}$** with or without rabeprazole. **Therefore, no axitinib dose adjustment is recommended.**"[176]

As is evident from the above excerpt, FDA's DDI analysis of axitinib is unique in that FDA separately took into account the influence on AUC, Cmax, tmax, and $t_{1/2}$.

e. Brivaracetam (Epilepsy) NDA 205-836, NDA 205-837, NDA 205-838

Brivaracetam is a small molecule drug for treating epilepsy. FDA's DDI analysis includes an account of another, second drug used for treating the same disease. FDA's review of brivaracetam is relevant to these situations:

- *First situation*. Where a given drug is only partially effective, and where the Sponsor wants to test if coadministering a second drug will better increase efficacy against the disease
- *Second situation*. Where a given drug is only partially effective and where the Sponsor wants to test if switching to a second drug will increase efficacy against the disease

Brivaracetam is for treating partial onset seizures in epilepsy. Brivaracetam is a ligand for SV2A. SV2A is a glycoprotein located in the presynaptic membrane. SV2A controls exocytosis of synaptic vesicles, where exocytosis results in concomitant neurotransmitter release. Binding of ligands to SV2A reduces seizures. The mechanism of action of SV2A is illustrated by the fact that a genetic deficiency in SV2A (in mice) increases seizures.[177,178]

[175] Cabozanitinib (thyroid cancer) NDA 203-756. Pages 23–25 of 106-page Clinical Pharmacology Review.

[176] Axitinib (renal cell carcinoma) NDA 202-324. Page 23 of 118-page Clinical Pharmacology Review.

[177] Coppola G, et al. New developments in the management of partial-onset epilepsy: role of brivaracetam. Drug Des. Dev. Ther. 2017;11:643–57.

[178] Menten-Dedoyart C, et al. Development and validation of a new mouse model to investigate the role of SV2A in epilepsy. PLoS One 2016;11:e0166525. doi: 10.1371/journal.pone.0166525.

This illustrates FDA's analysis for the two antiepileptic drugs, brivaracetam and levetiracetam. The structures are shown below. Brivaracetam differs in that it contains a propyl group (–CH_2–CH_2–CH_3).

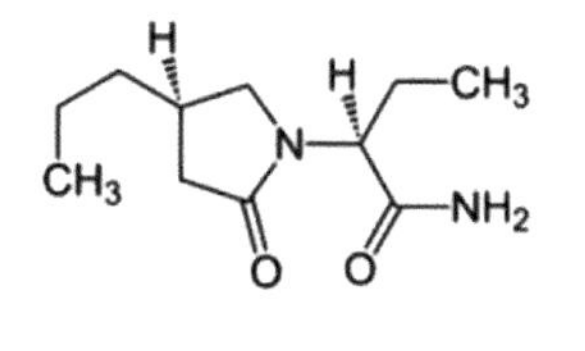

Brivaracetam 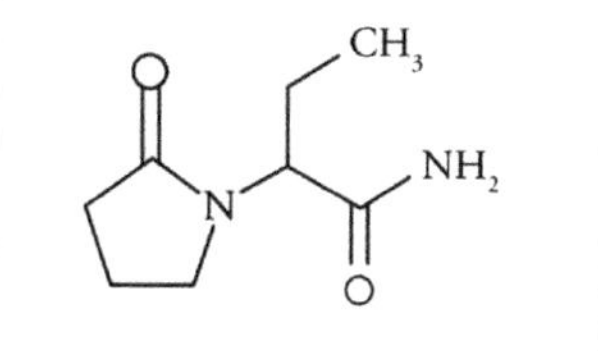 **Levetiracetam**

Partial-onset seizures are the most commonly encountered type of seizure in adults with epilepsy. The target of each of these drugs is a protein in nerves, namely, SV2A. SV2A controls presynaptic transmitter release. Brivaracetam has an affinity for SV2A that is 15–30 times greater than levetiracetam, as determined with human brain and rat brain.[179] SV2A coordinates synaptic vesicle exocytosis and neurotransmitter release.

i. Lipophilicity of Drugs that Act in the Brain

Where a small molecule drug needs to enter the brain, *greater lipophilicity enhances the rate of brain entry*. Brivaracetam is more lipophilic and levetiracetam is less lipophilic.[180] This difference in lipophilicity does not influence efficacy of levetiracetam in treating *chronic seizures*, in view of the fact that steady-state conditions are achieved after a day or so of drug treatment. On the other hand, for treating *acute seizures*, where a therapeutic response is needed within minutes in order to prevent permanent brain damage, low lipophilicity and consequent low rate of transport into the brain is a drawback for levetiracetam.

In addition to comparing lipophilicity, FDA compared other properties of the two drugs:

> Brivaracetam is pharmacologically similar to the anti-epileptic drug levetiracetam. The molecules are **structurally similar** where brivaracetam has a propyl moiety positioned on...the pyrrole ring that is not present on levetiracetam. Compared to levetiracetam, brivaracetam displays a markedly higher selectivity and **affinity for brain-specific binding site synaptic vesicle protein 2A** (SV2A). The safety characteristics of levetiracetam are likely to have similarity.[181]

ii. Coadministration of Two Drugs

FDA argued that coadministration of brivaracetam with another drug with a similar structure and similar physiological effects would not likely to have additive benefits. FDA's reasoning was that, "Levetiracetam and brivaracetam have the same target,

[179] Gao L, Li S. Emerging drugs for partial-onset epilepsy: a review of brivaracetam. Ther. Clin. Risk Manag. 2016;12:719–34.

[180] Nicolas JM. Brivaracetam, a selective high-affinity synaptic vesicle protein 2A (SV2A) ligand with preclinical evidence of high brain permeability and fast onset of action. Epilepsia 2016;57:201–9.

[181] Brivaracetam (epilepsy) NDA 205-836, NDA 205-837, NDA 205-838.

synaptic vesicle protein 2A (SV2A) in the brain, raising the possibility that treatment with either drug may saturate the available response and **no additional benefit** would result from **addition of the other drug**. . .[t]herefore, brivaracetam **should not be added** to the therapeutic regimen in patients already taking levetiracetam."[182]

FDA contemplated the possibility that coadministering two drugs that had the same biological target of the same class may actually *reduce efficacy of the first drug*. The Sponsor's clinical study showed that brivaracetam alone could reduce seizures by 30%, that levetiracetam alone could reduce seizures by 10%, but that the coadministration of both drugs could reduce seizures by only 10% (there was no additive effect):

> twenty percent of enrolled patients were permitted concomitant treatment with levetiracetam, an anti-epilepsy drug with an **overlapping mechanism of action** with brivaracetam. This allowed. . .examination of the potential for synergy between these agents. . .with concomitant levetiracetam treatment **there was a reduced therapeutic effect** when compared to those not on concomitant levetiracetam. This finding suggests there is no therapeutic synergy between levetiracetam and brivaracetam. This property of dual "acetam" treatment suggests that refractory patients who entered the study on concomitant levetiracetam may have captured most of the available benefit from levetiracetam and experience limited additional benefit from brivaracetam treatment.[183]

The Sponsor's clinical studies included subjects receiving various levels of brivaracetam with stepwise increases in levetiracetam. The steps were at 0 mg/day (left-most datapoint), 20, 50, 100, and 200 mg/day. Fig. 7.7 demonstrates that increasing dose of brivaracetam results in greater efficacy, i.e., fewer seizures (lower curve).

In detail, Fig. 7.7 (lower curve) shows that increasing brivaracetam doses resulted in increased plasma drug concentrations and also resulted in fewer seizures. Fig. 7.7 (upper curve) shows that coadministering levetiracetam prevented the use of increased birvaracetam from having any greater efficacy.[184,185,186] Based on these results, FDA made a recommendation for the package label, "Therefore, brivaracetam **should not be added** to the therapeutic regimen in patients already taking levetiracetam."[187] As shown below, this recommendation found a place on the Drug Interactions section of the package label.

[182] Brivaracetam (epilepsy) NDA 205-836, NDA 205-837, NDA 205-838. Pages 26–27 of 143-page Clinical Review. Pages 32–33 of 382-page pdf file for Medical Review.

[183] Brivaracetam (epilepsy) NDA 205-836, NDA 205-837, NDA 205-838. Page 17 of 382-page pdf file for Medical Review.

[184] Brivaracetam (epilepsy) NDA 205-836, NDA 205-837, NDA 205-838. Pages 32-33 of 382-page pdf file for Medical Review.

[185] Brivaracetam (epilepsy) NDA 205-836, NDA 205-837, NDA 205-838. Pages 307–311 of 405 pdf file of Clinical Pharmacology Review.

[186] Brivaracetam (epilepsy) NDA 205-836, NDA 205-837, NDA 205-838. Page 122 of 382-page pdf file for Medical Review.

[187] Brivaracetam (epilepsy) NDA 205-836, NDA 205-837, NDA 205-838. Page 33 of 382-page pdf file for Medical Review.

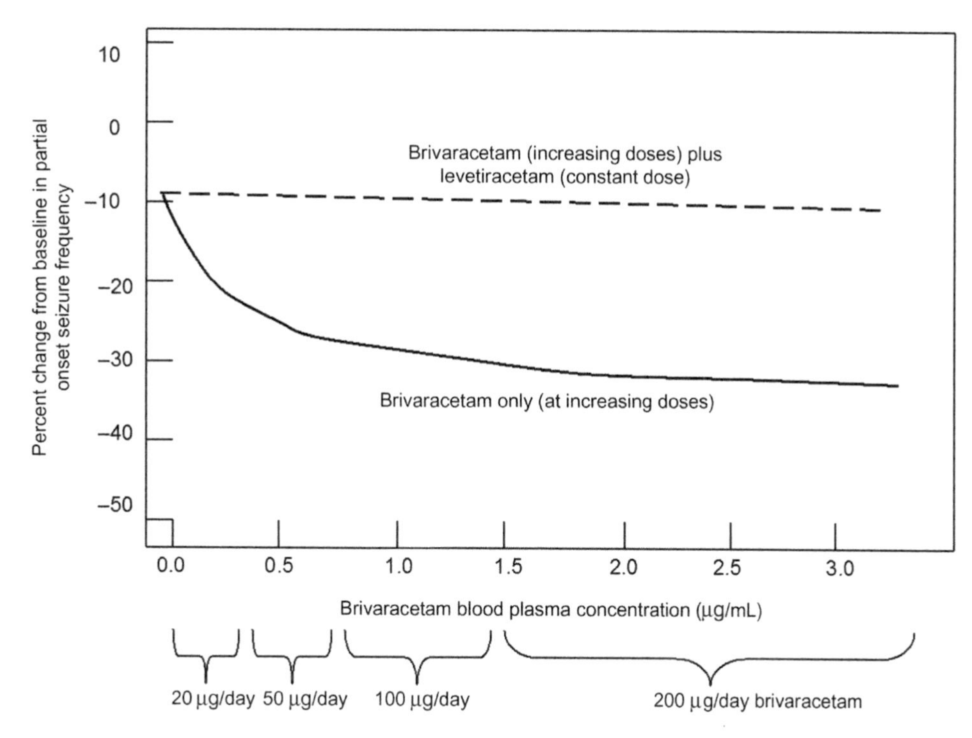

FIGURE 7.7 Reduction in seizure frequency versus brivaracetam plasma concentration (+/− coadministered levetiracetam). The data points are the result of combining results for spread of blood plasma brivaracetam, where brivaracetam doses were 20, 50, 100, or 200 mg/day. Roughly, the greater the dose, the lesser the seizure frequency (lower curve). With coadministered levetiracetam (upper curve), increasing brivaracetam dose did not result in improved seizure control.

Package label. The Drug Interactions section of the label warned that brivaracetam (Briviact®) has no benefit, where the patient is already being given levetiracetam:

> **DRUG INTERACTIONS**. . .Levetiracetam: BRIVIACT had no added therapeutic benefit when coadministered with levetiracetam.[188]

f. Cabozantinib (Thyroid Cancer) NDA 203-756

Cabozantinib is a small molecule for treating thyroid cancer. The drug inhibits several different tyrosine kinase enzymes.[189] The targeted tyrosine kinases include VEGFR-2, MET, and RET.[190] These abbreviations refer to different enzymes or genes. VEGFR-2

[188] Package label. BRIVIACT® (brivaracetam), for oral use, CV; June 2016 (15 pp.).

[189] Cabozanitinib (thyroid cancer) NDA 203-756. 106-page Clinical Pharmacology Review.

[190] Weitzman SP, Cabanillas ME. The treatment landscape in thyroid cancer: a focus on cabozantinib. Cancer Manag. Res. 2015;7:265–78.

means vascular endothelial growth factor receptor-2; MET means mesenchymal-epithelial transition; and RET means rearranged during transfections. Cabozantinib has the following structure:

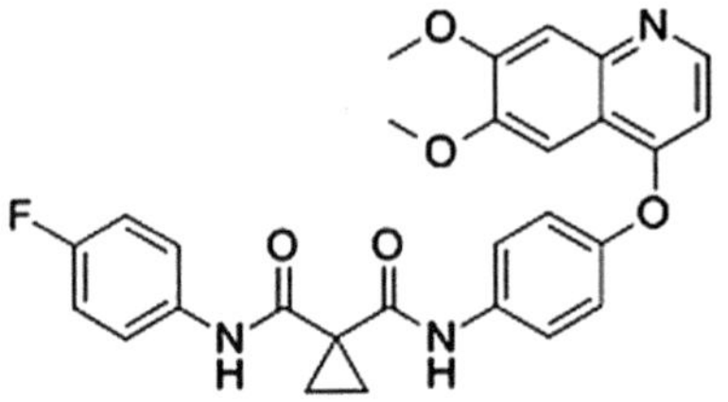

The following describes in vitro DDI studies occurring at the location of various CYP enzymes and at the location of a transporter (P-gp).

i. Influence of Cabozantinib on Exposure of a Second Drug (Exposure Predicted by Assays of CYP Inhibition or Induction)

The FDA reviewer assessed cabozantinib's ability to inhibit or to induce various CYP enzymes. FDA observed that, "Cabozantinib **inhibited** recombinant CYP2C8 and CYP2C19 isozymes with IC50 values of 5.0 micromolar and 8.3 micromolar, respectively… [u]sing human liver microsomal preparations, cabozantinib also inhibited…activities for CYP2C8 and CYP2C19, as well as CYP2C9, with IC50 values of 6.4, 6.2, and 6.1 micromolar, respectively[191]…[i]n vitro study data indicate that cabozantinib is an **inducer** of CYP1A1, but is not a potent inducer of CYP1A2, CYP2B6, CPY2C8, CYP2C9, CYP2C19, or CYP3A4…IC50 values >20 micromolar were observed for CYP1A2, CYP2D6, and CYP3A4 isozymes in both recombinant and human liver microsome assay systems."[192]

ii. Methodology for Induction Assays

FDA's Pharmacology Review provides a detail on methodology for measuring CYP enzyme induction. The detail was that induction can be measured by assays that detect messenger RNA or that detect protein:

> CYP induction following exposure to cabozantinib was studied in hepatic fractions of…rats administered doses of 1, 3, 10, 30, or 100 mg/kg/day for up to 8 days. CYP induction was measured by both total protein and mRNA content. CYP1A1 protein content increased by 1.5-fold in rats administered 100mg/kg, while CYP3A levels increased by 1.5 to 1.7-fold at doses ≥ 3 mg/kg. CYP2B was generally consistent, and unaffected. mRNA analysis indicated increased levels for CYP1A1 at 30 and 100 mg/kg, although induction was variable. CYP3A2 mRNA levels increased 5-fold at 100 mg/kg.[193]

[191] Cabozantinib (thyroid cancer) NDA 203-756. Pages 25–26 of 106-page Clinical Pharmacology Review.

[192] Cabozantinib (thyroid cancer) NDA 203-756. Pages 25–26 of 106-page Clinical Pharmacology Review.

[193] Cabozantinib (thyroid cancer) NDA 203-756. Page 46 of 122-page Pharmacology Review.

Zhang et al.[194] and Gerbal-Chaloin et al.[195] provide details on using antibodies and immunoblot assays for direct measuring of CYP1A1, CYP2B, CYP2C, and CYP3A proteins. Also, these references describe assays of messenger RNA encoding these CYP proteins.

iii. Influence of Cabozantinib on Exposure of a Second Drug (Exposure Predicted by P-Glycoprotein Efflux Assays)

Cultured cells (MDCK cells; Caco-2 cells) were used to assess DDIs occurring at the point of the P-gp transporter. The FDA reviewer observed that, "Cabozantinib was an inhibitor (IC50 = 7.0 micromolar), but not a substrate, of P-gp transport activities in a...system using...MDCK cells...[i]n a separate study, cabozantinib was observed to be a more potent P-gp inhibitor (IC50 = 0.5 micromolar) in a Caco-2 cell monolayer system."[196,197]

FDA's Medical Review commented on the clinical implications of the Caco-2 cell studies. The implication is that physicians should expect to find adverse effects caused by a coadministered drug, in the situation where the coadministered drug is a substrate for P-gp. FDA commented that, "P-glycoprotein Inhibition: Cabozantinib is an inhibitor (IC50 = 7.0 μM), but not a substrate, of P-gp transport activities in a bi-directional assay system using MDCK-MDR1 cells. Therefore, cabozantinib may have the potential to increase plasma concentrations of co-administered substrates of P-gp."[198]

Showing increased concern for P-gp mediated increases in exposure of coadministered drugs, FDA imposed a Postmarketing Requirement (PMR and required that the Sponsor conduct a clinical study using a probe substrate that assessed P-gp mediated transport. FDA's requirement was that the Sponsor, "Conduct a pharmacokinetic drug interaction trial in subjects administered an oral P-glycoprotein probe substrate with and without cabozantinib in accordance with the FDA draft Guidance for Industry: Drug Interaction Studies – Study Design, Data Analysis, and Implications for Dosing, and Labeling Recommendations."[199]

iv. FDA's Recommendations for the Package Label

FDA's ClinPharm Review provided a recommendation for the package label, which took the form of a verbatim account of the needed wording. One take-home lesson, regarding DDIs, is that where it is absolutely essential that the patient take a second drug (where the second drug can increase exposure of the first drug), then the label should

[194] Zhang QY, et al. Characterization of mouse small intestinal cytochrome P450 expression. Drug Metab. Dispos. 2003;31:1346–51.

[195] Gerbal-Chaloin S, et al. Induction of CYP2C genes in human hepatocytes in primary culture. Drug Metab. Dispos. 2001;29:242–51.

[196] Cabozantinib (thyroid cancer) NDA 203-756. Pages 25–26 of 106-page Clinical Pharmacology Review.

[197] Cabozantinib (thyroid cancer) NDA 203-756. Page 48 of 122-page Pharmacology Review.

[198] Cabozantinib (thyroid cancer) NDA 203-756. Page 113 of 120-page Medical Review.

[199] Cabozantinib (thyroid cancer) NDA 203-756. Page 4 of 106-page Clinical Pharmacology Review.

require reducing the dose of the first drug. FDA recommended that the package label state:

> "**CYP3A4 Inhibitors.** Avoid the use of concomitant strong CYP3A4 inhibitors (e.g., ketoconazole, itraconazole, clarithromycin, atazanavir, nefazodone, saquinavir, telithromycin, ritonavir, indinavir, nelfinavir, voriconazole) in patients receiving COMETRIQ...[f]or patients who require treatment with a strong CYP3A4 inhibitor: Reduce COMETRIQ dose by approximately 40%."[200]

Although FDA recommended, "Reduce COMETRIQ dose by approximately 40%," the package label was not so exact and did not mention anything about forty percent, and instead the label recommended, "Strong CYP3A4 inhibitors: Reduce the COMETRIQ dosage."[201]

FDA's Medical Review commented on the issue of greatest concern, that is, AEs resulting from excessive exposure to cabozantinib. FDA's comments were in the context of an account of AEs requiring dose reductions in clinical studies. FDA remarked, "The main adverse events leading to dose reduction were, in decreasing order, palmar-plantar erythrodysesthesia syndrome, weight decrease, decreased appetite, fatigue, diarrhea, stomatitis, asthenia and nausea."[202]

On the other side of the coin, FDA made the same sort of recommendations where patients needed to take a coadministered drug that induces CYP enzymes. Here, FDA recommended that the label requires increasing the cabozantinib dose, "**Strong CYP3A4 Inducers.** Avoid the use of concomitant strong CYP3A4 inducers (e.g., phenytoin, carbamazepine, rifampin, rifabutin, rifapentine, phenobarbital) if alternative therapy is available...[f]or patients who require treatment with a strong CYP3A4 inducer: Increase the dose of COMETRIQ in increments of 40 mg by only two weeks as tolerated."[203]

v. Clinical Studies on Ketoconazole's Influence on Exposure of Cabozantinib; and Rifampicin's Influence on Exposure of Cabozantinib

FDA commented that, "Cabozantinib is a CYP3A4 substrate. Administration of a strong CYP3A4 inhibitor, ketoconazole (400 mg daily for 27 days) to healthy subjects increased...plasma cabozantinib exposure (AUC_{0-inf}) by 38%. Administration of a strong CYP3A4 inducer, rifampin (600 mg daily for 31 days) to healthy subjects decreased...plasma cabozantinib exposure (AUC_{0-inf}) by 77%."[204] FDA's comments on CYP3A4 inhibitors and on CYP3A4 inducers found a place on the package label, as illustrated below.

vi. Clinical Studies on Cabozantinib on Exposure of a Second Drug (Rosiglitazone)

FDA's ClinPharm Review observed that, "Cabozantinib at steady-state plasma concentrations (≥100 mg/day daily for a minimum of 21 days) **has no effect on single-dose**

[200] Cabozantinib (thyroid cancer) NDA 203-756. Pages 34–35 of 106-page Clinical Pharmacology Review.

[201] Package label. COMETRIQ® (cabozantinib) capsules, for oral use; May 2016 (22 pp.).

[202] Cabozantinib (thyroid cancer) NDA 203-756. Pages 87–88 and 90 of 120-page Medical Review.

[203] Cabozantinib (thyroid cancer) NDA 203-756. Pages 34–35 of 106-page Clinical Pharmacology Review.

[204] Cabozanitinib (thyroid cancer) NDA 203-756. Pages 3, 7, and 27 of 106-page Clinical Pharmacology Review.

plasma exposure (Cmax and AUC) of rosiglitazone (a CYP2C8 substrate) in patients with solid tumors."[205]

The FDA reviewer concluded that the study drug (cabozantinib) is not an inducer of CYP2C8. A view of the Drug Interactions section of the package label reveals that it states this fact. What the label reads is, "Cabozantinib is an inducer of CYP1A1...but not of...CYP2C8."

Package label. Consistent with FDA's recommendations for cabozantinib (Cometriq®) the Drug Interactions section of the label reads:

> **DRUG INTERACTIONS.** Strong CYP3A4 inhibitors: Reduce the COMETRIQ dosage. Strong CYP3A4 inducers: Increase the COMETRIQ dosage.[206]

The Full Prescribing Information section of the package label provided these details:

> **DRUG INTERACTIONS. Effect of CYP3A4 Inhibitors.** Administration of a strong CYP3A4 inhibitor, ketoconazole to healthy subjects increased single dose plasma cabozantinib exposure by 38%. Avoid taking a strong CYP3A4 inhibitor (e.g., ketoconazole, itraconazole, clarithromycin, atazanavir, indinavir, nefazodone, nelfinavir, ritonavir, saquinavir, telithromycin, voriconazole) while taking COMETRIQ or reduce the dosage of COMETRIQ if concomitant use with strong CYP3A4 inhibitors cannot be avoided...[a]void ingestion of foods (e.g., grapefruit, grapefruit juice) or nutritional supplements that are known to inhibit cytochrome P450 while taking COMETRIQ.
>
> **Effect of CYP3A4 Inducers.** Administration of a strong CYP3A4 inducer, rifampin to healthy subjects decreased single-dose plasma cabozantinib exposure by 77%. Avoid chronic co-administration of strong CYP3A4 inducers (e.g., phenytoin, carbamazepine, rifampin, rifabutin, rifapentine, phenobarbital, St. John's Wort) with COMETRIQ or increase the dosage of COMETRIQ if concomitant use with strong CYP3A4 inducers cannot be avoided.
>
> **Effect of MRP2 Inhibitors.** Concomitant administration of MRP2 inhibitors may increase the exposure to cabozantinib. Monitor patients for increased toxicity when MRP2 inhibitors (e.g., abacavir, adefovir, cidofovir, furosemide, lamivudine, nevirapine, ritonavir, probenecid, saquinavir, and tenofovir) are coadministered with COMETRIQ.[207]

At the time of FDA approval, none of FDA's reviews disclosed any study with MRP2 inhibitors. MRP2 refers to the drug transporter, multidrug resistance protein-2. MRP2 is one of the ABC transporters. The package label (November 2012) published with FDA's Approval Letter did not mention MRP2, but the supplementary package label (May 2016) did mention MRP2, as quoted above. The reason for this was that following FDA approval, the Sponsor had conducted and published a DDI study assessing interactions at MRP2.[208] This was an in vitro study, where transport was measured across membrane vesicles, instead of the more usual technique of using cell monolayers.

[205] Cabozanitinib (thyroid cancer) NDA 203-756. Pages 3, 7, and 27 of 106-page Clinical Pharmacology Review.

[206] Package label. COMETRIQ® (cabozantinib) capsules, for oral use; May 2016 (22 pp.).

[207] Package label. COMETRIQ® (cabozantinib) capsules, for oral use; May 2016 (22 pp.).

[208] Lacy S, et al. Metabolism and disposition of cabozantinib in healthy male volunteers and pharmacologic characterization of its major metabolites.Drug Metab. Dispos. 2015;43:1190–207.

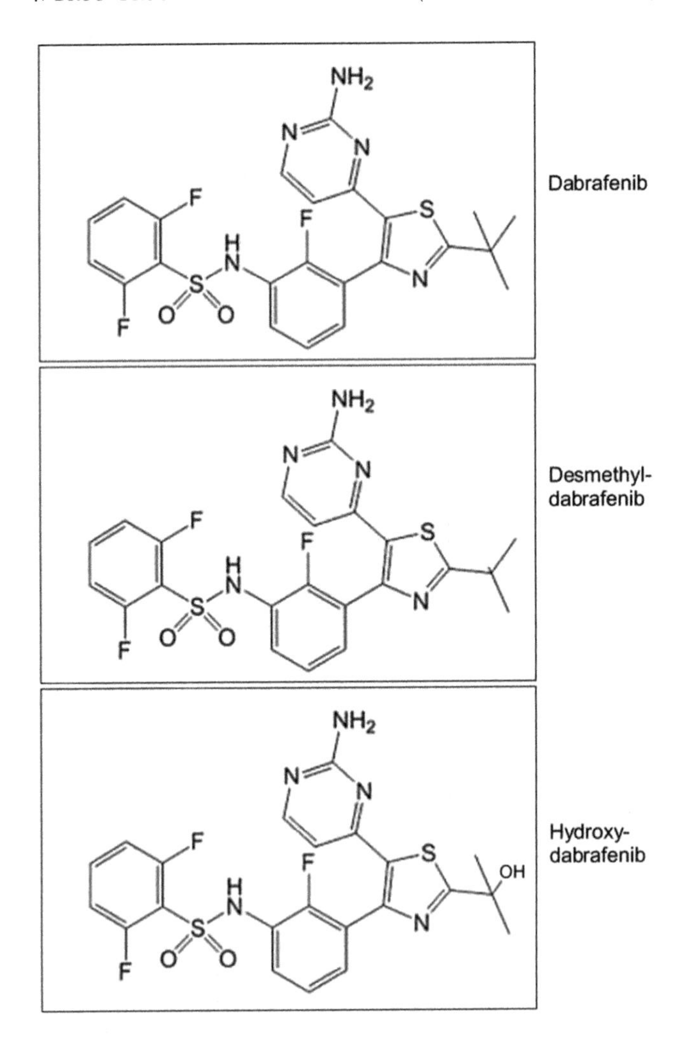

FIGURE 7.8 Structures of dabrafenib and its metabolites. Dabrafenib, desmethyl-dabrafenib, and hydroxy-dabrafenib are shown at the top, middle, and bottom, respectively.

g. Dabrafenib (Melanoma with BRAF V600E Mutation) NDA 202-806

Dabrafenib is a small molecule drug for treating melanoma. FDA's ClinPharm Review and Pharmacology Review focused on the metabolites, hydroxy-dabrafenib (M7), desmethyl-dabrafenib (M8), and carboxy-dabrafenib (M4). "M" symbols are standard in all FDA-submissions for identifying drug metabolites. Fig. 7.8 shows the structures of dabrafenib and its metabolites.

i. Exposure of Dabrafenib Metabolites

Regarding the metabolites, hydroxy-dabrafenib, carboxy-dabrafenib, and desmethyl-dabrafenib, FDA observed that, "Following administration of the recommended twice daily 150 mg dose of dabrafenib, these three metabolites were present at human plasma levels (AUC_{0-24}) of approximately 8, 100, and 6, respectively.[209] The unit of exposure for these AUC measurements was [(μg) (h)/(mL)].

ii. Efficacy of Dabrafenib Metabolites

After FDA's ClinPharm Review disclosed the concentrations of these metabolites, FDA further developed its argument, and commented on the CYP enzymes responsible for generating these metabolites, and speculated that the metabolites "are likely to contribute to the clinical activity of dabrafenib":

> The metabolism of dabrafenib is primarily mediated by CYP2C8 and CYP3A4 to form **hydroxy-dabrafenib**, which is further oxidized via CYP3A4 to form carboxy-dabrafenib and is excreted in bile and urine. Carboxy-dabrafenib can be decarboxylated via a non-enzymatic process in the gut to form **desmethyl-dabrafenib** and reabsorbed. **Desmethyl-dabrafenib** is metabolized by CYP3A4 to oxidative metabolites...[m]ean **metabolite to parent AUC ratios** following repeat-dose administration were 0.9, 11, and 0.7 for hydroxy-, carboxy-, and desmethyl-dabrafenib, respectively. Based on exposure, relative potency and pharmacokinetic properties, both **hydroxy- and desmethyl-dabrafenib are likely to contribute to the clinical activity of dabrafenib.**[210]

FDA's Pharmacology Review provided enzyme inhibition data that were used to support the speculative comment, "both hydroxy- and desmethyl-dabrafenib are likely to contribute to the clinical activity." Regarding inhibition and clinical activity, FDA stated that, "major human metabolites were identified during the development of dabrafenib...and were shown to have inhibitory activity against BRAF and the BRAFV600E, V600K, and V600D mutations. The desmethyl-dabrafenib metabolite had activity similar to that of the parent compound followed by...hydroxy-dabrafenib."[211]

Further detailing the enzyme inhibition results, FDA stated that, "Specifically, while desmethyl-dabrafenib (M8) showed similar activity than parent dabrafenib...1.5X less active than parent against...BRAF...hydroxy-dabrafenib (M7) and carboxy-dabrafenib (M4) demonstrated...20-50X lower activity than parent against...BRAF wild type enzyme."[212]

Further addressing the potential efficacy of the hydroxy-dabrafenib (M7) and desmethyl-dabrafenib (M8) metabolites, the Sponsor provided cell culture data, where cultured human melanoma cells expressing the BRAF V600E kinase were treated with dabrafenib, M7, M8, or carboxy-dabrafenib (M4). Effectiveness of the metabolites M7 and M8 in inhibiting the melanoma cells was in the same ballpark as dabrafenib, while carboxy-dabrafenib (M4) was not effective. Table 7.3 shows that the IC50 value for

[209] Dabrafenib (melanoma with BRAF V600E mutation) NDA 202-806. Page 10 of 135-page Pharmacology Review.

[210] Dabrafenib (melanoma with BRAF V600E mutation) NDA 202-806. Pages 19, 20, and 25 of 83-page Clinical Pharmacology Review.

[211] Dabrafenib (melanoma with BRAF V600E mutation) NDA 202-806. Page 4 of 135-page Pharmacology Review.

[212] Dabrafenib (melanoma with BRAF V600E mutation) NDA 202-806. Page 28 of 135-page Pharmacology Review.

TABLE 7.3 Ability of Dabrafenib and its Metabolites to Inhibit Proliferation of Cultured Melanoma Cells

	Dabrafenib	Hydroxy-dabrafenib (M7)	Desmethyl-dabrafenib (M8)	Carboxy-dabrafenib (M4)
Inhibition constant (IC50, nm)	6	17.7	9	223

IC50 is concentration of inhibitor, as present in the cell culture medium, required to inhibit proliferation by 50%.
Dabrafenib (melanoma with BRAF V600E mutation) NDA 202-806. Pages 29–30 of 135-page Pharmacology Review.

carboxy-dabrafenib (M4) is about 40-fold greater than that for the parent compound, and thus can be considered to be ineffective.

Note that, generally speaking, low values for IC50 mean that the inhibition is strong (here, low concentrations of inhibitors are sufficient to give 50% inhibition), while high values for IC50 mean that the inhibition is weak (here, high concentrations of inhibitors are needed to give 50% inhibition).

iii. Drug–Drug Interactions for Dabrafenib and Its Metabolites

The first step for most DDI studies is to determine if the study drug inhibits a variety of CYP enzymes. FDA commented on the Sponsor's assays using microsomes as the source of CYP enzymes, "In vitro studies in...human liver microsomes...indicated that dabrafenib inhibited CYPs 2C8, 2C9, 2C19, and 3A4, hydroxy-dabrafenib and desmethyl-dabrafenib inhibited CYP2C9.[213]"

FDA's ClinPharm Review provided a table of information on inhibition constants, where human liver microsomes were the source of CYP enzymes (Table 7.4):

Lawrence et al.[214] reiterated the Sponsor's finding that hydroxy-dabrafenib inhibits CYP1A2, 2C9, and 3A4, and that desmethyl-dabrafenib inhibits CYP2B6, 2C8, 2C9, 2C19, and 3A4, but concluded that there would be "no anticipated change of exposure" of various coadministered drugs, due to inhibition of CYP by dabrafenib, hydroxy-dabrafenib, and desmethyl-dabrafenib. This conclusion was based on the finding that, when the drug and drug metabolites are in the bloodstream they are mostly bound to plasma proteins. Thus the ability to inhibit CYP enzymes in human subjects was less than predicted from in vitro assays (in the situation where the in vitro assays do not include any added plasma proteins).

iv. FDA's Recommendations

FDA's Cross Discipline Team Leader Review provided a statement that likely was the basis for DDI information on the package label. FDA's comments reiterated the fact that

[213] Dabrafenib (melanoma with BRAF V600E mutation) NDA 202-806. Page 27 of 83-page Clinical Pharmacology Review.

[214] Lawrence SK, et al. The metabolic drug–drug interaction profile of dabrafenib: in vitro investigations and quantitative extrapolation of the P450-mediated DDI risk. Drug Metab. Dispos. 2014;42:1180–90.

TABLE 7.4 Inhibition Constants (IC50)

	Dabrafenib	Hydroxy-dabrafenib	Desmethyl-dabrafenib
CYP1A2	87	83	No inhibition
CYP2B6	No inhibition	No inhibition	78
CYP2C8	8.2	No inhibition	49.3
CYP2C9	7.2	28.6	6.3
CYP2C19	22	No inhibition	35.9
CYP3A4A	16	No inhibition	19.6

This shows IC50 values for the indicated cytochrome P450, where dabrafenib and its metabolites were tested for inhibitory properties. Activity of each CYP isozyme, and the degree of inhibition thereof, was determined with standard probe substrates specific for each CYP isozyme.
Dabrafenib (melanoma with BRAF V600E mutation) NDA 202-806. Page 27 of 83-page Clinical Pharmacology Review. FDA's review failed to disclose the unit for IC50.

dabrafenib's metabolites are active, thus implying that CYP enzyme inducers could destroy not only dabrafenib, but also destroy dabrafenib's active metabolites:

> Dabrafenib induces cytochrome P450 isoenzyme (CYP) 3A4-mediated metabolism and may induce other enzymes including CYP2B6, CYP2C8, CYP2C9, and CYP2C19. **Dabrafenib and its active metabolites** are primarily metabolized by CYP2C8 and CYP3A4. **Strong inhibitors or inducers of CYP3A4** or CYP2C8 **may increase or decrease systemic exposure to dabrafenib**, respectively. The effects of strong inhibitors or inducers of CYP3A4 or CYP2C8 on pharmacokinetics of dabrafenib in vivo will be studied under postmarketing requirements (PMR).[215]

Regarding drugs that increase stomach pH, FDA's reviews did not disclose any laboratory data. But FDA did impose a PMR that the Sponsor conduct a study on proton pump inhibitors and antacids. Corresponding information found a place in the package label. FDA's requirement took the form of this PMR. FDA's PMR requested that the Sponsor, "Conduct a clinical trial to evaluate if proton pump inhibitors, H_2 antagonists and antacids alter the bioavailability of dabrafenib. You may study the worst case scenario first, and then determine if further studies of other drugs are necessary. The study results should allow for a determination on how to dose dabrafenib with regard to concomitant gastric pH elevating agents."[216]

Package label. FDA review of dabrafenib (Tafinlar®) found a place in the Drug Interactions section of the package label:

> **DRUG INTERACTIONS. Concurrent administration of strong inhibitors of CYP3A4** or CYP2C8 is not recommended...[c]oncurrent administration of strong inducers of CYP3A4 or CYP2C8 is not recommended...**[d]rugs that increase gastric pH may decrease dabrafenib concentrations**...[c]oncomitant

[215] Dabrafenib (melanoma with BRAF V600E mutation) NDA 202-806. Page 17 of 39-page Cross Discipline Team Leader Review.

[216] Dabrafenib (melanoma with BRAF V600E mutation) NDA 202-806. Page 17 of 39-page Cross Discipline Team Leader Review.

use with agents that are sensitive substrates of CYP3A4, CYP2C8, CYP2C9, CYP2C19, or CYP2B6 may result in loss of efficacy of these agents.[217]

The package label admitted that no study on DDIs had been conducted with proton pump inhibitors or antacids:

> **DRUG INTERACTIONS**...However, no formal clinical trial has been conducted to evaluate the effect of gastric pH-altering agents on the systemic exposure of dabrafenib. When TAFINLAR is coadministered with a proton pump inhibitor, H_2-receptor antagonist, or antacid, systemic exposure of dabrafenib may be decreased and the effect on efficacy of TAFINLAR is unknown.[218]

h. Fingolimod (Multiple Sclerosis) NDA 022-527

Fingolimod is for treating multiple sclerosis. Fingolimod's structure is shown below. Metabolites of fingolimod include M27, M28, M29, and M30, each of which takes the form of fingolimod with a fatty acid group attached to fingolimod's amino group.[219]

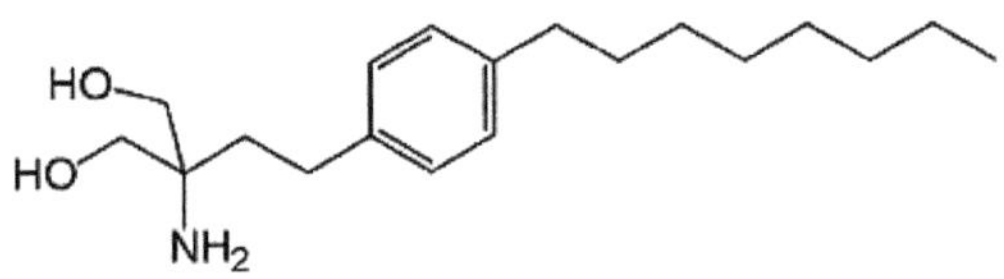

i. Influence of Fingolimod on Catabolism a Second Drug (Coumarin; Diclofenac) (CYP Enzyme Assays)

FDA used the inhibition constant (Ki) values obtained with assays using microsomes as the source of CYP enzymes and the blood plasma value (0.0325 μM) attained in human subjects, plugged these values into the [I]/Ki ratio, and used the calculated ratio for deciding if DDIs in human patients was likely.

Table 7.5 reveals the best probe substrate for measuring activity of the indicated CYP enzyme. Human microsomes was the source of enzymes. The microsomal preparation contains all of the indicated CYP enzymes, and hence there was a need to use probe substrates to ensure that the readout corresponded to only one of the CYP enzymes. Each of the assay mixtures contained only one of the probe substrates. Table 7.5 also provides the Sponsor's laboratory results, where the table discloses the inhibition constants (IC50). Each inhibition constant measures the ability of fingolimod to inhibit the indicated CYP enzyme.

FDA's attention then turned to the [I]/Ki ratio formula. The fingolimod dose proposed by the Sponsor results in human plasma levels of 0.0325 μM (this is the value for [I]). FDA stated how to interpret the [I]/Ki ratio algorithm, writing, "An estimated **[I]/Ki ratio of greater than 0.1** is considered positive and a follow-up in vivo evaluation is recommended."[220] This means that, if the ratio is greater than 0.1, then the Sponsor needs to conduct a drug–drug inhibition study with human subjects.

[217] Package label. TAFINLAR (dabrafenib) capsules for oral use; May 2013 (19 pp.).

[218] Package label. TAFINLAR (dabrafenib) capsules for oral use; May 2013 (19 pp.).

[219] Fingolimod (multiple sclerosis) NDA 022-527. Page 40 of 371-page Clinical Pharmacology Review.

[220] Fingolimod (multiple sclerosis) NDA 022-527. Page 87 of 371-page Clinical Pharmacology Review.

TABLE 7.5 Probe Substrates Specific for Assessing Activity of a Given CYP Enzyme

CYP Enzyme	Probe Substrate	Inhibition Constant (Ki)
CYP1A2	7-Ethoxyresorufin *O*-dealkylation	CYP1A2 (Ki = 89 μM)
CYP2A6	Coumarin 7-hydroxylation	CYP2A6 (Ki = 62 μM)
CYP2B6	7-Ethoxy-4-trifluoromethylcoumarin	CYP2B6 (Ki = 37 μM)
CYP2C9	Diclofenac 4′-hydroxylation	CYP2C9 (Ki = 55 μM)
CYP2C19	*S*-Mephenytoin 4′-hydroxylation	(No inhibition detected)
CYP2D6	Dextromethorphan *O*-demethylation	CYP2D6 (Ki = 53 μM)
CYP2E1	Chlorzoxazone 6-hydroxylation	CYP2E1 (Ki = 90 μM)
CYP3A4/5	Testosterone 6ß-hydroxylation	CYP3A4/5 (Ki = 9 μM)
CYP4A9/11	Lauric acid 12-hydroxylation	CYP4A9/11 (Ki = 170 μM)

Fingolimod (multiple sclerosis) NDA 022-527. Pages 85–87 of 371-page Clinical Pharmacology Review.

FDA plugged the [I] value and the Ki values from each of the different probe substrates into the algorithm, and where the result was that all of the calculated ratios were far below 0.1. Because the calculated ratios were far below 0.1, FDA concluded that, "A drug interaction is remote."[221]

Regarding CYP enzyme induction, the Sponsor used human hepatocytes for testing induction. Fingolimod was added to cultures of human hepatocytes, followed by a 72 hour incubation, and PCR assays to assess induction of mRNA encoding CYP3A. *Rifampicin* was used as a positive control for inducing CYP3A. The results demonstrated that *fingolimod did not induce CYP3A.*[222]

ii. Influence of a Second Drug (Ketoconazole) on Catabolism of Fingolimod (CYP Enzyme Assays)

The Sponsor conducted in vitro studies with human liver microsomes, with and without ketoconazole, an established inhibitor of CYP4F2 and CYP4F12. The results demonstrated that ketoconazole inhibits catabolism of fingolimod, where 50% inhibition of degradation occurred at 1.0 μM ketoconazole. CYP4F2-mediated catabolism of fingolimod is via hydroxylation at the methyl terminal of the octyl chain.[223] The octyl chain of fingolimod can be seen in its structure.

David et al.[224] provide an insightful remark that should be evaluated for possible relevance and applicability, as soon as any Sponsor identifies which CYP enzymes, for any

[221] Fingolimod (multiple sclerosis) NDA 022-527. Page 87 of 371-page Clinical Pharmacology Review.

[222] Fingolimod (multiple sclerosis) NDA 022-527. Page 102 of 371-page Clinical Pharmacology Review.

[223] Tanasescu R, Constantinescu CS. Pharmacokinetic evaluation of fingolimod for the treatment of multiple sclerosis. Expert Opin. Drug Metab. Toxicol. 2014;10:621–30.

[224] David OJ, et al. Clinical pharmacokinetics of fingolimod. Clin. Pharmacokinet. 2012;51:15–28.

drug, metabolize the study drug. The insightful remark was, "Fingolimod is largely cleared through metabolism by cytochrome P450 CYP4F2. Since few drugs are metabolized by CYP4F2, fingolimod would be expected to have a relatively low potential for drug–drug interactions."

In other words, if a Sponsor discovers that its study drug inhibits cytochrome P450 enzymes X, Y, and Z, then the Sponsor should determine if any commonly used drugs are metabolized by cytochrome P450 X, cytochrome P450 Y, and cytochrome P450 Z. In the situation where there are not any drugs in common use that are metabolized by cytochrome P450 enzymes X, Y, and Z, the Sponsor is justified in concluding that the study drug does not pose significant risk for DDIs. The terms X, Y, and Z are not real names for any cytochrome P450 enzymes, and they are used here to represent a hypothetical situation.

FDA's ClinPharm Review contained similar information, revealing that CYP4F2 was responsible for most of fingolimod's catabolism, "Fingolimod is metabolized...by recombinant human CYP4F2 (>80%), CYP2D6 (<10%), CYP2E1 (<10%), CYP2E1 (<10%), CYP3A4, CYP4F3B, and CYP4F12 (smaller contributions)...[k]etoconazole inhibited the biotransformation of fingolimod by recombinant CYP4F2 and CYP4F12 with IC50 of 1.6 μM and 0.6 μM, respectively."[225]

Regarding ketoconazole, Jin et al.[226] reported that ketoconazole inhibits CYP4F2 as well as CYP4F12, and that ketoconazole readily inhibits fingolimod catabolism by recombinant CYP4F2, and also readily inhibits fingolimod catabolism by recombinant CYP4F12. Please note that ketoconazole is also an inhibitor of CYP3A4.

Package label. Taken together, the above collection of facts explains why the package label refrained from warning about coadministering with any CYP4F2 inhibitors, aside from the CYP4F2 inhibitor, ketoconazole. The drug–drug information on the package label for fingolimod (Gilenya®), relating to CYP enzymes, was limited to:

> **DRUG INTERACTIONS.** Patients who use GILENYA and systemic ketoconazole concomitantly should be closely monitored, as the risk of adverse reactions is greater.[227]

> **CLINICAL PHARMACOLOGY**...The biotransformation of fingolimod in humans occurs...by oxidative biotransformation catalyzed mainly by the cytochrome P450 4F2 (CYP4F2)...[i]nhibitors or inducers of CYP4F2 and possibly other CYP4F isozymes might alter the exposure of fingolimod.[228]

iii. Influence of a Second Drug (Ketoconazole) on Exposure of Fingolimod (Clinical Study)

The Sponsor also conducted a clinical study of ketoconazole's influence on fingolimod exposure. FDA's ClinPharm Review reported the results as, "Co-administration of a single 5 mg dose of fingolimod with steady state ketoconazole...increased both fingolimod Cmax by 1.2-fold and AUC by 1.7-fold...[d]ose adjustment for fingolimod is recommended.

[225] Fingolimod (multiple sclerosis) NDA 022-527. Page 54 of 371-page Clinical Pharmacology Review.

[226] Jin Y, et al. CYP4F enzymes are responsible for the elimination of fingolimod (FTY720), a novel treatment of relapsing multiple sclerosis. Drug Metab. Dispos. 2011;39:191–8.

[227] Package label. GILENYA (fingolimod) capsules, for oral use; February 2016 (25 pp.).

[228] Package label. GILENYA (fingolimod) capsules, for oral use; February 2016 (25 pp.).

The reviewer recommends decrease the dose of fingolimod by 50% when it is co-administered with ketoconazole."[229]

Note that, although FDA recommended that the package label instructs the physician to reduce the fingolimod dose by 50% when coadministered with ketoconazole, the Drug Interaction section only recommended monitoring (and not any dose reduction):

> **DRUG INTERACTIONS**. Systemic ketoconazole: Monitor during concomitant use.[230]

iv. Influence of the Fingolimod on Efficacy of a Second Drug (Vaccine) (Assays that Measure the Vaccine's Generation of Antibodies)

The DDI in question was where fingolimod inhibits immune response, and where this inhibition prevents vaccines from working. The Sponsor's experiments involved a model vaccine. It was not a vaccine that is ever used in medical practice. The vaccine, which was used in human subjects, was with an antigen from molluscs (keyhole limpet hemocyanin, KLH). The vaccination response test measured the subject's generation of IgM antibodies and IgG antibodies.

Regarding the influence of fingolimod on a neoantigen's ability to stimulate antibody formation, FDA's Medical Review stated, "Immunologic evaluation in response to neoantigen, recall antigen and cellular immunity in a clinical...study...indicated a dose-related decrease in immune responses for fingolimod...IgM antibody response to KLH neoantigen was 0% for fingolimod 1.25 mg, 23% for fingolimod 0.5 mg and >90% for placebo. IgG antibody response was 57% for fingolimod 1.25 mg and >90% for fingolimod 0.5 mg and placebo...[t]hese results suggest that the **immunologic response to vaccination may be decreased during treatment with fingolimod**."[231]

FDA went a step further and recommended that fingolimod's impairment on vaccines should be on the package label, "However, vaccination prior to initiation of long-term fingolimod therapy should be considered, as well as wording to the effect that immunosuppressants may affect vaccination, and therefore, vaccination may be less effective during treatment with fingolimod."[232]

Package label. The Drug Interaction section warned about ketoconazole, and instructed physicians to monitor patients and to avoid vaccines:

> **DRUG INTERACTIONS. Systemic ketoconazole:** Monitor during concomitant use...Vaccines: Avoid live attenuated vaccines during, and for 2 months after stopping GILENYA treatment.[233]

[229] Fingolimod (multiple sclerosis) NDA 022-527. Page 11 of 371-page Clinical Pharmacology Review.

[230] Package label. GILENYA (fingolimod) capsules, for oral use; February 2016 (25 pp.).

[231] Fingolimod (multiple sclerosis) NDA 022-527. Pages 143, 170, 269, 348, and 351 of 519 page Medical Review.

[232] Fingolimod (multiple sclerosis) NDA 022-527. Pages 270 and 480 of 519 page Medical Review.

[233] Package label. GILENYA (fingolimod) capsules, for oral use; February 2016 (25 pp.).

The Clinical Pharmacology section warned about coadministered ketoconazole and recommended monitoring patients:

> **CLINICAL PHARMACOLOGY**...**Ketoconazole**. The co-administration of ketoconazole (a potent inhibitor of CYP3A and CYP4F) 200 mg twice-daily at steady-state and a single dose of fingolimod 5 mg led to a 70% increase in AUC of fingolimod and fingolimod-phosphate. Patients who use GILENYA and systemic ketoconazole concomitantly should be closely monitored, as the risk of adverse reactions is greater.[234]

i. Lumacaftor/Ivacaftor Combination (Cystic Fibrosis for Patients Age 12 Years or Older With F508del Mutation in CFTR Gene) NDA 206-038

The lumacaftor (LUM)/ivacaftor (IVA) combination (Orkambi®) is for cystic fibrosis, a disease of the lungs.[235] The structures of lumacaftor and ivacaftor are shown below. The main metabolite of lumacaftor is designated M28-LUM, while the main metabolites of ivacaftor are M1-IVA and M6-IVA. Lumacaftor is a strong inducer of CYP3A, and ivacaftor is a substrate of CYP3A.[236]

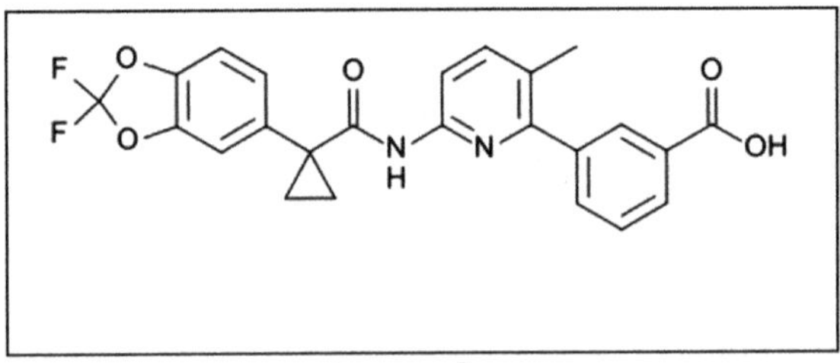

Lumacaftor 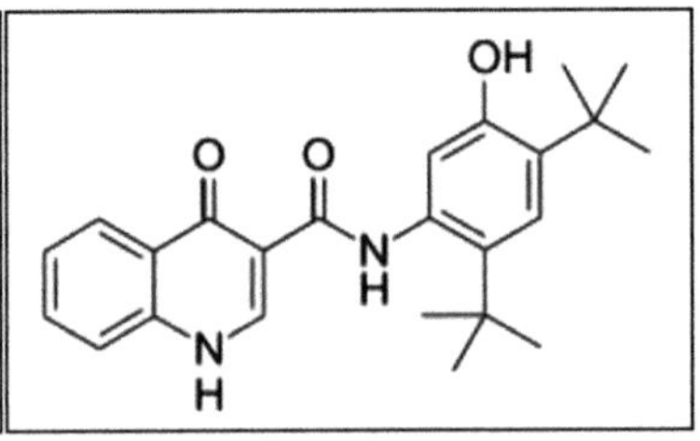 | Ivacaftor

FDA's comments provide take-home lessons applicable to monotherapy drugs and to combination therapy drugs. Where the study drug takes the form of a combination of two drugs, there is an increased risk for DDIs. At the time of the clinical trial on this combination, ivacaftor monotherapy had already been FDA-approved for at 150 mg doses. FDA realized that the one drug in the combination would stimulate CYP-mediated destruction of the other drug, writing that, "In a PK study in healthy volunteers, **LUM exposure** reduced **IVA exposure** by approximately 80%. Similar results were observed...in cystic fibrosis patients during the LUM/IVA dose-ranging study."[237]

To mitigate the DDI resulting in reduced ivacaftor exposure, the Sponsor jacked up the ivacaftor dose in the combination drug. This mitigation approach was described by FDA's Medical Review, "Because LUM is CYP3A inducer and IVA is a CYP3A substrate,

[234] Package label. GILENYA (fingolimod) capsules, for oral use; February 2016 (25 pp.).

[235] Lumacaftor/ivacaftor combination (cystic fibrosis for patients age 12 years or older with F508del mutation in CFTR gene) NDA 206-038. 99 page Medical Review.

[236] Davis PB. Another beginning for cystic fibrosis therapy. New Engl. J. Med. 2015;373:274–6.

[237] Lumacaftor/ivacaftor combination (cystic fibrosis for patients age 12 years or older with F508del mutation in CFTR gene) NDA 206-038. Page 24 of 99 page Medical Review.

co-administration of LUM with IVA results in substantially **lower IVA exposures**. As such, for LUM/IVA, compared to the approved IVA monotherapy dose of 150mgq12, the Applicant used a higher IVA dose of 250 mg q12 in the LUM/IVA dose-ranging trial and pivotal trials."[238]

In view of the fact that many fixed dose combination (FDC) drugs on the market (see 21 CFR §300.50), it could be argued that the Sponsor's decision to use IVA at a higher dose has a broadly applicable and practical value for the drug development process for many drugs. The Sponsor's efforts to mitigate the problem of CYP3A-mediated destruction of the second drug were partially effective. The FDA reviewer wrote, "Despite the nominally higher IVA dose in the LUM/IVA combination, IVA **exposure was still lower**. . .compared to IVA 150mg q12 monotherapy."[239]

As a reminder, the term "exposure" refers to a parameter of blood plasma concentration, such as AUC, Cmax, Cmin, and tmax.[240,241,242] FDA provided a histogram (Fig. 7.9) showing that greater lumacaftor doses result in progressively lower ivacaftor exposures. The *Y*-axis shows ivacaftor exposure (Cmin), and the *X*-axis shows lumacaftor dose.

The FDA provided a broader warning and referred to the potential for LUM to cause reduced exposure of any coadministered antibiotics, antifungals, proton pump inhibitors, and so on:

> As LUM is a strong CYP3A inducer, and because in vitro studies suggest that LUM has the potential to induce CYP2B6, CYP2C8, CYP2C9, CYP2C19, and inhibit CYP2C8 and CYP2C9, concomitant use of LUM/IVA may alter the exposure of many common medications. . .such as antibiotics, antifungals, proton pump inhibitors, ibuprofen, antidepressants, etc. As a result, concomitant use of LUM/IVA may require dose adjustments for some drugs.[243]

FDA chose this list of drugs from those that are commonly used by cystic fibrosis patients in general medical practice.

[238] Lumacaftor/ivacaftor combination (cystic fibrosis for patients age 12 years or older with F508del mutation in CFTR gene) NDA 206-038. Page 18 of 99 page Medical Review.

[239] Lumacaftor/ivacaftor combination (cystic fibrosis for patients age 12 years or older with F508del mutation in CFTR gene) NDA 206-038. Pages 17-18 of 81-page Clinical Review. 99-page Medical Review.

[240] U.S. Department of Health and Human Services. Food and Drug Administration. Center for Drug Evaluation and Research (CDER). Center for Biologics Evaluation and Research (CBER). Guidance for industry. Exposure-response relationships—study design, data analysis, and regulatory applications; 2003 (25 pp.).

[241] U.S. Department of Health and Human Services. Food and Drug Administration. Center for Drug Evaluation and Research (CDER). Guidance for industry. Bioequivalence studies with pharmacokinetic endpoints for drugs submitted under an ANDA; 2013 (20 pp.).

[242] U.S. Department of Health and Human Services. Food and Drug Administration. Center for Drug Evaluation and Research (CDER). Guidance for Industry Bioavailability and Bioequivalence Studies Submitted in NDAs or INDs—General Considerations; 2014 (26 pp.).

[243] Lumacaftor/ivacaftor combination (cystic fibrosis for patients age 12 years or older with F508del mutation in CFTR gene) NDA 206-038. Page 24 of 81-page Clinical Review. 99-page Medical Review.

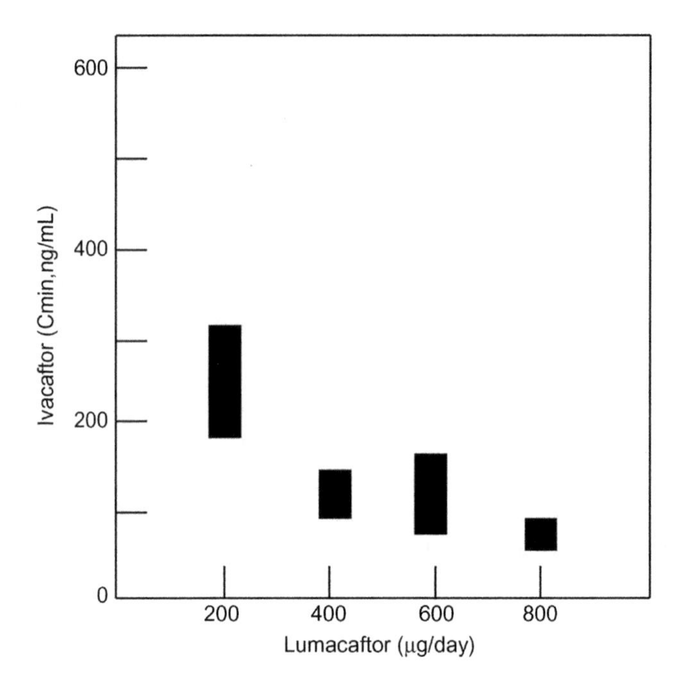

FIGURE 7.9 Histogram showing ivacaftor exposure versus lumacaftor dose. Lumacaftor induces CYP3A, where increased CYP3A results in increased CYP3A-mediated catabolism of ivacaftor.

Package label. Comments from FDA's review of lumacaftor (LUM)/ivacaftor (IVA) combination (Orkambi®) found a place in the Warnings and Precautions section of the package label:

> **WARNINGS AND PRECAUTIONS**...Drug interactions. Use with CYP3A substrates or CYP3A substrates with a narrow therapeutic index may decrease...exposure of the medicinal product and co-administration is not recommended. Hormonal contraceptives should not be relied upon as an effective method of contraception...[u]se with strong CYP3A inducers may diminish exposure of ivacaftor, which may dimninish its effectiveness, therefore, co-administration is not recommended.[244]

The Drug Interactions section provided an account of CYP3A inhibitors, such as itraconazole and ketoconazole:

> **DRUG INTERACTIONS**. Potential for Other Drugs to Affect Lumacaftor/Ivacaftor...Inhibitors of CYP3A. Co-administration of lumacaftor/ivacaftor with itraconazole, a **strong CYP3A inhibitor**, did not impact the exposure of lumacaftor, but increased ivacaftor exposure by 4.3-fold. Due to the induction effect of lumacaftor on CYP3A, at steady-state, the net exposure of ivacaftor is not expected to exceed that when given in the absence of lumacaftor at a dose of 150 mg every 12 hours (the approved dose of ivacaftor monotherapy). Therefore, no dose adjustment is necessary when CYP3A inhibitors are initiated in patients currently taking ORKAMBI. However, when initiating ORKAMBI in patients taking **strong CYP3A**

[244] Package label. ORKAMBI—lumacaftor and ivacaftor tablet, film coated; September 2016 (32 pp.).

inhibitors, reduce the ORKAMBI dose to 1 tablet daily (lumacaftor 200 mg/ivacaftor 125 mg total daily dose for patients aged 12 years and over; lumacaftor 100 mg/ ivacaftor 125 mg total daily dose for patients aged 6 through 11 years) for the first week of treatment to allow for the steady-state induction effect of lumacaftor...Examples of **strong CYP3A inhibitors** include: ketoconazole, itraconazole, posaconazole, and voriconazole • telithromycin, clarithromycin. No dose adjustment is recommended when used with **moderate or weak CYP3A inhibitors**.[245]

The same Drug Interactions section then provided an account of CYP3A inducers, such as rifampin:

DRUG INTERACTIONS...**Inducers of CYP3A**. Co-administration of lumacaftor/ivacaftor with **rifampin, a strong CYP3A inducer**, had minimal effect on the exposure of lumacaftor, but decreased ivacaftor exposure (AUC) by 57%. This may reduce the effectiveness of ORKAMBI. Therefore, co-administration with **strong CYP3A inducers**, such as rifampin, rifabutin, phenobarbital, carbamazepine, phenytoin, and St. John's Wort (Hypericum perforatum), is not recommended...No dose adjustment is recommended when used with **moderate or weak CYP3A inducers**.[246]

j. Ponatinib (Chronic Myeloid Leukemia) NDA 203-469

Ponatinib is for treating chronic myeloid leukemia (CML).[247] Ponatinib's structure is shown below. Ponatinib's metabolites include M14, M15, M23, and M29. The metabolites M15 and M29 are glucuronides of ponatinib.[248]

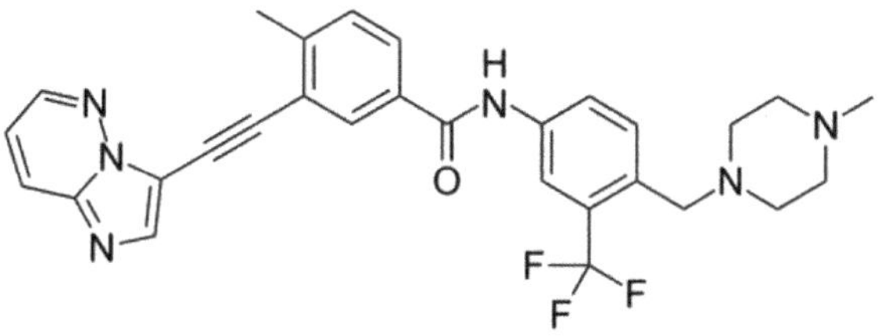

i. CYP Enzyme Inhibitors (In Vitro Study)

The Sponsor assessed the ability of various CYP enzymes to catabolize ponatinib by separate incubations with individual purified, recombinant CYP enzymes. As described by FDA, "What are the characteristics of drug metabolism? In vitro, ponatinib was incubated with...individual recombinant human CYP enzymes (CYP1A2, CYP2B6, CYP2C8, CYP2D6, CYP3A4/5, and CYP2C9...at a...concentration of 5-10 micromolar...[s]electively inhibiting ponatinib metabolism with CYP-specific inhibitors and monoclonal antibodies...suggested that **ponatinib was metabolized primarily by CYP3A4** and to a lesser extent by CYP2C8, CYP2D6, and CYP3A5."[249]

[245] Package label. ORKAMBI—lumacaftor and ivacaftor tablet, film coated; September 2016 (32 pp.).

[246] Package label. ORKAMBI—lumacaftor and ivacaftor tablet, film coated; September 2016 (32 pp.).

[247] Ponatinib (chronic myeloid leukemia) NDA 203-469. 89-page Clinical Pharmacology Review.

[248] Ponatinib (chronic myeloid leukemia) NDA 203-469. Page 21 of 89-page Clinical Pharmacology Review.

[249] Ponatinib (chronic myeloid leukemia) NDA 203-469. Page 19 of 89-page Clinical Pharmacology Review.

Identifying specific CYP enzymes as responsible for ponatinib's catabolism was by the way of:

1. using recombinant CYP enzymes of a known identity;
2. adding an inhibiting antibody with a known specificity;
3. adding a known inhibitor with a known specificity.

FDA's ClinPharm Review cited Sai et al.[250] as the source of cytochrome P450-specific inhibitors. However, Khojasteh et al.[251] provide a more comprehensive account of CYP-specific inhibitors suitable for use with human microsomes, as reproduced here in the bulletpoints (this list does not in any way imply that these inhibitors are acceptable for in vivo use with drug-drug interaction studies with human subjects):

- CYP3A4: ketoconazole
- CYP3A4: azamulin
- CYP1A2: furafylline
- CYP2B6: 2-phenyl-2-(1-piperidinyl)propane (PPP)
- CYP2C9: sulfaphenazole
- CYP2C19: (−)-*N*-3-benzyl-phenobarbital
- CYP2D6: quinidine
- CYP2A6: tranylcypromine
- CYP2A6: 3-(pyridin-3-yl)-1H-pyrazol-5-yl)methanamine (PPM)

ii. CYP Enzyme Inhibitors (Clinical Study)

The Sponsor conducted a clinical study to evaluate the influence of a CYP3A inhibitor (ketoconazole) on ponatinib exposure. The data showed that ketoconazole increased ponatinib's AUC and Cmax. The FDA reviewer stated, "The applicant conducted a...trial to evaluate the effects of...**ketoconazole, a strong CYP3A inhibitor** on the PK profile of...ponatinib administration in healthy subjects. Each subject received...15 mg ponatinib, once given alone and once co-administered with daily doses of 400 mg of ketoconazole for 5 days. Plasma concentrations for ponatinib...were sampled over a 96 hour period...The estimated mean ratios of $AUC_{0-\infty}$ and of Cmax for ponatinib increased by 78% and 47%, respectively."[252]

The Sponsor proposed that the package label merely requires "caution" while, in contrast, FDA disagreed and required that the package label includes instructions for dose reduction. FDA referred to the value of "78% reduction in AUC," mentioned above, and arrived at the conclusion, "the reviewer disagrees with the applicant's conclusion that a dose adjustment is not required and only Caution should be exercised with concurrent use of ponatinib with strong CYP3A inhibitors. As stated above, an average 78% increase in ponatinib exposure would be expected if ponatinib were co-administered with a strong

[250] Sai Y, et al. Assessment of specificity of eight chemical inhibitors using cDNA-expressed cytochromes P450. Xenobiotica 2000;30:327–43.

[251] Khojasteh SC, et al. Chemical inhibitors of cytochrome P450 isoforms in human liver microsomes: a re-evaluation of P450 isoform selectivity. Eur. J. Drug Metab. Pharmacokinet. 2011;36:1–16.

[252] Ponatinib (chronic myeloid leukemia) NDA 203-469. Page 27 of 89-page Clinical Pharmacology Review.

CYP3A4 inhibitor...[t]herefore, **the reviewer recommends a dose reduction to 30 mg daily**...if ponatinib is co-administered with a strong CYP3A4 inhibitor."[253]

Note the value of "30 mg daily" in FDA's recommendation, as this value should be compared to the warning that eventually materialized on the package label.

Package label. The package label did warn about strong CYP3A4 inhibitors, but the label's warning was not so exact as to require reducing the dose to only 30 mg daily. Instead, the label merely required, in the Full Prescribing Information section, that the "starting dose should be reduced":

> **DRUG INTERACTIONS**...[w]hen administering...[ponatinib] with **strong CYP3A inhibitors** (e.g., boceprevir, clarithromycin, conivaptan, grapefruit juice,...**ketoconazole**,...telaprevir, telithromycin, voriconazole), **the recommended starting dose should be reduced**.[254]

iii. CYP Enzyme Inducers

In addition to considering CYP enzyme inhibitors, FDA's review contemplated inducers of CYP enzymes, but on this point, the Sponsor had not provided any data. In FDA's words:

> it is anticipated that inducers of CYP3A will theoretically increase the clearance of ponatinib; however, the **magnitude is unknown**. The effect of CYP3A4 enzyme induction on the metabolism of ponatinib **was not specifically evaluated** by the applicant in vitro or in vivo. The applicant has submitted a proposed clinical trial protocol on...to evaluate the effect of rifampin (a strong CYP3A4 inducer) on the pharmacokinetics of ponatinib when administered concomitantly in healthy subjects...[c]ompletion of this trial should be a post marketing requirement.[255]

FDA agreed with the Sponsor's proposed clinical study, and instructed the Sponsor to conduct the rifampin study, "Conduct a dedicated drug interaction trial in humans to determine the effect of **co-administration of the strong CYP3A4 inducer rifampin** on the pharmacokinetics of ponatinib in healthy subjects."[256] As shown below, the package label succinctly stated, "Strong CYP3A Inducers: Avoid concurrent use."

iv. Influence of Ponatinib on Second Drug Exposure (CYP Enzyme Assays)

Ponatinib's inhibition of CYP enzymes was tested using probe substrates (Table 7.6).[257] EMA has published the same list of enzymes, in the same order, together with appropriate probe substrates for detecting CYP enzyme activity.[258] EMA's list is reproduced at the end of this chapter in Appendix A.

[253] Ponatinib (chronic myeloid leukemia) NDA 203-469. Page 27 of 89-page Clinical Pharmacology Review.

[254] Package label. ICLUSIG® (ponatinib) tablets for oral use; November 2016 (23 pp.).

[255] Ponatinib (chronic myeloid leukemia) NDA 203-469. Page 28 of 89-page Clinical Pharmacology Review.

[256] Ponatinib (chronic myeloid leukemia) NDA 203-469. Page 4 of 89-page Clinical Pharmacology Review.

[257] Ponatinib (chronic myeloid leukemia) NDA 203-469. Pages 28–29 of 89-page Clinical Pharmacology Review.

[258] European Medicines Agency. Guideline on the Investigation of Drug Interactions; 2012 (59 pp.).

TABLE 7.6 Prediction of Ponatinib's Ability to Inhibit CYP450 in Human Subjects

Cytochrome P450	Ki	[I]/Ki
1A2	6.65	0.021
2B6	2.8	0.049
2C8	3.05	0.045
2C9	5.4	0.025
2C19	2.6	0.053
2D6	5.8	0.024
3 A4/5	4.15	0.033

The value for [I] was taken from the value for Cmax in human subjects. Cmax = 0.137 μM at the therapeutic dose of ponatinib of 45 mg.
Ponatinib (chronic myeloid leukemia) NDA 203-469. Pages 28–29 of 89-page Clinical Pharmacology Review.

The Sponsor's inhibition assays used cultured human hepatocytes, where the readout was IC50 values. The Sponsor then converted each IC50 value to a corresponding inhibition constant (Ki) value which, in turn, was plugged into the [I]/Ki ratio equation. The value for the ([I]/Ki) ratio can optionally be plugged into the following equation. The equation is ([AUC**plus inhibitor**]/[AUC**no inhibitor**] = 1 + ([I]/[Ki]).[259,260] Although this equation is found in published articles on pharmacokinetics, it is more often that FDA's ClinPharm reviews utilize the simpler [I]/Ki ratio equation to arrive at its conclusions.

FDA explained the origin of the value for "[I]" explaining that, "Cmax of ponatinib was 73 ng/mL (0.137 micromolar) at a therapeutic dose of 45 mg using the to-be marketed tablet formulation in patients with advanced hematologic malignancies."[261]

FDA compared the experimentally determined values for the [I]/Ki ratio (Table 7.6) with *the cutoff value of 0.1*, and concluded that, "These studies demonstrated little or no inhibition of CYP1A2, CYP2B6, CYP2C8, CYP2C9, CYP2C19, CYP2D6 or CYP3A activities. . .[t]he range of **[Cmax]/Ki ratio was 0.020 – 0.053, which is less than 0.1**. . .[n]o additional in vivo trials are required."[262]

This provides the dramatic and practical take-home lesson that, where in vitro studies reveal a [Cmax]/Ki ratio that is less than 0.1, it is unlikely that the Sponsor will need to conduct a corresponding clinical study.

[259] Wienkers LC, Heath TG. Predicting in vivo drug interactions from in vitro drug discovery data. Nat. Drug Discov. 2005;4:825–33.

[260] McGinnity DF, et al. Prediction of CYP2C9-mediated drug–drug interactions: a comparison using data from recombinant enzymes and human hepatocytes. Drug Metab. Dispos. 2005;33:1700–7.

[261] Ponatinib (chronic myeloid leukemia) NDA 203-469. Page 28 of 89-page Clinical Pharmacology Review.

[262] Ponatinib (chronic myeloid leukemia) NDA 203-469. Page 28 of 89-page Clinical Pharmacology Review.

Package label. FDA's recommendations to avoid coadministering CYP3A4 inhibitors and CYP3A4 inducers found a place on the package label for ponatinib (Iclusig®):

> **DRUG INTERACTIONS. Strong CYP3A Inhibitors:** Avoid concurrent use, or reduce Iclusig dose if co-administration cannot be avoided...**Strong CYP3A Inducers:** Avoid concurrent use.[263]

The Full Prescribing Information section provided further details:

> **DRUG INTERACTIONS...Drugs That Are Strong Inhibitors of CYP3A Enzymes**. Based on in vitro studies, ponatinib is a substrate of CYP3A and to a lesser extent CYP2C8 and CYP2D6. In a drug interaction study in healthy volunteers, co-administration of Iclusig with ketoconazole increased plasma ponatinib $AUC_{0\text{-inf}}$ and Cmax by 78% and 47%, respectively...[w]hen administering Iclusig with strong CYP3A inhibitors (e.g., boceprevir, clarithromycin, conivaptan, grapefruit juice,...**ketoconazole**...telaprevir, telithromycin, voriconazole), the recommended starting dose should be reduced.[264]

> **DRUG INTERACTIONS...Drugs That Are Strong Inducers of CYP3A Enzymes**. Co-administration of strong CYP3A inducers (e.g., carbamazepine, phenytoin, **rifampin**, and St. John's Wort) with Iclusig should be avoided unless the benefit outweighs the risk of decreased ponatinib exposure...In a drug interaction study in healthy volunteers, co-administration of Iclusig following multiple doses of rifampin resulted in decreased ponatinib AUC0-inf and Cmax values by 62% and 42%, respectively.[265]

k. Regorafenib (metastatic colorectal cancer) NDA 203-085

Regorafenib, as well as its metabolites, inhibits several different tyrosine kinase enzymes. The published literature describes the variety of tyrosine kinases inhibited by regorafenib. These include VEGFR-1, VEGFR-2, VEGFR-3, TIE2, KIT, RET, RAF-1, and BRAF.[266] Regorafenib is initially catabolized by CYP3A4, resulting in oxidation of regorafenib's ring nitrogen to produce M-2. M-2 can be further metabolized to M-5 by a yet-unidentified enzyme which catalyzes demethylation of M-5's amide group.[267]

FDA has assessed the potencies of various tyrosine kinase inhibitor (TKI) drugs, and their metabolites, in FDA's reviews of the NDAs cited below. As can be seen from this list, the metabolites are significantly less potent than the parent compound:

- *Axitinib.* "The M12 and M7 metabolites show approximately 400-fold and 8000-fold less in vitro **potency**, respectively, against VEGFR-2 compared to axitinib."[268]
- *Bosutinib.* "Plasma samples from clinical trials were assessed for the parent drug (bosutinib) which is the active moiety. Several of the clinical trials assessed exposures to

[263] Package label. ICLUSIG® (ponatinib) tablets for oral use; November 2016 (23 pp.).

[264] Package label. ICLUSIG® (ponatinib) tablets for oral use; November 2016 (23 pp.).

[265] Package label. ICLUSIG® (ponatinib) tablets for oral use; November 2016 (23 pp.).

[266] Grothey A, et al. Optimizing treatment outcomes with regorafenib: personalized dosing and other strategies to support patient care. Oncologist 2014;19:6669–80.

[267] Regorafenib (metastatic colorectal cancer) NDA 203-085. Page 18 of 64-page pdf file of Clinical Pharmacology Review.

[268] Axitinib (renal cell carcinoma) NDA 202-324. Page 25 of 118-page Clinical Pharmacology Review.

the major metabolites, M2 and M5, which are **inactive**.[269]" "Most trials analyzed samples for the major metabolites M2 and M5. The M2 and M5 metabolites are **inactive**."[270]

- *Cabozantinib*. "metabolites (cabozantinib N-oxide and cabozantinib half-dimer) possess <1% of the. . .kinase inhibition **potency** of parent cabozantinib."[271]
- *Ponatinib*. "the AP24600 metabolite is inactive and the metabolite, AP24567, is approximately 4-fold less **potent** than ponatinib in vitro."[272]
- *Tofacitinib*. "All metabolites have less than <8% of total drug exposure and their potency was reported to be <10% of the **potency** of tofacitinib for JAK1/2 inhibition."[273]
- *Regorafenib*. In contrast to the situation for the metabolites of all of the kinase inhibitors disclosed above, regorafenib's metabolites M2 and M5 exhibit anticancer activity similar to that of regorafenib."[274]

Regorafenib's metabolites have a potency similar to that of the parent compound, for treating cancer. FDA focused on the metabolites called "M2" and "M5." As can be seen by FDA's comments, these two metabolites had a similar anticancer activity as the parent compound. FDA stated that, "Regorafenib inhibits multiple kinases, including VEGFR1, VEGFR2, VEGFR3, TIE2, KIT, RET, RAF-1, BRAF, BRAFV600E, PDGFR, and FGFR. . .[t]he **M2 and M5 metabolites of regorafenib inhibited some of the same kinases as regorafenib**, such as VEGFR2, TIE2, KIT (mutant and wild type), and BRAF (mutant) at IC50 values similar to regorafenib. These metabolites exhibited similar anticancer activity compared to regorafenib in tumor models of colorectal cancer."[275]

The issue of DDIs for regorafenib and its metabolites is elaborated upon below.

i. In Vitro Influence of Regorafenib on CYP Enzymes (Inhibition)

The Sponsor assessed the ability of the regorafenib and its metabolites to inhibit various cytochrome P450 enzymes, as revealed by FDA's comment that, "Regorafenib, M2, or M5 inhibited CYP2B6, CYP2C9, CYP2C8, CYP2C19, CYP2D6 or CYP3A4 in vitro."[276]

[269] Bosutininib (chronic myelogenous leukemia; CML) NDA 203-341. Page 12 of 89-page Clinical Pharmacology Review.

[270] Bosutininib (chronic myelogenous leukemia; CML) NDA 203-341. Page 40 of 89-page Clinical Pharmacology Review.

[271] Cabozanitinib (thyroid cancer) NDA 203-756. Page 12 of 106-page Clinical Pharmacology Review.

[272] Ponatinib (chronic myeloid leukemia) NDA 203-469. Page 21 of 89-page Clinical Pharmacology Review.

[273] Tofacitinib (rheumatoid arthritis) NDA 203-214. Page 6 of 181-page Clinical Pharmacology Review.

[274] Regorafenib (metastatic colorectal cancer) NDA 203-085. Page 11 of 64-page pdf file of Clinical Pharmacology Review.

[275] Regorafenib (metastatic colorectal cancer) NDA 203-085. Page 11 of 64-page pdf file of Clinical Pharmacology Review.

[276] Regorafenib (metastatic colorectal cancer) NDA 203-085. Page 29 of 64-page pdf file of Clinical Pharmacology Review.

TABLE 7.7 The Ki Value Determined and the Calculated R Value (ratio) Based on Steady-State Concentrations in the Bloodstream of Regorafenib, M2, and M5. $R = 1 + ([I]/Ki)$

	Regorafenib		M2		M5	
	Ki	Ratio	Ki	Ratio	Ki	Ratio
CYP2B6	5.2	2.6	–	-	–	–
CYP2C8	0.6	14.5	1.0	7.6	1.3	5.6
CYP2C9	4.7	2.7	0.8	9.2	–	–
CYP2C19	16.4	1.5	–	–	–	–
CYP2D6	–	–	7.8	1.8	–	–
CYP3A4	11.1	1.7	4.0	2.6	–	–

Regorafenib (metastatic colorectal cancer) NDA 203-085. Page 30 of 64-page pdf file of Clinical Pharmacology Review.

Table 7.7 reproduces a table from FDA's ClinPharm Review. The table shows that regorafenib and its metabolites inhibit the indicated CYP enzymes. Also, the table makes use of the concentration [I] of regorafenib used for human patients, and plugs this concentration and the Ki values into the ratio equation, $R = 1 + ([I]/Ki)$. This ratio equation is used to predict if there will be a significant DDI in human patients.

FDA's narrative on the data in Table 7.7 states that the table, "lists the **ratio values** calculated assuming a maximal steady-concentration of 8.1 μM (3.9 μg/mL), 6.6 μM (3.3 μg/mL), and 6.0 μM (2.9 μg/mL) for regorafenib, M2 and M5, respectively."[277]

The take-home lesson is that investigators must take care to plug in the blood plasma concentration of the metabolite (and not the blood plasma concentration of the study drug), when performing ratio calculations that plug in Ki values produced by in vitro enzyme studies where the metabolite (not the study drug) is added to incubation mixtures with human microsomes.

FDA observed that the high values for the ratios can be used to predict if there will be DDIs in human patients. For this reason. FDA asked the Sponsor to conduct a clinical DDI study as a PMR. FDA requested a study where regorafenib was coadministered with a probe substrate (but did not request a study where M2 or M5 was coadministered with a probe substrate).

FDA's request that the Sponsor conduct a postmarketing clinical DDI study took the form, "These ratio values suggest that a study to assess the effects of regorafenib on the PK of sensitive substrates of CYP2B6, CYP2C19, CYP2C8, CYP2C9, CYP2D6, and CYP3A4 **is likely warranted**. A study to assess the effects of regorafenib on the PK of a probe substrate of CYP2C8, CYP2C9, CYP2C19 and CYP3A4 is ongoing. The final study report will be submitted by November, 2012. **The applicant will also be asked to assess the effect of regorafenib on the PK of a probe substrate of CYP2D6 post marketing.**"[278]

[277] Regorafenib (metastatic colorectal cancer) NDA 203-085. Page 29 of 64-page pdf file of Clinical Pharmacology Review.

[278] Regorafenib (metastatic colorectal cancer) NDA 203-085. Page 29 of 64-page pdf file of Clinical Pharmacology Review.

ii. In Vitro Influence of Regorafenib on CYP Enzymes (Induction)

The Sponsor's in vitro CYP enzyme induction experiments provided these results on induction of CYP enzymes, "Regorafenib did not induce the enzyme activity of CYP1A2, CYP2B6, CYP2C19 and CYP3A4 in vitro…[h]uman hepatocytes were treated with regorafenib at concentrations of 5 ng/mL to 10,000 ng/mL starting 48 hr post-plating and continuing for 5 days until the determination of enzyme activity on day 8.. . .[n]o studies were conducted to determine the ability of the active metabolites M2 and M5 to induce these enzymes."[279]

FDA expressed interest on the ability of M2 or M5 to induce various cytochrome P450 enzymes, despite the fact that the Sponsor was not seeking FDA-approval to market the metabolites, and requested that the Sponsor conduct in vitro induction experiments with M2 and M5, "The applicant should assess the ability of regorafenib, M2 and M5 to induce mRNA expression levels of CYP1A2, CYP2B6 and CYP3A4 in accordance with the 2012 draft FDA Guidance for Industry regarding drug interaction studies."[280]

iii. Clinical Study on CYP3A4 Inducer (Rifampin) and Regorafenib

FDA's ClinPharm Review described the Sponsor's clinical study with rifampin, a CYP3A4 inducer, and recommended that the package label should warn against coadministering CYP3A4 inducers with regorafenib.

FDA's ClinPharm Review revealed that the Sponsor's clinical CYP enzyme induction study gave dramatic results, where FDA observed, "Twenty-four healthy men received a single 160 mg dose of **regorafenib**…[f]ollowing a washout of three weeks, they received **rifampin** at a dose of 600 mg daily…[t]he…exposure of regorafenib was decreased by 50% and the mean exposure of the M5 metabolite was increased by 264%. The mean AUC of the M2 metabolite appeared to be unchanged, but the mean Cmax was substantially increased…[g]iven this magnitude of change in the mean exposure, the **co-administration of strong CYP3A4 inducers with regorafenib should be avoided**."[281]

The design of the clinical study included a baseline assessment of study drug AUC and also an assessment of AUC after 2 weeks of daily treatment with the inducing agent (rifampin). The baseline assessment (absence of rifampin) provided a control value. Fig. 7.10 illustrates the study design, by the way of drawings of the anatomy. The figure shows a drawing of the liver at the start of the study, and another drawing of the same liver after 2 weeks of rifampin treatment.

In addition to the clinical study with human subjects, the Sponsor conducted a dog study on rifampin's induction of various CYP enzymes.[282] In the dog study, the control took the form of control dogs receiving only corn oil (vehicle), while the experimental

[279] Regorafenib (metastatic colorectal cancer) NDA 203-085. Page 29 of 64-page pdf file of Clinical Pharmacology Review.

[280] Regorafenib (metastatic colorectal cancer) NDA 203-085. Page 29 of 64-page pdf file of Clinical Pharmacology Review.

[281] Regorafenib (metastatic colorectal cancer) NDA 203-085. Page 28 of 64-page Clinical Pharmacology Review.

[282] Graham RA, et al. In vivo and in vitro induction of cytochrome P450 enzymes in beagle dogs. Drug Metab. Dispos. 2002;30:1206–13.

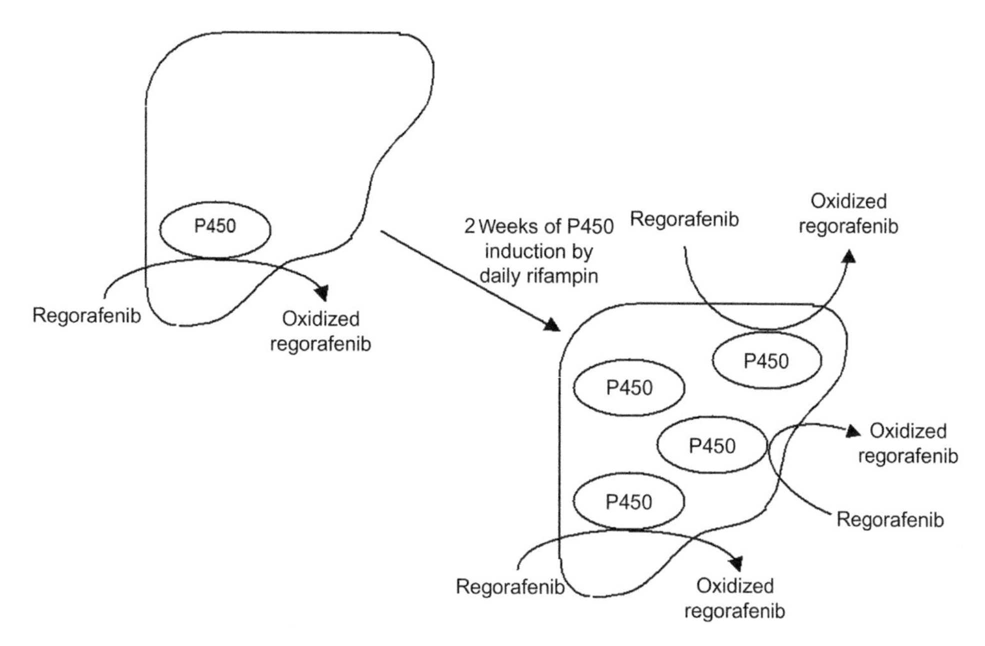

FIGURE 7.10 Increase in number of CYP3A4 enzymes in liver with rifampin treatment. The first picture of the liver is before rifampin treatment, and the second picture is the same liver after 2 weeks rifampin treatment, showing more CYP3A4 enzymes and greater ability to catabolize regorafenib.

dogs received either rifampin (daily for 4 days) or phenobarbital (daily for 2 weeks). The dog study concluded that, "dogs have CYP1A, CYP2B, CYP2E, and CYP3A enzymes and...the induction profile resembles the profile observed in humans more than in rats."

iv. Clinical Data on Coadministration of CYP3A4 Inhibitor (Ketoconazole) and Regorafenib

The Sponsor's data with human subjects included a CYP enzyme inhibitor study, demonstrating that ketoconazole coadministration increased exposure of regorafenib.

FDA warned against coadministration of CYP3A4 inhibitors, in an observation that, "The applicant evaluated the effect of a strong CYP3A4 inhibitor, ketoconazole, on the PK of regorafenib, M2 and M5...healthy men received...single dose of regorafenib...and ketoconazole at a dose of 400 mg daily starting day-4...[t]he...exposure of regorafenib increased by 33%, but the mean exposure of the M2 and M5 metabolites decreased each by 93%. The overall...exposure of regorafenib, M2 and M5 appears to be decreased by approximately 50%. Given this magnitude of change in...exposure, the **co-administration of strong CYP3A4 inhibitors with regorafenib should be avoided.**"[283]

The Sponsor's observation that ketoconazole did not increase exposure of M2 and M5, but instead decreased exposure M2 and M5, is explained by the fact that CYP3A4

[283] Regorafenib (metastatic colorectal cancer) NDA 203-085. Pages 9 and 27 of 64-page Clinical Pharmacology Review.

catalyzes the conversion of regorafenib to M2, where a yet-unidentified enzyme converts M2 to M5.[284]

v. Drug–Drug Interactions Mediated by UDP-Glucuronosyltransferase (UGT)

Regorafenib inhibits UDP-glucuronosyltransferase (UGT), an enzyme that connects a glucuronic acid moiety to various drugs, thereby increasing drug excretion. As stated near the beginning of this chapter, UGT catalyzes the glucuronication of drugs and other xenobiotics, as well as of bilirubin and steroid hormones. The glucuronic acid moiety can be attached to phenols, alcohols, carboxylic acids, and amines.[285]

The Sponsor chose to study DDIs between regorafenib and irinotecan, because irinotecan is commonly coadministered with regorafenib when treating colorectal cancer.[286] Schultheis et al.[287] described treating colorectal cancer with the combination of regorafenib and irinotecan.

Experiments from a journal article observed the DDI between regorafenib and irinotecan and proposed a type of study design for reducing this DDI. The journal article reported that, "The active metabolite of irinotecan, SN-38, is predominantly converted to an inactive metabolite by glucuronidation. As regorafenib is a strong inhibitor of the glucuronosyltransferases, UGT1A1 and UGT1A9, there was a potential for pharmacokinetic interaction. In anticipation of such an interaction, dosing of irinotecan was separated from regorafenib by 4 days."[288] Thus, the study design (dosing design) was to separate the two drugs by several days.

FDA's ClinPharm Review disclosed that regorafenib and its M2 and M5 metabolites inhibit certain UGT enzymes. FDA referred to the Ki values determined from in vitro experiments, and to the in vivo regorafenib concentration [I] in human subjects. A connection between *in vitro data* and *in vivo data* is established by the [I]/Ki ratio. FDA's ClinPharm Review provided the following table (Table 7.8).

Table 7.8 shows that the highest value for the ratio was with regorafenib and UGT1A9. FDA used the [I]/Ki ratio and predicted that regorafenib has the potential to inhibit UGT1A9 in patients. FDA's prediction took the form, "Regorafenib, M2, and M5 inhibited UGT1A1 and UGT1A9 in vitro. It does not appear that these compounds inhibit other UGT enzymes, including UGT1A4, UGT1A6, and UGT2B7. The fold difference in the Ki values and the steady-state concentrations **suggest that regorafenib, M2, and M5 has the**

[284] Regorafenib (metastatic colorectal cancer) NDA 203-085. Page 18 of 64-page Clinical Pharmacology Review.

[285] Milne AM, et al. A novel method for the immunoquantification of UDP-glucuronosyltransferases in human tissue. Drug Metab. Dispos. 2011;39:2258–63.

[286] Regorafenib (metastatic colorectal cancer) NDA 203-085. Page 31 of 64-page Clinical Pharmacology Review.

[287] Schulteis B, et al. Regorafenib in combination with FOLFOX or FOLFIRI as first- or second-line treatment of colorectal cancer: results of a multicenter, phase Ib study. Ann. Oncol. 2013;24:1560–7.

[288] Schulteis B, et al. Regorafenib in combination with FOLFOX or FOLFIRI as first- or second-line treatment of colorectal cancer: results of a multicenter, phase Ib study. Ann. Oncol. 2013;24:1560–7.

TABLE 7.8 Inhibition Constant (Ki) Values and Ratio of [I]/Ki

	Regorafenib		Metabolite M2		Metabolite M5	
	Ki (μM)	**Ratio**	**Ki (μM)**	**Ratio**	**Ki (μM)**	**Ratio**
UGT1A1	3.0	2.7	0.6	11.0	1.1	4.6
UGT1A9	2.1	3.9	4.3	1.5	7.9	0.6

[I] is the steady-state concentration when study drug is given to human subjects.
Regorafenib (metastatic colorectal cancer) NDA 203-085. Pages 30–31 of 64-page Clinical Pharmacology Review.

potential to inhibit UGT1A1 and regorafenib has the potential to inhibit UGT1A9 in humans."[289]

***Package label*:** The Sponsor's in vitro experiments on regorafenib's inhibition and induction of CYP enzymes found a place in the Clinical Pharmacology section:

> **CLINICAL PHARMACOLOGY**...Drug Interaction Studies. Effect of Regorafenib on Cytochrome P450 Substrates: In vitro studies suggested that **regorafenib is an inhibitor** of CYP2C8, CYP2C9, CYP2B6, CYP3A4 and CYP2C19; M-2 is an inhibitor of CYP2C9, CYP2C8, CYP3A4 and CYP2D6, and M-5 is an inhibitor of CYP2C8. In vitro studies suggested that **regorafenib is not an inducer** of CYP1A2, CYP2B6, CYP2C19, and CYP3A4 enzyme activity.[290]

The Sponsor's data that a drug coadministered with regorafenib can induce CYP3A4 and thus have an adverse influence on regorafenib exposure, and that a drug coadministered with regorafenib can inhibit CYP3A4 and thus have an adverse influence on regorafenib exposure, found a place in the Drug Interactions section:

> **DRUG INTERACTIONS. Strong CYP3A4 inducers:** Avoid strong CYP3A4 inducers...**Strong CYP3A4 inhibitors:** Avoid strong CYP3A4 inhibitors...BCRP substrates: Monitor patients closely for symptoms of increased exposure to BCRP substrates.[291]

Comments from FDA's ClinPharm Review on the study drug's active metabolites found a place on the package label for regorafenib (Stivarga®). The Drug Interaction section of the package label warned about CYP3A4 inducers and their influence on M-5 (but where there was no influence on M-2):

> **DRUG INTERACTIONS. Effect of Strong CYP3A4 Inducers on Regorafenib.** Co-administration of a strong CYP3A4 inducer (rifampin) with a single 160 mg dose of Stivarga decreased the mean exposure of regorafenib, increased...exposure of the active metabolite M-5, and resulted in no change in the mean exposure of the active metabolite M-2. Avoid concomitant use of strong CYP3A4 inducers (e.g. rifampin, phenytoin, carbamazepine, phenobarbital, and St. John's Wort).[292]

[289] Regorafenib (metastatic colorectal cancer) NDA 203-085. Pages 9 and 30–31 of 64-page Clinical Pharmacology Review.

[290] Package label. STIVARGA® (regorafenib) tablets, for oral use; August 2016 (21 pp.).

[291] Package label. STIVARGA® (regorafenib) tablets, for oral use; August 2016 (21 pp.).

[292] Package label. STIVARGA (regorafenib) tablets, oral; September 2012 (15 pp.).

The same Drug Interactions section warned against CYP3A4 inhibitors, and again warned that these inhibitors may decrease exposure to the active metabolites M-2 and M-5:

> **DRUG INTERACTIONS. Effect of Strong CYP3A4 Inhibitors on Regorafenib**. Co-administration of a strong CYP3A4 inhibitor (ketoconazole) with a single 160 mg dose of Stivarga increased the mean exposure of regorafenib and decreased. . .exposure of the active metabolites M-2 and M-5. Avoid concomitant use of strong inhibitors of CYP3A4 activity (e.g. clarithromycin, grapefruit juice, itraconazole, ketoconazole, posaconazole, telithromycin, and voriconazole).[293]

The package label for regorafenib (Stivarga®) referred to regorafenib's inhibition of UGT conjugating enzymes. The information was in the Clinical Pharmacology section of the package label:

> **CLINICAL PHARMACOLOGY**. . .Drug Interaction Studies. . .Effect of regorafenib on UGT1A1 substrates: In vitro studies showed that regorafenib, M-2, and M-5 competitively inhibit UGT1A9 and UGT1A1 at therapeutically relevant concentrations. Eleven patients received **irinotecan-containing combination chemotherapy with**. . .**regorafenib** at a dose of 160 mg. The mean AUC of SN-38 (SN-38 is a metabolite of irinotecan)[294] increased by 28% and the mean AUC of irinotecan increased by 44% when **irinotecan was administered 5 days after the last of 7 daily doses of regorafenib**.[295]

The implication of this information is concern that, with coadministration of regorafenib and irinotecan, the result could be increased toxicity from the irinotecan. The package label's information was based on the following steps of reasoning and logic:

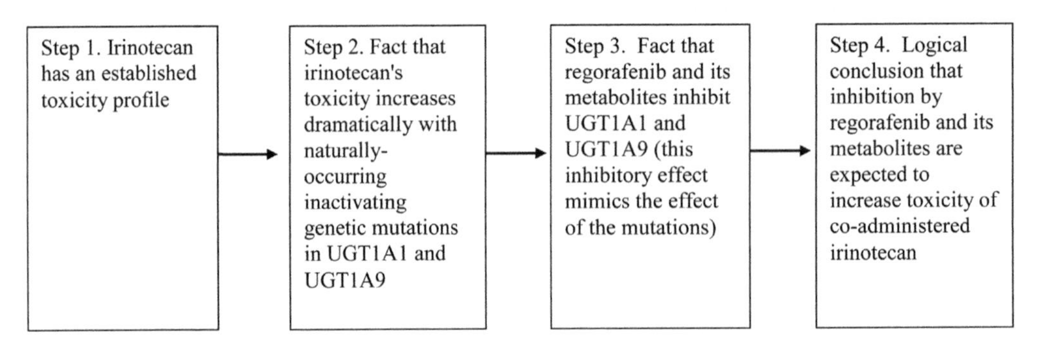

m. Tofacitinib (Rheumatoid Arthritis) NDA 203-214

Tofacitinib (Xeljanz®) is for treating rheumatoid arthritis. Tofacitinib's structure is shown below. The drug is metabolized to a number of compounds, including M8 and M9, which are oxidized on the pyrrolopyrimidine ring, M18, which is oxidized on the

[293] Package label. STIVARGA (regorafenib) tablets, oral; September 2012 (15 pp.).

[294] Regorafenib (colorectal cancer) NDA 203-085. Page 31 of 64-page ClinPharm Review.

[295] Package label. STIVARGA® (regorafenib) tablets, for oral use; August 2016 (21 pp.).

piperidine ring, and M20, which is tofacitinib glucuronide.[296] FDA's ClinPharm Review has extensive clinical data on DDIs, which include those for ketoconazole, fluconazole, rifampin, methotrexate, tacrolimus, cyclosporin, midazolam, and oral contraceptives.

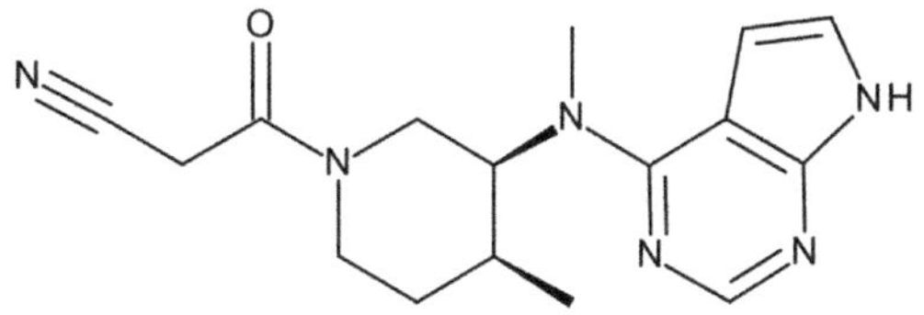

The Sponsor's DDI studies provided a basis for package label instructions to reduce tofacitinib's dose, where a coadministered drug was established as increasing tofacitinib's exposure. As a reference point, keep in mind the numerical value for the recommended dose of *5 mg twice daily*, in the absence of any coadministered drug, as shown by the following excerpt from the label:

> **DOSAGE AND ADMINISTRATION**. Rheumatoid Arthritis. The recommended dose of XELJANZ is 5 mg twice daily.[297]

i. Background on Azole Compounds as CYP Enzyme Inhibitors

A variety of azole compounds used for treating fungal infections have been tested for their contributions to DDIs in patients needing antifungal treatment. These azole compounds are fluconazole, itraconazole, ketoconazole, miconazole, posaconazole, and voriconazole.[298] The tests measured their inhibition of various CYP isozymes. Two of these, ketoconazole and fluconazole, are typically used for in vitro and in vivo DDI studies, as model compounds that are able to inhibit CYP3A4.

In a battery of tests assessing inhibition of CYP1A2, CYP2C9, CYP2C19, CYP2D6, and CYP3A4, ketoconazole was found to inhibit CYP3A4 the best, followed by a distant second, CYP2C19, and much more distant third, CYP2C9. Fluconazole was found to inhibit CYP2C19 the best, followed by a close second, CYP3A4, and a somewhat distant third, CYP2C9.[299]

ii. CYP3A4 Inhibitors Increase Tofacitinb Exposure (Study Published by the Sponsor)

A clinical study by Gupta et al.[300] assessed DDIs between ketoconazole and tofacitinib, and between fluconazole and tofacitinib, and determined that each of the coadministered

[296] Tofacitinib (rheumatoid arthritis) NDA 203-214. Pages 21–22 of 181-page Clinical Pharmacology Review.

[297] Package label. XELJANZ® (tofacitinib); November 2012 (27 pp.).

[298] Niwa T, et al. Drug interactions between nine antifungal agents and drugs metabolized by human cytochromes P450. Curr. Drug Metab. 2014;15:651–79.

[299] Lu C, et al. Quantitative prediction and clinical observation of a CYP3A inhibitor-based drug–drug interactions with MLN3897, a potent C–C chemokine receptor-1 antagonist. J. Pharmacol. Exp. Ther. 2010;332:562–8.

[300] Gupta P, et al. Evaluation of the effect of fluconazole and ketoconazole on the pharmacokinetics of tofacitinib in healthy adult subjects. Clin. Pharmacol. Drug Dev. 2014;3:72–7.

TABLE 7.9 Influence of Ketoconazole on Tofacitinib Exposure

Coadministered Drug	Study Drug	Percent Difference: [Tofacitinib Exposure Plus Coadministered Drug]/[Tofacitinib Exposure Minus Coadministered drug]	
		AUC Data	**Cmax Data**
Ketoconazole (inhibits both P-gp and CYP3A4)	Tofacitinib	203	116
Fluconazole (inhibits CYP2C19 and to a lesser extent CPY3A4)	Tofacitinib	179%	126%

Tofacitinib (rheumatoid arthritis) NDA 203-214. Page 35 (Table 19) and pages 124 and 126 (Table 57) and 180 of 181-page Clinical Pharmacology Review.

drugs increases tofacitinib exposure. Tofacitinib's AUC and Cmax were increased by 103% and 16%, respectively, with ketoconazole coadministration, and by 79% and 26%, respectively, with fluconazole coadministration. Please note that the statement, "Tofacitinib's AUC ... increased by 103%," means that tofacitinib's AUC was doubled.

This study was conducted by Pfizer, Inc. Pfizer, Inc. was also the Sponsor of NDA 203-214, and it can be seen that the data from Gupta et al.[301] and in FDA's ClinPharm Review are the same (the AUC and Cmax numbers are the same). Turning to the Sponsor's DDI study using ketoconazole and tofacibinib, the results in Table 7.9 demonstrate that there was an increase in tofacitinib exposure, where *tofacitinib's AUC increased by 103%* (doubling) and where tofacitinib's *Cmax increased by a lesser amount, 16%*.[302]

iii. Recommendation to Adjust Tofacitinib Does with Ketoconazole Coadministration

FDA's ClinPharm review took into account ketoconazole's potent inhibition of CYP3A4, acknowledged the Sponsor's recommendation to reduce tofacitinib's dose, and then expressly required that tofacitinib's dose be kept at a reduced level in the situation where there was coadministered ketoconazole.

FDA's recommendation for reducing tofacitinib took the form, "When given with ketoconazole (a potent CYP3A4 and P-gp inhibitor) tofacitinib AUC and Cmax increased by 103% and 16%, respectively...[s]ponsor recommended **keeping the maximum dose to 5 mg bid** in cases of co-administration with these drugs[303]...[i]f both 5 mg and 10 mg BID doses are approved, **tofacitinib dose should not exceed 5 mg BID** when it is coadministered with strong CYP3A4 inhibitors."[304]

[301] Gupta P, et al. Evaluation of the effect of fluconazole and ketoconazole on the pharmacokinetics of tofacitinib in healthy adult subjects. Clin. Pharmacol. Drug Dev. 2014;3:72–7.

[302] Tofacitinib (rheumatoid arthritis) NDA 203-214. Page 124 and Table 57 of 181-page Clinical Pharmacology Review.

[303] Tofacitinib (rheumatoid arthritis) NDA 203-214. Page 180 of 181-page Clinical Pharmacology Review.

[304] Tofacitinib (rheumatoid arthritis) NDA 203-214. Page 8 of 181-page Clinical Pharmacology Review.

This author points out an apparent discrepancy, in that FDA's review referred to reducing tofacitinib to 5 mg BID with coadministered ketoconazole while, in contrast, the package label required reducing tofacitinib to 5 mg once daily.

Package label. FDA's warnings about ketoconazole found a place on the Drug Interactions section:

> **DRUG INTERACTIONS**. Potent inhibitors of cytochrome P450 3 A4 (CYP3A4) (e.g., ketoconazole): **Reduce dose to 5 mg once daily**...[o]ne or more concomitant medications that result in both moderate inhibition of CYP3A4 and potent inhibition of CYP2C19 (e.g., fluconazole): **Reduce dose to 5 mg once daily.**[305]

iv. Recommendation to Adjust Tofacitinib Dose With Fluconazole Coadministration

FDA's ClinPharm review provided a rationale for the Sponsor's use of fluconazole, despite the fact that ketoconazole is more specific for CYP3A4, as compared to fluconazole (fluconazole inhibits both CYP3A4 and CYP2C19 to a similar extent).[306] The rationale was that, "The in-vitro and mass-balance suggest that tofacitinib is metabolized by CYP3A4 and to a lesser extent by CYP2C19. The sponsor recommends dose adjustment for patients receiving ***drug(s) that inhibit both CYP3A4 and CYP2C19*** (e.g., fluconazole) because of an approximate two-fold increase in exposure. However, tofacitinib dose adjustment is not warranted when co-administered with a CYP2C19 inhibitor."[307]

Package label. The Sponsor's idea that dose adjustments be made for patients receiving a drug that inhibits both CYP3A4 and CYP2C19 found a place on the package label, which instructed physicians to reduce the tofacitinib dose where patients are coadministered fluconazole. Note the unique and unusual concept described above, where the coadministered drug (fluconazole) was chosen because its catabolism by a variety of CYP enzymes (catabolism profile) closely matched the catabolism profile of the study drug (tofacitinib). The label reads:

> **DRUG INTERACTIONS**...One or more concomitant medications that result in both moderate inhibition of CYP3A4 and potent inhibition of CYP2C19 (e.g., fluconazole): Reduce dose to 5 mg once daily.[308]

v. Influence of a Second Drug (Rifampin) in Inducing CYP Enzymes

Rifampin's ability to induce CYP3A was shown with cultured hepatocytes, where the induction of CYP3A mRNA was determined with Northern blots. The Northern blot technique detected mRNA using a 46-mer nucleotide probe that corresponded to a 17-amino

[305] Package label. XELJANZ® (tofacitinib); November 2012 (27 pp.).

[306] Lu C, et al. Quantitative prediction and clinical observation of a CYP3A inhibitor-based drug–drug interactions with MLN3897, a potent C–C chemokine receptor-1 antagonist. J. Pharmacol. Exp. Ther. 2010;332:562–8.

[307] Package label. XELJANZ® (tofacitinib) tablets, for oral use; December 2015 (30 pp.).

[308] Package label. XELJANZ® (tofacitinib); November 2012 (27 pp.).

acid region in CYP3A. Also, the induction of CYP3A protein was determined by Western blots. Western blotting used an anti-CYP3A antibody.[309]

vi. Influence of a Second Drug (Rifampin) in Reducing Tofacitinib Exposure

In clinical studies, the Sponsor demonstrated that coadministering tofacitinib with rifampin resulted in severe reductions in tofacitinib exposure. With rifampin, tofacitinib's AUC decreased to only 16.1%, while Cmax decreased to 26.3%.[310] FDA concluded that rifampin should not be coadministered with the study drug, "because that will result in inefficacious concentrations of tofacitinib."

Package label. The label for tofacitinib (Xeljanz®) warned about the CYP-inducer rifampin in the Dosage and Administration section and also in the Drug Interactions section. The Full Prescribing Information section of the label provided the following information:

> **DOSAGE AND ADMINISTRATION**. Coadministration of potent inducers of CYP3A4 (e.g., rifampin) with XELJANZ may result in loss of or reduced clinical response to XELJANZ. Coadministration of potent inducers of CYP3A4 with XELJANZ is not recommended.[311]

> **DRUG INTERACTIONS**...Potent CYP3A4 Inducers. Tofacitinib exposure is decreased when XELJANZ is coadministered with potent CYP3A4 inducers (e.g., rifampin).[312]

vii. Drug–Drug Interactions at P-Glycoprotein

The Sponsor assessed tofacitinib's inhibition of P-gp mediated efflux of digoxin. Digoxin is transported by P-gp. The rate of digoxin transport is not influenced by metabolism by CYP enzymes, because digoxin is not a substrate for any of the CYP enzymes.[313] Cultured cell studies provided a value for the inhibition constant [IC50], while clinical studies provided the tofacitinib concentration [I]. IC50 is the concentration of inhibitor (tofacitinib) in the cell culture medium, required to inhibit a protein's transporting activity by 50%.[314] The Sponsor used the values of [I] and IC50 for calculating [I]/IC50.

viii. Hepatocyte Drug Transport Data

FDA observed that, "[I]/IC50 ratio is about 0.001...[which is] significantly below the level where a digoxin interaction study would be warranted."

[309] Hosagrahara VP, et al. Induction of the metabolism of midazolam by rifampin in cultured porcine hepatocytes: preliminary evidence for CYP3A isoforms in pigs. Drug Metab. Dispos. 1999;27:1512–8.

[310] Tofacitinib (rheumatoid arthritis) NDA 203-214. Pages 8, 35–36, 129–130 of 181-page Clinical Pharmacology Review.

[311] Package label. XELJANZ® (tofacitinib); December 2015 (30 pp.).

[312] Package label. XELJANZ® (tofacitinib); December 2015 (30 pp.).

[313] Fenner KS, et al. Drug–drug interactions mediated through P-glycoprotein: clinical relevance and in vitro–in vivo correlation using digoxin as a probe drug. Clin. Pharmacol. Ther. 2009;85:173–81.

[314] Lumen AA, et al. If the KI is defined by the free energy of binding to P-glycoprotein, which kinetic parameters define the IC50 for the Madin-Darby canine kidney II cell line overexpressing human multidrug resistance 1 confluent cell monolayer? Drug Metab. Dispos. 2010;38:260–9.

The concentration of study drug needed for 50% inhibition of digoxin transport by cultured hepatocytes was 1000 times greater than the concentration of study drug in blood plasma in human subjects. Because of this situation, FDA concluded that there was no need to conduct actual DDIs in human subjects, where the subjects were coadministered both study drug and digoxin.

ix. Enterocyte Drug Transport Data

Also, FDA assessed tofacitinib's influence on P-gp mediated drug transport in enterocytes. For both types of cells (liver and gut), FDA concluded that tofacitinib does not inhibit P-gp mediated transport, referring to two different cutoff values, one for use with enterocytes and one for use with the gut. FDA concluded that, "Tofacitinib has low potential to inhibit P-gp with an estimated IC50 value of 311 micromolar. At a steady-state unbound **Cmax of 310 nanomolar** and projected **gut concentration of 128 micromolar**. . .following a 10 mg BID dose, the systemic [I]/IC50 ratio is 0.001 and the gut [I]/IC50 ratio is 0.4. . .[b]oth of these ratios are significantly below the level where a digoxin interaction study would be warranted, i.e., >0.1 and >10, respectively."[315]

The phrase "Cmax of 310 namomolar" refers to tofacitinib's concentration in the bloodstream (systemic concentration). The phrase "128 micromolar" refers to a projected gut lumen concentration. The value of "128 micromolar" is that of 10 mg tofacitinib dissolved in 250 mL of solution.[316] Apparently the value of "250 mL solution" was based on the volume of a glass of water (240 mL) used by pharmaceutical companies who wish to file drug applications with drug regulatory agencies in the United States, Europe, and Japan. FDA's Guidance for Industry recommends that the drug product be ingested with 240 mL water.[317]

Another way of stating FDA's conclusion is that a clinical study would be justified, where the [I]/IC50 value is greater than 0.1 (plugging in blood plasma concentration) or where the [I]/IC50 value is greater than 10 (plugging in gut concentration).

Package label. FDA's account of P-gp found a place on the package label of tofacitinib (Xeljanz®) in the Clinical Pharmacology section. The label separately took into account tofacitinib's ability to inhibit transport of a coadministered drug, and a coadministered drug's ability to inhibit transport of tofacitinib. In both cases, the label stated that DDIs occurring at one or another of the drug transporters were low:

> **CLINICAL PHARMACOLOGY**. . .Potential for XELJANZ to Influence the PK of Other Drugs. . .In vitro data indicate that the potential for tofacitinib to inhibit transporters such as P-glycoprotein, organic anionic or cationic transporters at therapeutic concentrations is **low**. . .Potential for Other Drugs to Influence the PK of Tofacitinib. . .Inhibitors of. . .P-glycoprotein are **unlikely** to substantially alter the PK of tofacitinib.[318]

[315] Tofacitinib (rheumatoid arthritis) NDA 203-214. Pages 33 and 39 of 181-page Clinical Pharmacology Review.

[316] Tofacitinib (rheumatoid arthritis) NDA 203-214. Page 33 of 181-page Clinical Pharmacology Review.

[317] Mudie DM, et al. Quantification of gastrointestinal liquid volumes and distribution following a 240 mL dose of water in the fasted state. Mol. Pharm. 2014;11:3039–47.

[318] Package label. XELJANZ® (tofacitinib); November 2012 (27 pp.).

X. ORGANIC ANION TRANSPORTER POLYPEPTIDE (OATP) DATA

a. Background Information on Axitinib, Nilotinib, Pazopanib, and Sorafenib Transport

OAT polypeptide (OATP) family of transporters are expressed by hepatocytes. These are OATP1B1, OATP1B3, and OATP2B1.[319] These OATP transporters mediate uptake of drugs from the bloodstream through the basolateral membrane. Once inside the hepatocyte, the drugs can then be effluxed through the apical membrane via P-gp into the bile. Hu et al.[320] provide an account of various tyrosine kinase inhibitors (TKI) and their ability to inhibit OATP1B1. Axitinib, nilotinib, pazopanib, and sorafenib all inhibit human OATP1B1, where this inhibition blocks OATP1B1-mediated transport of another drug (docetaxel). Docetaxel transport is inhibited by more than 90%. The result of this inhibition is that docetaxel exposure increases by about 80%.

A view of the package label for one of these drugs (pazopanib) reveals that the label discloses this particular DDI and warns about increases in exposure of coadministered drugs:

> **CLINICAL PHARMACOLOGY**...In vitro studies also showed that **pazopanib** inhibits...organic anion-transporting polypeptide (OATP1PB) with IC50 of 0.79 μM...**[p]azopanib** may increase concentrations of drugs eliminated by...OATP1B1.[321]

Khurana et al.[322] describe the dramatic scenario where OATP-mediated transport of a drug into the hepatocyte can control not only the drug's fate of excretion into the bile, but also the drug's fate of destruction by the hepatocyte's cytochrome P450 enzymes. In the words of Khurana et al., "OATP1B1 plays a vital role in hepatic uptake of paclitaxel, making it vulnerable to metabolism by CYP3A4 and ultimately accelerating elimination by biliary secretion via P-glycoprotein (P-gp)."

The phrase, "ultimately accelerating elimination by biliary secretion" refers to the transport of paclitaxel via OATP1B1 through the basolateral membrane, followed by transport via P-gp through the apical membrane and into the bile duct.

b. Drug–Drug Interactions of Ponatinib Occurring at OATP

Note that the above commentary concerns *pazopanib*, while the following concerns a different drug, *ponatinib*. Regarding ponatinib, the Sponsor determined if ponatinib inhibited OATP-mediated uptake of a second drug, or if a second drug inhibited OATP-mediated

[319] Konig J, et al. Transporters and drug–drug interactions: important determinants of drug disposition and effects. Pharmacol. Rev. 2013;6:944–66.

[320] Hu S, et al. Inhibition of OATP1B1 by tyrosine kinase inhibitors: in vitro–in vivo correlations. Br. J. Cancer 2014;110:894–8.

[321] Package label. VOTRIENT® (pazopanib) tablets, for oral use; May 2017 (25 pp.).

[322] Khurana V, et al. Inhibition of OATP-1B1 and OATP-1B3 by tyrosine kinase inhibitors. Drug Metabol. Drug Interact. 2014;29:249–59.

uptake of ponatinib. Kidney epithelial cells were transfected with the OATP transporter of interest, thereby ensuring that transport tests with a given transfected cell line would be mediated by that specific OATP transporter. The individual cell lines expressed one of several types of OATP transporters. FDA's ClinPharm Review observed that

> Human...kidney epithelial cells...transfected with individual uptake transporter...were used to assess the substrate and inhibition potential of **ponatinib** toward the corresponding transporter. The cells were incubated with **ponatinib** (0.5, 1, and 2 μM) for 5, 10 and 20 min...[t]ransporter-specific positive controls were run in parallel and treated identically...[t]his study reports that **ponatinib** is not a substrate of OATP1B1/OATP1B3...[i]n addition, **ponatinib** was found not to be an inhibitor of OATP1B1, OATP1B3...OAT1, and OAT3...[t]he reviewer finds these results and the applicant's conclusion acceptable. No additional in vivo trials are required at this time.[323]

FDA's comments on ponatinib and OAT transporters found a place on the package label, as shown below.

Package label. The Clinical Pharmacology section of the package label provided this concise account of the OAT uptake transporter results:

> **CLINICAL PHARMACOLOGY**...Ponatinib is a weak substrate for both P-gp and ABCG2 in vitro. Ponatinib is not a substrate for organic anion transporting polypeptides (OATP1B1, OATP1B3) and organic cation transporter 1 (OCT1) in vitro.[324]

XI. A SMALL MOLECULE DRUG CAN STIMULATE THE CATABOLISM OF ITS OWN SELF

Small molecule drugs can stimulate their own catabolism. In other words, the drug induces expression of a cytochrome P450 enzyme that mediates catabolism of the same drug. The ability of small molecules to stimulate their own self-destruction is documented by the following list. Hence, DDI analysis sometimes includes an analysis of the study drug's ability to stimulate catabolism of its own self:

- carbamazepine[325]
- lamotrigene[326]
- rifampin[327]

[323] Ponatinib (chronic myeloid leukemia) NDA 203-469. Page 30 of 89-page Clinical Pharmacology Review.

[324] Package label. ICLUSIG® (ponatinib) tablets for oral use; November 2016 (23 pp.). NDA 203-469.

[325] Anderson GD. A mechanistic approach to antiepileptic drug interactions. Ann. Pharmacother. 1998;32:554–63.

[326] Anderson GD. A mechanistic approach to antiepileptic drug interactions. Annals Pharmacother. 1998;32:554–63.

[327] Blaschke TF, Skinner MH. The clinical pharmacokinetics of rifabutin. Clin. Infect. Dis. 1996;22 (Suppl. 1): S515–22.

- rifabutin[328]
- efavrenz[329]
- nevirapine[330]
- all-*trans*-retinoic acid[331,332]
- vitamin D[333]

XII. CONCLUDING REMARKS

Take-home lessons from this present chapter which are broadly applicable to a variety of FDA-submissions include:

- *Commonly used coadministered drugs.* For DDI studies, the second drug should be chosen from drugs that are commonly used in clinical practice with the first drug. This is illustrated for regorafenib (second drug: irinotecan) and lumacaftor/ivacaftor combination (second drug: antibiotics, antifungals, and antidepressants).
- *Fixed drug combinations.* For FDC, if there is a DDI between any two drugs in the FDC, and where the result is that one drug provokes increased exposure of another of the drugs in the FDC, the Sponsor should consider reducing the milligrams of the other drug in the FDC. This is illustrated for the lumacaftor/ivacaftor combination.
- *Instructions to reduce dose.* Package label recommendations to reduce the amount of the study drug should be considered, where a coadministered second drug increases exposure of the study drug. This type of package label instruction is illustrated for axitinib, cabozantinib, fingolimod, ponatinib, and tofacitinib.
- *Gastric interactions.* DDIs caused by events in the stomach may occur where the study drug or a coadministered drug is intended to influence gastric physiology. This situation can occur with drugs that are proton pump inhibitors, antacids, and drugs intended to slow gastric emptying. Gastric interactions are illustrated here by bosutinib and dabrafenib and, in the next chapter, by lixisenatide.[334]

[328] Blaschke TF, Skinner MH. The clinical pharmacokinetics of rifabutin. Clin. Infect. Dis. 1996;22 (Suppl. 1):S515–22.

[329] Gerber JG. Using pharmacokinetics to optimize antiretroviral drug–drug interactions in the treatment of human immunodeficiency virus infection. Clin. Infect. Dis. 2000;30 (Suppl. 2):S123-9.

[330] Gerber JG. Using pharmacokinetics to optimize antiretroviral drug–drug interactions in the treatment of human immunodeficiency virus infection. Clin. Infect. Dis. 2000;30 (Suppl. 2):S123-9.

[331] Krekels MD, et al. Induction of the oxidative catabolism of retinoic acid in MCF-7 cells. Br. J. Cancer 1997;75:1096–104.

[332] Marikar Y, et al. Retinoic acid receptors regulate expression of retinoic acid 4-hydroxylase that specifically inactivates all-*trans* retinoic acid in human keratinocyte HaCaT Cells. J. Investig. Dermatol. 1998;111:434–9.

[333] Chesdachai S, et al. The effects of first-line anti-tuberculosis drugs on the actions of vitamin D in human macrophages. J. Clin. Transl. Endocrinol. 2016;6:23–9.

[334] Lixisenatide (type-2 diabetes mellitus) NDA 208-471. Pages 13–14 of 198 Clinical Pharmacology Review.

- *Safety signals.* Wherever relevant, DDI studies should be designed with an eye to adverse events of increased concern, i.e., where increased exposure of the study drug or a coadministered drug is expected to provoke an adverse event. This is illustrated by bosutinib, where concern was for cardiac AEs.
- *Common efficacy target.* DDIs can be located at sites other than at one of the cytochrome P450 enzymes or one of the drug transporters. DDIs can occur in a biochemical pathway needed for efficacy of a coadministered drug. This is illustrated for fingolimod, where the coadministered drug is a vaccine. In detail, the goal of fingolimod is to inhibit T cells, and the mechanism of action of vaccines includes activating T cells. Yet another example of Drug A interfering with the mechanism of action of Drug B is that of evolocumab and statin drugs, as detailed in the next chapter.[335,336] This category of DDI is also illustrated for brivaracetam, where FDA assessed overlapping efficacy of a coadministered drug, levetiracetam.
- *Common toxicity target.* DDIs can also occur where a drug's toxicity inflicts the same cell or tissue as a coadministered drug's toxicity. This was the case for fingolimod, which has cardiac toxicity (may prolong the QT interval), where the label's Drug Interaction section warned that, "patients on QT prolonging drugs, e.g., citalopram, chlorpromazine, haloperidol, methadone, erythromycin, should be monitored overnight with continuous ECG in a medical facility."[337]

FDA reviews suggest the following multistep algorithm for arriving at the Drug Interactions section of the package label. The algorithm provided below takes into account only CYP enzymes and not other enzymes involved in drug catabolism, and takes into account only P-gp and not other transporters.

Step One. Assess if the study drug influences the rate of metabolism of a coadministered second drug, as determined by in vitro assays of CYP enzymes and in vitro P-gp efflux assays. Then assess the converse situation, that is, if a second drug influences the rate of metabolism of the study drug, as determined by in vitro assays of CYP enzymes and in vitro P-gp efflux assays. If there is no influence, stop and do not conduct any clinical DDI studies and do not include any information in the Drug Interaction section. But if there is an influence, proceed to Step Two.

Step Two. Acquire the value for the ratio [I]/Ki. This formula is described by papers from Ito et al.,[338,339] where one of these papers was cited by FDA's Guidance for

[335] Welder G, et al. High-dose atorvastatin causes a rapid sustained increase in human serum PCSK9 and disrupts its correlation with LDL cholesterol. J. Lipid Res. 2010;51:2714–21.

[336] Evolocumab (hyperlipidemia and hypercholesterolemia) BLA 125-522. Pages 51 and 57 of 148 page Clinical Pharmacology Review.

[337] Package label. GILENYA (fingolimod) capsules, for oral use; February 2016 (25 pp.).

[338] Ito K, et al. Which concentration of the inhibitor should be used to predict in vivo drug interactions from in vitro data? AAPS PharmSci. 2002;4:53–60.

[339] Ito K, et al. Database analysis for the prediction of in vivo drug–drug interactions from in vitro data. Br. J. Pharmacol. 2004;57:473–86.

Industry.[340] The value for [I] can be that of plasma total drug or it can be the concentration of the unbound (free and not bound to any plasma proteins) drug.[341] Where the values for [I] and Ki are plugged into the ratio formula [I]/Ki, and where the calculated value is less than 0.1, this means that the DDI is of low risk for patients, that the drug interaction inquiry may come to a halt, and that information should not be included in the Drug Interactions section, though the medical writer might want to consider including the in vitro DDI data in the Clinical Pharmacology section of the package label.

But where the value is between 0.1 and 1.0, thus showing medium risk, or where the value is greater than 1.0 showing high risk for DDIs in patients, the Sponsor should consider conducting a DDI clinical study.[342,343]

Step Three. If the ratio formula indicates medium or high risk, the Sponsor conducts clinical studies, where the study drug is coadministered with a second drug. The studies are designed to determine if the study drug influences exposure of the second drug, or the converse situation, testing if a second drug influences exposure of the study drug. If the clinical DDI studies reveal that drug interactions actually occur, these are provided in the Drug Interactions section.

APPENDIX A

Probe substrates that are specifically degraded by the indicated CYP enzyme, and inhibitors that specifically inhibit the indicated CYP enzyme. For in vitro CYP studies, the "probe drug" does not have to be a drug that is actually used in humans, it can be any organic molecule that is established to be metabolized by a specific CYP enzyme.

Cytochrome P450 enzyme	Probe drug for in vitro studies, to assess CYP enzyme activity§	Probe drug for human studies, to assess CYP enzyme activity	CYP enzyme inhibitors for in vitro studies
CYP1A2	Phenacetin *O*-deethylation	Theophylline, caffeine	Enoxacin
CYP2B6	Efavirenz hydroxylation; bupropion hydroxylation	Efavirenz, *S*-bupropion (hydroxylation)	Ticlopidine

[340] U.S. Dept of Health and Human Services. Food and Drug Administration. Center for Drug Evaluation and Research (CDER). Guidance for Industry. Drug interaction studies—study design, data analysis, implications for dosing and labeling recommendations; 2012 (Page 20 (Figure 2) of 75 pages).

[341] Ito K, et al. Database analysis for the prediction of in vivo drug–drug interactions from in vitro data. Br. J. Pharmacol. 2004;57:473–86.

[342] Ito K, et al. Database analysis for the prediction of in vivo drug–drug interactions from in vitro data. Br. J. Pharmacol. 2004;57:473–86.

[343] Williams JA, et al. Drug–drug interactions for UDP-glucuronosyltransferase substrates: a pharmacokinetic explanation for typically observed low exposure (AUCi/AUC) ratios. Drug Metab. Dispos. 2004;32:1201–8.

CYP2C8	Paclitaxel 6-hydroxylation, amodiaquine *N*-deethylation	Amodiaquine (*N*-deethylation), repaglinide	Gemfibrozil
CYP2C9	*S*-Warfarin 7-hydroxylation, diclofenac 4′-hydroxylation	*S*-Warfarin, tolbutamide	Fluconazole
CYP2C19	*S*-mephenytoin 4′-hydroxylation	Omeprazole	Omeprazole
CYP2D6	Bufarolol 1′-hydroxylation	Metoprolol, desipramine	Quinidine, paroxetine, fluoxetine
CYP3A	Midazolam 1-hydroxylation, testosterone 6β-hydroxylation	Midazolam	Ketoconazole, itraconazole, ritonavir, clarithromycin

European Medicines Agency. Guideline on the investigation of drug interactions; 2012 (59 pp.).
§ FDA has published essential the same list of probes for in vitro use. U.S. Food and Drug Administration. Drug development and drug interactions: table of substrates, inhibitors and inducers; September 2016 (FDA website accessed 03.02.17).

CHAPTER

8

Drug–Drug Interactions: Part Two (Therapeutic Proteins)

I. INTRODUCTION

When a submits a marketing application for a drug that is a biologic, the application is called a Biologics License Application (BLA).[1] Biologics include antibodies,[2] enzymes, vaccines,[3] recombinant products,[4] gene therapy,[5] blood products,[6] and

[1] U.S. Department of Health and Human Services. Food and Drug Administration. Center for Biologics Evaluation and Research (CBER). Guidance for industry. Providing regulatory submissions to the Center for Biologics Evaluation and Research (CBER) in electronic format—biologics marketing applications; November 1999 (63 pp.).

[2] U.S. Department of Health and Human Services. Food and Drug Administration. Center for Drug Evaluation and Research (CDER). Center for Biologics Evaluation and Research (CBER). Guidance for industry. Monoclonal antibodies used as reagents in drug manufacturing; 2001 (8 pp.).

[3] U.S. Department of Health and Human Services. Food and Drug Administration. Center for Biologics Evaluation and Research (CBER). Guidance for industry. Clinical considerations for therapeutic cancer vaccines; 2011 (16 pp.).

[4] U.S. Department of Health and Human Services. Food and Drug Administration. Center for Drug Evaluation and Research (CDER). Center for Biologics Evaluation and Research (CBER). Guidance for industry. Q6B specifications: test procedures and acceptance criteria for biotechnological/biological products; 1999 (21 pp.).

[5] U.S. Department of Health and Human Services. Food and Drug Administration. Center for Biologics Evaluation and Research (CBER). Guidance for industry. Gene therapy clinical trials—observing subjects for delayed adverse events; 2006 (23 pp.).

[6] U.S. Department of Health and Human Services. Food and Drug Administration. Center for Biologics Evaluation and Research (CBER). Guidance for industry. Bacterial risk control strategies for blood collection establishments and transfusion services to enhance the safety and availability of platelets for transfusion; 2016 (42 pp.).

FDA's Drug Review Process and the Package Label
DOI: https://doi.org/10.1016/B978-0-12-814647-7.00008-7

stem cells.[7] This chapter concerns drug–drug interactions (DDIs) involving biologics. These interactions include the following:

- Where the biologic is a cytokine, where the cytokine stimulates expression of one or more CYP enzymes, and where the increased CYP enzyme reduces exposure of a coadministered small molecule drug.
- Where the biologic is an anti-cytokine antibody, where this antibody blocks the cytokine's action resulting in reduced CYP enzyme expression, with the consequent increased exposure of a coadministered small molecule drug.
- Where the biologic is an antibody, and where a coadministered small molecule stimulates expression of the antibody's target, thus resulting in increases in formation of the antibody/target complex. In this way, the increase in target can deplete the antibody, i.e., reduce exposure of the antibody.
- Small molecule drug blocking the mechanism of action of a therapeutic protein.
- Therapeutic protein stimulating the patient's immune response to generate antibodies against the therapeutic protein that block the therapeutic protein's efficacy. (This is a frequent problem for biologics, though it appears not to have been classified as a DDI. But this topic fits best into this chapter.)

As stated earlier in this book, the term "exposure" is used in pharmacokinetics studies in referring to parameters of drug concentrations in the bloodstream, such as AUC, Cmax, tmax, and Cmin (Ctrough). Cmin is the minimal concentration residing in between two successive doses of the drug.

This concerns the universal technique of assessing if Drug A influences exposure of Drug B and for assessing the converse situation, that is, if Drug B influences exposure of Drug A. Regarding small molecule drugs that alter CYP enzyme activity or transporter activity, the possibility that a small molecule drug can influence CYP-mediated catabolism or transporter-mediated transport of a biologic is remote.

Unlike the case for small molecules, biologics are rarely or never substrates of cytochrome P450, bind to transcription factors that induce cytochrome P450, and interact with drug transporters such as P-glycoprotein (P-gp). Despite the fact that biologics, as a rule, do not interact with CYP enzymes, transcription factors, or drug transporters, FDA still contemplates and assesses these improbable scenarios.

Now, regarding the converse situation, what is a frequent concern is that a therapeutic protein can influence expression of CYP enzymes or of transporters, thus altering exposure of a coadministered small molecule drug.

a. Cytokine-Mediated Drug–Drug Interactions

Where a therapeutic antibody targets a cytokine, the consequent blocking of cytokine-mediated signaling can influence expression of one or more CYP enzymes.

[7] U.S. Department of Health and Human Services. Food and Drug Administration. Center for Biologics Evaluation and Research (CBER). Center for Devices and Radiological Health (CDRH). Office of Combination Products (OCP). Guidance for industry and FDA staff. Homologous use of human cells, tissues, and cellular and tissue-based products; 2015 (7 pp.).

The resulting changes in CYP enzyme activity, in turn, can influence the rate of catabolism of a coadministered small molecule drug.[8,9,10]

An exemplary situation is interleukin-6 (IL-6), a cytokine that reduces CYP3A4 activity. Hepatocyte studies demonstrate that IL-6 dramatically reduces expression of mRNA encoding CYP3A4, and as a consequence, also reduces the amount of residual CYP3A4 protein.[11] Simvastatin is one of the substrates of CYP3A4.[12] CYP3A4 catalyzes the conversion of simvastatin to simvastatin 3′-hydroxysimvastatin and 3′,5′-dihydrodiolsimvastatin.[13,14] Consistently, an antibody that blocks binding of IL-6 to its receptor, namely, tocilizumab, reduces simvastatin exposure when tocilizumab is coadministered with simvastatin.

b. Example of Ustekinumab Changing Exposure of Coadministered Warfarin

To provide an example, the Drug Interaction section of the package label for *ustekinumab* warns that ustekinumab blocks IL-12, where there is a consequent reduction in CYP enzymes. The package label warns that the reduced expression of CYP enzymes, could require dose-modifications of any coadministered warfarin or cyclosporine. The label refers to the scenario where chronic inflammation increases expression of inflammatory cytokines and where administering the antiinflammatory drug ustekinumab (Stelara®) counteracts the inflammatory cytokines:

> **DRUG INTERACTIONS...CYP450 Substrates.** The formation of CYP450 enzymes can be altered by increased levels of certain cytokines (e.g., IL-1, IL-6, IL-10, TNFα, IFN) during chronic inflammation. Thus, STELARA®, an antagonist of IL-12 and IL-23, could normalize the formation of CYP450 enzymes. Upon initiation of STELARA® in patients who are receiving concomitant CYP450 substrates, particularly those with a narrow therapeutic index, monitoring for therapeutic effect (e.g., for warfarin) or drug concentration (e.g., for cyclosporine) should be considered and the individual dose of the drug adjusted as needed.[15]

[8] Lee JI, et al. CYP-mediated therapeutic protein drug interactions. Clin. Pharmacokinet. 2010;49:295–310.

[9] Wang J, et al. Biological products for the treatment of psoriasis: therapeutic targets, pharmacodynamics and disease-drug-drug interaction implications. AAPS J. 2014;16:938–47.

[10] Evers R, et al. Critical review of preclinical approaches to investigate cytochrome p450-mediated therapeutic protein drug-drug interactions and recommendations for best practices: a white paper. Drug Metab. Dispos. 2013;41:1598–609.

[11] Evers R, et al. Critical review of preclinical approaches to investigate cytochrome p450-mediated therapeutic protein drug-drug interactions and recommendations for best practices: a white paper. Drug Metab. Dispos. 2013;41:1598–609.

[12] Stoll F, et al. Reduced exposure variability of the CYP3A substrate simvastatin by dose individualization to CYP3A activity. J. Clin. Pharmacol. 2013;53:1199–204.

[13] Prueksaritanont T, et al. The human hepatic metabolism of simvastatin hydroxy acid is mediated primarily by CYP3A, and not CYP2D6. Br. J. Clin. Pharmacol. 2003;56:120–4.

[14] Prueksaritanont T, et al. In vitro metabolism of simvastatin in humans [SBT] identification of metabolizing enzymes and effect of the drug on hepatic P450s. Drug Metab. Dispos. 1997;25:1191–99.

[15] Package label. STERLARA® (ustekinumab) injection, for subcutaneous or intravenous use; September 2016 (27 pp.).

Broadly viewed, the take-home lesson is that the onset of chronic inflammation could require dose modification for drugs such as warfarin, and that initiating ustekinumab therapy could also require dose modification for warfarin.

Ustekinumab is used for treating psoriasis, a disease characterized by chronic inflammation, where this chronic inflammation stimulates expression of CYP enzymes. The DDI of concern with ustekinumab treatment is that ustekinumab will have the desired effect of reducing psoriasis but at the same time, have the undesirable effect of reducing CYP enzyme levels back to a normal level. This effect is undesirable because it changes the expected exposure of any coadministered warfarin or cyclosporine.

c. Evolocumab—Example of Small Molecule Drug Reducing Exposure of Antibody Drug

Evolocumab provides a type of DDI mechanism, where Drug A interferes with, or modulates in some way, the mechanism of Drug B. This same category of DDI was revealed in the previous chapter, in the account of fingolimod (for treating multiple sclerosis), where fingolimod interfered with the mechanism of vaccines (for treating infections). The situation with evolocumab is that a coadministered drug (statin) stimulates expression of a plasma protein (PCSK9), where this plasma protein is targeted by evolocumab. The end-result in the DDI scenario is reduced evolocumab exposure.

Evolocumab binds to a protein (PCSK9) that circulates in the bloodstream. As evolocumab is used for treating high cholesterol and statin drugs are also for treating high cholesterol, evolocumab and statins are sometimes coadministered. The package label for evolocumab (Repatha®) reveals the need to coadminister evolocumab:

> **PATIENT INFORMATION**...REPATHA is used...along with diet and...statin therapy in adults with heterozygous familial hypercholesterolemia...or atherosclerotic heart or blood vessel problems, who need additional lowering of LDL cholesterol.[16]

But a problem with statins is that a coadministered statin drug stimulates an increase in plasma PCSK9. This stimulation occurs independently of evolocumab's binding to PCSK9. The consequence of this increase in PCSK9 is increased formation of antibody/PCSK9 complex and reduced exposure of the therapeutic antibody.[17,18]

FDA's ClinPharm Review described the scenario where coadministered statins provoke an increase in circulating PCSK9, and raised the concern that increased doses of evolocumab might be needed. FDA's comment on the DDI where statins reduced evolocumab

[16] Package label. REPATHA (evolocumab) injection, for subcutaneous use; August 2015 (34 pp., including the Patient Information section).

[17] Welder G, et al. High-dose atorvastatin causes a rapid sustained increase in human serum PCSK9 and disrupts its correlation with LDL cholesterol. J. Lipid Res. 2010;51:2714–21.

[18] Evolocumab (hyperlipidemia and hypercholesterolemia) BLA 125-522. Pages 51 and 57 of 148-page Clinical Pharmacology Review.

took the form, "It has been reported that statins upregulate PCSK9."[19,20] Dong et al.[21] reported that "**rosuvastatin** increased. . .liver expression of. . .HNF1α. . .a key transactivator for PCSK9 gene expression. Welder et al.[22] reported that **atorvastatin**. . .caused a rapid. . .increase in serum PCSK9. . .[t]hey put forth an explanation for why proportional LDL-cholesterol lowering was not achieved with increasing doses of statin. Based on the mechanism, it would be expected that statins by increasing circulating PCSK9 levels, would reduce the effectiveness of evolocumab."[23]

The Clinical Pharmacology section for evolocumab described the ability of statins to increase PCSK9, and where the consequence was reduced evolocumab exposure, but added that the reduced evolocumab exposure was not severe enough to require an increased evolocumab dose:

> **CLINICAL PHARMACOLOGY**. . .Drug Interaction Studies. An approximately **20% decrease in the Cmax and AUC of evolocumab** was observed in patients co-administered with a high-intensity statin regimen. This difference is not clinically meaningful and does not impact dosing recommendations.[24]

d. Evolocumab—Mechanism of Action of Evolocumab and PCSK9, and of Low Density Lipoprotein Receptor Destruction

The blocking of PCSK9 by evolocumab is shown immediately below, where the goal of evolocumab is to reduce plasma low density lipoprotein (LDL)-cholesterol and thus reduce risk for atherosclerosis:

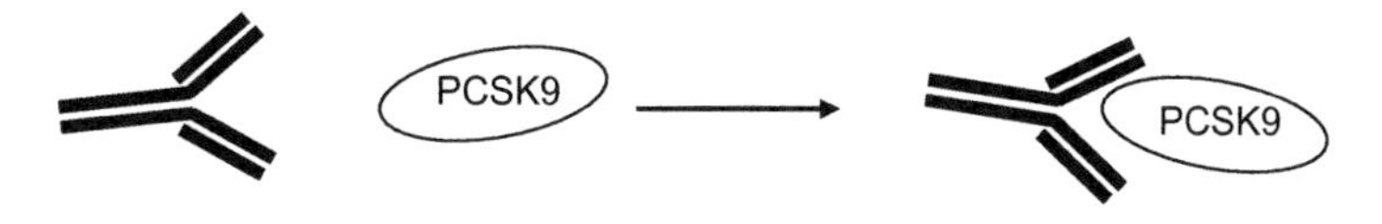

[19] Dubuc G, et al. Statins upregulate PCSK9, the gene encoding the proprotein convertase neural apoptosis-regulated convertase-1 implicated in familial hypercholesterolemia. Arteriosler. Thromb. Vasc. Biol. 2004;24:1454–9.

[20] Careskey HE, et al. Atorvastatin increases human serum levels of proprotein convertase subtilisin/kexin type 9. J. Lipid Res. 2008;49:394–8.

[21] Dong B, et al. Strong induction of PCSK9 gene expression through HNF1alpha and SREBP2: mechanism for the resistance to LDL-cholesterol lowering effect of statins in dyslipidemic hamsters. J. Lipid Res. 2010;51:1486–95.

[22] Welder G, et al. High-dose atorvastatin causes a rapid sustained increase in human serum PCSK9 and disrupts its correlation with LDL cholesterol. J. Lipid Res. 2010;51:2714–21.

[23] Evolocumab (hyperlipidemia and hypercholesterolemia) BLA 125-522. Page 51 of 148-page Clinical Pharmacology Review.

[24] Package label. REPATHA (evolocumab) injection, for subcutaneous use; August 2015 (15 pp.).

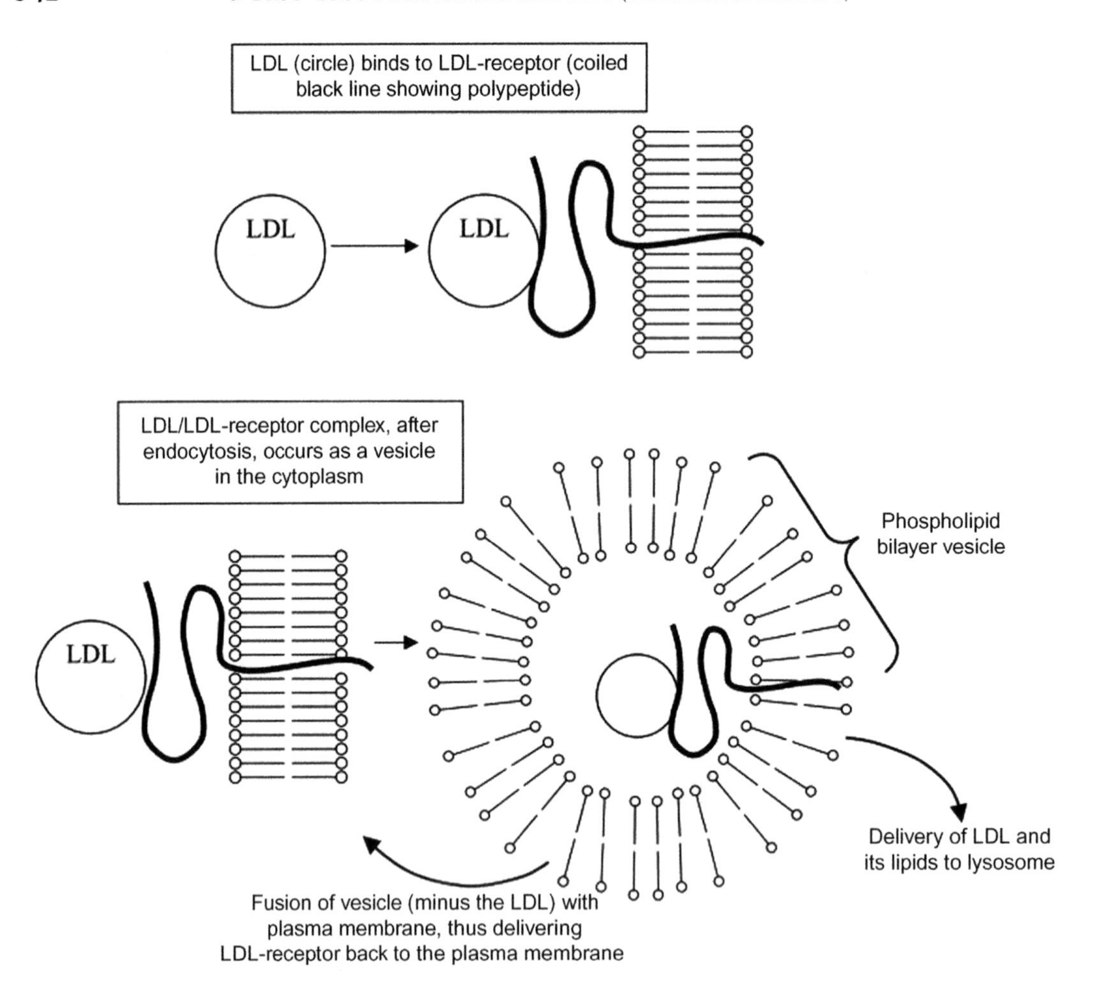

FIGURE 8.1 Fate of LDL-cholesterol, LDL, and LDL-receptor. The top picture shows binding of LDL to LDL-receptor. The lower picture shows invagination of the plasma membrane to create a phospholipid bilayer vesicle, followed by delivery of the LDL and its lipids to the lysosomes and recycling of LDL-receptor back to the plasma membrane.

Figs. 8.1 and 8.2 show the pathway for LDL, in its journey from the cytoplasm, with various fates.[25,26,27] Fig. 8.1 shows LDL metabolism in *absence of PCSK9* and Fig. 8.2 illustrates LDL metabolism in the *presence of PCSK9*.

[25] Reyes-Soffer G, et al. Effects of PCSK9 inhibition with alirocumab on lipoprotein metabolism in healthy humans. Circulation. 2017;135:352–62.

[26] Shapiro MD, Fazio S. PCSK9 and atherosclerosis—lipids and beyond. J. Atheroscler. Thromb. 2017;24:462–72.

[27] Wong ND, et al. Advances in dyslipidemia management for prevention of atherosclerosis: PCSK9 monoclonal antibody therapy and beyond. Cardiovasc. Diagn. Ther. 2017;7(Suppl. 1):S11–20.

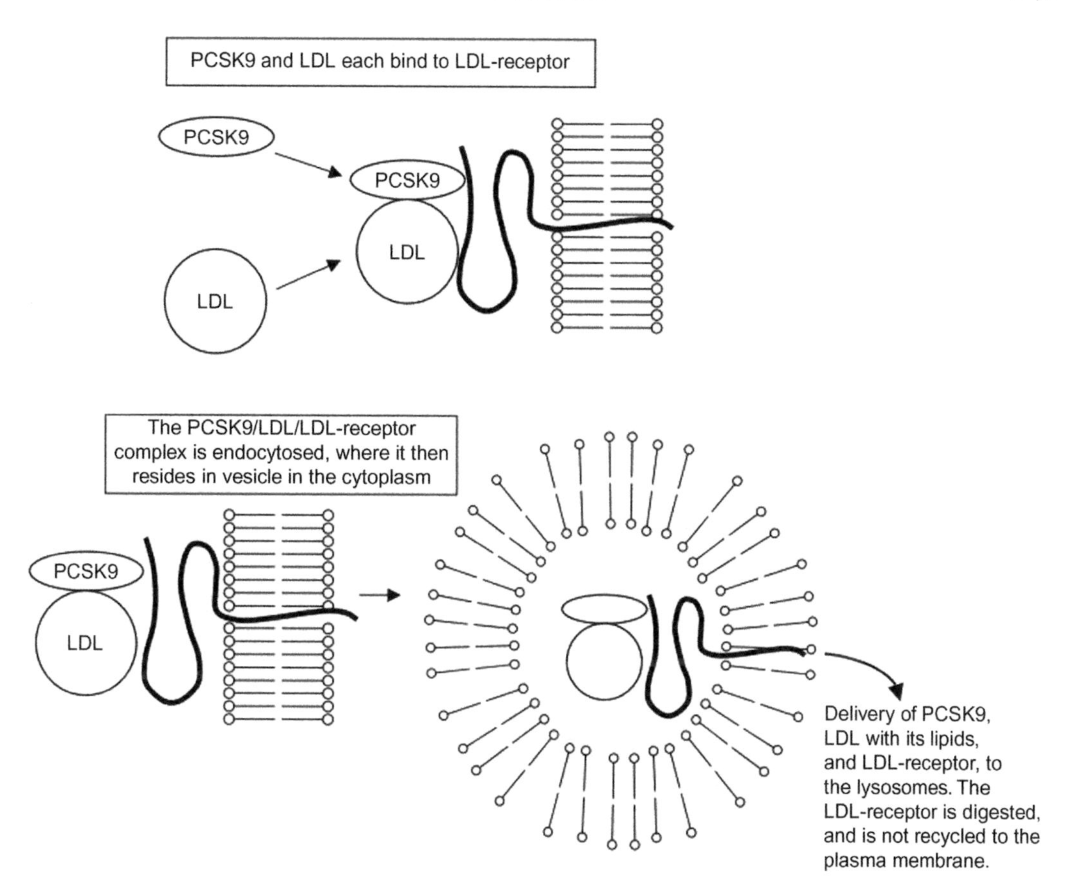

FIGURE 8.2 Fate of LDL-cholesterol, LDL, and LDL-receptor in presence of PCSK9. The top picture shows PCSK9 and LDL each binding to LDL-receptor. The lower picture shows invagination of the complex of PCSK9/LDL/LDL-receptor to produce a phospholipid bilayer vesicle, followed by delivery of PCSK9, LDL and its lipids, and LDL-receptor to the lysosomes.

i. Cycling of LDL/LDL-Receptor in Absence of PCSK9

LDL circulating in the plasma binds to LDL-receptor, to form the LDL/LDL-receptor complex. LDL-receptor is a membrane-bound protein, and when the complex is formed, the complex is also membrane bound. The complex is then endocytosed where it resides in a vesicle situated in the cytosol. This vesicle is shown in Fig. 8.1. The figure does not show the subsequent fate of the vesicle, but it is described here. The vesicle splits into two daughter vesicles. The first daughter vesicle contains only LDL-receptor and it returns to the plasma membrane, where it fuses with the plasma membrane, and delivers the LDL-receptor back to the plasma membrane. Once back in the plasma membrane, the LDL-receptor is able to bind more LDLs that are in the bloodstream. The second daughter vesicle contains only the LDL (plus lipids contained in the LDL, such as cholesterol), where this second vesicle delivers its contents to the lysosomes for digestion.[28]

ii. Cycling of LDL/LDL-Receptor in Presence of PCSK9

Fig. 8.2 shows PCSK9 in the circulation binding to LDL-receptor, and forming PCSK9/LDL-receptor complex. What also occurs is binding of LDL to LDL-receptor. The result is a complex of three proteins: (1) PCSK9, (2) LDL, and (3) LDL-receptor. The PCSK9/LDL/LDL-receptor complex is endocytosed to generate a vesicle that resides in the cytosol.

Fig. 8.2 shows this vesicle in the cytosol. The figure does not show subsequent steps, but these steps are described as follows. The vesicle in the cytosol then delivers its contents to the lysosome, where digestion in the lysosome destroys all three proteins of the complex (PCSK9, LDL, and LDL-receptor). In this scenario, which involves participation of PCSK9, the LDL-receptor is not delivered back to the plasma membrane, i.e., there is not any recycling of the LDL-receptor. In the words of Surdo et al.,[29] the low pH of the endosome vesicle enhances PCSK9/LDL-receptor affinity, ultimately leading to lysosomal LDL-receptor degradation. As LDL-receptor is not returned to the plasma membrane, there is less removal of circulating plasma LDL-cholesterol, where the consequence is higher levels of LDL-cholesterol and increased risk for atherosclerosis.[30] Descriptions of PCSK9's binding to LDL-receptor and of PCSK9's binding to LDL are in the cited references.[31,32]

II. BIOLOGY OF CYTOKINES AND OF TH1- AND TH2-TYPE IMMUNE RESPONSE

a. Introduction

This section introduces the biology of cytokines and their relation to cells of the immune system, such as T cells and dendritic cells (DCs). The cytokines include IL-12, tumor necrosis factor-alpha (TNFα), interferon-γ, IL-10, and IL-4. An understanding of cytokine biology enables an understanding of antibody drugs that bind to cytokines or that bind to cytokine receptors, such as certolizumab pegol (anti-TNFα), infliximab (anti-TNFα), siltuximab (anti-IL-6), tocilizumab (anti-IL-6 receptor), and ustekinumab (anti-IL-12).

[28] Reyes-Soffer G, et al. Effects of PCSK9 inhibition with alirocumab on lipoprotein metabolism in healthy humans. Circulation. 2017;135:352–62.

[29] Surdo P, et al. Mechanistic implications for LDL receptor degradation from the PCSK9/LDLR structure at neutral pH. EMBO Rep. 2011;12:1300–5.

[30] Reyes-Soffer G, et al. Effects of PCSK9 inhibition with alirocumab on lipoprotein metabolism in healthy humans. Circulation. 2017;135:352–62.

[31] Kosenko T, et al. Low density lipoprotein binds to proprotein convertase subtilisin/kexin type-9 (PCSK9) in human plasma and inhibits PCSK9-mediated low density lipoprotein receptor degradation. J. Biol. Chem. 2013;288:8279–88.

[32] Zhang Y, et al. Identification of a small peptide that inhibits PCSK9 protein binding to the low density lipoprotein receptor. J. Biol. Chem. 2014;289:942–55.

Cytokines are small proteins that transmit signals from one type of cell to another type of cell, in a manner similar to that of hormone-mediated signaling. For every cytokine, there is a corresponding cytokine receptor. In some cases, a given cytokine can bind to and activate more than one type of cytokine receptor, and in this situation the cytokine's binding activity is called promiscuous binding.[33,34]

Where the study drug is a cytokine, or where the study drug is an anticytokine antibody, a question that arises is whether the study drug can transmit a signal to the regulatory region of a cytokine P450 gene, with consequent change in CYP gene expression.

The influence of cytokines on CYP enzyme expression has been shown by the influence of IL-6 on the expression of CYP3A, CYP2C9, and CYP2C19,[35] and by the influence of IL-1beta, IL-6, and TNFα on expression of CYP3A4.[36] Also, the influence of cytokines on expression of various CYP enzymes has been shown in experiments where a mixture of various cytokines ("cytokine cocktail") altered expression of CYP3A4, CYP1A2, and CYP2C9.[37] DDIs involving IL-2, IL-6, TNFα, IFNα, and IFNγ have been reviewed.[38]

The immune network is best illustrated by the immunology of two diseases, where the behavior of the immune system for one disease is opposite that of the other. Cancer and psoriasis are suitable for this purpose. Cancer is distinguished by a dampened immune response, which allows unfettered proliferation of cancer cells and consequent death of the patient. Anticancer drugs intended to modulate the immune system work by reversing the dampening mechanism, in the case of some drugs, and by actively stimulating cancer cell killing, in the case of other drugs. Psoriasis and other autoimmune diseases are characterized by an increased immune response, where this immune response is pathological to the patient. In this case, antipsoriasis drugs act to cool down the immune system.

[33] Shahrara S, et al. IL-17 induces monocyte migration in rheumatoid arthritis. J. Immunol. 2009;182:3884–91.

[34] Lukacs NW, et al. Chemokine receptors in asthma: searching for the correct immune targets. J. Immunol. 2003;171:11–15.

[35] Zhuang Y, et al. Evaluation of disease-mediated therapeutic protein-drug interactions between an anti-interleukin-6 monoclonal antibody (sirukumab) and cytochrome P450 activities in a phase 1 study in patients with rheumatoid arthritis using a cocktail approach. J. Clin. Pharmacol. 2015;55:1386–94.

[36] Mimura H, et al. Effects of cytokines on CYP3A4 expression and reversal of the effects by anti-cytokine agents in the three-dimensionally cultured human hepatoma cell line FLC-4. Drug Metab. Pharmacokinet. 2015;30:105–10.

[37] Xu Y, et al. Physiologically based pharmacokinetic model to assess the influence of blinatumomab-mediated cytokine elevations on cytochrome P450 enzyme activity. CPT Pharmacometrics Syst. Pharmacol. 2015;4:507–15.

[38] Christensen H, Hermann M. Immunological response as a source to variability in drug metabolism and transport. Front. Pharmacol. 2012;3:Article 8 (10 pp.).

The field of cellular immunology is characterized by this dichotomy:

- *Th1-type immune response.* IL-12 is a "master regulator" for activating Th1-type immune response[39]
- *Th2-type immune response.* IL-10, or perhaps IL-4, are "master regulators" for activating Th2-type immune response[40,41,42]

In Th1-type immune response, the immune system gets "heated up." But in Th2-type immune response, the immune system gets "cooled down." Heating literally occurs in regions of the body where there are accumulations of lymphocytes, and where the lymphocytes engage in Th1-type immune response. This heating has been described as, "Local increases in temperature at sites of inflammation...are cardinal features of host responses to pathogenic stimuli. Proinflammatory cytokines such as IL-1b and TNF-alpha are **pyrogenic**."[43] Consistently, an account of arthritis states that, "IL-1...referred to [as]...endogenous **pyrogen**...is localized to the synovial pannus in rheumatoid arthritis...low concentrations of IL-1 have extraordinary potential to induce cartilage destruction and bone resorption."[44] The Th1/Th2 paradigm has been reviewed.[45]

b. Therapeutic Goal of Stimulating Immune Response (Immunology of Cancer)

Stimulation of Th1-type response is used to enhance immune attack against tumor cells. A number of anticancer drugs stimulate Th1-type response or inhibit Th2-type response. As reviewed by Disis et al.,[46] anticancer drugs that mediate immune response need to stimulate Th1-type response and also dampen Th2-type response. With optimal activation of Th1-type response and optimal dampening of Th2-type response, the outcome is more effective lysis of tumor cells by $CD8^+$ T cells that have infiltrated a tumor and reside near tumor cells. $CD8^+$ T cells that have been activated to kill their target cells, i.e., via release of perforin and granzyme, are called cytotoxic effector T cells.

[39] Bashyam H. Interleukin-12: a master regulator. J. Exp. Med. 2007;204:969.

[40] Couper KN et al (2008) IL-10: the master regulator of immunity to infection. J. Immunol. 180:5771–7.

[41] Onderdijk AJ, et al. IL-4 downregulates IL-1b and IL-6 and Induces GATA3 in psoriatic epidermal cells: route of action of a Th2 cytokine. J. Immunol. 2015;195:1744–52.

[42] Hsieh CS, et al. Development of TH1 $CD4^+$ T cells through IL-12 produced by Listeria-induced macrophages. Science. 1993;260:547–9.

[43] Wang WC, et al. Fever-range hyperthermia enhances L-selectin-dependent adhesion of lymphocytes to vascular endothelium. J. Immunol. 1998;160:961–9.

[44] Dayer JM. The pivotal role of interleukin-1 in the clinical manifestations of rheumatoid arthritis. Rheumatology (Oxford). 2003;42(Suppl. 2):ii3–10.

[45] Benson JM, et al. Discovery and mechanism of ustekinumab: a human monoclonal antibody targeting interleukin-12 and interleukin-23 for treatment of immune-mediated disorders. MAbs. 2011;3:535–45.

[46] Disis ML, et al. Th1 epitope selection for clinically effective cancer vaccines. Oncoimmunology. 2014;3: e954971.

c. Therapeutic Goal of Cooling Down Immune Response (Immunology of Psoriasis)

Psoriasis is caused by an overly active Th1-type immune response, where cells in psoriatic skin lesions express Th1-type cytokines, such as IL-6, IL-1beta, and IFN-γ.[47,48]

Psoriasis can be treated by administering an antibody that blocks Th1-type cytokines, such as ustekinumab, which targets and blocks IL-12.[49]

Alternatively, psoriasis can be treated by administering IL-4, which is a Th2-type cytokine.[50,51,52] In the words of one researcher, administration of "IL-4 improves psoriasis. . .and. . .alters the psoriatic skin phenotype towards a healthy skin phenotype."[53]

The therapeutic approaches of blocking Th1-type immune response and stimulating Th2-type immune response, in treating any autoimmune disorder, are mechanistically consistent with each other. IL-12 is a master switch that stimulates Th1-type response, while IL-4 is a master switch that stimulates Th2-type response.[54]

III. BACKGROUND FOR ACQUIRED IMMUNITY RESULTING IN IMMUNE RESPONSE AGAINST SPECIFIC ANTIGENS

Immune responses against specific antigens occur with immune responses against bacterial infections, viral infections, and cancer. Also, immune responses against specific antigens occur when a person acquires an immune response after vaccination, as with vaccination with a viral antigen, bacterial antigen, or tumor antigen. The mechanisms shown here should be kept in mind when contemplating disorders that do not involve acquired immunity, such as hypersensitivity reactions.

This shows relationships among antigens, DCs, MHC Class I, HLA subtypes, HLA polymorphisms, the immune synapse, and $CD8^+$ T cells.

[47] Onderdijk AJ, et al. IL-4 downregulates IL-1b and IL-6 and induces GATA3 in psoriatic epidermal cells: route of action of a Th2 cytokine. J. Immunol. 2015;195:1744–52.

[48] Eyerich S, et al. Mutual antagonism of T cells causing psoriasis and atopic eczema. New Engl. J. Med. 2011;365:231–8.

[49] Package label. STELARA® (ustekinumab) injection, for subcutaneous or intravenous use; September 2016 (27 pp.).

[50] Onderdijk AJ, et al. IL-4 downregulates IL-1b and IL-6 and induces GATA3 in psoriatic epidermal cells: route of action of a Th2 cytokine. J. Immunol. 2015;195:1744–52.

[51] Eberle FC, et al. Recent advances in understanding psoriasis. F1000Res. 2016;5:770. doi: 10.12688/f1000research.7927.1. eCollection 2016.

[52] Guenova E, et al. IL-4 abrogates T_H17 cell-mediated inflammation by selective silencing of IL-23 in antigen-presenting cells. Proc. Natl. Acad. Sci. 2015;112:2163–8.

[53] Onderdijk AJ, et al. IL-4 downregulates IL-1b and IL-6 and induces GATA3 in psoriatic epidermal cells: route of action of a Th2 cytokine. J. Immunol. 2015;195:1744–52.

[54] Onderdijk AJ, et al. IL-4 downregulates IL-1b and IL-6 and induces GATA3 in psoriatic epidermal cells: route of action of a Th2 cytokine. J. Immunol. 2015;195:1744–52.

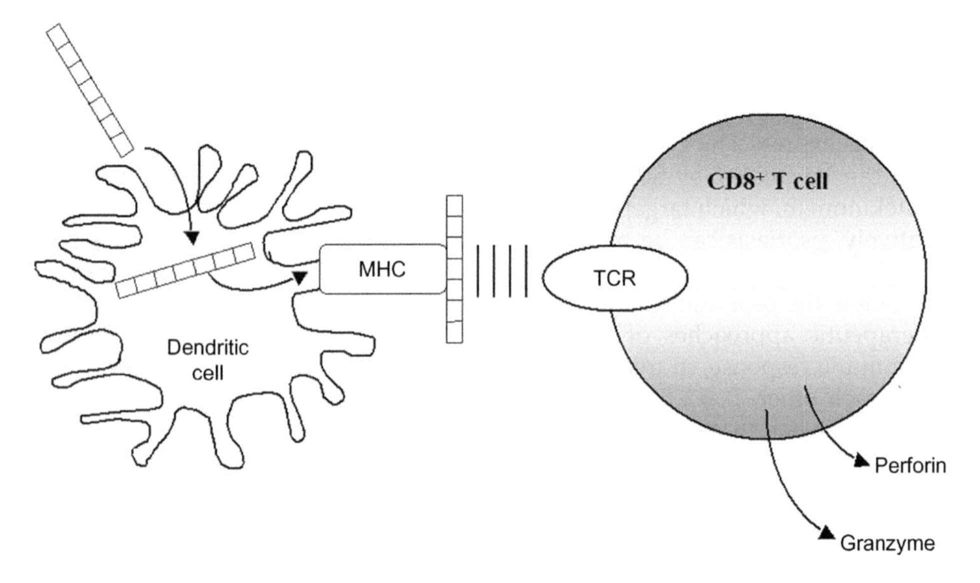

FIGURE 8.3 Events starting with antigen uptake and concluding with release of toxic proteins by the T cell. The dendritic cell (DC) takes up polypeptide antigen (segmented line), processes the polypeptide to short oligopeptides and presents them on MHC. MHC is presented to T cell receptor on the T cell, which stimulates the T cells to attack other cells that express the same oligopeptide antigen, where attack is by release of perforin and granzyme. The origin of the polypeptide antigen can also be a virus or bacterium infecting the DC, and in this case, the viral or bacterial antigen is also presented via MHC. Peptides presented by MHC Class I are derived mostly from proteins synthesized within the dendritic cell, while peptides presented by MHC Class II are derived mainly from proteins acquired by endocytosis.

Fig. 8.3 illustrates a polypeptide antigen being taken up by a DC. The polypeptide is subsequently processed by the DC, and presented to $CD8^+$ T cells, where the consequence is activation of the $CD8^+$ T cells to mount an immune response against any cell in the body that expresses the same antigen, where the $CD8^+$ T cell kills the target cell. In this context, "antigen expression" refers to any polypeptide or oligopeptide that is bound to or part of the extracellular face of the plasma membrane of any target cell, where the antigen is sensed by activated $CD8^+$ T cells. In other words, immune response takes the form of this pair of concepts: (1) First, an antigen stimulates DCs to activate T cells; and (2) Then, the activated T cells attempt to kill any cell that expresses the same antigen.

In Fig. 8.3, the segmented line refers to an oligopeptide. The parallel lines refer to binding between the oligopeptide and the T cell receptor (TCR). The parallel hatch marks are conventionally used to refer to hydrogen bonds. The arrows show an antigen being taken up by the DCs, where it is introduced to the MHC Class I protein, where the complex of MHC Class I protein plus antigen is expressed on the surface of the DC, and where the antigen is eventually presented to a $CD8^+$ T cell, resulting in the activation of the $CD8^+$ T cell.

Fig. 8.3 also shows release of granzyme and perforin from the activated $CD8^+$ T cell, which occurs when a $CD8^+$ T cell eventually contacts a cell infected with a virus or bacterium, or contacts a tumor cell. When released, granzyme and perforin kill the cell resulting in elimination of the infection or death of the cancer cell.[55,56]

DCs collect antigens, process and present them on MHC molecules to $CD8^+$ T cells.[57] MHC Class I molecules occur in three classes, known as HLA-A, HLA-B, and HLA-C. Another group of MHC molecules (MHC Class II) occurs as HLA-DR, HLA-DQ, and others.[58] DCs can acquire antigens by taking up proteins, oligopeptides, or entire cells from the surrounding medium.[59] Antigens that are the targets in autoimmune diseases, and antigens that are from tumors (thereby directing $CD8^+$ T cells to kill tumors) are acquired in this way.[60] Also, DCs can acquire antigens from within, i.e., where the DC is infected with a virus or a bacterium and where the virus or bacterium residing within the DC biosynthesizes the antigen.

MHC Class I is also expressed on the plasma membranes of other cells, in addition to DCs. When MHC Class I is expressed by a tumor cell, it can present tumor antigens, and when MHC Class I is expressed by host cells infected with a virus or bacterium, it can present viral antigens or bacterial antigens. This presentation allows circulating $CD8^+$ T cells to recognize and kill the tumor cell or the infected host cell, thereby eliminating the cancer or the infection.[61,62] (Fig. 8.3 does not illustrate the scenario where MHC Class I is expressed by host cells, and where the MHC Class I presents viral or tumor antigens, and where $CD8^+$ T cells kill these host cells.)

For any given class of HLA molecule, the HLA molecule naturally exists in the human population in many, many variant forms. From person to person, any given HLA molecule can differ in about 30 amino acids or in only a few amino acids. These differences or variations are called "polymorphisms." The polymorphisms exist in the human genome, and hence DNA analysis can easily be used to distinguish between different patients, according to their HLA polymorphisms. For example, certain polymorphisms are associated with resistance to HIV-1 virus infections.[63] Also, identifying polymorphisms in a patient's genome can determine if a given anticancer drug will be effective (or not be effective) in

[55] Willberg CB, et al. Protection of hepatocytes from cytotoxic T cell mediated killing by interferon-alpha. PLoS One. 2007;2:e791 (9 pp.).

[56] Martinez-Lostao L, et al. How do cytotoxic lymphocytes kill cancer cells? Clin. Cancer Res. 2015;21:5047–56.

[57] Shortman K. Ralph Steinman and dendritic cells. Immunobiol. Cell Biol. 2012;90:1–2.

[58] Tiercy JM. How to select the best available related or unrelated donor of hematopoietic stem cells? Haematologica. 2016;101:680–7.

[59] Inzkirweli N, et al. Antigen loading of dendritic cells with apoptotic tumor cell-preparations is superior to that using necrotic cells or tumor lysates. Anticancer Res. 2007;27:27:2121–9.

[60] Munz C. Live long and prosper for antigen cross-presentation. Immunity. 2015;43:1028–30.

[61] Reeves E, James E. Antigen processing and immune regulation in the response to tumours. Immunology. 2017;150:16–24.

[62] Albanese M, et al. Epstein-Barr virus microRNAs reduce immune surveillance by virus-specific $CD8^+$ T cells. Proc. Natl. Acad. Sci. 2016;113:E6467–75.

[63] Crawford H, et al. Evolution of HLA-B*5703 HIV-1 escape mutations in HLA-B*5703-positive individuals and their transmission recipients. J. Exp. Med. 2009;206:909–21.

that patient.[64] As shown elsewhere in this chapter, an identification of HLA-B polymorphisms can identify patients likely to suffer from Stevens–Johnson syndrome. The polymorphism in question is HLA-B*1502.[65]

IV. MATERIALIZATION OF ANTIBODIES AGAINST THERAPEUTIC PROTEINS, AS REVEALED BY FDA'S REVIEWS

FDA routinely assesses whether the Sponsor's therapeutic protein (antibody or enzyme) provokes the undesirable immune response where the patient develops antibodies against the therapeutic protein. FDA's uses this multistep analysis:

- *Step one*. Does the therapeutic protein induce the patient's immune system to generate antibodies against the therapeutic protein?
- *Step two*. Do the induced antibodies merely bind to the therapeutic protein, or do they also inactivate the activity of the therapeutic protein, as determined by vitro assays?
- *Step three*. Do the induced antibodies reduce exposure of the therapeutic protein, as determined by exposure parameters such as AUC, Cmax, tmax, and Cmin?[66]
- *Step four*. Do the induced antibodies reduce clinical efficacy of the therapeutic protein?

Various classes of AEs can result when a patient's immune system reacts against an antibody drug. One type of adverse event takes the form of an undesired laboratory value, namely, inactivation of the antibody drug (by inactivating antibodies).[67] The laboratory technique of "humanization" is used to reduce immune response against therapeutic proteins.[68,69,70] Humanization is performed when planning and engineering the amino acid sequence of the protein itself. But if the therapeutic protein is modified by the covalent modification of an organic molecule, such as polyethylene glycol (PEG), then humanization might not be applicable to that organic molecule, where the consequence is vigorous

[64] Kloth JS, et al. Genetic polymorphisms as predictive biomarker of survival in patients with gastrointestinal stromal tumors treated with sunitinib. Pharmacogenomics J. 2017. doi: 10.1038/tpj.2016.83.

[65] Package label. Tegretol®-XR (carbamezepine extended-release tablets); August 2015 (18 pp.).

[66] van Schouwenbug PA, et al. Functional analysis of the anti-adalimumab response using patient-derived monoclonal antibodies. J. Biol. Chem. 2014;289:34482–8.

[67] Krieckaert CL, et al. The effect of immunomodulators on the immunogenicity of TNF-blocking therapeutic monoclonal antibodies: a review. Arthritis Res. Ther. 2010;12:217 (6 pp.).

[68] Presta LG. Molecular engineering and design of therapeutic antibodies. Curr. Opin. Immunol. 2008;20:460–70.

[69] Apgar JR, et al. Beyond CDR-grafting: structure-guided humanization of framework and CDR regions of an anti-myostatin antibody. MAbs. 2016;8:1302–18.

[70] Frenzel A, et al. Phage display-derived human antibodies in clinical development and therapy. MAbs. 2016;8:1177–94.

immune response by the patient against the organic molecule. Moreover, where the therapeutic protein is an antibody, one particular region of the antibody (even with humanization) of enhanced immunogenicity is the antigen-binding site. Antibodies generated against the antigen-binding site are called, antiidiotypic antibodies.[71]

a. Atezolizumab (Urothelial Carcinoma) BLA 761-034

Atezolizumab is an antibody that binds to a PD-L1. When administered to a cancer patient, atezolizumab enhances immune response against cancer cells.

This outlines the antitumor mechanism of antibodies blocking PD-L1/PD-1 signaling. Immune response against cancer is mediated by cytotoxic T cells ($CD8^+$ T cells). In the optimal situation, these contact cancer cells, recognize them as being foreign, and kill them. However, cancer cells employ various strategies to evade immune response, such as expressing PD-L1. PD-L1 is a membrane-bound protein residing on the plasma membrane of tumor cells. PD-L1 and PD-1 form a ligand/receptor pair which transmits a signal that dampens immune response, where PD-L1 is expressed by tumor cells and PD-1 is expressed by T cells. The signaling pathway is illustrated by the following diagram, where the lightning represents the transmitted inhibitory signal:

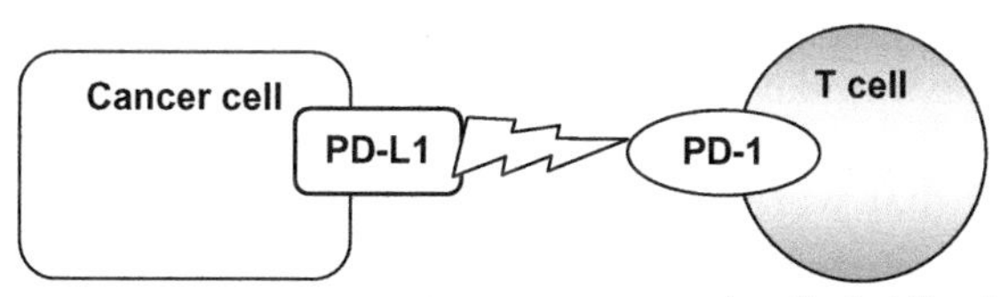

A variety of therapeutic proteins block this signaling pathway, thereby preventing cancer cells from transmitting a dampening signal to T cells. As a result of the therapeutic protein, T cells have a revived ability to kill the cancer cells.[72] Pembrolizumab[73] binds PD-1, nivolumab[74] also binds PD-1, while an antibody called "BMS-936559"[75] binds to PD-L1.

FDA's ClinPharm Review on atezolizumab addressed these questions:

- Does atezolizumab provoke the immune system to generate anti-atezolizumab antibodies?

[71] van Schouwenburg PA, et al. Adalimumab elicits a restricted anti-idiotypic antibody response in autoimmune patients resulting in functional neutralisation. Ann. Rheum. Dis. 2013;72:104–9.

[72] Liu X, Cho WC. Precision medicine in immune checkpoint blockade therapy for non-small cell lung cancer. Clin. Transl. Med. 2017;6:7. doi: 10.1186/s40169-017-0136-7 (4 pp.).

[73] Garon EB, et al. Pembrolizumab for the treatment of non-small-cell lung cancer. New Engl. J. Med. 2015;372:2018–28.

[74] Robert C, et al. Nivolumab in previously untreated melanoma without BRAF mutation. New Engl. J. Med. 2015;372:320–30.

[75] Brahmer JR, et al. Safety and activity of anti-PD-L1 antibody in patients with advanced cancer. New Engl. J. Med. 2012;366:2455–65.

TABLE 8.1 Incidence of Antibodies Against Atezolizumab and Association of These Anti-Atezolizumab Antibodies With Lack of Therapeutic Response

	Negative for Anti-Atezolizumab Antibodies	**Positive for Anti-Atezolizumab Antibodies**
Responders	30	27
Nonresponders	131	87
Percent of subjects that are responders	18.6%	23.7%

Atezolizumab (urothelial carcinoma) BLA 761-034. Page 23 of 53-page Clinical Pharmacology Review.

- In subjects producing anti-atezolizumab antibodies, was there a change in atezolizumab exposure?
- In subjects producing anti-atezolizumab antibodies, was atezolizumab's clinical efficacy impaired?

The FDA reviewer contemplated the influence of anti-therapeutic antibodies on the study drug's exposure, i.e., on AUC, Cmax, and Cmin. Here FDA observed that, "There was a trend to slightly lower Cmin concentrations in anti-therapeutic antibody-positive patients compared to anti-therapeutic negative patients...[p]ositive anti-therapeutic antibodies did not result in more than a 20% change in AUC, Cmax, or Cmin, from the typical patient."[76]

Table 8.1 reveals the number of patients where the immune system of the patient reacted to atezolizumab by generating anti-atezolizumab antibodies. Also, the table shows the number of patients that were "responders." "Responders" means patients where the administered atezolizumab resulted in shrinking tumors. As one can see from the data, the presence or absence of anti-atezolizumab antibodies did not reduce clinical response.

FDA's ClinPharm Review disclosed the levels of anti-atezolizumab antibodies in the subjects, as well as the association of these antibodies with failure of atezolizumab to have clinical efficacy. Clinical efficacy was determined by tumor size and number. Efficacy (tumors shrunk) was 18.6% in subjects not having anti-atezolizumab antibodies and 23.7% in subjects who had developed anti-atezolizumab antibodies.

FDA did not consider the values 18.6% and 23.7% (Table 8.1) to be significantly different from each other, and concluded, "Overall, anti-therapeutic antibody positivity did not seem to impact efficacy."

b. Atezolizumab Judged as Not Relevant to CYP Enzymes

The FDA reviewer contemplated atezolizumab's influence on CYP enzymes and on drug transporters, and dismissed these types of influences as being unlikely. FDA's checklist, which included dismissive judgments (with no data), took this form[77]:

> "Is there an in vitro basis to suspect in-vivo drug-drug interactions? No."
>
> "Is the drug a substrate of CYP enzymes? Unlikely."

[76] Atezolizumab (urothelial carcinoma) BLA 761-034. Page 24 of 53-page Clinical Pharmacology Review.

[77] Atezolizumab (urothelial carcinoma) BLA 761-034. Page 29 of 53-page Clinical Pharmacology Review.

"Is the drug an inhibitor and/or an inducer of CYP enzymes? Unlikely."

"Are there metabolic/transporter pathways that may be important? Unlikely."

"Metabolism studies are not generally performed for biological protein products. As proteins are degraded into amino acids that are subsequently recycled into other proteins, the classical biotransformation studies for small molecule drugs are not applicable."

Package label. The lack of significance of the anti-atezolizumab antibodies found a place in the Clinical Trials Experience section of atezolizumab's package label:

CLINICAL TRIALS EXPERIENCE...Immunogenicity. As with all therapeutic proteins, there is a potential for immunogenicity. Among 275 patients...114 patients (41.5%) tested positive for treatment-emergent...anti-therapeutic antibodies...at one or more post-dose time points...the presence of anti-therapeutic antibodies did not appear to have a clinically significant impact on pharmacokinetics, safety or efficacy.[78]

c. Certolizumab Pegol (Crohn's Disease) BLA 125-160

Certolizumab pegol is an antibody conjugated to PEG. Certolizumab pegol is for treating Crohn's disease. The antibody binds to TNFα. TNFα occurs as a soluble form in blood plasma, as well as a membrane-bound form in plasma membranes. Certolizumab pegol binds to both of these forms of TNFα.[79] Changes in TNFα expression can influence expression of CYP3A, and thus it might be expected that administering certolizumab pegol (which blocks TNFα) can influence CYP3A-mediated metabolism of any coadministered small molecule drugs.[80]

i. Certolizumab Pegol Generated Anti-Certolizumab Pegol Antibodies

FDA observed that antibodies were generated, and these antibodies reduced exposure. FDA's comments on reduced exposure referred to reduced Cmax, Cmin, and AUC, "The incidences of anti-certolizumab pegol antibodies and neutralizing anti-certolizumab pegol antibodies appear to be inversely proportional to certolizumab pegol dose...[w]hen antibodies occur, they have a significant effect on the pharmacokinetics...antibodies to certolizumab pegol increased the clearance of certolizumab by...four-fold...[i]ncreased clearance in antibody positive subjects can be expected to result in a 52% reduction in Cmax, 86% reduction in Cmin, and 72% reduction in AUC in a typical...subject with Crohn's disease."[81]

[78] Package label. TECENTRIQ® (atezolizumab) injection, for intravenous use; April 2017 (24 pp.).

[79] Certolizumab pegol (Crohn's disease) BLA 125-160. 102-page Clinical Pharmacology Review.

[80] Nyagode BA, et al. Selective effects of a therapeutic protein targeting tumor necrosis factor-alpha on cytochrome P450 regulation during infectious colitis: implications for disease-dependent drug-drug interactions. Pharmacol. Res. Perspect. 2014;2:e200027 (12 pp.).

[81] Certolizumab pegol (Crohn's disease) BLA 125-160. Pages 7, 10, 21, 27, and 36 of 102-page Clinical Pharmacology Review.

ii. FDA Recommended Dose-Adjustment

FDA contemplated the generation of neutralizing antibodies, and recommended that the package label contain dose-adjustment instructions. But despite FDA's recommendation, a view of the original package label (April 2008) and the most recent package label (January 2017), reveals that there are not any dose-adjustment instructions. The fact that dose-adjustment instructions were assessed when making the final draft of the label is demonstrated by the existence of dose-adjustment recommendations depending on the patient's weight. To this point, it can be seen that the package label says *not to adjust the dose.*[82]

FDA's recommendations for dose-adjustment took the form, "Eight percent of patients exposed to certolizumab pegol...developed anti-certolizumab pegol antibodies, of which 80% were neutralizing...dose adjustment is recommended for antibody positive patients. No other dose adjustments are recommended."[83]

Again, note that these were only FDA's recommendations, and not information that eventually found its way to the package label.

iii. Lupus-Like Syndrome

FDA's Office Director's Memo observed that the Sponsor's clinical studies detected an immune reaction to certolizumab pegol (Cimzia®), taking the form of lupus-like syndrome. The Office Director wrote, "Autoantibodies developed in 4% of CIMZIA-treated and in 2% of placebo-treated patients. One CIMZIA-treated patient developed symptoms of a lupus-like syndrome."[84]

The detected case of lupus-like syndrome found a place on the package label, as shown below. As detailed in the next chapter in this book, lupus-like syndrome has occurred with adalimumab, etanercept,[85,86,87] infliximab, as well as with small molecule drugs, such as hydralazine, procainamide, and sulfadiazine.[88,89,90,91]

[82] Package label. CIMZIA (certolizumab pegol) for injection, for subcutaneous use; January 2017 (33 pp.).

[83] Certolizumab pegol (Crohn's disease) BLA 125-160. Page 50 of 80-page Medical Review (first of three pdf files).

[84] Certolizumab pegol (Crohn's disease) BLA 125-160. Pages 4 and 7 of 7-page Office Director Memo.

[85] Williams VL, Cohen PR. TNF alpha antagonist-induced lupus-like syndrome: report and review of the literature with implications for treatment with alternative TNF alpha antagonists. Int. J. Dermatol. 2011;50:619–25.

[86] Package label. ERELZI (etanercept-szzs) injection, for subcutaneous use; August 2016 (39 pp.). BLA 761-042.

[87] Package label. ENBREL® (etanercept) injection, for subcutaneous use; November 2016 (32 pp.). BLA 103-795

[88] Lomicova I, et al. A case of lupus-like syndrome in a patient receiving adalimumab and a brief review of the literature on drug-induced lupus erythematosus. J. Clin. Pharm. Ther. 2017;42:363–6. doi: 10.1111/jcpt.12506.

[89] Hogan JJ, et al. Drug-induced glomerular disease: immune-mediated injury. Clin. J. Am. Soc. Nephrol. 2015;10:1300–10.

[90] Klapman JB, et al. A lupus-like syndrome associated with infliximab therapy. Inflamm. Bowel Dis. 2003;9:176–8.

[91] Beigel F, et al. Formation of antinuclear and double-strand DNA antibodies and frequency of lupus-like syndrome in anti-TNF-α antibody-treated patients with inflammatory bowel disease. Inflamm. Bowel Dis. 2011;17:91–8.

iv. In Vitro Drug–Drug Interactions Mediated by CYP Enzymes or Drug Transporters

FDA's ClinPharm Review for certolizumab pegol conducted an analysis that is almost always relevant to small molecule drugs, but rarely relevant to therapeutic proteins. Despite the unlikelihood of any relevance to certolizumab pegol, FDA proceeded with the analysis and observed that, "In vitro cytochrome P450 inhibition studies with human microsomes were not performed because proteins and immunoglobulin antibodies do not compete for the cytochrome P450...metabolism system. An in vitro P-glycoprotein (P-gp) interaction study showed that neither certolizumab nor its non-PEGylated Fab′ fraction were inhibitors of P-glycoprotein-mediated transport."[92]

In other words, despite the unlikely possibility that certolizumab would inhibit transport mediated by P-gp, the Sponsor still tested this possibility.

v. Clinical Drug–Drug Interactions Between Certolizumab Pegol and Methotrexate

Certolizumab pegol binds to soluble TNFα and also to membrane-bound TNFα.[93] The scientific literature teaches that changes in TNFα expression can influence expression of CYP3A, and thus it might be expected that administering certolizumab pegol can influence CYP3A-mediated metabolism of small molecule drugs.[94]

The DDI study involved human subjects. Methotrexate dosing was followed the next day by certolizumab pegol dosing and, as stated by FDA's ClinPharm Review, "On Day 1 of the study, subjects received their...methotrexate dose, on Day 2 subjects received a single subcutaneous 400 mg...dose of...certolizumab pegol."[95]

The Sponsor tested if certolizumab pegol influenced methotrexate exposure, and if methotrexate influenced exposure of certolizumab pegol. Neither influence was found. To this end, FDA wrote that, "[t]he study demonstrated the lack of a...drug interaction between certolizumab pegol and methotrexate."[96]

The reverse DDI was also examined, and it was found that there was no interaction, where FDA observed, "the similarity of the plasma concentration-time curves and PK parameters in...suggest that the concurrent administration of methotrexate had no effect on the pharmacokinetics of a single dose of certolizumab pegol."[97]

[92] Certolizumab pegol (Crohn's disease) BLA 125-160. Page 25 of 102-page Clinical Pharmacology Review.

[93] Certolizumab pegol (Crohn's disease) BLA 125-160. 102-page Clinical Pharmacology Review.

[94] Nyagode BA, et al. Selective effects of a therapeutic protein targeting tumor necrosis factor-alpha on cytochrome P450 regulation during infectious colitis: implications for disease-dependent drug-drug interactions. Pharmacol. Res. Perspect. 2014;2:e200027 (12 pp.).

[95] Certolizumab pegol (Crohn's disease) BLA 125-160. Page 28 of 102-page Clinical Pharmacology Review.

[96] Certolizumab pegol (Crohn's disease) BLA 125-160. Pages 28–29 of 102-page Clinical Pharmacology Review.

[97] Certolizumab pegol (Crohn's disease) BLA 125-160. Page 25 of 102-page Clinical Pharmacology Review.

Package label. The package label for certolizumab pegol (Cimizia®), in the Adverse Reactions section, disclosed the formation of anti-certolizumab pegol antibodies, and added that these antibodies did not influence efficacy or safety:

> **ADVERSE REACTIONS**...Patients with Crohn's disease were tested at multiple time points for antibodies to certolizumab pegol...[i]n patients continuously exposed to CIMZIA, the overall percentage of patients who were antibody positive to CIMZIA on at least one occasion was 8%; approximately 6% were neutralizing in vitro. **No apparent correlation of antibody development to adverse events or efficacy** was observed.[98]

FDA's observations on the materialization of anti-certolizumab antibodies also found a place in the Clinical Pharmacology section:

> **CLINICAL PHARMACOLOGY** ...The presence of anti-certolizumab antibodies was associated with a 3.6-fold increase in clearance.[99]

The Warnings and Precautions section of the package label warned about lupus-like syndrome. The warning on the label was copied, word for word, from FDA's Office Director's Memo.[100] The label read:

> **WARNINGS AND PRECAUTIONS**...Autoantibodies developed in 4% of CIMZIA-treated and in 2% of placebo-treated patients. One CIMZIA-treated patient developed symptoms of a lupus-like syndrome.[101]

The Adverse Reactions section of the package label also warned about lupus-like syndrome:

> **ADVERSE REACTIONS**...Autoimmunity. Treatment with CIMZIA may result in the formation of autoantibodies and rarely, in the development of a lupus-like syndrome. If a patient develops symptoms suggestive of a lupus-like syndrome following treatment with CIMZIA, discontinue treatment.[102]

d. Denosumab (Osteoporosis) BLA 125-320

Denosumab for treating osteoporosis is an antibody that binds to receptor activator of nuclear factor-κB ligand (RANKL). Binding of denosumab to RANKL (ligand) blocks its interaction with RANK (receptor). Densumab's blocking of RANKL/RANK signaling inhibits the development and activity of osteoclasts, decreases bone resorption, and increases bone density.[103]

[98] Package label. CIMZIA (certolizumab pegol) for injection, for subcutaneous use; January 2017 (33 pp.).

[99] Package label. CIMZIA (certolizumab pegol) for injection, for subcutaneous use; January 2017 (33 pp.).

[100] Certolizumab pegol (Crohn's disease) BLA 125-160. Pages 4 and 7 of 7-page Office Director Memo.

[101] Package label. CIMZIA (certolizumab pegol) for injection, for subcutaneous use; January 2017 (33 pp.).

[102] Package label. CIMZIA (certolizumab pegol) for injection, for subcutaneous use; January 2017 (33 pp.).

[103] Cummings SR, et al. Denosumab for prevention of fractures in postmenopausal women with osteoporosis. New Engl. J. Med. 2009;361:756–65.

RANKL is a cytokine. It is a membrane-bound cytokine, not a soluble cytokine. RANKL resides on the plasma membrane of osteoblasts. It recognizes and binds to its receptor (RANK) on bone marrow macrophages, and stimulates the macrophages to become osteoclasts.[104] In the diagram, the string of vertical lines is conventionally used to represent hydrogen bonds that mediate binding. Denosumab is shown binding to RANKL, thus preventing RANKL from binding to RANK that would occur in absence of antibody:

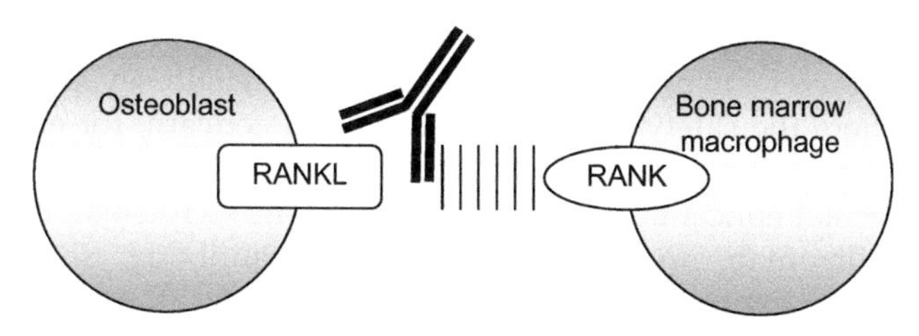

i. Argument that DDIs are Not Expected with Denosumab

FDA's ClinPharm Review focused on whether denosumab induces or inhibits any of the CYP enzymes.[105] FDA stated that this type of DDI is not expected, for two reasons. First, denosumab acts on cells that are not used in CYP-mediated catabolism (osteoblasts; macrophages). RANKL/RANK signaling involves osteoblasts and macrophages, and appears not to involve hepatocytes.[106] Second, the liver is the main location of CYP enzymes in the body, and that evidence suggests that the liver does not express RANKL. In other words, osteoblasts do not contain CYP enzymes, and hepatocytes are not involved in RANKL/RANK signaling. FDA reiterated the Sponsor's argument, "A role of RANKL in cytochrome P450...regulation has not been demonstrated and is unlikely, given ***lack of expression*** of its receptor RANK on...human hepatocytes. A RANKL inhibitor is thus unlikely to directly impact CYP expression or activity."[107]

Despite this argument, FDA still required that the Sponsor conduct experiments to detect possible DDIs between denosumab and small molecules.

ii. Arguments that DDIs Should be Evaluated for Denosumab

FDA required that the Sponsor consider DDIs for these reasons:

- *First reason*. Denosumab is the first drug in its class.
- *Second reason*. DDIs have, in fact, been documented between other antibodies and small molecule drugs that are metabolized by CYP enzymes
- *Third reason*. The mechanism of cytokines influence on CYP enzyme expression are unclear.

[104] Teitelbaum SL. Osteoclasts: what do they do and how do they do it? Am J. Pathol. 2007;170:427–35.

[105] Denosumab (osteoporosis) BLA 125-320. 79-page Clinical Pharmacology Review.

[106] Denosumab (osteoporosis) BLA 125-320. Page 3 of 79-page Clinical Pharmacology Review.

[107] Denosumab (osteoporosis) BLA 125-320. Page 3 of 79-page Clinical Pharmacology Review.

- *Fourth reason*. Published articles showing that RANKL is not expressed by the liver were not convincing to FDA.

First reason. FDA stated that further experiments were needed, based on the fact that denosumab is the first antibody drug for osteoporosis, writing, "Since denosumab is the **first drug that belongs to this class** for the treatment of osteoporosis...it would be informative and beneficial in terms of public health perspective if the sponsor conducts a **drug-drug interaction study** with a CYP3A4 substrate."[108]

Second reason. FDA pointed out that other antibody drugs exhibited significant DDIs mediated by cytochrome P450 enzymes, writing, "there are still several cases of significant **drug-interaction** where the safety of patients becomes a concern, for example, basiliximab increases tacrolimus [small molecule] concentration by 63%, muromonab doubled cyclosporine [small molecule] concentration in renal transplant recipients...[a]nti-cytokine antibodies such as tocilizumab, an anti-IL-6 monoclonal antibody, showed the alteration of CYP substrate drug exposure...[t]hus, denosumab may affect the exposure to CYP substrate drugs by altering the concentration of RANKL, a cytokine that affects B-cell and T-cell differentiation."[109]

Third reason. FDA observed that, "As the mechanism of other cytokines' effects on CYP450 is unclear, the possibility of anti-RANKL on CYP450 is unknown."[110]

Fourth reason. FDA was not convinced by the Sponsor's argument that denosumab's target (RANKL) is not expressed by hepatocytes (hepatocytes being the main location of CYP enzymes in the body). FDA's ClinPharm Review stated, "it is still uncertain...to conclude that a RANKL antagonist will not impact CYP expression. First, the sponsor asserts that RANK and RANKL are not...expressed by...human liver, referencing a literature where thousands of gene patterns were predicted by microarray analysis...[h]owever...the negative detection does not necessarily mean its non-existence. It could be because...the probe set may not properly interrogate the expression of the gene."[111]

iii. FDA Imposed a Requirement for DDI Studies

FDA required that the Sponsor conduct *in vitro DDI studies*, and raised the possibility that *clinical DDI* studies might also be needed, where the vitro tests turn out to be positive, "Therefore, the Clinical Pharmacology Review Team recommends the sponsor conduct an **in vitro study** to assess whether RANKL modulate expression of...CYP enzymes...[i]f, upon review, there is no...modulation of any of the major CYP enzymes...further exploration would not be necessary. If results of the in vitro study are positive, a **drug interaction study...will be needed...in patients**."[112]

Another part of FDA's ClinPharm Review set forth the requirement that the Sponsor conduct a clinical study, "Recommendations: The Clinical Pharmacology Review Team

[108] Denosumab (osteoporosis) BLA 125-320. Pages 4, 5, and 20 of 79-page Clinical Pharmacology Review.

[109] Denosumab (osteoporosis) BLA 125-320. Pages 4, 5, and 20 of 79-page Clinical Pharmacology Review.

[110] Denosumab (osteoporosis) BLA 125-320. Pages 4, 5, and 20 of 79-page Clinical Pharmacology Review.

[111] Denosumab (osteoporosis) BLA 125-320. Page 4 of 79-page Clinical Pharmacology Review.

[112] Denosumab (osteoporosis) BLA 125-320. Pages 4, 5, and 20 of 79-page Clinical Pharmacology Review.

recommends the sponsor conduct an **in vivo drug-drug interaction study** with CYP3A4 substrate, for example, midazolam, in postmenopausal female patients with osteoporosis as a postmarketing requirement."[113]

The fact that the Sponsor actually conducted this in vivo study in the postmarketing situation, is demonstrated by the fact that the package label published on the date of FDA's Approval Letter was silent as to this in vivo study while, in contrast, the revised package label dated several years later detailed the results of this in vivo study.

iv. Denosumab and Anti-Denosumab Antibodies

The Sponsor determined if denosumab induced the immune system of human subjects to generate anti-denosumab antibodies. The Sponsor was careful to assess the presence of antistudy drug antibodies at baseline (Day 1), as well as at various times during the clinical study (months 1, 6, 12, 18, 24, 30, and 36).[114] Denosumab did, in fact, induce antistudy drug antibodies, but only in a small percentage of subjects.

Regarding the low prevalence of anti-denosumab antibodies, FDA wrote, "The immunogenicity potential with denosumab is low. Less that 1% (43 out of 8113) of patients treated with denosumab tested positive for binding antibodies. No patients tested positive for neutralizing antibodies. No evidence of altered PK, PD, safety profile or clinical response has been observed in patients who tested positive for binding antibodies."[115]

Package label. As stated above, FDA recommended that, "the sponsor conduct an in vivo drug-drug interaction study with CYP3A4 substrate...**midazolam**...as a postmarketing requirement."[116] The results of this midazolam study found a place in the Drug Interactions section of the denosumab (Prolia®) package label. The result of this DDI study, which used midazolam as a probe substrate, was that denosumab (Prolia®) had no effect:

> **DRUG INTERACTIONS.** In subjects with postmenopausal osteoporosis, **Prolia...did not affect the pharmacokinetics of midazolam**, which is metabolized by cytochrome P450 3A4 (CYP3A4), indicating that it should not affect the pharmacokinetics of drugs metabolized by this enzyme in this population.[117]

v. 2016 package label versus 2010 package label (new CYP enzyme information)

This concerns the package label published at the time of FDA-approval (June 2010), which was followed by the Sponsor's postmarketing study assessing the influence of denosumab to influence midazolam exposure, and which was followed by publication of a supplemental label (August 2016). As one can see, the June 2010 label stated that the DDI study was not conducted while, in contrast, the August 2016 label stated the results of this DDI study. The June 2010 label merely disclosed:

> **DRUG INTERACTIONS.** No drug-drug interaction studies have been conducted with Prolia.[118]

[113] Denosumab (osteoporosis) BLA 125-320. Pages 4, 5, and 20 of 79-page Clinical Pharmacology Review.

[114] Denosumab (osteoporosis) BLA 125-320. Page 50 of 79-page Clinical Pharmacology Review.

[115] Denosumab (osteoporosis) BLA 125-320. Pages 23 and 50–52 of 79-page Clinical Pharmacology Review.

[116] Denosumab (osteoporosis) BLA 125-320. Pages 4, 5, and 20 of 79-page Clinical Pharmacology Review.

[117] Package label. Prolia® (denosumab) injection, for subcutaneous use; August 2016 (26 pp.).

[118] Package label. Prolia® (denosumab) injection, for subcutaneous use; June 2010 (17 pp.).

In striking contrast to the June 2010 label, the August 2016 label disclosed:

> **DRUG INTERACTIONS**. In subjects with postmenopausal osteoporosis, **Prolia...did not affect the pharmacokinetics of midazolam**, which is metabolized by cytochrome P450 3A4 (CYP3A4), indicating that it should not affect the pharmacokinetics of drugs metabolized by this enzyme in this population.[119]

Immune response against denosumab resulting in anti-denosumab antibodies was in the Adverse Reactions section:

> **ADVERSE REACTIONS**...Immunogenicity. Denosumab is a human monoclonal antibody. As with all therapeutic proteins, there is potential for immunogenicity...less than 1% (55 out of 8113) of patients treated with Prolia for up to 5 years tested positive for binding antibodies...[n]one of the patients tested positive for neutralizing antibodies...[n]o evidence of altered pharmacokinetic profile, toxicity profile, or clinical response was associated with binding antibody development.[120]

e. Evolocumab (Hyperlipidemia and Hypercholesterolemia) BLA 125-522

Evolocumab is an antibody that binds to PCSK9, where the antibody's therapeutic effect is to lower LDL-cholesterol and reduce risk for atherosclerosis. The antibody lowers LDL-cholesterol by about 60%.[121] Evolocumab's mechanism of action is detailed earlier in this chapter. To reiterate, evolocumab's mechanism of action involves[122,123,124,125]:

- Injected evolocumab circulates in the bloodstream
- Circulating evolocumab binds to PCSK9
- Binding of evolocumab prevents PCSK9 from binding to LDL-receptors, and prevents PCSK9 from targeting LDL-receptors for destruction in the lysosomes.
- By preventing LDL-receptor destruction, the evolocumab lowers LDL-cholesterol and reduces risk for atherosclerosis

i. Drug–Drug Interaction with Statin Drugs

FDA's analysis of evolocumab revealed a type of DDI, where a coadministered drug (statins) reduced the efficacy of evolocumab. The Sponsor compared monotherapy with

[119] Package label. Prolia® (denosumab) injection, for subcutaneous use; August 2016 (26 pp.).

[120] Package label. Prolia® (denosumab) Injection, for subcutaneous use; August 2016 (26 pp.).

[121] Sabatine MS, et al. Efficacy and safety of evolocumab in reducing lipids and cardiovascular events. New Engl. J. Med. 2015;372:1500–9.

[122] Chaudhary R, et al. PCSK9 inhibitors: a new era of lipid lowering therapy. World J. Cardiol. 2017;26:76–91.

[123] Le QT, et al. Plasma membrane tetraspanin CD81 complexes with proprotein convertase subtilisin/kexin type 9 (PCSK9) and low density lipoprotein receptor (LDLR), and its levels are reduced by PCSK9. J. Biol. Chem. 2015;290:23385–400.

[124] Romagnuolo R, et al. Lipoprotein(a) catabolism is regulated by proprotein convertase subtilisin/kexin type 9 through the low density lipoprotein receptor. J. Biol. Chem. 2015;290:11649–62.

[125] Lammi C, et al. Lupin peptides modulate the protein-protein interaction of PCSK9 with the low density lipoprotein receptor in HepG2 cells. SciRep. 2016;6:29931. doi: 10.1038/srep29931.

combination therapy in human subjects. The result was that, "comparable percent reduction...[of] LDL-cholesterol between **monotherapy** and **statin combination therapy** was observed."[126]

In other words, the speculated DDI taking the form of reduced evolocumab efficacy was not detected. FDA arrived at a package label recommendation, "The implication of this finding is that **no dose adjustment** is recommended for patients on a background therapy of statins."[127]

The consequence of FDA's recommendation to refrain from any dose adjustment instructions was that the statin-effect found a place on the package label in a section that was purely informational and not as a warning or instruction.

Package label. The package label of evolocumab (Repatha®) disclosed the drug-drug interaction where coadministered statins reduced evolocumab exposure. But the Highlights of Prescribing Information section and the Full Prescribing section did not have any Drug Interactions section. Instead, information on the drug-drug interaction appeared in a more "information neutral" section of the label, the Clinical Pharmacology section:

> **CLINICAL PHARMACOLOGY**...Drug Interaction Studies. An approximately **20% decrease in the Cmax and AUC** of evolocumab was observed in patients co-administered with a...statin regimen. This difference is not clinically meaningful and does not impact dosing recommendations.[128]

ii. Anti-Evolocumab Antibodies

FDA observed that evolocumab rarely induced anti-evolocumab antibodies in human subjects, and where antibodies were induced, they did not neutralize the study drug. FDA wrote, "The incidence of anti-evolocumab binding antibodies was low...[t]he incidence...was 0.1% (7 out of 4846 subjects)...[i]n addition, neutralizing antibodies were not detected in any subject."[129]

The consequence was that the fact of anti-evolocumab antibodies found a place on the package label in a section that was purely informational (and not as any warning or instruction).

Package label (anti-evolocumab antibodies). The package label of evolocumab (Repatha®) provided information on the anti-evolocumab antibodies in the Adverse Reactions section:

> **ADVERSE REACTIONS**...In a pool of placebo- and active-controlled clinical trials, 0.1% of patients treated with at least one dose of REPATHA tested positive for binding antibody development. Patients whose sera tested positive for binding antibodies were further evaluated for neutralizing antibodies; none

[126] Evolocumab (hyperlipidemia and hypercholesterolemia) BLA 125-522. Page 57 of 148-page Clinical Pharmacology Review.

[127] Evolocumab (hyperlipidemia and hypercholesterolemia) BLA 125-522. Page 59 of 148-page Clinical Pharmacology Review.

[128] Package label. REPATHA (evolocumab) injection, for subcutaneous use; August 2015 (15 pp.).

[129] Evolocumab (hyperlipidemia and hypercholesterolemia) BLA 125-522. Page 49 of 148-page Clinical Pharmacology Review.

of the patients tested positive for neutralizing antibodies. There was no evidence that the presence of anti-drug binding antibodies impacted the pharmacokinetic profile, clinical response, or safety of REPATHA, but the long-term consequences of continuing REPATHA treatment in the presence of anti-drug binding antibodies are unknown.[130]

f. Lixisenatide (Type-2 Diabetes Mellitus) NDA 208-471

FDA's ClinPharm review of lixisenatide is distinguished by its thorough account of DDIs, where there are separate clinical studies of acetaminophen, atorvastatin, digoxin, oral contraceptives, ramipril, and warfarin.

Lixisenatide improves glycemic control in type-2 diabetes mellitus. The drug promotes the release of insulin in response to hyperglycemia, and slows gastric emptying.[131]

Lixisenatide is an agonist of GLP-1 receptor. GLP-1 is glucagon-like peptide-1.

Lixisenatide delays gastric emptying, as measurable after a standard meal.[132] The consequent prolonged absorption of glucose from the meal results in a blunted increase in plasma glucose. This blunted increase in plasma glucose has the consequence of reducing plasma levels of glycosylated hemoglobin, and successful treatment of diabetes. Meier et al.[133] described the influence of lixisenatide in delaying gastric emptying. Various other drugs also delay gastric emptying, as described in the cited articles.[134,135,136,137] Lixisenatide is a 44-amino acid oligopeptide (Fig. 8.4).[138] The C-terminal amino acid is modified by an attached amino group, i.e., the C-terminal carboxyl group takes the form of an amide.

i. In Vitro Assays (Microsomes)

The Sponsor's in vitro DDI studies asked if the study drug impaired CYP enzyme-mediated catabolism of a second drug. The second drugs (probe substrates) were

[130] Package label. REPATHA (evolocumab) injection, for subcutaneous use; August 2015 (15 pp.).

[131] Pfeffer MA, et al. Lixisenatide in patients with type 2 diabetes and acute coronary syndrome. New Engl. J. Med. 2015;373:2247–57.

[132] Meier JJ, et al. Contrasting effects of lixisenatide and liraglutide on postprandial glycemic control, gastric emptying, and safety parameters in patients with type 2 diabetes on optimized insulin glargine with or without metformin: a randomized, open-label trial. Diabetes Care. 2015;38:1263–73.

[133] Meier JJ, et al. Contrasting effects of lixisenatide and liraglutide on postprandial glycemic control, gastric emptying, and safety parameters in patients with type 2 diabetes on optimized insulin glargine with or without metformin: a randomized, open-label trial. Diabetes Care. 2015;38:1263–73.

[134] Pleuvry BJ. Pharmacodynamic and pharmacokinetic drug interactions. Anaesth. Intensive Care Med. 2005;6:129–33.

[135] Greiff JM, Rowbotham D. Pharmacokinetic drug interactions with gastrointestinal motility modifying agents. Clin. Pharmacokin. 1994;27:447–61.

[136] Brown CK, Khanderia U. Use of metoclopramide, domperidone, and cisapride in the management of diabetic gastroparesis. Clin. Pharm. 1990;9:357–65.

[137] Barone JA. Domperidone: a peripherally acting dopamine$_2$-receptor agonist. Ann. Pharmacother. 1999;33:429–40.

[138] Package label. ADLYXIN (lixisenatide) injection, for subcutaneous use; June 2016 (32 pp.).

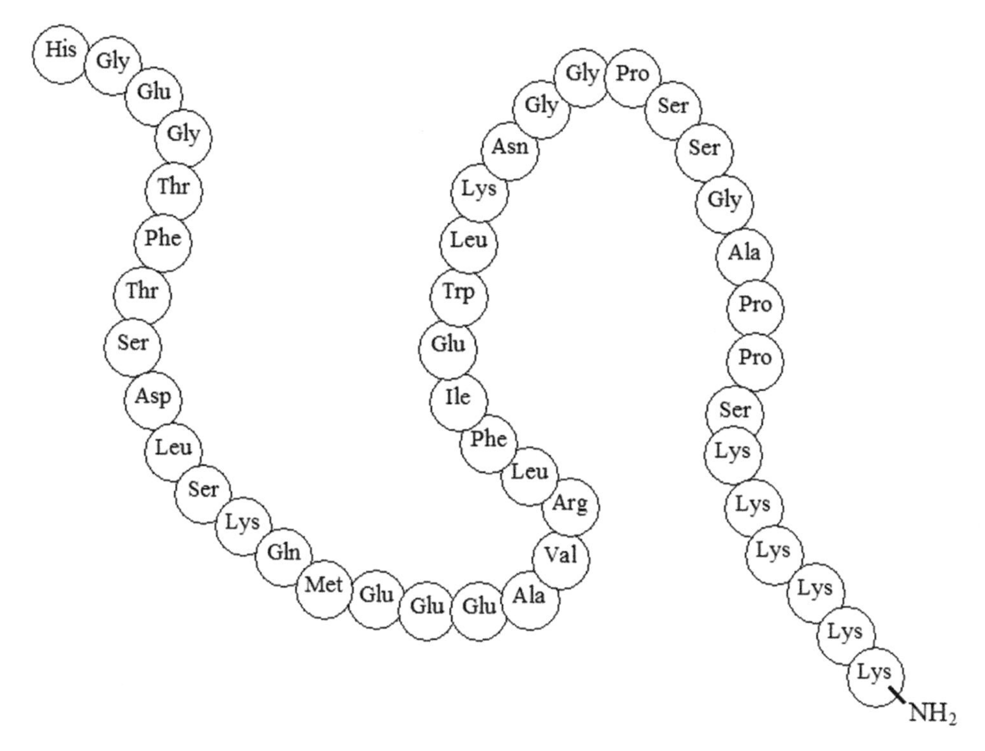

FIGURE 8.4 Lixisenatide. Lixisenatide is a 44-amino acid oligopeptide where the C-terminal amino acid is modified by an attached amino group. This amino group is attached to a lysine residue, as shown.

established to be specific for the CYP enzyme in Table 8.2.[139] Table 8.2 shows inhibition by lixisenatide of CYP enzyme-catalyzed processing of the probe substrate, where lixisenatide was 20 μM (97 μg/mL) in the enzyme incubation mixtures. Human microsomes were the source of enzyme. This concentration (97 μg/mL) was about one million times more than that of lixisenatide in blood plasma (50 pg/mL).

Low-to-moderate inhibitions of enzyme activity were found in some of the microsomal incubations. The extremely high concentration of lixisenatide (97 μg/mL) used for in vitro microsomal incubations should be compared with the much lower plasma levels of in blood plasma (only 50 pg/mL) when administered to humans.

Because of the *low-to-moderate inhibitions* and the *extremely high concentrations of study drug* in the microsomal assays, the Sponsor and FDA both concluded that lixisenatide would not be expected to result in any DDIs at the location of the CYP enzymes, when administered to human subjects. To this end, the FDA reviewer wrote, "Thus, lixisenatide is not expected to cause any drug-drug...inhibition of cytochrome P450s and we agree with the sponsor's conclusions."

[139] Lixisenatide (type 2-diabetes mellitus) NDA 208-471. Page 77 of 198-page Clinical Pharmacology Review.

TABLE 8.2 Tests for Lixisenatide-Mediated Inhibition of CYP Inhibition in Assays Measuring CYP-Catalyzed Catabolism of Probe Substrates

CYP Enzyme	Probe Substrate	Percent Inhibition at 20 μM (97 μg/mL) Lixisenatide
CYP1A2	Phenacetin	11
CYP2B6	Bupropion	9
CYP2C9	Diclofenac	0
CYP2C19	S-mephenytoin	32
CYP2D6	Dextromethorphan	4
CYP3A	Midazolam	10
CYP3A	Testosterone	20

Lixisenatide (type 2-diabetes mellitus) NDA 208-471. Page 77 of 198-page Clinical Pharmacology Review.

ii. In Vitro Assays (Hepatocytes)

Cultured hepatocytes were used to test induction of mRNA encoding CYP enzymes. Lixisenatide was added to the cell culture at concentrations of 180–180,000 pg/mL. The Sponsor reported that, "**No in vitro induction**...for CYP1A, CYP2B6, CYP2C9, and CYP3A was observed for lixisenatide...both at the enzyme activity and mRNA level."[140] FDA's analysis then turned to the clinical implications of the in vitro data, and stated that, "The sponsor proposed that in vivo drug interactions due to induction of CYP1A, CYP2B6, CYP2C9, and CYP3A by lixisenatide...**are unlikely**. This reviewer agrees with the sponsor."[141]

To summarize, the CYP inhibition tests (microsomes) and CYP induction tests (hepatocytes) showed that lixisenatide does not much influence CYP enzyme activity, and thus would not likely cause CYP-mediated DDIs in patients.

iii. Clinical Study of Drug–Drug Interactions

The Sponsor assessed interactions between lixisenatide and several different coadministered drugs, in studies with human subjects, where FDA's analysis of the following drugs are shown below:

- Acetaminophen
- Atorvastatin
- Oral contraceptives
- Warfarin

[140] Lixisenatide (type 2-diabetes mellitus) NDA 208-471. Page 77 of 198-page Clinical Pharmacology Review.

[141] Lixisenatide (type 2-diabetes mellitus) NDA 208-471. Page 77 of 198-page Clinical Pharmacology Review.

TABLE 8.3 Lixisenatide and Acetaminophen Treatments

Treatment	7:00 a.m.	8:00 a.m.	Breakfast at 8:30 a.m.	9:00 a.m.	12 noon
A	Acetaminophen	Placebo	Breakfast	–	–
B	Acetaminophen	Lixisenatide	Breakfast	–	–
C	–	Lixisenatide	Breakfast	Acetaminophen	–
D	–	Lixisenatide	Breakfast	–	Acetaminophen
E	–	Placebo	Breakfast	Acetaminophen	–

Lixisenatide (type 2-diabetes mellitus) NDA 208-471. Page 77 of 198-page Clinical Pharmacology Review.

Lixisenatide or placebo was administered to study subjects at only one time (8 o'clock in the morning). The coadministered drug (acetaminophen) was administered at either 7 a.m., 9 a.m., or at 12 noon. This was not a food-effect study. But food was included because food is necessarily a part of lixisenatide therapy (the mechanism of action of lixisenatide is to slow down the body's processing of food). Table 8.3 shows the experimental design.

FDA's ClinPharm Review provided graphs showing the time course of acetaminophen's peak in plasma levels, where the peak was followed by a tapering off until the 24-hour time point.

The value for each of the exposure parameters (AUC, Cmax, tmax), as they apply to oral drugs, is a function of the combined influences of stomach emptying rate, uptake by enterocytes, removal from blood plasma by hepatocyte drug transporters, and catabolism by CYP enzymes. But FDA's analysis, shown below, focused on absorption and gastrointestinal motility.

iv. Lixisenatide-Induced Drug–Drug Interactions (Acetaminophen)

FDA's ClinPharm Review focused on Cmax and tmax values, and warned:

> This drug interaction study was conducted to evaluate lixisenatide's effect on delaying GI motility. When administered 1 or 4 hours after 10 micrograms lixisenatide...Cmax of acetaminophen was decreased by 29% and 31% respectively, and median tmax was delayed by 2.0 and 1.75 hours respectively...[t]his study clearly demonstrated that a 10 microgram dose of lixisenatide can have a delaying effect of up to 4 hours on absorption of drugs such as acetaminophen which are absorbed rapidly with a short tmax (1-3 hours)...[o]ral medications that are **dependent on threshold concentrations for efficacy**, such as antibiotics, should be taken at least 1 hour before lixisenatide injection or with a meal or snack when lixisenatide is not administered.[142]

v. Lixisenatide-Induced Drug–Drug Interactions (Atorvastatin)

This concerns drug-drug interactions between lixisenatide with atorvastatin. FDA was concerned about failure of atorvastatin to reach a steady-state concentration to achieve efficacy, and recommended taking atorvastatin an hour before taking lixisenatide:

> The...**Cmax of atorvastatin was reduced by 31%**, AUC was unaltered, and tmax was prolonged (with change in tmax ranging from -1 to 9 hours) when atorvastatin was administered in morning 1 hour after

[142] Lixisenatide (type 2-diabetes mellitus) NDA 208-471. Pages 5, 80, and 81 of 198-page Clinical Pharmacology Review.

> lixisenatide injection...however, the significance of the effect of **prolongation in tmax by up to 9 hours** is not known...because **it may take longer to reach steady-state concentrations of atorvastatin** when administered with lixisenatide, patients taking atorvastatin may be advised to take atorvastatin preferably 1 hour before lixisenatide administration.[143]

This account provides the unique example, where FDA's DDI analysis focused on tmax.

vi. Lixisenatide-Induced Drug–Drug Interactions (Oral Contraceptives)

FDA's ClinPharm Review focused on Cmax and tmax values for the oral contraceptive, and observed that administering oral contraceptives (ethinylestradiol; levonorgestrel) at a time near to the time of administering lixisenatide resulted in substantial decrease in Cmax, as well as a delay of tmax. FDA warned that efficacy of the oral contraceptives depended on a "threshold concentrations," and recommended that when a dose of lixisenatide was taken, the time of taking oral contraceptives should be many hours later:

> Administration of an oral contraceptive 1 hour or 4 hours after lixisenatide did not affect AUC and $t_{1/2}$ of ethinylestradiol and levonorgestrel, whereas Cmax of ethinylestradiol was decreased by 52% and 39% respectively, and Cmax of levonorgestrel was decreased by 46% and 20%, respectively and median tmax was delayed by 1 to 3 hours. There was no change in AUC. The clinical implications of reduced peak exposure of ethinylestradiol and levonorgestrel when the oral contraceptive was administered 1 hour or 4 hours after lixisenatide injection, is unknown. **It is quite possible that oral contraceptives depend on threshold concentrations for efficacy**. Therefore, patients should be advised to take oral contraceptives at least 1 hour before lixisenatide administration or 11 hours after the morning dose of lixisenatide.[144]

This provides the unique example, as compared with FDA's drug–drug analyses for most other drugs, where FDA's analysis focused on threshold concentrations.

vii. Lixisenatide-Induced Drug–Drug Interactions (Warfarin)

This concerns drug-drug interactions between lixisenatide with warfarin. FDA expressed concern for the much-delayed tmax, and recommended monitoring of warfarin's desired efficacy reducing the tendency of blood to clot:

> While AUC [for warfarin] did not change between warfarin alone and warfarin plus lixisenatide...**Cmax** was reduced 19%...for S-warfarin...when warfarin was administered after lixisenatide. There was, however, a shift in **tmax** from 1 hour post warfarin alone treatment to 8.0 hours post warfarin plus lixisenatide treatment. The implication of the 7-hour delay in time to peak warfarin concentration is unknown. While no dose adjustment for warfarin is required when co-administered with

[143] Lixisenatide (type-2 diabetes mellitus) NDA 208-471. Pages 13–14 of 198 Clinical Pharmacology Review.

[144] Lixisenatide (type 2-diabetes mellitus) NDA 208-471. Pages 14–15 and 93–96 of 198-page Clinical Pharmacology Review.

lixisenatide based on lack of pharmacokinetic interaction, frequent monitoring of...[prothrombin time][145] in patients on warfarin...as well as...appropriate monitoring and dose adjustment is recommended at the time of...lixisenatide treatment.[146]

This provides another example of the unique situation, as compared to FDA's reviews of most other drugs, where FDA's DDI analysis focused on tmax.

viii. Antilixisenatide Antibodies Reduce Lixisenatide Efficacy in Human Subjects

Lixisenatide stimulates the immune system of subjects to generate antilixisenatide antibodies. Fig. 8.5, which is from FDA's ClinPharm Review, reveals that the antibodies are correlated with reduced efficacy.[147] The figure shows that concentration of antilixisenatide antibody after 24 weeks of lixisenatide treatment, with the associated reduction in lixisenatide efficacy.

Fig. 8.5 shows what might logically be expected, namely, that higher amounts of antilixisenatide result in failure of drug treatment to treat diabetes. With successful diabetes treatment glycosylated Hb levels will drop. But where diabetes treatment is interfered with by the antilixisenatide antibodies, then glycosylated Hb levels will be about the same as baseline glycosylated Hb levels. The long length of the X-axis takes into account the broad spread of plasma concentrations of antilixisenatide antibodies (the dots on the graph merely represent the median concentration of antilixisenatide antibodies).

FDA's account of the reduced efficacy resulting from the antilixisenatide antibodies, referred to efficacy measured by drops in glycosylated hemoglobin HbA1c. FDA's account stated, "What is the impact of anti-drug antibody on efficacy...[h]igher antibody concentrations resulted in attenuated HbA1c response...antibody concentration response analysis...showed a lower response in terms of change from baseline in HbA1c with increasing antibody concentrations."[148]

ix. Glycosylated Hemoglobin (HbA1c)

Lixisenatide's efficacy in treating diabetes was measured by the conventional technique of plasma glycosylated hemoglobin (HbA1c). Glycosylated hemoglobin occurs as a series of stable minor hemoglobin compounds that are formed by nonenzymatic condensation of glucose with hemoglobin. The rate of formation of glycosylated hemoglobin is directly proportional to glucose concentration.[149] Diabetes involves high plasma glucose. HBA1c has a sugar group linked to the N-terminal amino acid of the beta-globin chain of

[145] The normalized prothrombin time was expressed in terms of the International Normalized Ratio (INR), which is simply the prothrombin time for a given patient, normalized by a locally-determined normal prothrombin time.

[146] Lixisenatide (type-2 diabetes mellitus) NDA 208-471. Pages 13–14 of 198-page Clinical Pharmacology Review.

[147] Lixisenatide (type-2 diabetes mellitus) NDA 208-471. Figure 7B from page 131 of 198-page pdf file of Clinical Pharmacology Review.

[148] Lixisenatide (type-2 diabetes mellitus) NDA 208-471. Pages 129–130 of 198-page pdf file of Clinical Pharmacology Review.

[149] Goldstein DE, et al. Tests of glycemia in diabetes. Diabetes Care. 2004;27:1761–73.

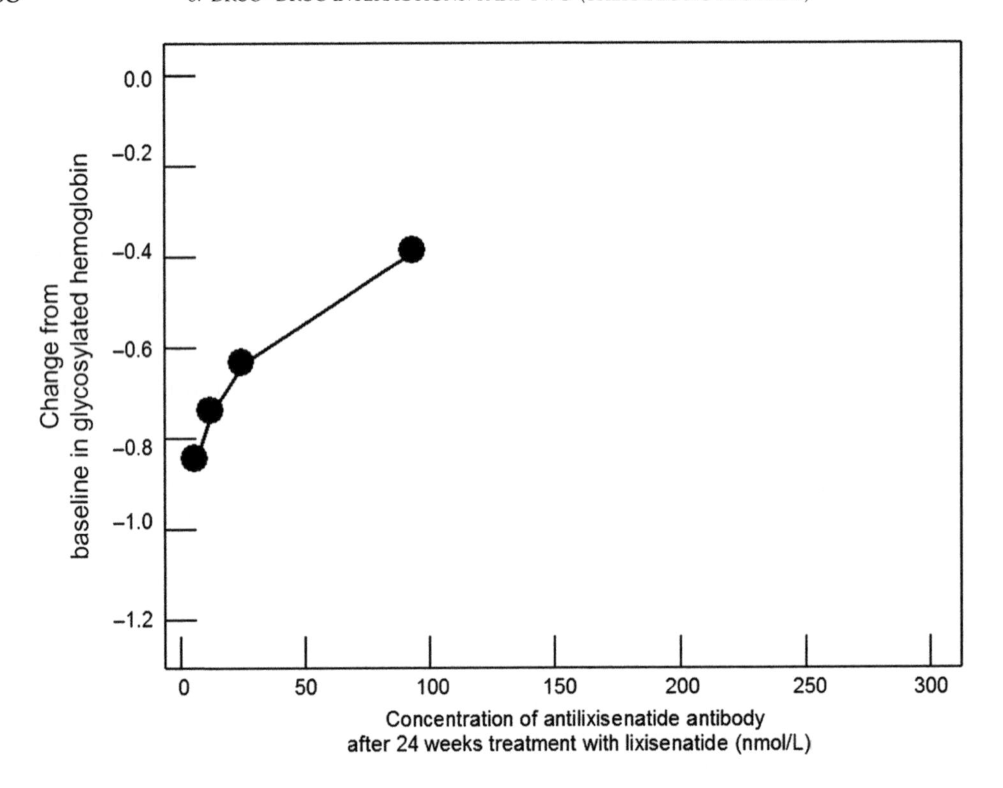

FIGURE 8.5 Antilixisenatide antibodies and reduced efficacy of lixisenatide. Efficacy of lixisenatide is measurable by levels of glycosylated Hb. With untreated diabetes, glycosylated Hb levels are high. But lixisenatide reduces glycosylated Hb. Antilixisenatide antibodies generated in patients block lixisenatide activity and reduce its ability to lower glycosylated Hb. The figure shows that the higher antilixisenatide antibody concentration, the higher is glycosylated Hb.

hemoglobin.[150] Levels of HBA1c increase in diabetes, a disorder where plasma glucose is chronically elevated. Where diabetes is treated, HBA1c levels get reduced. In the words of one researcher, "glycosylated haemoglobin is. . .a reliable yardstick of diabetic control that is simple, objective, and representative of average blood glucose concentrations over several weeks."[151]

Package label. FDA's DDI analyses for lixisenatide (Adlyxin®) and acetaminophen, and for lixisenatide and oral contraceptives, found a place in the Drug Interactions section. The label's mention of "threshold concentration" illustrates the fact that assessing if DDIs prevent a drug from reaching its threshold concentration should be considered for all DDI studies:

> **DRUG INTERACTIONS**. ADLYXIN delays gastric emptying which may impact absorption of concomitantly administered oral medications. Oral medications that are particularly dependent on **threshold concentrations for efficacy**, such as antibiotics, or medications for which a delay in effect is

[150] Peacock I. Glycosylated haemoglobin: measurement and clinical use. J. Clin. Pathol. 1984;37:841–51.

[151] Peacock I. Glycosylated haemoglobin: measurement and clinical use. J. Clin. Pathol. 1984;37:841–51.

> undesirable, such as **acetaminophen**, should be administered 1 hour before ADLYXIN. . .**[o]ral contraceptives** should be taken at least 1 hour before ADLYXIN administration or 11 hours after the dose of ADLYXIN.[152]

In the Full Prescribing Section, the Clinical Pharmacology section described the separate influences of lixisenatide (Adlyxin®) on the AUC, Cmax, and tmax, of coadministered acetaminophen, but the label did not go as far as to issue any warning or any dosage instructions:

> **CLINICAL PHARMACOLOGY**. . .ADLYXIN 10 micrograms did not change the overall exposure (AUC) of acetaminophen following administration of a single dose of acetaminophen 1000 mg, whether before or after ADLYXIN. No effects on acetaminophen Cmax and tmax were observed when acetaminophen was administered 1 hour before ADLYXIN. When administered 1 or 4 hours after 10 micrograms of ADLYXIN, Cmax of acetaminophen was decreased by 29% and 31% respectively and median tmax was delayed by 2.0 and 1.75 hours, respectively.[153]

FDA's observations on data antilixisenatide antibodies and consequent loss of clinical efficacy found a place in the Warnings and Precautions section:

> **WARNINGS AND PRECAUTIONS**. . .**Immunogenicity**: Patients may develop antibodies to lixisenatide. If there is worsening glycemic control or failure to achieve targeted glycemic control, significant injection site reactions or allergic reactions, alternative antidiabetic therapy should be considered.[154]

g. Nivolumab (Melanoma) BLA 125-554

Nivolumab is an antibody that binds to PD-1 and blocks signaling mediated by PD-1/PD-L1 interactions. Also, nivolumab blocks signaling mediated by PD-1/PD-L2 interactions. Nivolumab is used to treat various cancers such as melanoma,[155] Hodgkin's lymphoma,[156] and nonsmall-cell lung cancer (NSCLC).[157]

FDA's ClinPharm Review observed that antibodies against nivolumab were detected in human subjects, "A total of 24 out of the 281 evaluable patients (8.5%) who received nivolumab. . .tested positive for treatment emergent anti-nivolumab antibodies using an electrochemiluminescence. . .based assay."[158]

Going a step further from merely measuring the presence antibodies, FDA observed that these antibodies blocked nivolumab where blocking was determined with a cell-based

[152] Package label. ADLYXIN (lixisenatide) injection, for subcutaneous use; July 2016 (29 pp.).

[153] Package label. ADLYXIN (lixisenatide) injection, for subcutaneous use; July 2016 (29 pp.).

[154] Package label. ADLYXIN (lixisenatide) injection, for subcutaneous use; July 2016 (29 pp.).

[155] Larkin J, et al. Combined nivolumab and ipilimumab or monotherapy in untreated melanoma. New Engl. J. Med. 2015;373:23–34.

[156] Ansell SM, et al. PD-1 blockade with nivolumab in relapsed or refractory Hodgkin's lymphoma. New Engl. J. Med. 2015;372:311–9.

[157] Brahmer J, et al. Nivolumab versus docetaxel in advanced squamous-cell non–small-cell lung cancer. New Engl. J. Med. 2015; 373:123–35.

[158] Nivolumab (melanoma) BLA 125-554. Pages 5, 20, and 21 of 58-page Clinical Pharmacology Review.

functional assay, "Do the anti-product antibodies have neutralizing activity? Out of 12 anti-study drug antibody positive patients...2 patients (1 persistent positive, 1 other positive) each had 1 anti-study drug antibody positive sample with neutralizing antibodies...detected."[159]

Moving a step further from merely assessing if the antibodies were neutralizing, FDA contemplated if the antibodies resulted in lower nivolumab exposure. The result was that nivolumab exposure was not influenced, in the situation where the human subject developed anti-nivolumab antibodies, "Does the immunogenicity affect the PK...of the therapeutic protein? Nivolumab clearance for the patients whose samples **tested positive** for treatment emergent anti-study drug antibodies was in the same range of clearance for patients **tested negative** of anti-study drug antibodies treated with the same dose."[160]

Further on in FDA's analysis, FDA asked if the materialization of anti-nivolumab antibodies reduced nivolumab's clinical efficacy in treating melanoma. The result was there was no detectable influence. The FDA reviewer made the comments, "Neutralizing antibodies were detected in two patients (0.7%). No apparently altering or clinically meaningful difference in PK, safety and **efficacy profiles** were observed with the anti-nivolumab antibodies development[161]...[w]hat is the impact of anti-product antibodies on clinical efficacy? There was no evidence of apparent impact of treatment-emergent anti-product antibodies (APA) on the **clinical efficacy profile** for nivolumab due to lack of effect of immunogenicity on pharmacokinetic profile and flat exposure-response relationship[162]... [a] clear cause-effect relationship cannot be established between the presence of neutralizing antibodies and **loss of efficacy** and/or AEs."[163]

Package label. The observations in FDA's ClinPharm Review found a place on the Adverse Reactions section of the package label:

> **ADVERSE REACTIONS**...Of 2085 patients who were treated with OPDIVO as a single agent 3 mg/kg every 2 weeks and evaluable for the presence of anti-nivolumab antibodies, 233 patients (11.2%) tested positive for **treatment-emergent anti-nivolumab antibodies** by an electrochemiluminescent (ECL) assay and 15 patients (0.7%) had neutralizing antibodies against nivolumab. There was **no evidence of altered pharmacokinetic profile**...with anti-nivolumab antibody development.[164]

h. Pegloticase (Hyperuricemia and Gout) BLA 125-293

FDA's review for pegloticase provides a dramatic account of immune responses against therapeutic proteins, where these included the generation of antibodies against the therapeutic protein, as well as anaphylaxis. The best introduction to these issues might be

[159] Nivolumab (melanoma) BLA 125-554. Page 22 of 58-page Clinical Pharmacology Review.

[160] Nivolumab (melanoma) BLA 125-554. Page 22 of 58-page Clinical Pharmacology Review.

[161] Nivolumab (melanoma) BLA 125-554. Pages 5 and 20–21 of 58-page Clinical Pharmacology Review.

[162] Nivolumab (melanoma) BLA 125-554. Page 22 of 58-page Clinical Pharmacology Review.

[163] Nivolumab (melanoma) BLA 125-554. Page 22 of 58-page Clinical Pharmacology Review.

[164] Package label. OPDIVO (nivolumab) injection, for intravenous use; February 2017 (57 pp.).

FDA's Medical Review, which provides a bird's-eye view of the problems that had occurred with pegloticase, and which could occur with any therapeutic protein. FDA's account included the therapeutic strategy of coadministering drugs (corticosteroids) for reducing undesirable immune responses. In FDA's own words:

> Pegloticase was highly immunogenic with 88% of patients in the pegloticase...group...[h]igher titers of antibody to pegloticase were associated with higher rates of infusion reactions and decreases in urate-lowering effects of therapy. A review of all infusion reactions...revealed that approximately 5% of these patients met clinical criteria for anaphylaxis despite the mandated administration of antihistamines, acetaminophen, and corticosteroids prior to study drug infusions to attenuate infusion reactions.[165]

The enhanced immunogenicity of pegloticase arose from the fact that it was an animal enzyme, purified from pigs, and from the fact that it was not subject to "humanization." Humanization is a technique that is conventional when designing and manufacturing recombinant therapeutic proteins.[166,167] As stated by FDA's Medical Review, "Since pegloticase is a **non-human protein**, it is not surprising that more than 90% of all pegloticase-treated subjects developed anti-pegloticase antibodies...with higher rates of infusion reactions and decreases in urate-lowering effects of therapy."[168]

Pegloticase is an enzyme conjugated with an artificial polymer, methoxy-PEG. The enzyme is urate oxidase. According to FDA's Chemistry Review, each urate oxidase protein has nine strands of methoxy-PEG connected to it.[169,170] Urate oxidase catalyzes the conversion of uric acid to allantoin, thereby preventing the formation of uric acid deposits in the body, and consequent gout. In gout, crystals of uric acid can form in the joints and in the kidneys.[171] PEG is frequently conjugated to therapeutic proteins to enhance the lifetime of the drug in the bloodstream. Reagents for detecting anti-PEG antibodies are available.[172,173]

Patients treated with pegloticase generated antibodies against both portions of pegloticase, the enzyme (urate oxidase) portion and the PEG portion. A publication by the Sponsor (Savient, Inc.) revealed that a total of 69 out of 212 patients developed

[165] Pegloticase (hyperuricemia and gout) BLA 125-293. Page 3 of 238-page Medical Review.

[166] Pegloticase (hyperuricemia and gout) BLA 125-293. Page 16 of 28-page Chemistry Review.

[167] Pegloticase (hyperuricemia and gout) BLA 125-293. Page 56 of 238-page Medical Review.

[168] Pegloticase (hyperuricemia and gout) BLA 125-293. Page 4 of 238-page Medical Review.

[169] Pegloticase (hyperuricemia and gout) BLA 125-293. Page 4 of 28-page Chemistry Review.

[170] Lipsky PE, et al. Pegloticase immunogenicity: the relationship between efficacy and antibody development in patients treated for refractory chronic gout. Arthritis Res. Ther. 2014;16:R60. doi: 10.1186/ar4497 (8 pp.).

[171] Martillo MA, et al. The crystallization of monosodium urate. Curr. Rheumatol. Rep. 2014;16:400 (13 pp.).

[172] Krishna M, et al. Development and characterization of antibody reagents to assess anti-PEG IgG antibodies in clinical samples. Bioanalysis. 2015;7:1869–83.

[173] Dong H, et al. Development of a generic anti-PEG antibody assay using BioScale's acoustic membrane microparticle technology. AAPS J. 2015;17:1511–6.

anti-pegloticase antibodies that bound to the PEG part of pegloticase, and that a total of 24 out of these 212 patients developed anti-pegloticase antibodies that bound to the urate oxidase part of pegloticase.[174] Other publications have also described anti-pegloticase antibodies.[175,176]

An avenue for reducing the formation of antibodies generated in response to therapeutic proteins is coadministering an immunosuppressant.[177,178,179] Berhanu et al.,[180] confronted the problem of antibodies generated against pegloticase, and provided evidence that coadministering immunosuppressants such as azathioprine, cyclosporine, tacrolimus, or mycophenolate mofetil, can prevent the generating of antibodies against pegloticase, where patients receive pegloticase for gout therapy. For other therapeutic antibodies (infliximab; adalimumab), coadministering the immunosuppressants thiopurine or methotrexate prevented the generation of antidrug antibodies.[181]

For detecting anti-pegloticase antibodies in the study subjects, the Sponsor used the ELISA technique. Using this technique, the Sponsor was able to distinguish anti-pegloticase antibodies in the IgE, IgG, and IgM classes of immunoglobulins.[182]

FDA's ClinPharm Review for pegloticase disclosed that antibody data was available from 75 subjects in the treatment arm where each subject received 8 mg pegloticase once *every 2 weeks*, from 25 subjects in the treatment arm receiving 8 mg pegloticase *every 4 weeks*, and from 41 subjects in the *placebo group*.[183]

[174] Lipsky PE, et al. Pegloticase immunogenicity: the relationship between efficacy and antibody development in patients treated for refractory chronic gout. Arthritis Res. Ther. 2014;16:R60. doi: 10.1186/ar4497 (8 pp.).

[175] Abeles AM. PEG-ing down (and preventing?) the cause of pegloticase failure. Arthritis Res. Ther. 2014;16:112.

[176] Zhang P, et al. Anti-PEG antibodies in the clinic: current issues and beyond PEGylation. J. Control Release. 2016;244:184–93.

[177] Berhanu AA, et al. Pegloticase failure and a possible solution: immunosuppression to prevent intolerance and inefficacy in patients with gout. Semin. Arthritis Rheum. 2016. doi: 10.1016/j.semarthrit.2016.09.007.

[178] Krieckaert CL, et al. Methotrexate reduces immunogenicity in adalimumab treated rheumatoid arthritis patients in a dose dependent manner. Ann. Rheum. Dis. 2012;71:1914–5.

[179] Apgar JR, et al. Beyond CDR-grafting: structure-guided humanization of framework and CDR regions of an anti-myostatin antibody. MAbs. 2016;8:1302–18.

[180] Berhanu AA, et al. Pegloticase failure and a possible solution: immunosuppression to prevent intolerance and inefficacy in patients with gout. Semin. Arthritis Rheum. 2016. doi: 10.1016/j.semarthrit.2016.09.007.

[181] Strik AS, et al. Suppression of anti-drug antibodies to infliximab or adalimumab with the addition of an immunomodulator in patients with inflammatory bowel disease. Ailiment Pharmacol. Ther. 2017;45:1128–34.

[182] Pegloticase (hyperuricemia and gout) BLA 125-293. Pages 25 and 42 of 80-page Clinical Pharmacology Review.

[183] Pegloticase (hyperuricemia and gout) BLA 125-293. Pages 23–24 and 51-56 of 80-page Clinical Pharmacology Review.

In the 8 mg *every 2 weeks* group, 25 of the subjects had *high antibody levels*, and *none of these responded to the drug* by having lowered plasma uric acid. In other words, for these 25 subjects the drug had no efficacy. But for subjects receiving 8 mg *every 2 weeks* where the antibody levels were not detectable, or low, or moderate, the outcome was that pegloticase was effective in lowering plasma uric acid.

The FDA reviewer commented on the residual efficacy of pegloticase in subjects with anti-pegloticase antibodies, writing that the level of reduction in plasma uric acid, "was inversely related to the levels of circulating antibodies to pegloticase."[184]

Package label. The Sponsor's data on the generation of antibodies against pegloticase (Krystexxa®) found a place on the package label. The label also warned about variability in assays for detecting anti-pegloticase antibodies:

> **ADVERSE REACTIONS**...Immunogenicity. **Anti-pegloticase antibodies developed in 92% of patients** treated with KRYSTEXXA every 2 weeks, and 28% for placebo. Anti-PEG antibodies were also detected in 42% of patients treated with KRYSTEXXA. High anti-pegloticase antibody titer was associated with a **failure to maintain pegloticase-induced normalization of uric acid**...[t]he observed incidence of antibody positivity in an assay is highly dependent on several factors including assay sensitivity and specificity and assay methodology, sample handling, timing of sample collection, concomitant medications, and underlying disease. For these reasons, the comparison of the incidence of antibodies to pegloticase with the incidence of antibodies to other products may be misleading.[185]

Also, the package label also warned that any anti-pegloticase antibodies generated after pegloticase treatment had the potential of interfering with any coadministered drug having a PEG moiety:

> **DRUG INTERACTIONS**...Because anti-pegloticase antibodies **appear to bind to the PEG portion** of the drug, there may be potential for binding with other PEGylated products. The impact of anti-PEG antibodies on patients' responses to other PEG-containing therapeutics is unknown.[186]

i. Sebelipase Alfa (Deficiency in Lysosomal Acid Lipase) BLA 125-561

Sebelipase alfa is an enzyme for treating lysosomal lipase deficiency. Lipase deficiency takes two forms, infantile-lipase deficiency and late-onset lipase deficiency. Infantile-onset deficiency is usually fatal within the first 6 months of life. Sebelipase alfa is recombinant human lysosomal acid lipase, purified from the egg white of transgenic hens.[187] The recombinant lipase has the same amino acid sequence as the native human lipase.

This concerns the question of why human subjects generated anti-sebelipase alfa antibodies, despite the fact the enzyme was of human origin. The recombinant human sebelipase

[184] Pegloticase (hyperuricemia and gout) BLA 125-293. Page 60 of 80-page Clinical Pharmacology Review.

[185] Package label. KRYSTERXXA (pegloticase) injection, for intravenous infusion; September 2016 (14 pp.).

[186] Package label. KRYSTERXXA (pegloticase) injection, for intravenous infusion; September 2016 (14 pp.).

[187] Sebelipase alfa (deficiency in lysosomal acid lipase) BLA 125-561. 87-page Clinical Pharmacology Review.

alfa was manufactured by chicken eggs. Oligosaccharides from chicken glycoproteins have been proven to induce immune responses in mammals.[188] Sebelipase alfa contains attached oligosaccharides which were created by enzymes in the chicken pathway of oligosaccharide biosynthesis.[189] If these oligosaccharides were of chicken origin rather than of human origin, this could account for the anti-sebelipase alfa antibodies in the human subjects.

i. Anti-Sebelipase Alfa Antibodies

FDA commented on the anti-sebelipase alfa antibodies that were detected in the bloodstream following administration of the study drug, "Overall, 10 of 51 sebelipase alfa treated subjects (20%) in the clinical development program developed antibodies against sebelipase alfa (antidrug antibody-positive)."[190]

ii. Neutralization by Anti-Sebelipase Alfa Antibodies of Cellular Uptake (In Vitro Tests)

Neutralization was detected by enzyme activity and also by cellular uptake. FDA's ClinPharm Review stated that:

> in the pivotal study of infantile-onset...[lipase] deficiency...4 of 7 evaluable infants (57%) developed anti-drug antibodies during treatment with sebelipase alfa. Two patients were determined to be positive for neutralizing **antibodies that inhibit in vitro enzyme activity and cellular uptake of the enzyme**. In the...study of late-onset...[lipase] deficiency...5 of 35 evaluable children and adults (14%) treated with sebelipase alfa developed anti-drug antibodies during the 20-week double-blind period of the study.[191]

Regarding the time course for generating antidrug antibodies in the four infants, FDA observed, "Among infants...4 of 7 evaluable subjects (57%) had positive anti-drug antibody titers...[i]n these 4 subjects, **positive titers first occurred** at Week 5 (n = 1), Week 8 (n = 2), or Week 59 (n = 1)."[192]

In vitro assays for enzyme uptake, and for assessing the ability of anti-sebelipase alfa antibodies to block uptake, were as follows. As stated by the FDA reviewer, the oligosaccharide moiety of sebelipase alfa (as well as the oligosaccharide of the naturally occurring counterpart, lysosomal acid phosphatase), mediates uptake of the enzyme into cultured cells. The undesired blocking activity of the antibodies was tested by uptake assays with a cultured cell line.

[188] Hwang HS, et al. Cleavage of the terminal N-acetylglucosamine of egg-white ovalbumin N-glycans significantly reduces IgE production and Th2 cytokine secretion. Biochem. Biophys. Res. Commun. 2014;450:1247–54.

[189] Sebelipase alfa (deficiency in lysosomal acid lipase) BLA 125-561. Pages 13, 26, 38, 41, and 59 of 68-page Chemistry Review.

[190] Sebelipase alfa (deficiency in lysosomal acid lipase) BLA 125-561. Pages 6 and 21 of 87-page Clinical Pharmacology Review.

[191] Sebelipase alfa (deficiency in lysosomal acid lipase) BLA 125-561. Pages 6 and 22 of 87-page Clinical Pharmacology Review.

[192] Sebelipase alfa (deficiency in lysosomal acid lipase) BLA 125-561. Pages 6 and 22 of 87-page Clinical Pharmacology Review.

According to FDA's Pharmacology Review, "sebelipase alfa glycans consist of predominately N-acetylglucosamine (GlcNAc) and mannose terminated N-linked structures, as well as mannose-6-phosphate moieties. These glycans target uptake via receptors expressed on a number of cell types including Kupffer cells and hepatocytes...N-glycan structures...facilitate protein uptake into cells via the macrophage mannose or mannose-6-phosphate receptors (Stahl PD et al (1978) Proc. Natl. Acad. Sci. 75:1399-403; Coutinho MF et al (2012) Mol. Genet. Metab. 105:542)."[193]

Regarding the mechanism of sebelipase alfa uptake, FDA's ClinPharm Review stated that, "sebelipase alfa uptake by cells...is most likely mediated by binding of N-linked glycans."[194] FDA's review further disclosed that antibody blocking of sebelipase alfa uptake was tested using a rat macrophage cell line.[195] The result from this cell uptake assay, was that two out of four subjects with positive anti-sebelipase antibodies were found to be positive for antibodies that actually inhibited uptake of the enzyme by the rat macrophage cells.[196]

iii. Adverse Effects of Anti-Sebelipase Alfa Antibodies (Clinical Efficacy)

Efficacy, in terms of supporting growth of human infants, was reduced in some of the infants having anti-sebelipase alfa antibodies in their bloodstream, "**only 1 patient experienced loss of efficacy attributable to neutralizing anti-drug antibodies**. In this patient, the decreased effectiveness of sebelipase alfa was not associated with any life-threatening sequelae, and appeared to show signs of improvement following a dose escalation to 5 mg/kg."[197]

Further regarding impaired clinical efficacy, FDA observed, "For 1 of 4 infants...the **neutralizing antibodies** may have been a factor in the **continued suboptimal rate of growth**...[h]owever, other factors may have contributed to the observed suboptimal growth rate, such as infection, feeding difficulties, and chronic prednisolone premedication."[198]

FDA contemplated the influence of these antibodies on the study drug's efficacy and safety and decided that conclusions could not be drawn, writing, "Overall, it is not feasible to draw a definitive conclusion on the impact of anti-drug antibodies...on pharmacokinetics, pharmacodynamics, efficacy, and safety."[199]

[193] Sebelipase alfa (deficiency in lysosomal acid lipase) BLA 125-561. Page 6 of 141-page Pharmacology Review.

[194] Sebelipase alfa (deficiency in lysosomal acid lipase) BLA 125-561. Page 8 of 87-page Clinical Pharmacology Review.

[195] Sebelipase alfa (deficiency in lysosomal acid lipase) BLA 125-561. Pages 58, 65, and 73 of 87-page Clinical Pharmacology Review.

[196] Sebelipase alfa (deficiency in lysosomal acid lipase) BLA 125-561. Page 74 of 87-page Clinical Pharmacology Review.

[197] Sebelipase alfa (deficiency in lysosomal acid lipase) BLA 125-561. Page 12 of 229-page Medical Review.

[198] Sebelipase alfa (deficiency in lysosomal acid lipase) BLA 125-561. Page 26 of 87-page Clinical Pharmacology Review.

[199] Sebelipase alfa (deficiency in lysosomal acid lipase) BLA 125-561. Pages 6 and 22 of 87-page Clinical Pharmacology Review.

A published account of a clinical trial on 35 subjects receiving sebelipase alfa observed the formation of anti-sebelipase alfa, and arrived at the same type of conclusion, namely, that there was not any persuasive influence that the antibodies impaired clinical efficacy, "Antidrug-antibody titers developed in few patients and were generally low and transient, with **no apparent effect on the treatment response or safety profile**, including infusion associated reactions."[200]

iv. Validating Laboratory Tests

A recurring theme in many FDA-submissions is the need for the Sponsor to use validated assay methods. FDA's Guidance for Industry provides information for validating instruments and methods.[201,202] FDA review for another drug, denosumab,[203] recommended the DeSilva et al.[204] reference for guidance on assay validation. FDA mentioned the Sponsor's attempts to validate the tests, and then complained about the Sponsor's attempt, writing that, "The applicant submitted...validation reports that contain...an enzymatic activity-based neutralizing assay for the detection of the anti-sebelipase neutralizing antibodies...and a cell-based assay for the detection of neutralizing antibodies (inhibition of cellular uptake) in human serum."[205]

FDA complained, "The neutralizing...assay appears to be **insensitive**. See Product Quality Review...for more detailed information regarding immunogenicity assay validation."[206] The lack of assay validation did not prevent FDA from granting approval to sebelipase, but it did result in a warning on the package label, regarding the unreliable nature of the assays.

[200] Burton BK, et al. A phase 3 trial of sebelipase alfa in lysosomal acid lipase deficiency. New Engl. J. Med. 2015;373:1010–20.

[201] U.S. Department of Health and Human Services. Food and Drug Administration. Center for Drug Evaluation and Research (CDER). Center for Biologics Evaluation and Research (CBER). Center for Veterinary Medicine (CVM). Guidance for industry. Process validation: general principles and practices; 2011 (19 pp.).

[202] U.S. Department of Health and Human Services. Food and Drug Administration. Center for Drug Evaluation and Research (CDER). Center for Biologics Evaluation and Research (CBER). Guidance for industry. Analytical procedures and methods validation for drugs and biologics; 2015 (15 pp.).

[203] Denosumab (osteoporosis) BLA 125-320. Page 50 of 79-page Clinical Pharmacology Review.

[204] DeSilva B, et al. Recommendations for the bioanalytical method validation of ligand-binding assays to support pharmacokinetic assessments of macromolecules. Pharm. Res. 2003;20:1885–900.

[205] Sebelipase alfa (deficiency in lysosomal acid lipase) BLA 125-561. Page 58 of 87-page Clinical Pharmacology Review.

[206] Sebelipase alfa (deficiency in lysosomal acid lipase) BLA 125-561. Page 30 of 87-page Clinical Pharmacology Review.

Package label (anti-sebelipase alfa antibodies). FDA's complaint about the "insensitive" nature of the assays found a corresponding complaint in the Adverse Reactions section of the package label:

> **ADVERSE REACTIONS.** Immunogenicity assay results are highly dependent on the **sensitivity** and specificity of the assay and may be influenced by several factors such as: assay methodology, sample handling, timing of sample collection, concomitant medications, and underlying disease.[207]

In addition the Adverse Reactions section mentioned that the antibodies materializing in human subjects seemed not to result in reduced clinical efficacy:

> **ADVERSE REACTIONS**...Five of 35 (14%) KANUMA-treated pediatric and adult patients who completed the 20-week double-blind period of study treatment developed...anti-drug antibodies...anti-drug antibody titers decreased to undetectable levels during continued treatment. Two patients developed **in vitro neutralizing antibodies** during the open-label extension phase after 20 weeks and 52 weeks of treatment with KANUMA, respectively. There is **no clear association between the development of anti-drug antibodies and decreased efficacy in pediatric and adult patients** treated with KANUMA.[208]

j. Ustekinumab (Plaque Psoriasis) BLA 125-261

Ustekinumab is an antibody that binds to and inactivates two cytokines, IL-12 and IL-23.[209] Ustekinumab binds to a subunit that is used by both of these cytokines. The common subunit is "p40."[210] Ustekinumab is used to treat psoriasis. The mechanism of action of autoimmune diseases such as psoriasis includes abnormalities in cell-signaling mediated by cytokines and cytokine receptors.[211]

i. FDA Contemplated the Mechanism of Drug–Drug Interactions Common for Small Molecules (But Rarely Occurring for Therapeutic Proteins)

FDA considered and then rejected two hypothetical mechanisms for drug-drug interactions with ustekinumab. Figs. 8.6 and 8.7 illustrate these two hypothetical mechanisms. Although these two mechanisms are typical where DDIs occur between a first small molecule drug and a second small molecule drug, they rarely occur between an antibody drug and a small molecule drug. Despite the fact that these two mechanisms rarely occur between an antibody drug and small molecule drug, it is FDA's habit to contemplate the two mechanisms and then to dismisses them as unlikely.

[207] Package label. KAUMA (sebelipase alfa) injection, for intravenous use; December 2015 (13 pp.).

[208] Package label. KAUMA (sebelipase alfa) injection, for intravenous use; December 2015 (13 pp.).

[209] Ustekinumab (plaque psoriasis) BLA 125-261. 194-page Clinical Pharmacology Review.

[210] Tang C, et al. Interleukin-23: as a drug target for autoimmune inflammatory diseases. Immunology. 2012;135:112–24.

[211] Cho JH, Gregersen PK. Genomics and the multifactorial nature of human autoimmune disease. New Engl. J. Med. 2011;365:1612–23.

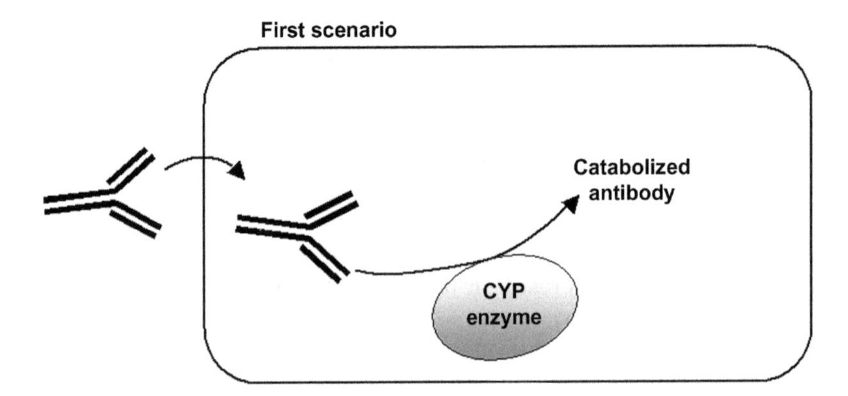

FIGURE 8.6 Therapeutic antibody or some other type of therapeutic protein enters the cell and is then catabolized by a cytochrome P450 enzyme. Drug–drug interaction is expected to occur where a small molecule inhibits or induces cytochrome P450, thereby changing the rate of antibody catabolism.

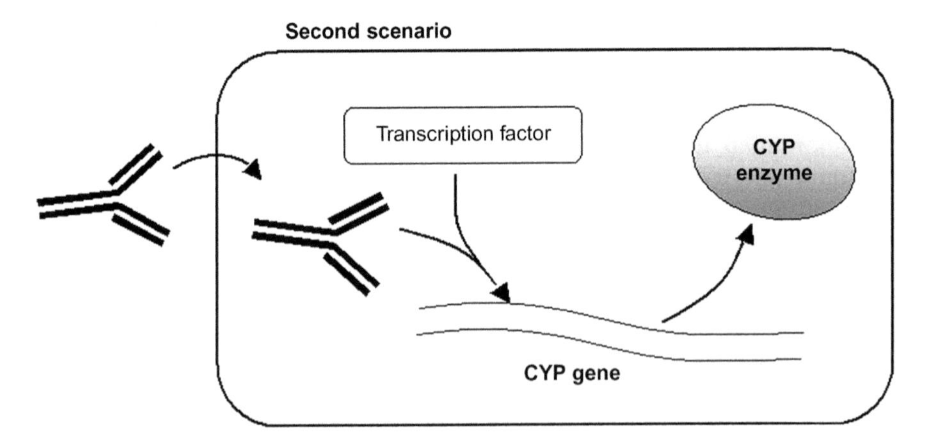

FIGURE 8.7 Therapeutic antibody or some other type of therapeutic protein enters the cell, interacts with a transcription factor controlling expression of cytochrome P450, resulting in a change in the rate of expression of cytochrome P450 expression. Drug–drug interaction is expected to occur where an administered therapeutic antibody changes cytochrome P450 expression, thereby changing the rate of catabolism of a coadministered small molecule drug.

ii. First Mechanism (Antibody Catabolized by CYP Enzyme)

Small molecule induces or inhibits CYP enzyme-mediated catabolism of antibody, with consequent change in antibody level in the bloodstream. The forked object in the figure is conventionally used to represent antibodies (Fig. 8.6).

iii. Second Mechanism (Antibody Induces or Inhibits CYP Enzyme)

Where an antibody induces or inhibits expression of CYP enzyme, where the result is an increase or decrease in CYP enzyme-mediated catabolism of a small molecule (Fig. 8.7).

In this mechanism, the antibody binds to a transcription factor to form a complex, where the complex then induces CYP gene expression. Another version of this mechanism (not in the picture) is where the antibody directly contacts CYP enzyme and inhibits it.

The first and second mechanisms are implausible, because each of these requires that ustekinumab enter the cell. Despite the implausibility of these two mechanisms, it is still the case that they are contemplated by FDA reviewers. The FDA reviewer discounted the possibility that a small molecule could induce or inhibit CYP expression with any consequent change in ustekinumab exposure, writing, "Antibodies are not metabolized by cytochrome P450 enzymes...[t]herefore, direct...interactions via the CYP pathway is not expected between ustekinumab and co-administered small molecule drugs."[212]

Despite the unlikelihood that a coadministered small molecule drug could induce or inhibit ustekinumab degradation, and result in a consequent change in ustekinumab exposure, the Sponsor did conduct relevant DDI studies. FDA's assessment of these studies was that, "Potential drug-drug interactions were evaluated...among the...most frequently used concomitant medications...including atorvastatin, metformin, acetylsalicylic acid, ibuprofen, and paracetamol...**none of these concomitant medications had...effect upon the...clearance of ustekinumab**."[213]

Regarding any drug's ability to induce or inhibit gene expression, it does so by way of binding to a transcription factor.[214,215,216] One would not expect any administered therapeutic antibody to enter any cell and gain access to any transcription factor, unless the antibody was a recombinant *intrabody*.[217] Also, even if the therapeutic antibody did reach the transcription factor, one would not expect the antibody to bind the transcription factor (unless the antibody was intentionally created so that it would bind that transcription factor).

iv. Third Mechanism (Antibody Blocks Cytokines in Plasma)

Therapeutic antibodies targeting cytokines or cytokine receptors can bind to cytokines in the plasma, or to cytokine receptors, thereby blocking cytokine-mediated signaling. In some cases, this blocking can induce or reduce CYP enzyme expression, depending on the normal biological activity of the cytokine itself. Liptrott et al.[218] provide an example where a cytokine regulates CYP enzyme expression. In this example, adding IL-12 to cultured

[212] Ustekinumab (plaque psoriasis) BLA 125-261. Page 13 of 194-page Clinical Pharmacology Review.

[213] Ustekinumab (plaque psoriasis) BLA 125-261. Page 13 of 194-page Clinical Pharmacology Review.

[214] Kublbeck J, et al. Up-regulation of CYP expression in hepatoma cells stably transfected by chimeric nuclear receptors. Eur. J. Pharm. Sci. 2010;40:263–72.

[215] Wilson TM, Kliewer SA. PXR, CAR and drug metabolism. Nat. Rev. Drug Discov. 2002;1:259–66.

[216] Tompkins LM, Wallace AD. Mechanisms of cytochrome P450 induction. J. Biochem. Mol. Toxicol. 2007;21:176–81.

[217] Marschall AL, Dubel S. Antibodies inside of a cell can change its outside: can intrabodies provide a new therapeutic paradigm? Comput. Struct. Biotechnol. J. 2016;14:304–8.

[218] Dallas S, et al. Interleukines-12 and -23 do not alter expression or activity of multiple cytochrome P450 enzymes in cryopreseved human hepatocytes. Drug Metab. Dispos. 2013;41:689–93.

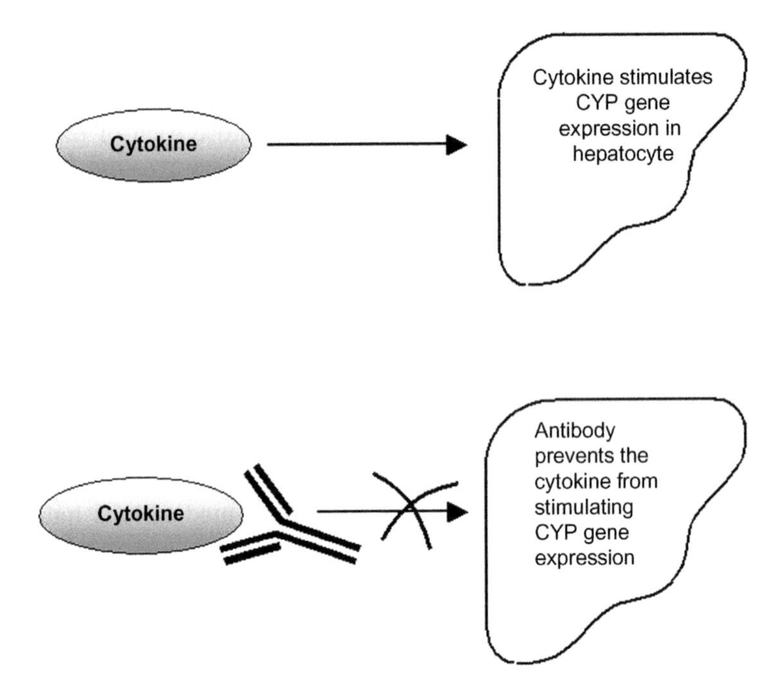

FIGURE 8.8 The upper drawing shows a cytokine stimulating a change in CYP enzyme expression and the lower drawing shows a therapeutic antibody blocking the cytokine's effect. The scenario where a therapeutic anticytokine antibody blocks the ability of the cytokine to module CYP enzyme expression is a scenario that has been demonstrated for various antibodies.

human white blood cells changed CYP2B6 expression.[219] To give another example of a cytokine regulating CYP enzyme expression, Dallas et al.[220] determined that tocilizumab, which binds IL-6 receptor (IL-6R), changed CYP3A4 expression. The resulting changes in CYP enzyme activity can, in turn, alter the catabolism of a coadministered small molecule drug.[221,222] Fig. 8.8 illustrates these types of scenarios. The wiggly shape at the right, which resembles a clam shell, is conventionally used to represent the liver, and in Fig. 8.8 it is used to represent a hepatocyte.

v. Ustekinumab-Mediated Drug–Drug Interactions With Cytokine Intermediary

In FDA's review of ustekinumab, FDA found it implausible that ustekinumab would have any *direct influence* on CYP regulation, or that any small molecule would have any

[219] Liptrott NJ, et al. The impact of cytokines on the expression of drug transporters, cytochrome P450 enzymes and chemokine receptors in human PBMC. Br. J. Pharmacol. 2009;156:497–508.

[220] Dallas S, et al. Interleukines-12 and -23 do not alter expression or activity of multiple cytochrome P450 enzymes in cryopreseved human hepatocytes. Drug Metab. Dispos. 2013;41:689–93.

[221] Lee JI, et al. CYP-mediated therapeutic protein drug interactions. Clin. Pharmacokin. 2010;49:295–310.

[222] Wang J, et al. Biological products for the treatment of psoriasis: therapeutic targets, pharmacodynamics and disease-drug-drug interaction implications. AAPS J. 2014;16:938–47.

direct influence on ustekinumab degradation by CYP enzymes, writing, "direct...interactions via the CYP pathway is not expected between ustekinumab and co-administered small molecular weight drugs."[223]

Leaving behind the topic of the unlikely *direct influence* of a therapeutic antibody on CYP enzyme expression, FDA then turned to the issue of the *indirect influence* of a therapeutic antibody on CYP enzyme expression, writing, "Ustekinumab, however, might indirectly influence the expression of CYP enzymes by antagonizing cytokine activities...because cytokines are known to reduce the expression level of multiple CYP enzymes. Therefore, **it is postulated that ustekinumab may increase CYP expression**...in patients leading to decreased exposure of drugs that are metabolized by CYP enzymes."[224] Please recall that ustekinumab binds to two different cytokines (IL-12 and IL-23) and that it blocks the signaling activity of each of these two cytokines.

FDA's ClinPharm Review of ustekinumab, together with FDA's ClinPharm Review of another antibody (nivolumab), provides a consistent account of DDI analysis of therapeutic antibodies. Addressing the possibility that nivolumab binds to any cytokine, FDA stated that nivolumab does not bind to any cytokine, writing, "Given...nivolumab is not a cytokine modulator, it is unlikely to have an effect on drug metabolizing enzymes or transporters, in terms of inhibition or induction."[225]

In reviewing ustekinumab, FDA required that the Sponsor consult package labels of other anti-cytokine antibodies for their warnings about the drug's influence on CYP enzyme expression and consequent influence on DDIs with co-administered small molecule drugs. FDA identified these other antibody drugs as rilonacept (anti-IL-1 antibody) and tocilizumab (anti-IL-6 antibody). The FDA reviewer moved a step forward and made a recommendation for ustekinumab's package label, writing, "The current recommendation is to ask the applicant to address this by including the suggested wording...[in FDA's] labeling recommendations based on the approved label for rilonacept (IL-1 antagonist) and the proposed label for tocilizumab (IL-6 antagonist)."[226]

This concerns the package label for rilonacept (not ustekinumab). A view of rilonacept's Drug Interactions section reveals that chronic inflammation in patients results in increased IL-1 which suppresses CYP enzymes, and that coadministering rilonacept reverses this suppression, resulting in CYP normalization:

> **DRUG INTERACTIONS**. Cytochrome P450 Substrates. The formation of CYP450 enzymes is suppressed by increased levels of cytokines (e.g., IL-1) during chronic inflammation. Thus it is expected that for a molecule that binds to IL-1, such as **rilonacept**, the formation of CYP450 enzymes could be normalized. This is clinically relevant for CYP450 substrates with a narrow therapeutic index, where the dose is individually adjusted (e.g., warfarin). Upon initiation of **rilonacept**, in patients being treated with these types of medicinal products, therapeutic monitoring of the effect or drug concentration should be performed and the individual dose of the medicinal product may need to be adjusted as needed.[227]

[223] Ustekinumab (plaque psoriasis) BLA 125-261. Page 13 of 194-page Clinical Pharmacology Review.

[224] Ustekinumab (plaque psoriasis) BLA 125-261. Page 13 of 194-page Clinical Pharmacology Review.

[225] Nivolumab (melanoma) BLA 125-554. Page 23 of 58-page Clinical Pharmacology Review.

[226] Ustekinumab (plaque psoriasis) BLA 125-261. Pages 13 and 41 of 194-page Clinical Pharmacology Review.

[227] Package label. ARCALYST™ (rilonacept) injection, for subcutaneous use; February 2008 (7 pp.). BLA 125-249.

FDA's ClinPharm Review for rilonacept (BLA 125-249) based its DDI analysis on a publication showing that IL-1 downregulates CYP2C8 and CYP3A4, and thus would be expected to change exposure of any coadministered small molecule drug that is metabolized by either of these CYP enzymes.[228] Rilonacept is a fusion protein, where what is fused together, to form a continuous polypeptide, is part of IL-1 receptor and the Fc portion of an IgG1 antibody.[229]

In view of FDA's recommendation that ustekinumab's Drug Interaction section be also based on tocilizumab's label, it is interesting to see that tocilizumab's Drug Interactions section teaches that inflammation increases expression of IL-6, where the increased IL-6 suppresses CYP enzymes, and that tocilizumab (Actemra®) reverses this phenomenon:

> **DRUG INTERACTIONS**. Interactions with CYP450 Substrates. Cytochrome P450s in the liver are down-regulated by infection and inflammation stimuli including cytokines such as IL-6. Inhibition of IL-6 signaling in rheumatoid arthritis patients treated with tocilizumab may restore CYP450 activities to higher levels than those in the absence of tocilizumab leading to increased metabolism of drugs that are CYP450 substrates. In vitro studies showed that tocilizumab has the potential to affect expression of multiple CYP enzymes including CYP1A2, CY2B6, CYP2C9, CYP2C19, CYP2D6 and CYP3A4. Its effects on CYP2C8 or transporters is unknown. In vivo studies with omeprazole, metabolized by CYP2C19 and CYP3A4, and simvastatin, metabolized by CYP3A4, showed up to a 28% and 57% decrease in exposure one week following a single dose of ACTEMRA, respectively. The effect of tocilizumab on CYP enzymes may be clinically relevant for CYP450 substrates with narrow therapeutic index, where the dose is individually adjusted. Upon initiation or discontinuation of ACTEMRA, in patients being treated with these types of medicinal products, therapeutic monitoring of effect (e.g., warfarin) or drug concentration (e.g., cyclosporine or theophylline) should be performed and the individual dose of the medicinal product adjusted as needed. Prescribers should exercise caution when ACTEMRA is coadministered with CYP3A4 substrate drugs where decrease in effectiveness is undesirable, e.g., oral contraceptives, lovastatin, atorvastatin, etc. The effect of tocilizumab on CYP450 enzyme activity may persist for several weeks after stopping therapy.[230]

FDA's ClinPharm review for tocilizumab (BLA 125-276) based its DDI analysis on in vitro incubations with hepatocytes and on a DDI study with human subjects, where the subjects were coadministered with tocilizumab plus omeprazole.[231] Tocilizumab is an antibody that binds to IL-6R. The Sponsor discovered that IL-6 is elevated in patients with rheumatoid arthritis (40 pg/mL) as compared to healthy people (4 pg/mL), and predicted that, if IL-6 downregulated cytochrome P450, then expression of CYP3A4 would be reduced, with the consequent increase in exposure of any coadministered drug that is metabolized by this CYP enzyme.

The Sponsor conducted hepatocyte incubations, showing that tocilizumab inhibits IL-6 mediated downregulation of cytochrome P450 CYP1A2, CYP2B6, CYP2D6, CYP2C9,

[228] Aitken AE, Morgan ET. Gene-specific effects of inflammatory cytokines on cytochrome P450 2C, 2B6 and 3A4 mRNA levels in human hepatocytes. Drug Metab. Dispos. 2007;35:1687–93.

[229] Rilonacept (cryopyrin associated periodic syndromes) BLA 125-249. Pages 7 and 22 of 82-page Clinical Pharmacology Review.

[230] Package label. ACTEMRA® (tocilizumab) injection, for intravenous infusion; January 2010 (20 pp.). BLA 125-276.

[231] Tocilizumab (rheumatoid arthritis) BLA 125-276. Pages 8, 21, and 28 of 136-page Clinical Pharmacology Review.

CYP2C19, and CYP3A4. Also, a study with human subjects showed that coadministering tocilizumab with omeprazole (CYP2C19 substrate) resulted in decreased exposure of omeprazole. To summarize, the previous few paragraphs provide background information on antibodies and antibody-like proteins and their potential for blocking cytokine-mediated signaling, and their consequent potential for influencing CYP enzyme expression. This background information concerned nivolumab (binds to PD-1; this is not a cytokine), rilonacept (binds the cytokine IL-1), and tocilizumab (binds IL-6 receptor). Now, the discussion returns to ustekinumab.

Package label. Ustekinumab's package label, as published on the date of FDA's Approval Letter (2009), warned about drug-drug interactions between ustekinumab and warfarin. The Drug Interactions section referred to the need to adjust warfarin dose, and instructed physicians to be aware of chronic inflammation resulting in a change in the patient's cytokine profile, and where this change in cytokine profile changed cytochrome P450-mediated warfarin catabolism. The Drug Interactions section further warned the physician that any administered ustekinumab could "normalize" the rate of cytochrome P450-mediated warfarin metabolism:

> **DRUG INTERACTIONS**...**CYP450 Substrates**. The formation of CYP450 enzymes can be altered by increased levels of certain cytokines (e.g., IL-1, IL-6, IL-10, TNFα, IFN) during chronic inflammation. Thus, ustekinumab could normalize the formation of CYP450 enzymes...upon initiation of ustekinumab in patients who are receiving concomitant CYP450 substrates, particularly...**warfarin**...should be considered and the individual dose of the drug adjusted as needed.[232]

vi. FDA Required More Experiments

FDA required that the Sponsor determine if IL-12 influences CYP enzyme expression. Please note that ustekinumab binds to IL-12 and to IL-23. FDA also designed an experiment for use with psoriasis patients that could make the same determination.

FDA set forth its requirement that the Sponsor conduct another experiment, writing, "Please conduct an in vitro study...to determine whether IL-12...modulates CYP enzyme expression and whether ustekinumab is able to reverse the effects of IL-12...on CYP expression, e.g., in vitro hepatocyte study. An alternative in vivo approach would be to determine the potential of ustekinumab for the alteration of CYP substrate metabolism in psoriasis patients, for example, a cocktail study with CYP probe drugs."[233]

FDA explained the scientific basis to suspect that ustekinumab (which blocks IL-12) could result in changes in CYP enzyme expression, with the consequent changes in catabolism of any coadministered drug. FDA's scientific basis was that, "Cytokines such as...IL-2, IL-6, and IL-10, are known to down-regulate the expression of cytochrome P450 enzymes in humans and inhibit the metabolism of CYP substrates. On the contrary, cytokine antagonists such as basiliximab (anti-IL-2 receptor antibody) and tocilizumab (anti-IL-6 antibody) are known to reverse the effect of the cytokines on CYP substrates, resulting in a normalization of CYP regulation."[234]

[232] Package label. STELARA™ (ustekinumab); September 2009 (12 pp.).

[233] Ustekinumab (plaque psoriasis) BLA 125-261. Page 8 of 194-page Clinical Pharmacology Review.

[234] Ustekinumab (plaque psoriasis) BLA 125-261. Page 8 of 194-page Clinical Pharmacology Review.

FDA continued with this narrative about cytokines, where FDA's goal was to lead up to its suggestion that the Sponsor should test if exposing human hepatocytes to IL-12 caused any changes in CYP enzyme expression. FDA's continued narrative was, "As a disease state, psoriatic patients have elevated cytokine levels. Ustekinumab as a IL-12...antagonist has the potential to reverse any IL-12...cytokine mediated CYP suppression. Thus, in psoriasis patients who have been stabilized on drugs with CYP mediated metabolism, ustekinumab has the potential, through this normalization of CYP activity, to require dose adjustment."[235]

Package label (2009). The package label for ustekinumab in its originally approved form (2009) did not contain the information required by FDA's requests to conduct more experiments.[236] The 2009 label stated that it was not known ("has not been reported") if IL-12 can modulate CYP enzyme expression:

> **DRUG INTERACTIONS**...CYP450 Substrates. The formation of CYP450 enzymes can be altered by increased levels of certain cytokines (e.g., IL-1, IL-6, IL-10, TNFα, IFN) during chronic inflammation. Thus, ustekinumab could normalize the formation of CYP450 enzymes. **A role for IL-12 or IL-23 in the regulation of CYP450 enzymes *has not been reported*.** However, upon initiation of ustekinumab in patients who are receiving concomitant CYP450 substrates, particularly those with a narrow therapeutic index, monitoring for therapeutic effect (e.g., for warfarin) or drug concentration (e.g., for cyclosporine) should be considered and the individual dose of the drug adjusted as needed.[237]

The package label referred to the fact that warfarin has a narrow therapeutic index. Warfarin's narrow therapeutic index is described in the cited references.[238,239] The problem with drugs with a narrow therapeutic index is that dosing mistakes can more easily result in toxicity to the patient, or in a dose too low to be effective—and a related problem for drugs with a narrow therapeutic index is that changes in expression of CYP enzymes, can more easily result in the situation in unpredictably high (or unpredictably low) concentrations of a coadministered drug.

Package label (2014). A supplementary package label for ustekinumab issued a few years later (2014) does, in fact, state that studies with cultured hepatocytes were conducted, where these studies demonstrated that *IL-12 does not change activity of CYP enzymes*:

> **CLINICAL PHARMACOLOGY**...**Drug-Drug Interactions**. The effects of IL-12 or IL-23 on the regulation of CYP450 enzymes were evaluated in an in vitro study using human hepatocytes, which showed that **IL-12 and/or IL-23 at levels of 10 ng/mL did not alter human CYP450 enzyme activities** (CYP1A2, 2B6, 2C9, 2C19, 2D6, or 3A4). However, the clinical relevance of in vitro data has not been established.[240]

[235] Ustekinumab (plaque psoriasis) BLA 125-261. Page 8 of 194-page Clinical Pharmacology Review.

[236] Package label. STELARA™ (ustekinumab) injection, for subcutaneous use; September 2009 (12 pp.).

[237] Package label. STELARA™ (ustekinumab) injection, for subcutaneous use; September 2009 (12 pp.).

[238] Fohner AE, et al. Variation in genes controlling warfarin disposition and response in American Indian and Alaska Native people: CYP2C9, VKORC1, CYP4F2, CYP4F11, GGCX. Pharmacogenet. Genomics. 2015;25:343–53.

[239] Kuruvilla M, Gurk-Turner C. A review of warfarin dosing and monitoring. BUMC Proceedings. 2001;14:305–6.

[240] Package label. STELARA™ (ustekinumab) injection, for subcutaneous use; March 2014 (20 pp.).

V. CONCLUDING REMARKS

FDA's reviews of therapeutic proteins assess whether the administered protein provokes human subjects to generate antibodies against the protein. FDA's reviews for atezolizumab, certolizumab pegol, denosumab, evolocumab, lixisenatide, nivolumab, pegloticase, and sebelipase alfa illustrate FDA's analysis of antibodies materializing in human subjects. FDA's reviews typically track these steps of analysis:

- *Step one*. Does the immune system of human subjects generate antibodies against the study drug?
- *Step two*. If antibodies against the study drug are formed, do they block any of the biochemical mechanisms of the study drug?
- *Step three*. If antibodies against the study drug are formed, is there a reduction in the concentration of the free study drug in the plasma of human subjects?
- *Step four*. If antibodies against the study drug are formed, is there a consequent reduction in clinical efficacy of the study drug?

But antibodies to many therapeutic proteins can be generated, in patients, in response to a part of the drug that is merely fused to or connected to the therapeutic protein. This textbook provides three examples of this scenario:

- PEG as the nontherapeutic portion (pegloticase)
- Oligosaccharide as the nontherapeutic portion (cetuximab)
- Constant region from immunoglobulin, also known as an "Ig portion" (abatacept)

Where the therapeutic protein takes the form of a covalent complex with PEG, an oligosaccharide, or an Ig portion, the Sponsor should determine if any antibodies that are generated bind to the therapeutic protein itself or, alternatively, bind to the PEG moiety, oligosaccharide moiety, or Ig portion. Where these antibodies materialize, they can prevent efficacy of any coadministered biological that also has a PEG moiety, oligosaccharide moiety, or Ig portion.

Jefferis[241] warned that the oligosaccharide group of recombinant antibodies are "potentially immunogenic and unacceptable as therapeutics." In particular, recombinant proteins with an oligosaccharide moiety can have undesired immunogenicity, where the oligosaccharide has α1,3-galactose residues, *N*-glycolylneuraminic acid residues, α1,3-fucose residues, or β1,3-xylose residues.[242,243,244]

Cetuximab administration can result in the patient's immune system generating immune responses against the oligosaccharide moiety of cetuximab. This immune response, which can result in severe anaphylaxis, resulting from the patient's immune

[241] Jefferis R. Glycosylation of recombinant antibody therapeutics. Biotechnol. Progr. 2005;21:11–16.

[242] Jefferis R. Criteria for selection of IgG isotype and glycoform of antibody therapeutics. BioProcess Int. 2006;4:40–43.

[243] Raju TS. Glycosylation variations with expression systems. BioProcess Int. 2003;1:44–53.

[244] Reusch D, Tejada ML. Fc glycans of therapeutic antibodies as critical quality attributes. Glycobiology. 2015;25:1325–34.

system generating antibodies that recognize galactose-α-1,3-galactose, which is a disaccharide residing in cetuximab's oligosaccharide moiety.[245] This problem resulted from the fact that cetuximab was manufactured by cultured mouse cells, which express an enzyme that transfers galactose residues, instead of being manufactured by cells that do not express this enzyme.

The problem of antibodies being generated to part of a therapeutic protein that does not have therapeutic activity, is also illustrated for abatacept. Abatacept has a portion having therapeutic activity (CTLA4 portion) and a portion not having therapeutic activity (Ig portion). This situation is detailed in the next chapter by FDA's comments that, "studies were carried out in serum from normal individuals. . .[t]hese studies determined that the reactivity was to the Ig portion of the molecule, and not to the CTLA4 portion. . .[o]ut of a total of 385 subjects receiving. . .abatacept. . .only two subjects. . .seroconverted for CTLA4 portion specific antibodies."[246]

FDA's ClinPharm Reviews addressed the possibility of DDIs occurring at the location of cytochrome P450 enzymes. This type of interaction is more likely to occur where the therapeutic protein inhibits cytokine-mediated signaling, and is not expected to occur where the therapeutic protein has no relevance to cytokine-mediated signaling. FDA reviewers assessed DDIs involving the cytochrome P450 enzymes for the following therapeutic proteins. The mechanism of action of each of these therapeutic proteins is shown:

- Atezolizumab (blocks PD-1/PD-L1 mediated signaling)
- Certolizumab (blocks TNFα)
- Denosumab (blocks RANK/RANKL mediated signaling)
- Lixisenatide (stimulates GLP-1 receptor)
- Rilonacept (blocks IL-1)
- Tocilizumab (blocks IL-6)
- Ustekinumab (blocks IL-12)

Another category of DDI is where there is an intersection in the mechanisms of action between the study drug and a coadministered drug. This category is illustrated in this chapter for evolocumab, and in the previous chapter for fingolimod.

[245] Chung CH, et al. Cetuximab-induced anaphylaxis and IgE specific for galactose-alpha-1,3-galactose. New Engl. J. Med. 2008;358:1109–17.

[246] Abatacept (CTLA-4 Ig) BLA 125-118. Page 179 of 237-page Medical Review.

CHAPTER

9

Immunosuppression, Drug-Induced Hypersensitivity Reactions, and Drug-Induced Autoimmune Reactions

I. INTRODUCTION

This chapter concerns a family of drug-induced adverse events, all related by the fact that they have an immune component. The term "immune-related AEs" is a recognized category of adverse events, as demonstrated by an account of AEs associated with nivolumab, pembrolizumab, and atezolizumab. As reported by Costa et al.,[1] these "immune-related AEs" include rash, pruritus, pneumonitis, and colitis. This chapter divides immune-related AEs into the following categories:

i. **Antibodies induced by the therapeutic protein against the therapeutic protein**. Where the administered therapeutic protein is a human antibody, the antibodies against it have been called, HAHAs. A more general term, which applies to antibodies against any type of drug, is antidrug antibodies.[2]

[1] Costa R, et al. Analyses of selected safety endpoints in phase 1 and late-phase clinical trials of anti-PD-1 and PD-L1 inhibitors: prediction of immune-related toxicities. Oncotarget. 2017;8(40):67782–9. doi: 10.18632/oncotarget.18847.

[2] Sandborn WJ, et al. Effects of transient and persistent anti-drug antibodies to certolizumab pegol: longitudinal data from a 7-year study in Crohn's disease. Inflamm. Bowel Dis. 2017;23:1047–56.

DOI: https://doi.org/10.1016/B978-0-12-814647-7.00009-9

ii. **Drug-induced hypersensitivity reactions**. This includes reactions such as anaphylaxis,[3] Stevens–Johnson syndrome (SJS), toxic epidermal necrolysis (TEN),[4] erythema multiforme,[5,6] and bullous pemphigoid.
iii. **Drug-induced autoimmunity**. This concerns autoimmunity against various tissues and organs in the body. This autoimmunity can materialize with drugs that interrupt naturally occurring mechanisms that dampen the immune system. Lupus-like syndrome is one type of drug-induced autoimmunity.[7,8]
iv. **Immunosuppressant-induced infections and cancer**. Immunosuppressants can suppress the immune system to the point where residual immunity is not sufficient to halt infections and cancers.

The term HAHA is used, on occasion, by FDA and in medical journals, to refer to a patient's immune response to an administered therapeutic human antibody, where the result is antibodies against the human antibody.[9,10,11] HAHA means "human anti-human antibody." Immune responses generated against therapeutic proteins can be minimized by the process of humanization, or alternatively, by using protein sequences that are fully of human origin.[12,13,14] Therapeutic antibodies

[3] Thong BY, Tan TC. Epidemiology and risk factors for drug allergy. Br. J. Clin. Pharmacol. 2011;71:684–700.

[4] Takahashi R, et al. Defective regulatory T cells in patients with severe drug eruptions: timing of the dysfunction is associated with the pathological phenotype and outcome. J. Immunol. 2009;182:8071–9.

[5] Gonzalez FJ, et al. Erythema multiforme to phenobarbital: involvement of eosinophils and T cells expressing the skin homing receptor. J. Allergy Clin. Immunol. 1997;100:135–7.

[6] Patterson JW, et al. Eosinophils in skin lesions of erythema multiforme. Arch. Pathol. Lab. Med. 1989;113:36–9.

[7] Moulis G, et al. Is the risk of tumour necrosis factor inhibitor-induced lupus or lupus-like syndrome the same with monoclonal antibodies and soluble receptor? A case/non-case study in a nationwide pharmacovigilance database. Rheumatology (Oxford) 2014;53:1864–71.

[8] Vedove CM, et al. Drug-induced lupus erythematosus with emphasis on skin manifestations and the role of anti-TNFα agents. J. Dtsch Dermatol. Ges. 2012;10:889–97.

[9] Mirick GR, et al. A review of human anti-globulin antibody (HAGA, HAMA, HACA, HAHA) responses to monoclonal antibodies. Not four letter words. Q. J. Nucl. Med. Mol. Imaging. 2004;48:251–7.

[10] Nechansky A. HAHA—nothing to laugh about. Measuring the immunogenicity (human anti-human antibody response) induced by humanized monoclonal antibodies applying ELISA and SPR technology. J. Pharm. Biomed. Anal. 2010;51:252–4.

[11] Hall B. HAHA antibodies—not such a funny story. J. Crohns Colitis. 2014;8:439–40.

[12] Presta LG. Molecular engineering and design of therapeutic antibodies. Curr. Opin. Immunol. 2008;20:460–70.

[13] Apgar JR, et al. Beyond CDR-grafting: structure-guided humanization of framework and CDR regions of an anti-myostatin antibody. MAbs. 2016;8:1302–18.

[14] Frenzel A, et al. Phage display-derived human antibodies in clinical development and therapy. MAbs. 2016;8:1177–94.

with sequences of fully human origin can be prepared from transgenic mice expressing repertoires of human antibody gene sequences.[15] One such mouse is the XenoMouse®.[16]

Drug-induced hypersensitivity reactions include anaphylaxis, SJS, TEN, and erythema multiforme. These are detailed later on in this chapter. Drug-induced hypersensitivity reactions arising from recombinant antibody drugs also include bullous pemphigoid, a disorder described in case reports for infliximab,[17] adalimumab,[18,19] nivolumab, and pembrolizumab.[20,21]

Drug-induced autoimmunity occurs where the goal of the drug is to enhance immune response, and where an unfortunate side-effect is enhanced immune response against healthy tissues and organs. An understanding of drug-induced autoimmunity might best be understood by a background in autoimmune diseases. The following lists some of the autoimmune diseases. Autoimmune diseases include rheumatoid arthritis, multiple sclerosis, psoriasis, Crohn's disease, ulcerative colitis, systemic lupus erythematosus (SLE), and graft-versus-host disease. In rheumatoid arthritis the immune system inflicts damage in joints,[22] in multiple sclerosis the immune system damages the brain,[23,24] and in psoriasis it damages the skin.[25] Autoimmune diseases involve recognition of and attack by the immune system against a **self-antigen**.

Drug-induced autoimmunity can occur where a therapeutic protein blocks a signaling pathway, where the normal and desired function of the pathway is to dampen immune response, that is, to prevent overly active immune responses. Therapeutic proteins that block this kind of signaling pathway includes drugs that block signaling mediated by PD-L1, PD-1, or CTLA-4. Where the desired goal is to block this type of signaling pathway, for example, when treating cancer, there is a risk for the adverse event of autoimmunity.

[15] Lonberg N. Human antibodies from transgenic animals. Nature Biotechnol. 2005;23:1117–25.

[16] Foltz IN, et al. Discovery and bio-optimization of human antibody therapeutics using the XenoMouse® transgenic mouse platform. Immunol. Rev. 2016;270:51–64.

[17] Hall B. HAHA antibodies—not such a funny story. J. Crohns Colitis. 2014;8:439–40.

[18] Altindago O, et al. A case with anti TNF-α induced bullous pemphigoid. Rheumatology. 2010;25:214–6.

[19] Stausbol-Gron B, et al. Development of bullous pemphigoid during treatment of psoriasis with adalimumab. Clin. Exp. Dermatol. 34:285–6.

[20] Jour G, et al. Autoimmune dermatologic toxicities from immune checkpoint blockade with anti-PD-1 antibody therapy: a report on bullous skin eruptions. J. Cutan. Pathol. 2016;43:688–96.

[21] Hwang SJ, et al. Bullous pemphigoid, an autoantibody-mediated disease, is a novel immune-related adverse event in patients treated with anti-programmed cell death 1 antibodies. Melanoma Res. 2016;26:413–6.

[22] Fox DA. Citrullination: a specific target for the autoimmune response in rheumatoid arthritis. J. Immunol. 2015;195:5–7.

[23] Seki SM, et al. Lineage-specific metabolic properties and vulnerabilities of T cells in the demyelinating central nervous system. J. Immunol. 2017;198:4607–17.

[24] Hohlfeld R and Steinman L. T cell-transfer experimental autoimmune encephalomyelitis: pillar of multiple sclerosis and autoimmunity. J. Immunol. 2017;198:3381–3.

[25] Cai Y, et al. New insights of T cells in the pathogenesis of psoriasis. Cell. Mol. Immunol. 2012;9:302–9.

Immunosuppressant drugs are used to reduce inflammation and to treat autoimmune diseases. But these drugs can also weaken the desired naturally-occurring immune responses against cancers and infections, with a consequent increased risk for cancers and infections. The naturally occurring and constitutive immune response against cancer cells is called immune surveillance.[26,27] But inhibiting immune surveillance by drugs can result in cancer. Immunosuppressants are used to prevent transplant rejections[28,29,30,31] and here there is an increased risk for squamous-cell carcinomas[32] and non-melanoma skin cancers.[33] Evidence suggests that rheumatoid arthritis patients treated with immunosuppressants are also at increased risk for cancer.[34,35]

Infections increased by immunosuppressants are classed as "opportunistic infections," and these include bacterial infections such as tuberculosis, and viral infections such as that caused by John Cunningham virus (JC virus). More specifically, opportunistic infections are infections caused by pathogens that usually do not cause disease in a host with a healthy immune system. A compromised immune system presents an opportunity for the pathogen to become infectious.[36] In an account of the immunosuppressive drugs used with organ transplants, Fishman[37] referred to the need to "determine an optimal immunosuppressive regimen while avoiding infection."

Opportunistic infections such as tuberculosis and fungal infections can occur with the tumor necrosis factor alpha (TNF-α) antagonist, infliximab. TNF-α antagonists also result

[26] Mohme M, et al. Circulating and disseminated tumour cells—mechanisms of immune surveillance and escape. Nature Rev. Clin. Oncol. 2017;14(3):155–67. doi: 10.1038/nrclinonc.

[27] Tu MM, et al. Licensed and unlicensed NK cells: differential roles in cancer and viral control. Front. Immunol. 2016;7:166 (11 pages).

[28] Curtis RE, et al. Impact of chronic GVHD therapy on the development of squamous-cell cancers after hematopoietic stem-cell transplantation: an international case-control study. Blood. 2005;105:3802–11.

[29] Navarro MD, et al. Cancer incidence and survival in kidney transplant patients. Transplant Proc. 2008;40:2936–40.

[30] Otley CC, Pittelkow MR. Skin cancer in liver transplant recipients. Liver Transpl. 2000;6:253–62.

[31] Khorsandi SE, Heaton N. Optimization of immunosuppressive medication upon liver transplantation against HCC recurrence. Transl. Gastroenterol. Hepatol. 2016;1:25. doi: 10.21037/tgh.2016.03.18.

[32] Curtis RE, et al. Impact of chronic GVHD therapy on the development of squamous-cell cancers after hematopoietic stem-cell transplantation: an international case-control study. Blood. 2005;105:3802–11.

[33] Navarro MD, et al. Cancer incidence and survival in kidney transplant patients. Transplant Proc. 2008;40:2936–40.

[34] Zogala RJ, et al. Management considerations in cancer patients with rheumatoid arthritis. Oncology (Williston Park) 2017;31:374–80.

[35] Lan JL, et al. Reduced risk of all-cancer and solid cancer in Taiwanese patients with rheumatoid arthritis treated with etanercept, a TNF-α inhibitor. Medicine (Baltimore) 2017;96:e6055. doi: 10.1097/MD.0000000000006055.

[36] Hersh CM. Infectious complications of multiple sclerosis therapies. Neurol. Rep. 2014;7:20–7.

[37] Fishman JA. Opportunistic infections—coming to the limits of immunosuppression? Cold Spring Harb. Perspect. Med. 2013;3:a015669. doi: 10.1101/cshperspect.a015669.

in aplastic anemia and hematological cancers.[38] Kim et al.[39] describes opportunistic viral infections that occur with the immunosuppressive drugs, infliximab, azathioprine, and natalizumab. Natalizumab, on rare occasions, can provoke the activation of John Cunningham virus in the brain, resulting in the disease, progressive multifocal leukoencephalopathy.[40]

The following bulletpoints introduce the topic of immunosuppressants:

- **Immunosuppressants can prevent the materialization of antibodies against therapeutic proteins.** This topic was described in Chapter 8, Drug–Drug Interactions—Part Two (Therapeutic Proteins), because the package labels needing to be detailed in the chapter had information on antibodies against the study drug. This topic is further developed here. Where the first drug is a therapeutic protein and the second drug is an immunosuppressant, the immunosuppressant can prevent the patient's immune system from generating antibodies against the therapeutic protein.
- **Immunosuppressants for treating autoimmune diseases.** Where the first drug is an immunosuppressant and the second drug is an immunosuppressant, enhanced efficacy in treating an autoimmune disease can result by administering two different immunosuppressants.
- **Immunosuppressants can increase risk for autoimmune damage to tissues and organs.** Where a drug blocks signaling mediated by PD-1, PD-L1, or CTLA-4, this increases risk for autoimmunity that damages the patient's own tissues and organs.
- **Immunosuppressants can increase risk for cancer and infections.** Where a drug blocks signaling mediated by PD-1, PD-L1, or CTLA-4, or blocks some other aspect of immune cell function, this increases risk for cancer and infections and can reduce efficacy of vaccines. Also, where the first drug is an immunosuppressant and the second drug is an immunosuppressant, this can increase risk for cancers and infections, and can reduce efficacy of vaccines.

a. Immunosuppressants Can Reduce Antibodies Against Therapeutic Proteins

This provides the examples of certolizumab pegol and pegloticase. Preventing antibodies against an administered therapeutic protein can occur where the coadministered

[38] Curtis JR, et al. Confirmation of administrative claims-identified opportunistic infections and other serious potential adverse events associated with tumor necrosis factor alpha antagonists and disease-modifying antirheumatic drugs. Arthritis Rheum. 2007;57:343–6.

[39] Kim SY, Solomon DH. Tumor necrosis factor blockade and the risk of viral infection. Nat. Rev. Rheumatol. 2010;6:165–74.

[40] Shirani A, Stuve O. Natalizumab for multiple sclerosis: a case in point for the impact of translational neuroimmunology. J. Immunol. 2017;198:1381–6.

immunosuppressant is azathioprine, mycophenolate mofetil, cyclosporine, or methotrexate (MTX).[41,42] The package label for certolizumab pegol informs the reader that, "Patients treated with concomitant immunosuppressants had a lower rate of antibody development than patients not taking immunosuppressants at baseline (3% and 11%, respectively).[43]" Another example is azathioprine to prevent antibodies against pegloticase, during pegloticase therapy for gout.[44] Pegloticase is a therapeutic enzyme.

b. Immunosuppressants for Treating Autoimmune Diseases

An example comes from use of MTX plus TNF antagonists for treating the autoimmune disease, rheumatoid arthritis. Greenberg et al.[45] describes coadministering two immunosuppresive drugs for treating arthritis, where the combination showed improved efficacy over monotherapy with either drug alone, and where fortunately, there was not any additive risk for infections. Greenberg et al. stated that:

> evidence that the combination of MTX and biological agents has **superior efficacy to monotherapy** with either agent alone also suggests that the potential for an additive risk of infections may be present for MTX and TNF antagonist combination therapy ... [w]e also observed that combination MTX and TNF antagonist therapy was **not associated with a synergistic risk of infection** compared with TNF antagonist monotherapy.[46]

c. Immunosuppressants Can Increase Risk for Autoimmune Damage to Tissues and Organs

Where the goal of a therapeutic agent, such as pembrolizumab (anti-PD-1) (Keytruda®), is to release naturally occurring inhibitions on immune response, thereby stimulating the immune system, a risk is autoimmunity that damages the patient's tissues and organs. In the Full Prescribing Information section of the package label for

[41] Berhanu AA, et al. Pegloticase failure and a possible solution: immunosuppression to prevent intolerance and inefficacy in patients with gout. Semin. Arthritis Rheum. 2017;46:754–8. doi: 10.1016/j.semarthrit.2016.09.007.

[42] Krieckaert CL, et al. Methotrexate reduces immunogenicity in adalimumab treated rheumatoid arthritis patients in a dose dependent manner. Ann. Rheum. Dis. 2012;71:1914–5.

[43] Package label. CIMZIA (certolizumab pegol) Lyophilized powder for subcutaneous injection. April 2008 (15 pages).

[44] Berhanu AA, et al. Pegloticase failure and a possible solution: immunosuppression to prevent intolerance and inefficacy in patients with gout. Semin. Arthritis Rheum. 2017;46:754–8. doi: 10.1016/j.semarthrit.2016.09.007.

[45] Greenberg JD, et al. Association of methotrexate and tumour necrosis factor antagonists with risk of infectious outcomes including opportunistic infections in the CORRONA registry. Ann. Rheum. Dis. 2012;69:380–6.

[46] Greenberg JD, et al. Association of methotrexate and tumour necrosis factor antagonists with risk of infectious outcomes including opportunistic infections in the CORRONA registry. Ann. Rheum. Dis. 2012;69:380–6.

pembrolizumab, the label warned about liver damage and provided physicians with information on treating the liver damage with immunosuppressive agents:

> **WARNINGS AND PRECAUTIONS** ... Immune-Mediated Hepatitis. KEYTRUDA can cause immune-mediated hepatitis. Monitor patients for changes in liver function. Administer **corticosteroids** (initial dose of 0.5 to 1 mg/kg/day [for Grade 2 hepatitis] and 1 to 2 mg/kg/day [for Grade 3 or greater hepatitis] **prednisone** or equivalent followed by a taper) and, based on severity of liver enzyme elevations, withhold or discontinue KEYTRUDA.[47]

Pembroluzumab's label also warned against immune-mediated colitis, immune-mediated pneumonitis, immune-mediated endocrinopathies, and immune-mediated nephritis. Similarly, the package label for another drug that disrupts the PD-1/PD-1 signaling pathway (atezolizumab; anti-PD-L1), the label warned against autoimmunity that damages the patient's tissues and organs, including autoimmunity taking the form of hepatitis.[48] This provides a general take-home lesson that may be applicable to most drugs. In drafting the package label, the medical writer might want to consider writing instructions for treating one or more of the expected adverse events.

d. Immunosuppressants Can Increase the Risk for Cancer and Infections

Regarding the goal of preventing the materialization of cancer and infections, the package label may warn about the increased risk for infections or cancer where one or more immunosuppressants are given. This is the situation with tofacitinib (Xeljanz®), an immunosuppressant for treating rheumatoid arthritis. The Black Box Warning warned about administering an additional, concomitant immunosuppressant:

> **BOXED WARNING**. SERIOUS INFECTIONS. Patients treated with XELJANZ are at increased risk for developing serious infections that may lead to hospitalization or death ... Most patients who developed these infections were **taking concomitant immunosuppressants such as methotrexate or corticosteroids.**[49]

Similarly, the package label for natalizumab (Tysabri®), which is an immunosuppressant, warns about increased risk for infections where natalizumab and a second immunosuppressant are both given. Natalizumab is used for treating Crohn's disease and for treating multiple sclerosis. The label instructs physicians to avoid giving natalizumab plus a second immunosuppressant:

> **DOSAGE AND ADMINISTRATION** ... For patients with Crohn's disease who start TYSABRI while on **chronic oral corticosteroids, commence steroid tapering** as soon as a therapeutic benefit of TYSABRI has occurred; if the patient with Crohn's disease cannot be tapered off of oral corticosteroids within six months of starting TYSABRI, discontinue TYSABRI.[50]

[47] Package label. KEYTRUDA® (pembrolizumab) for injection, for intravenous use. May 2017 (42 pages). BLA 125-514

[48] Package label. TECENTRIQ® (atezolizumab) injection, for intravenous use. April 2017 (23 pages).

[49] Package label. XELJANZ® (tofacitinib) tablets, for oral use. December 2015 (30 pages).

[50] Package label. TYSABRI (natalizumab) injection, for intravenous use. May 2016 (28 pages).

Regarding the increased risk for infections with a second, concomitant immunosuppressant, the package label for natalizumab further warns:

> **DRUG INTERACTIONS** ... Because of the potential for increased risk of PML and other infections, Crohn's disease patients receiving TYSABRI should not be treated with **concomitant immunosuppressants (e.g., mercaptopurine, azathioprine, cyclosporine, or methotrexate)** or inhibitors of TNFα, and corticosteroids should be tapered in those patients with Crohn's disease who are on chronic corticosteroids when they start TYSABRI therapy.[51]

II. AEs INVOLVING ACQUIRED IMMUNITY ILLUSTRATED BY FDA'S MEDICAL REVIEWS

a. Introduction

For distinguishing AEs involving acquired immunity from AEs that are hypersensitivity reactions, AEs involving acquired immunity will be defined as situations resulting from changes in responses by dendritic cells (DCs), T cells, T-cell activation to form cytotoxic T cells, and B cells and production of immunoglobulin G (IgG) antibodies. In contrast, hypersensitivity reactions will be defined as situations resulting from increased response by mast cells, eosinophils, and the production of IgE antibodies.[52,53] AEs with an acquired immunity component are shown for these drugs:

- Abatacept (AEs of sepsis, pneumonia and other infections, and AEs taking the form of the drug-drug interaction of, reduced efficacy of vaccines and immunizations. In this drug-drug interaction, the first drug is abatacept and the second drug is a vaccine)
- Adalimumab (AEs of tuberculosis, fungal infections, and lymphoma)
- Atezolizumab (AE of autoimmunity that damages various tissues and organs)
- Denosumab (AE of increased risk for infections, with co-administered immunosuppressant)
- Fingolimod (AE of viral infections)
- Infliximab (AEs of lymphoma, bacterial infections, fungal infections)
- Natalizumab (AE of John Cunningham virus)
- Tofacitinib (AEs of tuberculosis, fungal infections, lymphoma)

b. Abatacept (Rheumatoid Arthritis) BLA 125-118

Abatacept is a fusion protein for treating rheumatoid arthritis. The abatacept construct includes part of the constant region (Fc region) of an antibody, where this constant region resembles a bridge that resides between two different CTLA4 molecules. In abatacept, the entire CTLA4 protein is not used but only the extracellular portion of CTLA4. This type of construct is called an "Ig fusion protein" because one protein having a desired biological

[51] Package label. TYSABRI (natalizumab) injection, for intravenous use. May 2016 (28 pages).

[52] Marquardt DL, Wasserman SI. Mast cells in allergic diseases and mastocytosis. West. J. Med. 1982;137:195–212.

[53] Baldo BA. Adverse events to monoclonal antibodies used for cancer therapy: focus on hypersensitivity responses. OncoImmunology. 2013;2:e26333.

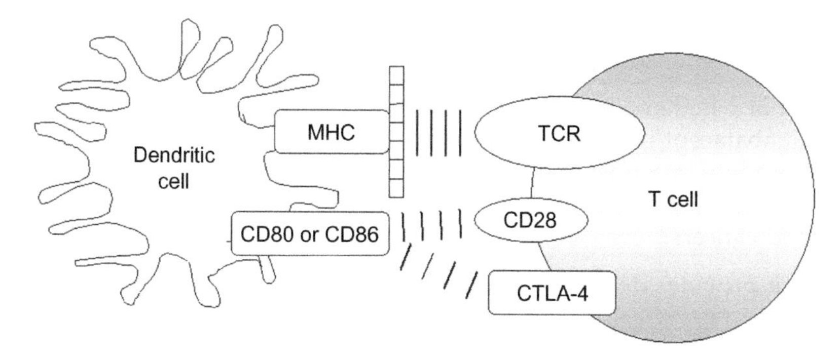

FIGURE 9.1 Interactions between a dendritic cell and a T cell, including CTLA-4. This shows binding of CTLA4 to CD80 of the dendritic cell, thereby preventing CD80 from transmitting an activating signal to CD28. The name "dendritic cell" comes from the word "dendrites." "Dendrites" is derived from a Greek word that means, resembling a tree with branches.

activity (CTLA4) is fused to a polypeptide derived from an immunoglobulin. For abatacept, the immunoglobulin is from human IgG1.[54,55] The terms "Ig fusion," "Fc fusion," "Ig chimera," and "immunoadhesion" are synonyms of each other.[56]

Fig. 9.1 shows that CTLA4 occurs as a membrane-bound protein on T cells. The figure illustrates the immune synapse. The goal of the dendritic cell in forming the immune synapse with the T cell is to present an antigen to the T cell, thereby activating the T cell. The dendritic cell holds and positions the antigen (shown by a segmented line representing amino acids), where the holding is by a membrane-bound protein called major histocompatibility complex (MHC). The dendritic cell presents the antigen to the T-cell receptor of the T cell. The T cell responds by being activated and by proliferating, and by killing cells in the body that express the same antigen on its surface.

Fig. 9.1 shows CTLA4 of the T cell transmitting a signal to the dendritic cell, where this signal results from binding of CTLA4 to the dendritic cell's CD80. Alternatively, CTLA4 binds to the dendritic cell's CD86. CD80 is also called B7-1; CD86 is also called, B7-2. CTLA4's interaction with CD80 prevents the CD80 from interacting with the T cell's CD28, thereby preventing CD80 from transmitting an activating signal to CD28.[57] The overall consequence of CTLA4's interactions with CD80 and CD86 is immunosuppression. Abatacept mimics one of the activities of the T cell, namely, the ability of the T cell to inhibit the dendritic cell's usual role in stimulating T cells. In this way, abatacept is an immunosuppressant and is thus suitable for treating rheumatoid arthritis.

[54] Abatacept (rheumatoid arthritis) BLA 125-118. Page 2 of 237-pageMedical Review and 25 pages, Chemistry Review.

[55] Genovese MC, et al. Abatacept for rheumatoid arthritis refractory to tumor necrosis factor alpha inhibition. N. Engl. J. Med. 2005;353:1114–23.

[56] U.S. Patent No. 8,329,867 (Ser. No. 13/032,491) of GA Lazar and MJ Bernett.

[57] Xu Z, et al. Affinity and cross-reactivity engineering of CTLA4-Ig to modulate T cell costimulation. J. Immunol. 2012;189:4470–7.

FDA's Medical Review and FDA's Pharmacology Review described these four immunological scenarios:

1. The question of whether abatacept induces the formation of anti-abatacept antibodies.
2. The fact that abatacept is an immunosuppressant.
3. Measuring residual immune system activity in abatacept-treated animals.
4. Is residual immune system activity in abatacept-treated patients sufficient to prevent infections and cancers?

The following provides details on these four immunological scenarios:

i. Does the Patient's Immune System Generate Anti-abatacept Antibodies?

The Sponsor determined if animals and humans treated with abatacept developed anti-abatacept antibodies. Where a patient develops antibodies against a therapeutic protein, this is classified as an AE.

The Sponsor measured anti-abatacept antibodies in mice, monkeys, and human subjects. Female mice by 14 weeks developed antibodies, but male mice did not develop antibodies until 33 weeks of abatacept treatment. FDA's Pharmacology Review revealed that, "Female mice in the low dose group developed ... anti-abatacept-antibody responses ... by week 14 during the 6-month dosing period. Male mice in the low dose group ... did not develop anti-abatacept antibody response until week 33.[58]"

However, in monkeys, generation of anti-abatacept antibodies was negligible, "No significant anti-abatacept antibodies ... developed during the treatment period in any group ... anti-abatacept antibodies ... were observed in one low and two intermediate-dose monkeys 6 to 9 weeks after completion of treatment, after abatacept serum levels had dropped below immunosuppressive levels.[59]"

In human subjects treated with abatacept, anti-abatacept antibodies were detected. However, these antibodies were preexisting in the subjects and were not to the CTLA-4 portion of abatacept, but instead were to the immunoglobulin portion (the "Ig portion"). FDA's Medical Review stated, "Immunogenicity of abatacept has been determined ... [i]t became apparent that human serum contained an endogenous, pre-existing reactivity to abatacept. Additional studies were carried out in serum from normal [healthy] individuals ... [t]hese studies determined that the reactivity was to the Ig portion of the molecule, and not to the CTLA4 portion ... [o]ut of a total of 385 subjects receiving ... abatacept ... only two subjects ... seroconverted for CTLA4 portion specific antibodies.[60]"

ii. Abatacept's Desired Function as an Immunosuppressant

Because abatacept is an immunosuppressant, the Sponsor tested if abatacept reduced T-cell activity, inflammation, or antibody formation. Where the goal is treating an

[58] Abatacept (rheumatoid arthritis) BLA 125-118. Pages 79, 85, and 87–89 of 127-pagePharmacology Review.

[59] Abatacept (rheumatoid arthritis) BLA 125-118. Pages 79, 85, and 87–89 of 127-pagePharmacology Review.

[60] Abatacept (rheumatoid arthritis) BLA 125-118. Page 179 of 237-pageMedical Review.

autoimmune disease, such as rheumatoid arthritis, what is needed is inhibiting one or more of these.

The Sponsor provided data from mice, showing that abatacept treatment decreased the **numbers of T cells** and **decreased the numbers of B cells**. In the body, antibodies are biosynthesized and secreted by B cells.[61] B cells can cause damage to bone (joint erosions) by mechanisms involving autoantibodies as well as by non-antibody-dependent mechanisms. The autoantibodies include those that recognize citrullinated proteins.[62,63]

The Sponsor withdrew white blood cells from mice that had been treated with or without abatacept and then conducted an in vitro stimulation test. Stimulation was with "B-cell mitogen." The Sponsor determined that abatacept treatment of mice reduced the ability of B cells to respond to B-cell mitogen, as found in this in vitro test. Regarding the **B-cell mitogen test**, FDA's Pharmacology Review stated that, "**A decrease (98% and 83%) in** ... **peak day 2 proliferation to B-cell mitogen** ... was observed in male mice in the intermediate and high dose groups at the end of the 5 month dosing period.[64]" Thus, abatacept reduced B-cell proliferation in mice. This shows abatacept's desired function as an immunosuppressant.

The Sponsor also conducted a **T-cell mitogen test**, where the mitogen was anti-CD3 antibody. Here, the Sponsor discovered that abatacept treatment resulted in reduced T-cell response. Referring to this reduced T-cell response, FDA wrote, "**peak day 2 proliferation to** ... **anti-CD3 was reduced by** ... **97% and 84%** ... in the intermediate and high dose groups." The Sponsor determined if abatacept's inhibitory influence was reversible, as determined after a recovery period, and here FDA's Pharmacology Review observed that, "Recovery of B-cell and T-cell activation was demonstrated at the end of the 4-month recovery period.[65] Thus, abatacept reduced T-cell proliferation in mice. This further shows abatacept's desired function as an immunosuppressant.

In human subjects with rheumatoid arthritis, FDA's Medical Review observed the efficacy of abatacept in cooling down the immune system, "In clinical trials with abatacept ... **inhibition of T cell activation**, decreases in products of macrophages, fibroblast-like synoviocytes, and B cells, and reductions in acute phase reactants of inflammation were observed. Decreases were seen in serum levels of soluble IL-2 receptor, a marker of T cell activation.[66]" Thus, abatacept reduced T-cell activation in humans. This additionally shows abatacept's desired function as an immunosuppressant.

iii. Measuring Residual Immune System Activity in Abatacept-treated Animals

The Sponsor determined if the immune system, despite treatment with the immunosuppressant (abatacept), was still intact and functional enough to mount immune responses

[61] Carter MJ, et al. The antibody-secreting cell response to infection: kinetics and clinical applications. Front. Immunol. 2017;8:630. doi: 10.3389/fimmu.2017.00630.

[62] Meednu N, et al. Production of RANKL by memory B cells: a link between B cells and bone erosion in rheumatoid arthritis. Arthritis Rheumatol. 2016;68:805–16.

[63] Bugatti S, et al. B cell autoimmunity and bone damage in rheumatoid arthritis. Reumatismo. 2016;68:117–25.

[64] Abatacept (rheumatoid arthritis) BLA 125-118. Pages 80–85 of 127-pagePharmacology Review.

[65] Abatacept (rheumatoid arthritis) BLA 125-118. Pages 80–85 of 127-pagePharmacology Review.

[66] Abatacept (rheumatoid arthritis) BLA 125-118. Page 47 of 237-pageMedical Review.

against foreign antigens. This question can be addressed by testing if abatacept influences the ability to generate immune responses, taking form of T-cell activation, T-cell proliferation, or the generation of antibodies.

This question of whether abatacept impairs the generation of antibodies is reasonable, in view of the report by Young et al. that, "The fusion protein cytotoxic T lymphocyte-associated protein 4-immunoglobulin (CTLA4-Ig) blocks T cell activation and consequently **inhibits T-dependent B cell antibody production.**[67]" The issue here is abatacept's ability to block B cells from expressing any kind of antibody.

Normal functioning of the immune system, and also any residual function of the immune system in patients treated with immunosuppressants, can be measured with a standard a vaccine (KLH antigen). KLH means, keyhole limpet hemocyanin. A limpet is a type of mollusc.

The Sponsor determined the extent the immune system was still intact, despite treatment with abatacept. Monkeys were treated with weekly abatacept at one of 10, 22, or 50 mg/kd/dose for a period of 52 weeks. In these monkeys, "Drug-related changes consisted of **decreases in IgG levels** at 50 mg/kg."

The Sponsor also tested the extent that the monkey's immune system was intact, by administering a vaccine of keyhole limpet hemocyanin (KLH). The KLH vaccination was given after completing abatacept administration, that is, it was given nine weeks into the recovery period. The Sponsor determined that, "an **antibody response to the** ... **KLH** ... was demonstrated ... showing a functional immune system.[68]" Thus, this shows the existence of residual immune system activity in abatacept-treated animals.

iv. Is Residual Immune System Activity in Abatacept-treated Patients Sufficient to Prevent Infections and Cancers?

This concerns whether the immune system of subjects treated with an immunosuppressant drug is still intact enough to resist infections. This question can be addressed by counting the number and intensity of infections in the abatacept treatment arm versus in the placebo treatment arm.

Abatacept immunosuppression permitted infections and cancers in mice. The infecting agent in mouse experiments was an oncovirus. FDA's Pharmacology Review described the materialization of lymphomas, mammary gland tumors, and adenocarcinomas in abatacept-treated mice, writing, "the increased lymphomas and mammary tumors in mice secondary to ... immunosuppression by abatacept ... [was] likely due to the activation of endogenous retroviruses ... [t]his conclusion is consistent with the increased incidences of neoplasms in humans and mice.[69]"

Consistent with this, FDA contemplated the Sponsor's KLH vaccination studies, and FDA decided that, "The ability of abatacept to suppress ... KLH antibody responses ...

[67] Young JS, et al. Delayed cytotoxic T lymphocyte-associated protein 4-immunoglobulin treatment reverses ongoing alloantibody responses and rescues allografts from acute rejection. Am. J. Transplant. 2016;16:2312–23.

[68] Abatacept (rheumatoid arthritis) BLA 125-118. Pages 90 and 94 of 127-pagePharmacology Review.

[69] Abatacept (rheumatoid arthritis) BLA 125-118. Pages 72, and 73, 105 of 127-pagePharmacology Review.

support the conclusion by the Sponsor that the increased **malignancies** ... were not a direct effect of the drug, but to long-term ... immunosuppression and the control of ... **oncoviruses.**[70]" In other words, this means that the cancer in abatacept-treated mice did not result from failure of T cells to kill cancer cells, but instead resulted from failure of T cells to kill a cancer-causing virus.

Regarding infections, FDA's Medical Review observed that, "Pneumonias were the most common bacterial infection and occurred at twice the rate in abatacept-treated subjects (2%) as compared to placebo-treated subjects (1%) ... [h]erpes simplex occurred at a higher frequency among abatacept-treated subjects (2%) compared to placebo-treated subjects (1%).[71]"

Regarding malignancies and infections, FDA's Medical Review observed that, "Overall, the rate of malignancy ... is not higher among abatacept-treated patients than controls ... [t]he rate of lung cancer was higher with abatacept ... than controls ... it is difficult to assess the significance of the ... higher rate of lung cancers ... [i]t could be a chance observation.[72]"

FDA's Medical Review turned to the lymphomas and, because of the very small number of lymphomas occurring in the study subjects, turned to a cancer database. The database, known as SEER, was used as a historic control. Regarding lymphomas, FDA's Medical Review stated:

> One way to analyze cancer incidence rates ... is by comparison to expected rates from epidemiological data. Rates can be compared ... to expected rates in patients with rheumatoid arthritis ... the Agency [FDA] has derived expected incidence rates from National Cancer Institute's Surveillance, Epidemiology, and End Results (SEER) database[73] ... [t]he overall malignancy ... incidence rates ... are similar between the abatacept group (0.59), placebo group (0.63), and the SEER database (0.47) ... studies had demonstrated an increased incidence of lymphoma ... in a murine model ... [c]onsequently, lymphoma ... were identified as possibly occurring at greater frequency than that of a normal population or rheumatoid arthritis patients not on abatacept ... the ability to reach firm conclusions is limited by the modest number of subjects and the ... short period of drug exposure.[74]

Package label. The package label for abatacept (Orencia®) provided warnings and other types of information regarding the above four scenarios. The relevant package label excerpts are shown below.

v. Does the Patient's Immune System Generate Anti-abatacept Antibodies?

The Adverse Reactions section described the formation of anti-abatacept antibodies in human subjects, though the writing did not distinctly state that some of the antibodies bound only to the CTLA4 portion or to the Ig portion:

> **ADVERSE REACTIONS** ... Antibodies directed against the entire abatacept molecule or to the CTLA-4 portion of abatacept were assessed by ELISA assays in rheumatoid arthritis patients for

[70] Abatacept (rheumatoid arthritis) BLA 125-118. Page 64 of 127-pagePharmacology Review.

[71] Abatacept (rheumatoid arthritis) BLA 125-118. Page 147 of 237-pageMedical Review.

[72] Abatacept (rheumatoid arthritis) BLA 125-118. Pages 15 and 16 and 32 of 237-pageMedical Review. See also, pages 182–194 and 208 and 209 of 237-pageMedical Review.

[73] Abatacept (rheumatoid arthritis) BLA 125-118. Page 183 of 237-pageMedical Review.

[74] Abatacept (rheumatoid arthritis) BLA 125-118. Page 180 of 237-pageMedical Review.

up to 2 years following repeated treatment with ORENCIA. Thirty-four of 1993 (1.7%) patients developed binding antibodies to the entire abatacept molecule or to the CTLA-4 portion of abatacept.[75]

vi. Abatacept's Desired Function as an Immunosuppressant

Consistent with the observations in FDA's Medical Reviews on mechanism of action of abatacept in reducing inflammation and reducing various biomarkers for inflammation, such as a drop in serum levels of soluble IL-2 receptor, the Clinical Pharmacology section read:

> CLINICAL PHARMACOLOGY ... In clinical trials with abatacept at doses approximating 10 mg/kg, **decreases were observed** in serum levels of soluble interleukin-2 receptor (sIL-2 R), interleukin-6 (IL-6), rheumatoid factor (RF), C-reactive protein (CRP), matrix metalloproteinase-3 (MMP3), and TNFα.[76]

vii. Measuring Residual Immune System Activity in Abatacept-treated Patients

The package label warned about abatacept's ability to impair desired immune responses, such as that desired with vaccinations. The Warnings and Precautions section instructed physicians not to administer concurrent live vaccines:

> WARNINGS AND PRECAUTIONS ... Live vaccines should not be given concurrently or within 3 months of discontinuation ... [p]atients ... should be brought up to date with all immunizations prior to ORENCIA therapy ... [b]ased on its mechanism of action, ORENCIA may blunt the effectiveness of some immunizations.[77]

The Patient Counseling Information section of the package label instructed the physician:

> PATIENT COUNSELING INFORMATION. Inform patients that live vaccines should not be given concurrently with ORENCIA or within 3 months of its discontinuation ... the patient should be brought up to date with all immunizations in agreement with current immunization guidelines prior to initiating ORENCIA therapy and to discuss with their healthcare provider how best to handle future immunizations once ORENCIA therapy has been initiated.[78]

viii. Is Residual Immune System Activity in Abatacept-treated Patients Sufficient to Prevent Infections and Cancers?

Regarding increased risk for infections, the Warnings and Precautions section warned about risk for tuberculosis infections:

> WARNINGS AND PRECAUTIONS ... Concomitant use with a TNF antagonist can increase the risk of infections and serious infections ... [p]atients with a history of recurrent infections or underlying

[75] Package label. ORENCIA (abatacept) for injection, for intravenous use. June 2017 (41 pages).

[76] Package label. ORENCIA (abatacept) for injection, for intravenous use. June 2017 (41 pages).

[77] Package label. ORENCIA (abatacept) for injection, for intravenous use. June 2017 (41 pages).

[78] Package label. ORENCIA (abatacept) for injection, for intravenous use. June 2017 (41 pages).

conditions predisposing to infections may experience more infections ... [d]iscontinue if a serious infection develops ... [s]creen for latent tuberculosis infection prior to initiating therapy. Patients testing positive should be treated prior to initiating ORENCIA.[79]

Regarding cancers, FDA's observations about lung cancer, and the fact that there was little or no difference in overall cancers in the abatacept and placebo treatment arms, found a place in the Adverse Reactions section of the package label:

ADVERSE REACTIONS ... In the placebo-controlled portions of the clinical trials ... the overall frequencies of malignancies were similar in the ORENCIA and placebo-treated patients (1.3% and 1.1%, respectively). However, more cases of **lung cancer** were observed in ORENCIA-treated patients (4, 0.2%) than placebo-treated patients (0). In the cumulative ORENCIA clinical trials (placebo-controlled and uncontrolled, open-label) a total of 8 cases of lung cancer (0.21 cases per 100 patient-years) and 4 lymphomas (0.10 cases per 100 patient-years) were observed in 2688 patients (3827 patient-years).[80]

As might be expected for every type of AE detected for all clinical studies for all drugs, FDA's Medical Review contemplated whether any of the AEs in the study subjects could have been caused by the underlying disease and not by the study drug. FDA's observations regarding the AE of lymphoma found a place in the Adverse Reactions section. The package label refers to the SEER database on cancer. The use of the SEER database has been described.[81,82] Regarding lymphoma, the label read:

ADVERSE REACTIONS ... The rate observed for lymphoma is approximately 3.5-fold higher than expected in an age- and gender-matched general population based on the National Cancer Institute's *Surveillance, Epidemiology, and End Results* Database. Patients with rheumatoid arthritis, particularly those with highly active disease, are at a higher risk for the development of lymphoma.[83]

The cited references describe the increased risk for lymphoma in patients with rheumatoid arthritis.[84,85,86,87,88]

[79] Package label. ORENCIA (abatacept) for injection, for intravenous use. June 2017 (41 pages).

[80] Package label. ORENCIA (abatacept) for injection, for intravenous use. June 2017 (41 pages).

[81] Slamon D, et al. Adjuvant trastuzumab in HER2-positive breast cancer. N. Engl. J. Med. 2011;365:1273–83.

[82] Wang W, et al. The demographic features, clinicopathological characteristics and cancer-specific outcomes for patients with microinvasive breast cancer: a SEER database analysis. Sci. Rep. 2017; 7:42045. doi: 10.1038/srep42045.

[83] Package label. ORENCIA (abatacept) for injection, for intravenous use. June 2017 (41 pages).

[84] Beyaert R, et al. Cancer risk in immune-mediated inflammatory diseases (IMID). Mol. Cancer. 2013;12:98. doi: 10.1186/1476-4598-12-98.

[85] Smitten AL, et al. A meta-analysis of the incidence of malignancy in adult patients with rheumatoid arthritis. Arthritis Res. Ther. 2008;10:R45.

[86] Smedby KE, et al. Malignant lymphomas in autoimmunity and inflammation: a review of risks, risk factors, and lymphoma characteristics. Cancer Epidemiol. Biomarkers Prev. 2006;15:2069–77.

[87] Zogala RJ, et al. Management considerations in cancer patients with rheumatoid arthritis. Oncology (Williston Park) 2017;31:374–80.

[88] Hellgren K, et al. Rheumatoid arthritis and risk of malignant lymphoma—is the risk still increased? Arthritis Rheumatol. 2017;69(4):700–8. doi: 10.1002/art.40017.

c. Adalimumab (Rheumatoid Arthritis) BLA 125-057

Adalimumab is a recombinant antibody that binds TNF-α. By inhibiting TNF-α, adalimumab reduces TNF-α's contributions to the pathology of rheumatoid arthritis. These contributions include TNF-α's ability to attract leukocytes into affected joints, TNF-α's influence on increasing synthesis of metalloproteinases by synovial macrophages, and TNF-α's inhibition of the synthesis of proteoglycans in cartilage.[89] Adalimumab has similar efficacy against rheumatoid arthritis as the small-molecule drug, tofacitinib.[90] Adalimumab is a fully human antibody, as it was genetically engineered through phage display technology and is indistinguishable in structure and function from natural human IgG1.[91]

Immune system-related AEs from adalimumab fell into these categories:

- **Anti-adalimumab antibodies**. Immune response to adalimumab by generation of antibodies against adalimumab with consequent impaired efficacy.
- **Autoimmunity and lupus-like syndrome**. Materialization of autoimmunity in study subjects. This involved materialization of antibodies against proteins encoded by the human genome and consequent lupus-like syndrome.
- **Infections**. Increased infections, such as tuberculosis, and need for screening and monitoring patients for tuberculosis.

i. Anti-adalimumab Antibodies

FDA's Pharmacology Review revealed that adalimumab administered to monkeys resulted in the generation of anti-adalimumab antibodies, and in some of the animals, in reduced adalimumab exposure. The FDA reviewer observed, "anti-adalimumab antibodies were not detected until day 21, at which time they were detected in one to three animals in each treatment group. On days 49 and 56, anti-adalimumab antibodies were detected in all animals. Sharp declines in adalimumab levels were observed in some of the animals.[92]"

FDA also observed, "432 adalimumab-treated patients ... was evaluated for HAHAs. Twelve percent of patients in this study developed HAHAs. Among the adalimumab-treated patients, ACR20 response is lower ... among the HAHA positive adalimumab-treated patients ... than among all the HAHA negative adalimumab treated patients.[93]"

Please note FDA's use of the term "ACR20." ACR is American College of Rheumatology. ACR20 is a set of criteria, which requires at least a 20% improvement in the core set measures for a patient to reach improvement. This core set includes number of

[89] Weinblatt ME, et al. Adalimumab, a fully human anti-tumor necrosis factor alpha monoclonal antibody, for the treatment of rheumatoid arthritis in patients taking concomitant methotrexate: the ARMADA trial. Arthritis Rheum. 2003;48:35–45.

[90] van Vollenhoven RF, et al. Tofacitinib or adalimumab versus placebo in rheumatoid arthritis. N. Engl. J. Med. 2012;367:508–19.

[91] Frenzel A, et al. Phage display-derived human antibodies in clinical development and therapy. MAbs. 2016;8:1177–94.

[92] Adalimumab (rheumatoid arthritis) BLA 125-057. 45 pages, Pharmacology Review.

[93] Adalimumab (rheumatoid arthritis) BLA 125-057. 25 pages, Medical Review.

tender joints, number of swollen joints, patient's assessment of pain, and so on. These criteria are used in clinical studies in rheumatoid arthritis.[94]

ii. Autoimmunity and Lupus-like Syndrome

Lupus-like syndrome does not involve antibodies against the study drug, and it is not classified as an allergy, but instead it falls into the category of resembling the autoimmune disease, systemic lupus erythematosus (SLE). Antagonists of TNF-α are used to treat various autoimmune diseases, but they can also result in increased risk for infections, cancer, and autoimmunity. This autoimmunity includes autoantibodies such as antinuclear antibody and anti–double-stranded DNA (dsDNA) antibodies, where some patients with these autoantibodies have lupus-like symptoms.[95]

FDA remarked that adalimumab stimulated the formation of antinuclear antibodies and antibodies against dsDNA, which are both part of the pathology of lupus. FDA wrote, "increases in anti-nuclear antibodies and anti-dsDNA titers were observed more frequently in adalimumab-treated patients than in placebo-treated patients. At week 24, 12% of adalimumab-treated patients and 7% of placebo-treated patients shifted from antinuclear antibody negative at baseline to positive ... [a] few cases of lupus-like syndromes with skin rash, serositis ... were seen.[96]"

FDA's review used the abbreviation, "ANA," which refers to "antinuclear antibodies." Antinuclear antibodies refers to a type of antibody that characterizes the autoimmune disease SLE, and it does not refer to antibodies against the study drug.[97] The term "lupus-like syndrome" is derived from a feature of SLE.

iii. Infections

FDA's Medical Review revealed that adalimumab was associated with an opportunistic infection (tuberculosis).[98] FDA observed that there was an increase in tuberculosis, that this tuberculosis may have been a "recrudescence of latent disease," and that the clinical study was in danger of being placed on a Clinical Hold. Also, FDA recommended that the package label require screening and monitoring:

> Nine cases of tuberculosis were observed ... [m]ost of the cases of tuberculosis occurred within the first few months after initiation of therapy and may reflect recrudescence of latent disease ... [o]ccurrence of ... tuberculosis ... early in the clinical trials prompted discussions between the Agency and the sponsor and **considered placing the clinical program on hold** ... analysis of those ... cases determined that 3/4 of the cases had baseline chest x-rays consistent with tuberculosis, suggesting that screening might be an effective

[94] Felson DT, LaValley MP. The ACR20 and defining a threshold for response in rheumatic diseases: too much of a good thing. Arthritis Res. Ther. 2014;16:101 (5 pages).

[95] Roginic S, et al. Autoimmune pitfalls of anti-tumor necrosis factor-alpha therapy. Isr. Med. Assoc. J. 2015;17:117–9.

[96] Adalimumab (rheumatoid arthritis) BLA 125-057. Page 28 of 42-pageMedical Review (first of three pdf files for Medical Review).

[97] Beigel F, et al. Formation of antinuclear and double-strand DNA antibodies and frequency of lupus-like syndrome in anti-TNF-α antibody-treated patients with inflammatory bowel disease. Inflamm. Bowel Dis. 2011;17:91–8.

[98] Adalimumab (rheumatoid arthritis) BLA 125-057. 25 pages, Medical Review.

> way to identify patients at risk ... [a]t the recommendation of FDA, the sponsor instituted measures for screening ... procedures consisting of chest X-ray ... and the proposed labeling supports these recommendations.[99]

Recrudescence of latent tuberculosis is a well-documented phenomenon.[100] A Clinical Hold is a notification issued by FDA to the Sponsor to delay a proposed clinical trial or to suspend an ongoing clinical trial. The Code of Federal Regulations (21 CFR §312.42) states that "[w]hen an ongoing study is placed on Clinical Hold, no new subjects may be recruited to the study ... patients already in the study should be taken off therapy ... unless specifically permitted by FDA in the interest of public safety." A Clinical Hold may also be imposed, for example, for repeated failures to administer informed consent forms, or carelessness of the Institutional Review Board (IRB).[101]

Package label. FDA's observations on the materialization of anti-adalimumab antibodies, autoimmunity and lupus-like syndrome, and increased risk for infections such as tuberculosis, all found a place on the package label for adalimumab (Humira®). FDA's comments about anti-adalimumab antibodies found a place in the Adverse Reactions section. As shown immediately below, the Adverse Reactions section discloses the materialization of the antibodies, and the fact that they blocked adalimumab activity as tested by in vitro experiments. In fact, the package label that is dated June 2016 disclosed the materialization of these antibodies in a small percentage of patients being treated for one of a variety of different autoimmune diseases (rheumatoid arthritis, juvenile ideopathic arthritis, ankylosing spondylitis, Crohn's disease, psoriasis, and uveitis). The package label issued at the time of the Approval Letter (December 2002) only mentioned materialization of antibodies in subjects with rheumatoic arthritis. But none of the package labels went a step further by commenting on the ability of the antibodies to reduce adalimumab exposure, and none went a step further by commenting on the ability of the antibodies to reduce adalimumab's clinical efficacy. The Adverse Reactions section read:

> **ADVERSE REACTIONS** ... Immunogenicity. Patients ... were tested at multiple time points for antibodies to adalimumab during the 6 to 12 month period. Approximately 5% (58 of 1,062) of adult rheumatoid arthritis patients ... developed low-titer antibodies to adalimumab ... which were neutralizing in vitro.[102]

FDA's comments about autoimmunity and lupus-like syndrome found a place in the Warnings and Precautions section and the Adverse Reactions section:

> **WARNINGS AND PRECAUTIONS** ... Autoimmunity. Treatment with HUMIRA may result in the formation of autoantibodies and, rarely, in the development of lupus-like syndrome. If a patient develops symptoms ... of a lupus-like syndrome ... discontinue treatment.[103]

[99] Adalimumab (rheumatoid arthritis) BLA 125-057. 25 pages, Medical Review.

[100] Ehlers, S. Lazy, dynamic or minimally recrudescent? On the elusive nature and location of the mycobacterium responsible for latent tuberculosis. Infection 2009;37:87–95.

[101] US Department of Health and Human Services; Food and Drug Administration. Guidance for Industry and Clinical Investigators. The use of clinical holds following clinical investigator misconduct; September 2004 (8 pages).

[102] Package label. HUMURA (adalimumab) injection, for subcutaneous use. June 2016 (51 pages).

[103] Package label. HUMURA (adalimumab) injection, for subcutaneous use. June 2016 (51 pages).

> **ADVERSE REACTIONS** ... Two patients out of 3046 treated with HUMIRA developed clinical signs ... of new-onset lupus-like syndrome. The patients improved following discontinuation of therapy. No patients developed lupus nephritis ... [t]he impact of long-term treatment with HUMIRA on the development of autoimmune diseases is unknown.[104]

FDA's comments about tuberculosis, and the need for screening and testing, found a place on the Black Box Warning. This provides the generally applicable take-home lesson for the situation where an expected AE is especially serious. The take-home lesson is that the medical writer should consider adding a requirement to monitor patients, and another requirement for treating patients who test positive. The Black Box Warning read:

> **BOXED WARNING**. SERIOUS INFECTIONS AND MALIGNANCY ... Increased risk of serious infections leading to hospitalization or death, including tuberculosis (TB), bacterial sepsis, invasive fungal infections ... and infections due to other **opportunistic pathogens**. Discontinue HUMIRA if a patient develops a serious infection ... [p]erform test for **latent tuberculosis**; if positive, **start treatment for tuberculosis** prior to starting HUMIRA. Monitor all patients for active TB during treatment, even if initial latent TB test is negative.[105]

iv. Background on Lupus-Like Syndrome Induced by Various Drugs

The package label for adalimumab warns of "lupus-like syndrome." The medical literature has documented cases of lupus-like syndrome occurring with adalimumab, with comments such as, "Drug-induced lupus erythematosus occurs with some drugs and resolves with their withdrawal.[106]" Lupus-like syndrome, which involves the formation of pathological antibodies, may occur with adalimumab, certolizumab pegol, etanercept, and infliximab.[107,108,109,110,111,112,113]

The generated antibodies that are responsible for lupus-like syndrome are not antibodies against the administered drug. The generated antibodies are specific for histones and

[104] Package label. HUMURA (adalimumab) injection, for subcutaneous use. June 2016 (51 pages).

[105] Package label. HUMURA (adalimumab) injection, for subcutaneous use. June 2016 (51 pages).

[106] Lomicova, et al. A case of lupus-like syndrome in a patient receiving adalimumab and a brief review of the literature on drug-induced lupus erythematosus. J. Clin. Pharm. Ther. 2017 42(3):363–6. doi: 10.1111/jcpt.12506.

[107] Williams VL, Cohen PR. TNF alpha antagonist-induced lupus-like syndrome: report and review of the literature with implications for treatment with alternative TNF alpha antagonists. Int. J. Dermatol. 2011;50:619–25.

[108] Mocci G, et al. Dermatological adverse reactions during anti-TNF treatments: focus on inflammatory bowel disease. J. Crohns Colitis. 2013;7:769–79.

[109] Package label. ERELZI (etanercept-szzs) injection, for subcutaneous use. August 2016 (39 pages). BLA 761-042.

[110] Package label. ENBREL® (etanercept) injection, for subcutaneous use. November 2016 (32 pages). BLA 103-795

[111] Hogan JJ, et al. Drug-induced glomerular disease: immune-mediated injury. Clin. J. Am. Soc. Nephrol. 2015;10:1300–10.

[112] Klapman JB, et al. A lupus-like syndrome associated with infliximab therapy. Inflamm. Bowel Dis. 2003;9:176–8.

[113] Beigel F, et al. Formation of antinuclear and double-strand DNA antibodies and frequency of lupus-like syndrome in anti-TNF-α antibody-treated patients with inflammatory bowel disease. Inflamm. Bowel Dis. 2011;17:91–8.

for dsDNA, as is the case for SLE.[114] In SLE, immune complexes (antibody plus antigen) deposit in the kidneys and cause renal inflammation.[115] Regarding the antibodies formed in lupus-like syndrome, these have been described as follows. The letter "H" is used to refer to one of the histone types:

> Antihistone antibodies are common in drug-induced lupus, with (H2A-H2B)-DNA subnucleosome being the predominant antigen in **procainamide** drug-induced lupus and H1 and the H3-H4 complex being the predominant antigens in **hydralazine** drug-induced lupus. Anti-dsDNA antibodies have been observed in patients with **TNFα inhibitor** drug-induced lupus, but are rare in procainamide and hydralazine drug-induced lupus.[116]

d. Atezolizumab (Urothelial Carcinoma) BLA 761-034

Atezolizumab is an antibody that binds to PD-L1, thereby blocking the PD-L1/PD-1 signaling pathway. When administered to a cancer patient, atezolizumab removes the naturally occurring inhibitions against immune response, that is, inhibitions imposed by PD-L1/PD-1 signaling. The result of the removed inhibitions is the desired consequence of enhanced immune response against cancer. But unfortunately, releasing the inhibitions also increases inflammation and autoimmunity in various tissues and organs in the body.

In a publication describing AEs from a clinical study on atezolizumab, pneumonitis was described as "an immune-mediated adverse event.[117]" Another publication on AEs resulting from drugs that block PD-L1/PD-1 signaling stated that "immune-related AEs" include colitis, pneumonitis, and rash.[118]

The design of the atezolizumab clinical trial excluded subjects already suffering from autoimmune disorders. FDA's Medical Review described this exclusion criterion as, "The trial excluded patients with a history of autoimmune disease with the exception of patients with a history of autoimmune hypothyroidism ... and patients with ... type 1 diabetes mellitus.[119]" Excluding this type of study subject prevented the obscuring of any autoimmune AEs arising from the study drug itself.

Comments by FDA reviewers on AEs of individual patients included remarks attributing the AE to inflammation or autoimmunity and relatedness to the study drug. Sub-ileus and volvulus are types of intestinal obstruction.[120] FDA's review

[114] Klapman JB, et al. A lupus-like syndrome associated with infliximab therapy. Inflamm. Bowel Dis. 2003;9:176–8.

[115] Pisetsky DS. The complex role of DNA, histones and HMGB1 in the pathogenesis of SLE. Autoimmunity 2014;47:487–93.

[116] Hogan JJ, et al. Drug-induced glomerular disease: immune-mediated injury. Clin. J. Am. Soc. Nephrol. 2015;10:1300–10.

[117] Rosenberg JE, et al. Atezolizumab in patients with locally advanced and metastatic urothelial carcinoma who have progressed following treatment with platinum-based chemotherapy: a single-arm, multicentre, phase 2 trial. Lancet. 2016;387:1909–20.

[118] De Velasco G, et al. Comprehensive meta-analysis of key immune-related adverse events from CTLA-4 and PD-1/PD-L1 inhibitors in cancer patients. Cancer Immunol. Res. 2017;5:312–8.

[119] Atezolizumab (urothelial carcinoma) BLA 761-034. Page 73 of 125-pageMedical Review.

[120] Ballantyne GH. The meaning of ileus. Its changing definition over three millennia. Am. J. Surg. 1984;148:252–6.

for atezolizumab defined sub-ileus as an immunological AE.[121] Regarding inflammatory disorders caused by the study drug, FDA's Medical Review stated:

Pneumonitis Patient 1087. Reviewer note: Pneumonitis is a known toxicity from anti-PD-1 and anti-PD-L1 drugs. The patient's death due to septic shock and respiratory failure, occurred in close temporal proximity to **pneumonitis**. Thus, it is plausible that this patient's death was attributable to toxicity from the study drug.[122]

Subileus Patient 8013. Reviewer note: Anti-PD-1/PD-L1 drugs can induce **colitis**, which may plausibly predispose to volvulus. Although this patient had other risk factors for volvulus, including prior abdominal surgery and peritoneal metastases, volvulus and interstinal that led to death may have been related to atezolizumab.[123]

In addition to documenting the materialization of drug-induced autoimmunity, FDA's comments also included a "relatedness analysis." As is evident above, the relatedness analysis commented on "close temporal proximity" (this argues in favor of relatedness) and another relatedness analysis commented on "other risk factors" (this argues against relatedness). FDA observed various inflammatory disorders associated with atezolizumab, including inflammatory disorders of the lungs (pneumonitis), liver (hepatitis), gut (colitis), and eye (optic neuritis). For subjects with pneumonitis, FDA's Medical Review also commented, "Pneumonitis. Six patients (1.9%) developed **pneumonitis** ... [t]he median day of onset for the first event for these six patients was day 81 (range: 14-127) ... [f]ive of these patients were treated with corticosteroids.[124]"

For subjects with hepatitis, a diagnosis of autoimmune hepatitis was based on the fact that the hepatitis could be resolved by treating with an antiinflammatory agent (prednisone). FDA's Medical Review described the autoimmune hepatitis as:

Three patients developed liver enzyme elevations that appeared to be immune-mediated. Patient 1141 developed flu-like symptoms on day 2 and Grade 1 AST/ALT increases on day 13 followed by delirium on day 15. AST/ALT worsened to Grade 2 and he developed Grade 1 hyperbilirubinemia ... [a]tezolizumab was not withheld. He received **prednisone for possible autoimmune hepatitis** and his liver enzymes and encephalopathy resolved by days 42 and 47 respectively. Patient 1115 developed AST/ALT on day 8 which worsened to Grade 3 by day 16. Atezolizumab was temporarily interrupted. He received **prednisone** with resolution of liver enzyme elevation by day 51.[125]

FDA's Medical Review further described colitis, "Patient 1090 experienced Grade 2 microscopic **colitis**. Biopsies of the left and right colon showed questionable wall thickening. Atezolizumab was withheld and no other treatment was reported. Atezolizumab was resumed on day 163 and the colitis resolved by day 203.[126]"

FDA took a reasonable approach for identifying study subjects with AEs having an immune component. This approach was to identify subjects receiving antiinflammatory

[121] Atezolizumab (urothelial carcinoma) BLA 761-034. Page 112 of 125-pageMedical Review.

[122] Atezolizumab (urothelial carcinoma) BLA 761-034. Pages 78 and 79 of 125-pageMedical Review.

[123] Atezolizumab (urothelial carcinoma) BLA 761-034. Pages 78 and 79 of 125-pageMedical Review.

[124] Atezolizumab (urothelial carcinoma) BLA 761-034. Pages 87 and 88 of 125-pageMedical Review.

[125] Atezolizumab (urothelial carcinoma) BLA 761-034. Page 89 of 125-pageMedical Review.

[126] Atezolizumab (urothelial carcinoma) BLA 761-034. Page 96 of 125-pageMedical Review.

drugs (hydrocortisone, dexamethasone, and cortisone). Subjects receiving hydrocortisone can be detected with absolute certainty whereas, in contrast, a Case Report Form with the notation, "patient has rash," is relatively ambiguous. Thus, the technique of identifying study subjects receiving antiinflammatory agents is a useful technique for AE analysis.

FDA's account of antiinflammatory drug use distinguished between cases where the AE was immune-related and where the AE was not immune-related. On this point, FDA stated, "Corticosteroid use. Fifty-seven patients (18.3%) received systemic corticosteroids within 30 days of an adverse event ... [t]wenty of these patients (6.4%) were considered to have **experienced an immunological AE** ... [o]ne patient died from subileus, which was **considered to be an immunological AE**.[127]"

The importance of correlating corticosteroid use with immune-related AEs, as a means for identifying safety signals, is highlighted by the fact that FDA's Cross Discipline Team Leader Review observation that, "Adverse events that were likely to be immune-mediated and were **treated with corticosteroids** occurred in 6% of patients.[128]"

Package label. FDA's observations on immune-related AEs with atezolizumab found a place on the package label for atezolizumab (Tecentriq®):

> **WARNINGS AND PRECAUTIONS. Immune-Related Pneumonitis**: Withhold for moderate and permanently discontinue for severe or life-threatening pneumonitis ... **Immune-Related Hepatitis**: Monitor for changes in liver function. Withhold for moderate and permanently discontinue for severe or life threatening transaminase or total bilirubin elevation ... **Immune-Related Colitis**: Withhold for moderate or severe, and permanently discontinue for life-threatening colitis ... **Ocular Inflammatory Toxicity**: Withhold for moderate and permanently discontinue for severe ocular inflammatory toxicity.[129]

e. Denosumab (Osteoporosis) BLA 125-320

Denosumab is a fully human antibody that binds RANK-ligand (RANKL). RANKL is a cytokine that interacts with its receptor RANK on osteoclasts and stops the osteoclasts from resorbing bone.[130] RANKL is expressed by osteoblasts and T cells. T cells express both soluble and membrane-bound forms of RANKL, and both forms are involved in bone erosion associated with arthritis.[131] RANKL-mediated signaling controls the fusion of osteoclast precursors into multinucleated cells, their differentiation into mature osteoclasts, their attachment to bone surfaces, and ultimately in their activation to resorb bone.

Although denosumab is an antibody for treating osteoporosis, a view of denosumab's mechanism of action suggests that it might have the effect of suppressing the immune system. RANKL is expressed by T cells. Also, RANK is expressed by dendritic cells. In vitro

[127] Atezolizumab (urothelial carcinoma) BLA 761-034. Page 112 of 125-pageMedical Review.

[128] Atezolizumab (urothelial carcinoma) BLA 761-034. Page 21 of 24-pageCross Discipline Team Leader Review.

[129] Package label. TECENTRIQ® (atezolizumab) injection, for intravenous use. April 2017 (24 pages).

[130] Cummings SR, et al. Denosumab for prevention of fractures in postmenopausal women with osteoporosis. N. Engl. J. Med. 2009;361:756–5.

[131] Kearns AE, et al. Receptor activator of nuclear factor kappaB ligand and osteoprotegerin regulation of bone remodeling in health and disease. Endocrin. Rev. 2008;29:155–92.

studies show that binding of RANKL from T cells to RANK on dendritic cells regulates dendritic cell function and survival.[132] Moreover, RANKL induces $CD8^+$ T cells and can enhance immune response against viral infections.[133]

FDA's Medical Review for denosumab warned against increased infections and hypersensitivity reactions. Regarding infections, FDA wrote, "RANKL is expressed on ... T and B lymphocytes ... [t]herefore, a RANKL inhibitor such as denosumab may increase risk for infection ... the incidence of non-fatal serious infections was 4.0% in the denosumab-treated subjects as compared to 3.3% in placebo-treated subjects ... [w]hile the over infection rates were similar, denosumab-treated subjects appeared to have **infections that were more serious.**[134]" This excerpt provides the take-home lesson that, in preparing the Listings and in drafting Clinical Study Reports, the Sponsor should consider disclosing the severity of each of the AEs, in addition to merely disclosing the names of AEs associated with each study subject.

The Sponsor detected a slight increase in incidence of infections, and also a slight increase in severity of infections in the denosumab treatment arm.[135] FDA required that the package label warn about additive effects with a coadministered immunosuppressant, leading to increased infections, "Regarding labeling, the ... label should include language that denosumab may cause serious infections ... **[p]atients on concomitant immunusuppressive therapy** may be at increased risk of infections.[136]" Regarding hypersensitivity reactions, FDA wrote, "**with any protein product**, **hypersensitivity** is a concern ... drug hypersensitivity was reported in 11(0.3%) of subjects in the placebo group and 15(0.4%) of subjects in the denosumab group ... there is no clear evidence of significant hypersensitivity reactions in subjects treated with denosumab.[137]"

Package label. The Warnings and Precautions section for denosumab (Prolia®) included various warnings about AEs having an immune component. These included anaphylactic reactions, infections, and dermatitis. The Warnings and Precautions section instructed physicans to be especially vigilant for infections materializing in patients taking concommitant immunosuppressants:

> **WARNINGS AND PRECAUTIONS** ... Hypersensitivity including **anaphylactic** reactions may occur. Discontinue permanently if a clinically significant reaction occurs ... Serious infections including **skin infections**: May occur, including those leading to hospitalization. Advise patients to seek prompt medical attention if they develop signs or symptoms of infection, including cellulitis ... [d]ermatologic reactions: **Dermatitis**, rashes, and eczema have been reported. Consider discontinuing Prolia if severe symptoms develop.[138]

[132] Kearns AE, et al. Receptor activator of nuclear factor kappaB ligand and osteoprotegerin regulation of bone remodeling in health and disease. Endocrin. Rev. 2008;29:155–92.

[133] Finsterbuch K, Piguet V. Down-RANKing the threat of HSV-1:RANKL upregulates MHC-Class-I-restricted anti-viral immunity in herpes simplex virus infection. J. Invest. Dermatol. 2015;135:2565–7.

[134] Denosumab (osteoporosis) BLA 125-320. Pages 145, 168, 178, and 354 of 710-pageMedical Review.

[135] Denosumab (osteoporosis) BLA 125-320. Pages 145, 168, 178, and 354 of 710-pageMedical Review.

[136] Denosumab (osteoporosis) BLA 125-320. Pages 145, 168, 178, and 354 of 710-pageMedical Review.

[137] Denosumab (osteoporosis) BLA 125-320. Pages 232, 233, 391 and 392 of 710-pageMedical Review.

[138] Package label. Prolia® (denosumab) Injection, for subcutaneous use. May 2017 (26 pages).

In the Full Prescribing Information section, the Warnings and Precautions section warned against concomitant immunosuppressants:

> **WARNINGS AND PRECAUTIONS** ... Patients on **concomitant immunosuppressant agents** or with impaired immune systems may be at increased risk for serious infections ... [i]n patients who develop serious infections while on Prolia, prescribers should assess the need for continued Prolia therapy.[139]

f. Fingolimod (Multiple Sclerosis) NDA 022-527

Fingolimod is a small molecule derived from a natural product made by the fungus, *Isaria sinclairii*. The natural product is myriocin.[140] Fingolimod acts on T cells that reside in lymph nodes and prevents these T cells from exiting the lymph nodes, where the end-effect is preventing them from migrating to the central nervous system (CNS).[141,142] Infections were observed during the Sponsor's clinical studies on fingolimod. FDA commented on two viral infections, herpes and varicella zoster:

> Dr. Villalba notes that the two cases of **herpes infections resulting in death** occurred in young patients (23 and 29 years old) who were taking fingolimod 1.25 mg/day ...[t]he case of disseminated **varicella zoster also had massive hepatic necrosis and multiorgan failure** that could be due to disseminated varicella zoster ... I agree with Dr. Villalba that these cases are of particular concern in light of the immunosuppression that might be expected based on the mechanism of action of fingolimod.[143]

FDA's Medical Review provided these concepts for clinical trial design for study drugs that suppress the immune system where there is a consequent increased risk for infections:

- Need to monitor for appearance of infections
- Need to assess baseline white blood cells before administering the drug
- Where any other immune-stimulating drugs are needed, such as vaccines, consider administering these before administering the drug
- Need to avoid concomitant administration of any other immunosuppressant drugs
- Need for patient to undergo washout of the immunosuppressant drug before administering any other (different) immunosuppressant drugs such as fingolimod

[139] Package label. Prolia® (denosumab) Injection, for subcutaneous use. May 2017 (26 pages).

[140] Adachi K, Chiba K. FTY720 story. Its discovery and the following accelerated development of sphingosine 1-phosphate receptor agonists as immunomodulators based on reverse pharmacology. Perspect. Medicin. Chem. 2007;1:11–23.

[141] Kappos L, et al. A placebo-controlled trial of oral fingolimod in relapsing multiple sclerosis. N. Engl. J. Med. 2010;362:387–401.

[142] Cohen JA, et al. Oral fingolimod or intramuscular interferon for relapsing multiple sclerosis. N. Engl. J. Med. 2010;362:402–15.

[143] Fingolimod (multiple sclerosis) NDA 022-527. Page 100 of 519-pageMedical Review.

FDA's path of reasoning and logic in arriving at the package label is shown by this excerpt:

> The risk of serious infection should be addressed in the labeling ... I agree with Dr. Villalba that labeling should recommend that **baseline white blood cell count** (WBC) should be obtained prior to starting therapy. I also agree with Dr. Cavaille-Coll's recommendations as follows ... fingolimod can modify signs and symptoms of infection, and physicians should be cautioned to maintain a higher degree of suspicion for infection ... **[i]mmunologic response to vaccination may be decreased**, and physicians should consider **vaccination prior to initiation** of fingolimod therapy ... **patients should undergo washout** before being treated with other immunosuppressants other than corticosteroids.[144]

FDA's concerns about infections, coadministered immunosuppressants, and coadministered vaccines found a place on the package label as shown below.

Package label. The Warnings and Precautions section for fingolimod (Gilenya®) warned against opportunistic infections and the need for vigilance in monitoring for infections. This package label provides the generally-applicable take-home lessons that, when drafting the package label the medical writer should consider requiring that the physician monitor for certain adverse events, and that the physician should consider dose modification (dose reduction, dose interruption, dose discontinuation) following the materialization of a given AE. The label read:

> **WARNINGS AND PRECAUTIONS** ... GILENYA may increase the risk of infections. A recent complete blood count (CBC) should be available before initiating treatment with GILENYA. **Monitor for signs and symptoms of infection** during treatment ... [d]o not start GILENYA treatment in patients with active acute or chronic infections ... [c]onsider suspending treatment with GILENYA if a patient develops a serious infection, and reassess the benefits and risks prior to re-initiation of therapy. Because the elimination of fingolimod after discontinuation may take up to two months, continue **monitoring for infections** throughout this period.[145]

The Drug Interactions section of the package label warned against coadministering immunosuppressive drugs such as natalizumab and mitoxantrone, and against coadministering vaccines. Regarding vaccines, the term "live attenuated" refers to vaccine taking the form of a virus or a bacterium that has been genetically or chemically modified, so that it does not cause any infection, and where some aspect of metabolism is still intact. The Drug Interaction section mentions natalizumab. Fingolimod and natalizumab, which are each used for treating multiple sclerosis, have similar mechanisms of action (preventing transit of white blood cells into the CNS):

> **DRUG INTERACTIONS** ... **Vaccination may be less effective** during and for up to 2 months after discontinuation of treatment with GILENYA ... [t]he use of live attenuated vaccines should be avoided during and for 2 months after treatment with GILENYA because of the risk of infection ... immunosuppressive ... modulating therapies are expected to increase the risk of immunosuppression. Use caution when switching patients from long-acting **therapies with immune effects** such as natalizumab or mitoxantrone.[146]

[144] Fingolimod (multiple sclerosis) NDA 022-527. Page 123 of 519-pageMedical Review.

[145] Package label. GILENYA (fingolimod) capsules. September 2010 (16 pages).

[146] Package label. GILENYA (fingolimod) capsules. September 2010 (16 pages).

g. Infliximab (Crohn's Disease, Ulcerative Colitis, and Rheumatoid Arthritis) BLA 125-544

Infliximab is an antibody that blocks the activity of TNF-α by binding to the soluble form of TNF-α or to the membrane form of TNF-α. Therapeutic proteins that bind to TNF-α and block TNF-α activity include infliximab, adalimumab, etanercept, golimumab, and certolizumab pegol. The literature has addressed the possibility that these TNF-α blocking agents increase risk for the AEs of infections and cancer.[147,148]

i. AE of Cancer

FDA's Medical Review of infliximab (Inflectra®), marketed by Celltrion, Inc., warned about the AE of lymphoma, where this warning was based on the fact that the same drug, marketed under the trade name Remicade®, was associated with increased lymphoma risk. Package labels are cited here for infliximab as previously marketed by Centocor, Inc.[149] and later by Janssen Biotech, Inc.[150]

FDA commented on the AE of cancer, referring to the package labels from the same drug (Remicade®) from other manufacturers, "Malignancies, including **lymphoma**, have been identified as potential risk with ... Remicade® and other TNF-inhibitors as described in the Warnings and Precautions section of ... Remicade's USPI. There was a small number of **malignancies** reported in the ... [study drug] which were balanced between the treatment arms ... [t]he incidence and types of these malignancies is expected for the study population and treatment.[151]"

ii. AE of Tuberculosis and Pneumonia

FDA's Cross Discipline Team Leader Review observed that the infliximab treatment arm had more infections than the comparator drug (Remicade®) treatment arm, writing, "A numerical imbalance in serious infections, driven by several cases of **tuberculosis and pneumonia**, was observed in the controlled studies. The differences were small, and serious infections, including tuberculosis, are well-recognized risks with TNF-inhibition as indicated in the Boxed Warning for this class of biological products.[152]"

[147] Hadam J, et al. Managing risks of TNF inhibitors: an update for the internist. Cleveland Clin. J. Med. 2014;81:115–27.

[148] Thompson AE, et al. Tumor necrosis factor therapy and the risk of serious infection and malignancy in patients with early rheumatoid arthritis: a meta-analysis of randomized controlled trials. Arthritis Rheum. 2011;63:1479–85.

[149] Package label. REMICADE™ Infliximab for IV Injection. August 1998 (12 pages).

[150] Package label. REMICADE (Infliximab) lyophilized for injection, for intravenous use. October 2015 (51 pages).

[151] Infliximab (Crohn's disease, ulcerative colitis, rheumatoid arthritis) BLA 125-544. Page 202 of 226-page Medical Review.

[152] Infliximab (Crohn's disease, ulcerative colitis, rheumatoid arthritis) BLA 125-544. Page 16 of 31-pageCross Discipline and Team Leader Review.

iii. AE of Fungal Infections

FDA's Medical Review took note of the AE of fungal infections, stating that only one study subject had a fungal infection. Despite the fact that only one subject had a fungal infection, it found a place on the package label, and it might be suggested that this was because the fungal infection was fatal. But real reason for inclusion on the package label was likely that the package label for the comparator drug (Remicade®) warned against fungal infections.

Regarding this single fungal infection, FDA stated, "By Week 30, 36 (17%) of 210 enrolled patients experienced adverse events ... [i]nfusion reactions and serious infections occurred in 7% and 6% of all patients, respectively. One patient developed invasive **fungal sepsis, which resulted in death.**[153]"

Package label. The package label for infliximab (Inflectra®) had a Black Box Warning section warning about infections and cancer:

> **BOXED WARNING.** Increased risk of serious infections leading to hospitalization or death, including **tuberculosis** (TB), bacterial sepsis, invasive **fungal infections** (such as histoplasmosis) and infections due to other **opportunistic pathogens** ... Discontinue INFLECTRA if a patient develops a serious infection ... Monitor all patients for active TB during treatment, even if initial latent TB test is negative ... **Lymphoma and other malignancies**, some fatal, have been reported in children and adolescent patients treated with tumor necrosis factor (TNF) blockers, including infliximab products.[154]

iv. Package Label for the Comparator Drug (Remicade®)

The Sponsor's clinical studies for infliximab (Inflectra®) used a study design where infliximab (Inflectra®) was administered in the study drug treatment arm, while a comparator drug, infliximab (Remicade®), was administered in the control treatment arm. Inflectra and Remicade are drugs that are very similar to each other, and they are called, "biosimilars." Although the Sponsor's own clinical studies observed the AEs of malignancies,[155] bacterial infections,[156] and fungal infections,[157] it is likely that package labeling for the comparator drug infliximab (Remicade®) played the dominant role for arriving at labeling for infliximab (Inflectra®). So much is evident, from the Black Box Warning for infliximab (Remicade®), reproduced below:

> **BOXED WARNING.** Increased risk of **serious infections** leading to hospitalization or death, including tuberculosis (TB), bacterial sepsis, invasive fungal infections (such as histoplasmosis) and infections due to other opportunistic pathogens. Discontinue REMICADE if a patient develops a **serious infection**. Perform test for latent tuberculosis; if positive, start treatment for tuberculosis prior to starting REMICADE. **Monitor all patients for active tuberculosis** during treatment ... **[l]ymphoma and other malignancies,**

[153] Infliximab (Crohn's disease, ulcerative colitis, rheumatoid arthritis) BLA 125-544. Page 144 of 226-pageMedical Review.

[154] Package label. INFLECTRA (infliximab-dyyb) for injection, for intravenous use. April 2016 (56 pages).

[155] Infliximab (Crohn's disease, ulcerative colitis, rheumatoid arthritis) BLA 125-544. Page 202 of 226-page Medical Review.

[156] Infliximab (Crohn's disease, ulcerative colitis, rheumatoid arthritis) BLA 125-544. Page 16 of 31-page Cross Discipline and Team Leader Review.

[157] Infliximab (Crohn's disease, ulcerative colitis, rheumatoid arthritis) BLA 125-544. Page 144 of 226-page Medical Review.

some fatal, have been reported in children and adolescent patients treated with tumor necrosis factor (TNF) blockers, including REMICADE. Postmarketing cases of fatal **hepatosplenic T-cell lymphoma** have been reported in patients treated with TNF blockers including REMICADE.[158]

h. Natalizumab (Multiple Sclerosis) BLA 125-104 and (Crohn's Disease) Supplemental BLA 125-104

Natalizumab is a recombinant antibody for treating two autoimmune diseases, multiple sclerosis and Crohn's disease. Regarding multiple sclerosis, the antibody prevents transit of T cells from the bloodstream into the CNS, thus reducing T-cell–induced inflammation of the brain. The antibody binds to an extracellular protein ($\alpha4\beta1$-integrin) of blood vessel endothelial cells and prevents binding of T cells to $\alpha4\beta1$-integrin to the endothelial cells, thus preventing passage of T cells out of the bloodstream and into surrounding tissues.[159]

i. Opportunistic Infection Resulting in Progressive Multifocal Leukoencephalopathy

Natalizumab is associated with the opportunistic infection, progressive multifocal leukoencephalopathy (PML), which is caused by JC virus. PML is a viral infection of the brain. PML results from latent viruses already residing in the brain.[160] Natalizumab, in rare instances, results in activation of JC virus in the brain, where the virus causes PML.

ii. PML and the November 2004 Package Label, the May 2006 Package Label, and the May 2016 Package Label

The package label that issued with FDA-approval (November 2004) did not mention PML.[161] The first mention of PML occurs in the supplemental package label dated May 2006, "There are no known interventions that can reliably prevent PML or adequately treat PML if it occurs. It is not known whether early detection of PML and discontinuation of TYSABRI will mitigate the disease.[162]" Unfortunately, the most recently published package label (May 2016) also admits that, "There are no known interventions that can reliably prevent PML or that can adequately treat PML if it occurs.[163]" According to Misbah,[164]

[158] Package label. REMICADE (infliximab) Lyophilized Concentrate for Injection, for Intravenous Use. October 2015 (51 pages). BLA 103-772

[159] Brody T. Multistep denaturation and hierarchy of disulfide bond cleavage of a monoclonal antibody. Analyt. Biochem. 1997;247:247–56.

[160] Kappos L, et al. Switching from natalizumab to fingolimod: A randomized, placebo-controlled study in RRMS. Neurology 2015;85:29–39.

[161] Package label. TYSABRI® (natalizumab) November 2004 (11 pages).

[162] Package label. TYSABRI (natalizumab) May 2006 (17 pages).

[163] Package label. TYSABRI (natalizumab) injection for intravenous use. May 2016 (28 pages).

[164] Misbah SA. Progressive multifocal leukoencephalopathy—driven from rarity to clinical mainstream by iatrogenic immunodeficiency. Clin. Exp. Immunol. 2017;188(3):342–52. doi: 10.1111/cei.12948.

potential approaches for eliminating PML include the antiviral drugs, mirtazapine, cytarabine, and cidofovir. Misbah added that, "Despite these different therapeutic approaches, the fatality rate in PML remains worryingly high.[165,166]"

iii. Steps in Natalizumab's Drug Approval Process

FDA granted approval to natalizumab in November 2004, but subsequently fatal opportunistic infections were reported in three patients in the general patient population. The drug was withdrawn from the market in February 2005.[167] FDA placed the Sponsor's program on a Clinical Hold.[168]

On June 5, 2006, FDA determined that natalizumab should be returned to the market, but only for the limited situation where patients were not able to tolerate other therapies for multiple sclerosis or where other therapies were not effective.[169] With the return of natalizumab to the market, the Sponsor agreed to a Risk Evaluation and Mitigation Strategy, which included the TOUCH program. The TOUCH program found a place on the package label, where it resides in the Black Box Warning, as shown below.

The drug returned to FDA for review and FDA issued a supplementary Medical Review, which concerned the indication of Crohn's disease, and which documented AEs arising during treatment of Crohn's disease as well as of multiple sclerosis. The package label was revised to warn against PML.

In February 2005, the Sponsor informed FDA of three cases of PML of about 3000 subjects that had been treated with natalizumab.[170] At a meeting held in March 2006, FDA agreed to permit remarketing of natalizumab but only to a subset of multiple sclerosis patients, namely, patients with relapsing multiple sclerosis. Also, FDA required that patients treated with natalizumab must not be receiving concomitant treatment with another immune-modulating drug (Avonex®).[171]

In FDA's Summary Review of June 2006, FDA imposed a requirement for an updated package label, where FDA required a Black Box Warning. FDA's requirement was stated as, "Specifically with regard to labeling we believe ... and we and the sponsor have agreed ... that labeling should contain a boxed warning describing the risks of PML.[172]"

[165] Misbah SA. Progressive multifocal leukoencephalopathy—driven from rarity to clinical mainstream by iatrogenic immunodeficiency. Clin. Exp. Immunol. 2017;188(3):342–52. doi: 10.1111/cei.12948.

[166] Pontillo G, et al. Brain susceptibility changes in a patient with natalizumab-related progressive multifocal leukoencephalopathy: a longitudinal quantitative susceptibility mapping and relaxometry study. Front. Neurol. 2017;8:294 (5 pages).

[167] Natalizumab (Crohn's disease) Supplemental BLA 125-104. Page 175 of 465-pageApproval Package.

[168] Natalizumab (Crohn's disease) Supplemental BLA 125-104. Page 183 of 465-pageApproval Package of January 14, 2008.

[169] Natalizumab (Crohn's disease) Supplemental BLA 125-104. Page 185 of 465-pageApproval Package of January 14, 2008.

[170] Natalizumab (multiple sclerosis) Supplemental BLA 125-104. 6-pageSummary Review of June 5, 2006.

[171] Natalizumab (multiple sclerosis) Supplemental BLA 125-104. 6-pageSummary Review of June 5, 2006.

[172] Natalizumab (multiple sclerosis) Supplemental BLA 125-104. 6-pageSummary Review of June 5, 2006.

FDA's Clinical Review dated January 2008 set forth observations and recommendations relating to immune responses:

- Immunosuppressants
- Opportunistic infections
- Anaphylaxis

FDA's Clinical Review recommended that concomitant immunosuppressants not be used, writing, "Because of the ... risk of PML and other infections ... patients receiving natalizumab should not be treated with concomitant immunosuppressants, e.g., azothiopr-ine, 6-mercaptopurine, or methotrexate, or inhibitors of TNFα, and corticosteroids should be tapered ... [o]rdinarily, multiple sclerosis patients receiving chronic immunosuppres-sant ... should not be treated with natalizumab.[173]"

FDA also observed that the study drug was associated with hypersensitivity reactions, including anaphylaxis. These AEs found a place on the package label, as shown below. FDA's observations on hypersensitivity reactions read, "**Hypersensitivity reactions** were strongly associated with the development of antibodies to natalizumab ... the most fre-quent serious adverse events associated with natalizumab were ... hypersensitivity reac-tions ... 1.3%, including **anaphylaxis.**[174]"

Package label. The package label for natalizumab (Tysabri®) contained a Black Box Warning regarding the opportunistic JC virus infection that causes PML:

> **BOXED WARNING** ... TYSABRI increases the risk of progressive multifocal leukoencephalopathy (PML), an **opportunistic viral infection** of the brain that usually leads to death or severe disability. Risk factors for the development of PML include duration of therapy, prior use of immunosuppressants, and presence of anti-JCV antibodies ... [h]ealthcare professionals should monitor patients ... for any new sign or symptom ... suggestive of PML. TYSABRI dosing should be withheld immediately at the first sign or symptom suggestive of PML ... [b]ecause of the risk of PML, TYSABRI is available only through a restricted program under a Risk Evaluation and Mitigation Strategy (REMS).[175]

The Warnings and Precautions section warned against concomitant treatment with other immunosuppressants (azathioprine) mirroring the recommendations of the FDA reviewer:

> **WARNINGS AND PRECAUTIONS** ... Progressive multifocal leukoencephalopathy ... caused by the JC virus ... has occurred in three patients who received TYSABRI ... **[a]ll three cases of PML occurred in patients who were concomitantly** exposed to immunomodulators ... or ... **with immunosuppressants, e.g., azathioprine** ... therefore, patients receiving ... immunosuppressant ... therapy ... should not be treated with TYSABRI.[176]

[173] Natalizumab (Crohn's disease) Supplemental BLA 125-104. Page 179 of 465-pageApproval Package of January 14, 2008.

[174] Natalizumab (Crohn's disease) Supplemental BLA 125-104. Pages 44 and 183 of 465-pageApproval Package of January 14, 2008.

[175] Package label. TYSABRI (natalizumab) injection. May 2016 (27 pages).

[176] Package label. TYSABRI (natalizumab) injection for intravenous use. January 2008 (22 pages).

Regarding anaphylaxis, the package label warned:

> **WARNINGS AND PRECAUTIONS** ... Hypersensitivity reactions: Serious hypersensitivity reactions (e.g., **anaphylaxis**) have occurred. Permanently discontinue TYSABRI if such a reaction occurs.[177]

i. Nivolumab (Melanoma) BLA 125-554

Nivolumab is an antibody that blocks the normal, constitutive signaling via the PD-1/PL-L1 signaling pathway, where the desired result is activation of T cells. Nivolumab enhances immune response against various cancers. But an unfortunate consequence of this activation is increased immune response against the patient's own tissues and organs.

Nivolumab-induced lung inflammation can be detected by computed tomography (CT scans), and nivolumab-induced liver inflammation can be detected by lab tests for liver enzymes. Imafuku et al.[178] reported this AE as, "A CT scan revealed widespread bilateral ground glass opacity of his lungs consistent with a diffuse **alveolar damage**." In a separate published case report, Imafuku et al.[179] reported a case of liver inflammation, "After approximately 34 weeks of nivolumab administration ... laboratory tests revealed a sudden elevation of liver enzymes (AST/ALT) ... [a]lthough many diseases induce liver dysfunction, in this case, we definitively excluded other causes of **hepatitis** such as viral infection (e.g., HBV, HCV, HSV, CMV, and EBV), medication other than nivolumab, and other forms of autoimmune hepatitis."

FDA's analysis of nivolumab focused on tissues and organ inflammation. FDA's Cross Discipline Team Leader Review provides a bird's eye view of this type of AE, and it refers to FDA's own recommendations for package labeling. To this end, FDA's account of inflammation-related risks were:

> The FDA clinical review of safety identified the following clinically significant immune-mediated adverse reactions of nivolumab, **as reflected in recommended labeling**: **pneumonitis (3.4%), colitis (2.2%), hepatitis (1.1%), and nephritis** and renal dysfunction (0.7%). Additional immune-mediated adverse reactions occurring in < 1% of patients ... were **pancreatitis, uveitis, demyelination, autoimmune neuropathy**, adrenal insufficiency, and facial and abducens nerve paresis. Grade 1 or 2 hypothyroidism and Grade 1 or 2 hyperthyroidism were reported in 7.8% (21/268) and 3% (8/268) of patients receiving nivolumab, respectively ... [t]he most serious risk with nivolumab appears to be **pneumonitis (3.4% all Grades)**, with fatal cases occurring in 0.9% (5/574) of patients treated with nivolumab at a range of doses from 1 mg/kg to 10 mg/kg. Of note, the incidence of fatal pneumonitis appeared to decrease in clinical trials following recognition of pneumonitis as an adverse reaction and implementation of management algorithms, including nivolumab dose delays or discontinuation.[180]

[177] Package label. TYSABRI (natalizumab) injection for intravenous use. January 2008 (22 pages).

[178] Imafuku K, et al. Two cases of nivolumab re-administration after pneumonitis as immune-related adverse events. Case Rep. Oncol. 2017;10:296–300.

[179] Imafuku K, et al. Successful treatment of sudden hepatitis induced by long-term nivolumab administration. Case Rep. Oncol. 2017;10:368–71.

[180] Nivolumab (melanoma) BLA 125-554. Pages 22–25 of 29-pageCross Discipline Team Leader Review.

FDA's Medical Review disclosed a few additional details on the pneumonitis, "Five (1.7%) patients died ... due to **pneumonitis**. Three patients ... developed **pneumonitis** within 100 days of last dose of nivolumab ... [t]he three patients died after developing subsequent AEs of sepsis or respiratory failure.[181]"

FDA provided additional details on colitis, "Five (1.9%) patients experienced Grade 3 **colitis** or diarrhea in the nivolumab group ... [o]f the five patients, two ... experienced **colitis** and three patients ... experienced diarrhea. Three of the AEs, two **colitis** and one diarrhea, were considered study drug related by the investigator ... [n]ivolumab dose was delayed in two patients, one each for **colitis**, and withdrawn for **autoimmune colitis** in one patient ... [a]ll five patients in the nivolumab group were treated with high dose corticosteroids.[182]"

Also, FDA's account of nivolumab-induced autoimmune reactions referred to other therapeutic proteins with a similar mechanism of action as nivolumab. AE analysis based on other drugs with a similar mechanism of action is always called, "drug class" analysis. FDA predicted nivolumab's AEs from AEs known to occur with pembrolizumab and ipilimumab, warning that, "Prescribers are familiar with management of **immune-mediated adverse reactions** based on the similar safety profiles of another anti-PD-1 monoclonal antibody, pembrolizumab, that received accelerated approval in September 2014, and of ipilimumab, an anti-CTLA monoclonal antibody which received regular approval in 2011.[183]"

Package label. The package label for nivolumab (Opdivo®) warned against autoimmune damage to various tissues and organs. The label warned against immune-mediated pneumonitis, immune-mediated colitis, immune-mediated hepatitis, and immune-mediated nephritis. Each of the warnings about pneumonitis, colitis, and hepatitis warned against autoimmunity resulting from **monotherapy with nivolumab alone**, and also from the **combination of nivolumab with a second immunosuppressant** (ipilimumab). The Full Prescribing Information section disclosed:

> **WARNINGS AND PRECAUTIONS** ... ***OPDIVO as a Single Agent.*** In patients receiving OPDIVO as a single agent, immune-mediated pneumonitis occurred in 3.1% (61/1994) of patients. The median time to onset of immune-mediated pneumonitis was 3.5 months ... [a]pproximately 89% of patients with pneumonitis received high-dose corticosteroids (at least 40 mg prednisone equivalents per day) for a median duration of 26 days (range: 1 day to 6 months).[184]

The warning pertaining to combination therapy was:

> **WARNINGS AND PRECAUTIONS** ... ***OPDIVO with Ipilimumab.*** In patients receiving OPDIVO with ipilimumab, immune-mediated pneumonitis occurred in 6% (25/407) of patients ... [a]pproximately 84% of patients with pneumonitis received high-dose corticosteroids (at least 40 mg prednisone equivalents per day) for a median duration of 30 days (range: 5 days to 11.8 months).[185]

[181] Nivolumab (melanoma) BLA 125-554. Pages 82 and 83 of 156-pageMedical Review.

[182] Nivolumab (melanoma) BLA 125-554. Pages 89 and 142 of 156-pageMedical Review.

[183] Nivolumab (melanoma) BLA 125-554. Pages 22–25 of 29-pageCross Discipline Team Leader Review.

[184] Package label. OPDIVO (nivolumab) injection, for intravenous use. February 2017 (58 pages).

[185] Package label. OPDIVO (nivolumab) injection, for intravenous use. February 2017 (58 pages).

j. Tofacitinib (Rheumatoid Arthritis) NDA 203-214

Tofacitinib, used to treat rheumatoid arthritis, is a small molecule that inhibits Janus kinases. The Janus kinases are protein tyrosine kinases, which are operably linked to cytokine receptors and which are necessary for cytokines to transmit signals to the cell.[186]

i. Mechanism of Action

Tofacitinib's mechanism of action establishes a context for tofacitinib's desired effect in reducing inflammation in joints and for tofacitinib's AE of increasing risk for infections and cancer.

Naive $CD4^+$ T cells can differentiate into two phenotypes, Th1-type T cells and Th2-type T cells. Th1 cells are proinflammatory and combat infections (a good thing) and cause autoimmune diseases (a bad thing), whereas Th2 cells are anti-inflammatory and cool down the immune system. In health synovial fibroblasts express lubricating molecules, such as hyaluronan.

In rheumatoid arthritis, synovial fibroblasts express inflammatory cytokines that stimulate T cells to damage the joints. In rheumatoid arthritis, synovial fluids have higher levels of IL-6, IL-8, RANTES, and other cytokines.[187] Where synovial fibroblasts are activated by T cells or by cytokines and where these activated synovial fibroblasts express IL-6 and IL-8, tofacitinib functions to prevent this expression of IL-6 and IL-8.[188] In addition, tests with $CD4^+$ T cells show that tofacitinib inhibits expression of the Th1-type cytokine, IFN-gamma, by these cells.[189] Tofacitinib inhibits the production of chemokines, such as IP-10 and RANTES.[190] Chemokines are cytokines that attract T cells, resulting in inflammation. Also, tofacitinib inhibits expression by T cells of RANKL, resulting in lower inflammation in joints.[191] RANKL (ligand), which is expressed by a first type of cell, binds to RANK (receptor), a receptor expressed by a second type of cell, where the result is transmission of a signal to the second cell.[192]

[186] Hodge JA, et al. The mechanism of action of tofacitinib—an oral Janus kinase inhibitor for the treatment of rheumatoid arthritis. Clin. Exp. Rheumatol. 2016;34:318–28.

[187] Jones DS, et al. Profiling drugs for rheumatoid arthritis that inhibit synovial fibroblast activation. Nat. Chem. Biol. 2017;13:38–45.

[188] Hodge JA, et al. The mechanism of action of tofacitinib—an oral Janus kinase inhibitor for the treatment of rheumatoid arthritis. Clin. Exp. Rheumatol. 2016;34:318–28.

[189] Maeshima K, et al. The JAK inhibitor tofacitinib regulates synovitis through inhibition of interferon-γ and interleukin-17 production by human $CD4^+$ T cells. Arthritis Rheum. 2012;64:1790–8.

[190] Rosengren S, et al. The JAK inhibitor CP-690,550 (tofacitinib) inhibits TNF-induced chemokine expression in fibroblast-like synoviocytes: autocrine role of type I interferon. Ann. Rheum. Dis. 2012;71:440–7.

[191] LaBranche TP, et al. JAK inhibition with tofacitinib suppresses arthritic joint structural damage through decreased RANKL production. Arthritis Rheum. 2012;64:3531–42.

[192] Liu C, et al. Structural and functional insights of RANKL–RANK interaction and signaling. J. Immunol. 2012;184:6910–9.

ii. AEs of Cancer and Infections (Animal Data)

FDA's Pharmacology Review observed that tofacitinib was associated with cancer and infections. Regarding cancer in monkeys, FDA observed:

> **Lymphomas** were observed in the 9-month general toxicology study in cynomolgus monkeys ... [t]hey occurred in 3 of 8 adult monkeys dosed orally with tofacitinib ... [t]hey were not present at the lower dose ... [t]he **lymphomas** in the adult monkey study were associated with lymphocyptovirus and thought to occur due to tofacitinib mediated immune suppression allowing for viral reactivation ... [t]hese **lymphoma** findings in the non-human primate support the human clinical trial occurrences of lymphoproliferative disease.[193]

FDA distinguished between cancer arising from the direct influence of tofacitinib on the genome, which is called "genetic toxicology" or "carcinogenicity," from cancer arising from tofacitinib's mechanism of action of reducing immune response against cancers. This careful distinction was wisely set forth in FDA's comment that, "Overall, results from both the genetic toxicology and carcinogenic studies indicate a low risk of **direct drug-induced carcinogenicity** for patients, due to the large exposure margins relative to the therapeutic dose. However, this cannot be applied for **immunosuppression-associated malignancies**, since they occurred in the monkey toxicology study and in clinical trials.[194]"

This concerns cancer and infections in rats. Regarding cancer in rats, FDA observed, "The most common cause of death ... was pituitary neoplasia in males and pituitary or mammary neoplasia in females.[195]"

Regarding infections in rats, FDA observed that, "The increased mortality in males given 75 mg/kg/day [tofacitinib] was attributed, partly to **bacterial infection** attributed to tofacitinib-related **immunosuppression** ... [b]acterial infections were also a cause of death in females at 100 mg/kg/day. Six females treated with 100 mg/kg/day developed signs of Clostridium piliforme infection ... and died between weeks 15 and 22. Lowering the dose level to 75 mg/kg/day was effective at preventing any further ... disease outbreaks.[196]"

Further regarding infections in rats, FDA contemplated the Sponsor's rat infection studies but discounted the results from some of these studies, because the tofacitinib doses were at a toxic level. FDA complained that, "[t]he highest dose should be at or near the estimated maximum tolerated dose for that species.[197]" FDA's Guidance for Industry

[193] Tofacitinib (rheumatoid arthritis) NDA 203-214. Page 34 of 534-pagePharmacology Review.

[194] Tofacitinib (rheumatoid arthritis) NDA 203-214. Pages 35 and 289-295 of 534-pagePharmacology Review.

[195] Tofacitinib (rheumatoid arthritis) NDA 203-214. Page 328 of 534-pagePharmacology Review.

[196] Tofacitinib (rheumatoid arthritis) NDA 203-214. Page 327 of 534-pagePharmacology Review.

[197] Tofacitinib (rheumatoid arthritis) NDA 203-214. Page 303 of 534-pagePharmacology Review.

describes maximum tolerated dose.[198,199] As a consequence, FDA decided to accept only safety data from rats treated at or below the maximum tolerated dose. The lower doses tofacitinib still resulted in bacterial infections in the rats:

> The bacterial infections were attributed to **immunosuppression**, an ... effect of tofacitinib, but also indicated **the maximum tolerated dose had been exceeded**. Therefore, the dose level for high-dose females was lowered to 75 mg/kg/day beginning in week 19 (day 133) of the dosing phase. After the dose was decreased to 75 mg/kg/day in week 19, a few females at 75 and 30 mg/kg/day still had bacterial infections resulting in death.[200]

iii. AEs of Cancer and Infections (Human Data)

FDA observed that, "There appears to be an increased risk of lymphoma in particular ... [s]even cases of lymphoproliferative disorder occurred ... in tofacitinib-treated patients. Two of the cases occurred in highly atypical locations (CNS and breast) ... [f]ive lymphoma cases in 218 (2.3%) renal transplant patients who had received 15 mg BID.[201]"

Also, FDA acknowledged tofacitinib's risk for cancer, especially when coadministered with another immunosuppressant:

> Committee Discussion: Given the overall safety profile of the 5 mg and 10 mg dose, the committee agreed that the safety data are more favorable for the 5 mg dose. The committee agreed that there was a concern for **over-immunosuppression** and **malignancy** with the 10 mg dose, especially if given with other immunologics. However, the committee agreed that the 10 mg dose might be a viable option for refractory patients who are willing to accept the risk of treatment.[202]

Moreover, FDA acknowledged tofacitinib's risk for infections in human subjects:

> An increased risk of serious infections, including opportunistic infections, identifying a profile of tofacitinib as a major immunosuppressant ... [s]erious infections associated with tofacitinib use were common in the rheumatoid arthritis program with **pneumonia** being the most common (occurring only in tofacitinib treated patients) ... [o]pportunistic infections were not uncommon and included cases of cryptococcal infections, Pneumocystis jiroveci pneumonia, and BK virus encephalitis.[203]

[198] US Department of Health and Human Services; Food and Drug Administration; Center for Drug Evaluation and Research (CDER); Center for Biologics Evaluation and Research (CBER). Guidance for Industry M3(R2) Nonclinical Safety Studies for the Conduct of Human Clinical Trials and Marketing Authorization for Pharmaceuticals; 2010 (25 pages).

[199] US Department of Health and Human Services; Food and Drug Administration; Center for Drug Evaluation and Research (CDER). Guidance for Industry Estimating the Maximum Safe Starting Dose in Initial Clinical Trials for Therapeutics in Adult Healthy Volunteers; 2005 (27 pages).

[200] Tofacitinib (rheumatoid arthritis) NDA 203-214. Page 327 of 534-pagePharmacology Review.

[201] Tofacitinib (rheumatoid arthritis) NDA 203-214. Page 4 of 303-pageMedical Review.

[202] Tofacitinib (rheumatoid arthritis) NDA 203-214. Pages 207 and 208 of 303-pageMedical Review.

[203] Tofacitinib (rheumatoid arthritis) NDA 203-214. Page 4 of 303-pageMedical Review.

Regarding the "BK virus" materializing with tofacitinib treatment, it is interesting that BK virus is genetically similar to JC virus, which is notorious as an opportunistic infection arising with natalizumab.[204,205]

iv. Tofacitinib's Extent of Impairing White Blood Cell Count and Function

The Sponsor measured tofacitinib's influence on various types of T cells, that is, on $CD4^+$ T cells, $CD8^+$ T cells, naive $CD8^+$ T cells, central $CD8^+$ T cells, effector memory $CD8^+$ T cells, B cells, and NK cells. $CD4^+$ T cells function to stimulate $CD8^+$ T cells, and in this way $CD4^+$ T cells are nature's adjuvant. $CD8^+$ T cells are the immune cells that actively kill infections and tumors. NK cells also actively kill infections and tumors, and they do so in combination with an antibody in a process called antibody-dependent cell cytotoxicity.[206] B cells are the immune cells that manufacture antibodies and release them into the bloodstream. Regarding the results from the monkey study, the Sponsor found that there were, "Decreased absolute total T cells and $CD4^+$ and $CD8^+$ T cell subsets occurred in weeks 13 through 39 of the dosing phase in both males and females of the 10 mg/kg/day dose group.[207]"

The Sponsor was careful to measure T-cell recovery, NK cell recovery, and B cell recovery in monkeys when tofacitinib treatment was stopped, writing, "At the end of the dosing period (day 270) ... [p]artial to complete recovery of both total T cell and $CD4^+$ T cell subsets occurred ... [p]artial to complete recovery of NK cells was observed for males and females in the 2 and 10 mg/kg/day dose groups during the recovery phase.[208]" FDA's Pharmacology Review also observed that, "There were no effects on B cells at any dose or gender throughout the dosing period." Thus, by the criteria used, the concept of B-cell recovery was not relevant.

The Sponsor used a standard test for lymphocyte function, namely, the ability of a cross-linking agent to stimulate the proliferation of white blood cells. The cross-linking agent was the lectin, concanavalin A (ConA). ConA has several oligosaccharide-binding sites, each of which can bind strongly to an oligosaccharide, thereby facilitating cross-linking. Any given ConA protein can simultaneously bind to several oligosaccharides on the surface of a T-cell and cross link the associated glycoproteins, where this cross-linking transmits a signal to the cell that results in proliferation.[209,210]

[204] Osborn JE, et al. Comparison of JC and BK human papovaviruses with simian virus 40: DNA homology studies. J. Virol. 1976;19:675–84.

[205] Pinto M, Dobson S. BK and JC virus: a review. J. Infect. 2014;68 (Suppl 1) S2–8.

[206] Tu MM, et al. Licensed and unlicensed NK cells: differential roles in cancer and viral control. Front. Immunol. 2016;7:166 (11 pages).

[207] Tofacitinib (rheumatoid arthritis) NDA 203-214. Pages 276–278 of 534-pagePharmacology Review.

[208] Tofacitinib (rheumatoid arthritis) NDA 203-214. Pages 276–278 of 534-pagePharmacology Review.

[209] Fillingame RH, et al. Increased cellular levels of spermidine or spermine are required for optimal DNA synthesis in lymphocytes activated by concanavalin A. Proc. Natl. Acad. Sci. 1975;72:4042–5.

[210] Henis YI. Mobility modulation by local concanavalin A binding. Selectivity toward different membrane proteins. J. Biol. Chem. 1984;259:1515-9.

TABLE 9.1 Influence of Tofacitinib on Antibody Production in Response to Vaccinating With KLH

Antibody Class	Gender of Monkey	Tofacitinib (10 mg/kg/day) (+) or Control Salt Solution (−)	Level of Anti-KLH Antibody Found on Day 14[a]
IgM	Male	+	252
		−	35
	Female	+	441
		−	14
IgG	Male	+	284
		−	−30
	Female	+	195
		−	−5

[a]The values were all corrected for values of anti-KLH antibody that were found on blood at baseline, that is, prior to vaccinating with KLH.
From Tofacitinib (rheumatoid arthritis) NDA 203-214. Pages 280–281 of 534-page Pharmacology Review.

The source of white blood cells was monkey blood. Tofacitinib did not influence function by this criterion and the Sponsor concluded, "**tofacitinib did not reduce functionality of the remaining T cells** when tested in vitro by proliferation in response to ConA stimulation."

v. Tofacitinib and Antibody Response to a Foreign Antigen

FDA's Pharmacology Review detailed tofacitinib's influence of antibody production, in response to administrating a foreign antigen to animals. The foreign antigen was keyhole limpet hemocyanin (KLH). Table 9.1 discloses the Sponsor's data. It shows that tofacitinib blocked the mouse's ability to generate anti-KLH antibodies. The profound drop in antibody production was found in the IgG immunoglobulin class and in the IgM immunoglobulin class.

vi. KLH Vaccination of Mice

Onda et al.[211] provides additional information on study design and KLH, where the goal is to measure AEs taking the form of immune system inhibition by any study drug. Onda et al. determined the influence of tofacitinib on immune response in mice. The measured parameters of immune response are outlined by the following points:

- **Wild-type mice and KLH vaccination.** Antibody response by wild-type mice to injected KLH.
- **JAK3 knockout mice and KLH vaccination.** Antibody response by mice to injected KLH, where the mice had been genetically deleted in tofacitinib's target. In other words, the mice were Janus kinase-3 (JAK3) knockout mice.

[211] Onda, et al. Tofacitinib suppresses antibody responses to protein therapeutics in murine hosts. J. Immunol. 2014;193:48–55.

- **Immunoglobulin class.** Determining antibody response to KLH, where response is determined for the various immunoglobulin classes (IgG1, IgG2a, IgG2b, IgG3, and IgM).
- **B-cell number.** Determining B-cell number in mice treated with or without tofacitinib.
- **B-cell development.** Determining the presence of B cells at various developmental stages, in mice treated with or without tofacitinib.
- **In vitro B0cell stimulatability.** Determining B-cell proliferation, with or without tofacitinib treatment, where proliferation was stimulated with lipopolysaccharide plus IL-4.

KLH is conventionally used to test immune response in patients receiving an immunosuppressant.[212] KLH antigen has a high immunogenicity in normal human subjects, where both antibody production and T-cell activation and proliferation occur. Also, the fact that KHL is a protein found in an inedible marine mollusc (the limpet), with consequent unlikelihood of prior immune exposure, has led to the use of immunization with KLH as a standard test of immune responsiveness in humans.[213,214]

FDA's comments regarding infections, lymphoma, immunosuppression, and lowered white blood cell counts found a place on the package label, as shown below.

Package label. The package label for tofacitinib (Xaljanz®) has a Black Box Warning that warns against infections and cancers:

> **BOXED WARNING**: SERIOUS INFECTIONS AND MALIGNANCY... Serious infections leading to hospitalization or death, including **tuberculosis** and bacterial, invasive fungal, viral, and other opportunistic infections, have occurred in patients receiving XELJANZ ... Prior to starting XELJANZ, perform a test for latent tuberculosis; if it is positive, start treatment for tuberculosis prior to starting XELJANZ ... **Lymphoma** and other malignancies have been observed in patients treated with XELJANZ.[215]

The Indications and Usage section recommended against coadministering immunosuppressants.

> **INDICATIONS AND USAGE** ... Limitations of Use: Use of XELJANZ in combination with ... potent immunosuppressants such as azathioprine and cyclosporine is not recommended.[216]

[212] McMahan ZH, Bingham O. Effects of biological and non-biological immunomodulatory therapies on the immunogenicity of vaccines in patients with rheumatic diseases. Arthritis Res. Ther. 2014;16:506 (10 pages).

[213] Swaminathan A, et al. Keyhole limpet haemocyanin - a model antigen for human immunotoxicological studies. Br. J. Clin. Pharmacol. 2014;78:1135–42.

[214] Moroz LA, et al. Normal human IgG with antibody activity for keyhole limpet haemocyanin. Immunology. 1973;25:441–9.

[215] Package label. XELJANZ® (tofacitinib) tablets for oral administration. November 2012 (27 pages). NDA 203-214.

[216] Package label. XELJANZ® (tofacitinib) tablets, for oral use. December 2015 (30 pages).

The Adverse Reactions section disclosed that tofacitinib (Xeljanz®) reduces white blood cell counts, and that lowered white blood cell counts increased risk for infections:

> **ADVERSE REACTIONS** ... Lymphopenia. In the controlled clinical trials, confirmed decreases in absolute lymphocyte counts below 500 cells/mm3 occurred in 0.04% of patients for the 5 mg twice daily and 10 mg twice daily XELJANZ groups combined during the first 3 months of exposure. Confirmed lymphocyte counts less than 500 cells/mm3 were associated with an increased incidence of treated and serious infections.[217]

III. DRUG-INDUCED HYPERSENSITIVITY REACTIONS

This discloses the basis, as set forth in FDA's Medical Reviews, for package label warnings against drug-induced hypersensitivity reactions. The reactions are described in this order:

1. Anaphylaxis
2. Stevens-Johnson syndrome (SJS) and toxic epidermal necrolysis (TEN)
3. Erythema multiforme

Generally, drug-induced hypersensitivity reactions involve the participation of eosinophils[218,219,220,221,222] and mast cells.[223,224,225,226,227]

[217] Package label. XELJANZ® (tofacitinib) tablets, for oral use. December 2015 (30 pages).

[218] Kobayashi T, et al. Human eosinophils recognize endogenous danger signal crystalline uric acid and produce proinflammatory cytokines mediated by autocrine ATP. J. Immunol. 2010;184:6350–8.

[219] Burnham ME, et al. Human airway eosinophils exhibit preferential reduction in STAT signaling capacity and increased CISH expression. J. Immunol. 2013;191:2900–6.

[220] Yawalkar N, et al. Evidence for a role for IL-5 and eotaxin in activating and recruiting eosinophils in drug-induced cutaneous eruptions. 2000;106:1171–6.

[221] Roujeau JC, et al. Clinical heterogeneity of drug hypersensitivity. Toxicology 2005;209:123–9.

[222] Pichler WJ, et al. Cellular and molecular pathophysiology of cutaneous drug reactions. Am. J. Clin. Dermatol. 2002;3:229–38.

[223] Shade KT, et al. A single glycan on IgE is indispensable for initiation of anaphylaxis. J. Exp. Med. 2015;212:457–67.

[224] McNeil BD, et al. Identification of a mast-cell-specific receptor crucial for pseudo-allergic drug reactions. Nature 2015;519:237–41.

[225] Schnyder B, Pichler WJ. Mechanisms of drug-induced allergy. Mayo Clin. Proc. 2009;84:268–72.

[226] Cernadas JR, et al. General considerations on rapid desensitization for drug hypersensitivity–a consensus statement. Allergy 2010;65:1357–66.

[227] Pichler WJ, et al. Drug hypersensitivity reactions: pathomechanism and clinical symptoms. Med. Clin. North Am. 2010;94:645–6.

a. Anaphylaxis Background

Anaphylaxis in humans is most commonly associated with production of IgE antibodies. Mast cells and basophils express, on the surface of their plasma membranes, the IgE receptor, FcεR1. Mast cells and basophils release mediators of anaphylaxis, including histamine, tryptase, carboxypeptidase A, prostaglandin D2, leukotrienes, and platelet-activating factor following activation.[228]

IgE antibodies mediate allergic diseases. IgE specific for innocuous environmental antigens, food proteins, and therapeutic proteins can bind to mast cells expressing the IgE receptor, FcεRI. In IgE-mediated anaphylaxis, the reaction requires an initial exposure to an antigen, a sensitization period, and subsequent reexposure to the antigen.[229]

Where the allergen, for example, a food protein or a therapeutic protein, comes in contact with a mast cell harboring IgE bound to the FcεRI of the plasma membrane of mast cells, the result is cross-linking of the membrane-bound IgE/FcεRI complex, with the consequent release of inflammatory mediators. Cross-linking by allergens activates the cells causing release of mediators that induce[230]:

- Vasodilation
- Vascular permeability
- Smooth muscle contractility

Criteria for diagnosing anaphylaxis include an acute onset of illness (within minutes to hours), skin or mucosal involvement (hives, pruritus, or flushing), and either respiratory symptoms (dyspnea, wheezing, or **bronchospasm**) or a circulatory component with reduced blood pressure.[231] An account of an anaphylaxis reaction to a therapeutic antibody (cetuximab) has been described.[232] To provide yet another list of events that are part of anaphylaxis, anaphylaxis presents as **angioedema**, hypotension, cardiac arrhythmias, **respiratory tract obstruction**, abdominal pain, vomiting, and diarrhea, each of which may occur singly or in combination.[233] It is thus the case that any warning on a package label about angioedema, bronchospasms, or variations in these medical terms, constitute warnings against anaphylaxis.

[228] Gleich GJ, Leiferman KM. Anaphylaxis: implications of monoclonal antibody use in oncology. Oncology (Williston Park). 2009;23 (2 Suppl 1):7–13.

[229] Marquardt DL, Wasserman SI. Mast cells in allergic diseases and mastocytosis. West. J. Med. 1982;137:195–212.

[230] Shade KT, et al. A single glycan on IgE is indispensable for initiation of anaphylaxis. J. Exp. Med. 2015;212:457–67.

[231] Gleich GJ, Leiferman KM. Anaphylaxis: implications of monoclonal antibody use in oncology. Oncology (Williston Park) 2009;23 (2 Suppl 1):7–13.

[232] Chung CH, et al. Cetuximab-induced anaphylaxis and IgE specific for galactose-alpha-1,3-galactose. N. Engl. J. Med. 2008;358:1109–17.

[233] Marquardt DL, Wasserman SI. Mast cells in allergic diseases and mastocytosis. West. J. Med. 1982;137:195–212.

b. Background on Stevens-Johnson Syndrome

Stevens-Johnson syndrome (SJS)[234] and a related disorder, TEN, are rare disorders that can be caused by allopurinol, trimethoprim-sulfamethaxole, aminopenicillins, cephalosporins, quinolones, carbamazepine, phenytoin, phenobarbitol, and nonsteroidal antiinflammatory drugs s of the oxicam type (meloxicam, piroxicam, and tenoxicam).[235,236,237]. SJS and TEN involve blisters, sloughing off of sheets of skin,[238] full-thickness epidermal necrolysis, and apoptosis of keratinocytes.[239] SJS has a mortality of 1%-5% and involves detachment of up to 10% of the epidermis (total body skin area).[240,241,242] TEN has a mortality of 25%–30% and can involve detachment of than 30% of the epidermis. A transitional SJS/TEN has been defined as an epidermal detachment between 10% and 30%.[243]

SJS and TEN can have consequences similar to those of severe burns, that is, fluid losses, electrolyte imbalances, and infections.[244] Photographs of skin lesions of SJS have been published for SJS caused by trimethoprim and sulfamethoxazole,[245]

[234] Stevens AM, Johnson FC. A new eruptive fever associated with stomatitis and ophthalmia. Am. J. Dis. Child. 1922;24:526–33.

[235] Harr T, French LE. Toxic epidermal necrolysis and Stevens-Johnson syndrome. Orphanet. J. Rare Dis. 2010;5:39 (11 pages).

[236] Ghislain PD, Roujeau JC. Treatment of severe drug reactions: Stevens-Johnson syndrome, toxic epidermal necrolysis and hypersensitivity syndrome. Dermatol. Online J. 2002;8:5.

[237] Mockenhaupt M, Viboud C, Dunant A, et al. Stevens-Johnson syndrome and toxic epidermal necrolysis: assessment of medication risks with emphasis on recently marketed drugs. The EuroSCAR-study. J. Invest. Dermatol. 2008;128:35–44.

[238] De Rojas MV, et al. The natural history of Stevens Johnson syndrome: patterns of chronic ocular disease and the role of systemic immunosuppressive therapy. Br. J. Ophthalmol. 2007;91:1048–53.

[239] Harr T, French LE. Toxic epidermal necrolysis and Stevens-Johnson syndrome. Orphanet. J. Rare Dis. 2010;5:39 (11 pages).

[240] Harr T, French LE. Toxic epidermal necrolysis and Stevens-Johnson syndrome. Orphanet. J. Rare Dis. 2010;5:39 (11 pages).

[241] Ghislain PD, Roujeau JC. Treatment of severe drug reactions: Stevens-Johnson syndrome, toxic epidermal necrolysis and hypersensitivity syndrome. Dermatol. Online J. 2002;8:5.

[242] Mockenhaupt M, Viboud C, Dunant A, et al. Stevens-Johnson syndrome and toxic epidermal necrolysis: assessment of medication risks with emphasis on recently marketed drugs. The EuroSCAR-study. J. Invest. Dermatol. 2008;128:35–44.

[243] Ghislain PD, Roujeau JC. Treatment of severe drug reactions: Stevens-Johnson syndrome, toxic epidermal necrolysis and hypersensitivity syndrome. Dermatol. Online J. 2002;8:5.

[244] Harr T, French LE. Toxic epidermal necrolysis and Stevens-Johnson syndrome. Orphanet. J. Rare Dis. 2010;5:39 (11 pages).

[245] Assaad D, et al. Toxic epidermal necrolysis in Stevens-Johnson syndrome. Can. Med. Assoc. J. 1978;118:154–6.

lamotrigine,[246] tetrazepam,[247] penicillin,[248] vaccines,[249] and by recombinant antibodies.[250,251,252] Eye involvement can include conjunctivitis, erosion of the cornea, and blindness.[253,254,255]

SJS and TEN have an immune component because of evidence that these disorders are triggered by activated T cells and by signaling via the CD40L to CD40 signaling pathway.[256] Further evidence of immune system involvement is from data showing the production by $CD8^+$ T cells from skin lesions from SJS patients and TEN patients of toxins that kill keratinocytes. These toxins from $CD8^+$ T cells are granulysin and perforin.[257]

An immune component of SJS and TEN was further shown by the association of specific HLA types in patients with these disorders. For example, associations have been found between carbamazepine-induced SJS/TEN with HLA-B*1502 allele, carbamazepine and HLA-A*3101 and HLA-B*1511, phenytoin and HLA-B*1502, allopurinol and HLA-B*5801.[258]

c. Erythema Multiform Background

Erythema multiform involves localized lesions of the skin and mucosal surfaces, in contrast to SJS, which occurs over wide areas of the skin. Erythema multiform has been found

[246] Wetter DA, Camilleri MJ. Clinical, etiologic, and histopathologic features of Stevens-Johnson syndrome during an 8-year period at Mayo Clinic. Mayo Clin. Proc. 2010;85:131–8.

[247] Torres MJ, et al. Nonimmediate allergic reactions induced by drugs: pathogenesis and diagnostic tests. J. Investig. Allergol. Clin. Immunol. 2009;19:80–90.

[248] Cram DL. Life-threatening dermatoses. Calif. Med. 1973;118:5–12.

[249] Chopra A, et al. Stevens-Johnson syndrome after immunization with smallpox, anthrax, and tetanus vaccines. Mayo Clin. Proc. 2004;79:1193–6.

[250] Brown BA, Torabi M. Incidence of infusion-associated reactions with rituximab for treating multiple sclerosis: a retrospective analysis of patients treated at a US centre. Drug Saf. 2011;34:117–123.

[251] Scheinfeld N. A review of rituximab in cutaneous medicine. Dermatol. Online J. 2006;12:3.

[252] Salama M, Lawrance IC. Stevens-Johnson syndrome complicating adalimumab therapy in Crohn's disease. World J Gastroenterol. 2009;15:4449–52.

[253] De Rojas MV, et al. The natural history of Stevens Johnson syndrome: patterns of chronic ocular disease and the role of systemic immunosuppressive therapy. Br. J. Ophthalmol. 2007;91:1048–53.

[254] Tsubota K, et al. Treatment of severe ocular-surface disorders with corneal epithelial stem-cell transplantation. N. Engl. J. Med. 1999;340:1697–703.

[255] De Rojas MV, et al. The natural history of Stevens Johnson syndrome: patterns of chronic ocular disease and the role of systemic immunosuppressive therapy. Br. J. Ophthalmol. 2007;91:1048–53.

[256] Caproni M, et al. The CD40/CD40 ligand system is expressed in the cutaneous lesions of erythema multiforme and Stevens-Johnson syndrome/toxic epidermal necrolysis spectrum. Br. J. Dermatol. 2006;154:319–24.

[257] Iwai S, et al. Distinguishing between erythema multiforme major and Stevens-Johnson syndrome/toxic epidermal necrolysis immunopathologically. J. Dermatol. 2012;39:781–6.

[258] Yacoub MR, et al. Drug induced exfoliative dermatitis: state of the art. Clin. Mol. Allergy. 2016;14:9 (12 pages).

with various drugs, such as adalimumab,[259] alectinib,[260] imatinib,[261] bupropion,[262] crizotinib,[263] as well as with rofecoxib, penicillin, ciprofloxacin, and metformin.[264] Immune contribution to erythema multiform includes infiltration by $CD8^+$ T cells, $CD4^+$ T cells, and macrophages.[265]

This quotation reveals that erythema multiforme has an immune component, "There is evidence for $CD8^+$ T cell involvement in morbilliform and bullous drug eruptions that may represent a $CD8^+$ T cell ... hypersensitivity response in the skin. The histology of drug induced **erythema multiforme** is very similar to graft-versus-host disease, which is known to be mediated by ... T cells. In both graft-versus-host disease and **erythema multiforme**, the infiltrating cells of the epidermis are predominantly $CD8^+$ T cells.[266]"

d. Cytokine Storm

Hypersensitivity reactions to therapeutic proteins have been reviewed.[267,268] A severe hypersensitivity reaction, which is provoked by some recombinant antibodies, is the "cytokine storm." When it occurs, the cytokine storm occurs within 2 hours of administering the antibody. The cytokine storm involves expression and release into the bloodstream of inflammatory cytokines, such as TNF-α, IFN-γ, and IL-2, where the consequent AEs include fever, chills, nausea, and life-threatening organ failure. Where a cytokine storm is anticipated, it can be mitigated by pretreating the patient with corticosteroids.[269,270,271]

[259] Baillis B, Maize JC. Treatment of recurrent erythema multiforme with adalimumab as monotherapy. JAAD Case Rep. 2017;3:95–7.

[260] Kimura T, et al. Alectinib-induced erythema multiforme and successful rechallenge with alectinib in a patient with anaplastic lymphoma kinase-rearranged lung cancer. Case Rep. Oncol. 2016;9:826–32.

[261] Lee MK, et al. Imatinib mesylate-induced erythema multiforme: recurrence after rechallenge with 200 mg/day imatinib. Ann. Dermatol. 2015;27:641–3.

[262] Evrensel A, Ceylan ME. Bupropion-induced erythema multiforme. Ann. Dermatol. 2015;27:334–5.

[263] Sawamura S, et al. Crizotinib-associated erythema multiforme in a lung cancer patient. Drug Discov. Ther. 2015;9:142–3.

[264] Lamoreux MR, et al. Erythema multiforme. Am. Fam. Physician. 2006;74:1883–8.

[265] Yacoub MR, et al. Drug induced exfoliative dermatitis: state of the art. Clin. Mol. Allergy. 2016;14:9 (12 pages).

[266] Kalish RS, Askenase PW. Molecular mechanisms of $CD8^+$ T cell-mediated delayed hypersensitivity: implications for allergies, asthma, and autoimmunity. J. Allergy Clin. Immunol. 1999;103:192–9.

[267] Song S, et al. Understanding the supersensitive anti-drug antibody assay: unexpected high anti-drug antibody incidence and its clinical relevance. J. Immunol. Res. 2016;2016:3072586. doi: 10.1155/2016/3072586.

[268] Baldo BA. Adverse events to monoclonal antibodies used for cancer therapy: focus on hypersensitivity responses. OncoImmunology. 2013;2:e26333 (15 pages).

[269] Vessiller S, et al. Cytokine release assays for the prediction of therapeutic mAb safety in first-in man trials--whole blood cytokine release assays are poorly predictive for TGN1412 cytokine storm. J. Immunol. Methods. 2015;424:43–52.

[270] Eastwood D, et al. Severity of the TGN1412 trial disaster cytokine storm correlated with IL-2 release. Br. J. Clin. Pharmacol. 2013;76:299–315.

[271] Brady JL, et al. Preclinical screening for acute toxicity of therapeutic monoclonal antibodies in a hu-SCID model. Clin. Trans. Immunology. 2014;3:e29 (7 pages).

TABLE 9.2 Classes of Hypersensitivity Reactions

Hypersensitivity Type	Manifested by	Mediated by
Type I	Anaphylaxis; angioedema	IgE
Type II	Hemolytic anemia	IgE/IgM cytotoxic reactions
Type III	Vasculitis	IgG/IgM immune complexes
Type IV	Psoriasis, SJS	T cells

From Baldo BA. Adverse events to monoclonal antibodies used for cancer therapy: focus on hypersensitivity responses. OncoImmunology. 2013;2:e26333; Rive CM, et al. Testing for drug hypersensitivity syndromes. Clin. Biochem. Rev. 2013;34:15–38.

e. Classes of Hypersensitivity Reactions

Small molecule drugs, such as penicillin or tamoxifen, and therapeutic proteins can result in AEs that are hypersensitivity reactions. Baldo[272] and Rive et al.[273] described four types of hypersensitivity reactions. The four classes are known as the Gell and Coombs classes: Type I, IgE antibody-mediated reactions; Type II, cytotoxic reactions; Type III, immune-complex mediated hypersensitivities; and Type IV, delayed, cell-mediated responses.[274] Table 9.2 shows the four types of hypersensitivity reactions, of which the first type includes anaphylaxis:

IV. FDA's ANALYSIS OF HYPERSENSITIVITY REACTIONS

a. Introduction

As part of the flow of the above narratives on AEs involving immunity, hypersensitivity reactions were shown to be caused by abatacept, denosumab, and natalizumab. The following continues the account of drug-induced hypersensitivity reactions for the drugs:

- Brentuximab vedotin
- Bupropion
- Carbamazepine
- Certolizumab pegol
- Eslicarbazepine
- Sebelipase alfa

[272] Baldo BA. Adverse events to monoclonal antibodies used for cancer therapy: focus on hypersensitivity responses. OncoImmunology. 2013;2:e26333.

[273] Rive CM, et al. Testing for drug hypersensitivity syndromes. Clin. Biochem. Rev. 2013;34:15–38.

[274] Coombs RRA, Gell PGH. Classification of allergic reactions responsible for clinical hypersensitivity and disease. In: Gell PGH, Coombs RRA, Lachmann PJ, eds. Clinical Aspects of Immunology. Oxford: Blackwells, 1975: 761–81.

b. Brentuximab Vedotin (Hodgkin's Lymphoma) BLA 125-388

FDA's Medical Review disclosed that one subject experienced the AE of Stevens-Johnson syndrome (SJS). FDA summarized the AEs, writing that, "The major safety issues identified by the applicant include peripheral neuropathy, neutropenia, infusion reactions, and one case of Stevens-Johnson syndrome.[275]" The patient narrative for this one subject referred to SJS and to TEN:

> Adverse Events Leading to Treatment Discontinuation. Of the 102 patients ... 21 (21%) experienced an adverse event ... that resulted in treatment discontinuation ... Patient 10006-0057 discontinued treatment due to **Stevens-Johnson syndrome (SJS)**. The 37-year old patient developed a maculopapular rash two weeks after receiving ... brentizumab. The rash eventually involved 90% of the skin, and was associated with 30% desquamation. Skin biopsy showed toxic epidermal necrolysis (TEN). Patient recovered but had to discontinue study treatment due to this event. This event was confounded by concomitant use of naproxen (an NSAID).[276]

Regarding infusion reactions such as erythema and anaphylaxis, FDA's Medical Review observed that, "Infusion reactions were noted in 14% of the patients. The most common of these were chills (5 patients), dyspnea, nausea, and pruritis (4 patients each), cough (3 patients), and erythema, flushing, pyrexia, and throat tightness (2 patients each) ... [a]ll infusion reactions were Grade 1 or 2 in severity. No instances of anaphylaxis occurred during the study.[277]"

Package label. FDA's comments about SJS found a place on the package label for brentuximab vedotin (Adcetris®). The Warnings and Precautions section read:

> **WARNINGS AND PRECAUTIONS** ... Anaphylaxis and infusion reactions: If an infusion reaction occurs, interrupt the infusion. If anaphylaxis occurs, immediately discontinue the infusion ... Serious dermatologic reactions: Discontinue if Stevens-Johnson syndrome or toxic epidermal necrolysis occurs.[278]

c. Bupropion (Depression) NDA 022-108 (Example of a Section 505(b)(2) submission)

The Sponsor's submission for bupropion was by way of NDA 022-108. The submission process provides the interesting situation where the Sponsor's submission, and the data reviewed by FDA during its drug-approval process, was based only partially on studies conducted by the Sponsor. According to FDA's Medical Review, "This is a 505(b)(2) application for an extended release (once daily) formulation of bupriopion.[279]"

Regarding 505(b)(2) applications, FDA's Guidance for Industry teaches that, "This provision expressly permits FDA to rely, for approval of an NDA, on data not developed by the applicant ... Section 505(b)(2) ... permits reliance for such approvals on literature or

[275] Brentuximab vedotin (Hodgkin's lymphoma) BLA 125-388. Pages 56 of 93-page Medical Review.

[276] Brentuximab vedotin (Hodgkin's lymphoma) BLA 125-388. Pages 68, 76, and 77 of 93-page Medical Review.

[277] Brentuximab vedotin (Hodgkin's lymphoma) BLA 125-388. Pages 75 and 76 of 93-page Medical Review.

[278] Package label. ADCETRIS™ (brentuximab vedotin) for injection, for intravenous use. September 2016 (26 pages).

[279] Bupropion (depression) NDA 022-108. Page 2 of 20-page Medical Review.

on an Agency finding of safety and/or effectiveness for an approved drug product ... [a] 505(b)(2) application is one for which one or more of the investigations relied upon by the applicant for approval were not conducted by or for the applicant and for which the applicant has not obtained a right of reference or use from the person by or for whom the investigations were conducted.[280]"

The above commentary appears to account for the fact that the amount of information on the package label is far beyond that which is disclosed in FDA's Medical Review. Turning to the information on immune-related AEs, as described in FDA's Medical Review for bupropion (Aplenzin®), we learn of only one type of AE with an immune component (rash). FDA observed that:

> Adverse events. No deaths or serious adverse events were reported ... [t]here were two discontinuations due to ... **rash** ... [t]he most commonly reported adverse events not leading to discontinuations were headache (6.3%), dizziness (5.6%), **rash** (2.1%), and puritus (2.1%). These adverse events were consistent with adverse events seen with the Reference Listed Drug (RLD) ... [b]ased on previous findings with bupropion, all of the adverse events listed above could be related to treatment with bupropion.[281]

Reference Listed Drug is used in FDA submissions where the Sponsor is seeking to gain FDA approval of a new drug formulation that is equivalent to another drug that had already been approved by FDA review.[282]

Package label. FDA's concern for hypersensitivity reactions for bupropion (Aplenzin®) found a place on the package label. The package label, as issued on the date of the Approval Letter (April 2008), contained more information about immune-related AEs than that disclosed by FDA's Medical Review. The Warnings and Precautions section recited:

> **WARNINGS AND PRECAUTIONS** ... Allergic Reactions. Anaphylactoid/ anaphylactic reactions characterized by symptoms such as **pruritus, urticaria, angioedema, and dyspnea** requiring medical treatment have been reported in clinical trials with bupropion. In addition, there have been rare spontaneous postmarketing reports of **erythema multiforme**, **Stevens-Johnson syndrome**, and **anaphylactic shock** associated with bupropion. A patient should stop taking APLENZIN and consult a doctor if experiencing allergic or anaphylactoid/ anaphylactic reactions (e.g., skin rash, pruritus, hives, chest pain, edema, and shortness of breath) during treatment. Arthralgia, myalgia, and fever with rash and other symptoms suggestive of **delayed hypersensitivity** have been reported in association with bupropion. These symptoms may resemble serum sickness.[283]

[280] US Department of Health and Human Services; Food and Drug Administration; Center for Drug Evaluation and Research (CDER). Guidance for Industry. Applications Covered by Section 505(b)(2); 1999 (12 pages).

[281] Bupropion (depression) NDA 022-108. 20-page Medical Review.

[282] US Department of Health and Human Services; Food and Drug Administration; Center for Drug Evaluation and Research (CDER). Guidance for Industry Bioequivalence Studies with Pharmacokinetic Endpoints for Drugs Submitted Under an ANDA; 2013 (20 pages).

[283] Package label. APLENZIN (bupropion hydrobromide) Tablet, Film Coated, Extended Release for oral use; April 2008 (27 pages).

d. Carbamazepine (Epilepsy) NDA 016-608 (Biomarker for Stevens-Johnson syndrome)

Carbamazepine is distinguished in that the package label's information on an immune-related adverse event (SJS) revealed increased risk for SJS for patients expressing a particular biomarker, that is, the HLA-B*1502 biomarker. This biomarker is used for assessing risk for certain populations in Asia. The published literature details the association of this and other HLA biomarkers with SJS.[284,285,286,287]

FDA's Medical Review for carbamazepine acquired a prediction of immune-related AEs from the package label of two related drugs, namely, opipramol and imipramine. The study drug (carbamazepine), opipramol, and imipramine, are all tricyclic antidepressants, and are thus related by structure and by biological function.[288]

FDA compared the AEs found with carbamazepine with those of opipramol and imipramine and observed that immune-related AEs for the related drugs included dermatitis and SJS, "The most important related drugs are Ensidon (Opipramol) and Imipramine (Tofranil). Ensidon, like Tegretol ... is a tranquilizer and mood elevator drug ... [t]he toxic reactions include dizziness, tremor, **dermatitis**, leucopenia, altered hepatic functions, alopecia, **Stevens-Johnson syndrome**, etc., reactions similar to some of those seen with Imipramine and Tegretol.[289]"

FDA's Medical Review was published in 1968, at the time FDA granted approval of the drug. As shown below, the 1968 package label does, in fact, disclose the risk for immune-related AEs, including erythema multiforme, erythematous rashes, and SJS. Also shown below is the corresponding package label information from 50 years later (the year 2015), a time when the field of genomic markers based on DNA sequences came into being.

Package label. The package label for carbamazepine (Tegretol®) published at the time of FDA approval (March 1968) contained information on immune-related AEs in the Adverse Reactions section:

> **ADVERSE REACTIONS.** If adverse reactions are of such severity that the drug must be discontinued, the physician must be aware that abrupt discontinuation of any anticonvulsant drug ... may lead to seizures ... Skin. Pruritic and **erythematous rashes**, urticaria, **Stevens Johnson syndrome**, photosensitivity reactions, alteration in skin pigmentation, exfoliative dermatitis ... **erythema multiforme** ... discontinuation may be necessary.[290]

[284] Ou GH, et al. A study of HLA-B*15:02 in 9 different Chinese ethnics: Implications for carbamazepine related SJS/TEN. HLA. 2017;89:225–9.

[285] Khor AH, et al. HLA-A*31: 01 and HLA-B*15:02 association with Stevens-Johnson syndrome and toxic epidermal necrolysis to carbamazepine in a multiethnic Malaysian population. Pharmacogenet. Genomics 2017;27:275–8.

[286] Maekawa K, et al. Development of a simple genotyping method for the HLA-A*31:01-tagging SNP in Japanese. Pharmacogenomics 2015;16:1689–99.

[287] Amstutz U, et al. Recommendations for HLA-B*15:02 and HLA-A*31:01 genetic testing to reduce the risk of carbamazepine-induced hypersensitivity reactions. Epilepsia 2014;55:496–506.

[288] Aherne GW, et al. The radioimmunoassay of tricyclic antidepressants. Br. J. Clin. Pharmac. 1976;3:561–5.

[289] Carbamazepine (epilepsy) NDA 016-608. Page 1 of first of two pdf files of Medical Review. The first pdf file is 13 pages, and the second pdf file is 6 pages.

[290] Package label. Tegretol® carbamazepine USP. Tablets of 200 mg. March 1968 (4 pages) NDA 016-608.

A more recent package label (August 2015) for carbamazepine (Tegretol®) has a Black Box Warning identifying SJS and by referring to a subgroup of patients at increased risk for SJS. The subgroup was identified by a biomarker that is a member of a family of proteins of the immune system (HLA protein family):

> **BOXED WARNING**. Serious and sometimes fatal dermatologic reactions, including **toxic epidermal necrolysis (TEN)** and **Stevens-Johnson syndrome** (SJS), have been reported during treatment with Tegretol®. These reactions have been estimated to occur in 1 to 6 per 10,000 new users in countries with mainly Caucasian populations, but the risk in some Asian countries is estimated to be about 10 times higher. Studies in patients of Chinese ancestry have found a strong association between the risk of developing SJS/TEN and the presence of HLA-B*1502, an inherited allelic variant of the HLA-B gene. HLA-B*1502 is found almost exclusively in patients with ancestry across broad areas of Asia.[291]

Although the Black Box Warning mentioned only the HLA-B*1502 biomarker, the Full Prescribing Information on the same package label provided further details on the HLA-B*1502 biomarker and on HLA-B*3101 biomarker. HLA-B*1502 is found in over 15% of people in Hong Kong, Thailand, and Malaysia, while HLA-B*3101 is found in over 15% of Japanese, Native Americans, and southern India.

e. Certolizumab Pegol (Crohn's Disease) BLA 125-160

Certolizumab pegol is an antibody conjugated to PEG. Certolizumab is for treating Crohn's disease. The antibody binds to TNF-α. FDA commented on hypersensitivity reactions associated with certolizumab pegol. One of the observations was based on drug class analysis, that is, on hypersensitivity AEs named on package labels of other drugs in the same class. The relevant drug class is TNF blockers.

FDA recommended that these AEs should be included on certolizumab pegol's package label, stating that, "A section describing the **skin reactions** seen in postmarketing experiences with other TNF blockers should be included as adverse event information.[292]"

In addition, FDA referred to data on hypersensitivity from the Sponsor's own clinical studies, writing, "Crohn's disease safety population was searched for events occurring within two hours of certolizumab pegol injection, that are commonly associated with **acute hypersensitivity**. A total of 35 subjects in the ... Crohn's disease population experienced events that fit in this category.[293]"

Package label. FDA's requirement to include hypersensitivity reactions on package label for certolizumab pegol (Cimzia®) found a place in the Warnings and Precautions section:

> **WARNINGS AND PRECAUTIONS** ... Hypersensitivity Reactions. The following symptoms that could be compatible with hypersensitivity reactions have been reported rarely following CIMZIA

[291] Package label. Tegretol®-XR (carbamazepine extended-release tablets); August 2015 (18 pages).

[292] Certolizumab pegol (Crohn's disease) BLA 125-160. Pages 20 and 44 of first of three 80 page Medical Reviews.

[293] Certolizumab pegol (Crohn's disease) BLA 125-160. Pages 20 and 44 of first of three 80 page Medical Reviews.

administration to patients: **angioedema, dyspnea, hypotension, rash, serum sickness, and urticaria**. Some of these reactions occurred after the first administration of CIMZIA. If such reactions occur, discontinue further administration of CIMZIA and institute appropriate therapy.[294]

f. Eslicarbazepine (Partial Onset Seizures) NDA 022-416

FDA's Medical Review for eslicarbazepine provides the following concepts, each of which found a place on the package label:

- **Skin and immune system disorders.** FDA's Medical Review detailed the AEs of Stevens- Johnson syndrome, hypersensitivity reactions, and eosinophilia, as they occurred with the study drug and with related drugs. Each of these found a place in the Warnings and Precautions section of the package label.
- **DRESS (Drug Reaction with Eosinophilia and Systemic Symptoms).** FDA's Medical Review devoted a few pages to the adverse event known as DRESS and to the RegiSCAR criteria for DRESS adverse events. DRESS found a corresponding place in the package label. DRESS is a rare type of adverse event occurring with drugs such as aromatic antiepileptic drugs, sulfonamides, and allopurinol.[295] DRESS is a type of hypersensitivity reaction, often presenting 2–8 weeks after starting the drug. RegiSCAR is a registry for severe cutaneous adverse reactions (SCAR), where these adverse events include SJS, TEN, and DRESS. The DRESS diagnosis scoring system has been described.[296]
- **AEs occurring with similar drugs.** FDA's Medical Review based its prediction of the AEs associated with eslicarbazepine on the Sponsor's own clinical data, as well as on AEs found from two similar drugs, carbamazepine and oxcarbazepine. Carbamazepine and oxcarbazepine are both named on the package label, for their use in predicting AEs expected from eslicarbazepine.
- **Genetic biomarkers.** FDA's Medical Review recommended that the Sponsor explore genetic biomarkers to predict dermatological reactions. This concept found a place in the package label, in the label's statement that, "Risk factors for development of serious dermatologic reactions with APTIOM use have not been identified."

Details on the above concepts are provided below:

i. Skin and Immune System Disorders

FDA's account of serious cutaneous disorders included patient narratives. For one patient, a serious skin reaction was described as, "skin dropping off hands, mouth, and feet," and for another patient, "life-threatening allergic exanthema with skin detachment on the entire body a few days after starting eslicarbazepine.[297]"

[294] Package label. CIMZIA (certolizumab pegol) for injection, for subcutaneous injection; January 2017 (32 pages).

[295] Kardaun SH, et al. Drug reaction with eosinophilia and systemic symptoms (DRESS): an original multisystem adverse drug reaction. Results from the prospective RegiSCAR study. Br. J. Dermatol. 2013;169:1071–80.

[296] Alkhateeb H, et al. DRESS syndrome following ciprofloxacin exposure: an unusual association. Am. J. Case Rep. 2013;14:526–8.

[297] Eslicarbazepine (partial onset seizures) NDA 022-416. Page 230 of 620-page Medical Review.

FDA's account of hypersensitivity reactions, in one patient narrative, stated that the patient, "developed an allergic reaction with urticaria, itching, dyspnea ... during the first hour after taking eslicarbazepine. Eslicarbazepine was discontinued and the patient recovered." And for another patient, "developed sore throat that progressed to pharyngospasm with complete anarthria on Day 17 ... eslicarbazine was discontinued and the patient recovered the next day.[298]"

FDA's Medical Review observed, "Dr. Doi identified one eslicarbazepine subject ... with an SAE of possible Stevens-Johnson syndrome (SJS) ... that included mucosal ulceration and skin exfoliation, but SJS was not confirmed by biopsy.[299]" The FDA reviewer was careful to take into account a concomitant medication, namely, lamotrigine, and to observe that lamotrigine does have a Black Box Warning for SJS. FDAs' conclusion regarding association of the study drug with SJS was, "Although confounded by lamotrigine, the role of eslicarbazepine cannot be ruled out.[300]" This provides an example of assessing relatedness by using a type argument resembling the burden-shifting arguments that are used in the legal profession. The phrase, "cannot be ruled out" used by the FDA reviewer enabled the most responsible type of conclusion that could have been made, when faced with the incomplete information on the relation between the study drug and the AE.

ii. DRESS (Drug Reaction With Eosinophilia and Systemic Symptoms)

FDA described the DRESS adverse events from the Sponsor's clinical studies, writing, "Dr. Doi reports that there are two eslicarbazepine subjects ... meeting RegiSCAR criteria for DRESS, occurring within 1 week to 1 month of beginning eslicarbazepine with positive dechallenge ... [i]n a third case, the event resolved while the patient continued on eslicarbazepine ... I agree with Dr. Doi that there are cases of DRESS associated with eslicarbazepine use and I agree with her recommendation that information regarding DRESS be included in the Warnings and Precautions section.[301]"

iii. AEs Occurring With Similar Drugs

FDA's Medical Review referred to AEs from drugs similar to eslicarbazepine, "Dr. Doi notes that serious skin reactions are included in the prescribing information for **carbamazepine** and **oxcarbazepine** in Warnings and Precautions, I agree with her recommendation to include similar Warnings and Precautions in the eslicarbazepine label.[302]"

iv. Genetic Biomarkers

FDA's Medical Review recommended that the Sponsor conduct additional studies that correlated severe cutaneous AEs with genetic biomarkers. The package label stated that, as yet, this type of study had not been performed. FDA's recommendation was, "The following safety issues should be further studied as Postmarketing Requirements:

[298] Eslicarbazepine (partial onset seizures) NDA 022-416. Page 230 of 620-page Medical Review.

[299] Eslicarbazepine (partial onset seizures) NDA 022-416. Pages 114 and 115 of 620-page Medical Review.

[300] Eslicarbazepine (partial onset seizures) NDA 022-416. Pages 114 and 115 of 620-page Medical Review.

[301] Eslicarbazepine (partial onset seizures) NDA 022-416. Pages 114, 115, 235 and 236 of 620-page Medical Review.

[302] Eslicarbazepine (partial onset seizures) NDA 022-416. Pages 114 and 115 of 620-page Medical Review.

Genetic risk factors for developing severe cutaneous adverse reactions, specifically the association with presence of HLA alleles (HLA-B*1502, HLA-A*3101).[303]"

Package label. The package label for eslicarbazepine (Aptiom®) warned against SJS. The warning was based on the Sponsor's own clinical studies, as well as the AEs established to occur with other similar drugs. Regarding these similar drugs, the package label referred to, "dermatologic reactions ... have been reported in patients using oxcarbazepine or carbamazepine which are chemically related to APTIOM." Regarding the common practice of predicting AEs for a study drug based on AEs associated with similar drugs, this practice is called, "drug class analysis." The Warnings and Precautions section read:

> **WARNINGS AND PRECAUTIONS** ... Serious Dermatologic Reactions, Drug Reaction with Eosinophilia and Systemic Symptoms (DRESS), Anaphylactic Reactions and Angioedema: Monitor and discontinue if another cause cannot be established.[304]

The Full Prescribing Information section of the label provided further information on each of the above topics. The further information on dermatological reactions included:

> **WARNINGS AND PRECAUTIONS** ... Serious dermatologic reactions including Stevens-Johnson syndrome (SJS) have been reported in association with APTIOM use. Serious and sometimes fatal dermatologic reactions, including toxic epidermal necrolysis (TEN) and SJS, have been reported in patients using **oxcarbazepine or carbamazepine which are chemically related to eslicarbazepine (Aptiom®).** The reporting rate of these reactions associated with oxcarbazepine use exceeds the background incidence rate estimates by a factor of 3- to 10-fold. Risk factors for development of serious dermatologic reactions with APTIOM use have not been identified.[305]

Further information on DRESS took this form:

> **WARNINGS AND PRECAUTIONS** ... Drug Reaction with Eosinophilia and Systemic Symptoms (DRESS)/ Multiorgan Hypersensitivity Drug Reaction with Eosinophilia and Systemic Symptoms (DRESS), also known as Multiorgan Hypersensitivity, has been reported in patients taking APTIOM. DRESS may be fatal or life-threatening. DRESS typically, although not exclusively, presents with fever, rash, and/or lymphadenopathy, in association with other organ system involvement, such as hepatitis, nephritis, hematological abnormalities, myocarditis, or myositis sometimes resembling an acute viral infection. Eosinophilia is often present. Because this disorder is variable in its expression, other organ systems not noted here may be involved. It is important to note that early manifestations of hypersensitivity, such as fever or lymphadenopathy, may be present even though rash is not evident. If such signs or symptoms are present, the patient should be evaluated immediately. APTIOM should be discontinued and not be resumed if an alternative etiology for the signs or symptoms cannot be established.[306]

g. Sebelipase Alfa (Deficiency in Lysosomal Acid Lipase) BLA 125-561

Sebelipase alfa is an enzyme for treating lysosomal acid lipase deficiency, a genetic disease. Sebelipase alfa is recombinant human lysosomal acid lipase, purified from the egg whites of eggs of transgenic hens.[307]

[303] Eslicarbazepine (partial onset seizures) NDA 022-416. Pages 139 and 153 of 620-page Medical Review.

[304] Package label. ATPTIOM® (eslicarbazepine acetate) tablets, for oral use; August 2015 (22 pages).

[305] Package label. ATPTIOM® (eslicarbazepine acetate) tablets, for oral use; August 2015 (22 pages).

[306] Package label. ATPTIOM® (eslicarbazepine acetate) tablets, for oral use; August 2015 (22 pages).

[307] Sebelipase alfa (deficiency in lysosomal acid lipase) BLA 125-561. 87-page Clinical Pharmacology Review.

FDA evaluated the hypersensitivity reactions experienced by subjects in the Sponsor's studies, where this evaluation used criteria set forth by National Institute of Allergy and Infectious Disease (NIAID) and Food Allergy and Anaphylaxis Network (FAAN).[308] According to these criteria, which are in the footnote, anaphylaxis is highly likely when any one of the three criteria is fulfilled.[309] In these criteria, PEF is Peak expiratory flow and BP is blood pressure.

FDA observed, "Hypersensitivity reactions occurred in 4 of the 6 (67%) ... patients who received more than 4 infusions of sebelipase alfa. None of the patients discontinued treatment due to adverse reactions, and hypersensitivity reactions have been generally mild and manageable with treatment interruption, adjustment of infusion rates, and standard medical intervention.[310]"

FDA's account of the Sponsor's clinical studies described anaphylaxis in one of the study subjects, "one patient experienced **anaphylaxis**, including signs and symptoms included chest discomfort, eyelid edema, dyspnea, urticaria, hyperemia, and pruritus; the patient recovered after treatment with an antihistamine and hydrocortisone; treatment with sebelipase alfa has been held pending further evaluation.[311]" In FDA's account of anaphylaxis in a different subject, the FDA reviewer wrote, "The event of **anaphylaxis** was considered as severe and serious; the infusion was interrupted and the one patient who experienced anaphylaxis recovered.[312]"

As a technique for arriving at package label information, FDA considered other drugs in the class of enzyme replacement therapies and remarked that the associated hypersensitivity reactions can be treated with antipyretics, antihistamines, or corticosteroids. On this point, FDA stated, "The greatest risks associated with the **class of enzyme replacement therapies** are hypersensitivity reactions, and these reactions have been manageable with infusion rate adjustments and treatment with antipyretics, antihistamines, and/or corticosteroids.[313]"

[308] Sampson HA, et al. Second symposium on the definition and management of anaphylaxis: summary report-Second National Institute of Allergy and Infectious Disease/Food Allergy and Anaphylaxis symposium. J. Allergy Clin. Immunol. 2006;117:391–7.

[309] **First criterion.** Acute onset of an illness ... with involvement of the skin, mucosal tissue, or both (e.g., generalized hives, pruritus or flushing, swollen lips-tongue-uvula) and at least one of the following: (1) Respiratory compromise (e.g., dyspnea, wheeze-bronchospasm, stridor, reduced PEF, hypoxemia); (2) Reduced BP or associated symptoms of end-organ dysfunction (e.g., hypotonia [collapse], syncope, incontinence). **Second criterion.** Two or more of the following that occur rapidly after exposure to a likely allergen for that patient (minutes to several hours): (1) Involvement of the skin-mucosal tissue (e.g., generalized hives, itch-flush, swollen lips-tongue-uvula); (2) Respiratory compromise (e.g., dyspnea, wheeze-bronchospasm, stridor, reduced PEF, hypoxemia); (3). Reduced BP or associated symptoms (e.g., hypotonia [collapse], syncope, incontinence); (4) Persistent gastrointestinal symptoms (e.g., crampy abdominal pain, vomiting). **Third criterion.** Reduced BP after exposure to known allergen for that patient ... (1) Infants and children: low systolic BP (age specific) or greater than 30% decrease in systolic BP*; (2). Adults: systolic BP of less than 90 mm Hg or greater than 30% decrease from that person's baseline.

[310] Sebelipase alfa (deficiency in lysosomal acid lipase) BLA 125-561. Page 73 of 229-page Medical Review.

[311] Sebelipase alfa (deficiency in lysosomal acid lipase) BLA 125-561. Page 209 of 229-page Medical Review.

[312] Sebelipase alfa (deficiency in lysosomal acid lipase) BLA 125-561. Page 209 of 229-page Medical Review.

[313] Sebelipase alfa (deficiency in lysosomal acid lipase) BLA 125-561. Page 12 of 229-page Medical Review.

FDA's attention then turned to the terminology for package labeling, that is, to the terms "infusion-related reactions" versus "hypersensitivity reactions." FDA recommended that the term "infusion-related reactions" should no longer be used. Also, FDA recommended that anaphylaxis be assessed by the criteria of NIAID and FAAN, as set forth by Sampson et al.[314]:

> For labeling of enzyme replacement therapies ... categorization of AE as ... "infusion-related reactions" is no longer recommended. Instead, AEs which are temporally related to these medications and are likely immunologically mediated should be categorized as "Hypersensitivity Reactions." Anaphylaxis represents a specific subgroup of hypersensitivity reactions which fulfill Sampson's criteria.[315]

Sampson's criteria can assess hypersensitivity to recombinant antibodies and to foods such as hen's eggs.[316,317,318] FDA's Cross Discipline Team Leader Review referred to the fact that sebelipase alfa was purified from eggs, raising the possibility that small quantities of contaminating proteins encoded by the chicken genome may cause hypersensitivity reactions. Details on chicken egg allergens are in the cited references.[319,320] FDA's account of eggs as the starting material was, "Sebelipase alfa ... is produced in the egg whites of eggs laid by genetically engineered chickens ... the ... manufacturing process begins with cracking of eggs and separation of egg yolk from the egg white, and the egg whites serve as the starting material for downstream purification.[321]"

Package label. FDA's narratives regarding hypersensitivity reactions, anaphylaxis, egg products, treatment with antipyretics or antihistamines, and halting sebelipase alfa treatment, all found a place on the package label. The Warnings and Precautions section of the package label recited:

> **WARNINGS AND PRECAUTIONS. Hypersensitivity Reactions** including Anaphylaxis: Observe patients during and after the infusion. Consider interrupting the infusion or lowering the infusion rate, based on the severity of the reaction. If a severe hypersensitivity reaction occurs, immediately stop the

[314] Sampson HA, et al. Second symposium on the definition and management of anaphylaxis: summary report—Second National Institute of Allergy and Infectious Disease/Food Allergy and Anaphylaxis symposium. J. Allergy Clin. Immunol. 2006;117:391–7.

[315] Sebelipase alfa (deficiency in lysosomal acid lipase) BLA 125-561. Page 75 of 229-page Medical Review.

[316] Sampson HA, et al. Second symposium on the definition and management of anaphylaxis: summary report—Second National Institute of Allergy and Infectious Disease/Food Allergy and Anaphylaxis symposium. J. Allergy Clin. Immunol. 2006;117:391–7.

[317] Chung CH, et al. Cetuximab-induced anaphylaxis and IgE specific for galactose-alpha-1,3-galactose. N. Engl. J. Med. 2008;358:1109–17.

[318] Tripodi S, et al. Predicting the outcome of oral food challenges with hen's egg through skin test end-point titration. Clin. Exp. Allergy. 2009;39:1225–33.

[319] Matsuo H, et al. Common food allergens and their IgE-binding epitopes. Allergol. Int. 2015;64:332–43.

[320] Lin YT, et al. Correlation of ovalbumin of egg white components with allergic diseases in children. J. Microbiol. Immunol. Infect. 2016;49:112–8.

[321] Sebelipase alfa (deficiency in lysosomal acid lipase) BLA 125-561. Page 10 of 42-page Cross Discipline Team Leader Review.

infusion and initiate appropriate treatment. Pre-treatment with antipyretics and/or antihistamines may prevent subsequent reactions in those cases where symptomatic treatment is required ... **Hypersensitivity to Eggs or Egg Products**: Consider the risks and benefits of treatment in patients with known systemic hypersensitivity reactions to eggs or egg products.[322]

V. CONCLUDING REMARKS

This chapter describes the problem of the materialization of antibodies against abatacept and adalimumab, following administering these antibodies to study subjects. Chapter 8, Drug–Drug Interactions—Part Two (Therapeutic Proteins), describes this same problem for atezolizumab, denosumab, evolocumab, sebelipase alfa, lixisenatide, and nivolumab. A related issue is the option of coadministering an immunosuppressant such as azathioprine to prevent materialization of the anti-drug antibodies. Other related problems caused by anti-drug antibodies include the ability of the anti-drug antibody:

- to bind the therapeutic protein,
- to prevent a biochemical function of the therapeutic protein,
- to reduce exposure (AUC, tmax, Cmax, Cmin) of the therapeutic protein, and
- to reduce clinical efficacy of the therapeutic protein.

Hypersensitivity reactions are described above for brentuximab, bupropion, carbamazepine, certolizumab pegol, denosumab, eslicarbazepine, natalizumab, and sebelipase alfa. The hypersensitivity reaction of bullous pemphigoid was briefly mentioned above for adalimumab, infliximab, nivolumab, and pembrolizumab. Autoimmunity and lupus-like syndrome were detailed above for adalimumab, atezolizumab, certolizumab, etanercept, infliximab, nivolumab, and pembrolizumab.

Increased risk for infections and cancers was described for drugs that are immunosuppressants, for the examples of abatacept, adalimumab, denosumab, fingolimod, infliximab, natalizumab, and tofacitinib. A related issue is the need to monitor patients for infections. Other related issues include the option of measuring residual activity of the immune system in study subjects treated with an immunosuppressant drug and the option of assessing if vaccines should be avoided when a patient is being treated with an immunosuppressant. FDA's review for atezolizumab, described above, provides a useful shortcut for identifying subjects who had experienced an adverse event. This shortcut takes the form of identifying subjects who had been treated with a drug to counteract the adverse event. In other words, although the detection and description of many AEs can be characterized by ambiguity, the fact that a subject was treated with a given drug can be absolutely unambiguous, where it is stated in the subject's record.

[322] Package label. KANUMA (sebelipase alfa) injection, for intravenous use; December 2015 (13 pages).

CHAPTER

10

Drug Class Analysis

I. INTRODUCTION

Drug class analysis is a tool for predicting adverse events that need to be drafted on the package label. The most extreme use of drug class analysis is where it serves as the only source for arriving at adverse event (AE) warnings on the package label. But more typically, package label warnings result from the combination of AE data the Sponsor's own clinical trial, the Sponsor's animal toxicity data, drug class analysis, and sometimes case histories published in the medical literature.

Examples of a drug classes are tyrosine kinase inhibitors (TKIs), proton pump inhibitors, calcium channel blockers, and corticosteroids. These names facilitate the understanding of AEs associated with a drug or with members of a drug class.[1]

FDA's Guidance for Industry on Pharmacologic Class defines drug class. FDA's definition is entirely consistent with FDA's drug class analyses for all of the drug-approvals outlined in this chapter. FDA's definition is as follows[2]:

> Pharmacologic class is defined on the basis of any one of the following three attributes of the drug:
>
> 1. **Mechanism of action.** Pharmacologic action at the receptor, membrane, or tissue level
> 2. **Physiologic effect.** Pharmacologic effect at the organ, system, or whole body level
> 3. **Chemical structure.**

[1] Lanthier M, et al. An improved approach to measuring drug innovation finds steady rates of first-in-class pharmaceuticals. Health Aff. (Millwood) 2013;32:1433–9.

[2] U.S. Department of Health and Human Services. Food and Drug Administration. Center for Drug Evaluation and Research (CDER). Center for Biologics Evaluation and Research (CBER). Guidance for industry and review staff labeling for human prescription drug and biological products—determining established pharmacologic class for use in the highlights of prescribing information good review practice; 2009 (9 pp.).

DOI: https://doi.org/10.1016/B978-0-12-814647-7.00010-5

FDA's Guidance for Industry on Labeling acknowledges that drug class analysis is a useful tool for drafting AE warning on package labels. FDA's Guidance states that, "In some cases, the labeling of all members of a class of drugs includes identical statements ... [t]hese class labeling statements describe a risk or effect that is typically associated with **members of the class**, based on what is known about the pharmacology or chemistry of the drugs."[3]

FDA's Guidance for Industry on Safety Reporting also recognizes the concept of drug class in a comment on defining drug class by pharmacological properties of the drugs, "the investigator brochure should list adverse events that commonly occur with the **class of drugs** or may be predicted to occur based on the pharmacological properties of the drug, even if not yet observed with the drug under investigation, to alert the investigator to the possibility of their occurrence."[4]

FDA's Guidance for Industry on the Target Product Profile recommends that the Sponsor be aware of the study drug's class, for use in meetings with FDA officials. This recommendation concerned drug class based on the mechanism of action of the drug. FDA's Guidance informs us that:

> A sponsor developed a new molecular entity whose therapeutic target was similar to several approved agents in a **drug class**. The sponsor noted in correspondence and in meetings ... that the product achieved its therapeutic effect via a **novel mechanism of action** on the target ... when the sponsor submitted an NDA, the sponsor prominently mentioned the novel **mechanism of action** in the drug label implying treatment benefit based upon this mechanism. The preclinical studies submitted as documentation did not provide adequate evidence to support this statement, and the data could not rule out the conventional **mechanism of action** shared by other **drugs in the class**.[5]

a. Defining First-in-Class

The term "first-in-class" may not have a distinct definition. The lack of distinct definitions is evident, for example, from FDA's review for vorapaxar. FDA's review for this drug stated that, "vorapaxar is a first-in-class PAR-1 antagonist."[6] But in assessing the same NDA for the same drug, FDA also referred to other drugs in the same class, stating

[3] U.S. Department of Health and Human Services. Food and Drug Administration. Center for Drug Evaluation and Research (CDER). Center for Biologics Evaluation and Research (CBER). Guidance for industry. Labeling for human prescription drug and biological products—implementing the PLR content and format requirements; 2013 (30 pp.).

[4] U.S. Department of Health and Human Services. Food and Drug Administration. Center for Drug Evaluation and Research (CDER). Center for Biologics Evaluation and Research (CBER). Guidance for industry and investigators safety reporting requirements for INDs and BA/BE studies; 2012 (29 pp.).

[5] U.S. Department of Health and Human Services. Food and Drug Administration. Center for Drug Evaluation and Research (CDER). Guidance for industry and review staff target product profile—a strategic development process tool; 2007 (22 pp.).

[6] Vorapaxar (reduction of atherothrombotic events in patients with a history of myocardial infarction) NDA 204-886. Page 82 of 126-page Clinical Pharmacology Review.

that, "The other products in this class of platelet inhibitor drugs were all approved.[7] The study drug (vorapaxar) and the drugs used in drug class analysis are all platelet inhibitors and, at the same time, vorapaxar is a first-in-class PAR-1 antagonist. While this appears to be a contradiction, this author prefers to characterize the apparent contradiction, instead, as an ambiguity. In fact, FDA admits to this ambiguity.

In response to the author's inquiry, FDA replied that, "We readily admit that we don't have a hard and fast definition for first-in-class, rather a general methodological approach ... but the reality is that where one might draw the line between a new class and a variant within an existing class is not always easily defined.[8] A response from another FDA official replied with a more distinct reply, "First-in-Class means drugs with a new and unique mechanism for treating a medical condition."[9]

That said, it is reasonable to conclude that any ambiguities are inconsequential. As demonstrated in this chapter, drug class analysis, without regard to being "first-in-class," uses drug class defined by:

- Similarities in chemical structure
- Used to inhibit the same target (enzyme or receptor)
- Used to inhibit the same signaling pathway
- Used for treating the same disease.

b. Defining Drug Class

As revealed by FDA's narratives quoted in this chapter, the Sponsor has the option to define drug class by one or more of the above techniques. Regarding drug class, comments by physicians from various medical schools inform us that:

> Because serious drug-related adverse events ... are more often a function of the pharmacologic **class of a drug** rather than specific medication features, most Black Box Warnings are typically **applied to all members of a given class** ... which is defined on the basis of the mechanism of action, e.g., fluoroquinolone antibiotics. Nevertheless, differences in the Black Box Warning acquisition have been reported, where addition of a Black Box Warning is not universal for all drugs in the category ... **[d]rugs within the same class** may have similar pharmacodynamic properties and chemical structures, but different pharmacokinetic properties, and these may justify discrepant Black Box Warning labeling.[10]

[7] Vorapaxar (reduction of atherothrombotic events in patients with a history of myocardial infarction) NDA 204-886. Page 3 of 8-page Risk Assessment Mitigation Review; and page 46 of 234-page Medical Review.

[8] e-mail from M. L. Lanthier, U.S. Food and Drug Administration (May 31, 2017).

[9] E-mail from HT, Pharmacist. Drug Information Specialist. Division of Drug Information. Center for Drug Evaluation and Research. Food and Drug Administration (June 1, 2017).

[10] Panagiotou OA, et al. Different black box warning labeling for same-class drugs. J. Gen. Intern. Med. 2011;26:603–10.

Drug class can be defined by the drug's *chemical structure*, where the chemical structure is somewhat different for all drugs in the same class. Regarding the use of structure to define a drug class, Prof. Furberg gives the example of ACE inhibitors, stating that:

> All ACE inhibitors share [the] ... feature [of] ... binding of the functional group to the zinc component of the ACE active site, and it is this commonality that divides the class into its three subgroups: the sulfhydryl-containing ACE inhibitors ... the carboxyl- or dicarboxyl-containing ACE inhibitors ... and the phosphorous-containing or phosphinyl ACE inhibitors.[11]

Prof. Furberg has warned that, "the concept of class effect is a term of convenience that has no universally accepted definition."[12] In addition, Furberg has correctly warned that, "All drugs have multiple mechanisms of action ... [t]he common actions—favorable or unfavorable—define the class. The actions that are specific to an individual agent within a class may add to, subtract from, or have a neutral effect on the efficacy or safety attributed to the common class actions."[13]

Sometimes, drug class is defined according to the *mechanism of action* of the drug, for example, that it is a tyrosine kinase inhibitor (TKI). Drug class can be based on a *mechanism subgroup*, for example, a subgroup taking the form of drugs that inhibit TKIs that are vascular endothelial growth factor (VEGF)-receptor inhibitors.[14] In this chapter, descriptions of kinases *always* refer to enzymes that catalyze phosphorylation of amino acid residues on a protein, and *never* refer to kinases that catalyze phosphorylation of a small molecule such as a sugar or a vitamin.

II. OFF-TARGET ADVERSE EVENTS

a. Introduction

Adverse events detected in study subjects during the course of a clinical study, as well as in patients in the postmarketing situation, can result from *on-target effects*. On-target effects is the situation where the drug's action on the intended target can result in both efficacy and toxicity. Another class of AEs is *off-target effects*. Off-target effects occurs when a drug acts at a site not intended by the Sponsor and results in toxicity.[15] Muller and Milton[16] describe the concept of on-target toxicities versus off-target toxicities:

[11] Furberg CD. Class effects and evidence-based medicine. Clin. Cardiol. 2000;23 (Suppl. IV):IV15–19.

[12] Furberg CD. Class effects and evidence-based medicine. Clin. Cardiol. 2000;23 (Suppl. IV):IV15-19.

[13] Furberg CD, Psaty BM. Should evidence-based proof of drug efficacy be extrapolated to a "class of agents"? Circulation 2003;108:2608–10.

[14] Morabito A, et al. Tyrosine kinase inhibitors of vascular endothelial growth factor receptors in clinical trials: current status and future directions. The Oncologist 2006;21:753–64.

[15] Guengerich FP. Mechanisms of drug toxicity and relevance to pharmaceutical development. Drug Metab. Pharmacokinet. 2011;26:3–14.

[16] Muller PY, Milton MN. The determination and interpretation of the therapeutic index in drug development. Nat. Rev. Drug Discov. 2012;11:751–61.

> In general, drugs can cause toxicity either through on-target pharmacology (effects mediated through the primary drug target ... or through off-target pharmacology (effects that are mediated through ... unintended targets). Off-target pharmacology typically begins to be considered through in vitro secondary pharmacology assays during lead identification and optimization.[17]

b. Statins

Statins, such as atorvastatin, inhibit hydroxymethylglutaryl-CoA reductase, thus inhibiting hepatic biosynthesis of cholesterol, with a consequent reduction in blood cholesterol and reduced risk for atherosclerosis. All statins are associated with myopathy, ranging in severity from asymptomatic increases in creatine kinase to muscle aches or weakness to fatal rhabdomyolysis.[18] Myopathy is defined as a serum creatine kinase level more than 10 times the upper limit of normal (ULN) with unexplained muscle weakness or pain. The mechanism of action where off-target action of statins cause myopathies has been identified as the Qo site of mitochondrial complex III.[19] Analysis of muscle biopsies from patients suffering from statin-induced myopathies revealed that CIII enzyme activity is reduced by 18%.

The following excerpt from the Warnings and Precautions section of a statin drug illustrates the AE of myopathy, resulting from off-target actions of the statin:

> **WARNINGS AND PRECAUTIONS.** Atorvastatin, like other statins, occasionally causes myopathy, defined as muscle aches or muscle weakness in conjunction with increases in creatine phosphokinase (CPK) values >10 times upper limit of normal (ULN) ... [p]atients should be advised to report promptly unexplained muscle pain, tenderness, or weakness ... LIPITOR therapy should be discontinued if markedly elevated CPK levels occur or myopathy is diagnosed or suspected.[20]

c. Torcetrapib

Cholesteryl ester transfer protein (CETP) is a protein that circulates in the bloodstream. It mediates transfer of cholesterol from high-density lipoproteins (HDLs) to low-density lipoproteins (LDLs). Inhibiting CETP, for example, by torcetrapib, has the desired effect of raising HDL-cholesterol ("good cholesterol") and lowering LDL-cholesterol ("bad cholesterol").[21] However, clinical studies with torcetrapib revealed that an off-target effect of torcetrapib is an increase in blood pressure. The description of increased blood pressure was that, 12 months into the study, systolic blood pressure

[17] Muller PY, Milton MN. The determination and interpretation of the therapeutic index in drug development. Nat. Rev. Drug Discov. 2012;11:751–61.

[18] Egan A, Colman E. Weighing the benefits of high-dose simvastatin against the risk of myopathy. New Engl. J. Med. 2011;365:285–7.

[19] Schirris TJ, et al. Statin-induced myopathy is associated with mitochondrial complex III inhibition. Cell Metab. 2015;22:399–407.

[20] Package label. LIPITOR® (atorvastatin calcium) Tablets for oral administration. NDA 020-702; March 2015 (22 pp.).

[21] Liu S, et al. Crystal structures of cholesteryl ester transfer protein in complex with inhibitors. J. Biol. Chem. 2012;287:37321–9.

increased by a mean of 5.4 mm Hg in the torcetrapib group as compared to baseline blood pressure.[22] In addition to hypertension, the adverse events associated with torcetrapib include heart failure, angina, and death.[23] In 2006 torcetrapib was withdrawn from the market because it was associated with excessive deaths.[24] Torcetrapib's off-target effect in increasing blood pressure (hypertension) appears to result from torcetrapib's action on the adrenal gland.[25,26]

d. Sunitinib

Sunitinib is a multitargeting TKI of VEGF receptors (VEGFRs 1–3) and platelet-derived growth factor (PDGF) receptors (PDGFRα and β). The fact that sunitinib is able to inhibit the activity of multiple tyroskine kinases is evident, from the fact that it inhibits VEGFRs and PDGFRs. Novello et al.[27] illustrates the concept where a drug that inhibits only one receptor, such as only VEGFR, may result in fewer off-target toxic effects while, in contrast, a drug that inhibits a multiplicity of related receptors, such as VEGFR and PDGFR, may have more off-target toxic effects. The example is from sunitinib, a drug that inhibits both VEGFR and PDGFR, and where both of these receptors are intended targets and are both considered to be "on-target." Novello et al.[28] illustrates the concept that a drug with multiple intended targets may have increased risk for off-target toxicities:

> Co-inhibition of VEGF and PDGF pathways potentially offers greater anti-angiogenic effect than inhibition of either pathway alone ... [h]owever, it is possible that broader anti-tumour activity may also translate into a less favourable safety profile due to **off-target toxicity**.[29]

[22] Barter PJ, et al. Effects of torcetrapib in patients at high risk for coronary events. New Engl. J. Med. 2007;357:2109–22.

[23] Tall AR, et al. The failure of torcetrapib: was it the molecule or the mechanism? Arterioscler. Thromb. Vasc. Biol. 2007;27:257–60.

[24] Dalvie D, et al. Pharmacokinetics, metabolism, and excretion of torcetrapib, a cholesteryl ester transfer protein inhibitor, in humans. Drug Metab. Dispos. 2008;36:2185–98.

[25] Joy T, Hegele RA. The end of the road for CETP inhibitors after torcetrapib? Curr. Opin. Cardiol. 2009;24:364–71.

[26] Vergeer M, et al. Cholesteryl ester transfer protein inhibitor torcetrapib and off-target toxicity: a pooled analysis of the rating atherosclerotic disease change by imaging with a new CETP inhibitor (RADIANCE) trials. Circulation 2008;118:2515–22.

[27] Novello S, et al. Phase II study of continuous daily sunitinib dosing in patients with previously treated advanced non-small cell lung cancer. Br. J. Cancer 2009;101:1543–8.

[28] Novello S, et al. Phase II study of continuous daily sunitinib dosing in patients with previously treated advanced non-small cell lung cancer. Br. J. Cancer 2009;101:1543–8.

[29] Novello S, et al. Phase II study of continuous daily sunitinib dosing in patients with previously treated advanced non-small cell lung cancer. Br. J. Cancer 2009;101:1543–8.

Kerkela et al.[30] articulated the same concept that was set forth by Novello et al.,[31] writing, "Although sunitinib makes logical sense from a cancer therapeutic standpoint, the targeting of multiple kinases by one drug leads to an inherent lack of selectivity, **increasing the risk of off-target toxicities** due to inhibition of additional kinases, the identity of which may or may not be known."

Regarding the off-target effect of sunitinib, Kerkela et al.[32] reveals that sunitinib's toxicity results from sunitinib's inhibition of 5′-adenosine monophosphate-activated protein kinase (AMPK), where this article states that, "Herein we present data suggesting that **off-target inhibition by sunitinib of AMPK**, a kinase that plays key roles in maintaining metabolic homeostasis in the heart, especially in the setting of energy stress, accounts, at least in part, for the **toxicity seen in cardiomyocytes** exposed to sunitinib. This, therefore, represents the first example of off-target inhibition of a kinase by a tyrosine kinase inhibitor (TKI) leading to cardiotoxicity."

Yet another example of the analysis of off-target effects comes from two related drugs for treating HIV-1 virus infections. Ray et al.[33] provide an analysis of two drugs in the same class (TAF, TDF). Both drugs contain tenofovir. The two drugs share the same off-target adverse event, namely, an effect on lipid metabolism.

Determining if a given AE is the result of an off-target interaction can be done concurrently with drug class analysis. Whether a given AE is the result of an off-target interaction does not necessarily preclude, nor compel, the conclusion that the off-target interaction was a drug class effect.

III. SHORT EXAMPLES OF DRUG CLASS ANALYSIS

a. Example of Eribulin

Drug class analysis for eribulin included drug class as defined by chemical structure ("halichondrin class of anti-neoplastic drugs"), mechanism of action ("exert an effect through alteration of microtubule function"[34]), and disease ("drugs used to treat patients with advanced cancer"[35]).

FDA's Medical Review stated that eribulin was the first drug in the halichondrin class, implying that predicting eribulin's AEs would not be possible from drug class analysis

[30] Kerkela R, et al. Sunitinib-induced cardiotoxicity is mediated by off-target inhibition of AMP-activated protein kinase. Clin. Transl. Sci. 2009;2:15–25.

[31] Novello S, et al. Phase II study of continuous daily sunitinib dosing in patients with previously treated advanced non-small cell lung cancer. Br. J. Cancer 2009;101:1543–8.

[32] Kerkela R, et al. Sunitinib-induced cardiotoxicity is mediated by off-target inhibition of AMP-activated protein kinase. Clin. Transl. Sci. 2009;2:15–25.

[33] Ray AS, et al. Tenofovir alafenamide: a novel prodrug of tenofovir for the treatment of Human Immunodeficiency Virus. Antiviral Res. 2016;125:63–70.

[34] Eribulin (breast cancer) NDA 201-532. Page 23 of 222-page Medical Review.

[35] Eribulin (breast cancer) NDA 201-532. Pages 109–110 and 116 of 222-page Medical Review.

based on structure. However, FDA observed that eribulin and other anti-neoplastic drugs had the shared mechanism of action of altering microtubule function. FDA's drug class analysis recited:

> Important Safety Issues with Consideration to Related Drugs. Eribulin is the **first drug in the halichondrin class** of anti-neoplastic drugs to be submitted for approval. However, there are several other approved chemotherapeutic agents that exert an effect through **alteration of microtubule function**. These drugs include the taxanes ... vinca alkaloids ... and the epithilone ixabepilone. **Peripheral neuropathy** and myelosuppression are common adverse effects associated with the taxanes and vinca alkaloids.[36]

FDA further observed that, "The incidence of **peripheral neuropathy** appears to be related to the cumulative eribulin dose."[37] The take-home lesson is that FDA can conduct drug class analysis using the narrow definition structure (halichondrin class), more broadly by function (alteration of microtubule function), or both.

b. Example of Rofecoxib

This illustrates predicting AEs of a given drug by mechanism of action where the mechanism of action is mechanism of toxicity (not mechanism of action of efficacy). The facts of the analysis narrowly concerned the fact that rofecoxib had unique toxic catabolites (anion and free radical) that are not generated by other drugs of the same drug class (cyclooxygenase (COX)-2 inhibitor drugs).

Fig. 10.1 depicts rofecoxib's structure, and the structures of two of its toxic breakdown products (anion and free radical). Rofecoxib's conversion to toxic compounds is responsible for the drug's cardiotoxicity. Other COX-2 inhibitor drugs, such as celecoxib and valdecoxib, do not participate in these reactions. In view of the above, Mason et al.[38] warned against the *indiscriminate* use of the terms "drug class" and "drug class effect" as tools for predicting adverse events of a given drug. In other words, the example of rofecoxib demonstrates that contemplating a drug's catabolites can be used to predict that a given drug will not have the same AEs as drugs of the same drug class, as defined by intended target (COX-2).

Mason et al.[39] provides the example of contemplating the mechanisms of action of a drug in a drug class that was defined by *common enzyme target (COX-2)*. The example is from the drug class of COX-2 inhibitors, where drugs in this class include *rofecoxib*, celecoxib, valdecoxib, and lumiracoxib. COX-2 inhibitor drugs and their toxic effects have been reviewed.[40]

[36] Eribulin (breast cancer) NDA 201-532. Page 23 of 222-page Medical Review.

[37] Eribulin (breast cancer) NDA 201-532. Page 141 of 222-page Medical Review.

[38] Mason P, et al. Rofecoxib increases susceptibility of human LDL and membrane lipids to oxidative damage: a mechanism of cardiotoxicity. J. Cardiovasc. Pharmacol. 2006;47:S7–14.

[39] Mason P, et al. Rofecoxib increases susceptibility of human LDL and membrane lipids to oxidative damage: a mechanism of cardiotoxicity. J. Cardiovasc. Pharmacol. 2006;47:S7–14.

[40] Ghosh R, et al. NSAIDs and cardiovascular diseases: role of reactive oxygen species. Oxidative Med. Cell. Longevity 2015;2015:Article ID 536962 (25 pp.).

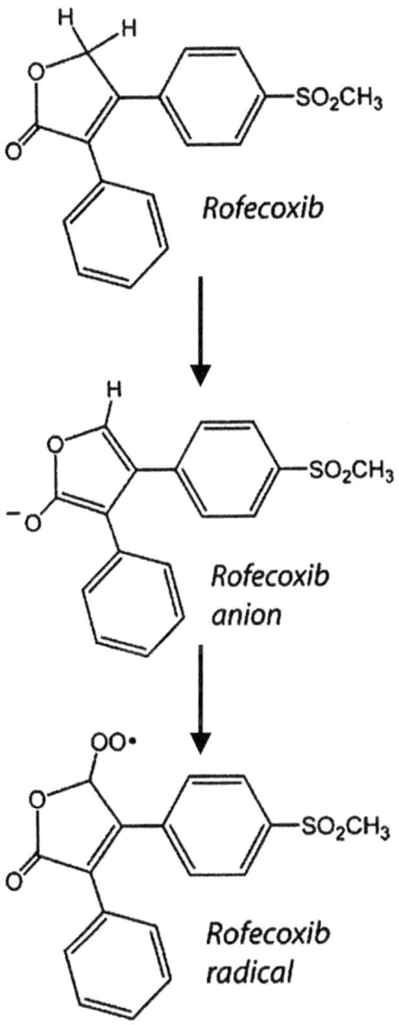

FIGURE 10.1 Rofecoxib and two of its metabolites. Rofecoxib contains two benzene rings and a five-membered ring. With metabolism, the five-membered ring is first converted to the "anion" metabolite, and then to the "radical" metabolites.

The warning against indiscriminate use of the term "drug class effect" was, "Experimental findings from independent laboratories now indicate that the cardiotoxicity of rofecoxib may not be a class effect but because of its intrinsic chemical properties.[41] In the case of rofecoxib, the intrinsic chemical property was breakdown to the anion and the free radical.

[41] Ghosh R, et al. NSAIDs and cardiovascular diseases: role of reactive oxygen species. Oxidative Med. Cell. Longevity 2015;2015:Article ID 536962 (25 pp.).

IV. FIRST-IN-CLASS AND INCREASED NEED FOR AN ADVISORY COMMITTEE

This concerns FDA's reviews of a drug that is "first-in-class" versus review of a drug where FDA had already approved other drugs of the same class. The existence of FDA-approved drugs in the same class can influence FDA's requirement that the Sponsor consult an advisory committee.

FDA's Approval Letter for regorafenib stated, "Your application for regorafenib was not referred to an FDA advisory committee because this application did not raise significant safety or efficacy issues that were unexpected for **a drug of this class** in the intended population.[42]

Similarly, FDA's Approval Letter for pomalidomide commented, "Your application for ... pomalidomide ... was not referred to an FDA advisory committee because this drug is not the **first in its class** and the safety profile is similar to that of other drugs or biologics approved for this indication.[43]

Also similarly, the potential to need an advisory committee is revealed by FDA's Approval Letter for macitentan, as well as in Approval Letters for other drugs, as cited.[44] FDA's Approval Letter for macitentan reads, "Your application for ... macitentan ... was not referred to an FDA advisory committee because this drug is not the **first in its class**, the safety profile is similar to that of other drugs approved for this indication, the clinical study designs are similar to that of previously approved products **in the class**, and the application did not raise significant safety or efficacy issues that were unexpected for **a drug of this class.**[45]

Moreover, for nebivolol, FDA wrote, "NDA 021-742 was not referred to an advisory committee for review because there are **several previously approved agents with the beta-blocker class of drugs**, evaluation of the safety data did not reveal particular safety issues that were unexpected for this class, and the design and results of the efficacy trials did not pose particular concerns."[46]

On the other hand, FDA's Approval Letter can also state that an advisory committee is not needed based on the fact that, "there were no controversial issues that would benefit from advisory committee discussion"[47] (without any mention of drug class analysis). Also,

[42] Regorafenib (metastatic colorectal cancer after prior treatment with various drugs) NDA 203-085. Approval Letter accompanying 76-page pdf file of Medical Review.

[43] Pomalidomide (multiple myeloma) NDA 204-026. Approval Letter accompanying FDA's Medical Review.

[44] Brivaracetam (epilepsy) NDA 205-836, NDA 205-837, NDA 205-838. Approval Letter accompanying Medical Review.

[45] Macitentan (pulmonary arterial hypertension) NDA 204-410. Approval Letter accompanying FDA's Medical Review.

[46] Nebivolol (hypertension) NDA 021-742. Approval Letter accompanying Medical Review.

[47] Bosutinib (chronic myeloid leukemia) NDA 203-341. Approval Letter accompanying FDA's Medical Review.

FDA's Approval Letter can state that an advisory committee is not needed based on the fact that, "the benefit/risk profile ... is clearly favorable for the proposed indication"[48] (without any mention of drug class analysis).

An all-encompassing and comprehensive statement on lack of need for any advisory committee was provided by FDA's Approval Letter for lenvatinib, which recited[49]:

> Your application for ... lenvatinib ... was not referred to an FDA advisory committee because:
>
> - the safety profile is acceptable for the treatment of patients with locally recurrent or metastatic, progressive, radioactive iodine-refractory differentiated thyroid cancer
> - the clinical trial design is acceptable
> - the application did not raise significant safety or efficacy issues that were unexpected for **a drug of this class**
> - the application did not raise significant public health questions on the role of the drug in the diagnosis, cure, mitigation, treatment, or prevention of a disease, and
> - there were no controversial issues that would benefit from advisory committee discussion

V. DRUGS FOR CANCER

a. Introduction

This provides FDA's drug class analysis for anti-cancer drugs, where the goal of the drug class analysis was to predict the study drug's AEs. The examples are:

- Bosutinib (drug class: drugs having the shared target of tyrosine kinases)
- Lenalidomide (drug class: similar structure to thalidomide; similar mechanism of action in killing cancer cells)
- Lenvatinib (drug class: drugs having the shared target of VEGFR-kinase inhibitors)
- Nivolumab (drug class: antibodies that inhibit PD-L1/PD-1 signaling pathway)
- Osimertinib (drug class: EGFR tyrosine kinase inhibitors (TKIs))
- Pomalidomide (drug class: similar structure to thalidomide and lenalidomide)
- Ponatinib (drug class: similar target of VEGFR-kinase inhibitors)
- Regorafenib (drug class: inhibitors of multiple kinases)
- Vemurafenib (drug class: serine/threonine protein kinase inhibitors and TKIs, but excluding VEGFR inhibitors).

b. Bosutinib (Chronic Myeloid Leukemia) NDA 203-341

Bosutinib is a small molecule drug that inhibits breakpoint cluster region (BCR)-Abelson oncogene locus (ABL) kinase as well as the Src-family of kinases including Src, Lyn, and Hck. All of these enzymes are tyrosine kinases. Thus bosutinib is a member of the drug class, TKIs.

[48] Vemurafenib (BRAF V600E mutation positive unresectable or metastatic melanoma) NDA 202-429. Approval Letter accompanying Medical Review.

[49] Lenvatinib (radioiodine-refractory differentiated thyroid cancer) NDA 206-947. Approval Letter accompanying Medical Review.

Chronic myeloid leukemia (CML) is characterized by a chromosomal abnormality that, when expressed, produces a BCR-ABL fusion protein. Thus the BCR-ABL kinase is an abnormal protein resulting from a chromosomal abnormality. The gene encoding this fusion results from chromosomal translocation generating the Philadelphia chromosome.[50] ABL, which normally encodes a tyrosine kinase, becomes constitutively active when fused to BCR. This constitutive tyrosine kinase activity causes the transforming ability of the BCR-ABL oncogene, where the result is that a normal human cell becomes a cancer cell.[51] BCR-ABL is a cytosolic protein that activates a number of cytosolic cell-signaling pathways.[52] ABL means Abelson oncogene locus, and BCR means breakpoint cluster region.

i. AEs From Package Labels of Other Tyrosine Kinase Inhibitors

FDA's Medical Review stated that the TKI drug class includes bosutinib, as well as imatinib, dasatinib, and nilotinib, as shown by the comment, "Bosutinib belongs to a pharmacologic class of drugs called TKIs. It inhibits the abnormal BCR-ABL kinase that promotes CML as well as the Src-family of kinases including Src, Lyn and Hck. Bosutinib causes minimal inhibition of PDGF receptor and c-Kit."[53]

FDA's review for bosutinib further defined the TKI drug class in the comment, "The Applicant considered the following 10 adverse events as events of special interest as they have been toxicities for **one or more other TKIs**: Prolongation of the QT interval (**nilotinib and dasatinib**); fluid retention and edema (**imatinib and dasatinib**); myelosuppression (**imatinib, dasatinib and nilotinib**), hepatotoxicity (**imatinib and nilotinib**), gastrointestinal irritation (**imatinib**), congestive heart failure, left ventricular dysfunction and myocardial infarction (**dasatinib**), hemorrhage (**imatinib and dasatinib**); and tumor lysis syndrome (**nilotinib**)."[54]

As part of the drug class analysis, FDA cited Giles et al.[55] which concerned class effects of TKIs. The very essence of any drug class analysis is a list of all of the AEs associated with all of the drugs of a given drug class.

[50] Cortes JE, et al. Bosutinib versus imatinib in newly diagnosed chronic-phase chronic myeloid leukemia: results from the BELA Trial. J. Clin. Oncol. 2012;30:3486–92.

[51] Hickey FB, Cotter TG. BCR-ABL regulates phosphatidylinositol 3-kinase-p110 transcription and activation and is required for proliferation and drug resistance. J. Biol. Chem. 2006;281:2441–50.

[52] Nammanapalli R, Bhalla K. Novel targeted therapies for Bcr-Abl positive acute leukemias: beyond STI571. Oncogene 2002;21:8584–90.

[53] Bosutinib (chronic myeloid leukemia) NDA 203-341. Page 12 of 74-page Clinical Review and 81-page pdf file of Medical Review.

[54] Bosutinib (chronic myeloid leukemia) NDA 203-341. Page 63 of 81-page pdf file of Medical Review.

[55] Giles F, et al. Class effects of tyrosine kinase inhibitors in the treatment of chronic myeloid leukemia. Leukemia 2009;23:1698–707.

ii. Toxicities Found in Animal Studies

But drug class analysis was not the only source of information on bosutinib's AEs. FDA's Pharmacology Review detailed the Sponsor's animal toxicity tests revealing gastrointestinal toxicity, myelosuppression, hepatic toxicity, and fluid retention:

- *Gastrointestinal irritation*: "Upon necropsy, animals ... had ... gastrointestinal toxicity (distention of the GI, and slight to moderate hypertrophy/hyperplasia of the large intestine) and discoloration of the mesenteric lymph nodes."[56]
- *Myelosuppression*: "Reduced red cell mass may be due to suppression of hematopoiesis secondary to inflammation; however, the magnitude of changes was small."[57]
- *Hepatotoxicity*: "hepatobiliary toxicity was of low incidence in animals and was mainly seen upon histopathology examination with no correlating changes in clinical chemistry parameters ... [h]epatobiliary findings were reported in rats ... but were of low incidence or of low severity. These findings included centrilobular hyperplasia. There were no changes in liver enzymes.[58]
- *Fluid retention*: The Pharmacology review noted that 2/2 male rats and 6/3 female rats had edema in the skin.[59]

iii. Sponsor's Adverse Event Data

The Sponsor's clinical study also revealed bosutinib's AEs, and these were quite similar to the animal toxicity findings:

- *Gastrointestinal irritation*: "Hemorrhage occurred in 79 patients (14%), was grade 3 or 4 in 13 patients (2%), and was a SAE in 18 patients (3%). The most frequent site of hemorrhage was the gastrointestinal tract."[60]
- *Myelosuppression*: "Thrombocytopenia, neutropenia and anemia occurred in 40%, 14%, and 23% of patients respectively. Grade 3 and 4 thrombocytopenia, neutropenia, and anemia occurred in 27%, 10%, and 12% of patients respectively."[61]
- *Hepatotoxicity*: "Liver Function. Twenty percent of patients experienced elevation of one or more liver enzymes during treatment with bosutinib. A third of the liver enzyme elevations were of grade 3 or 4 severity. Most patients were able to continue therapy after a dose hold or dose reduction. A total of 18 (3%) patients discontinued treatment due to liver enzyme elevations."[62]

[56] Bosutinib (chronic myeloid leukemia) NDA 203-341. Page 100 of 199-page Pharmacology Review.

[57] Bosutinib (chronic myeloid leukemia) NDA 203-341. Page 134 of 199-page Pharmacology Review.

[58] Bosutinib (chronic myeloid leukemia) NDA 203-341. Pages 4 and 53 of 199-page Pharmacology Review.

[59] Bosutinib (chronic myeloid leukemia) NDA 203-341. Page 193of 199-page Pharmacology Review.

[60] Bosutinib (chronic myeloid leukemia) NDA 203-341. Page 65 of 81-page Medical Review.

[61] Bosutinib (chronic myeloid leukemia) NDA 203-341. Page 65 of 81-page Medical Review.

[62] Bosutinib (chronic myeloid leukemia) NDA 203-341. Page 65 of 81-page Medical Review.

- *Fluid retention*: "Edema Localized and peripheral edema occurred in 87 (15%) patients, and was grade 1 or 2 severity in 95% of patients."[63]

Package label. The package label warnings for bosutinib (Bosulif®) were derived from drug class analysis as well as from the Sponsor's own data from humans and animals. The information includes a feature often found on package labels, namely, instructions for dose modification and instructions for monitoring:

> **WARNINGS AND PRECAUTIONS** ... Gastrointestinal toxicity: Monitor and manage as necessary. Withhold, dose reduce, or discontinue BOSULIF ... [d]iarrhea, nausea, vomiting, and abdominal pain occur ... [m]onitor and manage patients using standards of care, including antidiarrheals, antiemetics, and/or fluid replacement.[64]

The above warning was derived, in part, from FDA's drug class analysis using the package label from *imatinib*. The bosutinib label warned against myelosuppression:

> **WARNINGS AND PRECAUTIONS** ...... Myelosuppression: Monitor blood counts and manage as necessary.[65]

The above warning on bosutinib's label was derived, in part, from FDA's drug class analysis of the package labels from *imatinib*, *dasatinib*, and *nilotinib*. The bosutinib label also warned against hepatotoxicity, as shown below:

> **WARNINGS AND PRECAUTIONS.** Hepatic toxicity: Monitor liver enzymes at least monthly for the first three months and as needed. Withhold, dose reduce, or discontinue BOSULIF.[66]

This warning on the bosutinib label was derived, in part, from FDA's drug class analysis using package labels for *imatinib* and *nilotinib*. Moreover, the bosutinib package label also warned against the AE of fluid retention, as shown below:

> **WARNINGS AND PRECAUTIONS** ... Fluid retention: Monitor patients and manage using standard of care treatment. Withhold, dose reduce, or discontinue BOSULIF ... [a]dvise patients of the possibility of developing fluid retention (swelling, weight gain, or shortness of breath) and to seek medical attention promptly if these symptoms arise.[67]

The above warning on the bosutinib label was derived, in part, from FDA's drug class analysis that used package labels from *imatinib* and *dasatinib*.

[63] Bosutinib (chronic myeloid leukemia) NDA 203-341. Page 64 of 81-page Medical Review.

[64] Package label. BOSULIF (bosutinib) tablets, for oral use; September 2012 (13 pp.).

[65] Package label. BOSULIF (bosutinib) tablets, for oral use; September 2012 (13 pp.).

[66] Package label. BOSULIF (bosutinib) tablets for oral use; September 2012 (13 pp.).

[67] Package label. BOSULIF (bosutinib) tablets, for oral use; September 2012 (13 pp.).

c. Lenalidomide (Myelodysplastic Syndromes) NDA 021-880

This concerns lenalidomide (Revlimid®) for treating myelodysplastic syndromes (MDS).[68] MDS is group of hematological cancers.[69] Data on lenalidomide's teratogenicity from the human fetus or human infants were not available. However, data on human teratogenicity of a structurally related drug, thalidomide, were available and widely publicized.[70]

Drug class analysis was the only source of information for teratogenicity package label warnings for lenalidomide. FDA's drug class analysis took the following pathway of logic and reasoning (Steps 1-5):

Step 1. Drug class based on structure. The structures of the study drug (lenalidomide) and thalidomide are similar, thus supporting the utility of drug class analysis for arriving at warnings on lenalidomide's package label.
Step 2. Drug class based on mechanism of action for efficacy. In vitro cell culture data on the mechanism of action of lenalidomide and thalidomide support the rationale of drug class analysis for arriving at drug safety warnings. FDA's commentary on mechanisms stated that each of these drugs killed cancer cells and each stimulated T-cell proliferation.
Step 3. Metabolites. This concerns the logic in justifying drug class analysis on similarities in metabolites formed from lenalidomide, and metabolites formed from thalidomide. FDA concluded that *drug class could not be justified* on this basis, because of the fact that pathways of degradation for lenalidomide and thalidomide were not similar to each other.
Step 4. Animal toxicity studies. Regarding the logic in predicting lenalidomide's toxicity in humans from animal studies with lenalidomide and animal studies with thalidomide, FDA concluded that this type of logic was, in its present state, flawed. The Sponsor's logic was flawed, because of poor study design in the animal studies.
Step 5. FDA's conclusion. In view of the lack of animal studies on teratogenicity of the study drug, FDA concluded that the Sponsor must follow the requirements of a risk management plan.

Details on the above steps of logic and reasoning are as follows:

Step 1. Drug class based on structure. FDA observed that, "Lenalidomide and thalidomide have similarities in structure. Lenalidomide is an amide and bears an amino group in the aromatic ring, whereas thalidomide is an imide . . . [b]oth have an asymmetric center and both are manufactured as racemic mixtures."[71] The FDA reviewer reasoned that, "The structural similarity of lenalidomide to thalidomide, a

[68] Lenalidomide (myelodysplastic syndromes) NDA 021-880.

[69] Brody T. Chapter 18. Hematological cancers in CLINICAL TRIALS Study Design, Endpoints and Biomarkers, Drug Safety, and FDA and ICH Guidelines. 2nd ed. New York: Elsevier; 2016. pp. 331–376.

[70] Curran WJ. The thalidomide tragedy in Germany: the end of a historic medicolegal trial. New Engl. J. Med. 1971;284:481–2.

[71] Lenalidomide (myelodysplastic syndromes) NDA 021-880. Page 48 of 184-page pdf file Medical Review.

known teratogen, suggests developmental risk.[72] The structures of these two compounds are shown below[73]:

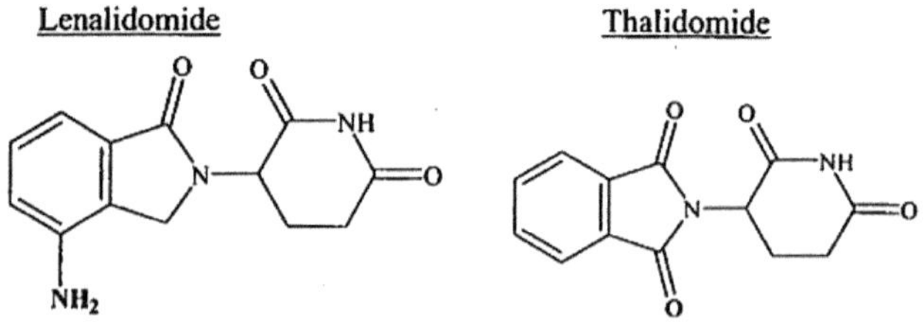

Lenalidomide   Thalidomide

Step 2. Drug class based on mechanism of action for efficacy. FDA's Medical Review observed similar mechanisms of action of the study drug and thalidomide, both of which are used to treat hematological cancers. Although thalidomide was first used to treat morning sickness in women, where the result was birth defects, it was later discovered that thalidomide could treat multiple myeloma.[74] Regarding similarity in **mechanism of action**, FDA stated that, "Both lenalidomide and thalidomide ... increase the secretion of anti-inflammatory interleukin-10 (IL-10) from lipopolysaccharide-stimulated PBMCs, stimulates T-cell proliferation, and production of IL-2 and IFN-gamma. Both inhibit the secretion of pro-inflammatory cytokines TNFα, IL-1β, and IL-6."[75]

This shows similarity of mechanism of action, supporting use of class analysis to aid in arriving at safety warnings on the package label. (Regarding the term "PBMC" used above, this term means peripheral blood mononuclear cell. PBMCs include T cells, B cells, NK cells, and monocytes, but not red blood cells and not neutrophils). Turning to similarity in **mechanism of action** in the desired effect of killing cancer cells, FDA observed, "In addition to these immune effects, there is evidence that thalidomide and its analogues [e.g., lenalidomide] ... act directly on tumor cells, via inducing apoptosis."[76]

FDA continued its observations on mechanisms of action that are shared (or diverge), between the study drug and thalidomide, writing, "Lenalidomide inhibited cell proliferation (IC50s) in some, but not all cell lines. Of cell lines tested, **lenalidomide but not thalidomide**, was effective in inhibiting growth of ... a human B cell lymphoma cell line."[77]

[72] Lenalidomide (myelodysplastic syndromes) NDA 021-880. Page 9 of 184-page pdf file Medical Review.

[73] Lenalidomide (myelodysplastic syndromes) NDA 021-880. Page 48 of 184-page pdf file Medical Review.

[74] Bartlett JB, et al. The evolution of thalidomide and its IMiD derivatives as anticancer agents. Nat. Rev. Cancer 2004;4:314–22.

[75] Lenalidomide (myelodysplastic syndromes) NDA 021-880. Page 9 of 184-page pdf file Medical Review.

[76] Lenalidomide (myelodysplastic syndromes) NDA 021-880. Page 9 of 184-page pdf file Medical Review.

[77] Lenalidomide (myelodysplastic syndromes) NDA 021-880. Page 22 of 184-page pdf file Medical Review.

Step 3. Metabolites. Regarding metabolism, FDA further observed:

> Lenalidomide and thalidomide are structurally related as they both possess piperidinedione and indoline moieties ... [b]ased upon the similarity in structure, one would predict that thalidomide and lenalidomide would be metabolized and degrade in a similar manner ... [t]heir degradative pathways, while apparently similar, have **not resulted in any common degradation products** in animals.[78]

Step 4. Animal toxicity studies. Regarding the most suitable animal for safety studies, FDA complained that the rat is not a suitable model. This complaint may have been based on publications stating that thalidomide is not teratogenic in rats.[79,80] FDA complained that:

> The rat, however is not an adequate species for the full assessment of lenalidomide's developmental effects, given the structural similarity to thalidomide. Historical data indicates that the **rat is not sensitive to the full range of thalidomide's teratogenic effects**.

After complaining about the Sponsor's rat toxicity studies, FDA complained about the Sponsor's rabbit toxicity studies, writing:

> An additional developmental study was conducted in the rabbit, with a concurrent thalidomide dose group. This study had a **confounding variable with some rabbits not eating** prior to the study and all these rabbits had a negative outcome in the study. Additionally, the highest dose tested did not meet the standard criteria for sufficient drug exposure.

Step 5. FDA's conclusions and requirement for monkey toxicity study. FDA required that the Sponsor conduct more animal studies, and that the Sponsor conduct a risk management plan. FDA concluded, "The embryo-fetal toxicity assessment of lenalidomide has not been adequately addressed. You will need to provide adequate information for this assessment in appropriate [animal] models ... [t]hese studies should be conducted in two different species that are appropriate to assess the full range of thalidomide embryo-fetal effects. The rat is not an acceptable model."[81]

As can be seen on the package label, the Sponsor used the monkey (not the rat) for assessing toxicity to the embryo and fetus.[82] Regarding the risk management plan, FDA wrote, "Due to lenalidomide's structural similarity to thalidomide and the inadequate developmental toxicity study, this reviewer recommends that a **risk management plan** very similar to that for thalidomide be instituted to prevent the risk of fetal exposure.[83]

[78] Lenalidomide (myelodysplastic syndromes) NDA 021-880. Page 21 of 184-page pdf file Medical Review.

[79] Scott WJ, et al. Non-confirmation of thalidomide induced teratogenesis in rats and mice. Teratology 1977;16:333–5.

[80] Chamberlain JG. Thalidomide and lack of teratogenesis in Long-Evans rats. Teratology 1979;19:129.

[81] Lenalidomide (myelodysplastic syndromes) NDA 021-880. Page 14 of 184-page pdf file Medical Review.

[82] Package label. REVILIMID® (lenalidomide) capsules, for oral use; February 2015 (36 pp.).

[83] Lenalidomide (myelodysplastic syndromes) NDA 021-880. Pages 20 and 30 of 184-page pdf file Medical Review.

Reiterating this requirement, FDA stated that, "this reviewer recommends full approval providing the sponsor agrees to ...Risk Management Plan: Due to **lenalidomide's structural similarity to thalidomide** and the **inadequate developmental toxicity study**, this reviewer recommends that a **risk management program** similar to that of thalidomide be instituted to prevent the risk of fetal exposure until developmental toxicity issues have been resolved."[84]

Regarding the package label, the FDA reviewer required, "A Black Box Warning should be placed in the label to include the unknown pregnancy risk and the recommendation to prevent fetal exposure ... [s]trong labeling: the labeling should include Black Box Warnings ... regarding prevention of fetal exposures."[85]

Package label. FDA's concerns about risk for fetal exposure found a place on the package label for lenalidomide (Revlimid®). The Black Box Warning is distinguished by its requirement for a Risk Evaluation and Mitigation Strategy (REMS) that included a "restricted distribution program":

> **BOXED WARNING**: EMBRYO-FETAL TOXICITY, HEMATOLOGIC TOXICITY, and VENOUS and ARTERIAL THROMBOEMBOLISM ... EMBRYO-FETAL TOXICITY. Lenalidomide, a thalidomide analogue, caused limb abnormalities in a developmental monkey study similar to birth defects caused by thalidomide in humans. If lenalidomide is used during pregnancy, it may cause birth defects or embryo-fetal death. Pregnancy must be excluded before start of treatment. Prevent pregnancy during treatment by the use of two reliable methods of contraception. REVLIMID is available only through a restricted distribution program called the REVLIMID **REMS program**.[86]

Risk Evaluation and Mitigation Strategy (REMS). FDA's mention of a "risk management plan" referred to the need for a Risk Evaluation and Mitigation Strategy (REMS). Many drugs are FDA-approved with the condition that the Sponsor conduct a REMS. The REMS is published on FDA's website along with the Approval Letter. The Sponsor's REMS for lenalidomide set forth requirements that applied to pharmacies supplying the drug and to patients taking the drug:

> To become a certified pharmacy, the pharmacy must agree to do the following before filling a REVLIMID prescription ... counsel females of reproductive potential on the ... potential for embryo-fetal toxicity with exposure to REVLIMID ... [u]sing 2 forms of effective birth control at the same time or abstaining from heterosexual sexual intercourse ... [o]btaining a pregnancy test weekly during the first 4 weeks of REVLIMID use, then a repeat pregnancy test every 4 weeks in females with regular menstrual cycles, and every 2 weeks in females with irregular menstrual cycles.[87]

[84] Lenalidomide (myelodysplastic syndromes) NDA 021-880. Page 4 of 184-page pdf file of Medical Review.

[85] Lenalidomide (myelodysplastic syndromes) NDA 021-880. Pages 4–5, 20, 39, and 42 of 184-page pdf file Medical Review.

[86] Package label. REVILIMID® (lenalidomide) capsules, for oral use; February 2015 (36 pp.).

[87] Lenalidomide (myelodysplastic syndromes) NDA 021-880. Risk Evaluation and Mitigation Strategy (REMS) (8 pp.).

The REMS further required that, "Each patient ... consents ... by acknowledging that he or she understands that ... severe birth defects or death to an unborn baby may occur if a female becomes pregnant while receiving REVLIMID ... they cannot donate blood while receiving REVLIMID ... they might be asked to participate in the REVLIMID Pregnancy Exposure Registry."[88]

d. Lenvatinib (Radioiodine-Refractory Differentiated Thyroid Cancer) NDA 206-947

FDA's review of lenvatinib defines drug class by the study drug's mechanism of action, where FDA's account of mechanism of action included the inhibition constant (IC50).[89] Each of the topics detailed below found a place on the package label.

Lenvatinib is a small molecule that inhibits various kinases, each taking the form of a different membrane-bound protein. Cicenas and Cicenas[90] reviewed the topic of multikinase inhibitors and single-kinase inhibitors. Lenvatinib is a multikinase inhibitor. Lenvatinib's targets are[91]:

- VEGFR
- FGFR
- PDGFα receptor
- RET/PTC

VEGFR (receptor) is the natural target of VEGF (ligand). The VEGF/VEGFR signaling pathway is used for inducing proliferation and sprouting of endothelial cells lining blood vessels and lymphatic vessels, and promotes angiogenesis and lymphangiogenesis in developing tissues and pathologies, such as cancer.[92]

Comments from FDA further detail lenvatinib's mechanism of action, "Lenvatinib is a receptor tyrosine kinase inhibitor that inhibits the kinase activities of vascular endothelial growth factor (VEGF) receptors VEGFR1 ... VEGFR2 ... and VEGFR3. Lenvatinib also inhibits other ... receptor tyrosine kinases that have been implicated in pathogenic angiogenesis, tumor growth, and cancer progression."[93] In addition to being inhibited by

[88] Lenalidomide (myelodysplastic syndromes) NDA 021-880. Risk Evaluation and Mitigation Strategy (REMS) (8 pp.).

[89] Lenvatinib (radioiodine-refractory differentiated thyroid cancer) NDA 206-947 (185-page pdf file Medical Review).

[90] Cicenas J, Cicenas E. Multi-kinase inhibitors, AURKs and cancer. Med. Oncol. 2016;33:43. doi:10.1007/s12032-016-0758-4.

[91] Lorusso L, et al. Lenvatinib and other tyrosine kinase inhibitors for the treatment of radioiodine refractory, advanced, and progressive thyroid cancer. OncoTargets Ther. 2016;9:6467–77.

[92] Davydova N, et al. Differential receptor binding and regulatory mechanisms for the lymphangiogenic growth factors vascular endothelial growth factor (VEGF)-C and -D. J. Biol. Chem. 2016;291:27265–78.

[93] Lenvatinib (radioiodine-refractory differentiated thyroid cancer) NDA 206-947. Pages 14 (Table 1) and page 30 of 185-page Clinical Review and 185-page pdf file Medical Review.

TABLE 10.1 IC50 for the Kinases Inhibited by Lenvatinib Compared to Other Multikinase Inhibitors

	Inhibition Constant (IC50) Unit of Nanomolar (nM)						
	Lenvatinib	Sorafenib	Motesanib	Axitinib	Pazopanib	Sunitinib	Vandetanib
VEGFR-1	22	26	2	0.1	10	10	–
VEGFR-2	4	90	3	0.2	30	10	40
VEGFR-3	5.2	20	6	0.29	47	10	110
PDGFRβ	39	57	84	2	84	39	–
c-KIT	–	68	8	1.7	74	1–10	–
RET	35	47	59	1.2	–	100	130
B-RAF	–	25	–	–	–	–	–

lenvatinib, VEGFR is also inhibited by sorafenib, pazopanib, sunitinib, motesanib, and axitinib.[94]

i. Defining the Drug Class by a List of Drugs, Including the Study Drug, That are Kinase Inhibitors

FDA's Medical Review provided a table of drugs in the same class as the study drug (lenvatinib). Table 10.1 reproduces, in part, the table from FDA's Medical Review.[95,96] IC50 is the concentration of inhibitor required to produce 50% inhibition of an enzymatic reaction at a specific substrate concentration.[97,98]

This author suggests that the reader devote a few moments to looking up and down the "lenvatinib" column, and contemplate the value for each inhibition constant, and then see if any of the other drugs inhibit the target, such as the target that is VEGFR-2, where the other drug has the same inhibition constant as found with lenvatinib. Generally, the lower the value for the inhibition constant, the more powerful is the drug as an inhibitor. As is evident from Table 10.1, lenvatinib was a powerful inhibitor of *VEGFR-2 and VEGFR-3* (As shown in this table, FDA's prediction of lenvatinib's toxicities was based on AEs seen with other

[94] Lorusso L, et al. Lenvatinib and other tyrosine kinase inhibitors for the treatment of radioiodine refractory, advanced, and progressive thyroid cancer. OncoTargets Ther. 2016;9:6467–77.

[95] Lenvatinib (radioiodine-refractory differentiated thyroid cancer) NDA 206-947. Page 14 of 185-page Clinical Review and 185-page pdf file of Medical Review.

[96] Stjepanovic N, Capdevila J. Multikinase inhibitors in the treatment of thyroid cancer: specific role of lenvatinib. Biol. Targets Therapy 2014;8:129–39.

[97] Yung-Chi C, Prusoff WH. Relationship between the inhibition constant (*KI*) and the concentration of inhibitor which causes 50 per cent inhibition (*I*50) of an enzymatic reaction. Biochem. Pharmacol. 1973;22:3099–108.

[98] Cortes A, et al. Relationships between inhibition constants, inhibitor concentrations for 50% inhibition and types of inhibition: new ways of analysing data. Biochem. J. 2001;357:263–8.

inhibitors of VEGFR-2 and VEGFR-3.) Axitinib was an even more powerful inhibitor of VEGFR-2 and VEGFR-3.

When defining a drug class by way of mechanism of action, and where the goal is to predict AEs for the study drug, it is best to acquire inspiration from AEs of other drugs in the same class, where inhibition by that other drug occurs at concentrations of drug that are achieved in the patient-care situation.

This concerns the scenario where the study drug and Drug X each inhibit only one target, namely, VEGFR-1. Where Drug X (another drug in the same drug class) inhibits VEGFR-1 with a very low IC, such as 10 nM, then the adverse events known to occur with Drug X can justifiably be predicted to occur with your drug. But if Drug X inhibits VEGFR-1 at a very high IC, such as 500 mM (a concentration obtainable in a test tube, but not in any way attainable in vivo when administering to patients), then the adverse events resulting from Drug X cannot justifiably be used to predict the adverse events of the study drug.

Once the drug class was established, FDA's Medical Review disclosed the AEs associated with each of the drugs in the drug class. Table 10.2 lists drugs in the same class as the study drug, together with some of the AEs from the corresponding package labels.[99]

FDA concluded that, "Based on its mechanism of action and IC50's, lenvatinib can be expected to produce toxicities related to VEGFR2/3 ... inhibition and FGFR2 inhibition [such] as **hypertension, proteinuria, and thrombotic events**."[100]

The published literature established that the drug class of VEGF inhibitors is associated with hypertension,[101] proteinuria,[102] and thromboembolism.[103]

ii. Package Labels for Various Kinase Inhibitor Drugs

Developing the drug class analysis even further, FDA pointed out that the drug class of multikinase inhibitors that share the same targets as the study drug (lenvatinib) identified the following AEs. FDA's list of AEs is reproduced as follows[104]:

- Cardiac dysfunction
- Hypertension
- Proteinuria
- Thromboembolitic events
- Hepatic impairment

[99] Lenvatinib (radioiodine-refractory differentiated thyroid cancer) NDA 206-947. Pages 18–20 of 185-page Medical Review.

[100] Lenvatinib (radioiodine-refractory differentiated thyroid cancer) NDA 206-947. Page 23 of 185-page Medical Review.

[101] Escalante CP, Zalpour A. Vascular endothelial growth factor inhibitor-induced hypertension: basics for primary care providers. Cardiol. Res. Practice 2011;Article ID 816897 (8 pp.).

[102] Izzedine H, et al. VEGF signalling inhibition-induced proteinuria: mechanisms, significance and management. Eur. J. Cancer 2010:46:439–48.

[103] Stortecky S, Suter TM. Insights into anticancer cardiovascular side-effects of modern treatment. Curr. Opin. Oncol. 2010;22:312–7.

[104] Lenvatinib (radioiodine-refractory differentiated thyroid cancer) NDA 206-947. Pages 81–82 of 185-page Clinical Review and pages 84–85 of 185-page pdf of Medical Review.

TABLE 10.2 Adverse Events of FDA-Approved Drugs in Same Drug Class as Lenvatinib

Approved Drug	Black Box Warning	Warnings and Precautions on Package Label
Cabozantinib	Perforation, fistula, hemorrhage	Thrombotic events, wound complications, hypertension, osteonecrosis of the jaw, proteinuria, embryofetal toxicity
Lapatinib	Hepatotoxicity	Decreases in left ventricular ejection fraction, diarrhea, ILD and pneumonitis, QT interval prolongation
Nilotinib	QT prolongation	Myelosuppression, sudden death, cardiac and vascular events, pancreatitis and elevated serum lipase, hepatotoxicity
Pazopanib	Hepatotoxicity	Increases in serum transaminase levels and bilirubin, prolonged QT intervals and torsades de pointes, cardiac dysfunction, fatal hemorrhagic events
Regorafenib	Hepatotoxicity	Hemorrhage, dermatological toxicology, hypertension, cardiac ischemia and infarction, gastrointestinal perforation or fistulae
Sunitinib	Hepatotoxicity	Cardiac toxicity, prolonged QT intervals, torsade de pointes, hypertension, hemorrhagic events, osteonecrosis of the jaw
Vandetanib	QT prolongation, sudden death, torsades	Severe skin reactions, ischemic cerebrovascular events, hemorrhage, heart failure, diarrhea
Ponatinib	Vascular occlusion, heart failure, hepatotoxicity	Hypertension, pancreatitis, neuropathy, ocular toxicity, hemorrhage, fluid retention, cardiac arrhythmias, myelosuppression

Lenvatinib (radioiodine-refractory differentiated thyroid cancer) NDA 206-947. Pages 21–23 of 185-page Medical Review.

- Renal impairment
- QTc prolongation
- Palmoplantar dysesthesia syndrome.

Further focusing on details of the drug class of multikinase inhibitors, FDA reproduced the names of several FDA-approved drugs in PKI drug class, together with the AEs listed on their package labels. FDA's presentation is reproduced in Table 10.2.[105] This presentation is truly exemplary and it provides a reasonable approach for conducting a drug class analysis for any kind of drug.

Regarding this table, FDA wrote, "As can be seen from the table ... many of the common toxicities of these agents can be attributed to specific kinase inhibition by the drugs, for example, VEGFR receptor inhibition and hypertension, proteinuria. Based on its mechanism of action and relative IC50s, lenvatinib can be expected to produce toxicities related to VEGFR2/3 inhibition and FGFR2 inhibition such as hypertension, proteinuria, and thrombotic events."[106]

If one contemplates the AEs from the labels for sorafenib, sunitinib, pazopanib, and vandetanib, it will be evident that some of these same AEs found a place on the package

[105] Lenvatinib (radioiodine-refractory differentiated thyroid cancer) NDA 206-947. Pages 18–22 of 185-page Medical Review.

[106] Lenvatinib (radioiodine-refractory differentiated thyroid cancer) NDA 206-947. Pages 21–23 of 185-page Medical Review.

label for lenvatinib. Taken together, FDA's review for lenvatinib shows the use of the following flow chart for arriving at the warnings on lenvatinib's label:

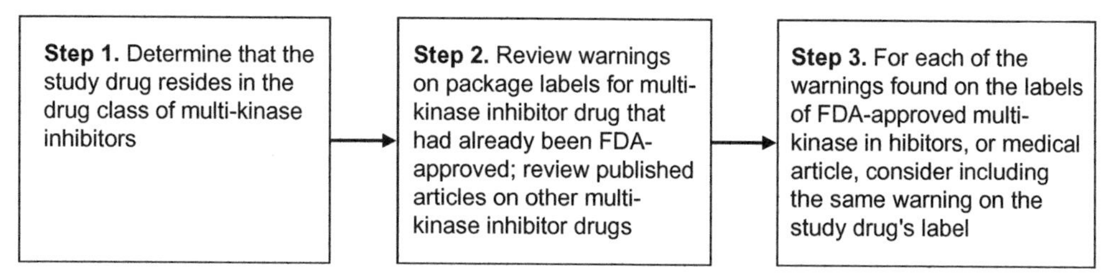

FDA's review of lenvatinib, for the AEs of hypertension, proteinuria, thromboembolic events, and QT prolongation, took into account AE warnings from package labels of other FDA-approved drugs of the same class (or from published medical articles). This feature of FDA's drug class analysis is revealed by these excerpts from FDA's review of lenvatinib:

- Regarding *hypertension*, FDA stated, "the patient had a convulsion related to elevated blood pressure ... [b]ased on these data, and consistent with the VEGF targeted effects, life threatening **hypertension** can occur ... [with] lenvatinib and hence ... is included in ... the label."[107]
- Similarly, for the AE of *proteinuria*, FDA stated that, "**Proteinuria** as an adverse event is expected considering the VEGF targeted mechanism of action of lenvatinib."[108]
- Regarding *thromboembolic events*, FDA stated that, "This reviewer notes that ... although the risk of ... **thromboembolic events** appears to be two-fold higher in patients treated with lenvatinib compared to placebo, it is comparable to the increased risk ... with other VEGF targeted tyrosine kinase inhibitors ... such as sorafenib and ... bevacizumab."[109]
- Regarding *hypocalcemia*, FDA stated that, "This reviewer ... concludes that most patients had ... **hypocalcemia** ... and that the hypocalcemia appeared to be exacerbated by lenvatinib [the study drug] ... [h]ypocalcemia led to QT prolongation ... [t]his reviewer hence concludes that hypocalcemia can occur following exposure to lenvatinib, and has been reported with drugs of similar class including sorafenib ... [t]his reviewer recommends addition of hypocalcemia to the Warnings and Precautions section of the label."[110]
- Regarding *QT prolongation*, FDA wrote that the, "safety concerns for treatment with lenvatinib were ... hypertension, proteinuria, arterial thromboembolic events, liver and

[107] Lenvatinib (radioiodine-refractory differentiated thyroid cancer) NDA 206-947. Pages 116–118 of 185-page Clinical Review and pages 119–121 of 185-page pdf file of Medical Review.

[108] Lenvatinib (radioiodine-refractory differentiated thyroid cancer) NDA 206-947. Pages 116–118 of 185-page Clinical Review and page 121 of 185-page pdf file of Medical Review.

[109] Lenvatinib (radioiodine-refractory differentiated thyroid cancer) NDA 206-947. Page 124 of 185-page pdf file of Medical Review.

[110] Lenvatinib (radioiodine-refractory differentiated thyroid cancer) NDA 206-947. Page 133–134 of 185-page Clinical Review and pages 136–137 of 185-page pdf file of Medical Review.

renal impairment, **QT prolongation** ... [t]hese events have been reported in the literature with other approved tyrosine kinase inhibitors targeting VEGF, such as sorafenib and cabozantinib."[111]

Package label. The results from FDA's drug class analysis found a place on the package label for lenvatinib (Lenvima®), where the AEs listed in FDA's analysis are highlighted in the following package label excerpts:

> **WARNINGS AND PRECAUTIONS. Hypertension**: Control blood pressure prior to treatment with LENVIMA. Withhold LENVIMA for Grade 3 hypertension despite optimal hypertensive therapy. Discontinue for life-threatening hypertension ... **Arterial Thromboembolic Events**: Discontinue LENVIMA following an arterial thromboembolic event ... **Proteinuria**: Monitor for proteinuria before initiation of, and periodically throughout, treatment with LENVIMA. Withhold LENVIMA for ≥2 grams of proteinuria for 24 hours. Discontinue for nephrotic syndrome ... **QT Interval Prolongation**: Monitor and correct electrolyte abnormalities in all patients. Withhold LENVIMA for the development of Grade 3 or greater QT interval prolongation ... **Hypocalcemia**: Monitor blood calcium levels at least monthly and replace calcium as necessary.[112]

e. Nivolumab (Metastatic Melanoma) BLA 125-554

FDA's review for nivolumab shows that drug class analysis was used to arrive at package label warnings, and also to formulate a suitable Risk Evaluation and Mitigation Strategy (REMS). Nivolumab is an antibody that activates T cells to mount an immune response against various cancers, including melanoma,[113,114] lymphoma,[115,116] nonsmall cell lung cancer (NSCLC),[117] renal cell carcinoma,[118] and ovarian cancer.[119]

Nivolumab binds to PD-1, thus blocking signaling mediated by binding of PD-1 to PD-L1, as well as blocking signaling mediated by binding of PD-1 to PD-L2.

[111] Lenvatinib (radioiodine-refractory differentiated thyroid cancer) NDA 206-947. Pages 133–133 of 185-page pdf file of Medical Review.

[112] Package label. LENVIMA (lenvatinib) capsules, for oral use; February 2015 (19 pp.).

[113] Kohn CG, et al. Cost-effectiveness of immune checkpoint inhibition in BRAF wild-type advanced melanoma. J. Clin. Oncol. 2017;35:1194–202.

[114] Weber JS, et al. Safety, efficacy, and biomarkers of nivolumab with vaccine in ipilimumab-refractory or -naive melanoma. J. Clin. Oncol. 2013;31:4311–8.

[115] Ansell SM, et al. PD-1 blockade with nivolumab in relapsed or refractory Hodgkin's lymphoma. New Engl. J. Med. 2015;372:311–9.

[116] Lesokhin AM, et al. Nivolumab in patients with relapsed or refractory hematologic malignancy: preliminary results of a phase Ib study. J. Clin. Oncol. 2016;34:2698–704.

[117] Gettinger S, et al. Nivolumab monotherapy for first-line treatment of advanced non-small-cell lung cancer. J. Clin. Oncol. 2016;34:2980–7.

[118] Motzer RJ, et al. Nivolumab versus everolimus in advanced renal-cell carcinoma. New Engl. J. Med. 2015;373:1803–13.

[119] Hamanishi J, et al. Safety and antitumor activity of anti-PD-1 antibody, nivolumab, in patients with platinum-resistant ovarian cancer. J. Clin. Oncol. 2015;33:3015–4022.

i. Nivolumab's Mechanism of Action

Cancer cells use various tactics to promote their own growth and survival. These include tactics for evading the immune system. In one tactic for immune evasion, cancer cells express PD-L1 on their plasma membrane, where PD-L1 (ligand) transmits signals to PD-1 (receptor) expressed by nearby immune cells (T cells). Transmission of the signal from the ligand (PD-L1) to the receptor (PD-1) inhibits the immune cell and, as a consequence, the cancer cell prevents the T cell from killing the cancer cell. PD-L1 expression by cancer cells facilitates escape from immune attack.[120] PD-1, the dominant receptor for PD-L1, is found on activated T cells, B cells, and NK cells in the tumor microenvironment.

ii. Therapeutic Protein Drug Class Versus Small Molecule Drug Class

Where a drug is an antibody, the definition of "drug class" encompasses much less than with small molecule drugs. Therapeutic antibodies are usually specific for only one target unlike, for example, the small molecule TKIs that are "multikinase inhibitors".[121] Also, defining drug class by chemical structure is almost impossible for therapeutic antibodies, because it would require comparing the three-dimensional structures of the binding sites for each of the antibodies that are candidates for being in the same drug class. In contrast, the structure of any small molecule drug can be described within the space of a minute, using a pencil and paper.

iii. Definition of Nivolumab's Drug Class

FDA defined the relevant drug class in terms of a list of antibody drug molecules (nivolumab, ipilimumab, pembrolizumab) with a shared mechanism of action. The shared mechanism took the form of cell-to-cell signaling pathways that prevent T cells from killing cancer cells. FDA's drug class definition took the form of the combination of nivolumab with, "other FDA-approved immune-modulating agents, ipilimumab and pembrolizumab."[122]

iv. Nivolumab's Mechanism Versus Ipilimumab's Mechanism

Although nivolumab and ipilimumab each block cell-to-cell signaling pathways that dampen T-cell activity, the targets of these two therapeutic antibodies are not the same, and the respective signaling pathways work differently from each other. Nivolumab binds to PD-1, whereas ipilimumab binds to cytotoxic T lymphocyte-associated protein-4 (CTLA-4). CTLA-4 binds to two different ligands, CD80 and CD86, which are also called B7.1 and B7.2, respectively.

When T cells infiltrate a tumor, for example, a melanoma tumor, the T cells are inhibited by both PD-L1 (expressed by tumor cells) and by CTLA-4 (expressed by suppressor

[120] Taube JM, et al. Differential expression of immune-regulatory genes associated with PD-L1 display in melanoma: implications for PD-1 pathway blockade. Clin. Cancer Res. 2015;21:3969–76.

[121] Gao L, et al. FGF19/FGFR4 signaling contributes to the resistance of hepatocellular carcinoma to sorafenib. J. Exp. Clin. Cancer Res. 2017;36:8 (10 pp.).

[122] Nivolumab (metastatic melanoma) BLA 125-554. Page 12 of 156-page Medical Review.

cells known as Tregs).[123] CTLA-4 is also expressed by activated effector T cells.[124] The way cells use their CTLA-4 to dampen immune response is that the CTLA-4 contacts CD80 and CD86 from an adjacent T cell, and yanks the CD80 and CD86 right out of the T cell, and then the cell degrades the CD80 and CD86.[125,126] This yanking out followed by degradation is called "trans-endocytosis."

The following concerns AEs that result from the above immune-modulating therapeutic proteins. Therapeutic anti-CTLA-4 antibodies have the AE of generating autoimmunity, resulting inflammation of various organs. Therapeutic anti-PD-L antibodies and therapeutic anti-PD-L1 antibodies also generate autoimmunity, but to a lesser extent than anti-CTLA-4 antibodies.[127]

v. Using Drug Class Analysis to Predict AEs for Nivolumab

FDA's Medical Review listed AEs associated with the earlier-approved drugs, ipilimumab, and pembrolizumab. Regarding the AEs for ipilimumab, FDA wrote, "Important Safety Issues With Consideration to Related Drugs. The **ipilimumab** prescribing information includes a boxed warning based on the risk of severe and fatal immune-mediated reactions due to T-cell activation and proliferation. The most common severe immune-mediated reactions included enterocolitis, hepatitis, dermatitis (including toxic epidermal necrolysis; TEN), neuropathy, and endocrinopathy."[128]

Regarding the AEs for pembrolizumab, FDA wrote, "The primary safety risks of **pembrolizumab** are immune-mediated adverse events which were managed with corticosteroids. The most common immune-mediated adverse reactions included pneumonitis, colitis, hepatitis, hypophysitis, nephritis, hyperthyroidism, and hypothyroidism."[129]

Regarding the sum of the AEs associated with other drugs that blocked the PD-1/PD-L1 signaling pathway, FDA's narrative was, "The reviewer identified additional AEs of special interest to identify immune-mediated adverse reactions of nivolumab based on **AEs observed with other anti-PD-1/PD-L1 approved drugs** or investigational agents. These included hypophysitis, hypopituitarism, myasthenia gravis, diabetic ketoacidosis, rhabdomyolysis, transverse myelitis, diabetes mellitus, and hypogonadism."[130]

[123] Curran MA, et al. PD-1 and CTLA-4 combination blockade expands infiltrating T cells and reduces regulatory T and myeloid cells within B16 melanoma tumors. Proc. Natl Acad. Sci. 2010;107:4275–80.

[124] Jeffrey LE, et al. Vitamin D antagonises the suppressive effect of inflammatory cytokines on CTLA-4 expression and regulatory function. PLoS One 2015;10:e0131539.

[125] Qureshi OS, et al. Trans-endocytosis of CD80 and CD86: a molecular basis for the cell-extrinsic function of CTLA-4. Science 2011;332:600–3.

[126] Hou TZ, et al. A transendocytosis model of CTLA-4 function predicts its suppressive behavior on regulatory T cells. J. Immunol. 2015;194:2148–59.

[127] Curran MA, et al. PD-1 and CTLA-4 combination blockade expands infiltrating T cells and reduces regulatory T and myeloid cells within B16 melanoma tumors. Proc. Natl Acad. Sci. 2010;107:4275–80.

[128] Nivolumab (metastatic melanoma) BLA 125-554. Page 16 of 156-page Medical Review.

[129] Nivolumab (metastatic melanoma) BLA 125-554. Page 16 of 156-page Medical Review.

[130] Nivolumab (metastatic melanoma) BLA 125-554. Page 109 of 156-page Medical Review.

FDA's recommendations for nivolumab's package label mentioned pneumonitis (pneumonitis is a risk from pembrolizumab), "Pneumonitis can have highly variable appearances on radiographic images. Despite the initiation of a pneumonitis management algorithm for early recognition and management of pneumonitis, an additional patient died in another study. The reviewer recommends a **boxed warning in the label for pneumonitis**."[131]

Package label. Most of the AEs identified by FDA's drug class analysis for nivolumab (Opdivo®) found a place on the package label. (The only exception was hypogonadism.) As is evident, many of the AEs took the form of autoimmunity, that is, AEs where the body's T cells inflicted damage on the body's own tissues. This type of indiscriminate response might be expected, in view of the mechanism of action of nivolumab. The Warnings and Precautions section read:

> **WARNINGS AND PRECAUTIONS. Immune-mediated pneumonitis**: Withhold for moderate and permanently discontinue for severe or life-threatening pneumonitis ... **Immune-mediated colitis**: Withhold OPDIVO when given as a single agent for moderate or severe and permanently discontinue for life-threatening colitis ... **Immune-mediated hepatitis**: Monitor for changes in liver function. Withhold for moderate and permanently discontinue for severe or life-threatening transaminase or total bilirubin elevation ... **Immune-mediated endocrinopathies**: Withhold for moderate or severe and permanently discontinue for life-threatening hypophysitis ... Withhold for severe and permanently discontinue for life-threatening hyperglycemia ... **Immune-mediated nephritis and renal dysfunction**: Monitor for changes in renal function. Withhold for moderate or severe and permanently discontinue for life-threatening serum creatinine elevation ... **Immune-mediated skin adverse reactions**: Withhold for severe and permanently discontinue for life-threatening rash.[132]

The package label warnings were based on drug class analysis, but also on AEs found in the Sponsor's own clinical studies. Some of the AEs in the following excerpt from FDA's Medical Review, correspond to AEs in the Warnings and Precautions section of the label, while others of these AEs instead found a place in the Adverse Reactions section of the label, as shown here:

> **ADVERSE REACTIONS** ... The primary safety risks of nivolumab are immune-mediated adverse events which occurred in 59% of patients in the nivolumab group. The most common immune mediated adverse events in more than 2% of patients were rash, diarrhea, pruritus, hepatotoxicity, hypothyroidism, vitiligo, pneumonitis, hyperthyroidism, and dermatitis. Seven percent of all immune-mediated events were Grade 3 or 4. Also observed at a lower frequency were colitis, nephritis, autoimmune thyroiditis ... autoimmune neuropathy, demyelination, pancreatitis, and ... abducens nerve paresis.[133]

vi. Drug Class Analysis and the REMS

FDA's drug class analysis was used to decide whether the Sponsor should plan a REMS, "This reviewer **does not recommend** a Risk Evaluation and Mitigation Strategy (REMS) given the current safety profile of nivolumab and the experience of the medical

[131] Nivolumab (metastatic melanoma) BLA 125-554. Page 84 of 156-page Medical Review.

[132] Package label. OPDIVO (nivolumab) injection, for intravenous use; February 2017 (58 pp.).

[133] Nivolumab (metastatic melanoma) BLA 125-554. Page 11 of 156-page Medical Review.

community in managing immune-mediated adverse reactions based on use of other FDA-approved immune modulating agents such as ipilimumab."[134]

FDA's Risk Assessment and Risk Mitigation Review concluded that there need not be any REMS. FDA's logic and reasoning was that oncologists are already familiar with safety risks for other antibody drugs (ipilimumab; pembrolizumab) that inhibit PD-L1/PD-1 signaling.

The FDA reviewer wrote, "The ... Division ... concurred that **nivolumab does not require a REMS** to ensure that the benefits outweigh the serious risks of immune-mediated skin rash, colitis ... pneumonitis, and nephritis ... concluded that oncology healthcare providers are informed on similar ... immune-mediated risks based use of **ipilimumab** (IPI), approved on March 25, 2011, for the treatment of ... melanoma. This same rationale was applied ... to not require a REMS for ... **pembrolizumab** ... approved September 4, 2014."[135]

Accordingly, FDA's Approval Letter for *nivolumab* did not impose any requirement for a REMS.[136]

But regarding another earlier-approved antibody (ipilimumab), FDA's Approval Letter dated March 2011 did require a REMS. FDA's Approval Letter insisted that the Sponsor devise a REMS, "we have determined that a **REMS is necessary for** ... **ipilimumab** to ensure the benefits of the drug outweigh the risks of severe and fatal immune-mediated adverse reactions such as fatal immune-mediated enterocolitis (including gastrointestinal perforation), fatal immune-mediated hepatitis (including hepatic failure), fatal immune-mediated toxicities of the skin."[137]

f. Osimertinib (Epidermal Growth Factor Receptor T790M Mutation Positive NSCLC) NDA 208-065

FDA's review of osimertinib analyzes an adverse event that is rare and unpredictable in its timing, and difficult to diagnose. This AE is interstitial lung disease (ILD). FDA's review for osimertinib showed that AEs of this kind can be predicted by the combination of:

- drug class analysis and
- subgroup analysis.

This concerns osimertinib, a small molecule that inhibits a mutated form of the signaling protein, epidermal growth factor receptor (EGFR). EGFR is part of signaling pathways that regulate tumorigenesis, invasion, and metastasis. In patients with NSCLC, the cancer may arise from mutations in EGFR gene. These mutations occur in 10%-35% of NSCLC patients.[138] Because these mutations contribute to the cause of the cancer, EGFR inhibitors

[134] Nivolumab (metastatic melanoma) BLA 125-554. Page 11 of 156-page Medical Review.

[135] Nivolumab (metastatic melanoma) BLA 125-554 (22-page Risk Assessment and Risk Mitigation Review).

[136] Nivolumab (metastatic melanoma) BLA 125-554. Approval Letter December 22, 2014.

[137] Ipilimumab (melanoma) BLA 125-377. Approval Letter March 25, 2011.

[138] Zhang H. Three generations of epidermal growth factor receptor tyrosine kinase inhibitors developed to revolutionize the therapy of lung cancer. Drug Des. Develop. Ther. 2016;10:3867–72.

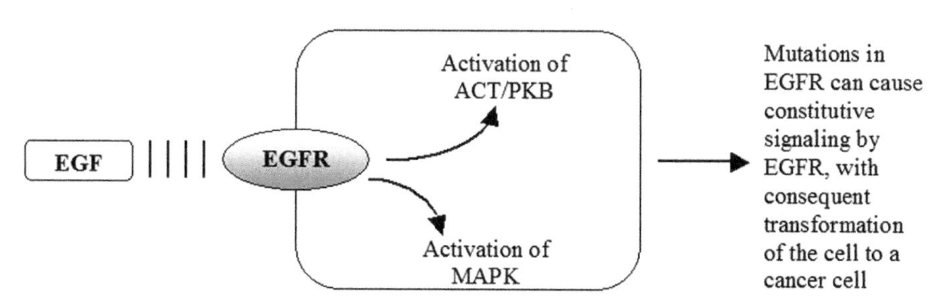

FIGURE 10.2 Pathway of EGF-mediated cell signaling. The picture shows EGF binding to EGFR, which provokes signaling events that activate ACT/PKB and MAPK. Some mutations in EGFR result in constitutively activated cell signaling by this pathway.

such as osimertinib, rociletinib, and olmutinib have been developed that inhibit the mutated EGFR (T790M EGFR) while sparing the wild-type EGFR.

EGFR is a tyrosine kinase, which catalyzes the phosphorylation of itself (autophosphorylation), and where this autophosphorylation activates the EGFR.[139]

Autophosphorylation occurs on specific tyrosine residues of EGFR. Once activated by autophosphorylation, EGFR transmits signals that activate downstream pathways in the cell. These downstream pathways are the ACT/PKB signaling pathway and the MAPK signaling pathway.

Fig. 10.2 illustrates the pathway, where the EGF (the ligand) is the starting point, EGF binds to EGFR (receptor), and transformation of the cell to a cancer cell is the end-result. The hatch marks in Fig. 10.2 are conventionally used to represent hydrogen bonds, for example, the hydrogen bonds that mediate EGF to EGFR binding.

i. Consistent Account of Drug Class

FDA's Medical Review of osimertinib provides a consistent account of the drug class, as shown by these excerpts. The drug class is EGFR TKIs. FDA's remarks describing this class include:

> The most frequently-observed treatment emergent adverse effects were consistent with expected toxicities based on preclinical data, and previous clinical experience with osimertinib, and class effects of other **EGFR TKIs**.[140]

> Based on toxicities identified in studies of other **EGFR TKIs**, there was an expectation that similar toxicities were possible with osimertinib. Known class toxicities include skin rash and nail changes, QT prolongation, diarrhea, ocular events, and interstitial lung disease (ILD).[141]

[139] Morgillo F, et al. Mechanisms of resistance to EGFR-targeted drugs: lung cancer. ESMO Open. 2016;1: e000060 (9 pp.).

[140] Osimertinib (epidermal growth factor receptor T790M mutation positive NSCLC) NDA 208-065. Page 17 of 137-page pdf file of Medical Review.

[141] Osimertinib (epidermal growth factor receptor T790M mutation positive NSCLC) NDA 208-065. Page 61 of 137-page pdf file of Medical Review.

> The most important rare ... adverse event that was thought to be the cause of 4 treatment-related deaths on study was interstitial lung disease. This is a known class effect of **EGFR TKIs**.[142]

What is consistent is FDA's connection of the drug class of EGFR TKIs with the "class toxicities" associated with EGFR TKIs. FDA's Medical Review defined another drug class, namely, a drug class taking the form of the combination of EGFR TKIs and ALK TKIs.[143] ALK means anaplastic lymphoma kinase.

ii. Sources of Information for Arriving at Safety Labeling

FDA's Medical Review provides the three sources of information used for drafting the package label. The sources were:

- Animal toxicity data
- AE data from the Sponsor's own clinical studies
- Drug class analysis

On these three sources, FDA wrote that, "A number of adverse events of special interest were prospectively identified by the Sponsor based on preclinical data, emerging safety data from clinical trials, and known class effects of EGFR TKIs."[144]

The AEs experienced by the study subjects were described by FDA as, "The general categories identified were skin effects, diarrhea, upper GI inflammation, interstitial lung disease (ILD)/pneumonitis, nail effects, ocular effects, and cardiac effects (including QT and cardiac contractility) ... [t]hese toxicities were in fact the most clinically relevant toxicities seen during safety review of osimertinib, with interstitial lung disease (ILD) being a cause of the only deaths related to osimertinib, and with diarrhea, skin, and nail effects being the most common toxicities, each affecting >20% of patients on-study."[145]

FDA's analysis reported that a particular type of AE is rare, unpredictable in its timing, and difficult to diagnose. This AE is interstitial lung disease. In FDA's own words, "Interstitial lung disease (ILD) and pneumonitis, while rare overall, occurred at rates comparable and even slightly higher than those reported with **other EGFR tyrosine kinase inhibitors** in NSCLC. There were four deaths in the phase 2 cohorts that occurred in patients who developed ILD/pneumonitis. Review of Case Report Forms (CRFs) for patients on-study showed that often, **ILD is difficult to diagnose** and occurs in the setting of overlapping lung infection ... [t]he onset of ILD with osimertinib is **not predictable in its timing**, with events occurring both early and late in the course of

[142] Osimertinib (epidermal growth factor receptor T790M mutation positive NSCLC) NDA 208-065. Page 19 of 137-page pdf file of Medical Review.

[143] Osimertinib (epidermal growth factor receptor T790M mutation positive NSCLC) NDA 208-065. Page 105 of 137-page Medical and Statistical Reviews.

[144] Osimertinib (epidermal growth factor receptor T790M mutation positive NSCLC) NDA 208-065. Page 105 of 137-page Medical and Statistical Reviews.

[145] Osimertinib (epidermal growth factor receptor T790M mutation positive NSCLC) NDA 208-065. Page 105 of 137-page Medical and Statistical Reviews.

therapy. Thus, vigilance about this toxicity is warranted by all practitioners prescribing osimertinib."[146]

FDA's analysis of the AE that was rare, unpredictable in its timing, and difficult to diagnose continued as follows. As stated by FDA, "While a relatively **uncommon toxicity** seen in patients on osimertinib, **ILD was cause of all osimertinib-related deaths**. This is not unexpected; fatal cases of ILD have been reported in all other trials of small molecule TKIs in non-small cell lung cancer (NSCLC). Consequently, ILD/pneumonitis appears in the Warnings and Precautions section of **labels for all other approved TKIs in NSCLC**, of which there are five. Table 44 is a comparison of the rates of occurrence of ILD/pneumonitis using numbers reported on FDA labels for each approved TKI in NSCLC. Please interpret cross TKI comparisons with great caution."[147]

The excerpt, "While a relatively uncommon toxicity seen in patients on osimertinib, **ILD was cause of all osimertinib-related deaths**," reveals a technique of subgroup analysis for detecting AEs that are of special interest. In other words, even though ILD is rare and unpredictable in its timing, and is difficult to diagnose, *subgroup analysis that looks only at subjects that died* brings ILD out of the shadows and into the limelight. This technique of subgroup analysis may or may not include a relatedness analysis. In other words, in using this technique of subgroup analysis, the Sponsor may or may not have conducted an inquiry as to whether the death resulted from the interstitial lung disease (or if death resulted from some other cause). In the situation where this relatedness analysis was conducted, and also in the situation where this relatedness was not conducted, the Sponsor has the option of submitting adverse event data by way of tables that disclose the degree of severity of each type of AE, where the tables discloses whether the subject died, and also where the tables include an opinion on relatedness of the death to the AE. Chapter 13 (Coding) reveals a format for submitting AE data by way of tables that disclose severity of each type of AE. This is illustrated for a clinical study on the trifluoridine/tipiracil combination (NDA 207–981).

As shown in Table 10.3, drug class analysis can be used as a basis for predictions whether osimertinib poses a risk for ILD, and as a basis for labeling. The table provides a consistent picture of ILD across various drugs in the same drug class (EGFR plus ALK drug class), where the consistent picture is that each of the drugs in this particular drug class shows a similar rate of the adverse event of ILD.[148]

Package label. The package label for osimertinib (Tagrisson®) reveals the outcome of subgroup analysis (all subjects that died had died of ILD) and drug class analysis (drug class results that ILD consistently occurs with all drugs that inhibit either EGFR or ALK). The outcome was this warning in the Warnings and Precautions section:

[146] Osimertinib (epidermal growth factor receptor T790M mutation positive NSCLC) NDA 208-065. Page 105 of 137-page Medical and Statistical Reviews.

[147] Osimertinib (epidermal growth factor receptor T790M mutation positive NSCLC) NDA 208-065. Page 105 of 137-page Medical and Statistical Reviews.

[148] Osimertinib (epidermal growth factor receptor T790M mutation positive NSCLC) NDA 208-065. Page 105 of 137-page Medical and Statistical Reviews.

TABLE 10.3 Interstitial Lung Disease (ILD) in FDA-Approved TKIs for Treating NSCLC

Drug	Target	Rate of ILD (%)
Erlotinib	EGFR	1.1
Afatinib	EGFR	1.5
Gefitinib	EGFR	1
Crizotinib	ALK	1.6
Ceritinib	ALK	4

WARNINGS AND PRECAUTIONS. Interstitial Lung Disease (ILD)/ Pneumonitis: Occurred in 3.3% of patients. Permanently discontinue TAGRISSO in patients diagnosed with ILD/Pneumonitis ... Withhold TAGRISSO and promptly investigate for ILD in any patient who presents with worsening of respiratory symptoms which may be indicative of ILD (e.g., dyspnea, cough and fever). Permanently discontinue TAGRISSO if ILD is confirmed.[149]

g. Pomalidomide (Multiple Myeloma) NDA 204-026

Pomalidomide is a small molecule drug for treating multiple myeloma. Although pomalidomide is the only drug present in the tablet, the package label states that pomalidomide is to be taken in combination with dexamethasone. The package label reads, "POMALYST is a thalidomide analogue indicated, in combination with dexamethasone, for patients with multiple myeloma."[150] Pomalidomide with low-dose dexamethasone acts synergistically in treating multiple myeloma.[151]

Structure-based drug class analysis, combined with animal toxicity studies, was used to arrive at the Black Box Warning on the package label.[152] FDA's Medical Review provided structures for pomalidomide and for two other drugs in the same drug class (Fig. 10.3). The circles in Fig. 10.3 highlight parts of the structures are moieties that confer individuality to drugs in this drug class.

Regarding safety analysis using the combination of structure-based drug class analysis and animal toxicity data, FDA wrote, "There are no ... studies in pregnant women. However, pomalidomide was found to be **teratogenic in both rats and rabbits**, when administered during the period of organogenesis. Since pomalidomide is an **analogue of thalidomide** ... no studies were conducted on pregnant women."[153]

[149] Package label. TAGRISSON™ (osimertinib) tablet, for oral use; November 2015 (16 pp.).

[150] Package label. POMALYST® (pomalidomide) capsules, for oral use; June 2016 (27 pp.).

[151] Richardson PG, et al. Pomalidomide alone or in combination with low-dose dexamethasone in relapsed and refractory multiple myeloma: a randomized phase 2 study. Blood 2014;123:1826–32.

[152] Pomalidomide (multiple myeloma) NDA 204-026 (94-page Clinical Review and 101-page pdf file of Medical Review).

[153] Pomalidomide (multiple myeloma) NDA 204-026. Page 14 of 94-page Clinical Review and 101-page pdf file of Medical Review.

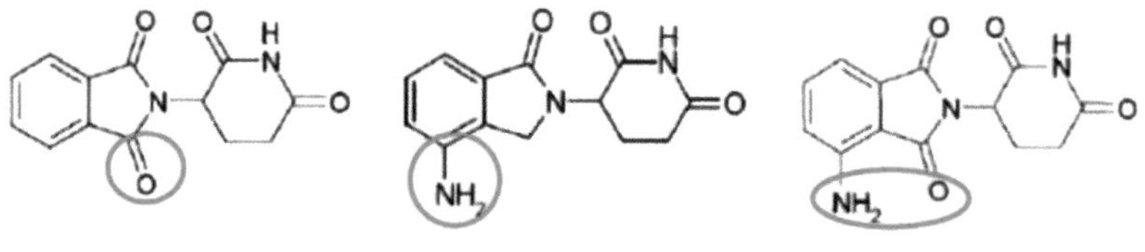

Thalidomide Lenalidomide Pomalidomide

FIGURE 10.3 Thalidomide and two other drugs in the same drug class. Thalidomide, lenalidomide, and pomalidomide are shown. The *circles* placed over each structure show the chemical moieties that are unique to each of these three drugs.

The FDA reviewer moved a step forward, and issued a requirement for the package label, "Therefore pomalidomide is contraindicated in pregnant women and women capable of becoming pregnant."[154]

FDA's drug class analysis extended into adverse events unrelated to teratogenicity, such as myelosuppression. Regarding these other AEs, FDA referred to warnings on package labels of other drugs in the same drug class, namely, thalidomide and lenalidomide, and observed that, "The ... toxicology review noted of immunomodulatory ... effects for pomalidomide similar to ... thalidomide and lenalidomide ... [t]he safety profile for pomalidomide is notable for the similarity to other [drugs] ... [t]hese similar features include **myelosuppression** (neutropenia, thrombocytopenia and anemia), infections, neuropathy, dizziness, GI toxicity, confusional state and fatigue."[155]

Package label. The package label for pomalidomide (Pomalyst®) warned against toxicity to the fetus, and stated that the pomalidomide is a thalidomide analogue. The label referred to the Risk Evaluation and Mitigation Strategy (REMS):

> **WARNING:** EMBRYO-FETAL TOXICITY ... POMALYST is contraindicated in pregnancy. POMALYST is a thalidomide analogue. Thalidomide is a known human teratogen that causes severe life-threatening birth defects ... For females of reproductive potential: Exclude pregnancy before start of treatment. Prevent pregnancy during treatment by the use of 2 reliable methods of contraception ... POMALYST is available only through a restricted program called POMALYST REMS.[156]

Comments from FDA reviewers about neutropenia found a place in the package label, in the Warnings and Precautions section and also in the Adverse Reactions section, as shown:

> **WARNINGS AND PRECAUTIONS.** Hematologic Toxicity: **Neutropenia** was the most frequently reported Grade 3/4 adverse event. Monitor patients for hematologic toxicities, especially neutropenia.[157]

[154] Pomalidomide (multiple myeloma) NDA 204-026. Page 91 of 94-page Clinical Review and 101-page pdf file of Medical Review.

[155] Pomalidomide (multiple myeloma) NDA 204-026. Page 69 of 94-page Clinical Review and page 70 of 101-page pdf file of Medical Review.

[156] Package label. POMALYST® (pomalidomide) capsules, for oral use; June 2016 (27 pp.).

[157] Package label. POMALYST® (pomalidomide) capsules, for oral use; June 2016 (27 pp.).

ADVERSE REACTIONS. Most common adverse reactions (≥30%) included fatigue and asthenia, **neutropenia**, anemia, constipation, nausea, diarrhea, dyspnea, upper respiratory tract infections, back pain, and pyrexia.[158]

h. Ponatinib (CP-CML; AP-CML, BP-CML; Ph + ALL) NDA 203-469

Ponatinib is a small molecule drug for treating various hematological cancers.[159] The abbreviations for these types of cancers are explained in the footnote.[160] CML is chronic myeloid leukemia. Ponatinib belongs to the drug class, TKIs. The pathway of research leading to the invention of ponatinib started with the realization that cancer patients developed resistance to another drug (imatinib). As reviewed by O'Hare et al.[161]:

> Resistance to imatinib ... involves point mutations in the kinase domain of BCR-ABL that impair inhibitor binding ... identification of BCR-ABL kinase domain mutations provides a potential explanation for imatinib resistance and suggests a clear treatment strategy ... [t]herefore, an optimal next-generation ... inhibitor capable of exerting ... disease control in CML ... we report on the design and ... of ponatinib, an orally active pan-inhibitor of BCR-ABL, including BCR-ABL T315I.

Ponatinib inhibits the enzyme activity of the mutated protein (the BCR-ABL fusion protein), which takes the form of the protein encoded by the Philadelphia chromosome, and it is also the case that ponatinib inhibits the enzyme activity of BCR-ABL fusion protein that has acquired a point mutation. The point mutation is T315I. The mechanism by which this mutation causes the enzyme to be resistant to anti-cancer drugs, such as imatinib, has been described as, "Substitution of the Bcr-Abl gatekeeper position with isoleucine (T315I) results in the loss of a direct hydrogen bond with the drug and also creates steric clash."[162] To summarize this situation, the formation of the gene encoding the BCR-ABL fusion protein is bad, because it causes cancer. And the materialization of the T315I point mutation in the BCR-ABL fusion protein is also bad, because it makes the BCR-ABL fusion protein resistant to the anti-cancer drug, imatinib.

i. Identifying the Drug Class by Mechanism of Action, Where Other Drugs Share the Same Mechanism

FDA's Medical Review identified the TKI drug class, "The treatment of CML and Ph + ALL have been revolutionized with the advent of tyrosine kinase inhibitors (TKI). The following TKIs are FDA-approved for the treatment of CML: imatinib (2001), dasatinib

[158] Package label. POMALYST® (pomalidomide) capsules, for oral use; June 2016 (27 pp.).

[159] Ponatinib (CP-CML; AP-CML, BP-CML; Ph + ALL) NDA 203-469.

[160] The abbreviations mean, chronic phase (CP), accelerated phase (AP), blast phase (BL), Philadelphia chromosome positive (Ph +). The indication for ponatinib is only for patients resistant or intolerant to prior tyrosine kinase inhibitor therapy.

[161] O'Hare T, et al. AP24534, a pan-BCR-ABL inhibitor for chronic myeloid leukemia, potently inhibits the T315I mutant and overcomes mutation-based resistance. Cancer Cell 2009;16:401–12.

[162] Panjarian S, et al. Structure and dynamic regulation of Abl kinases. J. Biol. Chem. 2013;288:5443–50.

(2006), nilotinib (2007), and bosutinib (2012). The following TKIs are FDA-approved for the treatment of Ph + ALL: imatinib (2001) and dasatinib (2006). FDA granted accelerated approval for omacetaxine mepesuccinate in 2012 for the treatment of adult patients with CP-CMP, AP-CML, BP-CML, and Ph + ALL."[163]

ii. List of AEs for Drugs in the Tyrosine Kinase Inhibitor (TKI) Drug Class

FDA provided a list of AEs associated with various drugs of the TKI drug class. FDA's list, reproduced in part below, was acquired by the FDA reviewer from the package labels from various marketed TKI drugs:

Imatinib
Fluid retention and edema
Hematological toxicity
Severe congestive heart failure and left ventricular dysfunction
Hepatotoxicity
Hemorrhage
Gastrointestinal disorders
Hypereosinophilic cardiac toxicity
Dermatological toxicities
Hypothyroidism
Toxicities from long-term se
Tumor lysis syndrome
Use in pregnancy
Growth retardation in children and adolescents
Dasatinib
Myelosuppression
Bleeding events associated with severe thrombocytopenia
Fluid retention
QT prolongation
Congestive heart failure, left ventricular dysfunction, and myocardial infarction
Pulmonary arterial hypertension
Use in pregnancy

iii. Drug Class That is a Subset of the TKI Drug Class

FDA developed its drug class analysis, and observed that within the TKI drug class, there resided a subset: *VEGFR-family of kinase inhibitors*.[164] Other *VEGFR-family of kinase inhibitors*, which were FDA-approved, were revealed by FDA to include, sorafenib, sunitinib, pazopanib, axitinib, regorafenib, bevacizumab, and ziv-aflibercept.

[163] Ponatinib (CP-CML; AP-CML, BP-CML; Ph + ALL) NDA 203-469. Page 14 of 108-page Medical Review.

[164] Ponatinib (CP-CML; AP-CML, BP-CML; Ph + ALL) NDA 203-469. Page 12 of Clinical Review. Evaluation for potential adverse events for similar drugs in drug class on pages 67–70 of Clinical Review. Clinical Review resides on pages 2–102 of 108-page Medical Review.

Regarding the AEs found on the package labels of this subset of TKIs, FDA observed, "a broad spectrum of kinase inhibition for ponatinib, which includes inhibition of the VEGFR-family of kinases. The safety profile for ponatinib is notable for similar features to kinase inhibitors active against the **VEGFR-kinases**. These similar features include arterial thromboembolic events, hypertension, gastrointestinal perforation, and compromised wound healing."[165]

FDAs explored the adverse events associated with various drugs of the *VEGFR-kinase inhibitor drug class* writing, "Possible mechanisms of action for the arterial ischemic events and arterial stenosis include the broad spectrum of kinase inhibition, VEGFR-kinase inhibition ... [p]onatinib was designated to target native and mutated BCR-ABL, however, in vitro testing revealed a broader spectrum of kinase inhibition, which included the VEGFR-family of kinases. Arterial ischemic events have been associated with **kinase inhibitors that target VEGFR** ... sorafenib, pazopanib, axitinib, and regorafenib ... [a]lso, the occurrence of other adverse events typically associated with ... **VEGFR-kinase inhibition** such as hypertension, proteinuria, oral mucositis, and gastrointestinal perforation further supports the above hypothesis."[166]

Package label. All of the AEs detected from FDA's drug class analysis of VEGFR-kinases, with the exception of the AE of proteinuria, found a place on the package label. In the excerpt appearing immediately above, these AEs are shown to be, arterial ischemic events, hypertension, proteinuria, oral mucositis, and gastrointestinal perforation. The Black Box Warning for ponatinib (Iclusig®) warned:

> **WARNING**: ARTERIAL THROMBOSIS and HEPATOTOXICITY ... Arterial Thrombosis: Cardiovascular, cerebrovascular, and peripheral vascular thrombosis, including fatal myocardial infarction and stroke have occurred in Iclusig-treated patients. In clinical trials, serious arterial thrombosis occurred in 8% of Iclusig-treated patients. Interrupt and consider discontinuation of Iclusig in patients who develop arterial thrombotic events.[167]

The Warnings and Precautions section warned against other AEs from the drug class analysis, namely, hypertension, gastrointestinal perforation, and impaired wound healing:

> **WARNINGS AND PRECAUTIONS** ... Hypertension: Monitor for high blood pressure and treat as clinically indicated ... Compromised Wound Healing and Gastrointestinal Perforation: Temporarily interrupt therapy in patients undergoing major surgical procedures.[168]

The package label's disclosure of oral mucositis found a place in the Adverse Reactions section:

> **ADVERSE REACTIONS** ... Adverse Reactions Occurring in >10% of Patients ... Mucositis.[169]

[165] Ponatinib (CP-CML; AP-CML, BP-CML; Ph + ALL) NDA 203-469. Page 12 of Clinical Review. Evaluation for potential adverse events for similar drugs in drug class on pages 67–70 of Clinical Review. Clinical Review resides on pages 2–102 of 108-page Medical Review.

[166] Ponatinib (CP-CML; AP-CML, BP-CML; Ph + ALL) NDA 203-469. Page 12 of Clinical Review. Pages 79–81 of Clinical Review. Clinical Review resides on pages 2–102 of 108-page Medical Review.

[167] Package label. ICLUSIG® (ponatinib) tablets for oral use; December 2012 (17 pp.).

[168] Package label. ICLUSIG® (ponatinib) tablets for oral use; December 2012 (17 pp.).

[169] Package label. ICLUSIG® (ponatinib) tablets for oral use; December 2012 (17 pp.).

The package label's disclosure of impaired wound healing appeared to rest entirely from drug class analysis, where the drug class as defined by mechanism of action. This much was stated in the Warnings and Precautions section:

> **WARNINGS AND PRECAUTIONS** ... No formal studies of the effect of Iclusig on wound healing have been conducted. Based on the mechanism of action ... Iclusig could compromise wound healing.[170]

i. Regorafenib (Metastatic Colorectal Cancer) NDA 203-085

Regorafenib inhibits various tyrosine kinases, including VEGFR, FGFR, and PDGFR. VEGFR, FGFR, and PDGFR are receptors, while the respective ligands are, VEGF, FGF, and PDGF. Stimulation of these receptors by their respective ligands transmits a signal to the cell that promotes angiogenesis (formation of new blood vessels).[171,172] For treating cancer, regorafenib blocks these signals, thereby preventing the angiogenesis that provides nutrients to growing tumor cells.

i. FDA Defines the Drug Class

FDA's Medical Review defined regorafenib's drug class in terms of mechanism of action against specific drug targets. The drug class is multi-kinase inhibitors:

> Regorafenib is a small molecular inhibitor of multiple kinases including BRAF, VEGFR 1/2/3, TIE2, PDGFR, FGFR, RAF-1, KIT and RET ... **[m]ulti-kinase agents that inhibit** at least 3 of the main tyrosine kinases targeted by regorafenib (VEGFR, PDGFR and KIT) include sorafenib ... sunitinib ... and pazopanib ... [t]he safety profile of this new molecular entity reflects its mechanism of action.[173]

FDA's review included a list of FDA-approved drugs of the same class, namely, sorafenib, sunitinib, and pazopanib, and then developed this topic by detailing the serious adverse events (SAEs) that were listed on the package labels for these particular drugs. FDA referred to AEs such as hepatotoxicity, cardiac ischemia, hemorrhage, hypertension, dermatologic toxicity, and gastrointestinal perforation, "The following ... serious adverse events have been described with the **multikinase inhibitors** noted above: hepatotoxicity, cardiac ischemia/infarction, ... hemorrhage, hypertension, dermatologic toxicity, GI

[170] Package label. ICLUSIG® (ponatinib) tablets for oral use; December 2012 (17 pp.).

[171] Wilhelm SM, et al. Regorafenib (BAY 73-4506): a new oral multikinase inhibitor of angiogenic, stromal and oncogenic receptor tyrosine kinases with potent preclinical antitumor activity. Int. J. Cancer 2011;129:245–55.

[172] Adenis A, et al. Survival, safety, and prognostic factors for outcome with regorafenib in patients with metastatic colorectal cancer refractory to standard therapies: results from a multicenter study (REBACCA) nested within a compassionate use program. BMC Cancer 2016;16:412 (8 pp.).

[173] Regorafenib (metastatic colorectal cancer after prior treatment with various drugs) NDA 203-085. Page 38 of 76-page pdf file of Medical Review.

perforation, elevation in INR when taking warfarin, wound healing complications, arterial and venous thrombotic events, RPLS, hypothyroidism, proteinuria, infection and fetal harm."[174]

ii. Combining Animal Data With Drug Class Analysis to Arrive at Information for the Package Label

The Sponsor's animal toxicity data were combined with drug class analysis, to arrive at predicted safety issues for the study drug. Regarding the animal study results, FDA observed that, "Findings of changes in **dentin and epiphyseal growth plates** were present in both species [rats; dogs]. These changes have been associated with many VEGF inhibitors and may be relevant to the pediatric population."[175]

Package label. FDA's tactic in combining drug class analysis with animal toxicity data resulted in corresponding information on the package label for regorafenib (Stivarga®). Although humans were not used for assessing the AE of dentin alteration, it was the case that dentin alteration in animals occurred, where this form of toxicity occurred at an exposure (AUC) similar to the exposure (AUC) achieved with the recommended therapeutic dose in humans:

> **USE IN SPECIFIC POPULATIONS**. Pediatric Use The safety and efficacy of Stivarga in pediatric patients less than18 years of age have not been established. In 28-day repeat dose studies in rats there were dose-dependent findings of **dentin alteration** and angiectasis. These findings were observed at regorafenib doses as low as 4 mg/kg (approximately 25% of the AUC in humans at the recommended dose). In 13-week repeat dose studies in dogs there were similar findings of **dentin alteration** at doses as low as 20 mg/kg (approximately 43% of the AUC in humans at the recommended dose). Administration of regorafenib in these animals also led to persistent growth and thickening of the femoral **epiphyseal growth plate**.[176]

FDA's comments in its drug class analysis, regarding hepatotoxicity, found a place in the Black Box Warning of regorafenib (Stivarga®):

> **BOXED WARNING.** Hepatotoxicity. Severe and sometimes fatal hepatotoxicity has been observed in clinical trials. Monitor hepatic function prior to and during treatment. Interrupt and then reduce or discontinue Stivarga for hepatotoxicity as manifested by elevated liver function tests or hepatocellular necrosis, depending upon severity and persistence.[177]

The other AEs brought to light in FDA's drug class analysis also found a place in the package label, as shown in the Warnings and Precautions section:

> **WARNINGS AND PRECAUTIONS**. Hemorrhage: Permanently discontinue Stivarga for severe or life-threatening hemorrhage ... **Dermatological toxicity**: Interrupt and then reduce or discontinue Stivarga

[174] Regorafenib (metastatic colorectal cancer after prior treatment with various drugs) NDA 203-085. Page 45 of 76-page pdf file of Medical Review.

[175] Regorafenib (metastatic colorectal cancer after prior treatment with various drugs) NDA 203-085. Pages 10–16 of 69-page Clinical Review.

[176] Package label. STIVARGA (regorafenib) tablets, oral; September 2012 (14 pp.).

[177] Package label. STIVARGA (regorafenib) tablets, oral; September 2012 (14 pp.).

depending on severity and persistence of dermatologic toxicity ... **Hypertension**: Temporarily or permanently discontinue Stivarga for severe or uncontrolled hypertension ... **Cardiac ischemia** and infarction: Withhold Stivarga for new or acute cardiac ischemia/infarction and resume only after resolution of acute ischemic events ... **Gastrointestinal perforation** or fistulae: Discontinue Stivarga.[178]

j. Vemurafenib (B-RAF V600E Mutation Positive Unresectable or Metastatic Melanoma) NDA 202-429

Vemurafenib is a small molecule drug for treating metastatic melanoma. The drug inhibits two kinds of protein kinases:

1. Serine-threonine protein kinases
2. Tyrosine protein kinases.

Vemurafenib's structure is shown below. Vemurafenib inhibits a serine-threonine protein kinase that is called B-RAF. The RAF kinases are serine/threonine protein kinases.[179] Vemurafenib also inhibits tyrosine kinases, such as KIT and c-RET, and thus this drug is also a tyrosine kinase inhibitor (TKI). Thus drug class analysis involved contemplating the AEs caused by two kinds of protein kinase inhibitors: (1) Serine/threonine protein kinases inhibitors and; (2) TKIs.

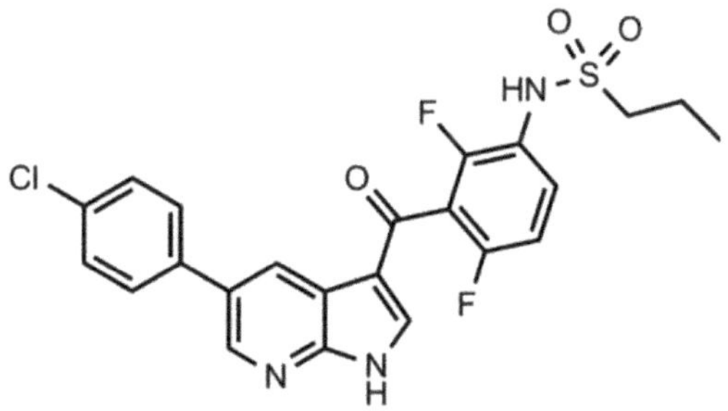

About half of metastatic cutaneous melanomas have a mutation at valine-670 in the B-RAF gene, where the mutations result in constitutive activation of the MAP-signaling pathway.[180] The full name for this pathway is the RAS-RAF-MEK-ERK MAPK pathway. There are three RAF genes, and these are called A-RAF, B-RAF, and C-RAF.[181]

Mutations of the RAF activator RAS are present in 30% of human cancers. RAS serves as an activator of RAF, where mutations in RAS or in RAF can transform normal mammalian cells to become cancer cells. The RAF protein, in particular B-RAF, is an attractive

[178] Package label. STIVARGA (regorafenib) tablets, oral; September 2012 (14 pp.).

[179] Wilhelm SM, et al. BAY 43-9006 exhibits broad spectrum oral antitumor activity and targets the RAF/MEK/ERK pathway and receptor tyrosine kinases involved in tumor progression and angiogenesis. Cancer Res. 2004;64:7099–109.

[180] Larkin J, et al. Combined vemurafenib and cobimetinib in BRAF-mutated melanoma. New Engl. J. Med. 2014;371:1867–76.

[181] Peng SB, et al. Inhibition of RAF isoforms and active dimers by LY3009120 leads to anti-tumor activities in RAS or BRAF mutant cancers. Cancer Cell 2015;28:384–98.

target for anti-cancer drugs. In melanoma resulting from B-RAF mutations, most of the mutations are B-RAF V600E, while a small percentage of the mutations are B-RAF V600K.[182]

This concerns wild-type B-RAF versus mutant B-RAF. In vitro enzyme assays demonstrated that vemurafenib inhibited B-RAF (wild-type), B-RAF V600E, and B-RAF V600E + G468A, with IC50 values of 32, 6.1, and 22 nM, respectively.[183]

FDA's Medical Review reiterated the established biochemistry for B-RAF mutations leading to melanoma, "mutations in B-RAF have been identified in melanoma ... [t]he most common alteration that occurs is the codon 600 valine to glutamate (V600E) mutation, which represents 90% of BRAF mutations. The next most common mutation is the codon 600 valine to lysine (V600K) followed by the valine to arginine mutation (V600R)."[184]

The following describes using enzyme inhibition data to narrow down the number of drugs used for drug class analysis, for the purpose of predicting AEs. FDA's Pharmacology Review provided a table (Table 10.4) of various kinases with vemurafenib inhibition data. The FDA reviewer stated that the table, "shows that vemurafenib inhibited numerous kinases at concentrations well below those achieved clinically." Table 10.4 includes serine/threonine protein kinases and tyrosine protein kinases, and reveals the fact that B-RAF V600E is the enzyme that is most susceptible to vemurafenib inhibition (B-RAF V600E has the lowest IC50). Vemurafenib does not inhibit VEGFR (the IC50 is extremely high) as indicated in the table[185] and by the references.[186,187]

The data teach that, for defining a drug class by way of mechanism of action, it is best to include only data guided by the study drug's targets *where inhibition occurs at concentrations of drug that are achieved in the patient-care situation*. In other words, because the study drug's greatest inhibition was on B-RAF V600E, C-RAF, A-RAF, and B-RAF, and because lesser inhibitions were found on FLT4, SRC, and other targets, use of drug class analysis to predict AEs should consider adverse events caused by drugs that inhibit B-RAF V600E, C-RAF, A-RAF, and B-RAF, and should ignore adverse events caused by kinase inhibitors that inhibit STK3, FLT4, SRC or VEGFR. To reiterate this point, a table showing the study

[182] McArthur GA, et al. Safety and efficacy of vemurafenib in BRAF(V600E) and BRAF(V600K) mutation-positive melanoma (BRIM-3): extended follow-up of a phase 3, randomised, open-label study. Lancet Oncol. 2014;15:323–32.

[183] Peng SB, et al. Inhibition of RAF isoforms and active dimers by LY3009120 leads to anti-tumor activities in RAS or BRAF mutant cancers. Cancer Cell 2015;28:384–98.

[184] Vemurafenib (BRAF V600E mutation positive unresectable or metastatic melanoma) NDA 202-429. Page 13 of 105-page pdf file of Medical Review.

[185] Vemurafenib (BRAF V600E mutation positive unresectable or metastatic melanoma) NDA 202-429. Page 30 (Table 6) of 115-page Pharmacology Review.

[186] EurekAlert! American Association Advancement Science. Drug targeting BRAF mutation slows thyroid cancer, too (quotation from Marcia Brose, MD, PhD); July 22, 2016.

[187] Cowey CL. A therapeutic renaissance: emergence of novel targeted agents for metastatic melanoma. Clin. Invest. 2012;2:883–93.

TABLE 10.4 Vemurafenib IC50 Values for Various Kinases

Enzyme Being Assayed	IC50 (nM)
B-RAF V600E	8
C-RAF	16
A-RAF	29
B-RAF (wild-type)	39
SRMS	18
ACK1	19
MAP4KS	51
FGR	63
BRK	209
KIT	538
STK3	891
FLT4	1920
SRC	2011
VEGFR	5300

Vemurafenib (B-RAF V600E mutation positive unresectable or metastatic melanoma) NDA 202-429. Page 30 of 115-page Pharmacology Review. The IC50 value for VEGFR is from page 29 of 115-page Pharmacology Review.

drug's inhibition of various enzymes, where the inhibition data was from in vitro experiments with purified enzymes, can show which enzymes are most susceptible to inhibition by the study drug, and which enzymes are least susceptible to inhibition by the study drug. A Sponsor interested in conducting a drug class analysis for predicting AEs expected from the study drug, will be guided towards the package labels for FDA-approved drugs that inhibit the susceptible enzymes (and will be dissuaded from using package labels for FDA-approved drugs that inhibit the least susceptible enzymes).

FDA's drug class analysis of vemurafenib focused on sorafenib, which was another B-RAF inhibitor.[188] In observing that sorafenib results in cutaneous squamous cell carcinomas, FDA cautioned that the study drug (vemurafenib) could also result in this same AE. The FDA reviewer realized that a confounding aspect of drug class analysis was that sorafenib inhibited many different enzymes (B-RAF, c-kit, PDGFR, VEGFR-2), and not just B-RAF. The fact that sorafenib inhibits many different enzymes is indicated by a term that is conventional in biochemistry, namely, promiscuous:

> To date, sorafenib is the only other approved agent that has demonstrated activity against B-RAF, but it is a promiscuous inhibitor that targets c-kit, PDGFR, and VEGFR-2. The adverse event profile of sorafenib includes hypertension, gastrointestinal perforation, and wound healing [sic, impaired wound healing],

[188] Vemurafenib (BRAF V600E mutation positive unresectable or metastatic melanoma) NDA 202-429.

> which is most likely related to the activity against VEGFR. An uncommon adverse reaction associated with sorafenib is the development of cutaneous squamous cell carcinomas ... the development of cutaneous squamous cell carcinomas appears to be related to targeting the B-RAF pathway.[189]

The following excerpt was used by FDA in two different ways, forming a symmetrical analysis. The excerpt predicted an AE expected for vemurafenib (cutaneous squamous cell carcinomas) and also predicted AEs that are not expected to be found with vemurafenib (hypertension, gastrointestinal perforation, wound healing). This is the excerpt with VEGFR-related adverse events highlighted:

> To date, sorafenib is the only other approved agent that has demonstrated activity against B-RAF, but it is a promiscuous inhibitor that targets c-kit, PDGFR, and VEGFR-2. The adverse event profile of sorafenib includes hypertension, gastrointestinal perforation, and wound healing [sic, impaired wound healing], which is most likely related to the activity against VEGFR. An uncommon adverse reaction associated with sorafenib is the development of cutaneous squamous cell carcinomas ... the development of cutaneous squamous cell carcinomas appears to be related to targeting the B-RAF pathway.[190]

Now, regarding this second part of the two-part drug class analysis, the package label shows another result from this two-part analysis, where the label *does not mention anything about hypertension (or high blood pressure), gastrointestinal perforation, or impaired wound healing.*[191]

FDA's reviews did not make any statement, to the effect that, "hypertension, gastrointestinal perforation, and impaired wound healing are not expected AEs of vemurafenib, because of the fact that vemurafenib is an extremely weak inhibitor of VEGFR, a drug target associated with these particular AEs." On the other hand, the data shown in Table 10.4 are consistent with the reasoning set forth, in this statement, for not including these particular AEs on vemurafenib's package label.

Package label. FDA's drug class analysis of vemurafenib (Zelboraf®), which involved contemplation of sorafenib's AE of cutaneous squamous cell carcinomas, found a place in the package label, as shown below. This warning was not based entirely on drug class analysis, and that it was mainly based on data from the Sponsor's own clinical studies:

> **WARNINGS AND PRECAUTIONS**. Cutaneous squamous cell carcinomas (cuSCC) occurred in 24% of patients. Perform dermatologic evaluations prior to initiation of therapy and every two months while on therapy. Manage with excision and continue treatment without dose adjustment.[192]

[189] Vemurafenib (BRAF V600E mutation positive unresectable or metastatic melanoma) NDA 202-429. Pages 12–13 of 105-page pdf file of Medical Review.

[190] Vemurafenib (BRAF V600E mutation positive unresectable or metastatic melanoma) NDA 202-429. Pages 12–13 of 105-page pdf file of Medical Review.

[191] ZELBORAF™ (vemuratenib) tablet, oral; August 2011 (13 pp.).

[192] ZELBORAF™ (vemuratenib) tablet, oral; August 2011 (13 pp.).

VI. DRUGS FOR INDICATIONS OTHER THAN CANCER

a. Introduction

This provides FDA's drug class analysis for indications other than cancer. The examples are from:

- Certolizumab pegol
- Hydromorphone
- Lixisenatide
- Lorcaserin
- Macitentan
- Vilazodone
- Vorapaxar

b. Certolizumab Pegol (Crohn's Disease) BLA 125-160

Certolizumab pegol is a pegylated antibody for treating Crohn's disease. The term "pegol" refers to a polyethylene glycol group that is covalently attached in order to increase the antibody's lifetime in the bloodstream. The antibody binds to tumor necrosis factor-alpha (TNFα) and inactivates it.

Signaling mediated by TNFα contributes to the pathology of Crohn's disease. Infliximab and adalimumab are other antibodies that bind to TNFα, and that are also used to treat Crohn's disease.

Certolizumab pegol contains the variable region derived from an anti-TNFα antibody, but it lacks the constant region (Fc region) of the original anti-TNFα antibody. Omission of the Fc region provides the therapeutic advantages of not inducing complement activation and of not inducing antibody-dependent cellular cytotoxicity.[193]

FDA conducted a drug class analysis with regard to these adverse events:

- Cancer
- Serious infections
- Stevens-Johnson syndrome.

i. Drug Class Analysis of the AE of Cancer

Drugs that impair immune response, such as drugs that block TNFα, often result in the risk of AEs taking the form of opportunistic infections and cancer.[194]

FDA's drug class analysis compared the study drug (certolizumab pegol, Cimzia®) with other anti-TNFα antibodies, infliximab (Remicade®) and adalimumab (Humira®). FDA recommended that the package label need not contain any warning against cancer:

[193] Sandborn WJ, et al. Certolizumab pegol for the treatment of Crohn's disease. New Engl. J. Med. 2007;357:228–38.

[194] Ali T, et al. Clinical use of anti-TNF therapy and increased risk of infections. Drug Healthcare Patient Saf. 2013;5:79–99.

> In the controlled studies in the Cimzia developmental program, no excess over placebo was seen; however, the concern is reasonably applicable to the class of TNF blockers as a whole, and it would be appropriate to include a warning regarding malignancies in the Cimzia labeling ... Remicade, which is approved for ... Crohn's disease, has a warning for the rare hepatosplenic T cell lymphoma ... [a]t present, a similar warning does not appear to be indicated for Humira. Until a clearer picture emerges as to the generalizability of this rare adverse reaction to the whole class of TNF blockers, Cimzia should not carry the warning either.[195]

Package label (cancer). FDA's recommendations to warn against malignancies (except for hepatosplenic T-cell lymphoma) found a place on the package label of certolizumab pegol (Cimzia®). The package label warned against "lymphoma and other malignancies" from other TNF blockers, but refrained from mentioning any risk for the rare hepatosplenic T-cell lymphoma. The Warnings and Precautions section read:

> **WARNINGS AND PRECAUTIONS** ... Cases of lymphoma and other malignancies have been observed among patients receiving TNF blockers.[196]

The Warnings and Precautions section admitted to the ambiguous area of the possible association between certolizumab pegol and cancer, where this ambiguous area took the form that cancer occurred in only two study subjects. The take-home lesson is that, where a serious adverse event, such as cancer, is detected in the study drug group, but where convincing data on relatedness to the study drug is not at hand, the medical writer has the option of disclosing the serious adverse event in the Warnings and Precautions section, but with the additonal comment that, "the potential role of the study drug in development of that serious adverse event is not known":

> **WARNINGS AND PRECAUTIONS** ... In controlled studies of CIMZIA for Crohn's disease and other investigational uses, there was one case of lymphoma among 2,657 Cimzia-treated patients and one case of Hodgkin lymphoma among 1,319 placebo-treated patients ... [t]he potential role of TNF blocker therapy in the development of malignancies is not known.[197]

ii. Drug Class Analysis of the AE of Infections

Regarding the potential for infections, FDA recommended that the package label should include a Black Box Warning about serious infections, writing, "The labeling should include warnings and precautions substantially similar to those in recently approved products in the **TNF blocker class**, namely warnings for **serious infections**, TB, hepatitis B

[195] Certolizumab (Crohn's disease) BLA 125-160. FDA divided the Medical Review into three parts: 1st part 80 pages, 2nd part 80 pages, 3rd part 56 pages.

[196] Package label. CIMZIA (certolizumab pegol). Lyophilized powder for solution for subcutaneous injection; April 2008 (16 pp.).

[197] Package label. CIMZIA (certolizumab pegol). Lyophilized powder for solution for subcutaneous injection. April 2008 (16 pp.).

reactivation, malignancies, hypersensitivity reactions, ... [t]he warnings regarding serious infections and TB should also appear in a boxed warning.[198]

Package label (infections). Consistent with FDA's recommendations, the package label contained a Black Box Warning about serious infections:

> **BOXED WARNING: RISK OF SERIOUS INFECTIONS.** Tuberculosis (TB), invasive fungal, and other opportunistic infections, some fatal, have occurred. Perform test for latent TB; if positive, start treatment for TB prior to starting CIMZIA. Monitor all patients for active TB during CIMZIA treatment, even if initial tuberculin skin test is negative.[199]

The Warnings and Precautions section of the package label also warned about infections. Please note that drug class analysis was invoked, in the recitation, "have been reported in patients receiving TNF blockers. Also, please note that the Warnings and Precautions section includes a contraindication. This contraindication reads, "Do not initiate treatment with CIMZIA in patients with active infections":

> **WARNINGS AND PRECAUTIONS.** Serious infections, sepsis, and cases of opportunistic infections, including fatalities, have been reported in patients receiving **TNF blockers**, including CIMZIA. Many of the serious infections reported have occurred in patients on concomitant immunosuppressive therapy that, in addition to their Crohn's disease, could predispose them to infections. In postmarketing experience with **TNF blockers**, infections have been observed with various pathogens including viral, bacterial, fungal, and protozoal organisms, and infections have been noted in all organ systems ... **[d]o not initiate treatment** with CIMZIA in patients with active infections, including chronic or localized infections.[200]

iii. Drug Class Analysis for the AE of Stevens-Johnson Syndrome

Regarding yet another AE associated with TNFα blockers, the FDA reviewer conducted a drug class analysis for the adverse event of Stevens-Johnson Syndrome, writing:

> Recently, reports have emerged of ... **Stevens-Johnson syndrome** ... in postmarketing experience with **TNF blockers**. Information about these reactions have been added to the Remicade labeling ... and DAARP [Division of Analgesia, Anesthesia, and Rheumatology Products] ... for other approved TNF blockers is viewing the reactions as a class effect ... [t]he same information also should be included in the ... Cimzia labeling.[201]

Package label (Stevens-Johnson syndrome). FDA's recommendation for Stevens-Johnson syndrome found a place on the package label. This warning was based entirely

[198] Certolizumab (Crohn's disease) BLA 125-160. FDA divided the Medical Review into three parts: 1st part 80-page pdf file, 2nd part 80-page pdf file, 3rd part 56-page pdf file.

[199] Package label. CIMZIA (certolizumab pegol). Lyophilized powder for solution for subcutaneous injection; April 2008 (16 pp.).

[200] Package label. CIMZIA (certolizumab pegol). Lyophilized powder for solution for subcutaneous injection; April 2008 (16 pp.).

[201] Certolizumab (Crohn's disease) BLA 125-160. FDA divided the Medical Review into three parts: 1st part 80 pages, 2nd part 80 pages, 3rd part 56 pages. Pages 18-19 of the first of the three 80-page pdf files for the Medical Review.

on drug class analysis, and not on any cases of Stevens-Johnson syndrome occurring in the Sponsor's clinical studies. The package label admitted to ambiguity in this particular warning, writing that it was not possible to "establish a causal relationship to drug exposure":

> **ADVERSE REACTIONS** ... Cases of severe skin reactions, including Stevens-Johnson syndrome, toxic epidermal necrolysis, and erythema multiforme, have been identified during post-approval use of **other TNF blockers**. Because these reactions are reported voluntarily from a population of uncertain size, it is not always possible to estimate reliably their frequency or establish a causal relationship to drug exposure.[202]

c. Hydromorphone (Acute and Chronic Pain) NDA 021-217

Hydromorphone is an opioid compound used for treating pain. Hydromorphone's structure is shown below along with the structure of another opioid, morphine:

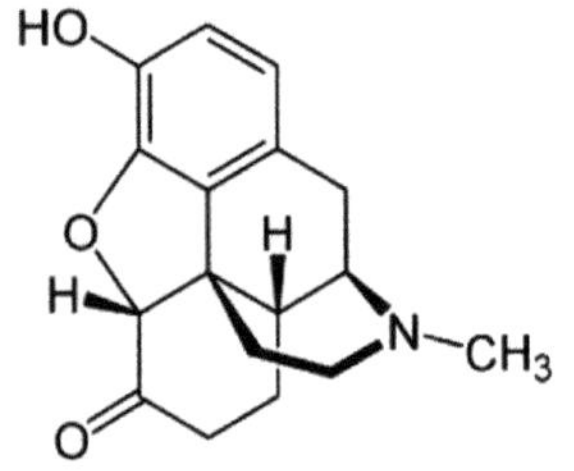

Hydromorphone (an opioid) 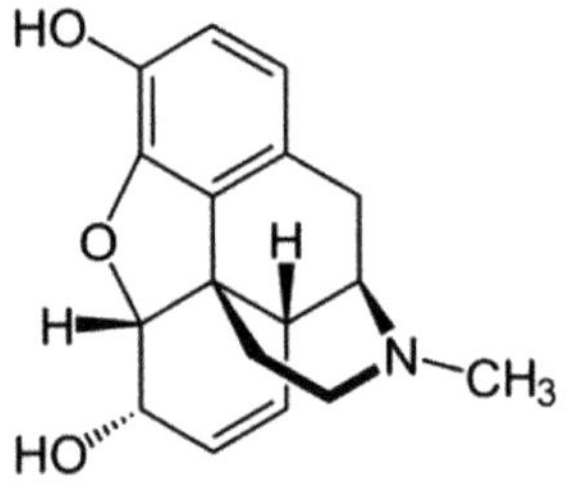 Morphine (an opioid)

FDA's review of hydromorphone used the following sources of information to arrive at package label information on adverse events:

- Drug class analysis
- Mechanism of action of the study drug's toxicity
- Animal toxicity data
- AE data from human subjects

High and low doses, concomitant disorders, and drug sensitivity were taken into account in FDA's AE analysis. Generally speaking, if a given clinical study resulted in a number of AEs that was lower than expected, the Sponsor should ask if that particular study had used a low dose of the study drug. Also, if a clinical study revealed AEs in more than the expected number, the Sponsor should ask if any subset of the study subject population had underlying condition resulting in increased risk for AEs. These issues are

[202] Package label. CIMZIA (certolizumab pegol). Lyophilized powder for solution for subcutaneous injection; April 2008 (16 pp.).

revealed in FDA's Pharmacology Review, which referred to high doses and which referred to patients with heightened sensitivity. The FDA reviewer observed:

> As is well known with **opioid drug products, respiratory depression** ... is the most prominent adverse effect of hydromorphone that is relevant to the proposed clinical use. Clinically significant respiratory depression rarely occurs with standard hydromorphone doses in the absence of **underlying pulmonary dysfunction**. However, at high doses or in sensitive patients, hydromorphone may produce **respiratory depression or irregular breathing** patterns.[203]

To summarize, in FDA's drug class analysis (drug class of opioid drugs), FDA's attention turned to the quantity of the dose and to the pre-existing condition in some patients of heightened sensitivity. In addition to using drug class analysis, the Sponsor's animal toxicity data also revealed respiratory depression, "The effects of hydromorphone on respiratory parameters were evaluated ... in beagle dogs ... hydromorphone initially increased respiratory rate from 0.5 to 2 hours after dosing followed by a **decrease in respiratory rate** from 3 to 6 hours after dosing."[204]

Although the Sponsor assessed toxicity on respiratory rate in rat studies, the rat data were not reproduced in FDA's review.[205] In comments on animal toxicology, the FDA reviewer referred to the transition from animal data to package label warnings for patients, "Respiratory depression, a known extension of the pharmacological [animal] action of hydromorphone, is the most prominent adverse effect of hydromorphone which would be **relevant to the proposed clinical use**."[206]

Finally, FDA's Medical Review utilized two other sources of AE information, namely, the Sponsor's own clinical data and drug class analysis, "Study DO-130 ... was stopped prematurely because of the number of patients who experienced **adverse events of decreased oxygen saturation** ... [a]t the time the study was halted, 50 of the 60 patients had been enrolled ... **[o]pioids are known to cause respiratory depression**. The patients who experienced decreased oxygenation is not unexpected.[207]

Package label. FDA's warnings about respiratory problems for hydromorphone (Exalgo®) found a place in the Black Box Warning of the package label. This referred to the AE of respiratory depression (but it did not refer to respiratory depression as a concomitant medical condition):

> **BOXED WARNING**: IMPORTANCE OF PROPER PATIENT SELECTION, LIMITATION OF USE, AND POTENTIAL FOR ABUSE ... EXALGO is indicated for opioid tolerant patients only ... **Fatal respiratory depression** could occur in patients who are not opioid tolerant.[208]

[203] Hydromorphone (acute and chronic pain) NDA 021-217. Page 11 of 207-page Pharmacology Review.

[204] Hydromorphone (acute and chronic pain) NDA 021-217. Page 31 of 207-page Pharmacology Review.

[205] Hydromorphone (acute and chronic pain) NDA 021-217. Page 25 of 207-page Pharmacology Review.

[206] Hydromorphone (acute and chronic pain) NDA 021-217. Page 16 of 241-page Medical Review.

[207] Hydromorphone (acute and chronic pain) NDA 021-217. Page 144 of 241-page Medical Review.

[208] Package label. EXALGO (hydromorphone hydrochloride) extended release tablets; March 2010 (30 pp.).

The Contraindications section referred to the situation where respiratory depression was a concomitant medical condition in patients:

> **CONTRAINDICATIONS**. Impaired Pulmonary Function. EXALGO is **contraindicated in patients with significant respiratory depression**, especially in the absence of resuscitative equipment or in unmonitored settings and in patients with acute or severe bronchial asthma or hypercarbia.[209]

The Clinical Pharmacology section referred to respiratory depression as an AE (but it did not refer to respiratory depression as a concomitant medical condition), where the information referred to mechanism of action of this form of toxicity:

> **CLINICAL TOXICOLOGY** ... Central Nervous System. Hydromorphone produces dose-related respiratory depression by direct action on brain stem respiratory centers. The respiratory depression involves a reduction in the responsiveness of the brain stem respiratory centers to increases in carbon dioxide tension and to electrical stimulation.[210]

The Sponsor's animal toxicity data of increased respiratory rate found a place on the package label. The package label referred to the rat data, but not the beagle dog data[211]:

> **USE IN SPECIFIC POPULATIONS** ... In the pre- and post-natal effects study in rats, neonatal viability was reduced at 6.25 mg/kg/day (~1.2) times the human exposure following 32 mg/day). Neonates born to mothers who have been taking opioids regularly prior to delivery will be physically dependent. The withdrawal signs include irritability and excessive crying, tremors ... increased respiratory rate.[212]

The package label information on rats compared the rat dose with the human dose. As can be seen, the dosing was about the same, when correcting for body weight. The fact that the dosing in rats and humans was the same, when correcting for weight, was stated in order to emphasize that implications for respiratory toxicity to human infants. To summarize, FDA's review of hydromorphone, is exemplary because of its integration of drug class analysis, animal toxicity data, and AE data from the Sponsor's own clinical study, for arriving at the package label. FDA's review of hydromorphone is also distinguished by the its analysis of a subgroup of patients, where this subgroup was defined as people with the pre-existing condition of, "not opioid tolerant" and the pre-existing condition of "significant respiratory depression." The Black Box Warning and the Contraindications section each referred to this subgroup.

[209] Package label. EXALGO (hydromorphone hydrochloride) extended release tablets. March 2010 (30 pp.).

[210] Package label. EXALGO (hydromorphone hydrochloride) extended release tablets. March 2010 (30 pp.).

[211] Hydromorphone (acute and chronic pain) NDA 021-217. Page 25 of 207-page Pharmacology Review.

[212] Package label. EXALGO (hydromorphone hydrochloride) extended release tablets; March 2010 (30 pp.).

d. Lixisenatide (Type-2 Diabetes) NDA 208-471

Lixisenatide is a small polypeptide for treating type-2 diabetes.[213] Pfeffer et al.[214] defined lixisenatide as belonging to a drug class definable by biochemical mechanism of action, physiological mechanism, and therapeutic effects, "Glucagon-like peptide-1 (GLP-1)-receptor agonists are a **class of parenteral glucose-lowering drugs** that activate the receptor for the endogenous incretin GLP-1. These drugs lower glucose levels by inhibiting the secretion of glucagon, promoting the release of insulin in response to hyperglycemia, slowing gastric emptying, and augmenting satiety."[215]

In other words, the above paragraph defines drug class by these methods:

- Biochemical mechanism (activating the receptor for one of the incretin hormones)
- Physiological effect (promoting insulin release; slowing gastric emptying)
- Therapeutic effect (glucose-lowering).

The relationship between naturally occurring glucagon-like peptide 1 (GLP-1) and lixisenatide is, "native GLP-1 is not suitable as a therapeutic agent because it is rapidly degraded by dipeptidyl peptidase-4 (DPP-4) and has a half-life of less than 2 minutes ... [t]hus, DPP-4-resistant GLP-1 receptor agonists with extended half-lives have been developed. Lixisenatide ... is a ... synthetic 44 amino acid ... GLP-1 receptor agonist modified C-terminally with six lysine residues and one proline deleted."[216]

i. Adverse Events Detected With Other Drugs, in the Drug Class Analysis of GLP-1 Receptor Agonists

The FDA reviewer acknowledged safety issues with other GLP-1 receptor agonists, namely anaphylaxis and hypersensitivity reactions, writing, "Anaphylaxis and serious hypersensitivity reactions were identified only in the post-marketing setting for the **other approved GLP-1 receptor agonists**."[217]

ii. Adverse Events Detected by Sponsor's Own Clinical Studies

FDA pointed out that anaphylaxis occurred in the Sponsor's own clinical studies, "There have been events of anaphylaxis ... that led to study drug discontinuation with lixisenatide which suggests that an association to study treatment was more

[213] Lixisenatide (type-2 diabetes) NDA 208-471. Medical Review contained three Clinical Reviews. Clinical Review for NDS 208-471 (pages 2–154 of the Medical Review); Clinical Review for NDA 204-961 (pages 156–251 of 690-page pdf file of Medical Review); and Clinical Review for NDA 204-961 (pages 426–676 of Medical Review). Medical Review is 690-page pdf file.

[214] Pfeffer MA, et al. Lixisenatide in patients with type 2 diabetes and acute coronary syndrome. New Engl. J. Med. 2015;373:2247–57.

[215] Pfeffer MA, et al. Lixisenatide in patients with type 2 diabetes and acute coronary syndrome. New Engl. J. Med. 2015;373:2247–57.

[216] Ratner RE, et al. Dose-dependent effects of the once-daily GLP-1 receptor agonist lixisenatide in patients with Type 2 diabetes inadequately controlled with metformin: a randomized, double-blind, placebo-controlled trial. Diabetes Med. 2010;27:1024–32.

[217] Lixisenatide (type-2 diabetes) NDA 208-471. Page 11 of 690-page Medical Review.

likely. In contrast, none of the placebo treated subjects with events discontinued study drug."[218]

iii. Subclasses Within the Drug Class of GLP-1 Receptor Agonists

FDA dissected the drug class of GLP-1 receptor agonists into *short-acting GLP-1 receptor agonists* and *long-acting GLP-1 receptor agonists*, where FDA's drug class analysis acknowledged the different types of AEs associated with each of these subclasses. FDA concluded that only the long-acting GLP-1 receptor agonists posed a risk for thyroid C-cell tumors in humans. FDA's Medical Review stated that, "the **long-acting GLP-1 receptor agonists** (exenatide LAR, liraglutide, dulaglutide and albiglutide) carry a boxed warning for the risk of thyroid C-cell tumors."[219]

A view of the package label for the *long-acting GLP-1 receptor agonist*, liraglutide (Victoza®), one finds a Black Box Warning that reads:

> **BOXED WARNING** ... Liraglutide (Victoza®) causes thyroid C-cell tumors at clinically relevant exposures in both genders of rats and mice. It is unknown whether VICTOZA causes thyroid C-cell tumors, including medullary thyroid carcinoma (MTC), in humans, as the human relevance of liraglutide-induced rodent thyroid C-cell tumors has not been determined.[220]

Turning to the package label for the *short-acting GLP-1 receptor agonist*, exenatide (Byetta®), it can be seen that the Nonclinical Toxicology section (but not in the Warnings and Precautions section) states that thyroid C-cell adenomas were found in animals:

> **NONCLINICAL TOXICOLOGY** ... Benign thyroid C-cell adenomas were observed in female rats at all exenatide (Byetta®) doses. The incidences in female rats were 8% and 5% in the two control groups and 14%, 11%, and 23% in the low-, medium-, and high-dose groups with systemic exposures of 5, 22, and 130 times, respectively, the human exposure resulting from the maximum recommended dose of 20 mcg/day, based on plasma area under the curve (AUC).[221]

Thus the nature of the package label warnings for the *long-acting GLP-1 receptor agonist*, liraglutide, and for the *short-acting GLP-1 receptor agonist*, exenatide, distinguishes the risk for these two subclasses of GlP-1 receptor agonists.

FDA's Medical Review compared lixisenatide with another *short-acting GLP-1 receptor agonist* (exenatide), and took note of the fact that lixisenatide at the recommended dose for humans (20 μg/day) was about 1000-times less than the dose (in terms of equivalent dose) that caused thyroid tumors in mice and rats. This recommended dose is in lixisenatide's package label.[222]

[218] Lixisenatide (type-2 diabetes) NDA 208-471. Page 11 of 690-page Medical Review.

[219] Lixisenatide (type-2 diabetes) NDA 208-471. Page 19 of 690-page Medical Review.

[220] Package label. VICTOZA (liraglutide) injection, for subcutaneous use. NDA 022-341; April 2016 (28 pp.).

[221] Package label. BYETTA® (exenatide) Injection. NDA 021-919; October 2009 (26 pp.).

[222] Package label. ADLYXIN (lixisenatide) injection, for subcutaneous use; July 2016 (30 pp.).

FDA's comparisons of the various GLP-1 receptor agonists read, "Thyroid C-Cell Proliferation: Based on the non-clinical carcinogenicity studies in mice and rats, the risk for thyroid C-cell proliferation with **lixisenatide** is expected to be similar to **exenatide** ... clinical exposure margins for lixisenatide at a dose of 20 micrograms/day were over 1,000-fold."[223]

FDA's Medical Review further distinguished the risk for AEs in *short-acting GLP-1 receptor agonists* [lixisenatide, exenatide (Byetta®)] from the risk for AEs in *long-acting GLP-1 agonists* (Victoza®, Bydureon®, Trulicity®, Tanzuem®), in an account of the need or lack thereof for a Risk Evaluation and Mitigation Strategy:

> While a Risk Evaluation Mitigation Strategy (REMS) is in place for the **long-acting GLP-1 receptor agonists** (i.e., VICTOZA, BYDUREON, TRULICITY, TANZUEM) to ensure that the benefits of the drug outweigh the potential risks of medullary thyroid carcinoma ... the **short-acting GLP-1 receptor agonist** (i.e., BYETTA) only carries a "Warning and precaution" statement in the label for pancreatitis. With lixisenatide, the ... reviewer's opinion was that the large clinical exposure margin of over 1000-fold for C-cell tumors indicates that the human risk for C-cell tumors is more similar to that of a **short-acting GLP-1 receptor agonist** (exenatide) than to **longer acting GLP-1 receptor agonists**. Therefore I do not think a REMS is required for lixisenatide.[224]

Package label. FDA's comments on anaphylaxis found a place on the package label for lixisenatide (Adlyxin®), in the Warnings and Precautions section and in the Contraindications section:

> **CONTRAINDICATIONS**. Hypersensitivity to ADLYXIN or any product components. Hypersensitivity reactions including anaphylaxis have occurred with ADLYXIN.[225]

> **WARNINGS AND PRECAUTIONS**. Anaphylaxis and Serious Hypersensitivity Reactions: Discontinue ADLYXIN and promptly seek medical advice.[226]

FDA's assessment that lixisenatide posed a low risk for thyroid tumors resulted in corresponding information in the Nonclinical Toxicology section, and not in the Warnings and Precautions section. The reason for placing this information in the Nonclinical Toxicology section is that the information was from animal studies:

> **NONCLINICAL TOXICOLOGY** ... Carcinogenicity studies of 2-years durations were conducted in ... mice and ... rats with twice daily subcutaneous doses of 40, 200, or 1,000 mcg/kg. A statistically significant increase in **thyroid C-cell adenomas** was observed in male mice at 2,000 mcg/kg/day, resulting in

[223] Lixisenatide (type-2 diabetes) NDA 208-471. Page 15 of 690-page Medical Review.

[224] Lixisenatide (type-2 diabetes) NDA 208-471. Page 15 of 690-page Medical Review.

[225] Package label. ADLYXIN (lixisenatide) injection, for subcutaneous use; July 2016 (30 pp.).

[226] Package label. ADLYXIN (lixisenatide) injection, for subcutaneous use; July 2016 (30 pp.).

systemic exposures that are $>$180-times the human exposure achieved at 20 mcg/day based on plasma AUC. Statistically significant increases in thyroid C-cell adenomas were seen at all doses in rats, resulting in systemic exposures that are $\geq$15-times the human exposure.[227,228]

e. Lorcaserin (Obesity) NDA 022-529

Lorcaserin is a small molecule that binds to one of the serotonin receptors, *5-HT2C receptor*.[229] Serotonin is also known as 5-hydroxytryptamine (5-HT). Lorcaserin's structure is shown below. Note the unusual seven-membered ring.

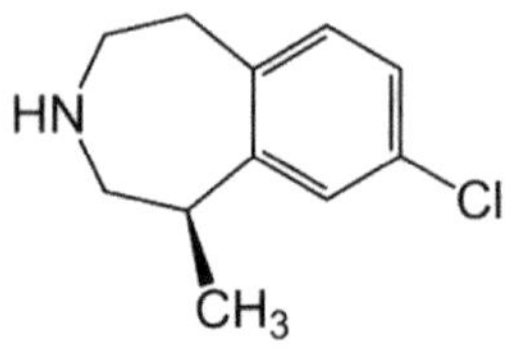

Lorcaserin is compared with another anti-obesity drug, dexfenfluramine. Dexfenfluramine binds to *5-HT2C serotonin receptor*, resulting in anorexia and weight loss (anorexia and weight loss are the desired effects of the drug).

But dexfenfluramine also binds to another *serotonin receptor*, *5-HT2B*. This drug's binding to the 5-HT2B receptor is responsible for dexfenfluramine's SAE of valvular heart disease.[230] Valvular heart disease involves thickening of the heart valve leaflets and regurgitation of blood. Any serotonin receptor agonist that does not activate 5-HT2B is not likely to result in valvular heart disease.[231,232] After discovering this serious adverse effect, dexfenfluramine was withdrawn from the market in 1977.[233]

[227] Package label. ADLYXIN (lixisenatide) injection, for subcutaneous use; July 2016 (30 pp.).

[228] This author detected an apparent discrepancy between FDA's reasoning and what is printed on lixisenatide's package label. FDA's Medical Review repeatedly refers to a huge safety margin, when comparing rat doses that led to tumors and the required dose for humans (see, "Based on the non-clinical carcinogenicity studies in mice and rats, the risk for thyroid C-cell proliferation with lixisenatide is expected to be similar to exenatide ... clinical exposure margins for lixisenatide at a dose of 20 micrograms/day were over 1,000-fold.") (also see, "With lixisenatide, the ... reviewer's opinion was that the large clinical exposure margin of over 1000-fold for C-cell tumors ..."). In contrast, lixisenatide's package label reads, "Statistically significant increases in thyroid C-cell adenomas were seen at all doses in rats, resulting in systemic exposures that are $\geq$15-times the human exposure."

[229] Smith SR, et al. Multicenter, placebo-controlled trial of lorcaserin for weight management. New Engl. J. Med. 2010;363:245–56.

[230] Lorcaserin (obesity) NDA 022-529. Page 56 of 425-page Medical Review.

[231] Rothman RB, Baumann MH. Serotonergic drugs and valvular heart disease. Expert Opin. Drug Saf. 2009;8:317–329.

[232] Rothman RB, et al. Evidence for possible involvement of 5-HT(2B) receptors in the cardiac valvulopathy associated with fenfluramine and other serotonergic medications. Circulation 2000;102:2836–41.

[233] Bello NT, Liang NC. The use of serotonergic drugs to treat obesity—is there any hope? Drug Des. Devel. Ther. 2011;5:95–109.

Drug class analysis of lorcaserin was based on mechanism of action, that is, whether the drug's target was serotonin receptor 5-HT1B, 5-HT2B, or 5-HT2C (one of these three different serotonin receptors). Regarding location in the body, 5-HT2A is located in the central nervous system (CNS), cardiac vessels, and heart valves. 5-HT2B is mainly in the cardiovascular system. 5-HT2C is in the CNS and participates in the process of controlling caloric balance. 5-HT2C is the target of drugs for anti-obesity treatment.[234] One can remember that 5-HT2C is the target of anti-obesity drugs by imagining that the letter "C" in 5-HT2C stands for "chubby."

FDA's analysis for identifying AEs of the study drug involved these following steps:

Step 1. Identify the receptors in a given family of receptors
Step 2. Identify drugs that bind to each of these receptors
Step 3. Identify AEs associated with each drug
Step 4. Use the above information to predict AEs expected from a new drug that is known to bind to one of the receptors.

FDA's analysis compared lorcaserin, which targets *5-HT2C* in the brain, with fenfluramine, which targets *5-HT2B*. Initially, FDA considered lorcaserin as a low risk for adverse events, in view of its specificity for binding to *5-HT2C* and low potency binding to *5-HT2B*. On this point, FDA wrote that, "lorcaserin preferentially activates **5HT2C**, with 8 to 15-fold greater potency compared to **5HT2A**, and 45 to 90-fold greater potency compared to **5HT2B**."[235] Embellishing upon lorcaserin's specificity, FDA added:

> maximal concentrations of lorcaserin ... observed in human plasma and anticipated in human brain tissue is notably lower than the EC50 for activation of 5HT2A and 2B, while remaining above the EC50 for activation of 5HT2C in vitro. Plasma concentrations of lorcaserin at the therapeutic dose are thus expected to remain within the selective range for activation of 5HT2C. Lorcaserin grouped with low-potency 5HT2B agonists [meaning that lorcaserin does not much bind to 5HT2B] that are not known to be associated with clinical valvulopathy in in vitro functional assays. Compounds known to cause clinical valvulopathy showed substantially higher 5HT2B receptor potency in these assays. The ... receptor potency data provides supportive evidence that ***off-target activation of the 5HT2A or 2B receptors is unlikely*** at the proposed clinical dose of lorcaserin.[236]

But then, FDA turned its attention to the risk of *any drug* that binds to 5-HT2B (even drugs that only weakly bind to 5-HT2B), writing, "Drugs that ... target 5HT receptors ... have been associated with an unusual cardiac valvular disease, characterized by fibrotic, regurgitant valves ... researchers have identified activation of the 5HT2B receptor as the ... mechanism of this adverse event."[237]

FDA then turned its attention to the fact that lorcaserin is a new and unique compound and, because of this, refused to use drug class analysis as the sole reason to characterize lorcaserin as a drug that poses no risk for valvular heart disease. FDA referred to a

[234] Bai B, Wang Y. The use of lorcaserin in the management of obesity: a critical appraisal. Drug Design Develop. Ther. 2011;5:1–7.

[235] Lorcaserin (obesity) NDA 022-529. Pages 204 and 273 of 425-page Medical Review.

[236] Lorcaserin (obesity) NDA 022-529. Page 23 of 425-page Medical Review.

[237] Lorcaserin (obesity) NDA 022-529. Page 65 of 425-page Medical Review.

program of adverse event screening, writing that, "Despite its relative 5HT2C specificity as compared to 5HT2B, lorcaserin is a novel 5HT2 agonist, and a ... program of echocardiographic screening ... was undertaken[238] ... [t]he ***clinical data as collected up to this point do not exonerate the drug for this potential risk*** [valvular heart disease].[239]

Please note the use of the peculiar set-point, which FDA used as a basis for requiring that the Warnings and Precautions section warned against valvular heart disease. This set-point was that, "data as collected up to this point do not exonerate the drug for this potential risk." This peculiar language is similar to the "burden of proof" concept that is used in legal arguments. As a consequence of FDA's review, a warning about valvular heart disease found a place in the Warnings and Precautions section of the package label.

Package label. The package label for lorcaserin (Belviq®) distinguished between 5-HT2B receptor agonists (which cause valvular heart disease) and 5-HT2C receptor agonists, such as lorcaserin, which had not yet been implicated in valvular heart disease. The Warnings and Precautions section read:

> **WARNINGS AND PRECAUTIONS** ... Valvular heart disease: If signs or symptoms develop consider BELVIQ discontinuation and evaluate the patient for possible valvulopathy.[240]

Further down in the same Warnings and Precautions section, the writing contemplated the notion that the AE of valvular heart disease could result from drugs that only weakly bind to 5-HT2B receptor (5-HT2B having already been implicated valvular heart disease caused by dexfenfluramine):

> **WARNINGS AND PRECAUTIONS** ... Regurgitant cardiac valvular disease, primarily affecting the mitral and/or aortic valves, has been reported in patients who took serotonergic drugs with **5-HT2B receptor** agonist activity. The etiology of the regurgitant valvular disease is thought to be activation of **5-HT2B receptors** on cardiac interstitial cells. At therapeutic concentrations, **BELVIQ is selective for 5-HT2C receptors** as compared to **5-HT2B receptors**. In clinical trials of 1-year duration, 2.4% of patients receiving BELVIQ and 2.0% of patients receiving placebo developed echocardiographic criteria for valvular regurgitation at one year (mild or greater aortic regurgitation and/or moderate or greater mitral regurgitation): none of these patients was symptomatic.[241]

FDA's Approval Letter required that the Sponsor conduct additional clinical studies for detecting cardiovascular AEs that were provoked by the study drug. FDA's requirement that the Sponsor conduct an additional clinical study took this form:

> Finally, there have been signals of a serious risk of major **adverse cardiovascular events** with some medications developed for the treatment of obesity, and available data **have not definitively excluded** the potential for this serious risk with ... lorcaserin ... We have determined that only a clinical trial ... will be sufficient to assess a signal of a serious risk of major adverse cardiovascular events with anti-obesity medications, including ... lorcaserin hydrochloride ... Therefore, based on appropriate scientific data, FDA has

[238] Lorcaserin (obesity) NDA 022-529. Page 66 of 425-page Medical Review.

[239] Lorcaserin (obesity) NDA 022-529. Page 227 of 425-page Medical Review.

[240] Package label. BELVIQ (lorcaserin hydrochloride) tablets, for oral use; June 2012 (23 pp.).

[241] Package label. BELVIQ® (lorcaserin hydrochloride) tablets, for oral use, CIV; May 2017 (24 pp.).

determined that **you are required to conduct** ... A randomized, double-blind, placebo-controlled trial to evaluate the effect of long-term treatment with Belviq on the incidence of major adverse cardiovascular events ... in obese and overweight subjects ... The timetable ... states that you will conduct this trial according to the following schedule: Final Protocol Submission: March 31, 2013 Trial Completion: December 31, 2017.[242]

f. Macitentan (Pulmonary Arterial Hypertension) NDA 204-410

This concerns FDA's review of macitentan for pulmonary arterial hypertension.[243] Pulmonary arterial hypertension is a chronic disease characterized by increased pulmonary vascular resistance, which leads to right ventricular failure and ultimately death if left untreated.[244]

i. Drug Class Defined by Chemical Structure

The structures of the study drugs, macitentan and bosentan, are shown below. Bosentan had been synthesized first and tested first, and bosentan was the basis for developing macitentan. Macitentan is about 10 times more potent than bosentan, as a result of longer receptor occupancy (time spent being bound to the receptor before dissociating) than for bosentan.[245] Macitentan has two heterocyclic bromine atoms. But in bosentan, one of these bromine atoms is replaced with a methyl group, and the aromatic ring bearing the other bromine group is completely missing.

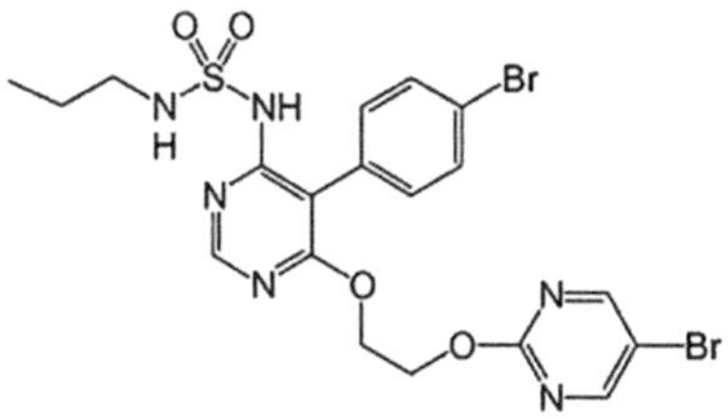

Macitentan

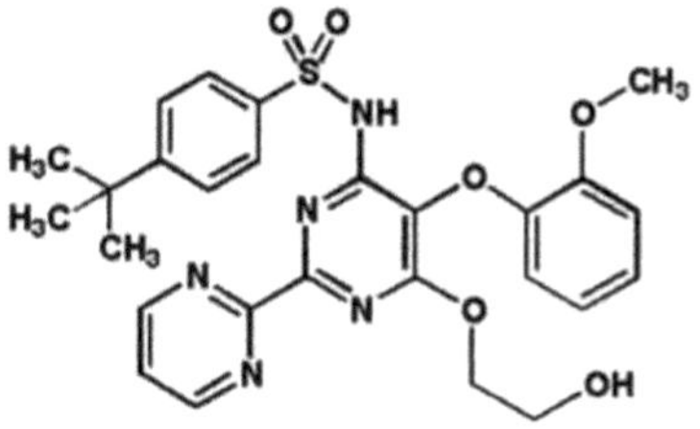

Bosentan

ii. Mechanism of Action of Macitentan

FDA's Medical Review revealed the mechanism of action of macitentan, stating that, "Macitentan is a ... dual ETA and ETB receptor antagonist that prevents the binding of ET-1 to its receptors. Endothelin (ET)-1 and its receptors (ETA and ETB) mediate a variety of deleterious effects such as vasoconstriction, fibrosis, proliferation, hypertrophy, and

[242] Lenvatinib (radioiodine-refractory differentiated thyroid cancer) NDA 206-947. Approval Letter; June 27, 2012.

[243] Macitentan (pulmonary arterial hypertension) NDA 204-410. 163-page pdf file of Medical Review.

[244] Sidharta PN, et al. Clinical pharmacokinetics and pharmacodynamics of the endothelin receptor antagonist macitentan. Clin. Pharmacokinet. 2015;54:457–471.

[245] Davenport AP, et al. Endothelin. Pharmacol. Rev. 2016;68:357–418.

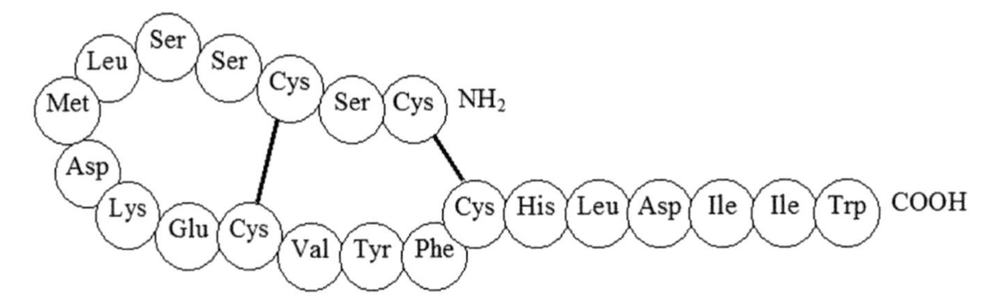

FIGURE 10.4 Amino acid sequence of endothelin. Endothelin is a ligand for the receptors, ETA and ETB. Macitentan prevents endothelin from binding to ETA and ETB.

inflammation. In disease conditions such as pulmonary vascular resistance (PAH), the local ET system is up-regulated and is involved in vascular hypertrophy and in organ damage."[246]

iii. Endothelin's Structure and Macitentan's Mechanism of Blocking of Endothelin

Endothelial cells form a single layer of cells lining every blood vessel in the cardiovascular system, including arteries and veins.[247] Endothelin is released by endothelial cells lining blood vessels. Fig. 10.4 shows the amino acid sequence of endothelin. Endothelin is an oligopeptide of 21-amino acids. The cysteine residues are connected via disulfide linkages.

The released endothelin binds to endothelin receptors in nearby smooth muscle cells, inducing contraction. The result of contraction is an increase in blood pressure. Fig. 10.5 shows the action of endothelin, in the absence (top drawing) and presence (lower drawing) of macitentan. The lightning bolt indicates macitentan's point of action where it blocks endothelin's interaction with its receptor. Macitentan, as well as the related drug bosentan, binds to endothelin receptor and prevents endothelin from binding to its own receptor.[248]

iv. Adverse Event Predicted From Drug Class Analysis Based on Chemical Structure

Referring to chemical structure, FDA wrote, "because this drug is ***chemically similar to bosentan, a drug associated with liver transaminase elevations***, I ... decided to re-examine the known effects of macitentan on the liver. In my re-review, I have found, at best, only a weak link between possible liver injury and the use of this agent [macitentan]."[249]

[246] Macitentan (pulmonary arterial hypertension) NDA 204-410. Page 18 of 163-page pdf file of Medical Review.

[247] Davenport AP, et al. Endothelin. Pharmacol. Rev. 2016;68:357–418.

[248] Gatfield J, et al. Distinct ETA receptor binding mode of macitentan as determined by site directed mutagenesis. PLoS One 2014;9:e107809 (15 pp.).

[249] Macitentan (pulmonary arterial hypertension) NDA 204-410. Liver Toxicity Review by Maryann Gordon, MD (3 pp.) occurring on page 6 of 163-page of Medical Review.

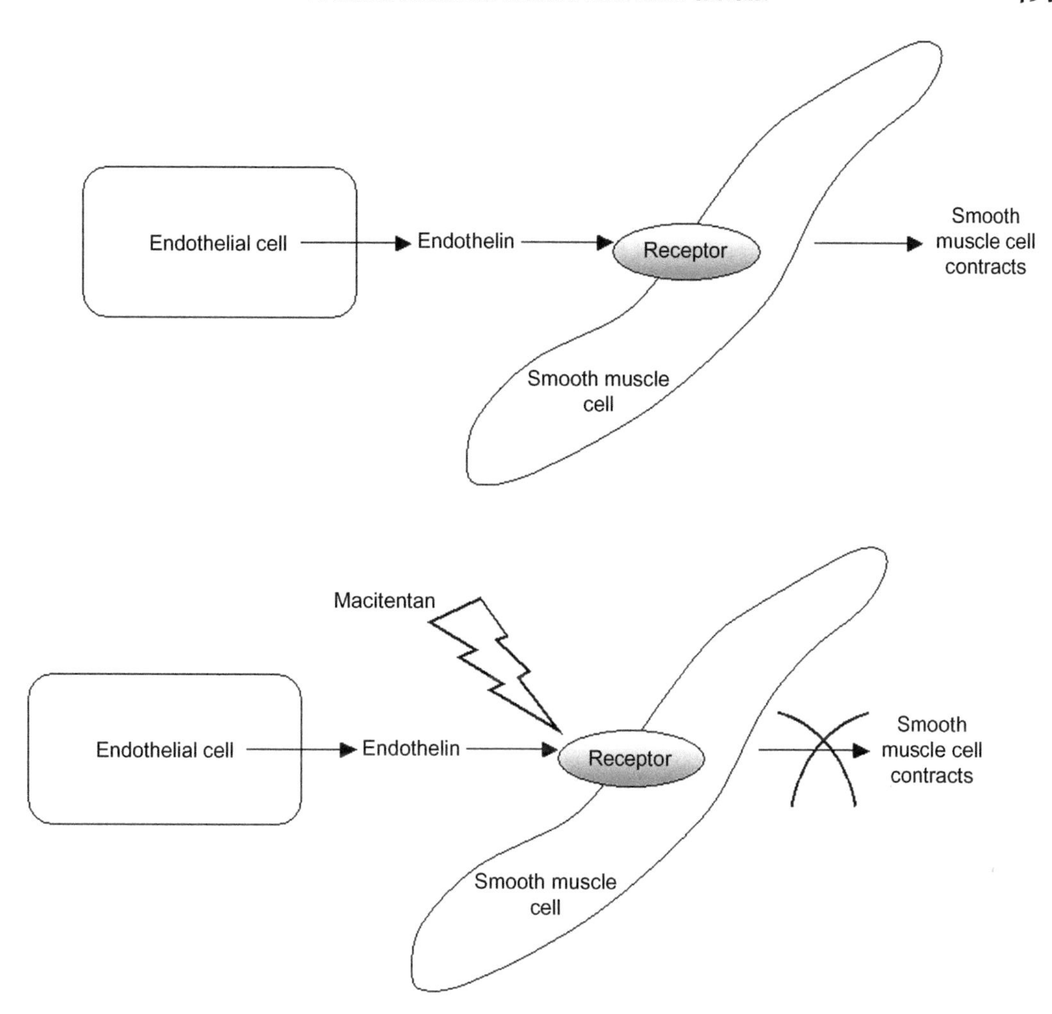

FIGURE 10.5 Endothelin signaling pathway. Endothelin is secreted by endothelial cells, where endothelin binds to its receptors on smooth muscle cells, where the result is contraction of the smooth muscle cells.

In other words, FDA's drug class analysis revealed that liver damage would be an expected AE but, in reviewing the Sponsor's own data on human subjects, there was only minimal liver damage. But despite FDA's judgment that there was "only a weak link" between macitentan and liver damage, the FDA reviewer realized that she was navigating in an ambiguous area which took the following form.

FDA realized that a firm conclusion on macitentan and liver damage could not yet be made, because *only low doses of macitentan had been used*, to date, and wrote, "However, there is **still need for vigilance** because ... the majority of macitentan doses studied has been 10 mg or less, so the safety of higher doses is unknown, and ... the total number of patients who have taken the drug is small."[250]

v. Drug Class Defined by Mechanism of Action

FDA's ClinPharm Review defined the drug class according to mechanism of action as consisting of four members: "Macitentan belongs to the class of ERAs used for the treatment of PAH. The other members of the ERA class are bosentan, ambrisentan and sitaxentan."[251] Embellishing on the members of this drug class, FDA added that, "Macitentan ... is a fourth-in-class, endothelin receptor antagonist (ERA). Other ERAs used for the treatment of pulmonary arterial hypertension (PAH) include bosentan (approved in 2001), ambrisentan (approved in 2007) and sitaxentan (not approved in US, and withdrawn from market worldwide)."[252]

vi. Adverse Events Predicted From Drug Class Analysis Based on Mechanism of Action

Using drug class analysis based on common mechanism of action, FDA's comments focused on these adverse events:

- Hepatic toxicity (resulting in release of liver enzymes to the bloodstream)
- Toxicity to seminiferous tubules and sperm production
- Toxicity to red blood cell production (resulting in low hematocrit and anemia)
- Toxicity to embryo (resulting in craniofacial abnormalities).

FDA's Medical Review outlined some of these AEs, using the terms "drug class," "class effect," and referring to the drug class of "endothelin receptor blockers."[253] FDA's drug class analysis that was used to predict AEs of the study drug included this writing:

> Important safety issues with consideration to related drugs:
>
> 1. Elevation of liver aminotransferases ... and liver failure have been reported with an endothelin receptor blocker (bosentan).
> 2. Based on animal data, endothelin receptor blockers are likely to cause major birth defects if used during pregnancy (class labeling).
> 3. Decreased sperm counts (class effect).

FDA's Medical Review also provided a narrative concerning the same drug class and another collection of AEs. Regarding the collection of other AEs, FDA wrote, "The non-

[250] Macitentan (pulmonary arterial hypertension) NDA 204-410. Liver Toxicity Review by Maryann Gordon, MD (3 pp.) occurring on page 6 of 163-page of Medical Review.

[251] Macitentan (pulmonary arterial hypertension) NDA 204-410. Page 49 of 91-page Clinical Pharmacology Review.

[252] Macitentan (pulmonary arterial hypertension) NDA 204-410. Page 52 of 91-page Clinical Pharmacology Review.

[253] Macitentan (pulmonary arterial hypertension) NDA 204-410. Page 19 of 163-page pdf file of Medical Review.

clinical work-up is consistent with macitentan being a mixed endothelin ETA and ETB receptor antagonist ... [t]ypical of these drugs, macitentan shows toxicity at seminiferous tubules, decreases hematocrit, and causes craniofacial abnormalities in fetuses exposed in utero. There is no identified off-target toxicity."[254]

Further regarding sperm, FDA referred to sperm-related AEs from bosentan, and imposed a requirement for macitentan's package label, "Effect on sperm. The testicular safety study (AC-055-113) conducted in healthy male subjects was poorly conducted and, therefore, uninformative. This agent will get a statement in the label that reflects the bosentan experience."[255]

The Sponsor's testicular safety study measured various parameters, as detailed by FDA's Medical Review, "AC-055-113 Investigation of the effect of macitentan on spermatogenesis, sperm quality and serum hormone concentrations of the hypothalamus pituitary-adrenal and gonadal axes."[256]

Further regarding hepatotoxicity, FDA referred to the package label for ambrisentan, and imposed a requirement for macitentan's labeling, "There is no hint of hepatotoxicity [for the study drug], but the baggage for endothelin receptors is such that the review team favors labeling similar to ambrisentan's, with additional clinical information obtained through a post-marketing commitment."[257]

Regarding hepatotoxicity, FDA's Office Director Memo stated, "Bosentan is hepatotoxic as was sitaxentan ... and dealing with its hepatotoxicity was a component of its limited distribution, but ambrisentan appears to be free of this toxicity ... [m]acitentan is clearly less hepatotoxic than bosentan, and may indeed prove not to be hepatotoxic, but it has not yet been shown to be entirely free of this risk."[258]

Regarding teratology, FDA's Office Director Memo utilized drug class analysis, which included the disclosure that bosentan was teratogenic. This comment on drug class actually referred to two drugs (bosentan, ambrisentan), and referred to teratogenicity and other AEs, "two other drugs, bosentan and ambrisentan, share macitentan's teratogenic effect, necessating restricted distribution, as well as its testicular toxicity and ability to cause anemia."[259]

In view of FDA's focus on bosentan and its proven teratogenic effects, a more dramatic account of FDA's drug class analysis can be provided by a view of the Black Box Warning on the package label of bosentan (Tracleer®):

[254] Macitentan (pulmonary arterial hypertension) NDA 204-410. Page 3 of 163-page pdf file of Medical Review.

[255] Macitentan (pulmonary arterial hypertension) NDA 204-410. Page 75 of 163-page pdf file of Medical Review.

[256] Macitentan (pulmonary arterial hypertension) NDA 204-410. Page 97 of 163-page pdf file of Medical Review.

[257] Macitentan (pulmonary arterial hypertension) NDA 204-410. Page 4 of 163-page pdf file of Medical Review.

[258] Macitentan (pulmonary arterial hypertension) NDA 204-410 (10-page Office Director Memo).

[259] Macitentan (pulmonary arterial hypertension) NDA 204-410 (10-page Office Director Memo).

Boxed Warning. Based on animal data, Tracleer is likely to cause major birth defects if used during pregnancy ... Must exclude pregnancy before and during treatment ... To prevent pregnancy, females of reproductive potential must use two reliable forms of contraception during treatment and for one month after stopping Tracleer.[260]

But FDA's recommendation for teratology warnings was not based entirely on drug class analysis. It was also based on the Sponsor's own animal toxicity studies. Regarding the animal toxicity tests, FDA wrote, "Macitentan was teratogenic in rabbits and rats, causing cardiovascular and mandibular arch fusion abnormalities at all doses tested ... [c]ontraindicating use of macitentan in pregnant women appears appropriate given the effects observed on fetal development."[261]

vii. FDA's Requirement for Risk Evaluation and Mitigation Strategy (REMS)

FDA's Approval Letter required a REMS and also pharmacovigilance. Regarding the REMS for teratogenicity, FDA's Approval Letter required:

we have also determined that ... macitentan ... can be approved only if elements necessary to assure safe use are required as part of a REMS to mitigate the risk of teratogenicity that is listed in the labeling. The elements to assure safe use will minimize the risk of fetal exposure and adverse fetal outcomes ... by certifying healthcare providers and pharmacies, and by documenting safe use conditions.[262]

Regarding active pharmacovigilance for hepatotoxicity, FDA's Approval Letter stated:

We have determined that an analysis of **spontaneous** postmarketing adverse events ... **will not be sufficient** to assess a signal of a serious risk of **hepatotoxicity** for ... macitentan[263] ... based on appropriate scientific data, FDA has determined that you are required to conduct the following: A long-term prospective observational study ... to evaluate potential serious **hepatic risks** related to the use of ... macitentan ... The registry will include a sample of patients prescribed ... macitentan ... and enroll at least 5000 patients. Patients should be followed for at least 1 year.[264]

Package label. Regarding teratogenicity, the results from FDA's drug class analysis, in combination with data from animal toxicity studies, found a place in the Black Box Warning for macitentan (Opsumit®).

BOXED WARNING: EMBRYO-FETAL TOXICITY. Do not administer OPSUMIT to a pregnant female because it may cause fetal harm ... Prevent pregnancy during treatment and for one month after stopping

[260] Package label. TRACLEER® (bosentan) tablets, for oral use; October 2016 (20 pp.).

[261] Macitentan (pulmonary arterial hypertension) NDA 204-410. Page 2 of 115-page Pharmacology Review.

[262] Macitentan (pulmonary arterial hypertension) NDA 204-410. Approval Letter accompanying 163-page pdf file of Medical Review.

[263] Macitentan (pulmonary arterial hypertension) NDA 204-410. Approval Letter accompanying 163-page pdf file of Medical Review.

[264] Macitentan (pulmonary arterial hypertension) NDA 204-410. Approval Letter accompanying 163-page pdf file of Medical Review.

treatment by using acceptable methods of contraception • For all female patients, OPSUMIT is available only through a restricted program called the OPSUMIT Risk Evaluation and Mitigation Strategy (REMS).[265]

The other AEs noted in FDA's Medical Review found a place on the package label, where they appear in the Warnings and Precautions section:

> **WARNINGS AND PRECAUTIONS**. Other endothelin receptor antagonists (ERAs) cause hepatotoxicity and liver failure. Obtain baseline liver enzymes and monitor as clinically indicated ... Decreases in hemoglobin ... Pulmonary edema in patients with pulmonary veno-occlusive disease. If confirmed, discontinue treatment ... Decreases in sperm count have been observed in patients taking endothelin receptor antagonists.[266]

g. Nebivolol (Hypertension) NDA 021-742

Nebivolol is a small molecule drug for treating hypertension. The relevant drug class is inhibitors of beta-adrenergic receptors. The literature provides an account of nebivolol's position in this drug class. A review article teaches that, "Nebivolol is a ... selective beta1-adrenergic receptor antagonist with a pharmacologic profile that differs from those of other drugs in its class. In addition to cardioselectivity mediated via beta1 receptor blockade, nebivolol induces nitric oxide-mediated vasodilation by stimulating endothelial nitric oxide synthase via beta3 agonism. This vasodilatory mechanism is distinct from those of other vasodilatory β-blockers (carvedilol, labetalol)."[267]

FDA's own drug class analysis did not contemplate the array of targeted receptors, but instead considered biochemical effects of influencing plasma levels of various hormones and on reproductive toxicity, where the account of reproductive toxicity utilized the safety margin concept. FDA provided a table, reproduced here as Table 10.5. The relevant drug class is, "beta-blockers". FDA referred to this table as, "A comparison of the effects of beta-adrenoreceptor antagonists on endocrine organs" and added, "As is readily apparent from the table ... inconsistent results on the hormonal system have been described for different beta-blockers and sometimes even for the same beta-blockers ... [o]ther than fetal growth restriction, little has been reported regarding adverse effects of beta-blocker treatment during pregnancy."[268]

Thus, although the hormone data in Table 10.5 are too sporadic and incomplete to compel a conclusion that nebivolol's AEs should be similar to those of other beta-blockers, or to compel a decision that nebivolol's AE are expected to differ from those of other beta-blockers, the existence of this table does provide an exemplary avenue for detecting subgroups in the drug class of any type of drug.

[265] Package label. OPSUMIT® (macitentan) tablets, for oral use; October 2013 (16 pp.).

[266] Package label. OPSUMIT® (macitentan) tablets, for oral use; October 2013 (16 pp.).

[267] Fongemie J, Felix-Getzik E. A review of nebivolol pharmacology and clinical evidence. Drugs 2015;75:1349–71.

[268] Nebivolol (hypertension) NDA 021-742. Page 12 of the eighth pdf file (80 pp.) of 12 pdf files in Medical Review.

TABLE 10.5 Effects of Beta-Adrenoreceptor Antagonists on Endocrine Organs

	Renin	Testosterone	Luteinizing Hormone	Follicle Stimulating Hormone	Estradiol
	Influence on Plasma Hormone Levels in Human Subjects				
Nebivolol	Down	No change	No change	No change	No change
Amosulalol	Down	–	–	–	–
Anapraline	–	Down	Up	Up	Up
Atenolol					
Bisoprolol	Down	–	–	–	–
Metoprolol	No change	Down	No change	–	–
Oxprenolol	Down	No change	–	–	–
Pindolol	No change	Down	–	–	–
Propranolol	Down	Up	Down	Down or no change	–

Nebivolol (hypertension) NDA 021-742. Pages 14–17 of the third pdf file (80 pages) of 12 pdf files in Medical Review.

FDA's analysis had the following features:

- Nebivolol results in AEs not found with other approved beta-blockers. These AEs are dystonia, cannibalism, and prolonged parturition.
- In addition to being characterized by unique reproductive AEs, nebivolol's reproductive adverse events occur at low doses while, in contrast, reproductive adverse events for other beta-blockers occur at much higher doses.
- As a consequence of FDA's drug class analysis, the package label for nebivolol warned against the reproductive AEs poor survival of newborns. In contrast, the adverse events of dystocia, reduced maternal care, and prolonged parturition, did not find any basis in drug class analysis, but instead had a basis in the Sponsor's animal toxicity studies. The package label used the euphemistic term "reduced maternal care" instead of cannibalism.

i. Safety Margin for Nebivolol is Much Smaller Than for Other Beta-Blockers

FDA's Medical Review dissected the drug class of beta-blockers as follows:

> There are other suggestions that nebivolol exerts endocrine-related effects during reproduction and differs from other beta-blockers. Nebivolol is unique among beta-blockers in that reproductive toxicity was observed at safety margins of ... **1.25-fold** the proposed human exposure. Safety margins for other beta-blockers is generally **10-fold** higher than the upper portion of their proposed dosing regimen.[269,270]

FDA referred to the fact that the AE of poor survival of pups "is common to beta antagonists" but added that, "However ... these reproductive findings occur at higher

[269] Nebivolol (hypertension) NDA 021-742. Page 12 of the eighth pdf file (80 pp.) of 12 pdf files in Medical Review.

[270] Nebivolol (hypertension) NDA 021-742. Pages 27–28 of the first pdf file (80 pp.) of seven pdf files in Pharmacology Review, and pages 20–21 of third pdf file (80 pp.) in Pharmacology Review.

exposure multiples with other antagonists in the class."[271] Further reiterating the point that other beta antagonists are characterized by a higher margin of safety, FDA added:

> while other beta adrenergic antagonists have reported non-clinical ... effects on reproduction, there are several distinctions that are ... unique to nebivolol ... the approved beta blockers showed in non-clinical studies at least **5X margin of safety** with the maximally recommended human doses.[272]

ii. Reproductive AEs for Nebivolol Differ From AEs of Beta-Blocker Drugs

FDA's Medical Review continued, "Dystocia, delayed parturition, and cannibalism were prominent features of reproductive toxicity studies with nebivolol, not described in current approved labeling for other beta-blockers."[273]

FDA stated that other drugs in the class included propranolol, atenolol, metoprolol, nadolol, and tolamolol, and observed that, "other than fetal growth restriction, little has been reported regarding adverse effects of beta-blocker treatment during pregnancy."[274]

Package label. FDA's comments on the Sponsor's animal toxicology data (dystocia, delayed parturition, and cannibalism) found a place on the package label in the Use in Specific Populations section. The term "pup" refers to a newborn animal:

> **USE IN SPECIFIC POPULATIONS** ... Pregnancy. Teratogenic Effects: Decreased pup body weights occurred at 1.25 and 2.5 mg/kg in rats, when exposed during the perinatal period ... [a]t 5 mg/kg and higher doses ... **prolonged gestation, dystocia and reduced maternal care** were produced with corresponding increases in late fetal deaths and stillbirths and decreased birth weight, live litter size and pup survival.[275]

h. Vilazodone (Antidepressant) NDA 022-567

Vilazodone is a drug for treating depression. The package label warnings for vilazodone were based only on drug class analysis, and not on animal toxicity studies or on AEs from clinical studies.[276] Vilazodone has the structure shown below:

[271] Nebivolol (hypertension) NDA 021-742. Pages 13–15, 34, and 66 of the second pdf file (80 pp.) of the seven Pharmacology Reviews.

[272] Nebivolol (hypertension) NDA 021-742. Page 21 of the third pdf file (80 pp.) of the seven Pharmacology Reviews.

[273] Nebivolol (hypertension) NDA 021-742. Page 12 of the eighth pdf file (80 pp.) of 12 pdf files in Medical Review.

[274] Nebivolol (hypertension) NDA 021-742. Pages 14–17 of the third pdf file (80 pp.) of 12 pdf files in Medical Review.

[275] Package label. BYSTOLIC® (nebivolol) tablets for oral use; December 2011 (14 pp.).

[276] Vilazodone (antidepressant) NDA 022-567 (126-page Clinical Review within the 259-page pdf file Medical Review).

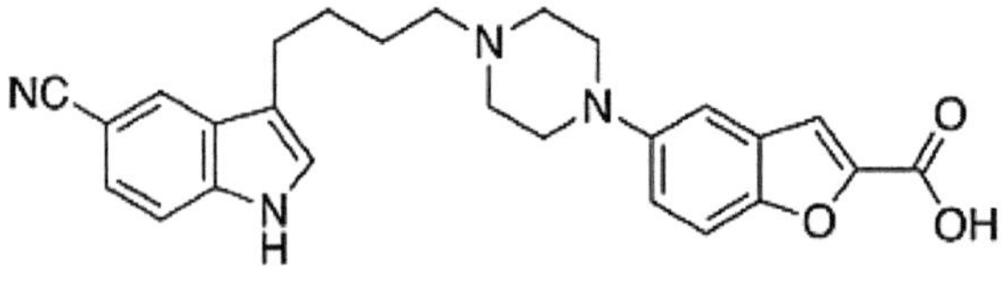

There are five different classes of antidepressant drugs. This list serves to break down the drug class of antidepressants into these five subclasses[277]:

1. Serotonin-specific reuptake inhibitors (SSRI)
2. Serotonin norepinephrine reuptake inhibitors (SNRI)
3. Tricyclic antidepressants (TCA)
4. Monoamine oxidase inhibitors (MAOI)
5. Multimodal antidepressants.

This provides a background on the physiology of reuptake inhibitors, as it applies to depression. The cause of depression, in part, is *depletion* of monoaminergic neurotransmitters [serotonin (5-hydroxytraptamine, 5-HT), norepinephrine, epinephrine, and dopamine]. Antidepressants act by *increasing* the concentration of serotonin (5-HT) in the extracellular space through the inhibition of serotonin reuptake transporters (SERTs).[278]

Vilazodone's mechanism of action is revealed by comments, which distinguish between SERTs from norepinephrine reuptake transporters:

> Vilazodone is an indolalkylamine that acts as both a potent serotonin-specific reuptake inhibitors (SSRI) (IC50 = 0.2 nM, Ki = 0.1 nM) and a 5-HT1A receptor partial agonist (IC50 = 0.5 nM). While vilazodone exhibits high affinity toward the 5-HT1A reuptake site, it does not bind to the norepinephrine (Ki = 56 nM) or dopamine (Ki = 37 nM) reuptake sites as avidly.[279]

FDA stated that drugs in the SSRI class include the study drug (vilazodone), fluoxetine, fluvoxamine, sertraline, paroxetine, citalopram, and escitalopram.[280] Moving in the path of reasoning used in most of FDA's drug class analyses, FDA contemplated AEs of other drugs in this same drug class (SSRI drug class):

> Safety Issues with Consideration to Related Drugs. Vilazodone is in the therapeutic class of serotonin specific reuptake inhibitors (SSRI). Although no specific safety issues related to the following items were identified in the vilazodone clinical development program, SSRIs as a class, have been associated with the

[277] Sahli ZT, et al. The preclinical and clinical effects of vilazodone for the treatment of major depressive disorder. Expert Opin. Drug Discov. 2016;11:515–23.

[278] Sahli ZT, et al. The preclinical and clinical effects of vilazodone for the treatment of major depressive disorder. Expert Opin. Drug Discov. 2016;11:515–23.

[279] Sahli ZT, et al. The preclinical and clinical effects of vilazodone for the treatment of major depressive disorder. Expert Opin. Drug Discov. 2016;11:515–23.

[280] Vilazodone (antidepressant) NDA 022-567. Page 8 of 126-page Clinical Review. Pages 2–127 of 259-page pdf file of Medical Review.

following safety concerns: • Suicidality • Serotonin syndrome • Seizures • Abnormal bleeding • Activation of mania or hypomania • Hyponatremia.[281]

FDA's Medical Review revealed that subjects from the Sponsor's clinical studies experienced the AEs in the list of SSRI drug class AEs. On the other hand, FDA's review disclosed that the rate of these particular AEs was either very low or that the AEs occurred in subjects not receiving vilazodone, as indicated by the bulletpoints. Please note the term, "listings". During the course of clinical studies, adverse events and other information on the study subjects are placed in a database called, "listings". Listings take the form of tables of data. When Clinical Study Reports are written, the medical writer bases the writing and analysis, in part, on the information in the listings. FDA's remarks on AEs of the SSRI drug class were:

- "**Suicidality**. There were two suicides during the clinical development program for vilazodone. Both occurred in Phase 2 studies; as indicated in the following narratives, neither patient was receiving vilazodone at the time of death."[282]
- "**Seizures**. Review of the entire vilazodone exposure database indicates one seizure."[283]
- "**Hyponatremia**. Review of the ... line listing for sodium values revealed only one result lower than 130 mmol/L ... [d]espite the known concern regarding hyponatremia/SIAH with this drug class, there was no evidence of hyponatremia in the ... database."[284]
- "**Activation of Mania or Hypomania**. Activation of mania or hypomania is a known complication of antidepressant use ... [r]eview of the ... line listing of AEs and narrative reports supports the applicant's assessment that the incidence of activation of manic/hypomanic symptoms associated with vilazodone use during treatment of MDD are within or below the expectations observed with other antidepressants."[285]
- "**Abnormal Bleeding**. The TEAEs related to bleeding occurred in <1% of subjects in the ... database ... [f]rom a clinical perspective, the bleeding events in the subjects taking vilazodone appear to be minor and not necessarily attributable to the drug."[286]
- "**Serotonin syndrome**. Because serotonin syndrome is a known complication of treatment with SSRIs, the AEs, ... were searched for preferred terms that might indicate serotonin syndrome ... [t]he ... line listings were reviewed to confirm the applicant's assessment. Two subjects were identified with probable vilazodone related serotonin toxicity."[287]

[281] Vilazodone (antidepressant) NDA 022-567. Page 8 of 126-page Clinical Review. Pages 2–127 of 259-page pdf file of Medical Review.

[282] Vilazodone (antidepressant) NDA 022-567. Page 120 of 259-page pdf file of Medical Review.

[283] Vilazodone (antidepressant) NDA 022-567. Pages 73–74 of 259-page pdf file of Medical Review.

[284] Vilazodone (antidepressant) NDA 022-567. Pages 73–74 of 259-page pdf file of Medical Review.

[285] Vilazodone (antidepressant) NDA 022-567. Pages 73 and 203 of 259-page pdf file of Medical Review.

[286] Vilazodone (antidepressant) NDA 022-567. Pages 73–74 of 259-page pdf file of Medical Review.

[287] Vilazodone (antidepressant) NDA 022-567. Page 202 of 259-page pdf file of Medical Review.

Package label. FDA's drug class analysis was based entirely on adverse events observed in the SSRI drug class. All of the AEs' characteristic of the SSRI drug class found a place on the package label for vilazodone (Viibryd®).[288] The dominant role of drug class analysis for predicting AEs is evident, in view of the fact that AE data from the Sponsor's own clinical studies showed, for example, that, "there was no evidence of hyponatremia in the ... database,"[289] "[r]eview of the entire vilazodone exposure database indicates one seizure,"[290] and "bleeding events in the subjects taking vilazodone **appear to be minor** and not necessarily attributable to the drug."[291] The warnings found a place in the Black Box Warning and in the Warnings and Precautions section:

> **WARNING: SUICIDAL THOUGHTS AND BEHAVIORS** ... Antidepressants increase the risk of suicidal thoughts and behaviors in patients aged 24 years and younger ... [m]onitor for clinical worsening and emergence of suicidal thoughts and behaviors ... [s]afety and effectiveness of VIIBRYD have not been established in pediatric patients.[292]

> **WARNINGS AND PRECAUTIONS** ... **Serotonin Syndrome**: Increased risk when co-administered with other serotonergic agents (e.g., SSRI, SNRI, triptans, amphetamines), but also when taken alone. If it occurs, discontinue VIIBRYD and initiate supportive treatment ... **Increased Risk of Bleeding**: Concomitant use of aspirin, nonsteroidal antiinflammatory drugs (NSAIDs), other antiplatelet drugs, warfarin, and other anticoagulants may increase this risk ... Activation of **Mania/Hypomania**: Screen patients for bipolar disorder ... **Seizures**: Can occur with treatment. Use with caution in patients with a seizure disorder ... **Hyponatremia** may occur as a result of treatment with SNRIs and SSRIs, including VIIBRYD. Cases of serum sodium lower than 110 mmol/L have been reported.[293]

i. Vorapaxar (Atherothrombotic Events in Patients With a History of Myocardial Infarction) NDA 204-886

Where a drug is "first-in-class," FDA may require the participation of an Advisory Committee, as is shown here for vorapaxar. Where a drug is first-in-class and where the Sponsor's clinical study shows that AEs result in an unfavorable ratio for risk/benefit, FDA is more likely to use of an Advisory Committee, as compared to the situation where the study drug is not first-in-class (where it is a member of drug class that is represented by drugs already approved by FDA).

An article in the journal, *Circulation*, provides details from the Advisory Committee meeting for vorapaxar. The article reveals interactions between the Advisory Committee,

[288] Package label. VIIBRYD™ (vilazodone hydrochloride) tablets; January 2017 (16 pp.).

[289] Vilazodone (antidepressant) NDA 022-567. Pages 73–74 of 259-page pdf file of Medical Review.

[290] Vilazodone (antidepressant) NDA 022-567. Pages 73–74 of 259-page pdf file of Medical Review.

[291] Vilazodone (antidepressant) NDA 022-567. Pages 73–74 of 259-page pdf file of Medical Review.

[292] Package label. VIIBRYD™ (vilazodone hydrochloride) tablets; January 2017 (16 pp.).

[293] Package label. VIIBRYD™ (vilazodone hydrochloride) tablets; January 2017 (16 pp.).

the Sponsor, and FDA, during the drug approval process.[294] The Advisory Committee was FDA's Cardiovascular and Renal Drugs Advisory Committee (CRDAC), and it included J. B.L. and S.K. The Sponsor was Merck, and the Sponsor's investigators at the meeting included K.M., D.M., and E.B. FDA's panel included P.S., M.R., R.T., Y.W., and M.J.K. The meeting was held on January 15, 2014.[295]

The Advisory Committee contemplated whether vorapaxar should be contraindicated in patients with these preexisting conditions:

1. Patients with history of stroke or transient ischemic attack (TIA)
2. Patients with low body weight
3. Elderly patients.

i. FDA's Drug Class Analysis Considers Patients With History of Stroke or Transient Ischemic Attack

The Advisory Committee decided that vorapaxar *should not be prescribed* to patients with a history of stroke or TIA, but rejected the notion that the drug should be contraindicated in low body weight patients. FDA's Medical Review revealed the recommendation of the Advisory Committee and the fact that the Sponsor agreed:

> After an Advisory Committee meeting where the majority of voting committee members supported approval and also voted to include patients with peripheral artery disease (PAD) without a history of stroke or TIA in the indicated population, the Applicant submitted proposed labeling with ... in patients with prior myocardial infarction (MI) or peripheral artery disease (PAD) without a history of stroke or TIA.[296]

The article published in *Circulation* described the logic used by the Advisory Committee, regarding the contraindications for patients with history of stroke or TIA. A decision made by the Advisory Committee was described as, "the increased risk of severe bleeding ... was substantially muted after the removal of patients with a history of stroke ... [t]his improved the benefit-to-risk ratio of vorapaxar. Given these findings, Daniel Bloomfield ... recommended to the FDA that vorapaxar should be contraindicated in patients with a history of stroke or TIA."[297]

The decision regarding stroke and TIA found a place in the Black Box Warning, as shown below. The Black Box Warning included the contraindication that voraxapar must not be given to patients with history of stroke, transient ischemic attack, or intracranial hemorrhage, etc.

[294] Baker NC, et al. Overview of the 2014 Food and Drug Administration Cardiovascular and Renal Drugs Advisory Committee meeting about vorapaxar. Circulation 2014;130:1287–94.

[295] Baker NC, et al. Overview of the 2014 Food and Drug Administration Cardiovascular and Renal Drugs Advisory Committee meeting about vorapaxar. Circulation 2014;130:1287–94.

[296] Vorapaxar (atherothrombotic events in patients with a history of myocardial infarction) NDA 204-886. Page 11 of 234-page Medical Review.

[297] Baker NC, et al. Overview of the 2014 Food and Drug Administration Cardiovascular and Renal Drugs Advisory Committee meeting about vorapaxar. Circulation 2014;130:1287–94.

Package label. The Black Box Warning for vorapaxar (Zontivity®) warns against risk for TIA and stroke:

> **WARNING: BLEEDING RISK.** Do not use ZONTIVITY in patients with a history of stroke, transient ischemic attack (TIA), or intracranial hemorrhage (ICH); or active pathological bleeding ... [a]ntiplatelet agents, including ZONTIVITY, increase the risk of bleeding, including ICH and fatal bleeding.[298]

ii. FDA's Drug Class Analysis Considers Low Body Weight Patients, and Decides That Warnings or Contraindications are not Needed for Low Body Weight Patients

FDA's Medical Review referred to the Advisory Committee, stating that, "An advisory committee meeting to consider this NDA is scheduled for January 15, 2014."[299] The results from this same Advisory Committee meeting are reported in Baker et al.,[300] which has a section of narrative with the title, "Should Vorapaxar Be Avoided in Patients With Low Body Weight?"

The publication of Baker et al. revealed the opinions from the Advisory Committee, for example, the opinion, "I do not find the analyses by **weight** particularly compelling, and other subgroup analyses by age and aspirin dose are less so. All need to be exposed in the label, but, I think, without advice as to interpretation, other than to be skeptical of subgroup analyses whether they show differences or not."[301] This same opinion, regarding the subgroup of low-weight patients, found a place on the package label.

Baker et al.[302] disclose that, "The panel spent a great deal of time discussing the labeling indication for patients with body weights <60 kg ... FDA asked the Sponsor to comment on whether vorapaxar should be contraindicated in this subgroup ... it is suspected the **patients with low body weights** are **also elderly and frail with multiple medical illnesses** that place them at higher risk for bleeding and reduce the clinical efficacy. Because of the small numbers in this subgroup ... the sponsor did not feel that a strong conclusion could be made. Furthermore, situations may arise in which a clinician may want to use vorapaxar in this subgroup. The sponsor prefers to keep this an individual clinician-patient decision, understanding the potential for increased bleeding, which therefore does not support a contraindication label."[303] As stated above, this same opinion about low-weight patients found a place on the package label.

The Clinical Pharmacology section of the package label mentioned *low body weight* patients, but only as information and not as any warning. Consistent with the

[298] Package label. ZONTIVITY™ (vorapaxar) Tablets 2.08 mg, for oral use; May 2014 (18 pp.).

[299] Vorapaxar (atherothrombotic events in patients with a history of myocardial infarction) NDA 204-886. Pages 11 and 191 of 234-page Medical Review.

[300] Baker NC, et al. Overview of the 2014 Food and Drug Administration Cardiovascular and Renal Drugs Advisory Committee meeting about vorapaxar. Circulation 2014;130:1287–94.

[301] Baker NC, et al. Overview of the 2014 Food and Drug Administration Cardiovascular and Renal Drugs Advisory Committee meeting about vorapaxar. Circulation 2014;130:1287–94.

[302] Baker NC, et al. Overview of the 2014 Food and Drug Administration Cardiovascular and Renal Drugs Advisory Committee meeting about vorapaxar. Circulation 2014;130:1287–94.

[303] Baker NC, et al. Overview of the 2014 Food and Drug Administration Cardiovascular and Renal Drugs Advisory Committee meeting about vorapaxar. Circulation 2014;130:1287–94.

recommendation of the Advisory Committee, the label stated that *low body weight did not much increase risk for AEs* and that *dose adjustments were not needed*:

> **CLINICAL PHARMACOLOGY** ... **Specific Populations**. In general, effects on the exposure of vorapaxar based on age, race, gender, **weight**, and moderate renal insufficiency were modest (20-40%). No dose adjustments are necessary based upon these factors.[304]

iii. FDA's Drug Class Analysis Considers use in Elderly Patients, and Recommends a Warning

FDA's drug class analysis focused on the safety issue of increased bleeding, as found in the drug class of antiplatelet drugs.[305] FDA's drug class analysis revealed increased bleeding risk in the elderly, stating that, "vorapaxar ... [is] less effective than placebo in the **elderly**, a finding consistent with the results for other antiplatelet drugs."[306] A warning regarding the elderly found a place in the package label, as shown below:

> **WARNINGS AND PRECAUTIONS** ... General **risk factors for bleeding** include **older age**, low body weight, reduced renal or hepatic function, history of bleeding disorders, and use of certain concomitant ... anticoagulants.[307]

The Use in Specific Populations section of the package label also disclosed increased bleeding risk in the elderly, reading:

> **USE IN SPECIFIC POPULATIONS** ... Geriatric Use ... ZONTIVITY increases the risk of bleeding in proportion to a patient's underlying risk. Because **older persons** are generally at a higher risk of bleeding, consider **patient age** before initiating ZONTIVITY.[308]

VII. CONCLUDING REMARKS

Determining safety information for the package label is best initiated by a review of FDA's Guidance for Industry, Warnings and Precautions, and Boxed Warnings Sections.[309] FDA's Guidance provides an introduction to drug safety and labeling, for example, that decisions to include adverse event information on the package label are a function of

[304] Package label. ZONTIVITY™ (vorapaxar) Tablets 2.08 mg, for oral use; May 2014 (18 pp.).

[305] Vorapaxar (reduction of atherothrombotic events in patients with a history of myocardial infarction) NDA 204-886 (234-page Medical Review).

[306] Vorapaxar (reduction of atherothrombotic events in patients with a history of myocardial infarction) NDA 204-886 (234-page Medical Review).

[307] Package label. ZONTIVITY™ (vorapaxar) Tablets 2.08 mg, for oral use; May 2014 (18 pp.).

[308] Package label. ZONTIVITY™ (vorapaxar) Tablets 2.08 mg, for oral use; May 2014 (18 pp.).

[309] U.S. Department of Health and Human Services. Food and Drug Administration. Center for Drug Evaluation and Research (CDER). Center for Biologics Evaluation and Research (CBER). Guidance for industry. Warnings and precautions, contraindications, and boxed warning sections of labeling for human prescription drug and biological products—content and format; 2011 (13 pp.).

whether the AE occurs to a greater extent in the study drug treatment arm or in the placebo treatment arm, and are a function of association of the AE with the drug as determinable by one or more of the Naranjo criteria.

Regarding drug class analysis, FDA's Guidance teaches that AEs that were not detected in the Sponsor's clinical studies can find a place in the Warnings and Precautions section, where the AEs are predictable from drug class analysis or predictable from animal toxicity data. Regarding drug class analysis, FDA's Guidance states that:

> The Warnings and Precautions section should include ... adverse reactions ... that are anticipated to occur with a drug if ... it appears likely that the adverse reaction will occur ... based on what is known about the pharmacology, chemistry, or class of the drug.[310]

FDA arrives at recommendations for the package label, by a review of AEs from the combination of the Sponsor's clinical studies, animal toxicity data, and drug class analysis. Drug class analysis can be dominant, where a similar drug has already been approved by FDA, and in this case, the Sponsor in collaboration with FDA reviewers contemplates package labeling from approved drugs, and determines which warnings are suitable for the study drug.

FDA's reviews for lixisenatide, lorcaserin, ponatinib, and vemurafenib illustrate techniques for improving accuracy of drug class analysis. Improved accuracy can be acquired by first tabulating all the AEs from all of the drugs in the drug class, then creating a subclass that reduces the number of drugs, and finally tabulating all the AEs from the subclass. The results from dissecting the drug class into its subclasses were:

- *Lixisenatide*. Drug class (GLP-1 receptor agonists). Subclass (short-acting GLP-1 receptor agonists)
- *Lorcaserin*. Drug class (agonist of serotonin receptors). Subclass (agonist of 5-HT2C receptor)
- *Ponatinib*. Drug class (tyrosine kinase inhibitors; TKIs). Subclass (VEGFR-family of kinase inhibitors)
- *Vemurafenib*. Drug class (serine-threonine protein kinases plus tyrosine kinases). Subclass (serine-threonine protein kinases plus tyrosine kinases, but excluding kinases with high value for IC50).

FDA's reviews for the following drugs illustrate how AEs from package labels from other drugs in the same drug class can be used to predict AEs for the study drug, where the available data on AEs from the Sponsor's own clinical studies are in short supply, totally absent, or ambiguous. When relying on package labels from another drug of the same drug class, the medical writer needs to be conscious of whether the AEs on the package label of the other drug were derived from human subjects or only from animal toxicity

[310] U.S. Department of Health and Human Services. Food and Drug Administration. Center for Drug Evaluation and Research (CDER). Center for Biologics Evaluation and Research (CBER). Guidance for industry. Warnings and precautions, contraindications, and boxed warning sections of labeling for human prescription drug and biological products—content and format; 2011 (13 pp.).

tests. This chapter shows how AEs from package labels from other drugs found a place on the package labels of the following drugs:

- Lenalidomide
- Pomalidomide
- Certolizumab pegol
- Macitentan
- Vilazodone

The special topic of pregnancy and teratogenicity to the embryo or fetus is illustrated in the accounts of lenalidomide,[311] pomalidomide,[312] and macitentan.[313] Adverse event labeling of lenalidomide and pomalidomide were based on previous human experiences with another drug in the same class (thalidomide). Adverse event labeling for macitentan was based on the previous package label for another drug in the same class (bosentan), where bosentan's package label warning was based on animal teratogenicity data. Ethical issues in conducting clinical studies on pregnant human subjects have been discussed.[314,315]

[311] Lenalidomide (myelodysplastic syndromes) NDA 021-880.

[312] Pomalidomide (multiple myeloma) NDA 204-026.

[313] Macitentan (pulmonary arterial hypertension) NDA 204-410.

[314] Chan M, et al. Prescription drug use in pregnancy: more evidence of safety is needed. Obstetric. Gynaecol. 2012;14:87–92.

[315] Merkatz RB, et al. Women in clinical trials of new drugs. A change in Food and Drug Administration policy. The Working Group on Women in Clinical Trials. New Engl. J. Med. 1993;329:292–6.

CHAPTER

11

Relatedness

I. INTRODUCTION

"Relatedness" refers to an association between administration of the study drug and a subsequent adverse event (AE), for example, as determined by one or more of the Naranjo criteria. During the course of a clinical trial, AEs experienced by study subjects are captured on the Case Report Form (CRF), where filling out this form includes assessing relatedness. After the data from the clinical study have been collected, relatedness analysis is used when drafting the package label.

Assessing "relatedness" of the study drug with a given AE can take into account particulars, such as the temporal relation between administering the drug and the AE, whether the study drug was administered orally, by subcutaneous injection, by intravenous injection, by infusion, and whether the study drug was administered on a daily basis, on a weekly basis, or only once as is the case with some vaccines. The need to take into account particulars is revealed by FDA's Guidance for Industry. FDA's Reviewer Guidance provides information on determining if a given AE is related to the study drug, stating that:

> For adverse events that seem drug related ... the reviewer should perform ... **[e]xplorations for dose dependency** ... [i]t may also be useful to evaluate safety as a function of weight-adjusted dose, body surface-adjusted dose, or cumulative dose ... [e]xplorations for demographic interactions (rates and comparisons with control for demographic and other subsets) for at least the more common and important adverse events.[1]

FDA's Guidance describes the influence of baseline characteristics on relatedness, writing that, "in examining whether ... **baseline risk factors are related to an adverse event**, the reviewer will either need to extract the baseline characteristics from case report

[1] US Department of Health and Human Services; Food and Drug Administration; Center for Drug Evaluation and Research (CDER). Reviewer Guidance Conducting a Clinical Safety Review of a New Product Application and Preparing a Report on the Review; 2005 (79 pages).

FDA's Drug Review Process and the Package Label
DOI: https://doi.org/10.1016/B978-0-12-814647-7.00011-7

tabulations ... it may be important to link individual safety observations with other on therapy data, such as dose, duration of treatment, concomitant therapy, other adverse events, lab data or effectiveness results."[2]

FDA provides guidance for the Adverse Reactions section of the package label, which discourages the disclosure of AEs not likely related to the study drug. Regarding unrelated AEs, FDA's Guidance states, "Exhaustive lists of every reported adverse event, including those that are infrequent and minor, commonly observed in the absence of drug therapy or **not plausibly related to drug therapy** should be avoided."[3]

Determining relatedness requires specialized experience and training, and because of this, relatedness is sometimes assessed by using an adjudication committee. Adjudication is detailed in Chapter 12, Adjudication of Clinical Data.

a. Case Report Form and MedWatch Form

In clinical trials, data on AEs are captured by the Case Report Form (CRF), while in the postmarketing situation, data on AEs may be captured and reported by FDA's MedWatch form (FDA Form 3500). For AEs that occur outside the United States, and where the AE is to be reported to the FDA, drug companies can use either an international form (CIOMS I form) or the MedWatch form.[4,5]

A page from a Case Report Form published by National Cancer Institute (NCI) is reproduced below in Fig. 11.1. The Case Report Form has a check box for writing the name of each type of adverse event, with a checklist for indicating relatedness (see checklist: Not related, Unlikely, Possible, Probable, Definite).

During the course of a clinical study, study subjects experience AEs, where these AEs are captured on a CRF (Fig. 11.1). The term "adverse event" does not imply that any person has assessed relatedness of the study drug with the AE, though the CRF may include check boxes to indicate an opinion on relatedness. A comment field may be provided on the CRF to elicit the investigator's description of the event and the rationale behind his causality opinion.[6] FDA's Guidance for Industry reveals that Sponsors have the option to draft the Clinical Study Protocol, so that it requires causality assessments, where FDA's Guidance states that, "non-serious events are recorded on the CRFs and are submitted to

[2] US Department of Health and Human Services; Food and Drug Administration; Center for Drug Evaluation and Research (CDER). Reviewer Guidance Conducting a Clinical Safety Review of a New Product Application and Preparing a Report on the Review; 2005 (79 pages).

[3] US Department of Health and Human Services; Food and Drug Administration; Center for Drug Evaluation and Research (CDER); Center for Biologics Evaluation and Research (CBER). Adverse Reactions Section of Labeling for Human Prescription Drug and Biological Products—Content and Format; 2006 (13 pages).

[4] US Department of Health and Human Services; Food and Drug Administration. Guidance for Industry. Postmarketing Safety Reporting for Human Drug and Biological Products Including Vaccines; 2001.

[5] Brody T. CLINICAL TRIALS. Study Design, Endpoints and Biomarkers, Drug Safety, and FDA and ICH Guidelines, 2nd ed., Elsevier, New York, NY; 2016 (pp. 541–6).

[6] Hsu PH, Stoll RW. Causality assessment of adverse events in clinical trials. II. An algorithm for drug causality assessment. Drug Inform. J. 1993;27:387–94.

U.S. Dept of Health and Human Services. National Institutes of Health. National Cancer Institute. Division of Cancer Prevention (March 2000) Guidelines for Designing and Completing Case Report Forms for Phase I and II Chemoprevention Trials (52 pages)

ADVERSE EVENTS

NCI Contact Number:		Study Title:				
Subject Number:		Subject Initials:				

Please enter visit date below:

Visit	Screen	Base-line	Visit 1	Visit 2	Visit 3	Visit 4	Visit 5	Visit 6
Date								

Adverse Event	Start Date MM/DD/YY	Stop Date MM/DD/YY	Event Recovery Status (Resolved or Not Resolved)	Check if Continu-ing	Relation-ship to Study Drug Not related Unlikely Possible Probable Definite	Toxicity Grade (1-4 CTCAE)§	Check if Serious Adverse Event (SAE)

§ CTCAE refers to Common Terminology Criteria for Adverse Events, which is a list of AE where that includes grades of severity for each AE. See, Dueck AC et al (2015) Validity and reliability of the US National Cancer Institute's Patient-Reported Outcomes version of the Common Terminology Criteria for Adverse Events (PRO-CTCAE). JAMA Oncol. 1:1051-1059.

FIGURE 11.1 Case Report Form with check boxes for opinions of relatedness of an AE to the study drug. This example of a CRF is from National Cancer Institute.

the sponsor ... investigator's assessment of causality is not required for non-serious adverse events by the regulations, although many sponsors may require it in the protocol."[7]

After the reproduction of the Case Report Form (CRF) (Fig. 11.1) is a reproduction of the MedWatch Form (Fig. 11.2). The Case Report Form and the MedWatch From each have check boxes for indicating relatedness. The relatedness inquiry is shown by a check box reading: Event Abated After Use Stopped or Dose Reduced? The relatedness is further shown by another check box reading: Event Reappeared After Reintroduction?

b. Naranjo Scale for Determining Relatedness

The Naranjo scale is an algorithm for determining relatedness.[8,9] The Naranjo scale, also known as the Adverse Drug Reaction (ADR) Probability Scale, asks the following questions, where the answers are then summed to generate a numerical value on relatedness:

- Are there previous conclusive reports of this reaction?
- Did the adverse event appear after the drug was given?
- Did the adverse reaction improve when the drug was discontinued or a specific antagonist was given?
- Did the adverse reaction reappear upon readministering the drug?
- Were there other possible causes for the reaction?
- Did the adverse reaction reappear upon administration of placebo?
- Was the drug detected in the blood or other fluids in toxic concentrations?
- Was the reaction worsened upon increasing the dose? Or, was the reaction lessened upon decreasing the dose?
- Did the patient have a similar reaction to the drug or a related agent in the past?
- Was the adverse event confirmed by any other objective evidence?

In actual practice, investigators may use only one or two of the criteria from the Naranjo scale. Also, it is typical to use only one, two, or perhaps three of the Naranjo criteria without mentioning or referring to the Naranjo scale.

c. Probability Scale Can Be Used With Naranjo scale

Following the use of one or more elements of the Naranjo scale, the result is compared to a probability scale, and the user determines the most appropriate level on the

[7] US Department of Health and Human Services; Food and Drug Administration; Center for Drug Evaluation and Research (CDER); Center for Biologics Evaluation and Research (CBER). Guidance for Industry and Investigators Safety Reporting Requirements for INDs and BA/BE Studies; 2012 (29 pages).

[8] Naranjo CA, et al. A method for estimating the probability of adverse drug reactions. Clin. Pharmacol. Ther. 1981;30:239–45.

[9] Seger D, et al. Misuse of the Naranjo adverse drug reaction probability scale in toxicology. Clin. Toxicol. (Phila.) 2013;51:461-6.

U.S. Department of Health and Human Services
Food and Drug Administration

Form Approved: OMB No. 09010-0291

MEDWATCH

The FDA Safety Information and Adverse Event Reporting Program

A. PATIENT INFORMATION

1. Patient Identifier In confidence	2. Age at Time of Event or Date of Birth:	3. Sex ☐ Female ☐ Male	4. Weight ___ lbs or ___kgs.

B. ADVERSE EVENT, PRODUCT PROBLEM OR ERROR
Check all that apply:
1. ☐ Adverse Event ☐ Product Problem (e.g., defects/malfunctions)
☐ Product Use Error ☐ Problem with Different Manufacturer of

2. Outcomes Attributed to Adverse Event
(Check all that apply)
☐ Death (mm/dd/yyyy) ☐ Disability or Permanent Damage
☐ Life-threatening ☐ Congenital Anomaly/Birth Defect
☐ Hospitalization ☐ Other Serious (Important Medical Events)
☐ Required Intervention to Prevent Permanent Impairment/Damage (Devices

3. Date of Event (mm/dd/yyyy)	4. Date of This Report (mm/dd/yyyy)

5. Describe Event or Problem or Produce Use Error

6. Relevant Tests/Laboratory Data, Including Dates

7. Other Relevant History, Including Preexisting Medical Conditions (e.g., allergies, race, pregnancy, smoking and alcohol use, hepatic/renal problems, etc.)

C. PRODUCT AVAILABILITY
Product Available for Evaluation?
☐ Yes ☐ No ☐ Returned to Manufacturer on: (mm/dd/yyyy)

D. SUSPECT PRODUCT(S)
1. Name, Strength, Manufacturer (from product label)
2. Dose or Amount Frequency Route
3. Dates of Use (If unknown, give duration) from/to (or best estimate).
4. Diagnosis or Reason for Use (Indication)
5. Event Abated After Use Stopped or Dose Reduced?
☐ Yes ☐ No ☐ Doesn't Apply
6. Lot #
7. Expiration Date
8. Event Reappeared After Reintroduction?
☐ Yes ☐ No ☐ Doesn't Apply
9. NDC# or Unique ID

E. OTHER (CONCOMITANT) MEDICAL PRODUCTS
Product names and therapy dates (exclude treatment of event)

F. REPORTER (See confidentiality section on back)
1. Name and Address
Name
Address
City State Zip

Phone #	E-mail
2. Health Professional? ☐ Yes ☐ No	3. Occupation

FORM FDA 3500 (1/09) Submission of a report does not constitute an admission that medical personnel or the product contributed to the event.

FIGURE 11.2 The MedWatch Form. It is provided by the US Food and Drug Administration.

probability scale. The probability scale can take the form of the WHO Causal Categories, which characterizes the relation between a given drug and a particular AE as one of:[10]

1. Certain
2. Probable
3. Possible
4. Unlikely
5. More data needed
6. Unassessable, report cannot be supplemented

Similarly, a Case Report Form used by Worldwide Antimalarial Research Network reveals similar check boxes for opinions for probability, where the check boxes are for Certain, Probable, Possible, Unlikely, Not related, and Unclassified.[11,12]

II. CODE OF FEDERAL REGULATIONS PROVIDES A BASIS FOR THE RELATEDNESS INQUIRY

The Code of Federal Regulations (CFR) provides guidance for assessing causality, stating that causality can be found where:

- "A single occurrence of an event that is uncommon and known to be strongly associated with drug exposure (e.g., angioedema, hepatic injury, Stevens-Johnson Syndrome)."[13]
- "An aggregate analysis of ... events observed in a clinical trial (such as known consequences of the underlying disease or condition under investigation or other events that commonly occur in the study population independent of drug therapy) that indicates those events occur more frequently in the drug treatment group than in a ... control group."[14]

The following excerpts also reveal the CFR's utilization of the concept of relatedness. According to 21 CFR §312.32, "**Adverse event** means any untoward medical occurrence associated with the use of a drug in humans, **whether or not considered drug related.**"[15]

[10] Meyboom RH, et al. Causal or casual? The role of causality assessment in pharmacovigilance. Drug Safety 1997;17:374–89.

[11] Case Report Form. Worldwise Antimalarial Research Network (WWARN); 2017.

[12] See also, The King's Health Partners Clinical Trials Office. Case Report Form Design. King's College, London, UK; 2017.

[13] 21 CFR §312.32. IND Safety Reporting.

[14] 21 CFR §312.32. IND Safety Reporting.

[15] 21 CFR §312.32. IND Safety Reporting.

Other rules in the CFR point out procedures where AEs must be reported, whether related or not to the study drug. For example, 21 CFR §312.64 requires that:

> An investigator must immediately report to the sponsor any serious adverse event, **whether or not considered drug related**, including those listed in the protocol or investigator brochure and must include an assessment of whether there is a reasonable possibility that the drug caused the event.[16]

Yet another part of the CFR invokes the concept of relatedness. Section §314.50 requires that CRFs need to include AEs, even if the AE is not believed to be related to the study drug:

> **Case report forms**. The application is required to contain copies of individual case report forms for each patient who died during a clinical study or who did not complete the study because of an adverse event, **whether believed to be drug related or not.**[17]

Section 312.32 requires that relatedness must be assessed in some situations (21 CFR §312.32). The CFR's requirements for causality and relatedness can be found in its definition for **suspected adverse reaction**:

> **Suspected adverse reaction** means any adverse event for which there is a **reasonable possibility that the drug caused the adverse event**. For the purposes of IND safety reporting, "reasonable possibility" means there is evidence to suggest a causal relationship between the drug and the adverse event. Suspected adverse reaction implies a lesser degree of certainty about causality than adverse reaction, which means any adverse event caused by a drug.[18]

Section 312.32 also contains FDA's most distinct and stringent requirement on causality and relatedness. This is in the CFR's definitions of AEs that are both serious and unexpected:

> Serious and unexpected suspected adverse reaction (SUSAR). The sponsor must report any suspected adverse reaction that is both serious and unexpected. The sponsor must report an adverse event as a suspected adverse reaction **only if there is evidence to suggest a causal relationship** between the drug and the adverse event.[19]

European Medicines Agency (EMA) distinguishes between the terms "adverse event" and "adverse drug reaction." An AE is, "Any untoward medical occurrence in a patient or clinical investigation subject administered a pharmaceutical product and which **does not necessarily have to have a causal relationship** with this treatment."[20] In contrast, EMA describes adverse drug reaction as requiring relatedness. EMA states that, "In the pre-approval clinical experience with a new medicinal product ... particularly as the

[16] 21 CFR §312.64. Investigator Reports.

[17] 21 CFR §314.50. Content and Format of an Application.

[18] 21 CFR §312.32. IND Safety Reporting.

[19] 21 CFR §312.32. IND Safety Reporting.

[20] European Medicines Agency. ICH E2A. Clinical Safety Data Management: Definitions and Standards for Expedited Reporting; 2005 (79 pages).

therapeutic dose(s) may not be established: all **noxious and unintended responses to a medicinal product related to any dose** should be considered adverse drug reactions."[21]

III. HOW TO RESOLVE AMBIGUITY IN THE RELATEDNESS INQUIRY

Attempts to assess relatedness are often characterized by ambiguous data and inability to arrive at a conclusion on relatedness. Ambiguity in assessing relatedness for a given AE can be mitigated by various techniques:[22]

- Adjudication., Adjudication is detailed in Chapter 12, Adjudication of Clinical Data.
- Pharmacovigilance after the drug enters the marketplace for acquiring additional safety data. Pharmacovigilance can involve a registry.
- Conducting additional clinical studies that assess AEs
- Disclosing the AE in a section of the package label that does not imply that the Sponsor believes that relatedness exists. To avoid this implication, an AE can be disclosed in the Clinical Trials Experience section or Adverse Reactions section, rather than in the Warnings and Precautions section, or the package label can expressly state that relatedness is not certain.

FDA's Medical Review for ustekinumab provides an example of ambiguity in the relatedness inquiry. This concerns the association of ustekinumab to the AEs of cancer and infections. Regarding ambiguity, FDA stated, "The reviewer agrees with the applicant that it is not clear that the findings in animals will apply to humans. The data from the applicant's 6-month monkey study were not adequate ... for definitive conclusions regarding malignancy risk of ustekinumab because the study did not provide treatment durations that cover ... a sufficiently long duration."[23]

Another reviewer expressed an opinion on ambiguity in relatedness, "The reviewer agrees that her presentation could have resulted from a combination of infectious events, and a contributory role for ustekinumab cannot be excluded."[24]

FDA's Approval Letter imposed the requirement that, in the postmarketing situation, the Sponsor should complete a 5-year clinical study for assessing relatedness of ustekinumab for cancer and other AEs. FDA's Approval Letter required:

> Finally, we have determined that **only clinical trials (rather than an observational study) will be sufficient to assess the known risk** of serious infection; or to identify unexpected serious risks of malignancy, tuberculosis, opportunistic infections, hypersensitivity reactions, autoimmune disease ... [t]herefore, based on appropriate scientific data, FDA has determined that you are required ... to ... [c]omplete the ongoing ... **C0743T08 trial for a total of 5 years** from initial enrollment ... [e]valuation of subjects should continue through 5 years (even if treatment is not continued for this duration).[25]

[21] European Medicines Agency. ICH E2A. Clinical Safety Data Management: Definitions and Standards for Expedited Reporting; 2005 (79 pages).

[22] Fingolimod (multiple sclerosis) NDA 022-527. Page 271 of 519-page Medical Review.

[23] Ustekinumab (psoriasis) BLA 125-261. Pages 216 and 217 of 246-page Medical Review.

[24] Ustekinumab (psoriasis) BLA 125-261. Page 151 of 246-page Medical Review.

[25] Approval Letter of September 25, 2009. Ustekinumab (psoriasis) BLA 125-261.

In addition, FDA's Approval Letter required that the Sponsor establish a registry for collecting information on AEs of ustekinumab-treated patients, and where the registry is to be active for 8 years. Regarding the registry, FDA's Approval Letter required:

> FDA has determined that you are required ... to conduct the following studies ... [e]nroll 4,000 ... ustekinumab-treated subjects into the Psoriasis Longitudinal Assessment and Registry ... and follow for 8 years from the time of enrollment. Subjects will be followed for the occurrence of serious infection, tuberculosis, opportunistic infections, malignancy, hypersensitivity reactions, autoimmune disease, neurologic or demyelinating disease, cardiovascular, gastrointestinal or hematologic adverse events.[26]

IV. ASSESSING RELATEDNESS OF AEs TO THE STUDY DRUG, AS ILLUSTRATED BY FDA'S MEDICAL REVIEWS

FDA's Medical Reviews provide guidance for assessing relatedness. The examples shown below fall into these analysis categories:

1. Relatedness established by temporal relationship
2. Relatedness established by comparing rate of AEs in study drug treatment arm versus control treatment arm
3. Relatedness established by mechanism of action of study drug
4. Relatedness established by taking into account concomitant drugs and other factors extrinsic to the patient, or taking into account the disease being treated by the study drug
5. Relatedness from drug class analysis
6. Relatedness inquiry that takes into account all AEs for one given tissue

a. Relatedness Established by Temporal Relationship

Examples of the relatedness inquiry by way of temporal relationships are shown for the following examples, as detailed below:

- Ceftazidime/avibactam combination
- Certolizumab
- Lixisenatide
- Lumacaftor/ivacaftor combination
- Simeprevir
- Vemurafenib

i. Ceftazidime/Avibactam (CAZ/AVI) Combination (Urinary Tract Infections) NDA 206-494

This concerns the temporal relationship of the study drug administration with the AE of urticaria. FDA's Medical Review for CAZ/AVI reveals the situation where relatedness

[26] Approval Letter of September 25, 2009. Ustekinumab (psoriasis) BLA 125-261.

was established because early discontinuation of CAZ/AVI resulted in prompt recovery from the AE. FDA's assessment of relatedness tracks one of the Naranjo criteria, namely, "Did the adverse event improve when the drug was discontinued?"[27]

The safety narrative provided by the Sponsor is reproduced as follows. Please note the observation that drug administration "was discontinued early and the event resolved after 1.5 hours." The narrative read, "Subject E0001038, a 26-year old White male, receiving an infusion of [study drug] ... experienced urticaria of moderate intensity 32 minutes after the start of the infusion ... [t]he infusion was discontinued early, and the event resolved after 1.5 hours ... [t]he event was assessed as causally related to the investigational product administration."[28]

Package label. Although urticaria was only a minor AE associated with ceftazidime/avibactam (AVYCAZ®), the AE of urticaria did find a place on the package label. It can reasonably be assumed that if there was not any causal relationship, then the package label would have been silent regarding urticaria. The Adverse Reactions section read:

> **ADVERSE REACTIONS** ... Other Adverse Reactions of AVYCAZ and Ceftazidime. Skin and subcutaneous tissue disorders. Rash, Rash maculo-papular, **Urticaria**, Pruritus.[29]

ii. Certolizumab (Crohn's Disease) BLA 125-160

Duration of study drug treatment was used to assess relatedness for the AE of rectal cancer. The study drug was certolizumab (Cimzia®). FDA quoted from the Sponsor's opinion that, "A 44-years old male was on Cimzia® for 30 days ... [p]atient had remained in the study for 30 days and the last dose had been administered 11 days prior to the event. Evaluation revealed a **rectal cancer** ... **the short time on Cimzia** makes it **unlikely to be related** to the study drug.[30]

To this, the FDA reviewer added, "I agree with the investigator that the ... **short duration of exposure to study drug makes it unlikely** that Cimzia is related to cause of death.[31]

Package label. The Warnings and Precautions section warned about cancer, but this particular warning was based on drug class analysis, and was not a result of cancer detected during clinical studies on certolizumab (Cimzia®). This warning read:

> **WARNINGS AND PRECAUTIONS** ... Cases of lymphoma and other malignancies have been observed among patients receiving TNF blockers.[32]

[27] Naranjo CA, et al. A method for estimating the probability of adverse drug reactions. Clin. Pharmacol. Ther. 1981;30:239–45.

[28] Ceftazidime/avibactam (CAZ/AVI) combination (urinary tract infections) NDA 206-494. Page 61 of 163-page Medical Review.

[29] Package label. AVYCAZ (ceftazidine and avibactam) for injection, for intravenous use; January 2017 (24 pages).

[30] Certolizumab (Crohn's disease) BLA 125-160. Page 28 of 80-page pdf file of Medical Review. This is the first of three separate pdf files containing the Medical Review.

[31] Certolizumab (Crohn's disease) BLA 125-160. Page 28 of 80-page pdf file of Medical Review. This is the first of three separate pdf files containing the Medical Review.

[32] Package label. CIMZIA (certolizumab pegol) Lyophilized powder or solution for subcutaneous injection; April 2008 (15 pages).

The Full Prescribing Section of the package label provided more information on the AE of cancer and concluded that relatedness with certolizumab was not established. This part of the package label referred to analysis by drug class (tumor necrosis factor (TNF) blockers) and to the Sponsor's own data on AEs:

> **WARNINGS AND PRECAUTIONS** ... In the controlled portions of clinical studies of some TNF blockers, more cases of malignancies have been observed among patients receiving TNF blockers compared to control patients. During controlled and open-labeled portions of CIMZIA studies ... malignancies ... were observed at a rate ... of 0.6 (0.4, 0.8) per 100 patient-years among 4,650 CIMZIA treated patients versus a rate of 0.6 ... per 100 patient-years among 1,319 placebo-treated patients. The size of the control group and limited duration of the controlled portions of the studies **precludes the ability to draw firm conclusions.**[33]

iii. Lixisenatide (Type-2 diabetes) NDA 208-471

This is about lixisenatide for treating Type-2 diabetes and the AEs of anaphylaxis and cancer. The take-home lesson is that a study subject's decision to discontinue the drug can argue for relatedness of the AE to the drug.

1. **Relatedness of lixisenatide to the AE of anaphylaxis**. This concerns relatedness analysis for anaphylaxis. In FDA's analysis, the argument for relatedness was strengthened by the fact that the AE led to the decision by the study subject to discontinue taking the study drug. Instructions for dose reduction, dose discontinuation, and dropping out of a clinical trial are conventional parts of the Clinical Study Protocol and of Clinical Study Reports, but these are not included in the Naranjo criteria. However, it is self-evident that a subject's decision for drug discontinuation can enhance arguments that the AE was related to the drug. FDA took drug discontinuation into account, writing, "there have been events of anaphylaxis ... that led to study drug discontinuation with lixisenatide, which **suggest that an association to study treatment was more likely.**"[34]

   FDA's Manual of Policies and Procedures expressly recommends that the Sponsor disclose, for each AE, whether that AE prompted the study subject to discontinue the drug, withdraw, or to drop out of the clinical study. Regarding withdrawing and dropping out, FDA recommends, "Where the review contains applicant-generated tables ... [a] table may identify one or more adverse events as having caused a particular subject to withdraw, in which case it would represent the actual incidence of specific adverse events that led to drop out. This approach is preferred. Alternatively, a table may list the adverse events that a subject experienced at the time of drop out and not identify any event ... as causing the drop out. This approach does not provide the actual incidence of adverse events associated with dropouts and is of less value."[35]

[33] Package label. CIMZIA (certolizumab pegol) Lyophilized powder or solution for subcutaneous injection; April 2008 (15 pages).

[34] Lixisenatide (type-2 diabetes) NDA 208-471. Page 15 of 690-page pdf file of Medical Review.

[35] Office of New Drugs; Center for Drug Evaluation and Research (CDER). Manual of Policies and Procedures. Attachment B: Clinical Safety Review of an NDA or BLA. MAPP6010.3 Rev. 1; December 2010 (82 pages).

2. **Background on relatedness to cancer**. Relatedness of a study drug to cancer is distinguished by the need for long-term exposure before the AE of cancer materializes. Long-term exposure to a chemical such as a drug or other xenobiotic, or to ultraviolet light, is needed before a tumor develop.

   Carcinogenesis is the result of a multistep process involving mutations that accumulate over long period of time. Short-term exposure studies, such as 90-day studies, are used to assess toxicity taking the form of low RBC counts, neutropenia, body weight, and sperm motility. In contrast, long-term exposure studies, such as those lasting 2 years and longer, are used to assess carcinogenicity.[36,37] FDA's Guidance for Industry refers to animal carcinogenicity studies as being "chronic," and provides a definition of "subchronic" as meaning about 90 days.[38]

   Long-term exposure is needed to detect the AE of cancer, because a given mutation caused by the study drug may not be sufficient (without additional mutations) to cause the cancer, and also because it takes time for a transformed cell to divide and multiply to the point where there exists a detectable tumor.
3. **Relatedness of lixisenatide to cancer**. FDA's Medical Review revealed the situation where relatedness to cancer was discounted because of the short exposure time, writing, "The five cases of **malignant melanoma** were reviewed ... [t]hree cases occurred after **very short exposure** to study drug, making association to lixisenatide unlikely ... [g]iven short exposures for three patients ... association to lixisenatide seems unlikely.[39]

Package label. Consistent with FDA's Medical Review, the package label for lixisenatide (Adlyxin®) warned against association of the drug with anaphylaxis:

> **CONTRAINDICATIONS**. Hypersensitivity to ADLYXIN or any product components. Hypersensitivity reactions including anaphylaxis have occurred with ADLYXIN.[40]

> **WARNINGS AND PRECAUTIONS**. Anaphylaxis and Serious Hypersensitivity Reactions: Discontinue ADLYXIN and promptly seek medical advice.[41]

[36] Halmes NC, et al. Reevaluating cancer risk estimates for short-term exposure scenarios. Toxicol. Sci. 2000;58:32–42.

[37] Chhabra RS, et al. An overview of prechronic and chronic toxicity/ carcinogenicity experimental study designs and criteria used by the National Toxicology Program. Environ. Health Perspect. 1990;86:313–21.

[38] US Department of Health and Human Services; Food and Drug Administration; Center for Drug Evaluation and Research (CDER); Center for Biologics Evaluation and Research (CBER). Guidance for Industry. S1C(R2) Dose Selection for Carcinogenicity Studies; 2008 (10 pages).

[39] Lixisenatide (type-2 diabetes) NDA 208-471. Pages 138 and 139 of 690-page pdf file containing Medical Review.

[40] Package label. ADLYXIN (lixisenatide) injection, for subcutaneous use; July 2016 (29 pages).

[41] Package label. ADLYXIN (lixisenatide) injection, for subcutaneous use; July 2016 (29 pages).

iv. Lumacaftor/Ivacaftor (LUM/IVA) Combination (Cystic Fibrosis) NDA 206-038

This concerns the lumacaftor/ivacaftor combination drug for cystic fibrosis. Duration of exposure was used to assess relatedness of lumacaftor/ivacaftor combination to the AE of respiratory syndromes. Short duration of time compelled the decision that the lumacaftor/ivacaftor combination was, in fact, related to the respiratory syndromes.[42] Regarding the short period of time, FDA wrote that, "Respiratory syndrome-related AEs occurred **sooner after dosing** and more commonly in LUM/IVA patients compared to placebo ... [t]hese data suggest that LUM/IVA exposure ***is associated with*** ... **respiratory syndrome-related AEs.**"[43]

The Sponsor stated that the respiratory syndromes took the forms of:[44]

1. Chest discomfort
2. Dyspnea
3. Respiratory abnormal

These three terms are Preferred Terms from the MedDRA dictionary. Consistent with FDA's review, the published literature states that respiratory syndrome occurs very shortly after dosing with lumacaftor/ivacaftor. One publication states that, "LUM/IVA was generally well tolerated; however, the incidence of certain respiratory adverse events (AEs), including dyspnea and chest tightness, was higher in LUM/IVA-treated patients than placebo-treated patients. These **AEs were often associated with initiation of therapy.**"[45]

Package label. FDA's observations regarding respiratory AEs for the lumacaftor/ivacaftor combination (Orkambi®) found a place on the package label. The term "FEV1" refers to "forced expiratory volume in one second." This is a standard marker for cystic fibrosis severity.[46]

> **WARNINGS AND PRECAUTIONS** ... Respiratory events: Chest discomfort, dyspnea, and respiration abnormal were observed more commonly during initiation of ORKAMBI. Clinical experience in patients with percent predicted FEV1.[47]

v. Simeprevir

Simeprevir is a small-molecule drug for treating hepatitis C virus. This provides another example of relatedness involving durations of time. The AE was photosensitivity

[42] Lumacaftor/ivacaftor combination (cystic fibrosis) NDA 206-038.

[43] Lumacaftor/ivacaftor combination (cystic fibrosis) NDA 206-038. Pages 10 and 57 of 99-page Medical Review.

[44] Lumacaftor/ivacaftor combination (cystic fibrosis) NDA 206-038. Page 71 of 99-page pdf file Medical Review.

[45] Marigowda G, et al. Effect of bronchodilators in healthy individuals receiving lumacaftor/ivacaftor combination therapy. J. Cystic Fibrosis. 2016;S1569–S1993. DOI: 10.1016.

[46] Szczesniak R, et al. Use of FEV1 in cystic fibrosis epidemiologic studies and clinical trials: a statistical perspective for the clinical researcher. J. Cyst. Fibros. 2017;16:318–26.

[47] Package label. ORKAMBI® (lumacaftor/ivacaftor) tablets, for oral use; September 2016 (13 pages).

rash and the relevant duration of time was of moderate length (54 days). The simeprevir analysis also revealed that the subject promptly discontinued the drug after the adverse event (rash) increased in severity:

> The subject developed a grade 1 AE photosensitivity reaction on Study Day 51 which was judged as **very likely related** to simeprevir by the investigator and reported as resolved the same day. However, on Study Day 54 the subject developed a grade 1 rash judged **possibly related** to the study drug. The rash persisted and was changed to grade 2 on Study Day 65 and judged **very likely related** to simeprevir ... simeprevir was discontinued on Study Day 67. On Study Day 114 the AE of rash improved in severity to grade 1 and was reported as resolved on Study Day 194. This reviewer concurs with the investigator's assessment of causality.[48]

The medical literature on simeprevir reports that skin rash occurs in the time frame of 2 weeks after starting simeprevir treatment.[49]

Package label. Consistent with FDA's relatedness analysis, the Warnings and Precautions section for simeprevir (Olysio®) warned about photosensitivity reactions and about rash:

> **WARNINGS AND PRECAUTIONS** ... Photosensitivity: Serious photosensitivity reactions have been observed during OLYSIO combination therapy. Use sun protection measures and limit sun exposure during OLYSIO combination therapy. Consider discontinuation if a photosensitivity reaction occurs ... Rash: Rash has been observed during OLYSIO combination therapy. Discontinue OLYSIO if severe rash occurs.[50]

The Full Prescribing Information section of the same package label provided information on temporality:

> **WARNINGS AND PRECAUTIONS** ... Rash has been observed with OLYSIO combination therapy ... **[r]ash occurred most frequently in the first 4 weeks of treatment**, but can occur at any time during treatment. Severe rash and rash requiring discontinuation of OLYSIO have been reported in subjects receiving OLYSIO in combination with Peg-IFN-alfa and RBV ... [p]atients should be monitored until the rash has resolved.[51]

vi. Vemurafenib (BRAF V600E Mutation Positive Melanoma) NDA 202-429

This further illustrates the use of temporality to assess relatedness. The AEs were Stevens–Johnson syndrome (SJS) and toxic epidermal necrolysis. In FDA's Medical Review for vemurafenib, FDA observed that, "The patient with Stevens-Johnson syndrome (SJS) that appeared 17 days after initiation of treatment, the patient with toxic epidermal necrolysis (TEN) that appeared after 26 days after initiation of treatment, and the patient ... who experienced the delayed hypersensitivity reaction, all demonstrate that vemurafenib has the potential to cause severe hypersensitivity reactions."[52]

[48] Simeprevir (hepatitis C virus infection) NDA 205-123. Page 102 of 173-page Medical Review.

[49] Eyre ZW, et al. Photo-induced drug eruption in a patient on combination simeprevir/sofosbuvir for hepatitis C. JAAD Case Rep. 2016;2:224–6.

[50] Package label. OLYSIO (simeprevir) capsules, for oral use; May 2017 (48 pages).

[51] Package label. OLYSIO (simeprevir) capsules, for oral use; May 2017 (48 pages).

[52] Vemurafenib (BRAF V600E mutation positive melanoma) NDA 202-429. 98-page Clinical Review, occurring on Pages 2–99 of 105-page Medical Review.

FDA conducted a separate analysis that only assessed AEs occurring within three days. FDA stated that, "The following adverse events ... were considered possibly related to immunogenicity: chills, drug hypersensitivity, hypotension, pruritis, rash ... swelling face, wheezing and Stevens Johnson syndrome ... **events that occurred within three days** of vemurafenib administration were reviewed. Thirteen patients experienced **events within three days of study drug administration.**"[53]

Package label. Consistent with FDA's review, the package label for vemurafenib (Zelboraf®) warned about Stevens Johnson syndrome, other hypersensitivity reactions, and rash:

> **WARNINGS AND PRECAUTIONS** ... Serious Hypersensitivity Reactions including anaphylaxis and Drug Reaction with Eosinophilia and Systemic Symptoms ... Discontinue ZELBORAF for severe hypersensitivity reactions ... Severe Dermatologic Reactions, including **Stevens-Johnson Syndrome** and Toxic Epidermal Necrolysis: Discontinue ZELBORAF for severe dermatologic reactions.[54]

> **ADVERSE REACTIONS**. Most common adverse reactions (≥30%) are arthralgia, rash, alopecia, fatigue, photosensitivity reaction, nausea, pruritus, and skin papilloma.[55]

b. Relatedness Established by Comparing Rate of AEs in Study Drug Treatment Arm Versus Control Treatment Arm

i. Ustekinumab (Psoriasis) BLA 125-261

In FDA's review of ustekinumab, the reviewer assessed relatedness of the AE of tumors to the study drug, by comparing the incidence of tumors with the incidence in a historic control.[56] The clinical trial concerned using ustekinumab for treating psoriasis. The AE of interest was tumors. Ustekinumab binds to interleukin-12 (IL-12) and blocks its activity.

IL-12, as it is naturally expressed in the body, facilitates $CD8^+$ T-cell responses against cancer, promotes Th1-type immune response, enhances the lytic activity of NK cells against tumor cells, and induces the secretion of IFN-γ by both T cells and NK cells. The activation of T cells by IL-12, in addition to the production of IFN-γ underlies the well-known antitumor effects of IL-12.[57] In addition to the antitumor effects of naturally expressed IL-12, treatment by administering IL-12 also promotes tumor elimination.[58,59] In

[53] Vemurafenib (BRAF V600E mutation positive melanoma) NDA 202-429. Page 89 of 105-page Medical Review

[54] Package label. ZELBORAF® (vemurafenib) tablet for oral use. April 2017 (18 pages).

[55] Package label. ZELBORAF® (vemurafenib) tablet for oral use. April 2017 (18 pages).

[56] Ustekinumab (plaque psoriasis) BLA 125-261. Page 77 of 246-page Medical Review.

[57] Salem ML, et al. Paracrine release of IL-12 stimulates IFN-gamma production and dramatically enhances the antigen-specific T cell response after vaccination with a novel peptide-based cancer vaccine. J. Immunol. 2004;172:5159–67.

[58] Zhu S, et al. IL-12 and IL-27 sequential gene therapy via intramuscular electroporation delivery for eliminating distal aggressive tumors. J. Immunol. 2010;184:2348–54.

[59] Jaime-Ramirez AC, et al. IL-12 enhances the antitumor actions of trastuzumab via NK cell IFN-γ production. J. Immunol. 2011;186:3401–9.

view of the fact that ustekinumab binds to and blocks IL-12, it might be expected that ustekinumab treatment can increase tumors.

FDA's Medical Review for ustekinumab revealed the use of a historic control for use in assessing relatedness. The historic control resided in a database, the SEER database. SEER database is provided by NCI and it allows gathering data by cancer site, e.g., bone, brain, stomach, pancreas, leukemia, and by race, gender, and age. SEER means Surveillance, Epidemiology, and End Results Program.

FDA commented that, "Eight solid malignancies ... prostate ... kidney, thyroid, breast ... were reported in 8 subjects, fewer than would be expected **by comparison with SEER database** (per subject year exposure, adjusted for age, gender, and race) ... [t]he rate and types of solid tumor malignancies, as well as the ratio of basal to squamous cell carcinomas of the skin, do not suggest a malignancy signal related to immunosuppression."[60]

In other words, to assess the ability of the study drug to cause immunosuppression, with the consequence of inducing tumors, the Sponsor compared incidence of tumors in the study with incidence in the SEER database. In finding that the incidences of tumors to be greater in this database than in the Sponsor's clinical study, FDA concluded that the Sponsor's data "do not suggest a malignancy signal related to immunosuppression."

However, because of the fact that ustekinumab suppresses immune response (it blocks Th1-type immune response), FDA required that the Sponsor collect data on malignancies from patients in the general population. In other words, the Sponsor was required to maintain a registry to actively collect all serious AEs, for example, malignancies, and where the registry will last 8 years.[61] The registry will collect AEs from about 4000 patients who were treated with ustekinumab. The name of the registry is PSOLAR. A published analysis of the data in the PSOLAR registry concluded that **ustekinumab is not associated with increased risk for malignancy.**[62]

Package label. The package label for ustekinumab (Stelara®) revealed that cancer was not detected in the Sponsor's clinical trials but did state that in the postmarketing situation, skin cancer was reported by patients. Also, the package label was careful to state that ustekinumab is an "immunosuppressant," thus raising the alarm that the drug poses an increased risk for both cancer and infections. FDA's Approval Letter was dated September 2009, and it is interesting to point out that the most recent package label dating from 5 years later (March 2014), refrains from stating that there is any clear-cut relationship between ustekinumab and cancer. The Warnings and Precautions section includes an interesting comment, which medical writers should consider including in all package label to drugs that are immunosuppressants. This interesting comment is a remark about

[60] Ustekinumab (plaque psoriasis) BLA 125-261. Page 77 of 246-page Medical Review.

[61] Ustekinumab (plaque psoriasis) BLA 125-261. Pages 9–11 of 246-page Medical Review.

[62] Papp K, et al. Safety surveillance for ustekinumab and other psoriasis treatments from the Psoriasis Longitudinal Assessment and Registry (PSOLAR). J. Drugs Dermatol. 2015;14:706–14.

patients having "pre-existing risk factors" for cancer. The Warnings and Precautions section read:

> **WARNINGS AND PRECAUTIONS** ... Malignancies STELARA® is an immunosuppressant and may increase the risk of malignancy. Malignancies were reported among subjects who received STELARA® in clinical studies ... [i]n rodent models, inhibition of IL-12/IL-23p40 increased the risk of malignancy ... [t]here have been post marketing reports of the rapid appearance of multiple cutaneous squamous cell carcinomas in patients receiving STELARA® who had pre-existing risk factors for developing non-melanoma skin cancer. All patients receiving STELARA® should be monitored for the appearance of non-melanoma skin cancer.[63]

c. Relatedness Established by Mechanism of Action of Study Drug

i. Certolizumab (Crohn's Disease) BLA 125-160

Certolizumab (Cimzia®) is an antibody for treating Crohn's disease. The antibody binds to TNF-α, resulting in the inactivation of TNF-α, with consequent reduction in the immune system's attack on bacterial infections.[64] Certolizumab and other drugs in the drug class of TNF antagonists result in increased risk for tuberculosis (TB).[65] The AE in question was TB.

FDA reported that, "This is the first fatal case of TB reported for a Cimzia treated patient."[66] Regarding relatedness and mechanism of action of the study drug, as it applies to TB, the FDA reviewer wrote, "A 50-year old female ... developed disseminated tuberculosis ... [d]espite intensive treatment, the subject died. The investigator assessed ... the event of disseminated tuberculosis as **definitely related** ... Cimzia and other TNF-α antagonists interfere with host immune function and may predispose the subject to infections. TB is a recognized risk factor with TNF-α antagonists and ten cases of TB have been reported in ... [this] program to date."[67]

Package label. The Black Box Warning for certolizumab (Cimzia®) provided detailed warnings against TB, which included instructions for discontinuation, TB testing, and TB monitoring:

> **BOXED WARNING**. Increased risk of serious infections leading to hospitalization or death including tuberculosis (TB), bacterial sepsis, invasive fungal infections ... and infections due to other opportunistic pathogens ... CIMZIA should be discontinued if a patient develops a serious infection or sepsis ... Perform test for latent TB; if positive, start treatment for TB prior to starting CIMZIA ... Monitor all patients for active TB during treatment, even if initial latent TB test is negative.[68]

[63] Package label. STELARA® (ustekinumab) injection, for subcutaneous use. March 2014 (19 pages).

[64] Suhreiber S, et al. Maintenance therapy with certolizumab pegol for Crohn's disease. N. Engl. J. Med. 2007;357:239–50.

[65] Zhang Z, et al. Risk of tuberculosis in patients treated with TNF-α antagonists: a systematic review and meta-analysis of randomised controlled trials. BMJ Open. 2017;7:e012567.

[66] Certolizumab (Crohn's disease) BLA 125-160. Page 27 of 80-page Medical Review. This is the first of three separate pdf files containing the Medical Review.

[67] Certolizumab (Crohn's disease) BLA 125-160. Page 27 of 80-page Medical Review. This is the first of three separate pdf files containing the Medical Review.

[68] Package label. CIMZIA (certolizumab pegol) for injection, for subcutaneous use. January 2017 (33 pages).

d. Relatedness Established by Taking Into Account Concomitant Drugs and Other Factors Extrinsic to the Patient or Taking Into Account the Disease Being Treated by the Study Drug

Introduction. The relatedness analysis for the following drugs attributed the AEs not to the study drug but instead to:

- concomitant disease (congestive heart failure),
- concomitant drugs (chemotherapy),
- concomitant disease (pancreatitis),
- concomitant drugs (oral contraceptives and pain drug),
- concomitant drug (cefazolin) and concomitant condition (bee sting), and
- the disease being treated by the study drug (multiple myeloma).

i. Alvimopan (Gastrointestinal Recovery Following Bowel Resection Surgery) NDA 021-775

The relatedness analysis attributed the AEs to a concomitant disease (congestive heart failure). Alvimopan is a μ-opioid receptor antagonist designed to reduce the antimotility effects of opioids. After intestinal surgery, the use of postoperative opioids for pain control acts on enteric nervous system μ receptors, slowing intestinal transit times. An option for avoiding the slowed intestinal transit is to avoid using opioids after surgery; however, many patients still require opioid pain control after abdominal surgery. Alvimopan is an oral, μ-opioid receptor antagonist that mitigates antimotility effects of opioids on the gastrointestinal tract without reducing opioid-based analgesia.[69]

This concerns the AE of cardiovascular death. The FDA reviewer distinguished between the drug as the cause of the AE and the underlying disease as the cause of the AE.[70] FDA observed the following features of the patient's concomitant diseases:

- Congestive heart failure
- Diabetes mellitus
- Chronic obstructive pulmonary disease

FDA's observations and analysis considered relatedness by temporal sequence, but then turned to concomitant diseases, and concluded that the AE was due to the concomitant diseases. FDA wrote, "the cardiovascular death in Patient 14CL302-22-0118 ... was **unlikely related** to alvimopan because this patient had an **improbable temporal sequence**

[69] Berger NG, et al. Delayed gastrointestinal recovery after abdominal operation—role of alvimopan. Clin. Exp. Gastroenterol. 2015;8:231–5.

[70] Alvimopan (GI recovery following bowel resection surgery) NDA 021-775. Pages 7–9 of 43-page Special Safety Review. The 383-page Medical Review contains three reviews. These are Special Safety Review (55 pages) on Pages 13–67 of Medical Review, Clinical Review (128 pages) on Pages 91–218 of Medical Review, and Clinical Review (124 pages) on Pages 257–380 of the 383-page Medical Review.

from the administration of alvimpan (he only received one dose of alvimopan and developed congestive heart failure 6 days later and died 13 days later) and his death can be reasonably explained by ... **his underlying congestive heart failure**."[71]

Regarding another patient, FDA commented that, "the cardiovascular death in Patient 14CL313-13-13015 (acute myocardial infarction) was **possibly related to alvimopan** because there was a reasonable temporal sequence and ... because of the existence of reports of myocardial infarctions associated with alvimopan use in the longer-term ... trials. However, other features (he was high risk for myocardial infarctions ... because of his underlying diabetes mellitus, increased age, and surgical stress) may be the only reason for his death."[72]

Regarding still another patient, FDA wrote that, "the death of Patient 13C304-003-001 was **unlikely related to alvimopan** because she had known COPD ... [t]his patient had COPD exacerbation in the past without being on alvimopan. In addition, she was off alvimopan when her symptoms of her COPD exacerbation started."[73]

ii. Certolizumab (Crohn's Disease) BLA 125-160

The relatedness analysis attributed the AEs to concomitant drugs (chemotherapy). Certolizumab (Cimzia®) is an antibody for treating Crohn's disease. The antibody binds to TNF-α, resulting in the inactivation of TNF-α, with consequent reduction in the immune system's attack on bacterial infections.[74]

Drugs in the TNF-α antagonist drug class can increase risk for cancer. Increased risk is especially evident for patients already predisposed to cancer or for patients taking a second immunosuppressant in addition to the TNF-α antagonist.[75] FDA issued a general warning about risk for cancer with drugs that block TNF-α, where FDA's publication stated, "Healthcare professionals should remain vigilant for cases of malignancy in patients treated with TNF blockers and report them to the FDA MedWatch program.[76]

FDA's Medical Review considered drugs that were administered concomitantly with the study drug. Note that the study drug (certolizumab) dampens immune response. The concomitant chemotherapy was not identified in FDA's review. Hence, it is possible that the chemotherapy was of a type that stimulated immune response (against the cancer) or that the chemotherapy was of a type that does not involve immune response, such as

[71] Alvimopan (GI recovery following bowel resection surgery) NDA 021-775. Page 163 of 383-page Medical Review.

[72] Alvimopan (GI recovery following bowel resection surgery) NDA 021-775. Pages 163 and 164 of 383-page Medical Review.

[73] Alvimopan (GI recovery following bowel resection surgery) NDA 021-775. Page 167 of 383-page Medical Review.

[74] Suhreiber S, et al. Maintenance therapy with certolizumab pegol for Crohn's disease. N. Engl. J. Med. 2003;357:239–50.

[75] Haynes K, et al. Tumor necrosis factor α inhibitor therapy and cancer risk in chronic immune-mediated diseases. 2013;Arthritis Rheum. 65:48–58.

[76] US Food and Drug Administration. FDA Drug Safety Communication: UPDATE on Tumor Necrosis Factor (TNF) blockers and risk for pediatric malignancy; 2011.

paclitaxel or cisplatin. FDA concluded that the AEs were not due to certolizumab but instead resulted from chemotherapy, "A 43-year old female ... had ... metastatic rectal tumor ...[s]ubject died from complications **related to chemotherapy**. Death considered by investigator to **not be related to Cimzia** ... [m]alignancy is a recognized safety concern with treatment with TNF-α antagonists.[77]

iii. Denosumab (Osteoporosis in Postmenopausal Women) BLA 125-320

The relatedness analysis attributed the AEs to the concomitant disease of pancreatitis.

Denosumab is an antibody for treating osteoporosis, the antibody that binds to RANKL (RANK ligand). FDA's relatedness analysis concerned the AEs of pancreatitis and dermatological AEs. FDA's relatedness analysis of pancreatitis established that this AE was not due to denosumab but was instead due to the fact that the study subjects were already at risk for pancreatitis. For help in relatedness analysis, the Sponsor chose to consult a team of experts (FDA's Division of Dermatology and Dental Products).

Regarding pancreatitis, FDA's Medical Review observed, "Pancreatitis. In trial 20030216, there was an imbalance in events of pancreatitis in subjects randomized to denosumab. A total of 4 subjects in the placebo group, and 8 subjects in the denosumab group reported an event of pancreatitis. ... **[m]any of these subjects had underlying risk factors for pancreatitis** ... [t]he temporal relationship between duration of denosumab exposure and the development of pancreatitis is highly variable. In addition, most cases in the denosumab group were **confounded by prior episodes of pancreatitis or risk factors for the development of pancreatitis.**[78]

Regarding dermatological AEs, FDA's observations and analysis was as follows. Please note the algorithm used in arriving at the conclusion, namely, "denosumab could not be clearly implicated as causative nor could it be definitely ruled out as the cause."[79] The algorithm took the form of a "burden-of-proof" style statement, of the type encountered in legal arguments. The FDA reviewer observed, "An imbalance in dermatologic adverse events ... was noted ... [a] total of 501 (12.4%) of placebo-treated subjects and 610 (15.1%) of denosumab-treated subjects reported an adverse event related to skin ... [which included] dermatitis, eczema, and rashes ... [t]he Division of Dermatology and Dental Products was consulted to assist in the evaluation of these cases and their relationship to denosumab ... [a]fter review, **denosumab could not be clearly implicated as causative nor could it be definitely ruled out as the cause."**[80]

[77] Certolizumab (Crohn's disease) BLA 125-160. Page 28 of 80-page Medical Review. This is the first of three separate pdf files containing the Medical Review.

[78] Denosumab (osteoporosis in postmenopausal women) BLA 125-320. Pages 166 and 167, 325 of 710-page pdf file of the Medical Review.

[79] Denosumab (osteoporosis in postmenopausal women) BLA 125-320. Pages 166 and 167 of 710-page pdf file of the Medical Review.

[80] Denosumab (osteoporosis in postmenopausal women) BLA 125-320. Pages 166 and 167 of 710-page pdf file of the Medical Review.

Package label. Despite FDA's unwillingness to implicate denosumab (Prolia®) as causative for dermatologic AEs, the Warnings and Precautions section did, in fact, warn about this type of AE:

> **WARNINGS AND PRECAUTIONS** ... Dermatologic reactions: Dermatitis, rashes, and eczema have been reported. Consider discontinuing Prolia if severe symptoms develop.[81]

FDA's reluctance to imply or state that denosumab was related to the AE of pancreatitis is mirrored in the package label, that is, by the fact that this AE was reported in the Adverse Reactions section and not in the Warnings and Precautions section:

> **ADVERSE REACTIONS** ... Pancreatitis has been reported in clinical trial.[82]

The above quoted package label excerpts were exactly the same for the package label, as first issued in June 2010[83] and in the package label from 7 years later, May 2017,[84] suggesting that further data for resolving the admitted ambiguities did not become available in the intervening years.

iv. Fingolimod (Multiple Sclerosis) NDA 022-527

The relatedness analysis attributed the AEs to concomitant oral contraceptives and pain drugs. This concerns the oral concomitant drugs of contraceptives, gabapentin, pregabalin, and the concomitant disease of idiopathic thrombocytopenia purpura (ITP). ITP is a disease that causes low platelet counts.

FDA's Medical Review of fingolimod for treating multiple sclerosis focuses on the AE of papilledema.[85] Papilledema occurred on Day 11 of the clinical study. Fingolimod was discontinued, but 30 days after discontinuation, the subject still had papilledema. FDA concluded that the AE was due to the concomitant drugs, "On a follow-up visit, 188 days after drug discontinued, an ophthalmologist said that papilledema was still present ... [t]he ... ophthalmologist thought it was most likely related to **papillophlebitis due to oral contraceptives use** and not related to drug.[86]

The published medical literature discloses the same association, as cited.[87] FDA's review contemplated attributing the AE of low platelets to concomitant drugs (gabapentin and pregabalin). One subject was receiving the concomitant medications of gabapentin and pregabalin for treating neuropathic pain. FDA observed that this AE was, "temporarily related to initiation of ... fingolimod therapy but the patient did not recover 9 months after drug discontinuation." FDA stated that the concomitant neuropathic pain drugs

[81] Package label. Prolia® (denosumab) injection, for subcutaneous use; May 2017 (26 pages).

[82] Package label. Prolia® (denosumab) injection, for subcutaneous use; May 2017 (26 pages).

[83] Package label. Prolia® (denosumab) injection, for subcutaneous use; June 2010 (17 pages).

[84] Package label. Prolia® (denosumab) injection, for subcutaneous use; May 2017 (26 pages).

[85] Fingolimod (multiple sclerosis) NDA 022-527. 519-page Medical Review.

[86] Fingolimod (multiple sclerosis) NDA 022-527. Page 212 of 519-page Medical Review.

[87] Eski YO, et al. Papillophlebitis associated with use of oral contraceptive: a case report. Gynecol. Endocrinol. 2015;31:601–3.

"may have played some role."[88] The published literature had documented an association of low platelets (thrombocytopenia) with gabapentin[89] and with pregabalin.[90]

v. Lixisenatide (Type-2 Diabetes) NDA 208-471

The relatedness analysis attributed the AEs to the concomitant drug (cefazolin) and a bee sting. Relatedness analysis can involve comparing the incidence of a given AE in the study drug treatment arm with that in the control arm, as shown by FDA's analysis of lixisenatide. FDA remarked on the greater number of anaphylaxis AEs in the study drug group, using the term "imbalance" to refer to the greater number of AEs found in the study drug group, writing, "analyzed by adjudicated events ... the imbalance for anaphylaxis events is clearly seen with lixisenatide."[91]

The term used by the FDA reviewer, "adjudicated events," refers to the fact that the AEs occurring in the clinical study was first processed by an adjudication committee to determine if the AE was definitely, probably, or unlikely related to the administered substance (drug or placebo), prior to comparing the number of AEs in the study drug arm versus in the control arm.

Now, referring to a concomitant drug in one study subject and to a bee sting in another study subject, FDA observed, "Of these 11 cases, nine cases were associated with lixisenatide treatment. The remaining two were attributed to cefazolin and bee sting hypersenstivity."[92] In other words, the conclusion was that two of the adverse events in the lixisenatide study were **not related to the study drug** but instead were due to a concomitant drug (cefazolin) or to a bee sting.

Package label. Although FDA's Medical Review for lixisenatide (Adlyxin®) determined that some of the instances of anaphylaxis were not related to the study drug, FDA's review did not eliminate all the instances of anaphylaxis as unrelated. As a result, the package label warned against this type of hypersensitivity reaction:

> **CONTRAINDICATIONS**. Hypersensitivity to ADLYXIN or any product components. Hypersensitivity reactions including anaphylaxis have occurred with ADLYXIN.[93]

> **WARNINGS AND PRECAUTIONS**. Anaphylaxis and Serious Hypersensitivity Reactions: Discontinue ADLYXIN and promptly seek medical advice.[94]

vi. Pomalidomide (Multiple Myeloma) NDA 204-026

FDA's relatedness analysis attributed the AEs to the disease being treated by the study drug (multiple myeloma). This concerns pomalidomide for treating multiple myeloma. Multiple myeloma is a cancer where plasma cells produce excessive amounts of antibodies

[88] Fingolimod (multiple sclerosis) NDA 022-527. Page 225 of 519-page Medical Review.

[89] Atakli D, et al. Thrombocytopenia with gabapentin usage. Ideggyogy Sz. 2015;68:270–2.

[90] Qu C, et al. Neuropsychiatric symptoms accompanying thrombocytopenia following pregabalin treatment for neuralgia: a case report. Int. J. Clin. Pharm. 2014;36:1138–40.

[91] Lixisenatide (Type-2 diabetes) NDA 208-471. Page 49 of 690-page pdf file of Medical Review.

[92] Lixisenatide (Type-2 diabetes) NDA 208-471. Pages 105 and 106 of 690-page pdf file of Medical Review.

[93] Package label. ADLYXIN (lixisenatide) injection, for subcutaneous use. July 2016 (12 pages).

[94] Package label. ADLYXIN (lixisenatide) injection, for subcutaneous use. July 2016 (12 pages).

TABLE 11.1 AEs Associated With Altered Renal function

	Pomalidomide Plus Dexamethasone	Pomalidomide Only
Acute renal failure	12 (10%)	16 (15%)
Chronic renal failure	1 (1%)	1 (1%)
Renal impairment	0	1 (1%)

From: Pomalidomide (multiple myeloma) NDA 204-026. Page 84 of 101-page Medical Review.

and where the antibodies are toxic to the kidneys. Multiple myeloma results in high concentrations of antibody fragments that result in nephropathy and renal failure.[95,96]

FDA's Medical Review provided a table showing AEs relating to renal function (Table 11.1).[97] Both treatment arms received the study drug (pomalidomide), potentially confounding the goal of attributing the AE to pomalidomide.

The FDA reviewer stated that, "The high incidence of ... renal failure is hard to attribute to treatment, because **renal failure may be related to the underlying disease of multiple myeloma**."

Apparently because of FDA's reluctance or reasoned unwillingness to attribute the AEs of renal failure and renal impairment to the study drug, the Warnings and Precautions section and the Adverse Reactions section of the package label were totally silent, as to renal AEs. Instead, warnings about patients with renal failure found a place in the Dosage and Administration section and the Use in Specific Populations section of the package label.

Package label. Consistent with FDA's remarks in FDA's Medical Review, the package label refrained from implying that pomalidomide (Pomalyst®) is associated with renal failure or renal impairment. Instead, the package label provided the physician with instructions when administering to renally impaired patients. The instructions resided in the Dosage and Administration section and in the Use in Specific Populations section:

> **DOSAGE AND ADMINISTRATION.** Multiple Myeloma: 4 mg per day taken orally on Days 1-21 of repeated 28 day cycles until disease progression ... Dosage Adjustment for Patients with Severe Renal Impairment on Hemodialysis. For patients with severe renal impairment requiring dialysis, the recommended starting dose is 3 mg daily (25% dose reduction). Take POMALYST after completion of dialysis procedure on hemodialysis days.[98]

> **USE IN SPECIFIC POPULATIONS** ... **Renal Impairment**. In patients with severe renal impairment requiring dialysis, the AUC of pomalidomide increased by 38% and the rate of serious adverse event (SAE) increased by 64% relative to patients with normal renal function; therefore, starting dose adjustment is recommended. For patients with severe renal impairment requiring dialysis, POMALYST should be

[95] Hutchison CA, et al. Treatment of acute myeloma failure secondary to multiple myeloma with chemotherapy and extended high cut-off hemodialysis. Clin. J. Am. Soc. Nephrol. 2009;4:745–54.

[96] Dimopoulos MA, et al. Pathogenesis and treatment of renal failure in multiple myeloma. Leukemia 2008;22:1485–93.

[97] Pomalidomide (multiple myeloma) NDA 204-026. Page 84 of 101-page Medical Review.

[98] Package label. POMALYST® (pomalidomide) capsules, for oral use; June 2016 (27 pages).

administered after the completion of hemodialysis on dialysis days because exposure of pomalidomide could be significantly decreased during dialysis.[99]

e. Relatedness From Drug Class Analysis

This concerns drug class analysis as part of the relatedness inquiry. The Naranjo criteria do not include drug class analysis. However, for any drug that is the first member in its class, Naranjo criteria will likely need to be applied for assessing relatedness. In other words, if the study drug is really and truly the first drug to exist in a class of drugs, then drug class analysis will be impossible.

The following concerns anticancer drugs that inhibit tyrosine kinase enzymes. For any group of members in a given drug class, the members may further be divided into subgroups by a number of techniques. One technique is to separate the drugs into those that inhibit at physiologically attained concentrations from those that inhibit only extremely weak at physiological concentrations. Another technique is to separate the drugs into those that are specific for inhibiting a given member of the group of the drug's potential targets (and do not inhibit all members of the group of the drug's potential targets) from drugs that are promiscuous and that inhibit several or all members of the drug's potential targets.

The connection between drug class analysis and any warnings that find a place on the package label is as follows. The connection is that the study drug is related to a drug previously approved by FDA, and that the AEs attributed to the previously approved drug are is used to predict AEs expected from the study drug.

i. Osimertinib (non–small-cell lung cancer) NDA 208-065

For the study drug, osimertinib, the relevant drug class, was a class of enzyme inhibitors, namely, epidermal growth factor receptor (EGFR) tyrosine kinase inhibitors. Osimertinib (Tagrisso®), as well as, erlotinib, gefitinib, lapatinib, and canertinib are each inhibitors of EGFR tyrosine kinase and thus are each members of the EGFR tyrosine kinase inhibitor drug class.[100,101,102]

The comments that follow concern a large drug class (tyrosine kinase inhibitors, TKIs) and a subset of this drug class (EGFR tyrosine kinase inhibitors). Regarding the drug class analysis used to predict AEs, FDA first acknowledged that the AE of most concern was interstitial lung disease, which occurred in the Sponsor's own clinical studies:

> Interstitial lung disease was the most concerning toxicity, causing 4 deaths on study and occurring in 2.7% of patients overall ... [o]verall, the safety profile of osimertinib is well-characterized. Its adverse event profile is consistent with **known class toxicities** ... [t]he most important rare but serious identified adverse event that was thought to be the cause of 4 treatment-related deaths on study was interstitial lung disease. **This is a known class effect of EGFR tyrosine kinase inhibitors**, and occurred at a rate of 2.7% overall in the study population.[103]

[99] Package label. POMALYST® (pomalidomide) capsules, for oral use; June 2016 (27 pages).

[100] Rocha-Lima CM, et al. EGFR targeting of solid tumors. Cancer Control. 2007; 14:295–304.

[101] Arora A, Scholar EM. Role of tyrosine kinase inhibitors in cancer therapy. J. Pharmacol. Exp. Ther. 2005;360:971–9.

[102] Su CM, et al. A novel application of E1A in combination therapy with EGFR-TKI treatment in breast cancer. Oncotarget. 2016;7:63924–36.

[103] Osimertinib (non-small-cell lung cancer) NDA 208-065. Pages 17 and 18 of 137-page pdf file of Medical Review.

Additional AEs from the EGFR tyrosine kinase inhibitors drug class were disclosed as, "Based on **toxicities identified in studies of other EGFR tyrosine kinase inhibitors**, there was an expectation that similar toxicities were possible with osimertinib. Known class toxicities include skin rash and nail changes, QT prolongation, diarrhea, ocular events, and interstitial lung disease (ILD)."[104]

FDA's drug class analysis then turned to AE warnings on package labels for other FDA-approved drugs in the same drug class, erlotinib, afatinib, gefitinib, crizotinib, and ceritinib, writing:

> While a relatively uncommon toxicity seen in patients on osimertinib, ILD was cause of all osimertinib-related deaths. **This is not unexpected; fatal cases of ILD have been reported in all other trials of small molecule tyrosine kinase inhibitors** in non-small cell lung cancer (NSCLC). Consequently, ILD ... appears in the Warnings and Precautions section of labels for all other approved tyrosine kinase inhibitors in NSCLC, of which there are five.[105]

As a consequence of FDA's drug class analysis, all of the AEs disclosed above found a place on the package label. The ocular events were listed as eye inflammation lacrimation, light sensitivity, blurred vision, eye pain, and red eye.[106] This may be called "relatedness analysis" because relatedness was established because the study drug was a member of a recognized drug class.

ii. Brivaracetam (epilepsy) NDA 205-836, NDA 205-837, NDA 205-838

FDA's review for brivaracetam used drug class analysis for arriving at the requirement that the package label warn against suicide.

FDA's Medical Review reveals that subject 1266-0002 and subject 1332-0009 had taken overdoses in suicide attempts and discloses that both of the suicide attempts were by subjects in the study drug group.[107] In addition, the FDA reviewer noted that suicidal behavior was a drug class effect for antiepileptic drugs. FDA required that the Warnings and Precautions section and also the Adverse Reactions section of the package label contain warnings against suicide:

> Cases of suicidality were observed and, as part of **class labeling** for anti-epileptic drugs, suicidality will be discussed in the Warnings and Precautions section. Common adverse events largely parallel those just described that will be included in the Warnings and Precautions section. They largely fall under the categories of nervous system and psychiatric disorders and will be described in the section on Adverse Reactions ... [s]uicide related events are believed to be, in part, **related to the class of drugs** ... [a]s such suicidal ideation anti-epileptic drug class labeling will be included in section 5 of the label.[108]

[104] Osimertinib (non-small-cell lung cancer) NDA 208-065. Pages 61, 105, and 114 of 137-page pdf file of Medical Review.

[105] Osimertinib (non-small-cell lung cancer) NDA 208-065. Page 105 of 137-page pdf file of Medical Review.

[106] Package label. TAGRISSO® (osimertinib) tablets, for oral use; March 30, 2017 (20 pages).

[107] Brivaracetam (epilepsy) NDA 205-836, NDA 205-837, NDA 205-838. Page 334 of 382-page Medical Review.

[108] Brivaracetam (epilepsy) NDA 205-836, NDA 205-837, NDA 205-838. Pages 2 and 21 of 38-page Cross-Discipline Team Leader Review.

FDA's Guidance for Industry Suicidal Behavior and Ideation and Antiepileptic Drugs describes this same drug class effect for antiepileptic drugs.[109] Thus, the package label's warning about suicide resulted from these sources:[110]

- AEs from the Sponsor's own clinical studies,
- Drug class effect (package labels from other antiepileptic drugs),
- FDA's Guidance for Industry Suicidal Behavior and Ideation and Antiepileptic Drugs.

f. Relatedness Inquiry That Takes Into Account All AEs for One Given Tissue

i. Vemurafenib (BRAF V600E mutation positive melanoma) NDA 202-429

Vemurafenib is a small-molecule drug for treating melanoma. The drug inhibits serine–threonine protein kinases and tyrosine protein kinases. FDA's Medical Review provided a relatedness inquiry for AEs inflicting the joints. These AEs were arthralgia, arthritis, and other joint-related AEs.

The study drug treatment arm received vemurafenib, while the control treatment arm received dacarbazine (comparator drug). FDA's review shows a valuable technique to increase the ability to detect and evaluate relatedness of AEs. The format is to make a table that lists all of the AEs, where the table separates all of the AEs by the tissue or organ in the body. In this case, FDA's review has a list of AEs inflicting the joints. Regarding AEs of the joints, the FDA observed, "The incidence of arthralgia and arthritis were both higher on the vemurafinib arm, as well as several other joint-related adverse events ... [t]he mechanism of arthralia is not known ... but is **likely related to vemurafenib treatment**, given the low rate of arthralgia and other joint-related AEs on the dacarbazine arm."[111]

Regarding other joint-related AEs, FDA observed that these were joint effusion, joint motion decreased, joint stiffness, and joint swelling. Table 11.2 below reveals the incidence of arthralgia and of the other joint-related AEs, for the study drug treatment arm and for the control treatment arm.

FDA's analysis and results provide the following take-home lesson. If FDA's review for vemurafenib had tabulated all of the AEs without any regard to location in the body, it is possible that little differences in AE rate would have been evident between study drug arm (vemurafenib) versus control arm (dacarbazine). For example, if the various AEs reported inflammation (without regard to location), pain (without regard to location), swelling (without regard to location), and so on, the result could have been that AEs for the joint would have been lost in the background noise. To give an example from Table 11.2, we see that one of the AEs is "joint effusion." The Sponsor had the option of

[109] US Department of Health and Human Services; Food and Drug Administration; Center for Drug Evaluation and Research (CDER). Guidance for Industry. Suicidal Ideation and Behavior: Prospective Assessment of Occurrence in Clinical Trials; 2012 (13 pages).

[110] US Department of Health and Human Services; Food and Drug Administration; Center for Drug Evaluation and Research (CDER). Guidance for Industry. Suicidal Ideation and Behavior: Prospective Assessment of Occurrence in Clinical Trials; 2012 (13 pages).

[111] Vemurafenib (BRAF V600E mutation positive unresectable or metastatic melanoma) NDA 202-429. Page 81 of 105-page pdf file Medical Review.

TABLE 11.2 Joint-Related AEs for Study Drug and for Active Comparator Drug

	Vemurafenib, $N = 336$	Dacarbazine, $N = 287$
Arthralgia	180 (53.6%)	19 (6.6%)
Arthritis	8 (2.4%)	0
Joint effusion	2 (<1%)	0
Joint range of motion decreased	2 (<1%)	0
Joint stiffness	8 (2.4%)	0
Joint swelling	12 (3.6%)	2 (<1%)

From: Vemurafenib (BRAF V600E mutation positive unresectable or metastatic melanoma) NDA 202-429. Page 82 (Table 37) of 105-page pdf file Medical Review.

recording this type of AE simply as "effusion" instead of "joint effusion." Effusion can occur in different locations in the body, for example, as "joint effusion," "pericardial effusion," and "pleural effusion."[112]

V. FORMATTING DISCLOSURES OF RELATEDNESS

a. Eribulin (Breast Cancer) NDA 201-532

Eribulin is a small-molecule drug that targets the cell's cytoskeleton, and more specifically, the microtubules. Microtubule-targeting anticancer drugs fall into two different drug classes, microtubule-stabilizing agents (taxanes such as paclitaxel and docetaxel) and microtubule-destabilizing agents. Microtubule-destabilizing agents include vinca alkaloids such as vincristine and vinblastine and the halichondrin analog, **eribulin.**[113]

This provides a useful format to assess relatedness, where the format is a table with one row per study subject and columns showing number of drug cycles (courses of treatment) before the subject experienced the AE. FDA's Medical Review for eribulin included this type of table, which is reproduced in part below (Table 11.3).[114] The numbers in **bold font** refer to the number of days from the ultimate dose (the last dose) to death. The table also discloses number of cycles of drug dosing, prior to the subjects' death, where the number of cycles enables the assessment of any cumulative effect of the dose on toxicity.

[112] Sedie AD, et al. Ultrasound imaging for the rheumatologist XXIV. Sonographic evaluation of wrist and hand joint and tendon involvement in systemic lupus erythematosus. Clin. Exp. Rheumatol. 2009;27:897–901.

[113] Eslamian G, et al. Efficacy of eribulin in breast cancer: a short report on the emerging new data. Onco Target. Ther. 2017;10:773–9.

[114] Eribulin (breast cancer) NDA 201-532. Pages 100–104 of 222-page Medical Review.

TABLE 11.3 Listing of Deaths Occurring Within 60 Days of Eribulin Therapy

Patient ID	Age	Number of Cycles	Days Since Last Dose	Probable Cause of Death
1006	68	8	**34**	11 days after C8, D1 general health deterioration reported, **deemed unrelated to study treatment** by investigator.
1005	53	1	**34**	Progressive disease noted as cause of death but also had ongoing pneumonia with pleural effusion that led to treatment discontinuation 13 days after C1 D8 eribulin dose; deemed unrelated by investigator but **may have been caused by study drug**.
1004	45	2	**8**	Type II diabetes mellitus. Non-neutropenic sepsis, presenting with hyperglycemia, hypotension. Developed renal failure. Reported by investigator as **unrelated** to eribulin, but there is a **reasonable possibility that sepsis was related** to chemotherapy.
1002 (at site 2303)	69	4	**25**	Patient developed dyspnea—**not related** per investigator.
1002 (at site 1601)	69	1	**0**	Although considered by the investigator to be **unrelated to study therapy**, pneumonia was diagnosed and death occurred on the same day as the Cycle 1, Day 4 dose; there is a **reasonable possibility that death was related to therapy**.
1001	54	3	**31**	Developed respiratory failure 23 days after dose. **Deemed not related** to study drug by investigator.
1011	49	9	**47**	Grade 2 sensory neuropathy developed after 3 months of therapy. Grade 3 paresthesia and lower extremity pain were reported after 6 months of therapy. One week later, 16 days after dose, paraparesis developed leading to discontinuation of eribulin. Paraparesis was ongoing at time of death 1 month later and considered fatal (47 days after last eribulin dose). Primary cause of death determined to be progressive (breast cancer), but there is a **possibility that neurologic toxicity due to eribulin contributed to paraparesis**.

Eribulin (breast cancer) NDA 201-532. Pages 102–104 of 222-page Medical Review. Data from, "Table 40: Tabular Listing of Deaths Occurring within 60 days of Eribulin Therapy. Study 305."

To summarize, relatedness analysis can be facilitated by a table with separate rows for individual study subjects, with separate columns disclosing the number of drug doses or cycles given to each subject and a comment on relatedness of the study drug to a given AE.

In addition to the table with rows for separate study subjects, as reproduced above, FDA's Medical Review for eribulin included the following exemplary relatedness narratives (see below). These are exemplary in that they illustrate the utility of taking into account "cumulative drug dose" for assessing relatedness and the utility of taking into account drug class analysis for assessing relatedness.

b. Relatedness of Study Drug to the AE of Peripheral Neuropathy (Cumulative Dose)

FDA stated that the AE of peripheral neuropathy was related to "cumulative" dose of the study drug, writing, "The incidence of peripheral neuropathy appears to be related to the **cumulative eribulin dose**. The ... analysis demonstrated that the incidence of peripheral neuropathy increased from thirteen percent in subjects with a **cumulative exposure** of $\leq 5.6 > 11.2\ \text{mg/m}^2$."[115]

The medical literature discloses that cumulative dose of chemotherapy drugs is a risk factor for neuropathies. For example, Grisold et al.[116] stated that treatment "length-dependent neuropathies ... develop after a typical **cumulative dose** ... although in at least 2 drugs (oxaliplatin and taxanes) immediate toxic effects occur."

c. Relatedness of Study Drug to the AE of Peripheral Neutropenia (Drug Class Analysis)

FDA's assessment of neutropenia was based on AEs of the study subjects and on drug class analysis. The drug class was "cytotoxic chemotherapy drugs." The drug class analysis is evident from FDA's statement that, "Neutropenia and severe infections are well recognized sequelae of cytotoxic chemotherapy drugs used to treat patients with advanced cancer ... [t]here is a clear association between eribulin therapy and neutropenia, which predisposes patients to serious infections. This information is described in the Warnings and Precautions and Adverse Reactions sections of the proposed label."[117]

Regarding cumulative dose as it applies to the AE of neutropenia, FDA concluded that the cumulative nature was not particularly relevant, "This figure indicates that the ... absolute neutrophil count (ANC) did not seem to be influenced by the number of cycles of eribulin therapy completed ... ANC toxicity does not appear to be cumulative."[118]

d. Relatedness of Study Drug to AE of Death (Temporal Relationship)

FDA's review for eribulin provides an example of relatedness analysis making use of temporal relationship between the last dose and the materialization of the AE. FDA's observations and comments were that, "One patient's death (patient number 20081018) was recorded as related to toxicity on Day 287; however, death occurred 238 days after her last eribulin dose and no adverse events leading to death were recorded. Comment: given the long time interval between the last dose of eribulin and the patient's death, it is unlikely that this patient's death was related to eribulin therapy."[119]

[115] Eribulin (breast cancer) NDA 201-532. Page 141 of 222-page Medical Review.

[116] Grisold W, et al. Peripheral neuropathies from chemotherapeutics and targeted agents: diagnosis, treatment, and prevention. Neuro-Oncology 2012;14:iv45–54.

[117] Eribulin (breast cancer) NDA 201-532. Pages 109 and 110 and 116 of 222-page Medical Review.

[118] Eribulin (breast cancer) NDA 201-532. Page 196 of 222-page Medical Review.

[119] Eribulin (breast cancer) NDA 201-532. Page 100 of 222-page Medical Review.

e. Relatedness of Study Drug to AE of Meningitis Listeria (Relationship Established by Mechanism of Action)

FDA's review for eribulin provides relatedness analysis by way of the mechanism of action of the study drug. The mechanism was immunosuppression. This eribulin-induced immunosuppression was neutropenia. A published account of the same clinical study (clinical study named "EMBRACE") reported that neutropenia occurred in 52% of subjects in the eribulin treatment arm and in fewer subjects (30%) in the active control treatment arm.[120] The AE was meningitis caused by listeria. It is a proven fact that neutrophils protect against listeria infections.[121] FDA's relatedness analysis was that:

> Patient 068007 experienced "meningitis listeria" and pyrexia eight days following receipt of Cycle 4, Day 15 therapy. Patient died 35 days later due to meningitis. There is a reasonable possibility that meningitis was due to immunosuppression caused by eribulin therapy ... **[t]here is a clear association** between eribulin therapy and neutropenia, which predisposes patients to serious infections. This information is described in the Warnings and Precautions and Adverse Reactions sections of the proposed label.[122]

f. Relatedness of Study Drug to AE of QT Prolongation (Electrocardiogram Data) Confounded by Concomitant Conditions

FDA's Medical Review for eribulin recommended a package label warning about QTc intervals, "A warning regarding increased QTc intervals following eribulin exposure was recommended by clinical pharmacology and clinical review team members. Definitive conclusions regarding the cause of the two reported cases of sudden death cannot be made due to the lack of ECG data recorded at the time of the events and confounding comorbid conditions and concomitant medications."[123]

g. Summary

FDA's Medical Review for eribulin illustrates these concepts of the relatedness inquiry:

- Formatting a table that correlates days of drug exposure prior to each type of AE
- Relatedness taking into account the cumulative dose
- Relatedness established (or rebutted) by temporal association
- Relatedness of drug to a disease established by mechanism of action (ability of drug to impair immune system)

[120] Cortes J, et al. Eribulin monotherapy versus treatment of physician's choice in patients with metastatic breast cancer (EMBRACE): a phase 3 open-label randomised study. Lancet 2011;377:914–23.

[121] Witter AR, et al. The essential role of neutrophils during infection with the intracellular bacterial pathogen Listeria monocytogenes. J. Immunol. 2016;197:1557–67.

[122] Eribulin (breast cancer) NDA 201-532. Pages 110 and 111 and 116 of 222-page Medical Review.

[123] Eribulin (breast cancer) NDA 201-532. Page 201 of 222-page Medical Review.

- Relatedness by drug class analysis (cytotoxic chemotherapy drugs)
- Relatedness inquiry confounded by concomitant medical conditions
- Relatedness inquiry confounded by concomitant drugs

VI. INCONCLUSIVE RELATEDNESS ANALYSIS FINDS A PLACE ON THE PACKAGE LABEL

The reveals the option to Sponsors and medical writers for stating, on the package label, that it cannot be determined with any degree of certainty that the study drug is related to any given AE.

a. Fingolimod (multiple sclerosis) NDA 022-527

This concerns FDA's relatedness analysis for fingolimod, a small-molecule drug for treating multiple sclerosis. FDA repeatedly stated that relatedness of fingolimod to the AE of lymphoma was unclear. Regarding lymphoma, FDA's Medical Review stated that its, "interpretation of the relationship to study drug is summarized ... Likely Related. 2 herpes viral infections. **Cannot be ruled out if related** ... possible T cell lymphoma ... lymphoproliferative disease ... Unlikely related. 1 traffic accident. 1 suicide."[124]

Ambiguities that were encountered in attempts to establish relatedness were that:

- Lymphoma was observed at the lower fingolimod dose, but not at the higher fingolimod dose. This author points out that, without further information, the occurrence of the AE at lower but not at higher doses is not logical.
- The database on lymphoma is too small for an adequate conclusion of relatedness.
- Because lymphoma is a type of cancer and because of the relatively long time generally required to induce cancer by chemicals, FDA stated that the "one year is too short for an immunosuppresive agent [the study drug] to induce a neoplasia."[125]

Regarding lymphoma, FDA complained that relatedness could not be established because of the fact that lymphoma occurred in the lower dose amount group but not in the higher dose amount group. FDA's complaint was that, "there was one case of B cell lymphoma, one T cell lymphoma, and one lymphoproliferative disorder among 456 patients in the fingolimod 2.5 mg group. No such cases were observed in the fingolimod 5 mg ... group ... [t]hese disorders were not observed in the higher fingolimod dose group. **It is unclear if these events are related to fingolimod.**"[126]

Further regarding lymphoma, FDA complained that the time frame of drug exposure was too short to establish relatedness and also that the database was too small to establish relatedness. FDA's complaint was, "**It is unclear whether the cases of lymphoma are related** to study drug. It appears that one year is too short for an immunosuppressive

[124] Fingolimod (multiple sclerosis) BLA 022-527. Pages 100 and 167 of 519-page Medical Review.

[125] Fingolimod (multiple sclerosis) BLA 022-527. Page 217 of 519-page Medical Review.

[126] Fingolimod (multiple sclerosis) BLA 022-527. Page 408 of 519-page Medical Review.

agent to induce a neoplasia. The database is relatively small and too short to adequately assess the risk of malignancy. The question whether fingolimod is associated with increased risk of malignancy needs to be addressed in a larger database, such as postmarketing registry, as proposed by the applicant."[127]

Package label. Because of the above ambiguities regarding lymphoma, FDA refrained from recommending that a warning be in the Warnings and Precautions section. Instead lymphoma was disclosed in the Clinical Trials Experience section. The Clinical Trials Experience section of the package label for fingolimod (Gilenya®) read:

> **CLINICAL TRIALS EXPERIENCE** ... Cases of lymphoma, including both T-cell and B-cell types and CNS lymphoma, have occurred in patients receiving GILENYA. The reporting rate of non-Hodgkin lymphoma with GILENYA is greater than that expected in the general population adjusted by age, gender, and region. **The relationship of lymphoma to GILENYA remains uncertain.**[128]

Information in the Clinical Trials Experience section is relatively neutral and do not rise to the point that it takes the form of any explicit warning or any instruction to the physician. The package label admits that "[t]he relationship of lymphoma to GILENYA remains uncertain."

b. Sofosbuvir/Velpatasvir Combination (Hepatitis C Virus) NDA 208-341

FDA's Medical Review of the sofosbuvir/velpatasvir (SOF/VEL) combination reveals ambiguity in relatedness analysis. FDA acknowledged that "there is no clear indication" that the study drug was related to neuropsychiatric events:

> There is **no clear indication for an increased risk** of neuropsychiatric events with SOF/VEL-containing treatment. However, depressive events have been observed in prior trials for Sovaldi [sofosbuvir] and Harvoni [sofosbuvir/ledipasvir], and these respective labels contain language pertaining to depression and suicidal events. For consistency, similar language is recommended for the SOF/VEL label as well.[129]

Package label. The package label for the SOF/VEL combination (Epclusa®) disclosed neuropsychiatric events (depression), where the disclosure was in the Adverse Reactions section and not in the Warnings and Precautions section. In addition to disclosing the AE of depression in the relatively neutral section of the Adverse Reactions section, the package label admitted that there was only a "potential" causal relationship:

> **ADVERSE REACTIONS** ... Less Common Adverse Reactions Reported in Clinical Trials. The following adverse reactions occurred in less than 5% of subjects ... treated with EPCLUSA for 12 weeks and are included because of a **potential causal relationship** ... Depression: In the ASTRAL-1 study, depressed mood occurred in 1% of subjects treated with EPCLUSA and was not reported by any subject taking placebo. No serious adverse reactions of depressed mood occurred and all events were mild or moderate in severity.[130]

[127] Fingolimod (multiple sclerosis) BLA 022-527. Page 217 of 519-page Medical Review.

[128] Package label. GILENYA (fingolimod) capsules, for oral use; February 2016 (19 pages).

[129] Sofosbuvir/velpatasvir combination (hepatitis C virus) NDA 208-341. Page 145 of 171-page Medical Review.

[130] Package label. EPCLUSA® (sofosbuvir and velpatasvir) tablets, for oral use. February 2017 (32 pages).

VII. RESPONSIBILITY FOR DETERMINING RELATEDNESS

a. Ticagrelor (Thrombic Events in Patients With Acute Coronary Syndromes) NDA 022-433

The responsibility of the Sponsor to determine relatedness is disclosed in FDA's Medical Review of ticagrelor. The terms "sponsor" and "investigator" have separate definitions (21 CFR §312.3). According to Section 312.3, "Sponsor means a person who takes responsibility for and initiates a clinical investigation. The sponsor may be an individual or pharmaceutical company, governmental agency, academic institution, private organization, or other organization. The sponsor does not actually conduct the investigation unless the sponsor is a sponsor-investigator."

Also, according to Section 312.3, "Investigator means an individual who actually conducts a clinical investigation (i.e., under whose immediate direction the drug is administered or dispensed to a subject). In the event an investigation is conducted by a team of individuals, the investigator is the responsible leader of the team."

FDA's Guidance for Industry also distinguishes between the Sponsor and investigator.[131,132] FDA's Medical Reviews use the terms "sponsor" and "applicant" interchangeably. Generally, the word "sponsor" is capitalized when referring to a specific sponsor, such as a specific pharmaceutical company. But if the subject matter does not concern any specific company, then "sponsor" may begin with a lower-case letter.

In FDA's review of ticagrelor, comments from T.A. Marciniak, MD, reveal that it is the responsibility of the Sponsor (or applicant), and not of the investigator, to determine relatedness. This concerns the serious adverse events (SAE) of seizures and headache leading to hospitalization. The comments mention the option of adjudication. One of the goals of adjudication is to determine relatedness of the study drug to AEs. The FDA reviewer wrote:

> For both patients the investigator discontinued study drug because of the SAE. Neither SAE was adjudicated. The applicant replied that the investigator had indicated that the SAE was unrelated to study drug and that it was its policy to use the investigator's determination of unrelatedness. Dr. Marciniak noted that it was ultimately the applicant's responsibility to determine unrelatedness, not just the investigator's ... I asserted that under both ICH E2A ... and the new reporting rule it was and is the **sponsor's ultimate responsibility** for determining relatedness.[133]

[131] US Department of Health and Human Services; Food and Drug Administration; Center for Drug Evaluation and Research (CDER); Center for Biologics Evaluation and Research (CBER). Guidance for Industry. Investigational New Drug Applications Prepared and Submitted by Sponsor-Investigators; 2015 (28 pages).

[132] US Department of Health and Human Services; Food and Drug Administration; Center for Drug Evaluation and Research (CDER); Center for Biologics Evaluation and Research (CBER); Center for Devices and Radiological Health (CDRH). Guidance for Industry Investigator Responsibilities—Protecting the Rights, Safety, and Welfare of Study Subjects; 2009 (18 pages).

[133] Ticagrelor (thrombic events in patients with acute coronary syndromes) NDA 022-433. Pages 3 and 4 of 640-page Medical Review.

The FDA reviewer admonished the Sponsor, stating that the Sponsor must comply with 21 CFR §312.32(c), which sets forth a requirement for determining causality (relatedness). The relevant parts of Section 312.32(c) include:

> Suspected adverse reaction means any adverse event for which there is a **reasonable possibility that the drug caused the adverse event**. For the purposes of IND safety reporting, "reasonable possibility" means there is **evidence to suggest a causal relationship** between the drug and the adverse event. Suspected adverse reaction implies a lesser degree of certainty about causality than adverse reaction, which means any adverse event caused by a drug[134] ... [t]he sponsor must report any suspected adverse reaction that is both serious and unexpected. The sponsor must report an adverse event as a suspected adverse reaction **only if there is evidence to suggest a causal relationship** between the drug and the adverse event.[135]

Section 312.32(c) provides examples on relatedness analysis. Although it might seem obvious to any biology student that relatedness analysis can be based on data showing that an AE occurs more in the study drug group than in the control group, this particular fact pattern is described in Section §312.32(c). Also, please note that certain AEs, such as SJS, may be automatically considered to be related to a study drug, because of the fact that this disorder, almost by definition, is a drug-induced AE.[136]

The examples in Section 321.32(c) are, "analysis of specific events observed in a clinical trial ... that indicates those events occur more frequently in the drug treatment group than in a concurrent or historical control group[137] ... [a] single occurrence of an event that is uncommon and known to be strongly associated with drug exposure (e.g., angioedema, hepatic injury, Stevens-Johnson Syndrome)."[138]

The FDA reviewer admonished the Sponsor on its responsibility to determine relatedness and complained about the Sponsor's home-grown policy to let the investigator determine relatedness. Note the option of utilizing adjudication to assess suspicious AEs:

> We discussed these cases with ... the Sponsor at a meeting and ... the Sponsor stated that, despite ticagrelor being discontinued in each case because of the SAE, ... the Sponsor did not report the cases as SUSARs because it was ... the Sponsor's policy to rely upon the investigators' determinations of relatedness to study drug ... [r]egardless, as I conveyed ... at the meeting, it is the sponsor's responsibility to make final determinations regarding the relatedness of SAEs for expedited reporting purposes.[139]

FDA's Medical Review emphasized the serious nature of the FDA reviewer's complaints about SAE reporting by suggesting that the Sponsor's study should be placed on a clinical hold. FDA's request for a clinical hold read, "We should consider placing on hold

[134] 21 CFR §312.32(c).

[135] 21 CFR §312.32(c).

[136] Brody T. CLINICAL TRIALS Study Design, Endpoints and Biomarkers, Drug Safety, and FDA and ICH Guidelines), 2nd ed., Elsevier, New York; 2016, pp. 514–9.

[137] 21 CFR §312.32(c).

[138] 21 CFR §312.32(c).

[139] Ticagrelor (thrombic events in patients with acute coronary syndromes) NDA 022-433. Page 34 of 640-page Medical Review.

ticagrelor clinical trials ... I judge the problems with AE reporting that I have documented in this review to be serious. I am not confident that ... the Sponsor is protecting adequately the safety of patients in its trials ... [p]lacing one or more trials on hold would force ... the Sponsor to address these problems.[140]

VIII. CONCLUDING REMARKS

AEs may present during the course of a clinical study on a given study drug, as well as in the postmarketing situation after physicians prescribe drugs to their patients. The definitions for AEs and SAEs do not require any assessments of relatedness. Section 312.32 defines AE as:

> Adverse event means any untoward medical occurrence associated with the use of a drug in humans, whether or not considered drug related.[141]

SAE is defined as:

> An adverse event ... is considered "serious" if, in the view of either the investigator or sponsor, it results in any of the following outcomes: Death, a life-threatening adverse event, inpatient hospitalization or prolongation of existing hospitalization, a persistent or significant incapacity or substantial disruption of the ability to conduct normal life functions, or a congenital anomaly/birth defect.[142]

Although the definitions for AEs and SAEs do not require any relatedness, the CRF and the MedWatch form (FDA form 3500) provide for inputting relatedness assessments. The relatedness inquiry is used when determining if a given AE is to be placed in the Black Box Warning, Warnings and Precautions section, Adverse Reactions section of the package label, or in some other section. As shown in this chapter, if relatedness is ambiguous, the AE can be left out of the Warnings and Precautions section and instead the AE can be placed in the Adverse Reactions section or in the Clinical Trials Experience section.

The inquiry for assessing relatedness includes one or more of:

- Was the time between drug dosing and the AE appropriate for the AE in question? Temporal assessments also ask if the severity of the AE was reduced after discontinuing the drug.
- Was the AE more prevalent in the study drug treatment arm versus in the placebo treatment arm?

[140] Ticagrelor (thrombic events in patients with acute coronary syndromes) NDA 022-433. Page 35 of 640-page Medical Review.

[141] 21 CFR §312.32(a) IND safety reporting. Definitions (2016).

[142] 21 CFR §312.32(a) IND safety reporting. Definitions (2016).

- Could the AE be attributed to the disease being treated, to a coadministered drug, or to a pre-existing risk factor such as renal impairment?
- Was the AE already established to be caused by the drug, as revealed by the package label of the same drug for a different formulation, previously approved by FDA? This relatedness assessment makes use of drug class analysis.

Another option is to determine that relatedness analysis was inconclusive. Where the Sponsor is concerned about an AE but where relatedness cannot be established, the Sponsor has the option to disclose the AE on the package label and to admit that relatedness was not established. This situation is shown for the AEs of lymphoma for fingolimod (Gilenya®) and for the AE of malignancies for certolizumab pegol (Cimzia®). The admissions on the package labels read:

- "Cases of lymphoma, including both T-cell and B-cell types and CNS lymphoma, have occurred in patients receiving GILENYA. The reporting rate of non-Hodgkin lymphoma with GILENYA is greater than that expected in the general population adjusted by age, gender, and region. The **relationship of lymphoma to GILENYA remains uncertain."**[143]
- "In ... clinical studies of some TNF blockers, more cases of **malignancies** have been observed among patients receiving TNF blockers compared to control patients. During CIMZIA studies ... malignancies ... were observed ... [t]he size of the control group and limited duration of the controlled portions of the studies **precludes the ability to draw firm conclusions."**[144]

[143] Package label. GILENYA (fingolimod) capsules, for oral use; February 2016 (19 pages).

[144] Package label. CIMZIA (certolizumab pegol) Lyophilized powder or solution for subcutaneous injection; April 2008 (15 pages).

CHAPTER

12

Adjudication of Clinical Data

I. INTRODUCTION

Adjudication assesses clinical data on efficacy and safety, for example, where the issue is whether the data meet the safety endpoints and efficacy endpoints for a given clinical trial. Endpoints on safety and efficacy are defined in the Clinical Study Protocol, and are used to determine if the data collected from clinical trial succeeded in showing that the drug is safe and effective.

Clinical Study Protocols are available from *New England Journal of Medicine*, where some examples are cited.[1,2] The components of the Clinical Study Protocol are detailed in the cited reference.[3]

Adjudication finds a basis in FDA's Guidance for Industry. FDA's Guidance refers to the need to compare the data collected with the endpoints defined in the Clinical Study Protocol, and to see if the collected data match or exceed the expectations set forth in the Clinical Study Protocol. FDA's Guidance states that:

> Sponsors may also choose to establish **an endpoint assessment/adjudication committee** ... in certain trials to review important endpoints reported by trial investigators to determine whether the

[1] See, for example, A Phase 3 Randomized Double-Blind Trial of Maintenance with Niraparib Versus Placebo in Patients with Platinum Sensitive Ovarian Cancer. Sponsor Protocol No. PR-30-5011-C. IND No. 100,996. Date of Protocol: March 21, 2013. Protocol published as supplement to Mirza MR, et al. Niraparib maintenance therapy in platinum-sensitive, recurrent ovarian cancer. New Engl. J. Med. 2016;375:2154–64.

[2] See also, for example, A Randomized Open-Label Phase III Trial of Pembrolizumab versus Platinum based Chemotherapy in 1L Subjects with PD-L1 Strong Metastatic Non-Small Cell Lung Cancer IND No: 116833. Date of Protocol: January 28, 2016. Protocol published as supplement to Reck M, et al. Pembrolizumab versus chemotherapy for PD-L1—positive non-small-cell lung cancer. New Engl. J. Med. 2016;375:1823–33.

[3] Brody T. CLINICAL TRIALS study design, endpoints and biomarkers, drug safety, and FDA and ICH guidelines. 2nd ed. New York: Elsevier; 2016.

FDA's Drug Review Process and the Package Label
DOI: https://doi.org/10.1016/B978-0-12-814647-7.00012-9

endpoints meet protocol-specified criteria. Information reviewed on each ... endpoint may include laboratory, pathology and/or imaging data, autopsy reports, physical descriptions, and any other data deemed relevant.[4]

FDA's Guidance for Industry further states that an advantage of adjudication is that it reduces subjectivity in assessing endpoints, and that adjudication is especially valuable where the adverse event in question has a complex definition. The account of ticagrelor, at a later point in this chapter, illustrates adjudication committee's ability to apply complex definitions.[5]

FDA's Guidance further states that, "These committees are typically **masked** to the assigned study arm when performing their assessments regardless of whether the trial itself is conducted in a blinded manner. Such committees are particularly valuable when endpoints are **subjective and/or require the application of a complex definition**, and when the intervention is not delivered in a blinded fashion."[6]

An adjudication committee is especially useful where there is an enhanced need to avoid bias.[7] The tasks of an adjudication committee overlap those of two other committees that are mandatory for regulated clinical trials, namely, the Data Safety Monitoring Committee (DSMC) and the Institutional Review Board (IRB). In addition to confirming or correcting the AEs that were reported by the investigator, an adjudication committee may assess compliance with the Clinical Study Protocol, that is, protocol deviations.[8] Excessive protocol deviations may provoke FDA to issue a Warning Letter to the Sponsor.[9]

A general account of the day-to-day activities of an adjudication committee reveals that, "The adjudication committee held an average of eight telephone conferences per clinical trial. During each conference, up to three to four cases were discussed for diagnostic adjudication alone. The adjudication committee would review up to 10 cases for diagnosis, outcome measures, or bleeding events in a given conference call."[10]

[4] U.S. Department of Health and Human Services. Food and Drug Administration. Center for Biologics Evaluation and Research (CBER). Center for Drug Evaluation and Research (CDER). Center for Devices and Radiological Health (CDRH). Guidance for clinical trial sponsors establishment and operation of clinical trial data monitoring committees; 2006 (34 pp.).

[5] Ticagrelor (thrombic events with acute coronary syndromes) NDA 022-433.

[6] U.S. Department of Health and Human Services. Food and Drug Administration. Center for Biologics Evaluation and Research (CBER). Center for Drug Evaluation and Research (CDER). Center for Devices and Radiological Health (CDRH). Guidance for clinical trial sponsors establishment and operation of clinical trial data monitoring committees; 2006 (34 pp.).

[7] Kradjian S, et al. Development of a charter for an endpoint assessment and adjudication committee. Drug Inform. J. 2005;39:53–61.

[8] McBee NA, et al. The importance of an independent oversight committee to preserve treatment fidelity, ensure protocol compliance, and adjudicate safety endpoints in the ATACH II trial. J. Vasc. Interv. Neurol. 2012;5:10–13.

[9] Brody T. Warning letters in CLINICAL TRIALS study design, endpoints and biomarkers, drug safety, and FDA and ICH guidelines. 2nd ed. New York: Elsevier; 2016. pp. 719–80.

[10] de Andrade J, et al. The idiopathic pulmonary fibrosis clinical research network (IPFnet). Diagnostic and adjudication processes. Chest 2015;148:1034–42.

Wittes et al.[11] inform us that, "The FDA has long had a rule (21 CFR §312.32) calling for prompt (within 15 days) reporting of any **serious unexpected** (i.e., not in the investigators' brochure or labeling) adverse experience 'associated with use of a drug' (i.e., if there was a reasonable possibility that the drug may have caused the event)." An adjudication committee can provide guidance on this "reasonable possibility."

Regarding the need for adjudication, Regev et al.[12] commented on the need for adjudication for one particular type of adverse event (drug-induced liver injury; DILI). The authors stated that, "Despite its many advantages, causality assessment based on expert opinion is largely subjective, and suffers from interobserver variance ... expert opinion was used by several authors as the 'gold standard' for development, validation and comparison of existing scoring systems ... authorities ... agree that consensus opinion among hepatologists with expertise in DILI **adjudication remains a gold standard**."[13]

Regev et al.[14] then outlined the nature of the adjudication committee, "When using expert opinion to assess hepatic cases ... it is ... advantageous to use three experts to increase the chances of a majority opinion ... each expert should perform causality assessment independently, and each expert should be blinded to treatment assignment where feasible. Ideally, during drug development, **causality assessment** should be performed prior to unblinding the treatment assignment within the company."

II. ACCOUNT OF ADJUDICATION FROM FDA'S BACKGROUND PACKAGE FOR OSTEOARTHRITIS

This is from FDA's Background Package, which was from an analysis of clinical studies on antinerve growth factor from various Sponsors. FDA's goal in writing this Background Package was, "to compare our adjudication results with those of the Sponsors."[15] The AE that was being adjudicated was joint destruction. FDA concluded that this AE was related to the drug, and was not a result of the disease (osteoarthritis):

> both sets of adjudications are in ... agreement with the adjudications conducted by the Sponsors. Differences in the adjudications may be due to differences in the adjudication processes and definitions used to determine the diagnoses, in addition to differences in the interpretation of the data by the adjudicators. There appears to be a safety signal of rapid joint destruction ... associated with both Anti-NGF agent monotherapy and Anti-NGF agent plus NSAID therapy. The incidence of this event is more pronounced in

[11] Wittes J, et al. The FDA's final rule on expedited safety reporting: statistical considerations. Stat. Biopharm. Res. 2015;7:174-90.

[12] Regev A, et al. Causality assessment for suspected DILI during clinical phases of drug development. Drug Safety 2014;37 (Suppl. 1):S47-56.

[13] Regev A, et al. Causality assessment for suspected DILI during clinical phases of drug development. Drug Safety 2014;37 (Suppl. 1):S47-56.

[14] Regev A, et al. Causality assessment for suspected DILI during clinical phases of drug development. Drug Safety 2014;37 (Suppl. 1):S47-56.

[15] FDA Center for Drug Evaluation and Research. Division of Anesthesia, Analgesia, and Addition Products (March 2, 2012) Background Package Addendum by Hertz S and Fields E (27 pp.).

patients receiving both the Anti-NGF agent and NSAID concurrently, but is clearly present in both treatment groups. The occurrence of these events was markedly disproportional, favoring **drug treatment over placebo** treatment, which supports that these **events of joint destruction are related to drug treatment**, and are not occurring as part of the natural history of osteoarthritis.[16]

A take-home lesson is that definitions used for various AEs and for their diagnosis need to be harmonized, before the Sponsor, the Sponsor's adjudication committee, and FDA's adjudication committee attempt to assess relatedness of the drug to adverse events.

III. ACCOUNTS OF ADJUDICATION FROM MEDICAL JOURNALS AND CLINICAL STUDY PROTOCOLS

a. Example of Asthma Adjudication

As reported in *New England Journal of Medicine*, a clinical study on asthma used an adjudication committee, "A joint adjudication committee ... was responsible for uniform determination of asthma-relatedness for study end points ... and ... [was] charged with ensuring responsible conduct of the trial and the safety of all the patients."[17]

The asthma adjudication included blinding as to treatment, where the adjudication committee assessed adverse events ("events"), and determined if the AEs were related to the disease being treated, "Events were reviewed by members of the joint adjudication committee who were unaware of the study-group assignments. All hospitalization events underwent initial screening by a member of the joint adjudication committee, and if the patient's condition was considered to be **potentially asthma-related**, a complete adjudication followed. All intubations and deaths were fully adjudicated."[18]

Intubation is defined as oro-tracheal or naso-tracheal placement of a tracheal tube. The fact that a patient needed intubation is a measure of severity of the asthma, in the same way that need for corticosteroids or mechanical ventilation are measures of severity.[19]

b. Example of Thrombosis Adjudication

In a clinical study of thrombosis reported in *New England Journal of Medicine*, an adjudication committee assessed an efficacy endpoint, which took the form of ability to reduce total venous thromboembolisms, "The primary efficacy outcome was the incidence of **adjudicated total venous thromboembolism**, which was a composite of

[16] FDA Center for Drug Evaluation and Research. Division of Anesthesia, Analgesia, and Addition Products (March 2, 2012) Background Package Addendum by Hertz S and Fields E (27 pp.).

[17] Stempel DA, et al. Serious asthma events with fluticasone plus salmeterol versus fluticasone alone. New Engl. J. Med. 2016;374:1822–30.

[18] Stempel DA, et al. Serious asthma events with fluticasone plus salmeterol versus fluticasone alone. New Engl. J. Med. 2016;374:1822–30.

[19] Barbers RG, et al. Near fatal asthma: clinical and airway biopsy characteristics. Pulm. Med. 2012; doi:10.1155/2012/829608.

asymptomatic deep-vein thrombosis ... symptomatic venous thromboembolism, fatal pulmonary embolism, or unexplained death for which pulmonary embolism could not be ruled out."[20]

The Clinical Study Protocol for the above thrombosis trial provided for the adjudication committee's responsibilities.[21,22] Regarding *safety*, the Protocol stated that, "All bleeding events will be reviewed by the CIAC. The procedures followed by the CIAC and definitions to classify bleeding are described in an **adjudication** operations manual." CIAC means Central Independent Adjudication Committee. Regarding *efficacy*, the Protocol referred to, "**Adjudicated** venogram between Day 8 and Day 12 post surgery period positive for DVT [deep vein thrombosis]." Moreover, the Protocol referred to two documents, the adjudication charter and the adjudication operations manual.

Further information on adjudication committees is provided by the cited articles from *New England Journal of Medicine*.[23,24,25,26,27]. These articles include, as a supplement, the Clinical Study Protocol used for the clinical study described by the corresponding journal article.

c. Example of Multiple Myeloma Adjudication

FDA's Medical Review of pomalidomide for multiple myeloma discloses adjudication of efficacy data. Analysis of efficacy data requires comparison with efficacy endpoints, as revealed in the excerpt. The FDA reviewer wrote, "The primary efficacy endpoint ... was progression free survival (PFS) defined as the time from date of randomization to the date of progression or death due to any cause, whichever occurred first ... [p]rimary efficacy analysis was based on response assessment by the Independent Response Adjudication

[20] Buller HR, et al. Factor XI antisense oligonucleotide for prevention of venous thrombosis. New Engl. J. Med. 2015;372:232–40.

[21] Buller HR, et al. Factor XI antisense oligonucleotide for prevention of venous thrombosis. New Engl. J. Med. 2015;372:232–40.

[22] ISIS Pharmaceuticals, Inc. An open-label, randomized, active comparator-controlled, adaptive parallel-group phase 2 study to assess the safety and efficacy of multiple doses of ISIS 416858 administered subcutaneously to patients undergoing total knee arthroplasty. Original Protocol July 24, 2012.

[23] Johnston SC, et al. Ticagrelor versus aspirin in acute stroke or transient ischemic attack. New Engl. J. Med. 2016;375:35–43.

[24] Wedzicha JA, et al. Indacaterol-glycopyrronium versus salmeterol-fluticasone for COPD. New Engl. J. Med. 2016;374:2222–2234.

[25] Agnelli G, et al. Oral apixaban for the treatment of acute venous thromboembolism. New Engl. J. Med. 2013;369:799–808.

[26] Henry DH, et al. Randomized, double-blind study of denosumab versus zoledronic acid in the treatment of bone metastases in patients with advanced cancer (excluding breast and prostate cancer) or multiple myeloma. J. Clin. Oncol. 2011;29:1125–32.

[27] Jiang W, et al. Safety and efficacy of sertraline for depression in patients with CHF (SADHART-CHF): a randomized, double-blind, placebo-controlled trial of sertraline for major depression with congestive heart failure. Am. Heart J. 2008;156:437–44.

Committee (IRAC) based on EBMT criteria. The IRAC was blinded as to which arm each patient was assigned."[28]

As is the case with other clinical trials using adjudication, persons conducting adjudication were blinded as to treatment arm and were independent of the Sponsor. EBMT refers to *European Group for Blood and Marrow Transplantation*. Use of EBMT criteria for assessing of drugs against multiple myeloma, and further details on adjudication of myeloma are detailed in the cited references.[29,30]

IV. TASKS OF THE ADJUDICATION COMMITTEE

The adjudication committee drafts an Adjudication Report. Briefing Documents, which had been submitted to FDA during FDA's review process, include descriptions of the Adjudication Report, as shown in the cited Briefing Documents.[31,32] The Adjudication Charter includes a flow chart showing how the adjudication committee fits into the stream of events[33]:

Step 1: Beginning with the occurrence of an AE.
Step 2: Followed by transmission of the AE report to a coordinator.
Step 3: Concluding with the Adjudication Report being entered into the Case Report Form.

Characteristics and tasks of adjudication committees include:

- The Sponsor has the option to choose an adjudication committee.
- The adjudication committee is typically blinded or masked.
- Adjudication has particular utility for the analysis of medical data that are subjective or complex.

[28] Pomalidomide/dexamethosone combination (multiple myeloma) NDA 204-026. 94-page Clinical Review and 101-page pdf file of Medical Review.

[29] Blade J, et al. Criteria for evaluating disease response and progression in patients with multiple myeloma treated by high-dose therapy and haemopoietic stem cell transplantation. Myeloma Subcommittee of the EBMT. European Group for Blood and Marrow Transplant. Br. J. Haematol. 1998;102:1115–23.

[30] Blade J, et al. Interpretation and application of the International Myeloma Working Group (IMWG) Criteria: proposal for uniform assessment and reporting in clinical trials based on the first study Independent Response Adjudication Committee (IRAC) Experience. Blood 2014;124:3460.

[31] Briefing Document. NDA 202-293. Dapagliflozin oral tablets, 5 and 10 mg. Sponsor: Bristol-Myers Squibb Advisory Committee Meeting; December 12, 2013.

[32] Briefing Material. NDA 22350: Saxagliptin (Onglyza) NDA 200678: Saxagliptin/Metformin (Kombiglyze XR). Applicant: AstraZeneca. Endocrinologic and Metabolic Drugs Advisory Committee Meeting; April 14, 2015.

[33] Briefing Material. NDA 22350: Saxagliptin (Onglyza) NDA 200678: Saxagliptin/Metformin (Kombiglyze XR). Applicant: AstraZeneca. Endocrinologic and Metabolic Drugs Advisory Committee Meeting; April 14, 2015. See page 92 of the 105-page Briefing Material.

- Adjudication committees assess safety data but may also assess efficacy data and Protocol deviations.
- Adjudication confirms or corrects the AEs that were reported by the investigator.
- An adjudication committee's review can take two steps, initial screening and then full adjudication.
- The Sponsor has the option of drafting an adjudication charter and an adjudication operations manual.

V. ACCOUNTS OF ADJUDICATION FROM FDA'S MEDICAL REVIEWS

a. Alvimopan (GI Recovery Following Bowel Resection Surgery) NDA 021-775

FDA's review for alvimopan describes the situation where there was a need for adjudicating case narratives.[34] The take-home lessons on adjudication were that:

- What is adjudicated is information in case narratives.
- Adjudication can be blinded or unblinded.
- The same adverse events can be analyzed by nonadjudicated analysis and also by an adjudicated analysis.
- The end-product of the adjudication included dividing one type of adverse event (myocardial infarction) into different severities of the adverse event, where this division was performed for each study subject.
- The end-product of the adjudication included pointing out discrepancies in a nonadjudicated review performed by the Sponsor.

The FDA reviewer explained how adjudication was used to divide AEs into various categories of AEs, "Medical Reviewer's Unblinded Adjudication of Cardiovascular Events in the Study 14: This medical officer ... reviewed the **narratives** of the 6 myocardial infarctions (MI) of the deaths ... in 6-month interim analysis of Study 14 ... [t]his medical officer ... **adjudicated events into** ... **three categories**: definite MI, likely MI, and unlikely MI. A definite MI was classified as having two of the following characteristics ... symptoms typical of a MI, ECG changes, and elevated cardiac enzymes."[35]

The FDA reviewer then contrasted the analysis from the adjudicated review with the Sponsor's nonadjudicated review, writing that, "This medical officer ... adjudicated a total of 6 MIs (5 nonfatal and 1 fatal). In contrast, the sponsor's non-adjudicated reports had a total of 7 MIs (6 nonfatal and 1 fatal). The discrepancy is with Patient #807 ... [t]his

[34] Alvimopan (GI recovery following bowel resection surgery) NDA 021-775. The 383-page Medical Review contains three reviews. These are Special Safety Review (55 pages) on pages 13–67 of Medical Review, Clinical Review (128 pages) on pages 91–218 of Medical Review, and Clinical Review (124 pages) on pages 257–380 of Medical Review.

[35] Alvimopan (GI recovery following bowel resection surgery) NDA 021-775. Page 180 of Medical Review. The 383-page Medical Review contains three reviews. These are Special Safety Review (55 pages) on pages 13–67 of Medical Review, Clinical Review (128 pages) on pages 91–218 of Medical Review, and Clinical Review (124 pages) on pages 257–380 of Medical Review.

medical officer ... thought this patient with a history of severe valvular disease and congestive heart failure (CHF) most likely developed CHF without MI."[36]

The results of the myocardial infarction (MI) adjudication apparently contributed to FDA's recommendation to limit study drug dosing to 15 doses. FDA's Summary Review focused on the rate of MI in short-term treatments versus in long-term (chronic) treatments. The enhanced importance of the adjudication committee's recommendations is apparent, because they resulted in a Black Box Warning. FDA's recommendation was, "The imbalance in ... myocardial infarction was seen in the chronic administration of alvimopan ... [t]his imbalance was not seen in the patients receiving ... alvimopan short-term, up to 15 doses. The **gastrointestinal advisory committee ... recommended that approval ... of this drug be limited to 15 doses administered in the hospital**."[37]

Package label. FDA's requirement for a limit to 15 doses of alvimopan found a place on the package label of alvimopan (Entereg®), that is, in the Black Box Warning:

> **BOXED WARNING**: FOR SHORT-TERM HOSPITAL USE ONLY. ENTEREG is available only for short-term (15 doses) use in hospitalized patients. Only hospitals that have registered in and met all of the requirements for the ENTEREG Access Support and Education (E.A.S.E.) program may use ENTEREG.[38]

b. Dapagliflozin (Type-2 Diabetes Mellitus) NDA 202-293

Dapagliflozin is a small molecule drug for reducing blood glucose and treating type-2 diabetes mellitus. During FDA's review of the drug, adjudication was used to assess relatedness to the AE of DILI. The result was that adjudication attributed the liver injury, not to the study drug, but instead to autoimmune hepatitis.[39]

i. Mechanism of Action for Dapagliflozin

Type-2 diabetes is characterized by reduced insulin secretion, development of insulin resistance, and increased hepatic glucose production. The disease is treated with metformin, sulfonylureas, thiazolidinediones, and with insulin injections. Dapagliflozin is another drug for type-2 diabetes mellitus. Its mechanism of action is to inhibit the glucose transporter (SGLT2) in the renal tubule, thus reducing resorption of glucose by this transporter, with the consequent reduced blood glucose.[40]

[36] Alvimopan (GI recovery following bowel resection surgery) NDA 021-775. Page 180 of Medical Review. The 383-page Medical Review contains three reviews. These are Special Safety Review (55 pages) on pages 13–67 of Medical Review, Clinical Review (128 pages) on pages 91–218 of Medical Review, and Clinical Review (124 pages) on pages 257–380 of Medical Review.

[37] Alvimopan (GI recovery following bowel resection surgery) NDA 021-775. 14-page Summary Review.

[38] Package label. ENTEREG (alvimopan) Capsules; May 2008 (13 pp.).

[39] Dapagliflozin (type-2 diabetes mellitus) NDA 202-293. Pages 13, 50, 76–77, 85–86, 89, 133, 161, and 428–9 of 494-page Medical Review.

[40] Cefalu WT. Paradoxical insights into whole body metabolic adaptations following SGLT2 inhibition. J. Clin. Invest. 2014;124:485–7.

ii. Distinguishing Between DILI Resulting from Dapagliflozin Versus from Autoimmune Hepatitis

A case report published after marketing the drug described DILI in a patient being treated with dapagliflozin. The liver injury was such that it required a liver transplant.[41] The case report was careful to exclude another cause of the DILI, autoimmune hepatitis. A liver biopsy from the patient was consistent with DILI and lacked the signs of *autoimmune hepatitis*.

The researchers considered autoimmune hepatitis, apparently, because the AE of autoimmune hepatitis had been encountered during the drug-approval process. In 2012 FDA delayed the approval of dapagliflozin partly because of concerns for liver toxicity as one study subject developed liver injury during the clinical trial. Analysis showed that the study subject suffered from *autoimmune hepatitis and not from DILI*. This diagnosis removed concerns for dapagliflozin-induced liver injury, and FDA subsequently approved dapagliflozin for use in early 2014.[42]

FDA's Medical Review for dapagliflozin reveals that this particular study subject, a 78-year-old man from India, was followed by the Sponsor for several years, where the goal was to assess the source of the liver injury.[43]

iii. The Sponsor's Adjudication Plan

FDA's Medical Review revealed that the Sponsor's adjudication committee consisted of three experts in liver diseases (hepatologists), and that the committee's goal was to determine relatedness, that is, if DILI was the cause of the observed liver toxicities. Procedural aspects of the adjudication committee are shown by these excerpts:

> The Applicant had an **adjudication plan** in place for assessment of possible liver-related abnormalities. This involved an independent Hepatic Adjudication Committee ... composed of three expert hepatologists, blinded to treatment assignments. The ... adjudication committee reviewed patients' cases in the dapagliflozin program to determine the likelihood that DILI was the cause of liver-related abnormalities. The ... adjudication committee created and maintained the ... Adjudication Committee Charter, and completed adjudication forms on which it summarized its assessment of liver-related cases.[44]

The following excerpt discloses the tasks performed by the adjudication committee:

> Each hepatologist submitted an assessment regarding the probability of DILI associated with study medication. This was followed by a consensus agreement on each case. The potential cases **adjudicated** were from the Applicant's ... 27 studies; N = 7303 dapagliflozin-treated patients and N = 4039 controls ... Criteria for referral to the **Hepatic Adjudication Committee** included at least one of the following four events:
>
> 1. AST and/or ALT >3X ULRR and total bilirubin (TBL) > 1.5X ULRR (within 14 days of the AST and/or ALT elevation)
> 2. AST and/or ALT >5X ULRR

[41] Levine JA, et al. Dapagliflozin-induced acute-on-chronic liver injury. ACG Case Rep. J. 2016;3:e169. doi: 10.14309.

[42] Levine JA, et al. Dapagliflozin-induced acute-on-chronic liver injury. ACG Case Rep. J. 2016;3:e169. doi: 10.14309.

[43] Dapagliflozin (type-2 diabetes mellitus) NDA 202-293. Pages 13, 50, 76–7, 85–6, 89, 133, 161, and 428–9 of 494-page Medical Review.

[44] Dapagliflozin (type-2 diabetes mellitus) NDA 202-293. Pages 79–84 of 494-page Medical Review.

TABLE 12.1 Hepatic Adjudication Causality Scale for Assessing Relatedness of Drug to Liver Injury

Causal Relation	Description of Causality Relation
Definite	The evidence for the study drug causing the injury is beyond a reasonable doubt
Highly likely	The evidence for the study drug causing the injury is clear and convincing but not definite
Probable	The preponderance of the evidence supports the link between the study drug and the liver injury
Possible	The evidence of the study drug causing the injury is equivocal but present
Unlikely	There is clear evidence that an etiological factor other than the study drug caused the injury

Dapagliflozin (type-2 diabetes mellitus) NDA 202-293. Pages 79–84 and 427 of 494-page Medical Review (Table 22. Hepatic Adjudication Causality Scale).

3. Liver-related serious or non-serious adverse event (SAE/AE) in subjects who prematurely discontinued study treatment due to any AE/SAE.
4. Liver-related SAE or liver-related AE in any patients who died.[45]

FDA's Medical Review described a causality scale, writing, "Causality for DILI events was assessed using a five-point numeric/descriptive likelihood causality scale, with causal relationship described as definite, highly likely, probably, and unlikely."[46] Table 12.1 shows the causality scale.

FDA observed that, "During the dapagliflozin clinical ... program ... cases from 80 patients met one or more criteria for ... the liver adjudication process across the 27 completed studies ... [n]one of the ... cases were assessed as 'definitely' or 'highly likely' associated with the ... study medication ... [s]eventeen events (10 dapagliflozin-treated patients vs. six controls) ... were assessed as 'possibly' related to study medication."[47]

iv. Reasons used to Conclude Lack of Relatedness

FDA's review disclosed the facts leading to the conclusion, by the adjudication committee, that hepatic adverse events (liver damage) in the adjudicated study subjects were *not related to dapagliflozin*. The conclusions of "not related" were based on these determinations:

- Concomitant infection (herpes zoster virus; hepatitis E virus)
- Concomitant alcohol binging
- Preexisting disorder (gallstones) where removing the gall bladder resolved the liver damage
- Autoimmune hepatitis

[45] Dapagliflozin (type-2 diabetes mellitus) NDA 202-293. Pages 79–84 of 494-page Medical Review.

[46] Dapagliflozin (type-2 diabetes mellitus) NDA 202-293. Pages 79–84 of 494-page Medical Review.

[47] Dapagliflozin (type-2 diabetes mellitus) NDA 202-293. Pages 79–84 of 494-page Medical Review.

TABLE 12.2 Facts Leading to Adjudication Committee's Conclusions That the Liver Injury Was Not Related to Dapagliflozin

Study Subject	Facts of the Case
D1691C00003-33-11. 64-year-old white female	Liver injury was attributed to a concomitant infection by herpes zoster and to a concomitant drug (acyclovir).
MM102077-321-771182. 60-year-old white female	Liver injury was attributed to excessive alcohol binging.
D1690C00018-203-4. 72-year-old white male	Liver injury was attributed to concomitant gallstones. After surgery to remove the gall bladder (cholecystectomy), the liver tests became normal.
MB10207788-70996. 57-year-old Asian man	Liver injury was attributed to concomitant hepatitis E virus.
D1690C00004-4402-6. 79-year-old man from India	Liver injury was attributed to autoimmune hepatitis. The same study subject was reevaluated at the age of 84, and it was confirmed that autoimmune hepatitis was the source of liver injury. FDA's Medical Review provided a graph showing spikes in liver function abnormalities over time (about 1300 days), with the conclusion that these spikes were characteristic of the pattern of inflammation in autoimmune hepatitis. The liver function tests were aminotransferases, bilirubin, and alkaline phosphatase.
D1690C00004-4916-16. 67-year-old white female	This study subject had been accidentally randomized to the placebo arm (not to the dapagliflozin arm), and had never received the study drug.
MB102029-4-276. 83-year-old white male	Liver injury was attributed to concomitant medications (niacin, pravastatin, levofloxacin).

Dapagliflozin (type-2 diabetes mellitus) NDA 202-293. Pages 80–90 of 494-page Medical Review.

- Study subject was found to be inadvertently assigned to the placebo arm
- Concomitant medications (niacin, pravastatin, levofloxacin).

Table 12.2 provides a distinct picture of the tasks and outcomes from the adjudication committee's review of the liver injury AEs presenting in the dapagliflozin clinical studies.[48]

FDA's Medical Review provided an overview of the adjudication process, and the conclusion that dapagliflozin was not related and not associated with the AEs of serious liver injury that had materialized during the clinical study:

> Based on the information provided in the briefing package and the presentations at today's meeting, discuss your level of concern with regard to dapagliflozin use and drug-induced liver injury. Specifically comment on whether you believe use of dapagliflozin is associated with an increased risk of drug-induced liver injury and explain your rationale ... [m]uch of the discussion focused on a relative **lack of evidence** across the Applicant's entire clinical program to suggest an association of serious liver injury with dapagliflozin.[49]

[48] Dapagliflozin (type-2 diabetes mellitus) NDA 202-293. Pages 80–90 of 494-page Medical Review.

[49] Dapagliflozin (type-2 diabetes mellitus) NDA 202-293. Pages 132–133 of 494-page Medical Review.

v. FDA Arrives at Package Label Recommendation

FDA's Medical Review revealed FDA's recommendations on the Black Box Warning, Warning and Precautions section, and on monitoring recommendations on the package label, "It was felt that the available data on hepatic safety did not support a Boxed Warning, however, it was questioned whether the Warnings and Precautions section of product labeling should include information related to potential hepatotoxicity with dapagliflozin ... [t]he hepatologist on the panel recommended against routine monitoring of liver laboratory tests."[50]

FDA's recommendations against any Black Box Warning, Warnings and Precautions section, and monitoring of liver function are reflected in the package label for dapagliflozin (Farxiga®). In other words, the package label does not include any of these things. The Warnings and Precautions section warns against hypotension, ketoacidosis, and urinary tract infections, but there is nothing about liver function. Also, although the package label does require monitoring, there does not exist any statement about monitoring liver function.[51] FDA's review for dapagliflozin is likely to provide the most detailed and thorough account of relatedness analysis of all of FDA's Medical Reviews for all NDAs submitted to the FDA, to date.

c. Denosumab (Postmenopausal Osteoporosis) BLA 125-320, BLA 125-331

FDA's Medical Review for denosumab provides an account of adjudication of various adverse events, where the AEs were identified by terms in the MedDRA dictionary. The utility and practical use of the adjudication process was that the adverse event under review, osteonecrosis of the jaw (ONJ), was determined by the adjudication committee actually to not be this adverse event at all. FDA wrote, "A ... list of MedDRA preferred terms was used to identify cases of potential ONJ cases to be **adjudicated by an expert committee**. The committee did not consider any of the 12 cases sent for adjudication to be a bona fide case of ONJ."[52]

FDA's review for denosumab describes the adjudication process, as used for ONJ and also for cardiovascular AEs.[53]

i. Osteonecrosis of the Jaw

The Sponsor's adjudication committee sets up MedDRA terms that would trigger cases of potential ONJ to be reviewed by the committee. As is evident from FDA's Medical Review, the term "osteonecrosis of the jaw (ONJ)" was a verbatim term that did not exist in the MedDRA dictionary. The Sponsor provided guidance to its adjudication committee by defining ONJ as, "Area of exposed alveolar ... bone where ... mucosa is normally found associated with non-healing after appropriate care by 8 weeks in a patient."

[50] Dapagliflozin (type-2 diabetes mellitus) NDA 202-293. Pages 132–133 of 494-page Medical Review.

[51] Package label. FARXIGA (dapagliflozin) tablets, for oral use; March 2017 (28 pp.).

[52] Denosumab (osteonecrosis in postmenopausal women) BLA 125-320, BLA 125-331. Page 117 of 710-page Medical Review.

[53] Denosumab (post-menopausal osteoporosis) BLA 125-320, BLA 125-331. Pages 90–6, 104–6, 202 of 369-page Clinical Review. This Clinical Review is on pages 253–621 of 710-page pdf file of Medical Review.

TABLE 12.3 List of MedDRA Terms Used to Query the Sponsor's Database of Adverse Events to Identify Potential Cases of ONJ

Abscess jaw	Osteomyelitis acute	Gingival ulceration
Alveolar osteitis	Osteomyelitis drainage	Jaw operation
Bone erosion	Pain in jaw	Maxillofacial operation
Bone infarction	Periodontal infection	Oral cavity fistula
Dental necrosis	Primary sequestrum	Oroantral fistula
Gingival erosion	Sequestrectomy	Osteomyelitis
Jaw lesion excision	Abscess oral	Osteomyelitis chronic
Loose tooth	Bone debridement	Osteonecrosis
Necrosis	Bone fistula	Periodontal destruction
Oral surgery	Dental fistula	Periodontal operation
Osteitis	Gingival abscess	Secondary sequestrum

Denosumab (postmenopausal osteoporosis) BLA 125-320, BLA 125-331. Page 357 of 710-page Medical Review.
This table (Table 43) is from page 105 of 369-page Clinical Review within the Medical Review.

After establishing this definition, the adjudication committee formulated a number of query terms, where the query terms were used to detect adverse events in the Sponsor's database of adverse events, and hopefully detect study subjects who had experienced the adverse event of ONJ. Regarding the formulation of query terms, FDA's Medical Review stated, "The true incidence and risk of ONJ related to ... denosumab is unknown ... [a]s a result, the applicant included ... a plan to ... evaluate patients ... for ONJ signs and symptoms. This was accomplished through formation of an adjudication committee, the Osteonecrosis of the Jaw Adjudication Committee ... **setting up MedDRA terms which would trigger cases of potential ONJ** to be reviewed by the committee."[54]

The chosen MedDRA terms that were used to query the safety database are shown in Table 12.3. FDA's Medical Review distinguished between verbatim terms and preferred terms, and scrutinized the validity of the chosen MedDRA terms:

> All adverse events were coded using ... MedDRA ... [a]dverse events were graded on a five-point intensity scale: mild, moderate, severe, life-threatening or fatal ... [a] **comparison of verbatim term to MedDRA Preferred Term** was performed ... to verify the precision and accuracy of the medical coding of adverse events. The coding was considered acceptable in the majority of events reviewed. In those cases where the FDA reviewers would have chosen a different preferred term, the differences were not clinically meaningful and would not have had a significant impact on the ... assessment of safety.[55]

[54] Denosumab (post-menopausal osteoporosis) BLA 125-320, BLA 125-331. Page 356 of 710-page Medical Review.

[55] Denosumab (post-menopausal osteoporosis) BLA 125-320, BLA 125-331. Pages 327–8 of 710-page Medical Review.

Table 12.3, reproduced from FDA's Medical Review, reveals a long list of terms naming various AEs, where the goal of using many terms was to ensure maximal detection of AEs that served as a signal for ONJ.

In addition to using MedDRA's terms for detecting study subjects, for subsequent analysis by the adjudication committee, the Sponsor devised some home-grown terms that could be used to detect additional cases of ONJ.[56] These home-grown terms included ischemic necrosis, aseptic necrosis bone, and osteomyelitis. Regarding the ability of the Sponsor to create new preferred terms to be part of the MedDRA dictionary, please note that creating new terms is not allowed, unless permission is acquired, "Companies are not allowed to add new terms but can suggest new terms—or alternate placing in the hierarchy—which will then be considered for the biannual update."[57]

Use of the query terms resulted in detecting 21 study subjects who had potentially experienced ONJ. FDA commented on these 21 potential cases of ONJ. The outcome of this review was that none of the AEs were actually ONJ. FDA stated that the adjudication committee had, "reviewed all 21 cases and concluded that **none were positive** for meeting the criteria."[58]

ii. Cardiovascular AEs

This concerns cardiovascular AEs associated with denosumab. First, the adjudication committee members were cardiologists, not otherwise associated with the trial. Second, FDA's review distinguished AEs that were adjudicated from those that were not adjudicated, by including a separate analysis entitled, "Unadjudicated Adverse Event Analysis." Third, adjudication was for AEs that were serious adverse events (SAEs). Fourth, before setting out to adjudicate the SAEs, the committee categorized the SAEs into:

1. Acute coronary disease
2. Congestive heart failure
3. Stroke/transient ischemic attacks
4. Cardiac arrhythmia
5. Other vascular disorders.

The adjudication committee analyzed deaths separately from SAEs, and divided deaths into cardiovascular versus noncardiovascular. The committee used an Event-Adjudication Manual. This type of manual has been described as useful for a variety of clinical trials.[59] Regarding the results of the adjudication process, FDA wrote that, "Adjudication of cardiovascular SAEs was done ... [t]he number of events submitted for adjudication was 526 in the placebo group and 572 in the denosumab group."[60]

[56] Denosumab (post-menopausal osteoporosis) BLA 125-320, BLA 125-331. Page 357 of 710-page Medical Review.

[57] Schroll JB, et al. Challenges in coding adverse events in clinical trials: a systematic review. PLoS One 2012;7:e41174.

[58] Denosumab (post-menopausal osteoporosis) BLA 125-320, BLA 125-331. Page 358 of 710-page Medical Review.

[59] Abbott. Advisory Committee Briefing Document for Meridia® (Sibutramine hydrochloride monohydrate). Endocrinologic and Metabolic Drugs Advisory Committee; 2010 (206 pp.).

[60] Denosumab (post-menopausal osteoporosis) BLA 125-320, BLA 125-331. Pages 146 and 164–5 of 710-page Medical Review.

After conducting its analysis, the adjudication committee arrived at the following conclusion. The final output of the adjudication process was that denosumab was not related to cardiovascular AEs. FDA concluded that, "Time to first any adjudicated cardiovascular event ... does **not suggest worsening cardiovascular outcomes** over time ... [t]he incidence of any adjudicated cardiovascular serious event ... cardiovascular death ... and other vascular disorder was similar in the 2 treatment arms."[61]

Package label. ONJ adverse event information, apparently with the aid of adjudication, found a place on the Warnings and Precautions section of the package label for denosumab (Prolia®):

> **WARNINGS AND PRECAUTIONS** ... ONJ ... is generally associated with tooth extraction and/or local infection with delayed healing. ONJ has been reported in patients receiving denosumab ... [a] routine oral exam should be performed ... prior to initiation of Prolia treatment ... [t]he risk of ONJ may increase with duration of exposure to Prolia ... [p]atients who are suspected of having or who develop ONJ while on Prolia should receive care by a dentist or an oral surgeon ... [d]iscontinuation of Prolia therapy should be considered based on individual benefit-risk assessment.[62]

Note the statement in the Warnings and Precautions section that ONJ can result from "tooth extraction" and with "local infection" (things other than denosumab), and that the terms used by the Sponsor to query the safety database included "loose tooth" and "periodontal infection." Hence, the text in this package label warning can reasonably be suspected to be a result from adjudication, where adjudication determined that a loose tooth, tooth extraction, and local infection are causes of ONJ (causes other than denosumab).

d. Lixisenatide (Type-2 Diabetes) NDA 208-471, NDA 204-961, NDA 204-961

Lixisenatide is an oligopeptide used for treating diabetes.[63] FDA's Medical Review described the Sponsor's use of an Allergic Reactions Adjudication Committee (ARAC).[64] The composition and responsibilities of the adjudication committee were, "An allergic reactions adjudication committee (ARAC) was established in April 2007 during the Phase 2 Study ... and it reviewed and adjudicated all potential allergic treatment-emergent adverse events (TEAEs) reported by the investigators in this study ... ARAC members

[61] Denosumab (post-menopausal osteoporosis) BLA 125-320, BLA 125-331. Pages 146 and 164–5 of 710-page Medical Review.

[62] Package label. Prolia® (denosumab) Injection, for subcutaneous use. August 2016 (26 pp.).

[63] Pfeffer MA, et al. Lixisenatide in patients with type 2 diabetes and acute coronary syndrome. New Engl. J. Med. 2015;373:2247–2257.

[64] Lixisenatide (type-2 diabetes) NDA 208-471 (Medical Review contained three Clinical Reviews). Clinical Review for NDS 208-471 (pages 2–154 pages of the Medical Review); Clinical Review for NDA 204-961 (pages 156–251 of 690-page pdf file of Medical Review); and Clinical Review for NDA 204-961 (pages 426–676 of Medical Review). Medical Review is 690-page pdf file. Adjudication committee information on pages 104–20 of 154-page Clinical Review.

were independent from the applicant and the investigators, and were blinded on the actual investigational product received by the patients."[65]

i. Association of Anaphylaxis with Lixisenatide

FDA's Medical Review observed that there was "an imbalance" in anaphylaxis, in comparing the study drug treatment arm with the control arm, meaning that anaphylaxis was associated with the study drug. The FDA reviewer observed, "An **imbalance in hypersensitivity reactions** adjudicated by the ARAC was noted ... [t]he ARAC adjudicated eight cases as anaphylactic reactions and three as anaphylactic shock ... [o]f these 11 cases, nine cases were associated with lixisenatide treatment. The remaining two were attributed to cefazolin and bee sting hypersensitivity."[66]

ii. Adjudication Committee Decided that the Associated Anaphylaxis was not Due to the Study Drug

FDA's Medical Review acknowledged the fact that anaphylaxis was "temporally associated" with the study drug, and revealed the useful and practical decision from the adjudication committee that anaphylaxis was a result of an "alternate" cause, namely bee stings and other drugs that were taken.

Regarding this useful decision regarding the "alternate" cause and nonrelatedness to lixisenatide, the FDA reviewer wrote, "The Applicant's independent ARAC identified ... 17 cases of anaphylactic reaction/anaphylactic shock **temporally associated** with lixisenatide treatment in the clinical development program ... five cases were discounted as not lixisenatide-related because an alternative likely cause was apparent (2 bee stings and 3 alternate drugs—cefazolin, mecillinam, protamine) and a sixth case submitted by the investigator as 'anaphylactic shock' was deemed unrelated by ARAC because the event occurred after 1 year of lixisenatide treatment."[67]

Package label. The package label for lixisenatide (Adlyxin®) warned against anaphylaxis. As revealed by FDA's Medical Review, 17 cases of anaphylaxis were detected during the Sponsor's clinical studies. Six of these cases (but not all) were discounted by the adjudication committee as being irrelevant to the study drug. As a consequence, the package label included warnings against anaphylaxis:

> **CONTRAINDICATIONS**. Hypersensitivity to ADLYXIN or any product components. Hypersensitivity reactions including anaphylaxis have occurred with ADLYXIN.[68]
>
> **WARNINGS AND PRECAUTIONS**. Anaphylaxis and Serious Hypersensitivity Reactions: Discontinue ADLYXIN and promptly seek medical advice.[69]

[65] Lixisenatide (type-2 diabetes) NDA 208-471, NDA 204-961, NDA 204-961. Pages 105, 245, and 679 of 690-page Medical Review.

[66] Lixisenatide (type-2 diabetes) NDA 208-471, NDA 204-961, NDA 204-961. Pages 105, 248, 250, 415, 518, 520, 684, 687, and 688 of 690-page Medical Review.

[67] Lixisenatide (type-2 diabetes) NDA 208-471, NDA 204-961, NDA 204-961. Page 415 of 690-page Medical Review.

[68] Package label. ADLYXIN (lixisenatide) injection, for subcutaneous use; July 2016 (33 pp.).

[69] Package label. ADLYXIN (lixisenatide) injection, for subcutaneous use; July 2016 (33 pp.).

e. Nivolumab (Metastatic Melanoma) BLA 125-554

Nivolumab is a recombinant antibody for treating melanoma. As is the case for most anticancer drugs, efficacy of nivolumab for treating melanoma was assessed by objective response and by clinical parameters such as progression-free survival (PFS) and overall survival (OS).

The term, objective response, as it is applied to solid tumors, refers to measurements of tumor size and number. Clinical studies on drugs against solid tumors, including studies on nivolumab for melanoma, use a method of objective response known as the RECIST criteria.[70,71] The RECIST criteria are the set of criteria most often used, and these are used with measurements of tumor size and number.[72] Data from computed tomography scans (CT scans) and from X-rays are interpreted by radiologists.

Eisenhauer et al.[73] warn about a source of variability in interpreting RECIST data, namely that, "RECIST measurements may be performed at most clinically obtained slice thicknesses. It is recommended that CT scans be performed at 5 mm contiguous slice thickness or less ... **variations in slice thickness can have an impact on lesion measurement and on detection of new lesions**." A publication by Holihan et al.[74] reveals that CT scans can be misinterpreted by radiologists. The warnings from Eisenhauer et al.[75] and from Holihan et al. raise the issue that adjudication may be needed where data on tumors are acquired from CT scans.

FDA's Medical Review disclosed use of adjudication, and that adjudication was used to resolve disagreements between readings from two different radiologists, "A prospectively defined blinded **independent** central review (BICR) which consisted of three radiologists (2 readers plus an **adjudicator** if needed) was used for the primary analysis of objective response rate (ORR) as well as the analysis of progression-free survival (PFS) (see below for further details regarding the BICR)."[76]

In the words of the FDA reviewer, "If any of the response evaluations after the global radiology analysis, including best overall response, confirmed best overall response, date of first response, or date of progression were not identical between the two readers, an adjudication by a third radiologist was performed. The **adjudicator reviewed all the**

[70] Robert C, et al. Nivolumab in previously untreated melanoma without BRAF mutation. New Engl. J. Med. 2015;372:320–30.

[71] Postow MA, et al. Nivolumab and ipilimumab versus ipilimumab in untreated melanoma. New Engl. J. Med. 2015;372:2006–17.

[72] Eisenhauer EA, et al. New response evaluation criteria in solid tumours: revised RECIST guideline (version 1.1). Eur. J. Cancer 2009;45:228–47.

[73] Eisenhauer EA, et al. New response evaluation criteria in solid tumours: revised RECIST guideline (version 1.1). Eur. J. Cancer 2009;45:228–47.

[74] Holihan JL, et al. Use of computed tomography in diagnosing ventral hernia recurrence: a blinded, prospective, multispecialty evaluation. JAMA Surg. 2016;151:7–13.

[75] Eisenhauer EA, et al. New response evaluation criteria in solid tumours: revised RECIST guideline (version 1.1). Eur. J. Cancer 2009;45:228–47.

[76] Nivolumab (metastatic melanoma) BLA 125-554. Page 34 of 156-page Medical Review.

results from both initial readers and indicated which primary reviewer's results they believe most accurately represents the above variables for the case."[77]

Package label. The Sponsor's use of independent review of radiological data, using the RECIST criteria, found a place on the package label of nivolumab (Opdivo®). The Clinical Review section of the package label revealed that an adjudication committee was used to provide a firm conclusion that nivolumab actually resulted in a reduction of tumor size and number:

> **CLINICAL REVIEW** ... Major efficacy outcome measures included confirmed objective response rate (ORR) as assessed by **independent radiographic review** committee ... using Response Evaluation Criteria in Solid Tumors (RECIST v1.1).[78]

f. Sofosbuvir/Velpatasvir Combination (Hepatitis C Virus) NDA 208-341

Sofosbuvir and velpatasvir are small molecule drugs for treating hepatitis C virus infections. The results from one of the clinical trials (ASTRAL-4) described in FDA's Medical Review were published.[79] The Clinical Study Protocol reveals that cirrhosis was determined by the Child-Pugh score.[80] An adjudication committee assessed relatedness between the sofosbuvir/velpatasvir combination and adverse events. Adjudication was needed to distinguish between these things:

- Need to distinguish between AEs from the study drug versus AEs caused by concomitant drugs.
- Need to distinguish AEs from the study drug versus AEs caused by the concomitant condition of cirrhosis.

i. Relatedness Analysis Confounded by Cirrhosis

One consequence of hepatitis C virus infections is cirrhosis, a form of liver injury.[81] But cirrhosis caused by hepatitis C virus is not a type of DILI. The AE needing adjudication was DILI. A system is used for scoring DILI and involved scores of, "definite, very likely, probable, possible, and unlikely." FDA's Medical Review stated that the problem was:

> Hepatic safety signals can be difficult to detect in HCV trials, especially among subjects with **advanced cirrhosis**. To facilitate detection of possible safety concerns, the Applicant convened an ... adjudication

[77] Nivolumab (metastatic melanoma) BLA 125-554. Pages 34–5 of 156-page Medical Review.

[78] Package label. OPDIVO (nivolumab) injection, for intravenous use; February 2017 (57 pp.).

[79] Curry MP, et al. Sofosbuvir and velpatasvir for HCV in patients with decompensated cirrhosis. New Engl. J. Med. 2015;373:2618–28.

[80] A Phase 3, Multicenter, Open-Label Study to Investigate the Efficacy and Safety of Sofosbuvir/GS-5816 Fixed-Dose Combination in Subjects with Chronic HCV Infection and Child-Pugh Class B Cirrhosis. Protocol ID: GS-US-342-1137. IND No. 118605. The Clinical Study Protocol was published as a supplement to Curry MP, et al. Sofosbuvir and velpatasvir for HCV in patients with decompensated cirrhosis. New Engl. J. Med. 2015;373:2618–28.

[81] Sofosbuvir/velpatasvir combination (hepatitis C virus) NDA 208-341. Pages 119–27 of 164-page Clinical Review and 171-page Medical Review.

committee to review possible cases of DILI. The panel reviewed all cases of pre-specified liver related laboratory abnormalities, treatment-emergent deaths, liver transplants, hepatic failure events, and hepatic events leading to discontinuation of study drug.[82]

ii. Relatedness Analysis Confounded by Concomittant Drugs

The adjudication committee observed that the relatedness inquiry was confounded by concomitant drugs, as well as by concurrent illnesses, writing:

> **adjudication committee** believed that the current drug-induced liver injury ... causality scoring system ... was not applicable to this subject population with advanced liver disease co-morbidities who are **often receiving numerous concomitant medications** ... assessment is **confounded** by the initiation of several medications, many of which were discontinued at the same time[83] ... [o]f the 55 cases screened for DILI evaluation by the ... **adjudication committee**, there was only one case in which DILI **could not be definitively excluded**; causality was **confounded** by concomitant medications and concurrent illness in this subject[84] ... [t]wo cases identified as potential DILI cases by the IAC were confounded by concomitant medications, cholelithiasis and/or viral illness. The remaining nine cases meeting ... **adjudication committee** screening criteria for potential DILI were **unlikely related** to the study drug (SOF/VEL) use due confounding events, alternative explanations and/or isolated liver laboratory elevations which improved while HCV treatment was continued.[85]

iii. Utility in Consulting Package Label of the Concomitant Drug

The adjudication committee assessing whether a concomitant drug is responsible for a given AE, by consulting the package label of the concomitant drug. FDA's Medical Review referred to measures of liver injury taking the form of high serum ALT and high serum bilirubin. FDA referred to the information on the eplerenone package label. On this point, FDA wrote, "The ... **adjudication committee** evaluation of this case is considered thorough ... this case may be confounded by concomitant eplerenone use. In the eplerenone label ... increases of ALT greater than 120 U/L and bilirubin greater than 1.2 mg/dL were reported 1/2259 patients administered eplerenone tablets and 0/351 placebo-treated patients."[86]

The adjudication committee provided a strategy for assessing relatedness, for the situation where relatedness was confounded and unclear, "the ... **adjudication committee** determined a more meaningful approach would be to categorize subjects as those for

[82] Sofosbuvir/velpatasvir combination (hepatitis C virus) NDA 208-341. Page 88 of 171-page Medical Review.

[83] Sofosbuvir/velpatasvir combination (hepatitis C virus) NDA 208-341. Page 126 of 171-page Medical Review.

[84] Sofosbuvir/velpatasvir combination (hepatitis C virus) NDA 208-341. Page 124 of 171-page Medical Review.

[85] Sofosbuvir/velpatasvir combination (hepatitis C virus) NDA 208-341. Page 124 of 171-page Medical Review.

[86] Sofosbuvir/velpatasvir combination (hepatitis C virus) NDA 208-341. Page 132 of 171-page Medical Review.

whom DILI could be excluded, those for whom DILI could not be excluded, and those with insufficient data to make a determination."[87]

Package label. Comments in FDA's review of sofosbuvir/velpatasvir combination (Epclusa®) and relatedness to adverse events found a place on the package label. The Use in Specific Populations section of the package label acknowledged that it was acceptable to administer the drug to patients with even severe hepatic impairment, but warned that where the patient has *cirrhosis then monitoring hepatic laboratory values is needed*:

> **USE IN SPECIFIC POPULATIONS** ... Hepatic Impairment. No dosage adjustment of EPCLUSA is required for patients with mild, moderate, or severe hepatic impairment (Child-Pugh Class A, B, or C) ... **hepatic laboratory monitoring** (including direct bilirubin) ... is recommended for patients with decompensated **cirrhosis** receiving treatment with EPCLUSA.[88]

g. Ticagrelor (Thrombotic Events With Acute Coronary Syndromes) NDA 022-433

FDA's review of ticagrelor described adjudication of two different adverse events:

1. **AE of stroke.** The goal of ticagrelor was to prevent stroke, where adjudication was used to improve accuracy in diagnosing the incidence of stroke, and in measuring ticagrelor's efficacy in preventing stroke.
2. **AE of bleeding.** An adverse event associated with ticagrelor is bleeding. Adjudication was used to acquire a more accurate accounting of the number of bleeding adverse events. Bleeding was an established AE related to ticagrelor, and here, adjudication was not used for any relatedness analysis, but instead to acquire a more accurate measure of the incidence of this type of AE.

Ticagrelor is for treating thrombotic events. Ticagrelor inhibits the P2Y12 adenosine diphosphate receptor, where the drug's effect is to stimulate the adenosine-induced increase in coronary blood flow velocity and to increase oxygen supply, in patients with acute coronary syndrome.[89]

FDA's comments referred to SUSARs (Serious Unexpected Adverse Reactions)[90,91] where FDA complained that the Sponsor failed to report AEs as SUSARs. Also, FDA complained that the Sponsor had failed to submit the SUSAR for adjudication. The SUSARs included syncope, ventricular tachycardia, and seizure. FDA complained, "On day 12 he was rehospitalized for syncope ... ventricular tachycardia ... seizure after hospitalization,

[87] Sofosbuvir/velpatasvir combination (hepatitis C virus) NDA 208-341. Page 126 of 171-page Medical Review.

[88] Package label. EPCLUSA® (sofosbuvir and velpatasvir) tablet, for oral use; June 2016 (32 pp.).

[89] Pelletier-Galarneau M, et al. Randomized trial comparing the effects of ticagrelor versus clopidogrel on myocardial perfusion in patients with coronary artery disease. J. Am. Heart Assoc. 2017;6:e005894. doi:10.1161.

[90] U.S. Department of Health and Human Services. Food and Drug Administration. Center for Drug Evaluation and Research (CDER). Center for Biologics Evaluation and Research (CBER). Guidance for industry and investigators safety reporting requirements for INDs and BA/BE studies; 2012 (29 pp.).

[91] Wallace S, et al. Serious adverse event reporting in investigator-initiated clinical trials. Med. J. Aust. 2016;204:231–3.

ticagrelor was discontinued ... [the Sponsor] did not report this event as a SUSAR **nor was it submitted for adjudication** as a possible cardiac ischemic event."[92]

FDA complained that the Sponsor had failed to use adjudication to determine relatedness of suspicious AEs:

> did not submit the thrombocytopenia SAE ... despite the investigator reporting it as related to study drug ... it is the sponsor's responsibility to make final determinations regarding the relatedness of SAEs for expedited reporting purposes ... I find the limited data collection highly disturbing for these two patients who had SUSARs ... [t]his **pattern of failing to report and submit for adjudication suspicious adverse events** is the precise pattern I have seen with other problematic NDA submissions.[93]

Emphasizing the Sponsor's failure to use adjudication, FDA further complained that the Sponsor had, "failed to submit potential endpoint events for **adjudication** ... [a] ticagrelor patient was hospitalized on day 22 with a coronary thrombosis ... [n]o other information on symptoms ... of this event ... are listed ... [the Sponsor] unblinded the patient as a SUSAR but did not distribute the SUSAR because the investigator indicated the event was unrelated to the study drug ... [the Sponsor] did not submit the event for **adjudication**."[94]

FDA's complaints ramped up to the point where FDA alleged that the Sponsor had concealed negative data. FDA learned about the AEs from MedWatch forms. The Sponsor had failed to spontaneously submit the MedWatch forms to FDA, and instead, the FDA reviewer requested the MedWatch forms:

> A ticagrelor patient was hospitalized on day 22 with a coronary thrombosis. A troponin T was reported as >5x and an echo showed hypokinesis of the inferior wall with ejection fraction 55%. ... [the Sponsor] unblinded the patient as a SUSAR but did not distribute the SUSAR because the investigator indicated the event was unrelated to study drug ... **[the Sponsor] did not submit the event for adjudication** ... I detected this case after requesting ... [the Sponsor] to supply Medwatch forms for 26 ticagrelor cases ... but for whom [the Sponsor] had not submitted Medwatch forms to the IND or the NDA.[95]

FDA further complained about not submitting Case Report Forms, and SAE forms, as part of the FDA-submission. FDA complained that the Sponsor, "initially submitted to the NDA only Medwatch forms for the 201 patients for whom Medwatch forms had previously been reported expeditedly to the IND ... [t]his case also suggests that [Sponsor's] apparent reluctance to provide complete CRFs, including SAE forms and other clinical records, is based on a desire to conceal negative data regarding ticagrelor."[96]

[92] Ticagrelor (thrombic events with acute coronary syndromes) NDA 022-433. Pages 6 and 33 of 640-page Medical Review.

[93] Ticagrelor (thrombic events with acute coronary syndromes) NDA 022-433. Page 34 of 640-page Medical Review.

[94] Division of Cardiovascular and Renal Products. Review of Complete Response. NDA 022-433. Drug: ticagrelor. Reviewer Marcianiak, TA (47 pp.). This Review of Complete Response is part of FDA's 640-page Medical Review.

[95] Ticagrelor (thrombic events with acute coronary syndromes) NDA 022-433. Page 44 of 640-page pdf file containing Medical Review (pages 1–158) and other reviews.

[96] Ticagrelor (thrombic events with acute coronary syndromes) NDA 022-433. Page 44 of 640-page pdf file containing Medical Review (pages 1–158) and other reviews.

Expanding the complaint to encompass clinical trials from a variety of Sponsors, FDA complained that other Sponsors had failed to submit SAEs to the adjudication committee, "Not submitting suspicious events for **adjudication** is a pattern I have seen in other problematic submissions ... [i]t can enable a sponsor to manipulate the endpoints while proclaiming that a blue ribbon academic research organization **adjudicated** events fairly blinded to treatment assignment."[97]

i. Outcome of Stroke Adjudication

FDA's Medical Review for ticagrelor describes adjudication of the AE of stroke, where the result of adjudication was that some of the stroke AEs were down-classified to "No Event":

> In this analysis, 17 ticagrelor [study drug] and 15 clopidogrel [control] subjects with ... "strokes" were down-classified to "No Event" ... after **adjudication** ... [i]n general, the cases where suspected strokes were downgraded appear to have been appropriately and reasonable **adjudicated** ... [i]t appeared that down-classification from "stroke" to "no event" was due to insufficient information on the clinical presentation and outcome, which ranged from inadequate to entirely non-existent.[98]

ii. Adjudicating Efficacy Endpoints

As demonstrated by the footnoted references,[99,100,101] the Sponsor adjudicated the adverse event of stroke in the context of efficacy analysis of ticagrelor. The goal was not to detect relatedness of ticagrelor to the AE of stroke, but instead the goal was to use adjudication to acquire a more accurate determination of ticagrelor's efficacy in preventing stroke. The Sponsor's Medical Review referred to a clinical study with ticagrelor that was still in the planning stage, where this study was named PEGASUS.[102] A publication on the results of the PEGASUS study referred to the use of adjudication for assessing ticagrelor's

[97] Ticagrelor (thrombic events with acute coronary syndromes) NDA 022-433. Page 45 of 640-page pdf file containing Medical Review (pages 1–158) and other reviews.

[98] Ticagrelor (thrombic events with acute coronary syndromes) NDA 022-433. Page 529 of 640-page pdf file containing Medical Review (pages 1–158) and other reviews.

[99] Bonaca MP, et al. Prevention of stroke with ticagrelor in patients with prior myocardial infarction: insights from PEGASUS-TIMI 54 (Prevention of cardiovascular events in patients with prior heart attack using ticagrelor compared to placebo on a background of aspirin-thrombolysis in myocardial infarction 54). Circulation 2016;134:861–71.

[100] Johnston SC, et al. Ticagrelor versus aspirin in acute stroke or transient ischemic attack. New Engl. J. Med. 2016;375:35–43.

[101] AstraZeneca. Ticagrelor NDA 22-433 briefing document for cardiovascular and renal drugs advisory committee meeting; June 2010; page 102 of 344-page Briefing Document.

[102] Ticagrelor (thrombic events with acute coronary syndromes) NDA 022-433. Page 182 of 640-page pdf file containing Medical Review (pages 1–158) and other reviews.

efficacy in preventing stroke, "We investigated the incidence of stroke ... and the efficacy of ticagrelor focusing on the approved 60 mg twice daily dose for reducing stroke in this population ... [p]atients were followed for ... 33 months. Stroke events were adjudicated by a central committee."[103]

iii. Adjudication of Bleeding

FDA's Medical Review demonstrates that the incidence of the AE of bleeding would be different, depending on the definition, that is, whether bleeding was adjudicated by the PLATO definition or by the TIMI definition:

> Bleeding was the major safety concern when the sponsor was designing PLATO. The primary safety endpoint designated in PLATO was time to first major bleeding event ... PLATO used an ... **adjudication committee** ... **to adjudicate bleeding events**. The ... **adjudication committee** judged each bleeding event against a set of definitions to maintain consistency and quality ... [t]he PLATO categories consider certain bleeds that are likely to be severe, such as intrapericardial bleed with tamponade and intracranial hemorrhage, to be unconditionally major/life-threatening while according to the TIMI definition these two types of bleeds are counted as minor, minimal or not at all unless they are symptomatic or are accompanied by a hemoglobin decrease of >5 gm/dL.[104]

Discrepancies in the definitions for AEs can occur between the investigator and the adjudication committee. FDA's Medical Review stated that, "**Adjudication committee** determined that some events reported by Investigators **did not qualify as bleeding events**. On occasion, Adjudication Committee identified additional events and directed the sponsor to query a site to register the events for official adjudication. If the Investigator agreed, the event was registered and processed by the adjudication committee."[105]

Package label. The importance of assessing relatedness, number, and severity of the AE of bleeding is evident from the fact that the package label for ticagrelor (Brilinta®) has a Black Box Warning on bleeding:

> **BOXED WARNING**: BLEEDING RISK ... BRILINTA, like other antiplatelet agents, can cause significant, sometimes fatal bleeding. Do not use BRILINTA in patients with active pathological bleeding or a history of intracranial hemorrhage.[106]

[103] Bonaca MP, et al. Prevention of stroke with ticagrelor in patients with prior myocardial Infarction: insights from PEGASUS-TIMI 54 (Prevention of cardiovascular events in patients with prior heart attack using ticagrelor compared to placebo on a background of aspirin-thrombolysis in myocardial infarction 54). Circulation 2016;134:861–71.

[104] Ticagrelor (thrombic events with acute coronary syndromes) NDA 022-433. Page 232–3 of 640-page Medical Review.

[105] Ticagrelor (thrombic events with acute coronary syndromes) NDA 022-433. Page 410 of 640-page Medical Review.

[106] Package label. BRILINTA® (ticagrelor) tablets, for oral use; September 2016 (23 pp.).

VI. CONCLUDING REMARKS

Commentaries from FDA's Medical Reviews have defined the structure and tasks of an adjudication committee. Adjudication is distinguished as follows:

- *Independence from Sponsor*. Adjudication is independent from the Sponsor and investigator. The Sponsor and investigator may be one and the same, or they may be separate entities.[107] The "sponsor" and the "investigator" have separate definitions, according to 21 CFR §312.3, and also according to FDA's Guidance for Industry.[108,109]
- *Blinding*. Members of the adjudication committee may be blinded as to whether the study subject experiencing an AE was in the study drug group or control group. Also, the members may be blinded as to the identity of the study drug.[110]
- *Appropriate terms for AEs*. Before engaging in adjudication, the terms used to query the Sponsor's database, for detecting study subjects of interest, need to be defined.
- *Diagnosing the AE*. An adjudication committee may evaluate the appropriateness of diagnostic procedures for adverse events.[111] Adjudication may be used to resolve disagreements in diagnosis that use radiology,[112] for diagnosing joint destruction in osteoarthritis,[113] and for diagnosing the etiology of liver injury.[114] The need for a consistent definition of the AE of interest is illustrated by FDA's review of ticagrelor.[115] The need for a consistent definition of the AE of ONJ was an issue in FDA's review of denosumab.[116]

[107] American Academy of Pediatrics (2014) Off-label use of drugs in children. Pediatrics 133:563–7.

[108] U.S. Department of Health and Human Services. Food and Drug Administration. Center for Drug Evaluation and Research (CDER). Center for Biologics Evaluation and Research (CBER). Guidance for industry. Investigational new drug applications prepared and submitted by sponsor-investigators; May 2015 (28 pp.).

[109] U.S. Department of Health and Human Services. Food and Drug Administration. Center for Drug Evaluation and Research (CDER). Center for Biologics Evaluation and Research (CBER). Center for Devices and Radiological Health (CDRH). Guidance for industry investigator responsibilities—protecting the rights, safety, and welfare of study subjects; October 2009 (18 pp.).

[110] Lixisenatide (type-2 diabetes) NDA 208-471, NDA 204-961, NDA 204-961. Pages 105, 245, and 679 of 690-page Medical Review.

[111] de Andrade J, et al. The idiopathic pulmonary fibrosis clinical research network (IPFnet). Diagnostic and adjudication processes. Chest 148:1034–42.

[112] Nivolumab (metastatic melanoma) BLA 125-554. Page 34 of 156-page Medical Review.

[113] FDA Center for Drug Evaluation and Research. Division of Anesthesia, Analgesia, and Addition Products. Background Package Addendum by Hertz S and Fields E; March 2, 2012 (27 pp.).

[114] Dapagliflozin (type-2 diabetes mellitus) NDA 202-293. Pages 13, 50, 76–7, 85–6, 89, 133, 161, and 428–9 of 494-page Medical Review.

[115] Ticagrelor (thrombic events with acute coronary syndromes) NDA 022-433. Page 232–3 of 640-page Medical Review.

[116] Denosumab (post-menopausal osteoporosis) BLA 125-320. Page 232 of 710-page Medical Review.

- *Relatedness*. An adjudication committee may determine relatedness of a given study drug with various AEs. Relatedness analysis by an adjudication committee may use conventional techniques, such as observing imbalance in rate or severity of AEs in the study drug arm versus control arm, or observing temporal relation between administering the drug and materialization of the AE.
- *Structure of relatedness inquiry when faced with confounding factors*. When faced with confounding factors, such as concomitant drugs, or concomitant conditions and disorders, the relatedness inquiry can be facilitated by determining study subjects where drug-induced AEs can be excluded, those for whom drug-induced AEs cannot be excluded, and those with insufficient data to make a determination.[117] FDA's review for ticagrelor revealed the situation where clinical data on the AE were inadequate or entirely nonexistent.[118]
- *Documents needed*. This chapter describes the potential need for an adjudication operations manual, adjudication charter, and adjudication instructions in the Clinical Study Protocol.

[117] Sofosbuvir/velpatasvir combination (hepatitis C virus) NDA 208-341. Page 126 of 171-page Medical Review.

[118] Ticagrelor (thrombic events with acute coronary syndromes) NDA 022-433. Page 529 of 640-page Medical Review.

CHAPTER

13

Coding

I. INTRODUCTION

Adverse events (AEs), as they are recorded on the Case Report Form (CRF), can take the form of observations jotted down by the healthcare worker using verbatim terms, histology data, photographs of skin lesions, electrocardiograms, and so on. The raw data on AEs are converted to a standard system of AE terms by an activity called coding. Converting raw data such as verbatim terms and photographs to standard AE terms is performed before the Sponsor analyzes the AEs and before the Sponsor drafts the Clinical Study Report.

Coding, without more, does not provide any information on the relatedness of the study drug to any given AE. But coding is a necessary step before relatedness analysis can occur, and thus before the AE finds a place in the Black Box Warning, Warnings and Precautions section, and Contraindications section of the draft package label.

An exemplary comment summarizing the physical act of capturing AEs, that is, where verbatim terms are written on the CRF, followed by coding, and the end-result of storing coded AEs in the Sponsor's database, is provided by a comment in one of FDA's Medical Reviews. FDA's comment about all of these activities refers to the time of signing the consent form (an early step in any clinical study) to the very end of the study:

> The applicant states that all AEs, regardless of their seriousness or relationship to the investigational product . . . were collected by open questioning at the time of signing of the informed consent through the end of the study and were recorded in the AE pages of the Case Report Form (CRF) ... [a]ll AEs were recoded in accordance with MedDRA version 14.1 in the integrated safety database.[1]

[1] Lixisenatide (type-2 diabetes) NDA 208-471. The Medical Review contains three Clinical Reviews. Clinical Review for NDS 208-471 (pages 2–154 of the Medical Review); Clinical Review for NDA 204-961 (pages 156–251 of 690-page pdf file Medical Review); and Clinical Review for NDA 204-961 (pages 426–676 of Medical Review). Medical Review is on a 690-page pdf file.

FDA's Drug Review Process and the Package Label
DOI: https://doi.org/10.1016/B978-0-12-814647-7.00013-0

a. Verbatim Terms

Verbatim terms for a given AE are recorded on the study subject's Case Report Form. These terms are then coded to a corresponding term (often a different term than the verbatim term). The coded term is used in documents submitted to the FDA. Specifically, FDA reviewers learn about AEs from the Listings and Clinical Study Reports. Coding is most frequently done with the MedDRA® dictionary.[2] Listings take the form of tables of data without explanations of how the data were collected and interpreted, and without any written analysis. Clinical Study Reports are more like publications in medical journals, in that they contain tables, figures, text, and conclusions.

Regarding verbatim terms, healthcare workers do their best to capture the primary data clearly, completely, and in as natural clinical language as possible.[3] But different sorts of natural clinical language may be used by different healthcare workers to describe the same AE. It is necessary for all coders to use the same medical dictionary for coding verbatim terms. The dictionary most often used in the pharmaceutical industry is Medical Dictionary for Regulatory Activities (MedDRA®). Other coding dictionaries can be used, such as Systematized Nomenclature of Medicine-Clinical Terms (SNOMED-CT). Coding AEs into a standard nomenclature should be done by trained experts to ensure accuracy and consistency. Where a coding dictionary is revised during the course of a clinical trial, the Sponsor may need to redo some of the coding to ensure that all AE terms adhere to the more recent dictionary.

Further regarding verbatim terms, ICH requires that the "MedDRA lowest level term(s) ... that most accurately reflects the reported verbatim information should be selected." ICH warns that even "a single letter difference in a reported verbatim text can impact the meaning of the word and consequently term selection."[4] The following flow charts show examples on the step where a verbatim term is coded to a MedDRA term. Note that the verbatim term "lip sore" has the closest MedDRA term "lip pain," whereas the verbatim term "lip sores" (note plural word, sores) has the closest MedDRA term, "cheilitis."[5]

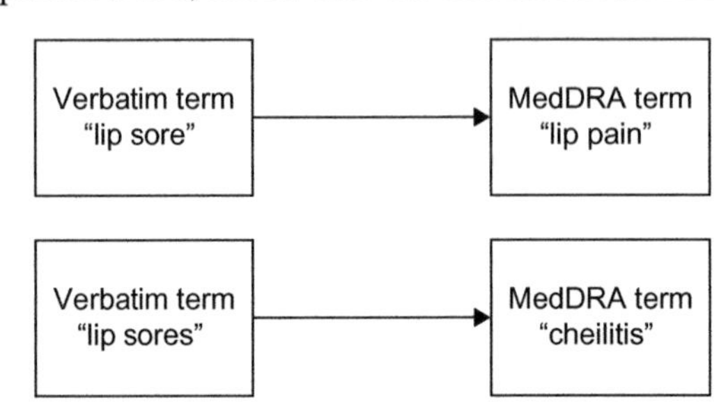

[2] Brown EG, et al. The Medical Dictionary for Regulatory Activities (MedDRA). Drug Saf. 1999;20:109–17.

[3] Gliklich AE, et al. Registries for evaluating patient outcomes: a user's guide. third ed. Rockville, MD: Agency for Healthcare Research and Quality (AHRQ), U.S. Dept. of Health and Human Services (HHS); 2014.

[4] MedDRA MSSO. MedDRA® term selection: points to consider. ICH-Endorsed Guide for MedDRA Users. McLean, VA: MedDRA MSSO. Sept. 2016 (57 pp.).

[5] MedDRA MSSO. MedDRA® term selection: points to consider. ICH-Endorsed Guide for MedDRA Users. McLean, VA: MedDRA MSSO. Sept. 2016 (57 pp.).

The following flow chart reveals the situation where there is a choice in the coding step, where a verbatim term can be coded into one of two different MedDRA terms[6]:

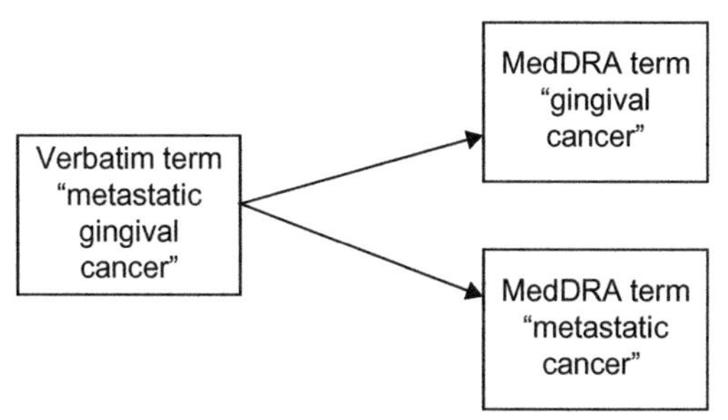

Where MedDRA appears not to have a corresponding Preferred Term, ICH recommends using the closest reasonable Preferred Term from MedDRA. For example, if the verbatim term for an AE is "brittle hair" (this term is not in MedDRA), ICH suggests coding this AE as "hair breakage" (this is in MedDRA). In Europe and Japan, pharmaceutical companies are required to use MedDRA, while in the United States MedDRA is standard but not required.[7]

As part of FDA's review of an NDA or BLA, FDA reviewers may scrutinize every verbatim term to ensure appropriate coding to a corresponding Preferred Term. This practice of manual term-by-term checking is described in FDA's Medical Review for atezolizumab:

> The clinical review of safety assessed the adequacy of the Applicant's mapping of AE verbatim terms to MedDRA Preferred Terms for 100% of the ... raw AE data set. The review used manual matching of all verbatim and MedDRA Preferred Terms to assess the acceptability of the Applicant mapping from the verbatim term to MedDRA Preferred Term. The Preferred Terms listed in the dataset adequately represented the investigator-recorded term and did not raise any significant issues.[8]

b. Devising New Preferred Terms

MedDRA's "Term Selection: Points to Consider" describes the Sponsor's ability to add new terms to the MedDRA dictionary, by way of the change request process: "When a specific medical concept is not represented by a single MedDRA term, consider requesting a new term through the change request process."[9]

[6] MedDRA MSSO. MedDRA® term selection: points to consider. ICH-Endorsed Guide for MedDRA Users. McLean, VA: MedDRA MSSO. Sept. 2016 (57 pp.).

[7] International Conference on Harmonisation of Technical Requirements for Registration of Pharmaceuticals for Human Use. Understanding MedDRA The Medical Dictionary for Regulatory Activities. 2013 (20 pp.).

[8] Atezolizumab (urothelial carcinoma) BLA 761-034. Page 68 of 125-page Medical Review.

[9] MedDRA MSSO. MedDRA® term selection: points to consider. ICH-Endorsed Guide for MedDRA Users. McLean, VA: MedDRA MSSO. Sept. 2016 (57 pp.).

MedDRA's Subscriber Portfolio describes the change request process as "change requests ... allow MedDRA subscribers to affect the ... content of the terminology. Simple change requests are changes affecting the Preferred Term and Lowest Level Term (LLT) levels of the MedDRA terminology ... [a]ll requests will be compared to the currently released version of MedDRA and to changes already accepted for the next release of MedDRA."[10]

FDA's Medical Review for simeprevir reveals the situation where the Sponsor had devised a new Preferred Term (though the Medical Review does not mention if the Sponsor had submitted any change request). FDA observed that "the ... term 'rash' also includes the MedDRA Preferred Terms that comprise the 'photosensitivity' grouped variable. In order to also assess 'rash' in the absence of the photosensitivity related Preferred Terms, **a new variable termed 'rash excluding photosensitivity' was created**. 'Rash excluding photosensitivity' occurred in 25% of subjects in the simeprevir group and 19% of subjects in the Control group during the first 12 weeks of treatment."[11]

c. Remapping

Remapping is a concept similar to that where the Sponsor proposes to add a new term to the MedDRA dictionary. Remapping is illustrated by FDA's Medical Review of brivaracetam. As stated by FDA, the Sponsor's goal in remapping was to allow easier assessment of a particular safety concern:

> the Applicant performed some **modifications to the MedDRA classification** ... [t]hese modifications (presented in the individual CSRs) included the **remapping** of some Preferred Terms to a different SOC (named "UCB SOC") than the automatically assigned primary MedDRA SOC, to "more appropriately grouped Preferred Terms to allow easier assessment of a particular safety concern" (e.g., Preferred Term Abnormal ECG was moved from the SOC Investigations to the SOC Cardiac disorders). Furthermore, MedDRA Preferred Terms were assigned to "UCB grouping terms" (e.g., the UCB SOC of "Eye disorders" was further classified into the UCB grouping terms of "Other eye disorders" and "Vision disorders").[12]

The term "UCB" as it occurs in the System Organ Class (SOC) called "UCB SOC" refers to the name of the Sponsor, UCB, Inc. UCB, Inc. is a pharmaceutical company that develops drugs for neurological diseases, including epilepsy.

d. Ambiguous Verbatim Terms

ICH MedDRA provides examples of ambiguous verbatim terms, such as "Turned green" and "His glucose was 40." ICH MedDRA states that "Turned green" is ambiguous because "Turned green reported alone is vague; this could refer to a patient condition or ... even to

[10] TRW, Inc. MedDRA. Subscriber Portfolio. 2002 (19 pp.).

[11] Simeprevir (hepatitis C virus) NDA 205-123. Page 101 of 173-page Medical Review.

[12] Brivaracetam (partial onset seizures) NDA 205-836, NDA 205-837, NDA 205-838. Page 193 of 383-page Medical Review.

a product, e.g., pills."[13] The verbatim term "His glucose was 40" is ambiguous, because it fails to give any unit for glucose concentration, such as millimolar (mM) or mg/L.[14]

e. Other Coding Dictionaries

In addition to MedDRA, other AE dictionaries are available, such as Systemized Nomenclature of Medical Clinical Terms (SNOMED-CT),[15,16,17] WHO-ART,[18,19] COSTART,[20,21,22,23] ADROIT,[24] International Classification of Diseases (ICD),[25] Rheumatology Common Toxicity Criteria,[26] and Oncology Common Toxicity Criteria.[27] MedDRA is distinguished in that it groups AEs by SOC.[28,29]

[13] MedDRA MSSO. MedDRA® term selection: points to consider. ICH-Endorsed Guide for MedDRA Users. McLean, VA: MedDRA MSSO. Sept. 2016 (57 pp.).

[14] MedDRA MSSO. MedDRA® term selection: points to consider. ICH-Endorsed Guide for MedDRA Users. McLean, VA: MedDRA MSSO. Sept. 2016 (57 pp.).

[15] Alecu I, et al. A case report: using SNOMED CT for grouping Adverse Drug Reactions Terms. BMC Med. Inform. Decis. Mak. 2008;8 (Suppl. 1):54. DOI:10.1186.

[16] Richesson RL. Heterogeneous but "standard" coding systems for adverse events: issues in achieving interoperability between apples and oranges. Contemp. Clin. Trials. 2008;29:635–45.

[17] Bodenreider O. Using SNOMED CT in combination with MedDRA for reporting signal detection and adverse drug reactions reporting. AMIA 2009 Symposium Proceedings. 2009. Pp. 45–9.

[18] Alecu I, et al. A case report: using SNOMED CT for grouping Adverse Drug Reactions Terms. BMC Med. Inform. Decis. Mak. 2008;8(Suppl. 1):54. DOI:10.1186.

[19] Belknap SM, et al. Quality of methods for assessing and reporting serious adverse events in clinical trials of cancer drugs. Clin. Pharmacol. Ther. 2010;88:231–6.

[20] Varricchio F, et al. Understanding vaccine safety information from the Vaccine Adverse Event Reporting System. Pediatr. Infect. Dis. J. 2004;23:287–94.

[21] Zhou W, et al. A potential signal of Bell's palsy after parenteral inactivated influenza vaccines: reports to the Vaccine Adverse Event Reporting System (VAERS)--United States, 1991-2001. Pharmacoepidemiol. Drug Saf. 2004;13:505–10.

[22] Belknap SM, et al. Quality of methods for assessing and reporting serious adverse events in clinical trials of cancer drugs. Clin. Pharmacol. Ther. 2010;88:231–6.

[23] Brown EG, et al. The Medical Dictionary for Regulatory Activities (MedDRA). Drug Saf. 1999;20:109–17.

[24] Belknap SM, et al. Quality of methods for assessing and reporting serious adverse events in clinical trials of cancer drugs. Clin. Pharmacol. Ther. 2010;88:231–6.

[25] Brown EG, et al. The Medical Dictionary for Regulatory Activities (MedDRA). Drug Saf. 1999;20:109–17.

[26] Woodworth T, et al. Standardizing assessment and reporting of adverse effects in rheumatology clinical trials II: the Rheumatology Common Toxicity Criteria v.2.0. J. Rheumatol. 2007;34:1401–14.

[27] Woodworth T, et al. Standardizing assessment and reporting of adverse effects in rheumatology clinical trials II: the Rheumatology Common Toxicity Criteria v.2.0. J. Rheumatol. 2007;34:1401–14.

[28] Bousquet C, et al. Appraisal of the MedDRA conceptual structure for describing and grouping adverse drug reactions. Drug Saf. 2005;28:19–34.

[29] Bodenreider O. Using SNOMED CT in combination with MedDRA for reporting signal detection and adverse drug reactions reporting. AMIA 2009 Symposium Proceedings. 2009. pp. 45–9.

f. Grading Adverse Events Versus Coding Adverse Events

In AE reporting, coding AEs is a different activity from grading AEs. Grading any given AE is an option. CTCAE provides a grading scale for use in capturing the **severity** of AEs.[30] CTCAE states that "A grading (severity) scale is provided for each AE term."[31] The MedDRA[32] scheme for AE classifications does ***not take into account AE severity***. Regarding lack of severity distinctions, MedDRA states that "Descriptors of severity are not included in the terminology. Descriptors such as 'severe' and 'mild' are used only when pertinent to the specificity of the term (e.g., severe vs. mild mental retardation)."

CTCAE also refers to the MedDRA classification, in its commentary on SOC. The fact that CTCAE refers to MedDRA in this way demonstrates that the two dictionaries (MedDRA; CTCAE) are compatible with each other: "System Organ Class, the highest level of the MedDRA hierarchy, is identified by anatomical or physiological system, etiology, or purpose … CTCAE terms are grouped by MedDRA Primary SOCs. Within each SOC, AEs are listed and accompanied by descriptions of severity (Grade)."[33]

An excerpt from FDA's Clinical Review on an anticancer drug further reveals that these two dictionaries are compatible with each other and both can be used for AE analysis. The excerpt, which is from FDA's review of bosutinib, is:

> Adverse events (AEs) were coded using Medical Dictionary for Regulatory Activities (MedDRA) v. 14. AEs were summarized by MedDRA system organ class (SOC) and Preferred Term. AEs and SAEs were to be graded according to NCI-CTCAE version 3.0.[34]

g. System Organ Class (SOC)

Schroll et al.[35] reveals how SOC fits into MedDRA's scheme of naming Lowest Level Terms (LLTs), Preferred Terms, and SOC terms:

> MedDRA is a … hierarchy with Lowest Level Terms at the bottom, followed by Preferred Terms, and with System Organ Class (SOC) at the top … [e]vents are initially coded with Lowest Level Terms and they consist of thousands of synonyms … Preferred Terms are unique medical entities. Companies are not allowed to add new terms but can suggest new terms … which will then be considered for the biannual update.[36]

[30] Common Terminology Criteria for Adverse Events (CTCAE) Version 4.0. U.S. Dept. of Health and Human Services. National Institutes of Health National Cancer Institute. May 28, 2009 (v4.03: June 14, 2010) (195 pp.).

[31] Common Terminology Criteria for Adverse Events (CTCAE) Version 4.0. U.S. Dept. of Health and Human Services. National Institutes of Health National Cancer Institute. May 28, 2009 (v4.03: June 14, 2010) (195 pp.).

[32] MedDRA MSSO. Introductory Guide MedDRA Version 14.0. McLean, VA: MedDRA MSSO. 2011 (77 pp.).

[33] Common Terminology Criteria for Adverse Events (CTCAE) Version 4.0. U.S. Dept. of Health and Human Services. National Institutes of Health National Cancer Institute. May 28, 2009 (v4.03: June 14, 2010) (195 pp.).

[34] Bosutinib (chronic myeloid leukemia) NDA 203-341. Page 55 of 74-page Clinical Review. 81-page pdf file for Medical Review.

[35] Schroll JB, et al. Challenges in coding adverse events in clinical trials: a systematic review. PLoS One. 2012;7:e41174 (7 pp.).

[36] Schroll JB, et al. Challenges in coding adverse events in clinical trials: a systematic review. PLoS One. 2012;7:e41174 (7 pp.).

The MedDRA dictionary provides names for AEs that can designate the AE by LLT, Preferred Term, and SOC. SOC terms are broad terms, while Preferred Terms are more specific, and LLT are the most specific. Each SOC encompasses many Preferred Terms. Instructions from MedDRA warn about double counting and provide guidance on avoiding double counting. To this end, MedDRA states that after detecting an AE, capturing it by way of a verbatim term on the CRF and followed by coding of this AE into a Preferred Term, the Sponsor must take care to record this particular AE under only one SOC (and not under a multiplicity of different SOCs). In MedDRA's own words:

> To avoid **double counting** while retrieving information from all SOCs, each Preferred Term is assigned a **primary SOC**. This is required because Preferred Terms can be represented in more than one SOC (multi-axiality). It prevents an individual Preferred Term from being displayed more than once in cumulative SOC-by-SOC data outputs, which would result in **over-counting** of terms.[37]

The above excerpt refers to a "primary System Organ Class." This refers to the fact that, when using MedDRA, any given Preferred Term may be linked to more than one SOC and that any given MedDRA term may be attributed to more than one SOCs.[38] Du et al.[39] warned of this option, referring to the ability of any given Preferred Term to fall into more than one SOC, and of the need to designate a **primary SOC** where the goal is to avoid double counting. On this point, Du et al.[40] stated that "SOC is the highest level of the MedDRA terms, which comprises grouping by etiology, manifestation site or purpose ... each Preferred Term is linked to at least one SOC ... [t]o avoid **double counting**, we will need to identify the primary SOC for each term."

h. Listings

Clinical Study Reports include a title, narratives, tables, figures, and about 20 appendices, where one of the appendices takes the form of listings. The listings concern various topics, such as demographics, subject disposition, preexisting medical conditions, AEs, hematology, blood chemistry values, and electrocardiogram (ECG) data.[41] For each of these topics, the listing discloses the subject number, subject gender, and treatment.

The demographics information in the listings discloses gender, race, date of birth, weight, height, and body mass index. The disposition information in the listings states

[37] MedDRA MSSO. Introductory Guide MedDRA Version 14.0. McLean, VA: MedDRA MSSO. 2011 (77 pp.).

[38] van Hunsel F, et al. Comparing patients' and healthcare professionals' ADR reports after media attention. Br. J. Clin. Pharmacol. 2009;6:558–64.

[39] Du J, et al. Trivalent influenza vaccine adverse symptoms analysis based on MedDRA terminology using VAERS data in 2011. J. Biomed. Semant. 2016;7:13–19.

[40] Du J, et al. Trivalent influenza vaccine adverse symptoms analysis based on MedDRA terminology using VAERS data in 2011. J. Biomed. Semant. 2016;7:13–19.

[41] Maund E, et al. Benefits and harms in clinical trials of duloxetine for treatment of major depressive disorder: comparison of clinical study reports, trial registries, and publications. Br. Med. J. 2014;348:g3510. DOI: 10.1136/bmj.g3510.

whether or not the subject completed the study, date the subject left the study, and reason why the subject left the study, for example, where the reason was that the subject experienced an AE such as leucopenia, nausea, elevated serum aminotransferase, or papular rash. The treatment is whether the subject received the study drug or placebo, and the number of milligrams of study drug. The hematology values can include hematocrit, red blood cells, white blood cells, platelets, neutrophils, lymphocytes, monocytes, eosinophils, and basophils. The blood chemistry values in the listings can include time of blood withdrawal and values for sodium, potassium, urea, creatinine, lactate dehydrogenase, creatinine phosphokinase, and C-reactive protein.

Table 13.1 provides an excerpt of adverse events information in a listing. The listing is from a clinical trial on a small molecule drug for treating a platelet disorder. As can be seen, the listing discloses the subject number, treatment, name of AE (verbatim term, lowest-level term, Preferred Term, and SOC term), and other types of information. The purpose in reproducing this excerpt of the listing is to demonstrate that the MedDRA terms, "low-level term," "Preferred Term," and "System Organ Class," do not remain hidden as part of the Sponsor's database of AEs, but may all be used by the medical writer in drafting the Clinical Study Report.

II. CODING EXAMPLES FROM FDA'S MEDICAL REVIEWS

a. Denosumab (Postmenopausal Osteoporosis) BLA 125-320

This concerns FDA's review of denosumab for the indication of postmenopausal osteoporosis.[42] Denosumab is an antibody that binds to RANK ligand, resulting in suppressed osteoclast function and reduced bone resorption.

i. Quality Control Recommended by FDA's Guidance for Industry

FDA's Guidance for Industry suggests that the Sponsor should conduct quality control on its coding process:

> One reason to review the details of individual cases is to determine whether the event was coded to the correct preferred term ... [a] case might be incorrectly included in the numerator of a rate calculation if the event is incorrectly coded to a specific preferred term. Events may be incorrectly coded to preferred term by the applicant when they summarize the data or because an **investigator used a verbatim term incorrectly when recording the event in the case report form**. An example of incorrect coding would be if an investigator used the verbatim term acute liver failure for a case of increased ALT and the applicant coded the event to acute liver failure.[43]

[42] Denosumab (post-menopausal osteoporosis) BLA 125-320. 710-page Medical Review.

[43] U.S. Department of Health and Human Services. Food and Drug Administration. Center for Drug Evaluation and Research (CDER). Reviewer guidance conducting a clinical safety review of a new product application and preparing a report on the review. 2005 (79 pp.).

TABLE 13.1 Format of Adverse Events as They Occur in the Listings

Subject Number	Treatment (Study Drug/Placebo)	Adverse Event (Verbatim Term)	Low-Level Term	Preferred Term	System Organ Class	Start Date	Duration (hours: minutes)	Doses so far
15	Placebo	Cannula site reaction	Procedural site reaction	Procedural site reaction	Injury, poisoning, and procedural complications	Sept. 16, 2015	1:30	18
16	200 mg drug	Cough	Cough	Cough	Respiratory, thoracic, and mediastinal disorders	Sept. 28, 2015	Unavailable	Unavailable
17	200 mg drug	Papular rash	Papular rash	Rash papular	Skin and subcutaneous tissue disorders	Oct. 07, 2015	Unavailable	30
18	200 mg drug	Intermittent dizziness	Dizziness	Dizziness	Nervous system disorders	Sept. 09, 2015	6:35	4
		Breathlessness	Breathlessness	Dyspnea	Respiratory, thoracic, and mediastinal disorders	Sept. 09, 2015	6:35	4
		Loose stools	Loose stools	Diarrhea	Gastrointestinal disorders	Sept. 13, 2015	7:00	10

ALT refers to the value for serum alanine aminotransferase activity, which is an indicator of mild-to-severe liver damage. Liver damage results in increased leakage of this enzyme into the bloodstream.

ii. Quality Control of Adverse Event Coding for Denosumab

The steps used by the Sponsor for collecting raw data, coding the data, and conferring with FDA regarding the acceptability of the coding are described in FDA's Medical Review. The fact that quality control was used is demonstrated by the phrase "to verify the precision and accuracy." The term ONJ refers to one of the AEs of concern in the clinical study:

> The Applicant collected adverse event information ... [i]n addition, hematology, serum chemistry, electrocardiogram ... were assessed at regular intervals ... [c]ases involving specific safety issues of interest, that is, ... ONJ ... were reviewed by an adjudication committee. All adverse events were coded using ... MedDRA and the submitted application was coded in MedDRA version 11.0. Adverse events were graded on a five-point intensity scale: mild, moderate, severe, life-threatening, or fatal [CTCAE Criteria] ... [a] **comparison of verbatim term to MedDRA Preferred Term was performed on a random sampling of adverse event terms to verify the precision and accuracy of the medical coding** of adverse events.[44]

FDA's Medical Review stated that the disagreements expressed by FDA reviewers were only slight and inconsequential: "The coding was considered acceptable in the majority of events reviewed. In those cases where the FDA reviewers would have chosen different preferred terms, the differences were not clinically meaningful and would not have had a significant impact on the overall assessment of safety for the trial."[45]

iii. MedDRA's System Organ Class (SOC) Finds Use in Adjudication of AE of Death

This shows how one particular AE (death) can be classified by SOC. Assessing whether death was due to the AE of neoplasms, cardiac AEs, nervous system AEs, and so on was determined separately by the Sponsor and by FDA. Determination was by adjudication, as indicated in Table 13.2. The first column in the table contains names for various SOCs.

The FDA reviewer observed that, "Across the ... denosumab clinical ... program, the cause of death did not differ greatly between the denosumab or placebo groups, by either adjudication. However, there were more fatal vascular events in the denosumab group, **primarily due to hemorrhage**. The majority of these patients had risk factors for the event ... history of varices ... [t]his reviewer agrees ... [that] increased number of fatal **hemorrhages** in the denosumab group appears to be due to underlying risk factors."[46]

The practical outcome of the presentation of the AE in question (death) by way of SOC was the conclusion that increased risk on the denosumab group was due not to

[44] Denosumab (post-menopausal osteoporosis) BLA 125-320, BLA 125-331. Clinical Review (pages 75–6 of 91-page Clinical Review).

[45] Denosumab (post-menopausal osteoporosis) BLA 125-320, BLA 125-331. Clinical Review (pages 75–6 of 91-page Clinical Review).

[46] Denosumab (post-menopausal osteoporosis) BLA 125-320. Pages 335–6 of 710-page Medical Review.

TABLE 13.2 Cause of Death by System Organ Class (SOC) for the Entire Denosumab Clinical Development Program

	Applicant Adjudication		FDA Adjudication	
	Placebo	Denosumab	Placebo	Denosumab
SOC	Number of Deaths and Percent			
Neoplasms	37 (27%)	54 (27.6%)	37 (27%)	82 (41.8%)
Cardiac	36 (26.3%)	39 (19.9%)	27 (19.7%)	26 (13.3%)
Nervous system	14 (10.2%)	19 (9.7%)	15 (11%)	16 (8.2%)
Respiratory	13 (9.5%)	18 (9.2%)	10 (7.3%)	8 (4.1%)
Infections and infestations	12 (8.8%)	14 (7.1%)	13 (9.5%)	17 (8.7%)
Metabolism and nutrition	1 (0.7%)	9 (4.6%)	2 (1.5%)	2 (1%)
Hepatobiliary	1 (0.7%)	7 (3.6%)	1 (0.7%)	2 (1%)
Gastrointestinal	3 (2.2%)	5 (2.6%)	5 (3.7%)	6 (3.1%)
Blood and lymphatics	0	2 (1%)	0	2 (1%)
Renal and urinary	1 (0.7%)	2 (1%)	0	3 (1.5%)

Denosumab (post-menopausal osteoporosis) BLA 125-320. Pages 335–6 of 710-page Medical Review.

drug-related AEs, but mostly to underlying risk factors such as hemorrhage. In detail, the take-home lesson seems to be as follows. If the Sponsor detects an underlying, concomitant disorder in a subgroup of the population of study subjects, the Sponsor should consider performing the following analysis. The example of denosumab was with the underlying disorder of "history of varices." The fact that some of the AEs detected in the Sponsor's clinical study found a place in the SOC of "Cardiac," makes it easy for the Sponsor to downgrade all of these AEs as being not likely related to denosumab, for the subjects in the Cardiac SOC group that suffered from "history of varices."

b. Eslicarbazepine (Partial Onset Seizures) NDA 022-416

This illustrates the problem where splitting can result in underestimation of a particular AE. FDA noticed that several closely related AEs, during the process of coding, ended up contributing to AEs in different SOCs. These SOCs were somewhat disparate from each other. FDA complained that "However, there were instances where the MedDRA coding process resulted in splitting likely related AEs into separate SOCs, leading to an underestimation of the true incidence for a particular event or syndrome."[47]

FDA's implication was that all of the AEs could very well have been the same type of AE. The AEs were initially coded as the MedDRA Preferred Terms, as shown below.

[47] Eslicarbazepine (partial onset seizures) NDA 022-416. Page 158 of 620-page Medical Review.

During the coding process, these Preferred Terms were pigeon-holed to the indicated SOC. As is evident from the bulletpoint list, each of these AEs, which could have been the same kind or a closely related AE, found its way into a different SOC:

- "Gait disturbance" (SOC of General disorders)
- "Confusional state" (SOC of Psychiatric disorders)
- "Vertigo" (SOC of Ear/labyrinth disorders)
- "Ataxia and dizziness" (SOC of Nervous system disorders)

FDA's solution to this problem was to suggest that all of these AEs be pigeonholed into the SOC of Nervous system disorders. FDA's suggestion to this effect can be found where the FDA reviewer used the phrase "grouped together within the SOC Nervous system disorders."[48] The meaning of this phrase is more evident from a more complete quotation:

> After reviewing the analysis AE dataset ... to assess the coding of the verbatim terms to the MedDRA preferred terms, the coding process overall seemed appropriate and allowed for reliable estimates of AE risks. However, there were instances where the **MedDRA coding process resulted in splitting likely related AEs into separate SOCs** leading to an underestimation of the true incidence for a particular event or syndrome. For example, the MedDRA Preferred Term, gait disturbance, was coded only under the primary ***SOC of General disorders, administration site conditions*** which provided less precise information than the secondary ***SOC of Nervous system disorders***. Other preferred terms that described similar symptoms were also coded to other primary SOCs instead of **grouped together within the *SOC Nervous system disorders*** ... [t]herefore, in order to account for the splitting of the preferred terms into different system organ classes, additional analyses were performed by the reviewer ... to group these preferred terms across SOCs to provide more accurate estimates of adverse event syndromes.[49]

c. Hydroxybutyrate (Cataplexy in Narcolepsy; Excessive Daytime Sleepiness in Narcolepsy) NDA 021-196

This concerns FDA's review for a drug with psychiatric AEs. There were many mismatches between verbatim terms and Preferred Terms. The COSTART (Coding Symbols for a Thesaurus of Adverse Reaction Terms) dictionary was used for coding, not the more frequently used MedDRA dictionary.

Coding using either the MedDRA or COSTART dictionary results in conversion of verbatim terms to "preferred terms," as is evident from comments by Labenz et al. that "the data were coded using the MedDRA terminology and stored by the FDA in the Adverse Event Reporting System ... FDA provided one-to-one mapping from COSTART preferred terms to MedDRA preferred terms. This was necessary to produce summary frequency

[48] Eslicarbazepine (partial onset seizures) NDA 022-416. Page 158 of 620-page Medical Review.

[49] Eslicarbazepine (partial onset seizures) NDA 022-416. Page 158 of 620-page Medical Review.

TABLE 13.3 Verbatim Terms and Preferred Terms from Clinical Study of 4-Hydroxybutyrate

Verbatim Term	Preferred Term
"Lethargy"	**Somnolence**
"Paranoia"	Paranoid reaction
"Felt drunk"	Stupor
"Heart ache"	Emotional lability
"Chest pressure"	Pain chest
"Down in the dumps"	Depression
"Poor concentration"	Thinking abnormal
"Intoxicated feeling"	Stupor
"Difficulty breathing"	Dyspnea
"Increased sleepiness"	**Somnolence**
"Restless leg increased"	Hyperkinesia
"Itching of extremities"	Pruritis
"Phlegm knot in throat"	Pharyngitis
"Laughing continuously"	Emotional lability
"Finding fault with everything"	Emotional lability
"Patient lost bowel control while asleep"	Incontinence fecal
"Bit tongue due to falling faster to ground"	Convulsion
"Hit temple against furniture due to falling faster to ground"	Convulsion

4-Hydroxybutyrate (cataplexy in narcolepsy; excessive daytime sleepiness in narcolepsy) NDA 021-196. Pages 25–27 of 41-page pdf file (14th pdf file of 15 pdf files for Medical Review). Pages 4–6 and 15 of 41-page pdf file (15th pdf file of 15 pdf files in Medical Review). Also, page 8 of 41-page pdf file (10th pdf file of 15 pdf files in Medical Review).

tables in the MedDRA format."[50] Other researchers also refer to "preferred terms" in describing the COSTART dictionary.[51,52]

Although many of the verbatim terms used by the Sponsor were exactly the same as the Preferred Terms, Table 13.3 illustrates that there existed differences between verbatim terms and Preferred Terms. Table 13.3 reveals the phenomenon of lumping, where both "lethargy"

[50] Labenz J, et al. A summary of Food and Drug Administration-reported adverse events and drug interactions occurring during therapy with omeprazole, lansoprazole and pantoprazole. Aliment. Pharmacol. Ther. 2003;17:1015–9.

[51] Martin-Mola E, et al. Sustained efficacy and safety, including patient-reported outcomes, with etanercept treatment over 5 years in patients with ankylosing spondylitis. Clin. Exp. Rheumatol. 2010;28:238–45.

[52] Chan AT, et al. Multicenter, phase II study of cetuximab in combination with carboplatin in patients with recurrent or metastatic nasopharyngeal carcinoma. J. Clin. Oncol. 2005;23:3568–76.

and "increased sleepiness" are coded to "somnolence" (highlighted in **bold font**). Table 13.3 also reveals verbatim terms that might seem informal, for example, "drugged feeling" and "down in the dumps." Moreover, Table 13.3 reveals use of verbatim terms that take the form of small narratives, as with "hit temple against furniture due to falling faster to ground."

FDA commented on an ambiguity inherent in capturing AEs taking the form of undesired mental states. The issue is whether the Sponsor should examine the subject's mental status or, alternatively, infer the subject's mental state from the symptoms of the undesired mental state. The patient narrative read:

> Patient 01-251. This 65 year old man with narcolepsy and sleep apnea was taking a nightly dose of 7.5 grams when 2 months after study entry, he reportedly was feeling "drunk," confused and unsteady ... [o] ther adverse events noted during the study included "sleep walking," "feelings of shakiness," "an upset stomach," "felt like he was on a drug binge," "interrupted breathing," and "dry heaves."[53]

Regarding this study subject, FDA commented, "As contemporaneous formal mental status examinations were not carried out in patients with 'confusion,' it is unclear if any patients coded as having this adverse event were really confused ... [t]his adverse event appears to have been recorded largely, if not entirely, on patients' symptoms. Nevertheless, 'confusion' as it pertains to these patients and other associated symptoms, e.g., unsteadiness, are not unexpected with a drug that has sedative properties."[54]

FDA compares the CRFs with tables submitted in the Sponsor's NDA submission, as revealed by the following excerpt. First, the FDA reviewer described a table that the Sponsor had included in the NDA submission: "The table provides the following data ... patient ID#, sex, age, dosage at onset of confusion, start and stop date for adverse event, investigator term [verbatim term], whether serious or not, frequency, relationship to study drug, and relevant medical history."[55]

Then, the FDA reviewer complained about a discrepancy between a CRF and the table: "It is unclear what the term 'heart aches' coded as 'emotional lability' refers to. It could mean periods of depressed mood rather than emotional lability (she had a previous history of depression and melancholia). The Case Report Form for this patient does not list this adverse event at all."[56]

The take-home lesson is that FDA's scrutiny of coding can include a review of the verbatim terms on the CRFs.

[53] 4-Hydroxybutyrate (cataplexy in narcolepsy; excessive daytime sleepiness in narcolepsy) NDA 021-196. Page 23 of 41-page pdf file (14th pdf file of 15 pdf files of Medical Review).

[54] 4-Hydroxybutyrate (cataplexy in narcolepsy; excessive daytime sleepiness in narcolepsy) NDA 021-196. Page 23 of 41-page pdf file (14th pdf file of 15 pdf files of Medical Review).

[55] 4-Hydroxybutyrate (cataplexy in narcolepsy; excessive daytime sleepiness in narcolepsy) NDA 021-196. Pages 25–7 of 41-page pdf file (14th pdf file of 15 pdf files of Medical Review).

[56] 4-Hydroxybutyrate (cataplexy in narcolepsy; excessive daytime sleepiness in narcolepsy) NDA 021-196. Pages 25–7 of 41-page pdf file (14th pdf file of 15 pdf files of Medical Review).

d. Ponatinib (CP-CML; AP-CML, BP-CML; Ph + ALL) NDA 203-469

This concerns pooling of data from a multiplicity of clinical studies and reveals the situation where recoding might be needed. Pooling data are used where there is a need to acquire conclusions of greater statistical significance. FDA's Medical Review revealed the vexing problem, where coding in one clinical study used one version of the MedDRA dictionary, while coding in another clinical study used another version of the MedDRA dictionary. Another problem was different versions of the CTCAE dictionary. The FDA reviewer described the problem and a solution. The solution was to recode the AEs using the more recent version of the MedDRA dictionary. However, the Sponsor was not able to perform recoding according to the more recent CTCAE dictionary:

> Adverse events were coded using the Medical Dictionary for Regulatory Activities (**MedDRA version 13.0**) AE coding system for purposes of summarization. AEs were graded according to the National Cancer Institute Common Terminology Criteria for Adverse Events (NCI-CTCAE, v. 4.0) ... [p]ooled analysis of the adverse events in Study 07-101 and 10-201 was performed to increase the sensitivity of detecting the adverse events. The Applicant had recoded the adverse events in both Study 07-101 and Study 10-201 to **MedDRA version 15.0**. The Applicant was not able to recode CTCAE grading to the same scale due to inherent differences in versions of the CTCAE versions.[57]

e. Simeprevir (Hepatitis C Virus) NDA 205-123

FDA complained about coding in a clinical study on simeprevir, a drug for hepatitis C virus infections. This concerns the coding of a verbatim term "sunburn" which resulted in coding into two different SOCs. Also, this concerns the choice of Preferred Terms, with the coding of verbatim terms into one of two different Preferred Terms and where the identical verbatim term ("severe photosensitivity") was coded into two different Preferred Terms.

The first complaint concerned a decision tree during the coding process. The source of FDA's complaint, apparently, was coding the verbatim term of "photosensitivity" under the SOC of ***Skin and Subcutaneous Disorders SOC*** or under the ***Injury, Poisoning, and Procedural Complications SOC*** (see flow chart below). FDA wrote, "Events suggesting photosensitivity were not limited to the Skin and subcutaneous disorders system organ class (SOC), e.g. 'sunburn' is coded under Injury, poisoning and procedural complications."

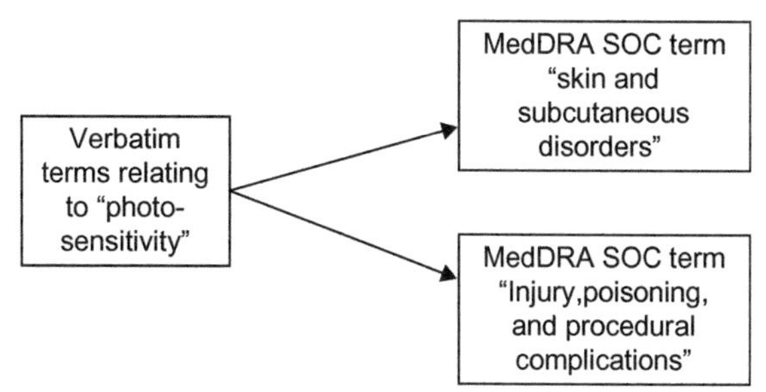

Similarly, FDA complained about the Sponsor's inconsistent behavior in using both arms of the following decision tree for coding the primary data into both of two different Preferred Terms (see flow chart below). The primary data took the form of photographs.

[57] Ponatinib (CP-CML; AP-CML, BP-CML; Ph + ALL) NDA 203-469. Page 61 of 108-page Medical Review.

The FDA reviewer complained, "For example, the cutaneous adverse event for subject C206-0292 was coded as 'drug eruption,' and the reaction for subject C208-0243 was coded as 'rash.' However, review of the photographs for these two subjects suggests that they both suffered severe photosensitivity reactions. Therefore, photosensitivity events may have been underreported."[58]

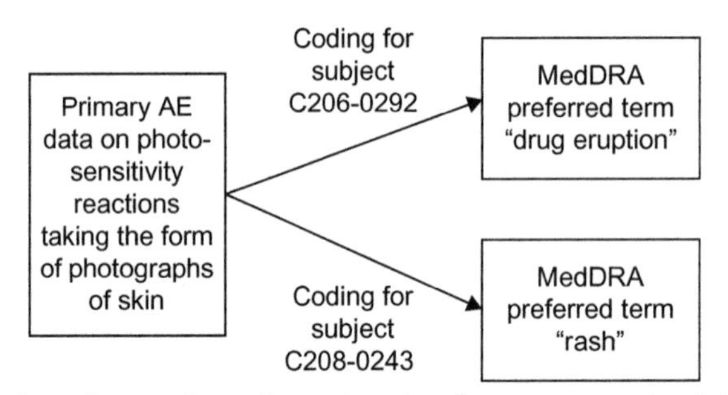

FDA complained that the coding done by the Sponsor resulted in underreporting of photosensitivity reaction and also that the Sponsor's coding of this same reaction as either "drug eruption" or as "rash" had the unfortunate result of diluting the safety signal of this event. In the table, the reason for highlighting the terms, "rash," "photosensitivity reaction," and "drug eruption," is to show how reporting the same type of AE by one of three different terms (a different term, depending on the study subject being analyzed), can result in dilution of the signal and under-reporting of the same type of AE. Table 13.4, which is from FDA's Medical Review, shows the existence of the Preferred Terms, rash, photosensitivity reaction, psoriasis, and drug eruption.[59]

i. Example of Incorrect Coding

Regarding one of the photographs of a "rash," FDA examined the photographs of cutaneous AEs from study subjects and found that the Sponsor had incorrectly identified the AE as "rash" and the FDA reviewer implied that the correct Preferred Term was "herpes simplex virus infection." The FDA reviewer wrote, "the cutaneous adverse event for subject C206-0426 was coded as 'rash.' However, photographs of the subject appear to reveal a single cluster of vesiculopustular lesions on an erythematous base ... and suggest a herpes simplex virus infection."[60]

ii. An Unexpected Type of Adverse Event Can Arouse Suspicion of Incorrect Coding (Example of Psoriasis from Simeprevir)

The FDA reviewer also complained about the large number of psoriasis AEs in the simeprevir group (11 subjects) as compared to the placebo group (two subjects). This difference raised doubts in the reviewer's mind as to whether the AE was really psoriasis,

[58] Simeprevir (hepatitis C virus) NDA 205-123. Pages 4–5 of 173-page Medical Review.

[59] Simeprevir (hepatitis C virus) NDA 205-123. Pages 17–18 of 173-page Medical Review.

[60] Simeprevir (hepatitis C virus) NDA 205-123. Pages 17–18 of 173-page Medical Review.

TABLE 13.4 Preferred Terms That Were Fit into the SOC of Skin and Subcutaneous Tissue Disorders

	Placebo Arm		Simeprevir Arm (150 mg)	
Preferred Term	**No. of Subjects**	**% of Subjects**	**No. of Subjects**	**% of Subjects**
Pruritus	92	23.2	203	26.0
Rash	**64**	**16.1**	**139**	**17.8**
Erythema	15	3.8	29	3.7
Photosensitivity reaction	**2**	**0.5**	**24**	**3.1**
Rash macular	5	1.3	11	1.4
Psoriasis	2	0.5	11	1.4
Rash papular	8	2.0	8	1.0
Rash erythematous	4	1.0	7	0.9
Rash pruritic	1	0.3	7	0.9
Drug eruption	**1**	**0.3**	**2**	**0.3**

Simeprevir (hepatitis C virus) NDA 205-123. Pages 17–18 of 173-page Medical Review.

and the reviewer believed that some of the AEs were really some kind of rash. FDA complained that, "The differential in the frequency of reports of psoriasis between the simeprevir and placebo treatment groups (1.2% and 0.3%, respectively) raises a question of whether these events truly capture the condition of psoriasis or whether some of the events represent a rash (of some sort) that was coded as psoriasis."[61]

FDA went a step forward and recommended that the package label include warnings about rash and photosensitivity:

> A clear signal for photosensitivity with simeprevir use has been identified ... [t]hat risk may be mitigated by **advisement in the label of photoprotective measures** ... manifestations of photosensitivity may range in severity and include burning and stinging to erythema with edema, blistering, pain and constitutional symptoms (e.g. nausea, fever, chills, tachycardia) ... [t]wo subjects in the ... trials were hospitalized for photosensitivity reactions that were considered to be serious adverse events[62] ... **[t]his Reviewer recommends that a discussion of rash and photosensitivity events be included in the Warnings and Precautions section of the product label**. This would include a recommendation that sun protection measures ... be initiated in all patients receiving simeprevir.[63]

[61] Simeprevir (hepatitis C virus) NDA 205-123. Pages 17–18 of 173-page Medical Review.

[62] Simeprevir (hepatitis C virus) NDA 205-123. Pages 3–4 of 173-page Medical Review.

[63] Simeprevir (hepatitis C virus) NDA 205-123. Page 109 of 173-page Medical Review.

Package label. The package label for simeprevir (Olysio®) included warnings about photosensitivity reactions and about rash. As stated above, the Sponsor used the Preferred Terms rash, rash macular, rash papular, rash erythematous, and rash pruritic. The label warned about photosensitivity and against rash:

> **WARNINGS AND PRECAUTIONS** ... Photosensitivity: Serious photosensitivity reactions have been observed during OLYSIO combination therapy. Use sun protection measures and limit sun exposure during OLYSIO combination therapy. Consider discontinuation if a photosensitivity reaction occurs. Rash: Rash has been observed during OLYSIO combination therapy. Discontinue OLYSIO if severe rash occurs.[64]

III. SPLITTING AND LUMPING

a. Introduction

FDA's Medical Reviews refer to the linguistic development of verbatim terms where, during the coding process, the verbatim terms are transmogrified into other terms by way of splitting and lumping.[65] Lumping is also called "combining." Published journal articles provide little information on splitting and lumping. Fortunately, FDA's Medical Reviews provide more expansive coverage on this topic, and for this reason, generous quotations from FDA's Medical Reviews are provided here.

The need to modify or correct verbatim terms can arise during routine quality control by the Sponsor, or with scrutiny by FDA reviewers, or during adjudication, or where the version of MedDRA or the version of CTCAE changed during the course of the Sponsor's activities in coding.

Ultimately, the strategies taken during coding can influence the AEs that seem to be of greatest risk to patients ("safety signals") and consequent warnings on the package label. Inappropriate splitting or lumping can result in underemphasis or overemphasis of a given safety signal. Any coding that uses splitting or lumping can inspire increased scrutiny from FDA reviewers. For example, in FDA's review of ivacaftor, FDA's scrutiny took the form of the remark "There was no evidence of splitting or lumping in the individual coding noted, and it was appropriate."[66]

[64] Package label. OLYSIO (simeprevir) capsules, for oral use. February 2017 (48 pp.).

[65] The author thanks Dr. Yuliya Yasinskaya, MD of the Food and Drug Administration, for confirming the accuracy of the author's account of splitting. Dr. Yasinskaya responded, "Your splitting examples are spot on: the like or identical verbatim terms are split between different Preferred Terms ... [f]rom the clinical reviewer's perspective, we look at the dictionary of Preferred Terms vs verbatim for mismatches and do group like Preferred Terms together when assessing the rate of certain events of interest." (e-mail of May 16, 2017).

[66] Ivacaftor (cystic fibrosis) NDA 203-188. Page 97 of 130-page Medical Review.

b. Published Journal Articles on Splitting and Lumping

Brajovic[67] warned that splitting can result in failure of the Sponsor to recognize that a particular AE should be of increased concern, that is, that the particular AE was a safety signal, writing: "**Splitting similar terms.** Splitting results in lower incidence. May minimize or mask a safety signal."

Consistently, Cobert[68] warns that "Splitting may ... falsely decrease the incidence of AEs." "Fluid retention" may give a better signal and truer incidence than dividing the AEs into "pedal edema, general edema, peripheral edema." Toneatti et al.[69] warned about the same problem where "splitting of narrow verbatim/events in too many distinct Preferred Terms which might result in artificially small incidence event rates providing possible misleading information."

This concerns a nondesirable type of splitting, where the coding procedure applied to many subjects results in the same Preferred Term being coded into one type of SOC for one subject, but in a different SOC for the next subject. Wittes et al.[70] issued the same type of warning about splitting. This is from the situation where a coder reviews all of the Preferred Terms from AEs captured during a clinical study, where a Preferred Term derived from an AE experienced by one subject is coded into one SOC and where for the exact same Preferred Term from another study subject it is coded into a different SOC. The example is from the Preferred Term "pulmonary edema." Wittes et al.[71] warned that "For example, 'pulmonary edema' will be classified in the Respiratory SOC and 'heart failure' in the Cardiovascular SOC even though they may represent the same medical condition. This type of splitting, while allowing precise description of a given event, often makes it difficult to identify all of the cases that reflect a medical concept of interest."

On the other hand, in favor of splitting, Cobert[72] advocates, "Splitting: Using multiple terms may at times be more useful than combinations [lumping] ... dyspnea, cough, wheezing, and pleuritis is more sensitive and useful than pulmonary toxicity." Babre also reveals an advantage of splitting, referring to the problem where "multiple medical concepts are recorded together. To code we need to split the terms."[73]

[67] Brajovic S. MedDRA Use at FDA. MedDRA Workshop. March 19, 2010. International Conference of Harmonisation (ICH). 2010 (37 pp.).

[68] Cobert B. Cobert's manual of drug safety and pharmacovigilance. 2nd ed. Sudbury, MA: Jones and Bartlett Learning. 2012. p. 203.

[69] Toneatti C, et al. Experience using MedDRA for global events coding in HIV clinical trials. Contemp. Clin. Trials. 2006;27:13–22.

[70] Wittes J, et al, The FDA's final rule on expedited safety reporting: statistical considerations. Stat. Biopharm. Res. 2015;7:174–90.

[71] Wittes J, et al. The FDA's final rule on expedited safety reporting: statistical considerations. Stat. Biopharm. Res. 2015;7:174–190.

[72] Cobert B. Cobert's manual of drug safety and pharmacovigilance. 2nd ed. Sudbury, MA: Jones and Bartlett Learning. 2012. p. 203.

[73] Babre D. Medical coding in clinical trials. Perspect. Clin. Res. 2010;1:29–32.

A disadvantage of lumping has been observed. Brajovic[74] pointed out a problem with lumping different terms together, namely, the fact that a unique type of AE can be obscured because it is lumped together with other AEs that are actually different AEs. Brajovic et al.[75] warned that "**Lumping dissimilar terms.** Specific AEs all coded under an 'umbrella' term. May obscure a safety signal under the lumped term."

Brajovic[76] provided the example of lumping of four different types of edema that had all been lumped together under the MedDRA Preferred Term of oedema, "Face edema. Lip edema. Eyelid edema. Edema of hands. Foot edema. Coding issue: all lumped to Preferred Term Oedema."[77]

Consistently, Cobert has warned about the disadvantage of lumping, where the disadvantage is hiding of certain AEs. Cobert commented that "coding may be done at a less specific level, coding 'edema' instead of 'facial edema' or 'lung disease' instead of a more specific ... 'pneumococcal pneumonia.' This type of lumping can mask or hide certain AEs."[78]

Disadvantages of lumping were also noted by Wittes et al.[79] who provided the warning that lumping can hide AEs of real clinical concern. The warning was that:

> However, **overaggressive lumping** may commingle adverse events of real clinical concern with many others that are not, or events with no common pathophysiology. Often, the least clinically relevant items can be the most frequent in the higher order classification. Thus, noise can overwhelm important signals of harm and so reviewers of these tables may miss the important signals.[80]

c. Advantage of Lumping in the Situation Where FDA Requires a Cutoff Point for AE Reporting

FDA as well as some journals may require that for AE reporting only AEs that occur over a specific frequency be reported. To provide a hypothetical example, this means that if the AEs were "hallucinations," "abnormal dreams," and "mental status changes," and where each of these occurred at a frequency of only 4%, then these AEs would never be

[74] Brajovic S. MedDRA Use at FDA. MedDRA Workshop. March 19, 2010. International Conference of Harmonisation (ICH). 2010 (37 pp.).

[75] Brajovic S. MedDRA Use at FDA. MedDRA Workshop. March 19, 2010. International Conference of Harmonisation (ICH). 2010 (37 pp.).

[76] Brajovic S. MedDRA Use at FDA. MedDRA Workshop. March 19, 2010. International Conference of Harmonisation (ICH). 2010 (37 pp.).

[77] Brajovic S. MedDRA Use at FDA. MedDRA Workshop. March 19, 2010. International Conference of Harmonisation (ICH). 2010 (37 pp.).

[78] Cobert B. Cobert's manual of drug safety and pharmacovigilance. 2nd ed. Sudbury, MA: Jones and Bartlett Learning. 2012. p. 203.

[79] Wittes J, et al. The FDA's final rule on expedited safety reporting: statistical considerations. Stat. Biopharm. Res. 2015;7:174–90.

[80] Wittes J, et al. The FDA's final rule on expedited safety reporting: statistical considerations. Stat. Biopharm. Res. 2015;7:174–90.

reported. On the other hand, if all three of these AEs were lumped together and reported as the term "mental status changes," then the frequency would be above the cutoff point, and in this case all three AEs would be reported. Wittes et al.[81] describes this advantage of lumping, "To simplify the tables, some sponsors ... committees, publications, and FDA advisory panels, list only those events that occur with a frequency of more than 1% (or 5% or 10%). Often the FDA for drug labels and journals require such simplification ... **we ... recommend that sponsors ... devise systems to lump similar events**. Below we suggest some approaches to such lumping."

The lumping technique advocated by Wittes et al.[82] involves, "Looking only at higher order term classifications (i.e., lumping) produces tables with higher counts for many listed events and thus more stability in the estimated event rates." In other words, the lumping technique suggested by Wittes et al. is to refrain from using a MedDRA Lowest Level Term of one of the MedDRA Preferred Terms, and instead to use one of MedDRA's High Level terms or one of MedDRA's SOC terms.

IV. FDA'S MEDICAL REVIEWS, EXAMPLES OF SPLITTING AND LUMPING

a. Splitting (Example of Apremilast)

Apremilast is a small molecule drug for psoriatic arthritis. Splitting can refer to the situation where exactly the same AE, as it occurs in several different study subjects, is written down in each CRF with a different verbatim term for each of the study subjects. Splitting can also refer to the situation where the same verbatim term was written down for each CRF, but where coding resulted in use of different Preferred Terms for each different study subject.

The FDA reviewer highlighted the value and goal of not splitting the verbatim terms, during the coding process, into different Preferred Terms, such as "nightmares" and "abnormal dreams." FDA wrote, "a subject in the trial experienced 'nightmares each night' which was coded to the MedDRA Preferred Term 'nightmares.' Although this is not objectionable on face, other subjects had events coded to the Preferred Term 'abnormal dreams,' some of which may have been nightmares. Thus, in order to avoid minimizing the incidence of these events by splitting, this event should be coded to 'abnormal dreams' as well."[83]

b. Splitting (Example of Dimethyl Fumarate)

Dimethyl fumarate is a small molecule for treating multiple sclerosis.[84] FDA's Medical Review revealed the situation where splitting can result in dispersion, during coding, of

[81] Wittes J, et al. The FDA's final rule on expedited safety reporting: statistical considerations. Stat. Biopharm. Res. 2015;7:174–90.

[82] Wittes J, et al. The FDA's final rule on expedited safety reporting: statistical considerations. Stat. Biopharm. Res. 2015;7:174–90.

[83] Apremilast (psoriatic arthritis) NDA 206-088. Page 119 of 132-page Medical Review.

[84] Dimethyl fumarate (multiple sclerosis). Page 61 of 288-page Medical Review.

the AEs into a plethora of different Preferred Terms, with consequent weakening of the safety signal. This undesirable process is illustrated by the flow chart:

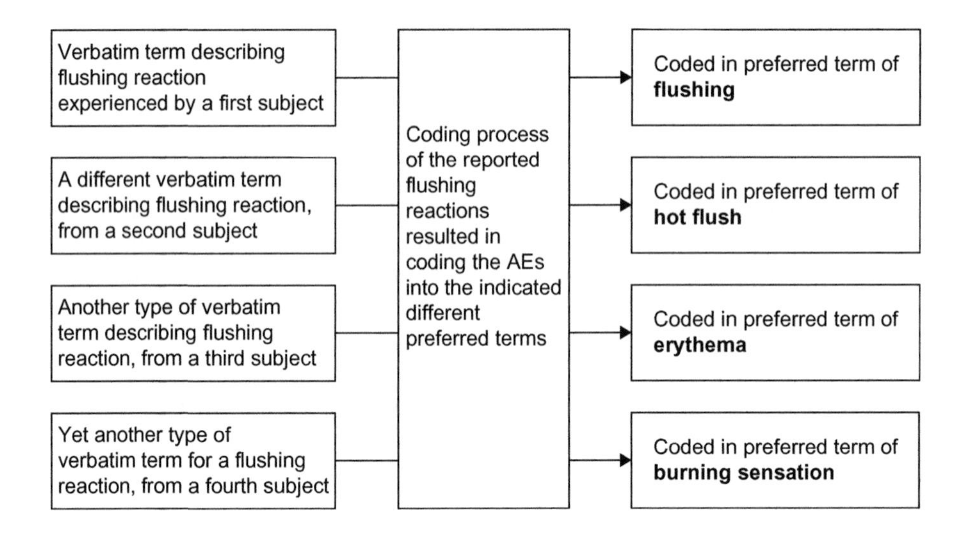

The above fact-pattern is described by the reviewer's comments that "There were … instances where the coding process could have led to **splitting of likely related events into separate Preferred Terms** … [f]or example, the coding process resulted in verbatim terms describing flushing reactions to be mapped to a number of different Preferred Terms, including flushing, hot flush, erythema, generalized erythema, burning sensation, etc."[85]

c. Splitting (Example of Linaclotide)

FDA's Medical Review for linaclotide reveals decisions in coding that might be utilized by an experienced coder, when faced with the task of translating verbatim terms to one or more MedDRA Preferred Terms. This excerpt provides no other take-home lesson, apart from the fact that coding can be an arduous fact-dependent process:

> The Investigator initially reported the event as "symptomatic ischemic colitis," but changed it to "ileus secondary to ischemic ulcerative colitis" one day later. This Investigator Term was split into two terms for coding, initially as Preferred Terms "ileus" and "colitis ulcerative" but subsequently, based on the clinical picture, to Preferred Terms "ileus" and "colitis ischemic."[86]

[85] Dimethyl fumarate (multiple sclerosis). Page 61 of 288-page Medical Review.

[86] Linaclotide (irritable bowel syndrome) NDA 202-811. Page 418 of 489-page Medical Review.

This reveals the situation where similar or identical AEs, as reported by two different subjects, were coded as two different Preferred Terms. FDA stated that:

> There was some **splitting** of the adverse events terms in the datasets. For example, the applicant's decode contained the terms abdominal pain, abdominal pain lower, abdominal pain upper, abdominal tenderness and abdominal discomfort. It is not clear how the distinction in the coding of verbatim terms to these decoded terms was made. "Black stools" and "dark stools" were coded as "faeces discolored" yet "black tarry stools" was coded as "melaena." Likewise, "abdominal spasms" were coded to "abdominal rigidity" yet "abdominal cramping" was coded to "abdominal pain" while "stomach cramps" and "stomach cramping" were coded to "upper abdominal pain." Another example would be the **splitting** of terms for "diarrhea" vs. "frequent bowel movements" and "gastrointestinal motility disorder" vs. "ileus."[87]

FDA's Medical Review for linaclotide revealed that the Sponsor's coding resulted in inappropriate splitting and that FDA stepped in and repaired this inappropriate splitting, before assessing the study drug's AEs. This repair took the form of (1) detecting the splitting which resulted in different Preferred Terms for the same AE and (2) combining the Preferred Terms to create one Preferred Term that encompassed all of the previously split AEs. FDA's account of this need for repairing was as follows:

> The appropriateness of the applicant's coding was assessed by examining the ... datasets submitted with the application and comparing the preferred terms in the "AE-DECOD" column to the verbatim terms reported by investigators and recorded in the "AE-TERM" column. In general the coding appeared to be adequate for review. There was some **splitting of preferred terms**. When appropriate, the medical officer combined preferred terms (e.g. abdominal pain vs. abdominal pain upper or abdominal pain lower; diarrhea vs. frequent bowel movements; anemia vs. hematocrit decreased vs. red blood cell count decreased).[88]

FDA's review for linaclotide also revealed a cutoff point, where the goal of the cutoff point was to guide writing on the package label. In short, AEs below the cutoff point do not go on the package label, but AEs above the cutoff point should go on the package label. The issue was that splitting of AEs could have resulted in AEs occurring at a rate below the cutoff point, where the outcome was silence on the package label regarding this AE, versus lumping of the AEs where the result was a rate of AEs occurring at a rate above the cutoff point, where the outcome was presence on the package label. In comments about labeling, FDA described the raw data and referred to the cutoff point:

> The applicant proposed to include in labeling those **adverse events that occurred in greater than 2% of the population** and occurring at a greater incidence in the Linaclotide treated group relative to the placebo group. The treatment emergent adverse events (TEAEs) experienced by at least 2% of all Linaclotide patients in the Phase 3 Double Blind trials and at an incidence of at least 1% more than placebo patients were diarrhea (15.1% vs. 4.7%), abdominal pain (4.2% vs. 3.1%), and abdominal distension (3.5% vs. 2.4%).[89]

[87] Linaclotide (irritable bowel syndrome) NDA 202-811. Page 42 of 489-page Medical Review.

[88] Linaclotide (irritable bowel syndrome) NDA 202-811. Page 165 of 489-page Medical Review.

[89] Linaclotide (irritable bowel syndrome) NDA 202-811. Page 165 of 489-page Medical Review.

Then, FDA described how splitting and lumping can influence what is written on the package label. In particular, FDA noticed that the AE of "abdominal discomfort" had somehow been split away from similar Preferred Terms that had been combined, where these combined similar Preferred Terms were "abdominal pain," "upper abdominal pain," and "lower abdominal pain." FDA's comment about the option of either splitting or lumping terms for AEs was "There was some **splitting of the preferred terms that may represent abdominal pain**. When the clinical reviewer combined the preferred terms 'abdominal pain,' 'upper abdominal pain,' and 'lower abdominal pain' into one group, the incidence was 6% in placebo treated patients, 7% in patients treated with Linaclotide (145 μg) and 6% in patients treated with Linaclotide (290 μg). It is possible that this incidence would have been higher if abdominal discomfort was also included in the analysis."[90]

d. Splitting (Example of Lurasidone and the AE of Dystonias)

This is from FDA's review of lurasidone for schizophrenia. FDA complained that splitting of, what was likely the same or very similar AE presenting in many different subjects, resulted in undercounting of this particular AE. The issue was that AEs taking the form of various types of dystonias were coded by the Sponsor into a variety of different Preferred Terms, where the concern was that people reviewing only the Preferred Terms would not be aware of the true frequency of dystonias. FDA complained:

> Adverse events occurring in > 5% of patients treated with lurasidone and with an incidence > placebo included akathisia (15%), nausea (12%), sedation (11.9%), somnolence (10.7%), insomnia (8.4%), dyspepsia (7.6%), agitation (6.4%), and anxiety (6.3%). Due to **issues of coding adverse events (splitting/lumping)**, frequencies for ... dystonias were difficult to determine. Dystonias (as a preferred term) occurred in 3.5% of patients in the lurasidone groups, 0.7% of patients in the placebo group and 12.5% of patients in the comparator haloperidol 10 mg group. However, rates of dystonia are higher if other dystonia-related preferred terms are included such as **oculogyric crisis**, **oromandibular dystonia** and **torticollis.**[91]

The variety of types of dystonias was described in a publication on drugs for schizophrenia. A take-home lesson broadly applicable to all people involved in coding is that the same AE can have a different term, depending on which part of the body is involved.

Regarding the issue of the location in the body of the AE, Yamamoto et al.[92] wrote that "**dystonic reactions** ... typically consist of bizarre movements involving tonic contractions of skeletal muscles ... [t]hese dystonic reactions are most often localized in the face, neck

[90] Linaclotide (irritable bowel syndrome) NDA 202-811. Page 185 of 489-page Medical Review.

[91] Luradisone (schizophrenia) NDA 200-603. Page 93 of 210-page Medical Review.

[92] Yamamoto N, Inada T. Dystonia secondary to use of antipsychotic agents, dystonia - the many facets. R. Rosales (Ed.), ISBN: 978-953-51-0329-5. InTech Europe. University Campus STeP. Rijeka, Croatia. InTech China. Yan An Road (West), Shanghai, China. 2012 (page 57 of 64 pages).

and upper part of the body ... [l]aryngeal dystonia occurs only rarely, but may be life-threatening. The name of the reaction is derived from the anatomic region that is affected. Hence, the terms '**torticollis**,' 'laryngospasm,' '**oculogyric crisis**,' and 'opisthotonos' are used to describe dystonic reactions in the specific body regions."

Similarly, the variety of types of dystonias was described as encompassing "isolated **spasmodic torticollis**, isolated blepharospasm, isolated **oromandibular dystonia**, cranial dystonia ... isolated spasmodic dysphonia, and isolated writer's cramp."[93]

e. Splitting (Example of Lurasidone and Coding to a Preferred Term That Could Cause the AE to Avoid Detection by Persons Reviewing the AE Data)

This provides an example of coding of a dystonia AE, where the coding was accurate, but where the Preferred Term used in coding could have resulted in the AE in question being unnoticeable or invisible by any person reviewing the Preferred Terms and interested in calculating the frequency of dystonia-related AEs. The problem was in the choice of the Preferred Term, "eye rolling." FDA articulated this problem as

> The ... file for adverse events was reviewed with an emphasis on the verbatim to preferred term coding. In general, it appeared that most verbatim terms were appropriately coded to preferred terms. Some terms potentially relating to dystonic-type events **may not have been captured in appropriate preferred terms**. One example is the verbatim term "upgoing eyes (EPS)" which was mapped to the preferred term eye rolling.[94]

f. Splitting (Example of Lurasidone and the AE of Parkinson's Disease)

FDA's review of lurasidone for schizophrenia complained that the exact same AE presenting in different study subjects could be coded in a variety of different Preferred Terms. These different Preferred Terms included tremor, cogwheel rigidity, bradykinesia, and drooling. The problem with this type of splitting was that a person reviewing the coded AEs might not be aware of the true extent of "parkinsonian adverse events." FDA wrote:

> Similarly, parkinsonian adverse events occurred in 4.9% of patients in the lurasidone groups, 0.4% of patients in the placebo group and 0 patients in the haloperidol 10 mg group (though 18% experienced "extrapyramidal disorder" in the haloperidol group, coding differences). Determining the frequency of parkinsonian-related adverse events was difficult since there was potential splitting for preferred terms tremor, cogwheel rigidity, bradykinesia, drooling, etc.[95]

The potential problem with splitting and lumping, as articulated by the FDA reviewer, was inaccurate relatedness determinations for the study drug. Regarding confounding of

[93] Soland VL, et al. Sex prevalence of focal dystonias. J. Neurol. Neurosurg. Psychiatry. 1996;60:204–5.

[94] Luradisone (schizophrenia) NDA 200-603. Pages 97–8 of 210-page Medical Review.

[95] Luradisone (schizophrenia) NDA 200-603. Pages 93 and 165 of 210-page Medical Review.

accurate relatedness determinations, FDA's Medical Review warned, "Due to difficulties in mapping of adverse event terms related to parkinsonian adverse events (e.g. **splitting and lumping**), it is more difficult to determine a dose-relationship."[96]

g. Splitting (Example of Brivaracetam)

This reveals the situation where a MedDRA Preferred Term was split into different MedDRA SOC terms. Brivaracetam is a small molecule drug used for treating seizures.[97] FDA's Medical Review stated that the coding involved the AEs of somnolescence, dizziness, and fatigue. The medical literature reveals that treatment-emergent adverse events (TEAEs) for brivaracetam include these same AEs, namely, 12.4% for somnolence, 10.4% for headache, 9.6% for dizziness, 7.7% for fatigue, 4.9% for nausea, 4.2% for nasopharyngitis, 2.8% for irritability, and 2.5% for insomnia.[98]

The FDA reviewer complained that the Sponsor's coding procedures included splitting, where the result was underestimation of the true incidence for particular AEs. FDA's complaint was as follows: "there were also instances where the coding process resulted in **splitting** likely related AEs into separate SOCs leading to an underestimation of the true incidence for a particular event or syndrome. Therefore, in order to account for the **splitting** of the Preferred Terms into different System Organ Classes (SOCs) in this NDA, additional analyses were performed by the reviewer ... to group these Preferred Terms across SOCs to provide more accurate estimates of adverse event syndromes."[99]

The following reveals the situation where a verbatim term was split into a multiplicity of Preferred Terms. This excerpt illustrates the fact that FDA reviewers scrutinize and reanalyze the Sponsor's coding activities. FDA's Medical Review stated that "Please refer to Dr. Dinsmore's review of efficacy for details regarding worsening seizures, status epilepticus, and rebound epilepsy. To address the issue of the **splitting** of potentially similar neurological events into multiple Preferred Terms, I performed additional analyses in order to pool together related events ... I reanalyzed the AEs in the following main groups: Somnolence and fatigue along with dizziness and gait disturbance."[100]

h. Lumping (Example of Droxidopa)

Droxidopa is a small molecule drug for treating neurogenic orthostatic hypertension. FDA's Medical Review referred to the AEs of "hallucination," "abnormal dreams," and "mental status changes," and two different ways of coding these AEs.

[96] Luradisone (schizophrenia) NDA 200-603. Page 120 of 210-page Medical Review.

[97] Brivaracetam (epilepsy) NDA 205-836, NDA 205-837, NDA 205-838. 382-page pdf file Medical Review.

[98] Coppola G, et al. New developments in the management of partial-onset epilepsy: role of brivaracetam. Drug Des. Dev. Ther. 2017;11:643–57.

[99] Brivaracetam (partial onset seizures) NDA 205-836, NDA 205-837, NDA 205-838. Pages 192–3 of 382-page Medical Review.

[100] Brivaracetam (partial onset seizures) NDA 205-836, NDA 205-837, NDA 205-838. Pages 239–40 and 253–4 of 382 pages.

TABLE 13.5 Adverse Events Leading to Discontinuation

Patient No.	Treatment	Preferred Term	Study Day	Dose
115004	Droxidopa	**Worsening of hallucination**	10	200 mg
131005	Droxidopa	Hypertension	20	600 mg
141004	Droxidopa	**Worsening of vivid dreams**	19	100 mg
156007	Droxidopa	**Altered mental status**	20	400 mg
16003	Droxidopa	Elevated blood pressure	20	100 mg
16003	Droxidopa	Worsening of Parkinson's	5	500 mg
146006	Placebo	Syncopal episode	22	600 mg

Droxidopa (neurogenic orthostatic hypotension) NDA 203-202. Page 66 of 436-page Medical Review.

FDA's comments did not take the form of any complaint, but were set forth in a neutral manner. FDA's Medical Review also disclosed that these three different AEs occurred in the droxidopa group, not in the placebo group.

The comments referred to coding these AEs separately for each of the three subjects, as "hallucination," "abnormal dreams," and "mental status changes" or, alternatively, as lumping these three different AEs under the umbrella of one name, for each of the three subjects. The FDA reviewer stated that "The most common AE leading to discontinuation was hypertension ... followed by hallucination, abnormal dreams, or mental status changes (1 each on droxidopa, or 3 patients if these AEs are lumped). Since there were more discontinuations in droxidopa-treated patients compared to those on placebo, one can speculate that there were additional ... tolerability issues."[101]

A table in FDA's Medical Review discloses the AEs by the Preferred Terms of "hallucination" (patient 115004), "abnormal dreams" (patient 141004), and "altered mental status" (patient 156007). Because each of these AEs were memorialized as Preferred Terms, it is apparent that they were not all lumped into one Preferred Term. FDA's review table revealed the use of three different names rather than one name encompassing all three of these AEs. The Preferred Terms for these three different names are highlighted in the table. An excerpt from this table is shown in Table 13.5. The CRF (patient 156007) described this AE by the verbatim term, "While sitting at the breakfast table, the patient was observed by wife to be unresponsive, like in a daze and pale in color. The patient was not aware of his surroundings."[102]

This verbatim term was coded to the Preferred Term, "altered mental status." Table 13.5 shows the three different terms (without any lumping).

FDA's comments on "altered mental status" and "hallucinations" found a place on the package label, but the package label did not have any warning about vivid dreams. FDA's comments about the verbatim term, "While sitting at the breakfast table ..." and its coding to the Preferred Term "altered mental status," and FDA's comments about the option of

[101] Droxidopa (neurogenic orthostatic hypotension) NDA 203-202. Pages 38 and 66 of 436-page Medical Review.

[102] Droxidopa (neurogenic orthostatic hypotension) NDA 203-202. Pages 38 and 66 of 436-page Medical Review.

lumping together the AEs described by the Preferred Terms of "hallucination," "abnormal dreams," and "altered mental status" provide a standalone example of coding and lumping (but does not provide any lesson for package label warnings).

Package label. The Warnings and Precautions section of the package label for droxidopa (Northera®) reveals warnings about altered mental status:

> **WARNINGS AND PRECAUTIONS**. Postmarketing cases of a symptom complex resembling neuroleptic malignant syndrome (NMS) have been reported with NORTHERA use during postmarketing surveillance ... NMS is an uncommon but life-threatening syndrome characterized by fever or hyperthermia, muscle rigidity, involuntary movements, **altered consciousness, and mental status changes.**[103]

The Adverse Reactions section of the package label for droxidopa (Northera®) reveals a warning about "hallucinations."

> **ADVERSE REACTIONS** ... Postmarketing Experience. The following adverse reactions have been identified during post-approval use of NORTHERA ... Cardiac Disorders ... Blurred vision ... Fatigue Nervous System Disorders: Cerebrovascular accident Psychiatric Disorders: **Psychosis, hallucination**, delirium, agitation, memory disorder.[104]

i. Ambiguity in Study Design and Ambiguity in Lumping

FDA's Medical Review of linaclotide provides the situation where a set group of verbatim terms were available for choosing by study subjects, but where the available terms were ambiguous. Please note the verbatim terms "abdominal discomfort" and "abdominal pain," These terms inspired the FDA reviewer to describe how lumping (combining), or refraining from lumping (not combining), would have influenced whether or not a given AE was reported on the package label:

> patients answered a series of questions about the severity of their abdominal symptoms. Using a 5 point ordinal scale, patients rated their "**abdominal pain**" "**abdominal discomfort**" and "bloating" as "None" (1), "Mild" (2), "Moderate" (3), "Severe" (4) or "Very Severe" (5) over the preceding 24 hours ... [t]he difference in "abdominal pain" and "abdominal" discomfort **may not be completely clear and patients were not given instructions on how to differentiate between the two**. In addition, for questions such as this, one can not be sure that all respondents have the same understanding of what constitutes each symptom assessed and one is unable to adequately objectively quantify differences between each of the ordinal categories.[105]

j. Miscoding and Coding Omissions. Examples from Brivaracetam and Lurasidone

FDA's Medical Review for brivaracetam illustrates the concepts of coding omissions and miscoding. This example is where the verbatim term "broken brace after fall" had not been

[103] Package label. NORTHERA® (droxidopa) capsules, for oral use. February 2017 (10 pp.).

[104] Package label. NORTHERA® (droxidopa) capsules, for oral use. February 2017 (10 pp.).

[105] Linaclotide (irritable bowel syndrome) NDA 202-811. Page 76 of 489-page Medical Review.

coded to "fall" and where "tingling and numbness in her feet" were miscoded to "unevaluable event." The FDA reviewer complained: "However, there were rare cases that appeared to be **coding omissions and miscoding**. For example, the verbatim term of 'broken brace after fall' was only coded to the Preferred Term device breakage (and not also to fall)."[106]

Cobert[107] used a similar example, which involved the AE of falling, in a general account of coding omissions and miscoding. Cobert's example was that of the primary AE of dizziness and of consequent secondary events of fall, shoulder fracture, and skin abrasions. More generally stated, Cobert highlighted the need, in coding, to decide that coding should refer to one or both of the primary AE and the secondary AE.

This is from FDA's Medical Review of lurasidone, a drug for schizophrenia. With coding, the verbatim term was coded to the Preferred Term of "fall." However, the FDA Reviewer determined that the term "fall" did not accurately capture the nature of the AE, in view of the fact that the verbatim term was "jump from a freeway overpass." FDA complained that "Adverse Event Coding. A number of potential issues in coding verbatim terms to preferred terms were noted during the review. For example, one SAE coded 'fall' was, according to the narrative, a jump from a freeway overpass. Though it is unclear what the motivation for the jump was (suicidal thoughts denied, auditory hallucinations denied), 'fall' **does not seem to capture the seriousness of the event**."[108]

k. Summary of Problems Caused by Careless Splitting and Lumping

FDA's Medical Review for lurasidone provides the best account available of the situation where carelessness in splitting and lumping can have the eventual consequence of inaccurate writing on the package label. The potential problems are inaccurate assessment of incidence of a given AE and of inaccurate relatedness analysis. FDA stated that:

> As previously stated, it is **difficult to ascertain the incidence of parkinsonian symptoms in these clinical trials due to lumping and splitting of terms**. Some terms seem to be lumping—e.g. parkinsonism, extrapyramidal disorder, while others seemed to split—e.g. tremor, bradykinesia, drooling. The Sponsor did not capture "salivary hypersecretion" as an EPS-related term, though this seems similar to "drooling" to this reviewer and the latter term was captured as an EPS-related term. Gait disturbance was also not necessarily captured adequately when mapping from verbatim terms—some cases where the gait was parkinsonian-like (e.g. "decrease arm swing during walk" mapped to preferred term "gait disturbance" with other adverse events consistent with EPS, "shuffling gait"). In the Parkinsonism-related terms, the most frequently reported adverse events were "parkinsonism", "tremor", "salivary hypersecretion" and "extrapyramidal disorder."[109]

The term "EPS," as used in the above excerpt, refers to "extrapyramidal signs and symptoms."[110] Regarding relatedness determination, FDA stated, "Due to difficulties in

[106] Brivaracetam (partial onset seizures) NDA 205-836, NDA 205-837, NDA 205-838. Pages 192–3 of 382 pages.

[107] Cobert B. Cobert's manual of drug safety and pharmacovigilance. 2nd ed. Sudbury, MA: Jones and Bartlett Learning. 2012. p. 81.

[108] Lurasidone (schizophrenia) NDA 200-603. Page 19 of 210-page Medical Review.

[109] Luradisone (schizophrenia) NDA 200-603. Page 165 of 210-page Medical Review.

[110] Package label. LATUDA (lurasidone hydrochloride) tablets for oral use. February 2017 (55 pp.).

mapping of adverse event terms related to parkinsonian adverse events (e.g. **splitting and lumping**), it is more difficult to determine a dose-relationship."[111]

Splitting or lumping can be desired features of the coding process, where the desired outcome is a more precise detection of safety signals from the study drug (advantage provided by splitting) or where they can result in the detection of safety signals that are otherwise undetectable because of background noise (advantage provided by lumping).

V. FORMATTING ADVERSE EVENTS

FDA's Manual of Policies and Procedures provides some guidance for AE tables, pointing out that information on duration of treatment should not be overlooked. FDA's advice is that "Most applicants will construct adverse event tables by compiling and presenting the numbers and/or percentages of subjects experiencing an adverse event in a clinical trial ... without regard to the duration of treatment received. This approach is often satisfactory for relatively short-term clinical trials. If clinical trials of significantly different durations are pooled, however, or if there is a different discontinuation rate in the treatment arms and the risk of the adverse reaction persists over time, one must consider these durations to understand the real occurrence rate that patients will experience."[112] Shown below is an example of formatting AEs from FDA's review of the trifluridine/tipiracil combination.

a. Trifluridine/Tipiracil Combination (Metastatic Colorectal Cancer) NDA 207-981

FDA's Medical Review for the trifluridine/tipiracil combination provides a format for disclosing AEs, where the format encompasses both MedDRA's Preferred Terms and MedDRA's SOCs. This format shows the names for some of the Preferred Terms and for some of the SOCs and reveals how the Preferred Terms fit into the various SOCs.

The name of one of the System Organ Classes is "Investigations." This term refers to adverse events taking the form of abnormal laboratory tests. The SOC term "Investigations" refers to data, such as ECG data, body weight, blood pressure, and blood test data, such as red blood cell counts, bilirubin, glucose, aminotransferases, and triglycerides.[113] MedDRA's Introductory Guide defines Investigations as, "a clinical laboratory test concept (including biopsies), radiologic test concept, physical examination parameter, and physiologic test concept (e.g., pulmonary function test)."[114]

Within each SOC, Table 13.6 arranges the AEs that are identified by MedDRA's Preferred Terms in descending order of frequency. The columns reveal the AEs from the study drug treatment arm (533 subjects) and the placebo arm (265 subjects). The formatting in Table 13.6

[111] Luradisone (schizophrenia) NDA 200-603. Page 120 of 210-page Medical Review.

[112] Director, Office of New Drugs. Center for Drug Evaluation and Research (CDER). Manual of Policies and Procedures. Attachment B: Clinical Safety Review of an NDA or BLA. MAPP6010.3 Rev. 1. December 2010 (82 pp.).

[113] Some of these types of investigations are disclosed in FDA's Medical Reviews for lurasidone (NDA 200-603) 210-page Medical Review, perampanel (NDA 202-834) 401-page Medical Review, and brivaracetam (NDA 205-836) 382-page Medical Review.

[114] MedDRA MSSO. Introductory Guide MedDRA Version 14.0. MedDRA MSSO: McLean, VA. 2011 (77 pp.).

TABLE 13.6 Adverse Events in Trifluridine/Tipiracil Combination Arm and in Placebo Arm

System Organ Class (SOC)/ Preferred Term	Trifluridine/Tipiracil Combination Arm All Grades of AEs	Trifluridine/Tipiracil Combination Arm Grades 3/4/5 of AEs	PLACEBO ARM All Grades of AEs	PLACEBO ARM Grades 3/4/5 of AEs
BLOOD AND LYMPHATIC SYSTEM DISORDERS				
Anemia	207 (38.8%)	83 (15.6%)	22 (8.3%)	7 (2.6%)
Neutropenia	156 (29.3%)	107 (20.1%)	0 (0%)	0 (0%)
Thrombocytopenia	37 (6.9%)	11 (2.1%)	1 (0.4%)	1 (0.4%)
Leukopenia	29 (5.4%)	13 (2.4%)	0 (0%)	0 (0%)
GASTROINTESTINAL DISORDERS				
Nausea	256 (48.0%)	10 (1.9%)	63 (23.8%)	3 (1.1%)
Diarrhea	170 (31.9%)	16 (3.0%)	33 (12.5%)	1 (0.4%)
Vomiting	147 (27.6%)	11 (2.1%)	38 (14.3%)	1 (0.4%)
Constipation	80 (15.0%)	1 (0.2%)	40 (15.1%)	3 (1.1%)
Abdominal pain	79 (14.8%)	11 (2.1%)	34 (12.8%)	9 (3.4%)
Stomatitis	42 (7.9%)	2 (0.4%)	16 (16.0%)	0 (0%)
Ascites	21 (3.9%)	5 (0.9%)	14 (5.3%)	8 (3.0%)
INVESTIGATIONS				
Neutrophil count decreased	148 (27.8%)	85 (15.9%)	1 (0.4%)	0 (0%)
White blood cell count decreased	146 (27.4%)	55 (10.3%)	1 (0.4%)	0 (0%)
Platelet count decreased	81 (15.2%)	13 (2.4%)	6 (2.3%)	0
Blood alkaline phosphatase increased	46 (8.6%)	17 (3.2%)	24 (9.1%)	11 (4.2%)
Blood bilirubin increased	44 (8.3%)	20 (3.8%)	19 (7.2%)	9 (3.4%)
Weight decreased	41 (7.7%)	1 (0.2%)	27 (10.2%)	0 (0%)
Aspartate aminotransferase increased	28 (5.3%)	6 (1.1%)	22 (8.3%)	7 (2.6%)
Gamma-glutamyltransferase increased	24 (4.5%)	16 (3.0%)	13 (4.9%)	10 (3.8%)
METABOLISM AND NUTRITION DISORDERS				
Decreased appetite	206 (38.6%)	19 (3.6%)	76 (28.7%)	12 (4.5%)
Hypoatremia	16 (3.0%)	7 (1.3%)	14 (5.3%)	4 (1.5%)
MUSCULOSKELTAL AND CONNECTIVE TISSUE DISORDERS				
Back pain	42 (7.9%)	9 (1.7%)	18 (6.8%)	2 (0.8%)
NERVOUS SYSTEM DISORDERS				
Dysgeusia	36 (6.8%)	0 (0%)	6 (2.3%)	0 (0%)
Headache	29 (5.4%)	0 (0%)	13 (4.9%)	0 (0%)

Trifluridine/tipiracil combination (metastatic colorectal cancer) NDA 207-981. Pages 85–7 of 121-page Medical Review.

also reveals the option of breaking down the frequency of each AE according to severity of that AE, and where frequency is disclosed in terms of number of study subjects and percentage.

VI. CONCLUDING REMARKS

Package labels disclose AEs, where the primary basis is AEs detected in the course of the Sponsor's clinical studies.

It is conceivable that the package label can be drafted by a review of the raw data, that is, AE information that is captured by way of CRFs. On the other hand, the huge number of study subjects for most clinical trials prevents this approach. Moreover, information on CRFs usually takes the form of verbatim terms. Verbatim terms are distinguished in that the terms chosen for any given AE may be somewhat different, depending on the health-care worker filling out the CRF. Also, verbatim terms may be ambiguous, or be somewhat informal, or may include grammatical errors, such as the verbatim term, "His glucose was 40"[115] (the error was that the unit of concentration was not stated).

Moreover, the CRF may include raw data on AEs as acquired directly from devices, such as a blood pressure measuring device or a glucometer (the information taking the form of numbers or photographs). On this point, FDA's Guidance for Industry reveals that an electronic CRF (eCRF) can accept data directly from devices.[116,117] Verbatim terms as well as data from devices must be coded using a uniform dictionary. The dictionary most commonly used is Medical Dictionary for Regulatory Activities (MedDRA).

An intermediary between the verbatim terms on the CRF and the package label is the process of coding the verbatim terms into MedDRA Preferred Terms and into MedDRA SOC terms. Splitting and lumping are frequent issues in coding. Splitting and lumping can be problematic and can result in complaints from FDA reviewers.

Exemplary accounts of splitting, lumping, and miscoding, as provided by this chapter, include the following:

- **This concerns splitting.** Verbatim terms relating to dystonia, as captured on CRFs from different study subjects, suffered from splitting because of the fact that coding resulted in their translation into the Preferred Terms of dystonia, oculogyric crisis, oromandibular dystonia, and torticollis (instead of their all being coded to the Preferred Term of dystonia).

[115] MedDRA MSSO. MedDRA® term selection: points to consider. ICH-Endorsed Guide for MedDRA Users. McLean, VA: MedDRA MSSO. Sept. 2016 (57 pp.).

[116] FDA's Guidance for Industry includes this narrative: "Automatic Transmission of Data From Devices or Instruments Directly to the eCRF ... a device or instrument is the data originator (e.g., blood pressure monitoring device or glucometer) and data are automatically transmitted directly to the eCRF."

[117] U.S. Department of Health and Human Services. Food and Drug Administration. Center for Drug Evaluation and Research (CDER). Center for Biologics Evaluation and Research (CBER). Center for Devices and Radiological Health (CDRH). Guidance for Industry. Electronic source data in clinical investigations. 2013 (11 pp.).

- **This concerns lumping.** A desirable consequence of lumping was shown by FDA's review of 3-hydroxybutyrate. The verbatim terms "lethargy" and "increased sleepiness" were lumped together as the Preferred Term "somnolence" instead of to differing Preferred Terms. This lumping prevented undercounting of the AE, that is, lumping prevented this AE from becoming invisible to persons evaluating the AEs arising during the clinical study.
- **This concerns miscoding.** The verbatim term of "upgoing eyes" was coded to the Preferred Term of "eye rolling," instead of to a term that is more capable of capturing the AE associated with the study drug. The more capable Preferred Term is dystonia.

CHAPTER

14

Pooling

I. INTRODUCTION

The term "pooling" refers to the technique of combining data on efficacy or safety from separate clinical studies, where the goal is to perform data analysis on the pooled data. Pooled data can provide conclusions that are statistically more significant than when data are from only one of the Sponsor's clinical studies.

a. Justification and Advantages of Pooling

A comment from FDA's Guidance for Industry refers to criteria to justify pooling, as well as to an advantage of pooling, "When significant heterogeneity is not observed, pooling can reveal subpopulation differences that could not be detected in individual studies."[1]

Referring to one aspect of "heterogeneity," a published account from pooling of two different clinical studies referred to lack of heterogeneity in "baseline characteristics." The publication, which concerned maraviroc for treating HIV-1, stated the justification for pooling, "A pooled analysis was performed because the studies had identical designs, with similar baseline characteristics for each study population."[2]

A published study of colorectal cancer and fluoropyrimidine therapy stated that an advantage of pooling was to test hypotheses that would have been impossible with

[1] U.S. Department of Health and Human Services. Food and Drug Administration. Center for Drug Evaluation and Research (CDER). Center for Biologics Evaluation and Research (CBER). Integrated summary of effectiveness guidance for industry; 2015 (17 pp.).

[2] Fatkenheuer G, et al. Subgroup analyses of maraviroc in previously treated R5 HIV-1 infection. New Engl. J. Med. 2008;359:1442–55.

FDA's Drug Review Process and the Package Label
DOI: https://doi.org/10.1016/B978-0-12-814647-7.00014-2

analysis of individual clinical studies. The name "ACCENT" was used to refer to all of the clinical studies that were pooled. The term "power" refers to statistical power. The publication stated that:

> The ACCENT database provides the advantage of pooling large, mature clinical trials to test hypotheses that are **difficult or impossible to test** within individual trials. Given the fact that none of the included trials had over one quarter of the patients age equal or greater than 70 years, subset analyses by age in the individual studies have limited power.[3]

A published study on colorectal cancer treated with bevacizumab stated that the advantage of pooling was better ability to detect real differences, "Pooling of the data from the three studies allows evaluation of these three efficacy end points with greater statistical power **to detect real differences** between the groups of patients who were and were not treated with bevacizumab."[4]

A published study on depression treated with levomilnacipran revealed that **subgroup analysis** can be facilitated by pooling data from separate clinical studies. This concerns the subgroup of men enrolled in the clinical study. The publication stated that, "One advantage of pooling data from 5 studies was that it provided a robust sample of men (n = 941) in which to evaluate the effects of levomilnacipran on depressive symptoms. **This subgroup was of particular interest** because of the **lower prevalence of major depressive disorder in men** than women and potentially lower response to antidepressant treatment in men."[5]

b. Details from FDA's Guidance for Industry on Pooling

FDA's Guidance for Industry cautions that study designs should be similar in the clinical studies to be pooled. The take-home lesson is that, when contemplating pooling, the Sponsor should consider differences in study design, in baseline disease characteristics of the study subjects, and differences in subgroups of the study subjects, such as the subgroups of gender, age, prior therapy, concomittant conditions such as renal impairment, race, genetic biomarkers, and so on. Excerpts from FDA's Guidance for Industry teach that:

> When pooling data, sponsors should consider the possibility that various sources of systematic differences can interfere with interpretation of a pooled result. To ensure that pooling is appropriate, sponsors should confirm that study designs, as well as ascertainment and measurement strategies employed in the studies that are pooled, are **reasonably similar**.[6]

[3] McCleary NJ, et al. Impact of age on the efficacy of newer adjuvant therapies in patients with stage II/III colon cancer: findings from the ACCENT database. J. Clin. Oncol. 2013;31:2600–6.

[4] Kabbinavar FF, et al. Combined analysis of efficacy: the addition of bevacizumab to fluorouracil/leucovorin improves survival for patients with metastatic colorectal cancer. J. Clin. Oncol. 2005;23:3706–12.

[5] Montgomery SA, et al. Efficacy of levomilnacipran extended-release in major depressive disorder: pooled analysis of 5 double-blind, placebo-controlled trials. CNS Spectr. 2015;20:148–56.

[6] U.S. Department of Health and Human Services. Food and Drug Administration. Center for Drug Evaluation and Research (CDER). Center for Biologics Evaluation and Research (CBER). Guidance for industry. Premarketing risk assessment; 2005 (26 pp.).

FDA's Guidance for Industry cautions that differences in study design, in the patient populations, or in data analysis, can mean that pooling is not appropriate:

> Some issues for consideration in deciding whether pooling is appropriate include possible differences in the duration of studies, heterogeneity of patient populations, and case ascertainment differences across studies (i.e., different methods for detecting the safety outcomes of interest, such as differences in the intensities of patient follow-up).[7]

FDA's Guidance for Industry warns that:

> **Differences among individual studies** can affect the validity and interpretability of pooled analyses ... caution is warranted when the studies differ with respect to ... demographic or disease characteristics (e.g., duration, severity, specific signs and symptoms, previous treatment, concomitant diseases and treatments, prognostic or predictive biomarkers) ... [t]reatment practices, including methods of assessing effectiveness, specific test procedures ... [s]tudy design features (e.g., study duration, study size, doses studied, allocation ratio to treatment and control arms, visit frequency) ... [s]uch differences can make study results heterogeneous, requiring cautious interpretation. Thus, **consistency of populations and treatment practices across studies** should be examined before pooling.[8]

c. Publications Comment on Dissimilarities and Heterogeneities Among Clinical Studies to be Pooled

In contemplating whether data should be pooled, researchers sometimes express published doubts on whether pooling is appropriate:

- A publication on methotrexate for treating giant cell arteritis stated, "data pooling was inappropriate because of dissimilarities between studies."[9]
- A publication on a vaccine for respiratory syncytial virus stated, "Absence of homogeneity indicated pooling was inappropriate and no definitive conclusions can be drawn from the data."[10]
- A study on ipilimumab for melanoma stated, "a limitation of the pooled overall survival analysis is the different follow-up times of the studies."[11] This comment referred to one of the features of clinical study design, namely, the follow-up phase of

[7] U.S. Department of Health and Human Services. Food and Drug Administration. Center for Drug Evaluation and Research (CDER). Center for Biologics Evaluation and Research (CBER). Guidance for industry. Premarketing risk assessment; 2005 (26 pp.).

[8] U.S. Department of Health and Human Services. Food and Drug Administration. Center for Drug Evaluation and Research (CDER). Center for Biologics Evaluation and Research (CBER). Integrated summary of effectiveness guidance for industry; 2015 (17 pp.).

[9] Mahr AD, et al. Adjunctive methotrexate for treatment of giant cell arteritis: an individual patient data meta-analysis. Arthr. Rheum. 2007;56:2789–97.

[10] Glenn GM, et al. A randomized, blinded, controlled, dose-ranging study of a respiratory syncytial virus recombinant fusion (F) nanoparticle vaccine in healthy women of childbearing age. J. Infect. Dis. 2016;213:411–22.

[11] Schadendorf D, et al. Pooled analysis of long-term survival data from phase II and phase III trials of ipilimumab in unresectable or metastatic melanoma. J. Clin. Oncol. 2015;33:1889–94.

the clinical studies. Data from 12 clinical studies were pooled. From study to study, the follow-up time varied from about 1 year to about 10 years. The term "overall survival" is a standard measure of efficacy used in cancer clinical trials. "Overall survival analysis" refers to comparing overall survival of subjects in the study drug arm verus overall survival of subjects in the control arm.

- A study on ascorbate for treating pneumonia stated, "We planned that if a number of trials were available with sufficient uniformity in settings and outcome definitions, we would pool the data; but, if the trials were heterogeneous ... we would present them separately ... [but] the studies are clinically so divergent that pooling was inappropriate."[12]

d. Arguments to Justify Pooling Can be Made on Information Available at Study Initiation and Also Based on Information at Study Conclusion

Sponsors wanting to know if pooling is appropriate should compare the degree of balance in each of the several clinical trials at their *initiation*. Also, Sponsors interested in pooling data should compare the degree of balance in each of the several clinical studies at their *completion*. Imbalance typically can result when subjects die, withdraw, drop out, or are lost during the follow-up period, to a greater extent in the study drug arm than in the control treatment arm. The general fact that FDA scrutinizes data on efficacy and safety among the various subgroups in any given clinical study is exemplified by a comment on the vemurafenib clinical trial for melanoma, "Reviewer's Comment: A treatment effect in favor of vemurafenib treatment was observed across subgroups."[13]

II. POOLING EXAMPLES FROM FDA'S MEDICAL REVIEWS

a. Imbalance Existing at Baseline (Example of Vemurafenib)

This concerns subgroups of study subjects defined by a genetic biomarker taking the form of the V600E mutation, that is, whether a subject has or does not have this mutation. The V600E mutation occurs in the BRAF kinase gene in a modest percentage of cancer patients. Vemurafenib is a small molecule drug that inhibits BRAF kinase and is used for treating various types of cancer.[14]

In FDA's Statistical Analysis Review for vemurafenib, FDA complained about an imbalance in one particular subgroup, where the baseline ratio of:

> [Number of vemurafinib treatment V600E subgroup subjects]/[number of control treatment V600E subgroup subjects]

[12] Hemila H, Louhiala P. Vitamin C for preventing and treating pneumonia. Cochrane Database of Systematic Reviews. 2013;Issue 8. Art. No.: CD005532. doi: 10.1002/14651858.CD005532.pub3.

[13] Vemurafenib (melanoma). Pages 27 and 37 of 105-page Medical Review.

[14] Kopetz S, et al. Phase II pilot study of vemurafenib in patients with metastatic BRAF-mutated colorectal cancer. J. Clin. Oncol. 2015;33:4032–8.

did *not have the same value* as the baseline ratio for:

> [Total number of vemurafenib treatment subjects]/[total number of control treatment subjects]

FDA complained that the resulting data were of doubtful use, and could only be considered to be exploratory, "Among the randomized patients, [in] a subgroup of ... 111 vemurafenib patients, 109 dacarbazine [control treatment] patients ... 164 were detected as V600E ... **[t]hese analyses were considered exploratory since the subgroup patients may not preserve the balance** in baseline characteristics as randomized patients."[15]

b. Imbalance Caused by Discontinuations (Example of Ceftolozane/Tazobactam Combination Drug)

FDA's Statistical Review for ceftolozane/tazobactam for infections complained that discontinuations by study subjects, during the course of the clinical trials, resulted in an imbalance, "This imbalance was driven primarily by 21 (5.4%) of subjects in the ceftolozane/tazobactam ... arm compared to 11 (2.6%) of subjects in the ... [control treatment] arm **prematurely discontinuing study drug**."[16]

c. Issue of Imbalances in Inclusion/Exclusion Criteria (Example of Ceftolozane/Tazobactam Combination)

Where data are pooled, FDA scrutinizes whether there were differences in inclusion/exclusion criteria in the various studies that are to be pooled. Inclusion criteria and exclusion criteria, as well as efficacy endpoints, treatment duration, and so on, are always set forth in the Clinical Study Protocol, and any differences (in Clinical Study Protocols for different clinical studies) will necessarily result in differences in the characteristics of subjects enrolling in the different clinical studies. FDA's scrutiny of inclusion/exclusion criteria in different studies to be pooled is shown by FDA's remarks. FDA's remarks referred to the Clinical Study Protocols used for the different clinical studies, "During the design stage of Trial 08 and Trial 09, the Division considered **pooling to be acceptable** due to the similarity of Protocol 08 and Protocol 09. These protocols had **identical entry criteria**, primary efficacy variables, trial drug dosing regimen, comparator, treatment duration, outcome and safety assessments."[17]

Further regarding pooling of data from the various clinical studies on ceftolozane/tazobactam combination, FDA complained about imbalances among some of the subgroups and raised the possibility that these imbalances would result in "confounding" of the data interpretation. The subgroups included age group and prior therapy. FDA articulated its

[15] Vemurafenib (melanoma) 202-429. 41-page Statistical Review.

[16] Ceftolozane/tazobactam combination (infections) 206-829. Pages 22 and 33 of 118-page Statistical Review.

[17] Ceftolozane/tazobactam combination (infections) 206-829. Pages 22 and 33 of 118-page Statistical Review.

concern for possible errors arising from, "differences between protocols for 3 risk factors including use of prior antibiotics ... and age ≥ 65 ... such differences can ... **result in imbalances and confounding** in the ... analysis."[18]

Regarding the differences in clinical study design, as revealed by Clinical Study Protocols for each study, FDA decided that these differences were small and would have only "minimal effect" on the analysis:

> While **comparisons failed to show strong heterogeneity across protocols that would prohibit pooling**, they did show differences between protocols for 3 risk factors including use of prior antibiotics (47.2% vs. 68.1%) ... and age ≥ 65 (19.5% vs. 26.6%). Since such differences can potentially result in imbalances and confounding in the primary analysis, the Reviewer conducted further sensitivity analyses. These analyses showed that adjusting for the protocol had a **minimal effect on** ... **analysis findings.**[19]

d. Study Design Differences as a Reason to Refrain From Pooling (Example of Denosumab)

This concerns pooling of data from a multiplicity of clinical studies on denosumab, a drug for osteoporosis.[20] FDA's Medical Review warned against pooling of data from different clinical studies, because the studies differed in study design. The differences in study design were use of a *double-blind clinical study* versus an *open label clinical study*. FDA observed that there were differences in the rates of two adverse events (musculoskeletal AEs; gastrointestinal AEs), when comparing the differently designed studies, but FDA was not able to determine if differences in study design resulted in the differences in the detected AEs. FDA disclosed the reason for refraining from pooling:

> Data in this CR Safety Update have **not been pooled across trials** ...because of **significant differences in study population and design** ... [i]t is not clear whether the change from ***double-blind*** to ***open label*** would affect reporting of AEs ... overall AEs especially musculoskeletal and gastrointestinal (GI) ... were reported less frequently.[21]

e. Differences in Subject Demographics and in Drug Formulations as Reasons to Refrain From Pooling (Example of Lixisenatide)

FDA's Medical Review of lixisenatide for type-2 diabetes reveals that the Sponsor had conducted 20 clinical studies where the AE data were pooled, and additional nine clinical

[18] Ceftolozane/tazobactam combination (infections) 206-829. Pages 22 and 33 of 118-page Statistical Review.

[19] Ceftolozane/tazobactam combination (infections) 206-829. Pages 22 and 33 of 118-page Statistical Review.

[20] Denosumab (post-menopausal osteoporosis) BLA 125-320. 710-page Medical Review.

[21] Denosumab (osteoporosis) BLA 125-320. Page 79 of 710-page Medical Review.

studies where no pooling was done.[22] FDA described the absence of any pooling, referring to nine "stand-alone" studies that were distinct from each other and distinct from all of the other studies, and five other clinical studies that were distinguished by their different formulations. FDA stated that the reason for not pooling included the fact that one of the studies had special patients (suffering from renal impairment), and that other studies had special formulations of the study drug:

> There are nine completed clinical pharmacology studies presented as stand-alone studies in the NDA. The **applicant's rationale for not including these studies in the integrated safety database** is as follows: Four stand-alone studies in special/different populations with longer treatment duration ... [s]pecial population [was] PEP6053 (patients with renal impairment), POP11814 (elderly patients), and POP11320 (healthy Chinese subjects) ... [d]ifferent population and longer treatment duration: TDR11215 (spermatogenesis study in overweight or obese subjects ... studies with different formulations ... premixed or simultaneous injections of lixisenatide and insulin glargine ... prolonged release formulation of lixisenatide.[23]

The following bullet points list the four stand-alone studies and the five studies with "different formulations." The abbreviations are the names of each study[24]:

- POP11320. Pharmacokinetics study on healthy Chinese subjects
- POP11814. Pharmacokinetics study on elderly subjects
- POP6053. Pharmacokinetics study on subjects with renal impairment
- TDR11215. Spermatogenesis effects study with obese subjects
- BDR10880. Injection with formulation of study drug plus insulin glargine
- BDR11038. Injection with formulation of study drug plus insulin glargine
- BDR11540. Injection with formulation of study drug plus insulin glargine
- BDR11578. Injection with formulation of study drug plus insulin glargine
- TDU10121. Prolonged release formulation

f. Differences in Study Design as a Reason to Avoid Pooling (Example of Lenvatinib)

FDA's review of lenvatinib for thyroid cancer provides guidance of general interest, regarding pooling efficacy data.[25] FDA's Medical Review of lenvatinib reveals the situation where pooling efficacy data should not be performed where differences in study design included eligibility criteria, prior drug treatment, and endpoints.[26]

[22] Lixisenatide (type-2 diabetes) NDA 208-471. Pages 47, 56, 176, 237, and 253 of 690-page Medical Review.

[23] Lixisenatide (type-2 diabetes) NDA 208-471. Pages 47, 56, 176, 237, and 253 of 690-page Medical Review.

[24] Lixisenatide (type-2 diabetes) NDA 208-471. Pages 47, 56, 176, 237, and 253 of 690-page Medical Review.

[25] Lenvatinib (radioiodine-refractory differentiated thyroid cancer) NDA 206-947. Pages 55 of 185-page Clinical Review and page 58 of 185-page pdf file of Medical Review.

[26] Lenvatinib (radioiodine-refractory differentiated thyroid cancer) NDA 206-947. Pages 32 and 55 of 185-page Clinical Review and pages 35 and 58 of 185-page pdf file of Medical Review.

In FDA's review of lenvatinib, FDA referred to three of the Sponsor's clinical studies, and pointed out differences in study design for these three studies. FDA concluded that the Sponsor should not pool data from the three studies and should, instead, draft individual side-by-side summaries of the efficacy data. FDA wrote:

> Studies 303, 201, and 208 **differed with respect to many aspects including study design and endpoints** ... **eligibility criteria** ... **region of study conduct**, tumor progression assessment criteria, criteria for study drug discontinuation ... [t]his reviewer hence recommends that the readers **use caution in performing cross trial comparisons and interpreting pooled data for efficacy**. The applicant has recognized this in the submission and has summarized each trail individually in the Integrated Summary of Efficacy (ISE) and has compared the three trials side-by-side which is a reasonable approach to analyzing the data.[27]

g. Differences in Subject Baseline Characteristics as a Reason to Avoid Pooling (Example of Sofosbuvir/Velpatasvir Combination)

This concerned four different clinical studies, ASTRAL-1, ASTRAL-2, ASTRAL-3, and ASTRAL-4. FDA observed that data from three of these studies, ASTRAL-1, ASTRAL-2, and ASTRAL-3, were pooled to form the integrated safety population. However, data from ASTRAL-4 were not pooled.

The reason to refrain from pooling was that the *dose levels and durations of the study drug were lower* in the ASTRAL-4 clinical trial. Another reason to refrain from pooling was that, "Serious adverse events and Grade 3 and 4 adverse events **in these lower-dose/duration populations** were considered to be **significant predictors of potential drug-related toxicity** and were therefore reviewed but were not pooled with other trials."[28]

FDA further stated the reason to avoid pooling and the need to analyze ASTRAL-4 data separately, "Data from ASTRAL-4 were analyzed separately because we anticipated that the ***frequency and severity of AEs may differ in this population*** of decompensated cirrhotics compared to the integrated safety population, and that pooling of the data may confound interpretation of the safety results."[29]

Compensated study subjects were enrolled in the ASTRAL-1, ASTRAL-2, and ASTRAL-3 clinical studies[30] while decompensated study subjects were enrolled in the ASTRAL-4 clinical study.[31] The ASTRAL-1 clinical trial for subjects infected with hepatitis C virus

[27] Lenvatinib (radioiodine-refractory differentiated thyroid cancer) NDA 206947. Pages 55 of 185-page Clinical Review and page 58 of 185-page pdf file of Medical Review.

[28] Sofosbuvir/velpatasvir combination (hepatitis C virus) NDA 208-341. Page 39 of 171-page Medical Review.

[29] Sofosbuvir/velpatasvir combination (hepatitis C virus) NDA 208-341. Pages 88–89 of 171-page Medical Review.

[30] Sofosbuvir/velpatasvir combination (hepatitis C virus) NDA 208-341. Pages 3, 17, 77, and 127 of 171-page Medical Review.

[31] Sofosbuvir/velpatasvir combination (hepatitis C virus) NDA 208-341. Pages 64, 65, 74–76, 78, 85, 97, 107, 113, and 126–127 of 171-page Medical Review.

(HCV), and having *compensated* cirrhosis, was published.[32] The terms "compensated" and "decompensated" refer to different stages of cirrhosis, where ascites and esophageal bleeding are the first complications to occur and to mark the transition to decompensated cirrosis and of eventual liver-related death.[33] The ASTRAL-2 and ASTRAL-3 studies on subjects with *compensated* cirrhosis were published in one paper.[34] The ASTRAL-4 clinical trial for subjects infected with HCV, and having *decompensated* cirrhosis, was published in a separate paper.[35]

h. Similarity in Study Design to Justify Pooling (Example of Fingolimod)

FDA's Medical Review for fingolimod reveals use of a Pre-NDA Meeting with FDA to agree upon an appropriate pooling strategy. The proposed pooling was for several clinical studies with durations of 6 months, 12 months, and 24 months. The goal of pooling was, "Safety data in the multiple sclerosis population were pooled in 5 different groups, to better assess and compare risk/rate of events."[36] FDA approved this pooling strategy because differences in duration of the various clinical studies were not great, writing, "The pooling strategy was discussed with the Agency prior to the NDA submission. Given the difference in the duration of these studies, using several pools to evaluate safety was considered appropriate. All five safety pools are relevant."[37]

FDA referred to that fact that the Sponsor and FDA had discussed and agreed upon the pooling strategy before submitting the NDA. Pre-NDA and Pre-BLA meetings find a basis in 21 CFR §312.47, which states that:

> FDA has found that delays associated with the initial review of a marketing application may be reduced by exchanges of information about a proposed marketing application. The primary purpose of this kind of exchange is to uncover any major unresolved problems.

i. Differences in the Disease to be Treated as a Reason to Avoid Pooling (Example of Ponatinib)

This concerns FDA's Clinical Review of ponatinib, a drug for various hematological cancers.[38] Ponatinib is a protein kinase inhibitor used for treating, for example, chronic

[32] Feld JJ, et al. Sofosbuvir and velpatasvir for HCV genotype 1, 2, 4, 5, and 6 infection. New Engl. J. Med. 2015;373:2599–607.

[33] Benvegnu L, et al. Natural history of compensated viral cirrhosis: a prospective study on the incidence and hierarchy of major complications. Gut 2004;53:744–9.

[34] Foster GR, et al. Sofosbuvir and velpatasvir for HCV genotype 2 and 3 infection. New Engl. J. Med. 2015;373:2608–17.

[35] Curry MP, et al. Sofosbuvir and velpatasvir for HCV in patients with decompensated cirrhosis. New Engl. J. Med. 2015;373:2618–28.

[36] Fingolimod (multiple sclerosis) NDA 022-527. Page 151 of 519-page Medical Review.

[37] Fingolimod (multiple sclerosis) NDA 022-527. Page 151 of 519-page Medical Review.

[38] Ponatinib (CP-CML; AP-CML, BP-CML; Ph + ALL) NDA 203-469.

myeloid leukemia (CML) and Philadelphia chromosome positive acute lymphoblastic leukemia (Ph + ALL).[39] The concepts shown below are:

- Pooling of data from two different clinical studies (Study 07-101 and Study 10-201)
- Separating out two of the subgroups (Ph + ALL and BP-CML) for their own analysis, from a clinical study involving four different types of leukemia (CP-CML, AP-CML, BP-CML, and Ph + ALL).

The goal of pooling was to increase sensitivity for detecting AEs in the safety data from two separate clinical studies. FDA observed that, "Pooled analysis of the adverse events in Study 07-101 and 10-201 was performed to increase the sensitivity of detecting the adverse events."[40]

Regarding separating out one or more subgroups, subjects from these two clinical studies suffered from a variety of hematological cancers, CP-CML, AP-CML, BP-CML, and Ph + ALL. The biological characteristics and treatment history of two of these cancers (BP-CML and Ph + ALL) were so different from each other, that data from these two subject populations were not combined with each other. FDA's Medical Review wrote:

> The safety analyses for patients with BP-CML were separated from those patients with Ph + ALL due to **differences in the underlying disease biology, treatment history** ... [i]n addition, 82% of the patients (51/62) with BP-CML had myeloid blast phase, 18% had lymphoid blast phase, **which further supports the separated safety analyses** of patients with Ph + ALL from patients with BP-CML.[41]

j. Similarities in Adverse Events in Different Studies to Justify Pooling (Example of Alirocumab)

The Sponsor had pooled AE data from two different clinical studies, each using a different dosing scheme. One clinical study used 150 mg alirocumab only while the second clinical study used a titrated dose (starting with 75 mg, then increasing to 150 mg). AEs from these two groups were compared with AEs from the placebo arm. FDA commented on pooling of data from the constant dose group (150 mg Q2W) with data from the titrated dose group (75/150 mg Q2W):

> The safety analyses combined the two dose regimens of alirocumab (75/150 mg Q2W and 150 mg Q2W) and compared this alirocumab-treated group against the comparator ... this **pooling of alirocumab doses is acceptable** based on the following analyses which did not demonstrate a meaningful difference in adverse events by dose or treatment regimen.[42]

[39] Shamroe CL, Comeau JM. Ponatinib: a new tyrosine kinase inhibitor for the treatment of chronic myeloid leukemia and Philadelphia chromosome-positive acute lymphoblastic leukemia. Ann. Pharmacother. 2013;47:1540–6.

[40] Ponatinib (CP-CML; AP-CML, BP-CML; Ph + ALL) NDA 203-469. Pages 61–62 of 108-page Medical Review.

[41] Ponatinib (CP-CML; AP-CML, BP-CML; Ph + ALL) NDA 203-469. Pages 61–62 of 108-page Medical Review.

[42] Alirocumab (hyperlipidemia) BLA 125-559. Page 288 of 372-page Medical Review.

TABLE 14.1 Comparing AE Frequency in 150 mg Alirocumab Group (Constant Dose) With 75 mg/150 mg Alirocumab Group (Titrated Dose)

	Placebo (%)	Alirocumab (150 mg) (%)	Placebo (%)	Alirocumab (75 mg/150 mg) (%)
Treatment emergent AEs (TEAEs)	80.2	77.8	75.5	75.1
TEAEs leading to death	1.0	0.4	0.9	0.9
TEAEs leading to treatment discontinuation	5.3	6.1	4.8	4.0

Alirocumab (hyperlipidemia) BLA 125-559. Page 291 of 372-page Medical Review.

FDA's Medical Review included a table comparing AEs from the 150 mg dose group with the AEs from the 75 mg/150 mg dose titration group.[43] The table, reproduced below (Table 14.1), does reveal differences in AEs by the dosing schedule but these are extremely small. According to the FDA reviewer, the data "did not demonstrate a meaningful difference."

k. Differences in Racial Subgroups, and in Concomitant Medications, in Different Studies as Reason to Avoid Pooling the Studies (Example of Alirocumab and Example of Tofacitinib)

The following account provides the take-home lesson that one acceptable strategy for pooling, is to pool data on safety or efficacy from a multiplicity of studies, but to exclude data from certain subgroups. This concerns tofacitinib for rheumatoid arthritis. An advantage of pooling was stated in terms of improved detection of "safety signals" for rarely occurring AEs:

> **Pooling data from the 12 month safety data from these clinical trials** was used to **improve the precision of an incidence estimate**. This was considered particularly important for detecting a potential safety signal for rare events, such as malignancy, serious infections, gastrointestinal perforations, which were events of special interest based on the immunosuppressive properties of tofacitinib [the study drug].[44]

Also, FDA's Medical Review focused on reasons to exclude data from certain studies from pooling, namely, that one particular study consisted of subjects from a different racial group, and because of the fact that in a particular clinical study, subjects were taking background medications. FDA stated that "more accurate" comparison would result by excluding Japanese studies from pooling:

> Long-term extension study A3921041 included patients who completed Japanese Phase 2 studies which were **not included in the pooled analyses** of the controlled studies. Therefore, excluding these patients in the open label extension analyses may allow for more accurate comparison with the analyses from the controlled studies.[45]

[43] Alirocumab (hyperlipidemia) BLA 125-559. Page 291 of 372-page Medical Review.

[44] Tofacitinib (rheumatoid arthritis) NDA 203-214. Page 117 of 303-page Medical Review.

[45] Tofacitinib (rheumatoid arthritis) NDA 203-214. Page 30 of 303-page Medical Review.

The fact that the Sponsor had originally planned to include data from Japanese clinical studies, and that FDA subsequently complained about this pooling strategy, is shown by, "FDA Response: We acknowledge your methodological considerations, but we do not agree with your proposal. One difference is the studies for inclusion in the safety analysis. Studies 1025 and 1035 are of sufficiently similar design and patient population to the Phase 3 studies. Therefore, we request that they are included in the integrated safety analyses. **We do not agree with inclusion of the Japanese studies due to differences in patient population, background medication use.**"[46]

FDA defined background medications as follows. Note that some background medications (NSAIDs, opioids) were permitted throughout the clinical study, while other background medications (intravenous corticosteroids) were forbidden:

> **Concomitant Medications**. Patients continued on their stable **background arthritis therapy** which may have included any of the traditional . . .anti-rheumatic drugs (DMARDs) at the dosage range and intervals mentioned in the inclusion criteria. All patients were required to follow the standard of care monitoring required for such background therapies. Stable **background arthritis therapy** could include nonsteroidal antiinflammatory drugs (NSAIDs), selective cyclooxygenase-2 (COX-2) inhibitors . . . throughout the study. Intravenous or intramuscular corticosteroids . . . were not allowed during this study.[47]

l. Normalization to Justify Pooling of Safety Data From Different Clinical Studies (Example of Denosumab)

The following provides the take-home lesson that, an acceptable pooling strategy is to adjust for subject-years of exposure, before pooling the data. This reveals FDA's justification of pooling of safety data from two different clinical studies, Trial 2003-0216 and Trial 2006-0289. The first clinical trial was Trial 2003-0216. These two trials are related to each other, according to the Sponsor's Briefing Document, but the first study lasted 3 years and the second study was an "extension study" lasting 7 years. As described in the Briefing Document, "Study 2006-0289: This is an ongoing, multinational, multicenter, open-label, single-arm . . . extension study to evaluate long term safety and sustained efficacy of denosumab in the treatment of postmenopausal osteoporosis; enrolled 4550 subjects who completed Study 2003-0216."[48,49]

The normalization procedure enabling pooling the AE data from the two studies was, "in order to investigate possible changes in denosumab AEs over time, the Applicant compared AE rates in trial 2003-0216 and by **adjusting for subject-years of exposure** for each treatment group and each prior-treatment group, respectively. Thus, in the group exposed

[46] Tofacitinib (rheumatoid arthritis) NDA 203-214. Pages 33 and 80 of 303-page Medical Review.

[47] Tofacitinib (rheumatoid arthritis) NDA 203-214. Page 262 of 303-page Medical Review.

[48] Briefing Document. Background document for meeting of advisory committee for reproductive health drugs (August 13, 2009), Denosumab (Proposed trade name: PROLIA) Amgen, Inc. (page 22 of 87 pp.).

[49] Briefing Document. Background information for the meeting of the advisory committee for reproductive health drugs; August 13, 2009 (page 32 of 147 pp.).

to continuous denosumab for 4–5 years, the rate of each AE during the first 3 years could be compared to its rate in the subsequent ... years."[50] FDA further commented that, "This method of comparison is ... valid because AEs were assessed at the same intervals ... in both parent and follow-up study."

m. Definition of the Unit "Subject-Years"

The unit "subject-years" is defined by this formula[51,52]:

Subject-years = [number of subjects with the AE]/[sum of AEs for all subjects for time from baseline to AE]

The numerical value for subject-years is a function of only subjects that actually experience the AE in question. If a given subject did not experience an AE, then data from that subject is not used to calculate the value of subject-years. This illustrates the use, in FDA's Medical Review, of the unit "subject-years adjusted AEs."[53] Turning to data using the unit, "AEs per 100 subject-years," FDA wrote, "When adjusted for exposure, the rates of ONJ were **similar** ... **across the studies** ... the cumulative rate of ONJ at year 1 was approximately 1 event per subject-years, and at years 2 and 3, was approximately 2 events, per 100 subject years."[54]

III. SIMILAR VALUES FOR ALLOCATION RATIOS, AS A CONDITION FOR APPROPRIATE POOLING

a. Introduction

Allocation and allocation ratios need to be considered before pooling data. Study design for most clinical trials begins with the enrollment, randomization, and allocation of study subjects. Randomization refers to the assignment of a number to each subject, where the number is then assigned to the study drug group or to the control treatment group. Allocation ratio refers to the ratio:

[Number of subjects in study drug arm]/[number of subjects in control treatment arm]

A one-to-one allocation ratio is the most common form of allocation. Sometimes different allocation ratios are chosen, for example, where the investigational treatment is

[50] Denosumab (osteoporosis) BLA 125-320. Page 79 of 710-page Medical Review.

[51] Windeler J, Lange S. Events per person year—a dubious concept. Br. Med. J. 1995;310:454–6.

[52] Vandenbroucke JP. Farr's "on prognosis." A history of epidemiological methods and concepts (A. Morabia, editor). Basel: Birkhauser-Verlag; 2004. p. 191–193.

[53] Amgen. Background information for the meeting of the oncology drugs advisory committee; February 8, 2012 (124 pp.).

[54] Denosumab (osteoporosis) BLA 125-320. Page 79 of 710-page Medical Review.

unusually expensive or complicated to administer.[55] Another reason to choose a ratio other than [1]/[1] is the situation where the study drug is thought to be greatly superior to the control treatment and where, for example, a ratio of study drug/control of [3]/[1] is used.[56]

FDA's Guidance warns against an imbalance taking the form of different allocation ratios occurring in different clinical studies, writing, "Differing allocation ratios across studies need to be taken into account in pooling."[57]

Nested within the process of allocation is the process of stratification. With allocation of study subjects to the study drug arm and control treatment arm, stratification can be defined by subgroups, such as, gender, disease severity, race (black, white), geographic region, severity of concomitant disorders (renal impairment), and so on.[58,59]

Stratification ensures, for example, that for each male subject allocated to the study drug arm, another male subject is allocated to the control arm, or that for each subject with Stage IIIC cancer allocated to the study drug arm, another subject with Stage IIIC cancer is allocated to the control arm, or that for each subject suffering from renal impairment allocated to the study drug arm, another subject with renal impairment is allocated to the control arm.

b. Imbalance in Allocation (Example of Vemurafenib)

If study design requires an allocation ratio of [1]/[1], then this same ratio is imposed for each subgroup. This provides an example from a clinical study on melanoma that compared vemurafenib (study drug arm) with dacarbazine (control treatment arm). Referring to allocation by various subgroups (geographic region, disease severity, and a serum enzyme level), FDA's Medical Review described the study design as:

> The treatment allocation was based on ... the following balancing factors: Geographic region (North America, Western Europe, Australia/New Zealand), ... Metastatic classification (unresectable Stage IIIC, M1a, M1b, and M1c), and Serum lactate dehydrogenase (LDH) normal vs. LDH elevated.[60]

[55] Kirby A, et al. Determining the sample size in a clinical trial. Med. J. Aust. 2002;177:256–7.

[56] Kirby A, et al. Can unequal be more fair? Ethics, subject allocation, and randomised clinical trials. J. Med. Ethics 1998;24:401–8.

[57] U.S. Department of Health and Human Services Food and Drug Administration Center for Drug Evaluation and Research (CDER) Center for Biologics Evaluation and Research (CBER) (2015) Integrated Summary of Effectiveness Guidance for Industry (17 pp.).

[58] Kundt G, Glass A. Evaluation of imbalance in stratified blocked randomization. Methods Inf. Med. 2012;51:55–62.

[59] Vemurafenib (melanoma) NDA 202-429.

[60] Vemurafenib (melanoma). Pages 27 and 37 of 105-page Medical Review.

FDA assessed whether allocation into these various subgroups was balanced in the study drug arm and control treatment arm, at baseline (at the start of the clinical trial), writing, "There was no substantial imbalance between treatment arms with respect to the demographic characteristics of age, gender, and race."[61]

IV. SUMMARY OF CODING AND POOLING

Data on adverse events are captured on the Case Report Form (CRF)[62] where the healthcare worker can use verbatim terms or, alternatively, AE terms that are predefined on the Case Report Form. Verbatim terms are those chosen by the healthcare worker when filling out the Case Report Form. Examples of verbatim terms include, turned green and his glucose was 40,[63] sunburn,[64] symptomatic ischemic colitis,[65] upgoing eyes,[66] broken brace after fall,[67] and restless leg increased.[68] Although most verbatim terms chosen by coders in the pharmaceutical industry are likely to be appropriate, this short list of verbatim terms is from FDA's Medical Reviews, where FDA complained about the choice of the verbatim term.

Where data on safety are captured on the Case Report Form by way of verbatim terms or by readout from a diagnostic machine, the data need to be translated by coding. Coding is to Preferred Terms (PT) and to System Organ Class (SOC) terms found in the MedDRA dictionary. The end result of the coding process is that the terms for AEs are saved in the Sponsor's safety database.

The detection of adverse events of increased concern (also known as "safety signals") can be enhanced, and sometimes obscured, by the processes of splitting and lumping. Lumping occurs where similar AEs from several different study subjects are captured on the Case Report Forms by several different terms, but where coding results in these inputted into the database by exactly the same Preferred Term. Splitting occurs where the same verbatim term, as it occurs on Case Report Forms for several different study subjects, is

[61] Vemurafenib (melanoma). Pages 27 and 37 of 105-page Medical Review.

[62] Schroll JB, et al. Assessment of adverse events in protocols, clinical study reports, and published papers of trials of orlistat: a document analysis. PLoS Med. 2016;13:e1002101. doi: 10.1371/journal.pmed.1002101. eCollection.

[63] MedDRA MSSO. MedDRA® term selection: points to consider. ICH-endorsed guide for MedDRA users. McLean, VA: MedDRA MSSO; 2016 (57 pp.).

[64] Simeprevir (hepatitis C virus) NDA 205-123. Pages 4–5 of 173-page Medical Review.

[65] Linaclotide (irritable bowel syndrome) NDA 202-811. Page 418 of 489-page Medical Review.

[66] Lurasidone (schizophrenia) NDA 200-603. Pages 97–98 of 210-page Medical Review.

[67] Brivaracetam (partial onset seizures) NDA 205-836, NDA 205-837, NDA 205-838. Pages 192–193 of 382 pp.

[68] 4-Hydroxybutyrate (cataplexy in narcolepsy; excessive daytime sleepiness in narcolepsy) NDA 021-196. Pages 25-27 of 41-page pdf file (14th pdf file of 15 pdf files in Medical Review) and pages 4–6 and 15 of 41-page pdf file (15th pdf file of 15 pdf files in Medical Review). Also, page 8 of 41-page pdf file (10th pdf file of 15 pdf files in Medical Review).

inputted into the database by a variety of different Preferred Terms. Variations on these particular examples of splitting and lumping occur.

Pooling data results in greater quantities of analyzable data, as acquired from a multiplicity of clinical studies, for example, from Phase II trials, Phase IIb trials, Phase III trials, and Phase IIIb trials[69,70] or from replicated clinical trials.[71] Data from various studies should not be pooled where the study design differed markedly from the studies proposed to be pooled. Dramatic differences in study design can take the form of drug dosage amount, months of drug dosing, severity of the disease experienced by the study subjects, and the particular biomarkers required for enrolling subjects in the clinical study. Where studies to be pooled were conducted for different periods of time, AE data may be normalized as shown in the denosumab study.[72]

Another parameter for determining if pooling is appropriate is whether the subgroups within the patient population were balanced in the same way (when comparing the different studies proposed to be pooled). Balancing refers to the ratio of subjects in the [study drug group]/[control group]. The Sponsor has the option to assess balancing at the time of enrollment, and also to assess balancing at the time of completion.[73,74]

The issue of establishing and then maintaining balancing in the various subgroups in any population of study subjects was illustrated by FDA's Guidance for Industry,[75] as well as by FDA's Medical Reviews for vemurafenib[76] and for the ceftolozane/tazobactam combination.[77]

Where the Sponsor pools data from a multiplicity of clinical studies, and where the Sponsor's data analysis meets with FDA approval, the conclusions may find a place on the package label.

[69] Manns MP, et al. Simeprevir with peginterferon/ribavirin for treatment of chronic hepatitis C virus genotype 1 infection: pooled safety analysis from Phase IIb and III studies. J. Viral Hepat. 2015;22:366–75.

[70] Williams-Herman D, et al. Safety and tolerability of sitagliptin in patients with type 2 diabetes: a pooled analysis. BMC Endocr. Disord. 2008;8:14. doi: 10.1186/1472-6823-8-14.

[71] Hanania NA, et al. Lebrikizumab in moderate-to-severe asthma: pooled data from two randomised placebo-controlled studies. Thorax 2015;70:748–56.

[72] Denosumab (osteoporosis) BLA 125-320. Page 79 of 710-page Medical Review.

[73] Vemurafenib (melanoma). Pages 27 and 37 of 105-page Medical Review.

[74] Ceftolozane/tazobactam combination (infections) 206-829. Pages 22 and 33 of 118-page Statistical Review.

[75] U.S. Department of Health and Human Services. Food and Drug Administration. Center for Drug Evaluation and Research (CDER). Center for Biologics Evaluation and Research (CBER). Integrated summary of effectiveness guidance for industry; 2015 (17 pp.).

[76] Vemurafenib (melanoma) 202-429. 41-page Statistical Review.

[77] Ceftolozane/tazobactam combination (infections) 206-829. Pages 22 and 33 of 118-page Statistical Review.

Index

Note: Page numbers followed by "*f*" and "*t*" refer to figures and tables, respectively.

E

G

H

I

M

P

Q

R

T

U

V

W

X

Z

Edwards Brothers Inc.
Ann Arbor MI. USA
January 10, 2018